PROGRESS IN HETEROCYCLIC CHEMISTRY

Volume 15

Related Titles of Interest

Books

CARRUTHERS: Cycloaddition Reactions in Organic Synthesis
CLARIDGE: High-Resolution NMR Techniques in Organic Chemistry
FINET: Ligand Coupling Reactions with Heteroatomic Compounds
GAWLEY & AUBÉ: Principles of Asymmetric Synthesis
HASSNER & STUMER: Organic Syntheses Based on Name Reactions
KATRITZKY: Advances in Heterocyclic Chemistry
KATRITZKY & POZHARSKII: Handbook of Heterocyclic Chemistry, 2nd Edition
LEVY & TANG: The Chemistry of C-Glycosides
MATHEY: Phosphorus-Carbon Heterocyclic Chemistry: The Rise of a New Domain
McKILLOP: Advanced Problems in Organic Reaction Mechanisms
OBRECHT: Solid Supported Combinatorial and Parallel Synthesis of Small-Molecular-Weight Compound Libraries
OSBORN: Best Synthetic Methods – Carbohydrates
PELLETIER: Alkaloids; Chemical and Biological Perspectives
SESSLER & WEGHORN: Expanded Contracted and Isomeric Porphyrins
WONG & WHITESIDES: Enzymes in Synthetic Organic Chemistry

Major Reference Works

BARTON, NAKANISHI, METH-COHN: Comprehensive Natural Products Chemistry
BARTON & OLLIS: Comprehensive Organic Chemistry
KATRITZKY & REES: Comprehensive Heterocyclic Chemistry I CD-Rom
KATRITZKY, REES & SCRIVEN: Comprehensive Heterocyclic Chemistry II
KATRITZKY, METH-COHN & REES: Comprehensive Organic Functional Group Transformations
SAINSBURY: Rodd's Chemistry of Carbon Compounds
TROST & FLEMING: Comprehensive Organic Synthesis

Journals

BIOORGANIC & MEDICINAL CHEMISTRY
BIOORGANIC & MEDICINAL CHEMISTRY LETTERS
CARBOHYDRATE RESEARCH
HETEROCYCLES (distributed by Elsevier)
PHYTOCHEMISTRY
TETRAHEDRON
TETRAHEDRON: ASYMMETRY
TETRAHEDRON LETTERS

Full details of all Elsevier Science publications, and a free specimen copy of any Elsevier Science journal, are available on request at www.elsevier.com or from your nearest Elsevier Science office.

PROGRESS IN HETEROCYCLIC CHEMISTRY

Volume 15

A critical review of the 2002 literature preceded by three chapters on current heterocyclic topics

Editors

GORDON W. GRIBBLE
*Department of Chemistry, Dartmouth College,
Hanover, New Hampshire, USA*

and

JOHN A. JOULE
*Department of Chemistry, University of Manchester,
Manchester, UK*

2003

PERGAMON
An Imprint of Elsevier Science

Amsterdam – Boston – London – New York – Oxford – Paris
San Diego - San Francisco – Singapore – Sydney - Tokyo

ELSEVIER SCIENCE Ltd.
The Boulevard, Langford Lane
Kidlington, Oxford OX5 1GB, UK

© 2003 Elsevier Science Ltd. All rights reserved.

This work and the individual contributions contained in it are protected under copyright by Elsevier Science Ltd, and the following terms and conditions apply to its use:

Photocopying
Single photocopies of single chapters may be made for personal use as allowed by national copyright laws. Permission of the Publisher and payment of a fee is required for all other photocopying, including multiple or systematic copying, copying for advertising or promotional purposes, resale, and all forms of document delivery. Special rates are available for educational institutions that wish to make photocopies for non-profit educational classroom use.

Permissions may be sought directly from Elsevier via their homepage (http://www.elsevier.com) by selecting 'Customer Support' and then 'Obtaining Permissions'. Alternatively you can send an email to: permissions@elsevier.co.uk or fax: +44 1865 853333.

In the USA, users may clear permissions and make payments through the Copyright Clearance Center, Inc., 222 Rosewood Drive, Danvers, MA 01923, USA; phone: (+1) (978) 7508400, fax: (+1) (978) 7504744, and in the UK through the Copyright Licensing Agency Rapid Clearance Service (CLARCS), 90 Tottenham Court Road, London W1P 0LP, UK; phone: (+44) 20 7631 5555; fax: (+44) 20 7631 5500. Other countries may have a local reprographic rights agency for payments.

Derivative Works
Tables of contents may be reproduced for internal circulation but permission of Elsevier Science is required for external resale or distribution of such material.
Permission of the Publisher is required for all other derivative works, including compilations and translations.

Electronic Storage or Usage
Permission of the Publisher is required to store or use electronically any material contained in this work, including any chapter or part of a chapter.

Except as outlined above, no part of this work may be reproduced, stored in a retrieval system or transmitted in any form or by any means, electronic, mechanical, photocopying, recording or otherwise, without prior written permission of the Publisher. Address permissions requests to: Elsevier Science Global Rights Department, at the mail, fax and e-mail addresses noted above.

Notice
No responsibility is assumed by the Publisher for any injury and/or damage to persons or property as a matter of products liability, negligence or otherwise, or from any use or operation of any methods, products, instructions or ideas contained in the material herein. Because of rapid advances in the medical sciences, in particular, independent verification of diagnoses and drug dosages should be made.

Although all advertising material is expected to conform to ethical (medical) standards, inclusion in this publication does not constitute a guarantee or endorsement of the quality or value of such product or of the claims made of it by its manufacturer.

First edition 2003

Library of Congress Cataloging in Publication Data
A catalog record from the Library of Congress has been applied for.

British Library of Cataloguing in Publication Data
A catalogue record from the British Library has been applied for.

ISBN: 008 0442870   Hardcover
ISBN: 008 0442854   (ISHC members edition)

∞ The paper used in this publication meets the requirements of ANSI/NISO Z39.48-1992 (Permanence of Paper).
Printed in The Netherlands

Contents

Foreword   *vii*

Editorial Advisory Board Members   *viii*

Chapter 1: Recent Advances in the Synthesis of Heterocycles via Ring-Closing Metathesis   *1*
Michael A. Walters, *Pfizer Global Research and Development, Ann Arbor Laboratories, Ann Arbor, USA*

Chapter 2: Photochemical Isomerizations of Some Five-Membered Heteroaromatic Azoles   *37*
James W. Pavlik, *Department of Chemistry and Biochemistry, Worcester Polytechnic Institute, Worcester, MA, USA*

Chapter 3: Naturally Occurring Halogenated Pyrroles and Indoles   *58*
Gordon W. Gribble, *Department of Chemistry, Dartmouth College, Dartmouth, NH, USA*

Chapter 4: Three- and Four-Membered Ring Systems

Part 1.   **Three-membered ring systems**   *75*
Albert Padwa, *Emory University, Atlanta, GA, USA* and Shaun Murphree
Allegheny College, Meadville, PA, USA

Part 2.   **Four-membered ring systems**   *100*
Benito Alcaide, *Departamento de Química Orgánica I. Facultad de Química. Universidad Complutense de Madrid, Madrid, Spain* and Pedro Almendros,
Instituto de Química Orgánica General, CSIC, Madrid, Spain

Chapter 5: Five-Membered Ring Systems

Part 1.   **Thiophenes & Se, Te Analogs**   *116*
Erin T. Pelkey, *Hobart and William Smith Colleges, Geneva, NY, USA*

Part 2.   **Pyrroles and Benzo Derivatives**   *140*
Tomasz Janosik, *Department of Chemistry, Dartmouth College, Hanover NH, USA* and Jan Bergman, *Department of Biosciences at Novum, Karolinska Institute, Novum Research Park, Huddinge, Sweden, and Södertörn University College, Huddinge, Sweden*

Part 3.   Furans and Benzofurans   *167*
Xue-Long Hou, *Shanghai-Hong Kong Joint Laboratory in Chemical Synthesis and State Key Laboratory of Organometallic Chemistry, Shanghai Institute of Organic Chemistry, Chinese Academy of Sciences, Shanghai, China*, Zhen Yang, *Key Laboratory of Bioorganic Chemistry and Molecular Engineering of the Ministry of Education, Department of Chemical Biology, College of Chemistry, Peking University, Beijing, China* and Henry N. C. Wong, *Department of Chemistry, Institute of Chinese Medicine and Central Laboratory of the Institute of Molecular Technology for Drug Discovery and Synthesis, The Chinese University of Hong Kong, Hong Kong, China and Shanghai-Hong Kong Joint Laboratory in Chemical Synthesis, Shanghai Institute of Organic Chemistry, The Chinese Academy of Sciences, Shanghai, China*

Part 4.   With More than One N Atom   *206*
Larry Yet, *Albany Molecular Research, Inc., Albany, NY, USA*

Part 5.   With N & S (Se) Atoms   *230*
David J. Wilkins, *Key Organics Ltd., Camelford, UK* and Paul A. Bradley, *Pfizer Global Research & Development, Sandwich, UK*

Part 6.   With O & S (Se, Te) Atoms   *249*
R. Alan Aitken and Stephen J. Costello, *University of St Andrews, UK*

Part 7.   With O & N Atoms   *261*
Stefano Cicchi, Franca M. Cordero and Donatella Giomi, *Università di Firenze, Italy*

Chapter 6: Six-Membered Ring Systems

Part 1.   Pyridines and Benzo Derivatives   *284*
D. Scott Coffey, Stanley P. Kolis and Scott A. May, *Eli Lilly & Company, Indianapolis, IN, USA*

Part 2.   Diazines and Benzo Derivatives   *306*
Michael P. Groziak , *San José State University, San José, CA, USA*

Part 3.   Triazines, Tetrazines and Fused Ring Polyaza Systems   *339*
Carmen Ochoa and Pilar Goya, *Instituto de Química Médica (CSIC), Madrid, Spain*

Part 4.   With O and/or S Atoms   *360*
John D. Hepworth, *James Robinson Ltd., Huddersfield, UK* and B. Mark Heron, *Department of Colour Chemistry, University of Leeds, Leeds, UK*

Chapter 7: Seven-Membered Rings   *385*
John D. Bremner, *University of Wollongong, NSW, Australia*

Chapter 8: Eight-Membered and Larger Rings   *431*
George R. Newkome, *The University of Akron, Akron, OH, USA*

Index   *450*

Foreword

This is the fifteenth annual volume of *Progress in Heterocyclic Chemistry*, which covers the literature published during 2002 on most of the important heterocyclic ring systems. References are incorporated into the text using the journal codes adopted by *Comprehensive Heterocyclic Chemistry*, and are listed in full at the end of each chapter. This volume opens with three specialized reviews. The first, by Michael Walters covers Recent Advances in the Synthesis of Heterocycles via Ring-Closing Metathesis. The second, by James Pavlik, discusses Photochemical Isomerisations of Pyrazoles, Imidazoles, Thiazoles, and Isothiazoles. The third, contributed by Gordon Gribble, at short notice to replace a review which an author failed to produce, surveys Naturally Occurring Halogenated Pyrroles and Indoles.

The remaining chapters examine the recent literature on the common heterocycles in order of increasing ring size and the heteroatoms present. Last year's gap in coverage of seven-membered systems is made good with a review (chapter 7) covering the literature of two years (2001 and 2002) in this area. We are delighted to welcome some new contributors to this volume and we continue to be indebted to the veteran cadre of authors for their expert and conscientious coverage. In particular we thank Daniel Ketcha who is giving up his coverage of pyrroles and indoles after many years' excellent reviews. We are also grateful to Adrian Shell and Derek Coleman of Elsevier Science for some proof reading and for supervising the publication of the volume.

We hope that our readers find this series to be a useful guide to modern heterocyclic chemistry. As always, we encourage both suggestions for improvements and ideas for review topics.

Gordon W. Gribble
John A. Joule

Editorial Advisory Board Members
Progress in Heterocyclic Chemistry

2002 - 2003

PROFESSOR Y. YAMAMOTO (CHAIRMAN)
Tokyo University, Sendai, Japan

PROFESSOR D. P. CURRAN
University of Pittsburgh
USA

PROFESSOR A. DONDONI
University of Ferrara
Italy

PROFESSOR K. FUJI
Kyoto University
Japan

PROFESSOR T.C. GALLAGHER
University of Bristol
UK

PROFESSOR A.D. HAMILTON
Yale University
USA

PROFESSOR M. IHARA
Tohoku University
Japan

PROFESSOR G.R. NEWKOME
University of Akron
USA

PROFESSOR R. PRAGER
Flinders University
Australia

PROFESSOR R.R. SCHMIDT
University of Konstanz, Germany

PROFESSOR L. TIETZE
Georg-August University
Germany

PROFESSOR S.M. WEINREB
Pennsylvania State University
USA

Information about membership and activities of the International Society of Heterocyclic Chemistry (ISCH) can be found on the World Wide Web at http://euch6f.chem.emory.edu/hetsoc.html

Chapter 1

Recent Advances in the Synthesis of Heterocycles via Ring-Closing Metathesis

Michael A. Walters
Pfizer Global Research and Development, Ann Arbor Laboratories
Michael.Walters@pfizer.com

1.1 INTRODUCTION

The ring-closing metathesis (RCM) of heteroatom-functionalized dienes is a powerful method for the construction of an ever-broadening range of heterocycles. Several authoritative reviews on the use of metathesis in synthesis have been published <95AG(E)1833, 97AG(E)2036, 97S792, 98T4413, 98JMOC29, 98MI275, 99MI211, 99ACA75, 00MI565, 00MI112, 00MI91, 02MI1> and informative, general discussions have also appeared <02CEN(51)29, 02CEN(51)34>. This review presents advances in the field of RCM as applied to the synthesis of heterocyclic compounds subsequent to the most recent comprehensive review on metathesis <00AG(E)3012> up until the end of 2002. Emphasis is placed on the synthetic aspects of RCM. Details on the mechanism of this process can be found in the aforementioned reviews as can information on catalyst preparation and reactivity <01ACR18, 01MI155>. Given the volume of papers published on the use of RCM in the last two years (~800) some areas of intense activity in this area could not be covered and apologies to those working in those realms are offered in advance.

1.2 CATALYSTS

The most widely employed RCM catalysts remain those developed by the groups of Grubbs and Schrock (Figure 1). In this review the abbreviations SMC (Schrock metathesis catalyst <94JA3414>), GMC (Grubbs metathesis catalyst <96JA100>), and MC2 (second generation, unsaturated <99JA2674, 99TL2247> or saturated <99OL953>, metathesis catalyst) will be used. The SMC has excellent reactivity but must be handled in rigorously dried solvents in an inert atmosphere. The GMC is much more stable to solvent impurities and moisture, and the so-called second generation catalysts (MC2) appear to combine the best features of both. Along with these catalysts that are generally useful for RCM, Schrock, Hoveyda and co-workers have developed a variety of imido-catalysts for asymmetric RCM and ROM (ring-opening metathesis), ARCM and AROM <01MI945, 03JA2591>. For simplicity, these will be generically referred to as S-HMC in the text. In all cases, the original synthetic articles should be consulted for the catalyst employed in a given reaction. This is especially true in the case of the asymmetric reactions where the successful outcome of the process appears to be more dependent on correct catalyst selection.

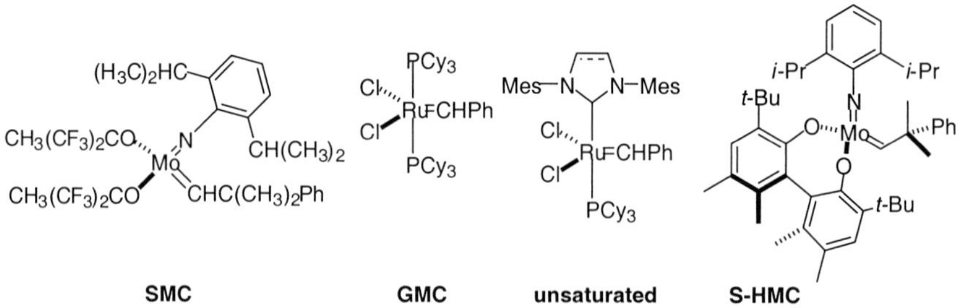

SMC **GMC** **unsaturated or saturated MC2** **S-HMC**

Figure 1

1.3 KEY SYNTHETIC ADVANCES

This section presents several of the most important advances in the synthesis of heterocycles via RCM that were made in the time period covered by this review.

In a process that is complementary to Lindlar reduction of cyclic alkynes, Fürstner and Radkowski reported the chemo- and stereoselective reduction of cycloalkynes (formed by ring-closing alkyne metathesis) to (*E*)-cycloalkenes using ruthenium-catalyzed *trans*-selective hydrosilylation followed by desilylation by AgF (dark, rt) (Scheme 1) <02CC2182>. Employing this process, the 21-membered macrocycle is formed in 85% yield with almost complete *E*-stereoselectivity.

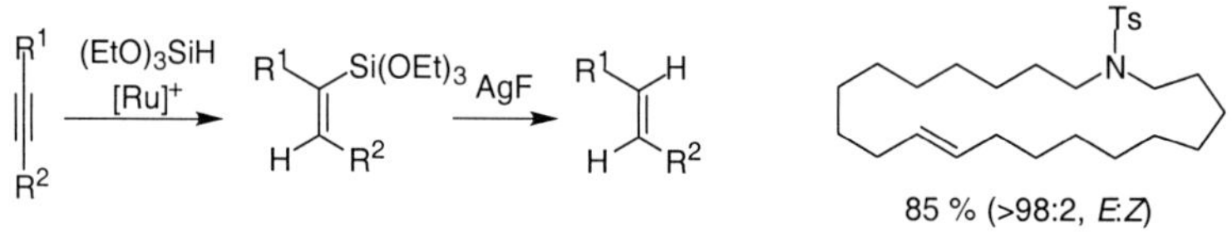

Scheme 1

Major advances were also made in the field of ARCM. Notable examples of this process are shown in Scheme 2 and include AROM <02JA10779>, and the desymmetrization reactions of acetals <00TL9553> and ethers <02JA2868>. Especially noteworthy are the ARCM to yield medium-sized, cyclic amines that occur in high yield and enantiomeric excess in the absence of solvent <01JA6991> [---*m* indicates the site of RCM].

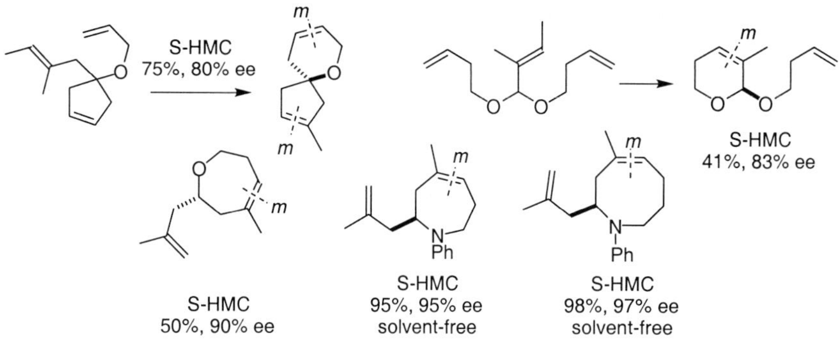

Scheme 2

Another area of intense activity was the use of unstrained cycloalkenes in combination reactions. These transformations are generically referred to as ring rearrangement metatheses (RRM). Blechert and co-workers have published extensively in this area and representative examples of the strategies developed are shown in Schemes 4, 5, and 6. Although not shown, the formation of O-heterocycles by many of these processes has also been reported. As with all metathesis reactions, these transformations are equilibrium processes. Typically, ROM occurs with strained cycloalkenes due to alleviation of ring strain. RCM reactions are often driven to completion by the loss of ethylene. In the case of the ROM-RC-enyne metathesis (Scheme 4) the formation of product is driven by the production of a 1,3-butadiene that cannot undergo further metathesis reaction <01TL5245, 02MI631> [---*rom* indicates the site of ROM]. Mori and Kitamura developed similar chemistry and employed it in combination with the Diels-Alder reaction to prepare tricycle **1** <01OL1161>.

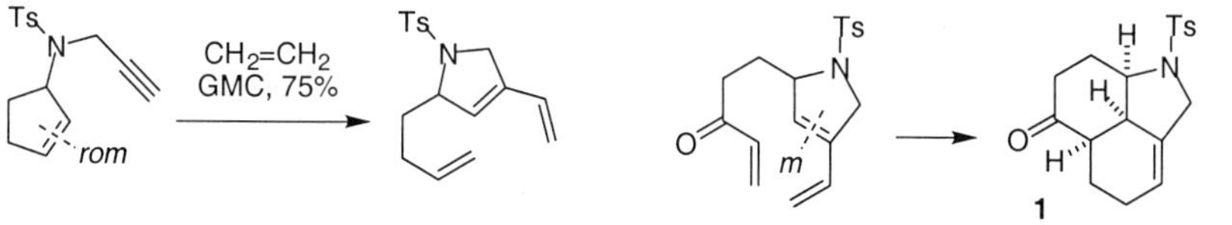

Scheme 4

Blechert and co-workers have also shown that suitably functionalized unstrained cyclopentenes participate in ROM-RCM (Scheme 5). The processes are especially attractive given the ready availability of chiral, non-racemic cyclopentenes. Cyclopentene **2** (R = TBDMS) is transformed almost quantitatively into **3** under the influence of GMC. When R = H, the ratio **2:3** is ~1:2, hinting at the delicate equilibrium that exists in these reactions. Compound **3** was transformed into the indolizidine **4** in several steps <00OL3971>. The utility of this process has been demonstrated by the preparation of (-)-swainsonine (**5**) <02JOC4325> (note that this strategy is amenable to the RCM of either the 5- or 6-membered ring in these indolizidines), the indolizidine **6** <00S893>, and (-)-indolizidine 167B <02TL6739, 00CC1501>. Ethylene is used in these reactions to increase the efficiency of the rearrangement process.

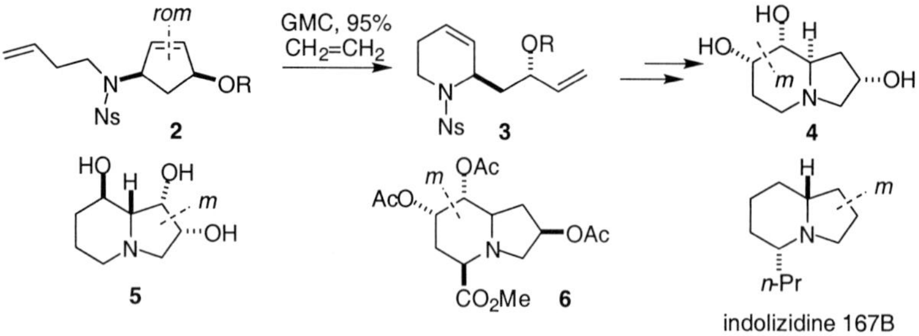

Scheme 5

Blechert and co-workers have also explored the use of cycloheptenes in RRM in processes where the thermodynamic driving force is the final RCM (Scheme 6). In one example of this novel process, ROM-RCM-RCM of **7** leads to an excellent yield of the substituted tetrahydrooxepin **8** <02T7503>. This strategy also featured significantly in the preparation of (+)-dihydrocuscohygrine **9** <02JOC6456>. The more common sequential reactions of strained cyclic alkenes have been reviewed <03EJO611>.

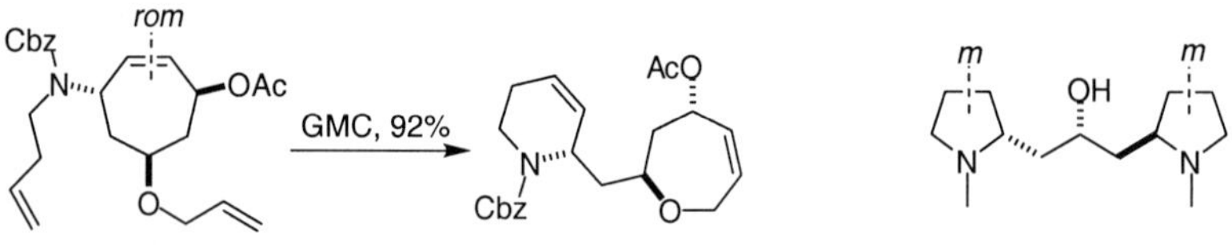

Scheme 6

In chemistry that should prove very useful in the asymmetric synthesis of azabicyclic alkaloids, Beak and co-workers reported the syntheses of 5-, 6-, and 7-membered fused bicyclic lactams <01JOC9056>. Pyrrolidine lactams are formed by asymmetric, anionic cyclization of **10**, followed by reaction of **11** with an unsaturated acylating agent and RCM. Asymmetric hydrogenation using the Noyori catalyst is the hallmark of the preparation of the homologous lactams **13** and **14**. Yields of the RCM to form these compounds are generally in the range of 60-90% and are typically effected with MC2. The [5,5]-fused system could not be formed from **12**. A similar observation was made by Nakagawa <00CPB1593>.

The control of alkene geometry in RCM reactions has been an area of intense research and interest since the process was first developed. While a general solution to this challenge has not yet been developed, intriguing observations of $E:Z$ control in macrocyclizations continue to be reported. For example, in the course of their studies on the synthesis of herbarumin I and II, Fürstner and co-workers reported the selective formation of either of the two isomeric alkene products **16** or **17** via RCM of diene **15** <02JA7061> (Scheme 8). The diene **15** was transformed into the E-alkene **17** using the ruthenium indenylidene catalyst (Fürstner Metathesis Catalyst; FMC, <01MI4811>) while use of the MC2 led to clean formation of the Z-isomer **16**.

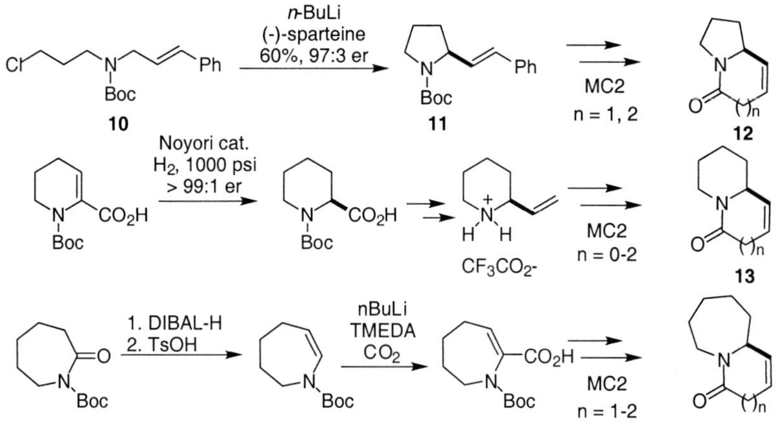

Scheme 7

Studies show that the formation of the *E*-alkene is likely the result of kinetic control while the MC2, probably as a reflection of its higher overall activity, leads to the thermodynamically favored *Z*-product.

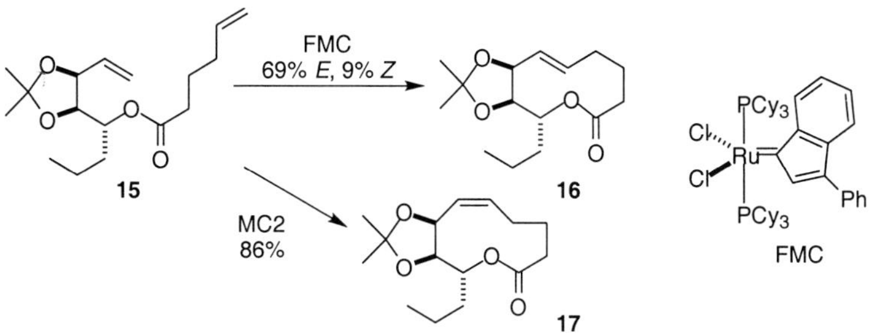

Scheme 8

In a related observation, Fürstner and co-workers reported that while both the FMC and the MC2 lead to the same *E*:*Z* alkene ratio in the case of **18**-RCM, the use of the same conditions with **19** leads to product enriched selectively in either of the two isomers <02MI657> depending on catalyst selection (Scheme 9). Fürstner notes that this "illustrates the subtle influence of remote substituents on the stereochemical outcome of RCM in the macrocyclic series." Studies concerning the effects of chelation of the reactivity of ruthenium carbene complexes were reported by Fürstner and co-workers <02OM331>. Paquette and co-workers observed that the use of either the GMC or the MC2 led to different outcomes in the macrocyclization RCM of 1,2-amino-alcohol-templated ene-dienes <02HCA3033, 02MI615> hinting that catalyst selection is also an important consideration in those processes.

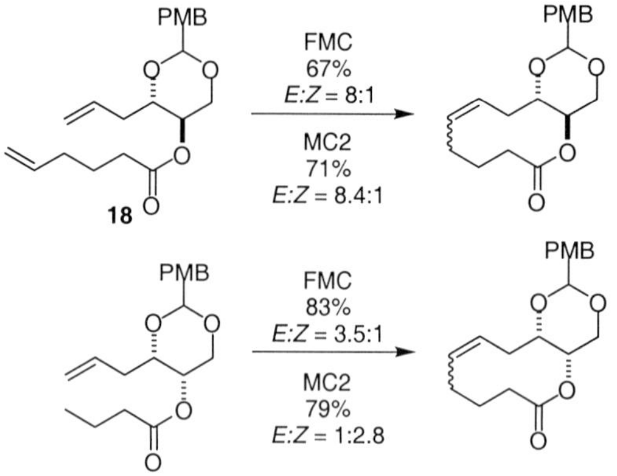

Scheme 9

1.4 NITROGEN-CONTAINING HETEROCYCLES

The following section contains highlights from the use of RCM in the preparation of nitrogen-containing heterocycles. Although reactions are categorized by ring size of the final products, some of these processes can be used to form larger, smaller, or other heterocyclic rings and the placement of these methods into one section or another is somewhat arbitrary. In most cases the reaction shown is illustrative of several that may have been reported.

1.4.1 Five-Membered Ring N-Heterocycles

Enamide-alkene RCM was developed as a convenient method for the preparation of cyclic enamides <01OL2045>. In an example of this methodology, deprotonation and N-alkylation of imine **20** leads to a moderate yield of **21** (Scheme 10). RCM is smoothly effected using the GMC in DCM (dichloromethane) at reflux. Six-membered rings could also be formed using this same reaction process and some of these reactions could be carried out more effectively using the MC2. Rings larger than 6-atoms could not be formed.

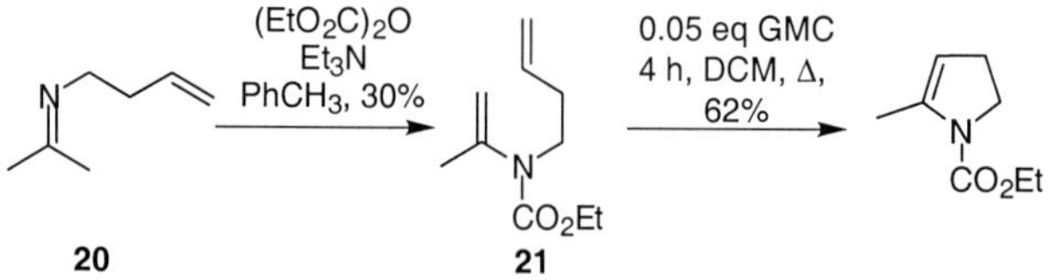

Scheme 10

Begtrup and co-workers employed iron-catalyzed coupling to prepare pyrrolidine RCM-precursors <02SL1889> (Scheme 11) while Grigg and co-workers employed a palladium-catalyzed allenylation process <01TL8673> (Scheme 12). Both of these processes appear fairly general, the latter procedure also being useful in the preparation of 3-substituted piperidines.

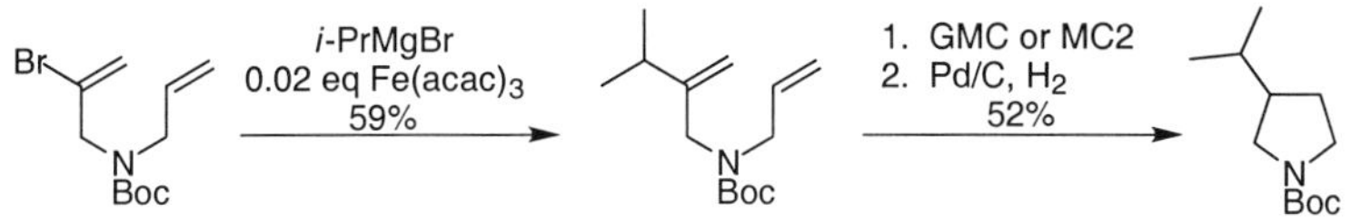

Scheme 11

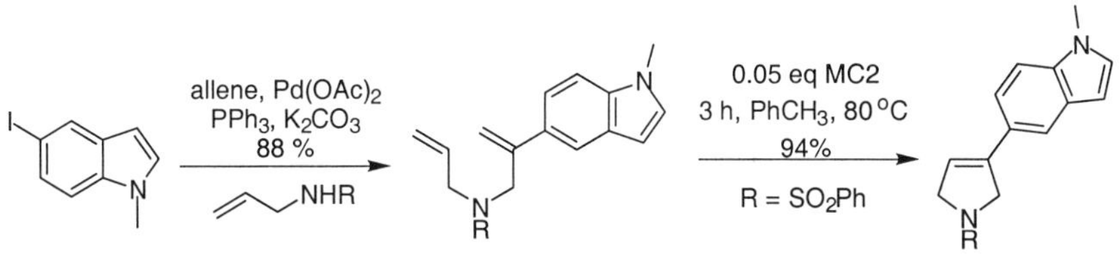

Scheme 12

Alkynes and ynamines have also been developed as useful participants in RCM approaches to pyrrolidine derivatives (Scheme 13). Mori and co-workers reported the formation of a cyclic dienamide via ene-ynamide RCM <02OL803>. They also studied substituent effects on the course of RCM of enynes and reported the formation of both pyrrolidines and piperidines <02MI678>. The syntheses of 5- and 6-ring carbocycles and O-containing heterocycles from appropriate precursors were also reported in that work. Synthetically-useful dienylboronic acids have been prepared from acetylenic boronates <00AG(E)3101>.

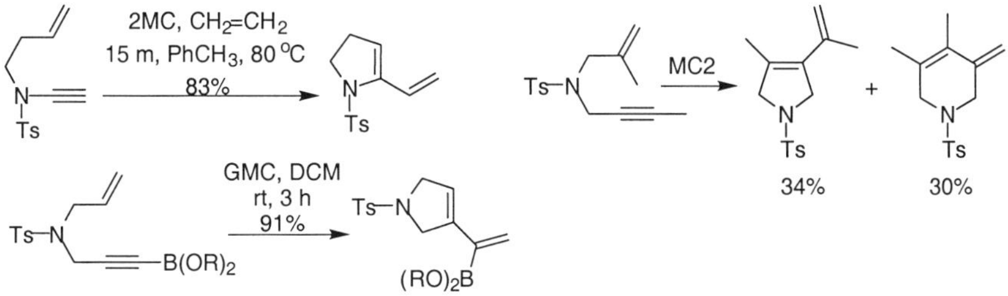

Scheme 13

The preparation of chiral, non-racemic pyrrolidines from enantiomerically-enriched vinyl epoxides has been reported <02SL731, 02JOC7774> (Scheme 14). In like fashion, the addition of allylamine to the enantiomerically-enriched epoxide **22** gives diol **23** which can be transformed into the proline analog **24** by a sequence of steps including RCM <02JOC6896>.

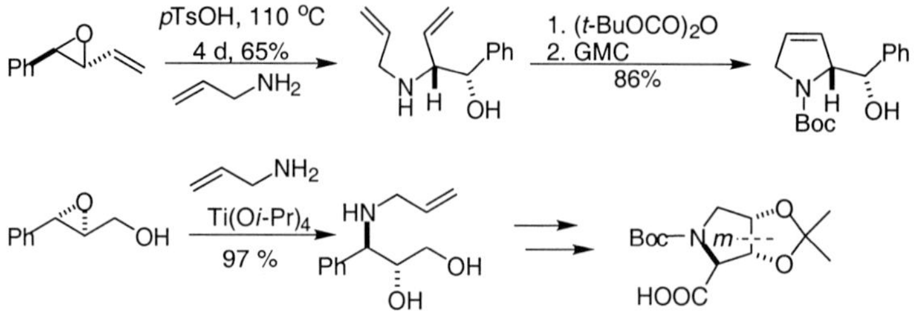

Scheme 14

1.4.2 Six-Membered Ring *N*-Heterocycles

Perhaps the most exciting advance in the preparation of *N*-heterocycles came in the synthesis of chiral, non-racemic piperidines via processes involving RCM. These processes generally involve the use of enantiomerically-enriched imines or templates as platforms for diene construction. The general strategies for the use of imines in RCM are shown in Scheme 15. These strategies usually involve the addition of an allyl organometallic (route A) to an imine, followed by *N*-allylation and RCM, or the addition of a vinyl organometallic (route B) followed by butenylation and RCM.

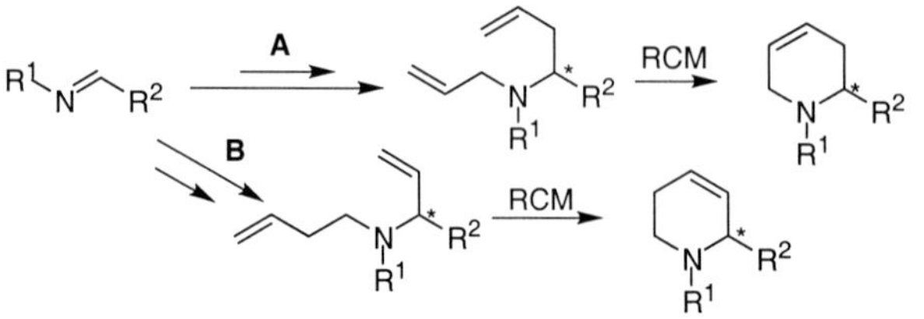

Scheme 15

A variety of imine derivatives have been used in these processes including those derived from norephedrine (**25**) <00TA4639>, Garner's aldehyde (**26**) <03TL527> (in this case the required

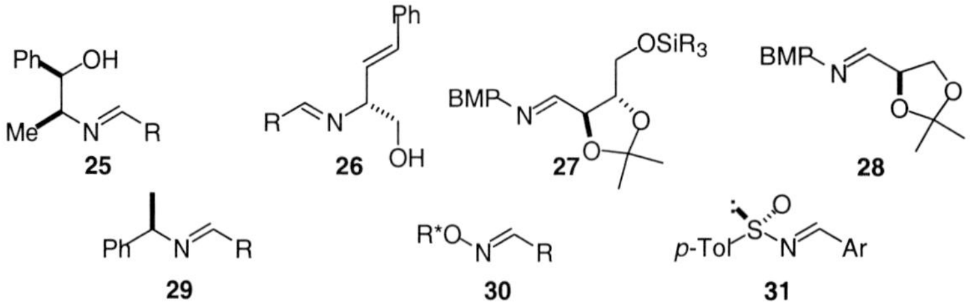

Scheme 16

unsaturation is Wittig-derived), (+)-dimethyl tartrate (**27**) and (+)-glyceraldehyde acetonide (**28**) <00CC1771, 02JCS(P1)2378>, and *N*-sulfinylimines (**31**) <00TL8157> have also been used as <01EJO2841>, and α-methyl benzylamine (**29**) <01JOM359> (Scheme 16). Oxime ethers (**30**) substrates. The asymmetric addition of allylmetals to aldehydes followed by *O*- to *N*-allyl interconversion has also been employed to prepare piperidines by RCM <00S1646, 02JOC1982>.

Several enantiomerically-enriched templates have also been employed in the preparation of chiral, non-racemic piperidines. The RCM of *N,O*- and *O,O*-acetals have been developed by Rutjes and co-workers into a powerful method for the preparation of a wide-variety of heterocycles <02MI736> (Scheme 17). Here the reaction of an enantiopure Ts-protected allyl glycine is employed to prepare a chiral, non-racemic, cyclic amino acid derivative. These *N,O*-acetals can be readily transformed into synthetically useful *N*-sulfonyliminium ions by treatment with BF$_3$•OEt$_2$ <00CC699>.

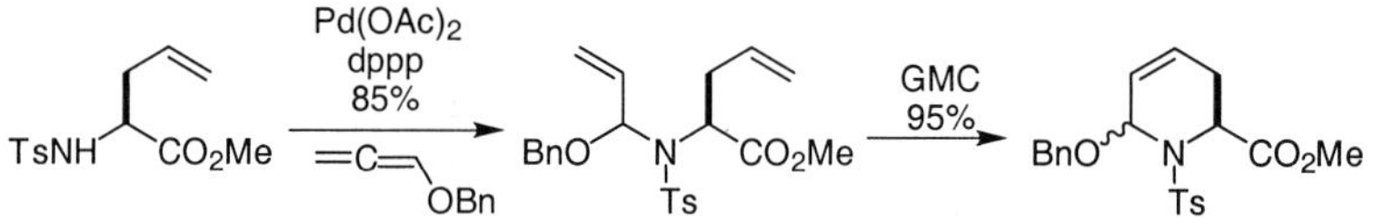

Scheme 17

A variety of amino-alcohol derivatives have been used as substrates for RCM approaches to piperidines. Representative examples of these methods are shown in Scheme 18 (**32**, <01TL1209>; **33**, <01TA817>; **34**, <02OL4499>). Application of this chemistry to the synthesis of deoxy-azasugars and hydroxypyrrolizidines has been reported <01TL4079>.

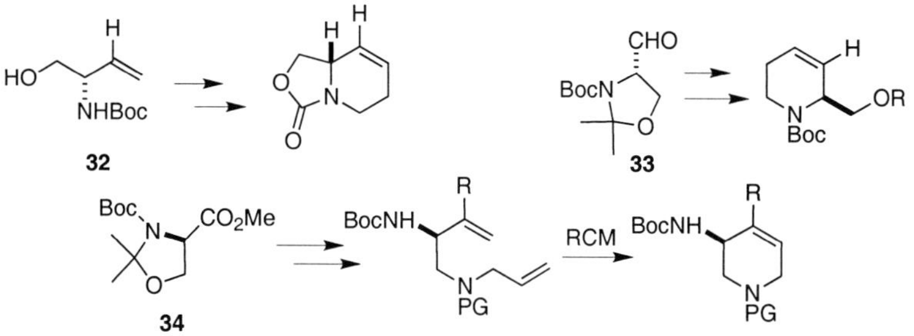

Scheme 18

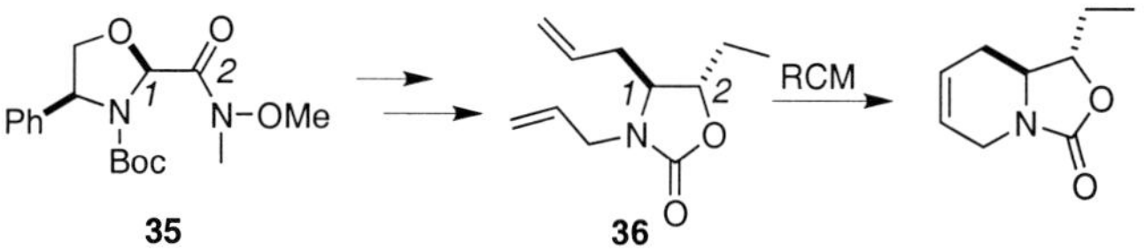

Scheme 19

In related work, the glycinol-derived Weinreb amide **35** was converted to the RCM precursor **36** by a series of steps including EtMgBr addition at C-2, allyl silane addition at C-1, and N-allylation <01T5393> (Scheme 19). Homologous alkenes have been used to prepare larger N-rings by this process. Use of this chemistry to prepare (-)-β-conhydrine <00TL4113> and tropanes <01TL4633> has been reported.

The addition of lithium (S)-N-allyl-N-α-methylbenzylamide to unsaturated esters was shown to be an effective method for producing RCM precursors <02SL1146> for piperidine synthesis, as was the addition of allylamine to the enantiomerically-enriched allylic epoxide **37** <02TL779> (Scheme 20).

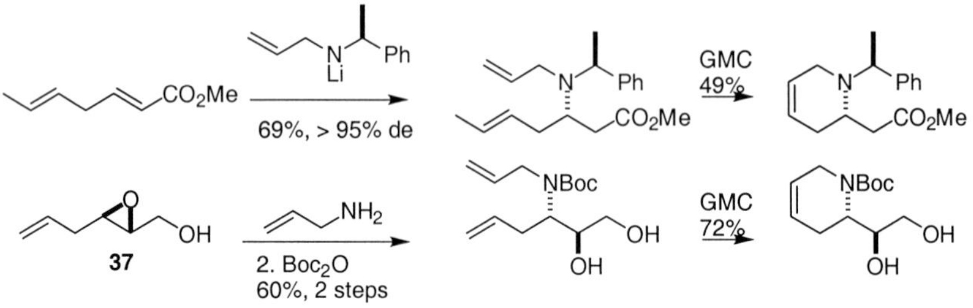

Scheme 20

The preparation of spirocyclic piperidines has also been reported. In an approach to the spirocyclic core of halichlorine and pinnaic acid, the annulation of 6-membered, N-containing heterocycles onto ketones was reported by Wright and co-workers <00OL1847> (Scheme 21). The annulation of α-tetralone by this method is illustrative of the general procedure. Condensation of allyl amine with the carbonyl group gives the imine that is treated, *in situ*, with allyl magnesium bromide. This addition was observed to be effective only in the case of allyl Grignard reagents: both vinyl and butenyl systems led to metalloenamine formation. RCM in the presence of *p*TSA (to effect amine protonation) was observed to be sluggish, but could be pushed to completion by the portionwise addition of catalyst. Use of the more stable MC2 (also in the presence of *p*TSA) leads to shorter reaction times and higher yields, as did N-protection prior to the RCM.

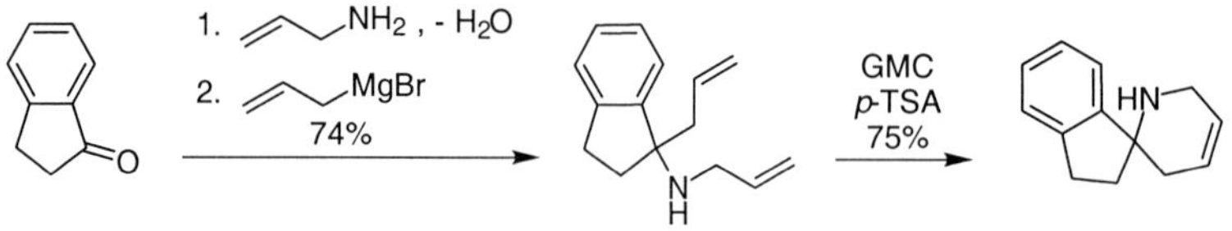

Scheme 21

An extension and expansion upon this process was reported by Marco and co-workers in their preparation of conjugated δ-lactams via the RCM of unsaturated amides <02T1185> (Scheme 22). Excellent yields are observed when the reacting alkene is unsubstituted. No RCM is observed, however, when the reacting alkene is internally substituted (isobutenyl), even with

MC2. Spirocycle **38** is formed in good yield under nearly identical conditions. This suggests that the lack of reactivity in the case of the acrylamide is probably due to disruption of the RCM catalytic cycle by disadvantageous complexation of a Ru-intermediate by the amide oxygen.

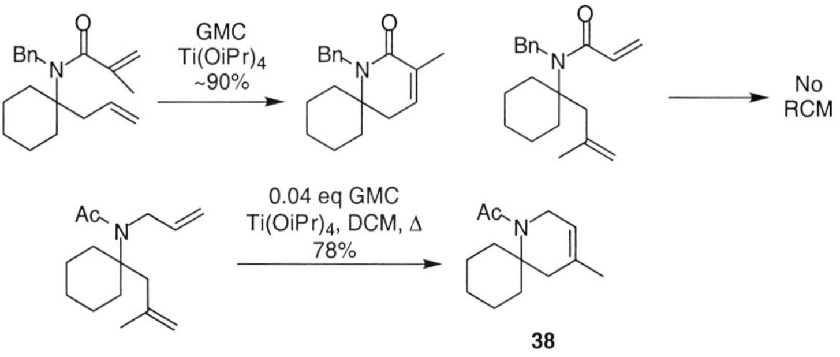

Scheme 22

Spiropiperidines can also be prepared by tandem RCM <02CC1542> (Scheme 23).

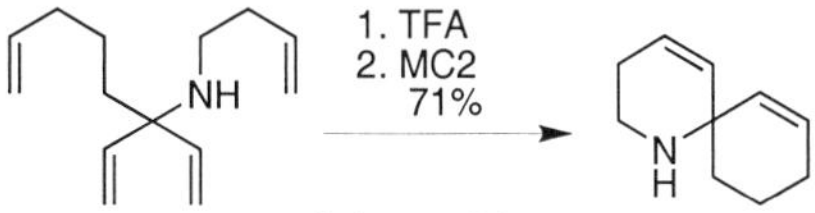

Scheme 23

1.4.3 Seven-Membered Ring *N*-Heterocycles

A novel, general route to 2,3,6,7-tetrahydroazepines via RCM was reported by Pearson and Aponick <01OL1327> (Scheme 24). Double-allylation of the amine α,α'-dication equivalent **39**, followed by *in situ* trapping with ClCO$_2$Ph gives rise to diene **40** which is efficiently converted to the 7-membered ring heterocycle **41** by treatment with the GMC in DCM. This process was effectively applied to a variety of substituted (2-azaallyl)stannanes.

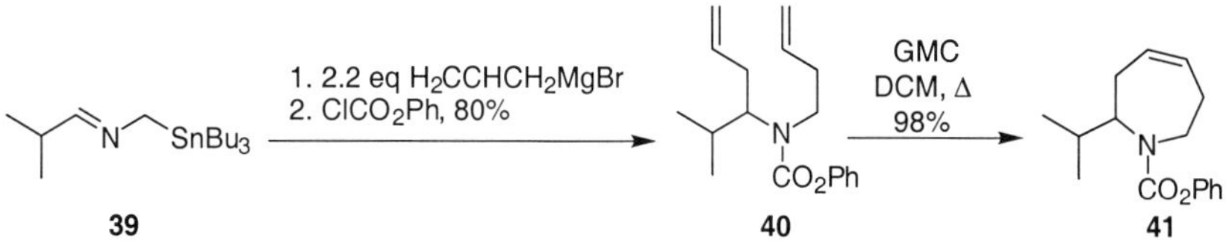

Scheme 24

Another example of the use of RCM to form azepines is shown in Scheme 25. Compound **42** is a precursor to novel, dual inhibitors of acetylcholinesterase and the serotonin transporter <02OL3359>.

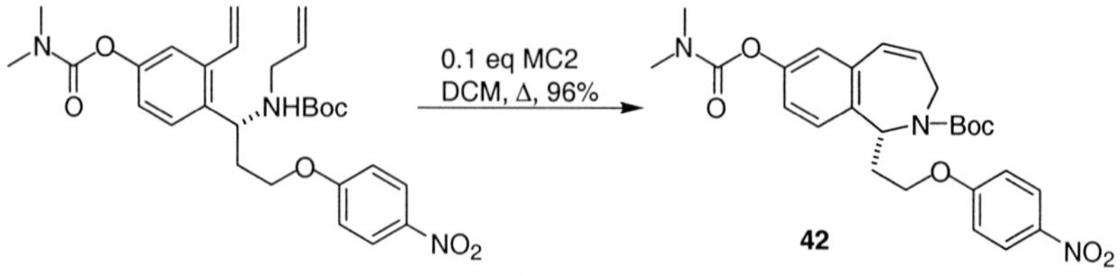

Scheme 25

1.4.4 Large Ring *N*-Containing Heterocycles

Ghosh and co-workers reported the preparation of the novel cyclourethane-derived HIV protease inhibitor **43**, along with 15- and 16-membered analogs <02BMCL1993> (Figure 2). RCM in the presence of Ti(iPrO)$_4$ gave the 14-membered 6-nor-fluvirucinin B$_1$ precursor **44** in 72% yield <02SL1724>. In studies directed toward the antitumor antibiotic geldanamycin, Bach and Lemarchand reported the formation of 18-20 membered ring lactams **45** by RCM <02SL1302>.

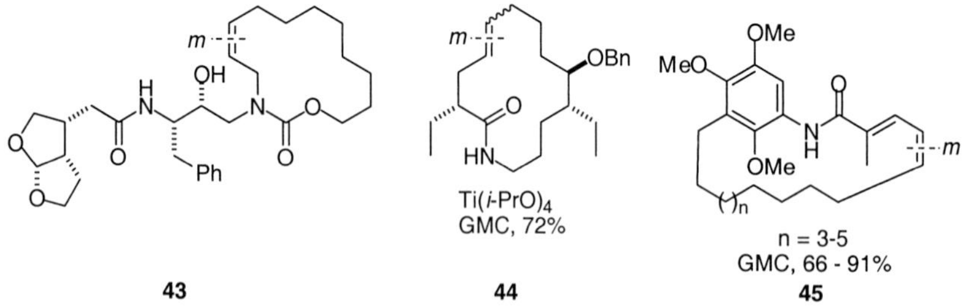

Figure 2

The preparation of simple, cyclic 7-9 membered lactams via RCM was also reported <01SL37> (Figure 3). Organ and co-workers reported the preparation of some 8-membered lactams in their enantiospecific synthesis of inhibitors of factor Xa <02TL8177>.

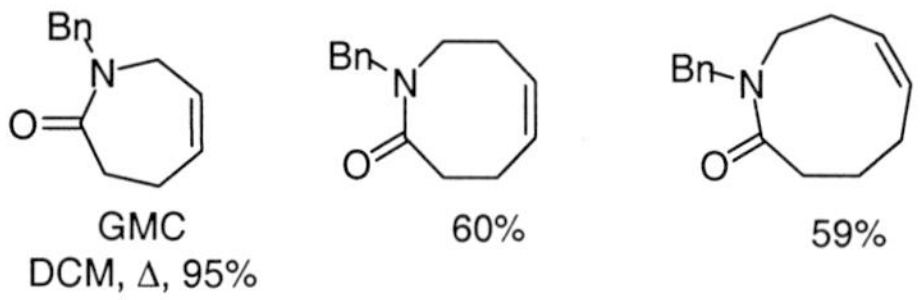

Figure 3

1.4.5 Indoles

Nishida and co-workers reported the preparation of indole **47** from *N*-allyl aniline **46** via alkene isomerization using vinyloxytrimethylsilane (VOT/CH$_2$Cl$_2$, 50 °C) in the presence of

MC2 followed by RCM at higher temperature (Scheme 26). The formation of a new catalyst in the isomerization process was inferred by spectroscopic techniques <02AG(E)4732>.

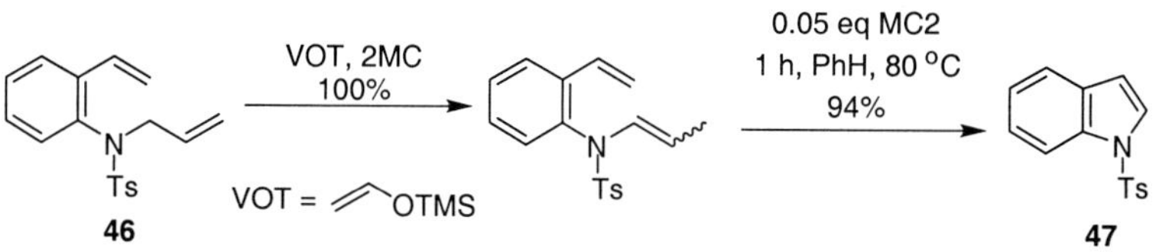

Scheme 26

The mitosene skeleton was readily formed by Wittig olefination of the indole-aldehyde **48** followed by RCM <02TL4765> (Scheme 27).

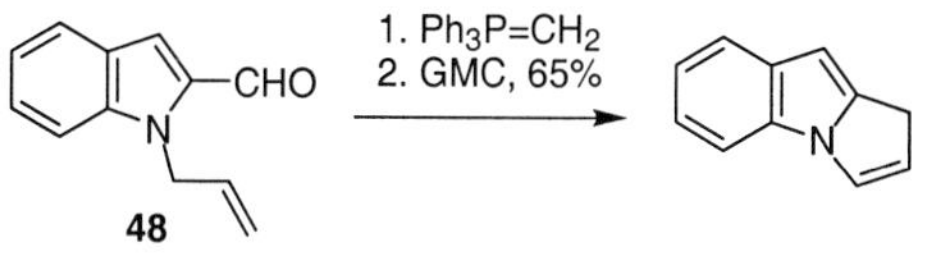

Scheme 27

The preparation of a variety of indole-lactams by the process shown in Scheme 28 was reported <02T10181>.

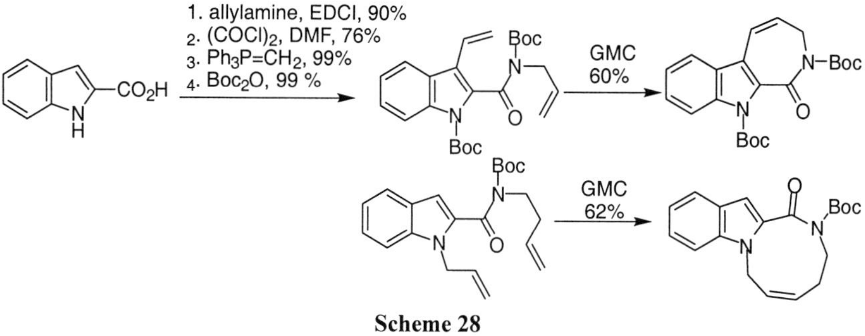

Scheme 28

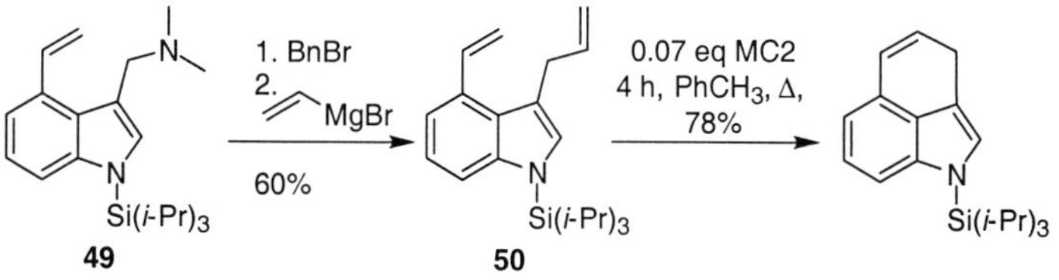

Scheme 29

The use of RCM in the synthesis of 3,4-carbocyclic indoles was reported by Pérez-Castells and co-workers as is exemplified in Scheme 29 <02T5407>. Vinyl indole **49** was quaternized with BnBr and quaternary salt underwent smooth addition of vinyl magnesium bromide to give the indole diene **50**. RCM was accomplished with the MC2 in good yields.

1.4.6 Fused Rings *N*-Heterocycles

A direct route to 5,8-disubstituted indolizidines and 1,4-disubstituted quinolizidines was reported by Barluenga and co-workers <02OL1971> (Scheme 30). Hetero-Diels-Alder reaction followed by reduction and desilylation gave good yields of the key piperidine core **51**. Further elaboration of the hydroxymethyl side chain followed by RCM and hydrogenation gave the targets in moderate yields.

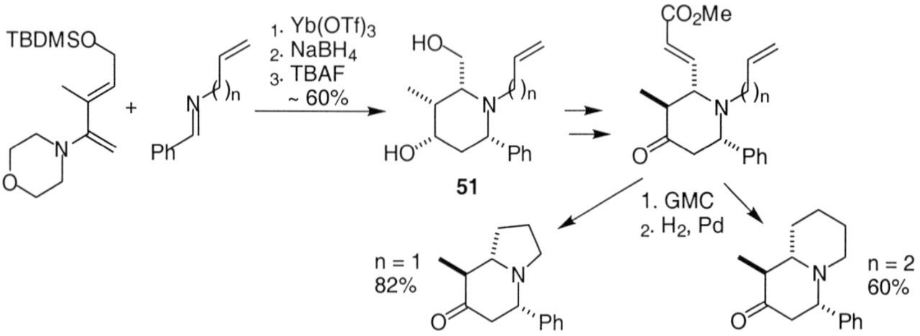

Scheme 30

Pilli and co-workers reported the use of acyliminium ions as key synthetic intermediates in the construction of compounds for RCM (Scheme 31) <00TA753>. Application of this methodology to other hydroxylated indolizidines <01TL5605> and trans-fused decahydroquinolines <00TL7843> was reported, as was the synthesis of larger rings <00SL319>. Preparation of 2,5,6-trisubstituted piperidines via *N*-sulfonyliminium ion intermediates was reported by Padwa and co-workers <02OL2029>.

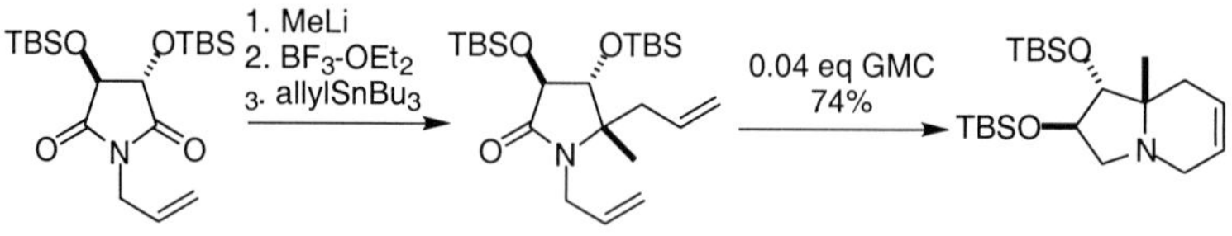

Scheme 31

A dramatic example of double RCM was reported by Ma and Ni (Scheme 32) <02OL639>. The direction of the RCM in these systems is dependent on the nature of the R-substituent as shown. A homologous derivative gave almost exclusively the [7,7]-fused system **52**.

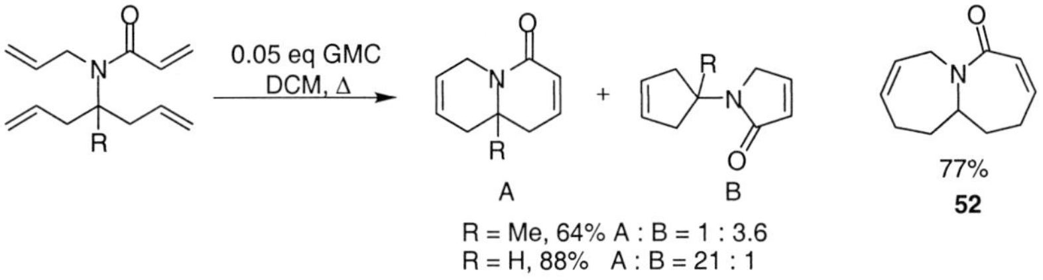

Scheme 32

R = Me, 64% A : B = 1 : 3.6
R = H, 88% A : B = 21 : 1

The RCM of isoquinoline and β-carboline enamines was reported by Grigg and co-workers <00TL3967> (Scheme 33).

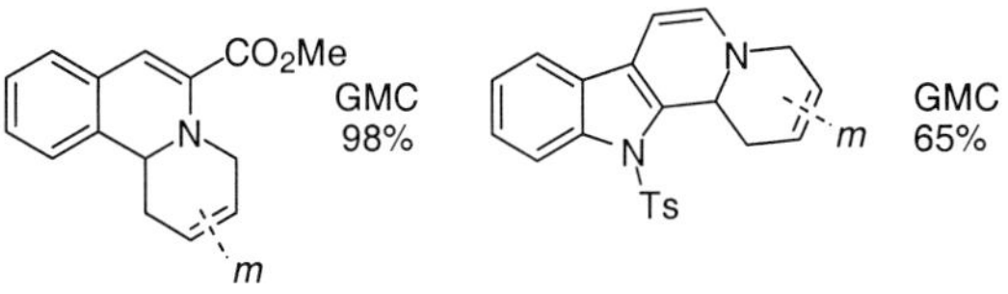

Scheme 33

Dihydroquinolines and quinolines are prepared in excellent yield using RCM (Scheme 34) <01TL8029>. The preparation of the 7- and 8-membered, and 4-oxygenated analogs (**53-55**) occurs in 95-100% yield with MC2.

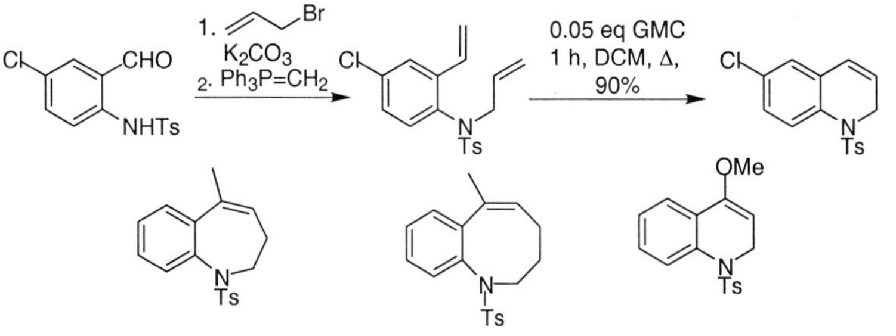

Scheme 34

1.4.7 Bridged Ring *N*-Heterocycles

Martin and Neipp developed a flexible and efficient synthesis of azabicyclo[n.3.1]alkenes starting from either glutarimide or 4-methoxypyridine (Scheme 35) <02T1779>. These reactions proceed in higher yield and at lower temperature with the MC2.

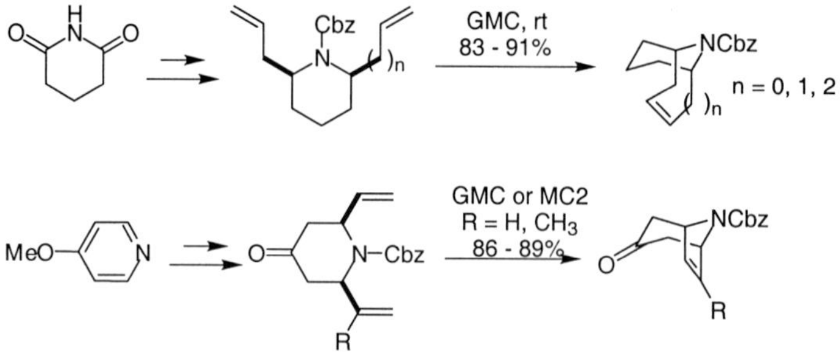

Scheme 35

A novel, two-step RCM-Heck approach to bridged N-heterocycles was reported by Grigg and York, one example of this methodology being shown in Scheme 36 <00TL7255>. In the case shown, the yield is 69% when the reactions are run separately, 37% when the reactions are run in combination using Pd(OAc)$_2$, and 73% when PS-Pd is employed in a two-step, one-pot procedure.

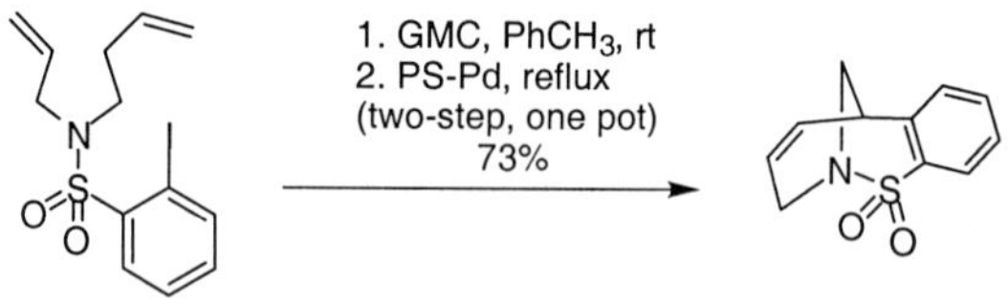

Scheme 36

1.5 OXYGEN-CONTAINING HETEROCYCLES

The following section contains highlights from the use of RCM in the preparation of oxygen-containing heterocycles. Following a section that covers processes that have been applied to a variety of ring sizes, the sections are organized by the ring size of the final products. As in the preceding section, many of these processes can be used to form larger, smaller, or other heterocyclic rings and the placement of these methods into one section or another is somewhat arbitrary. In most cases the reaction shown is illustrative of several that may have been reported.

1.5.1 General Oxygen-Containing Heterocycles

A tandem RCM-alkene isomerization sequence to form 5-, 6-, and 7-membered enol ethers was reported by Snapper and co-workers <02JA13390> (Scheme 36). In this process the RCM reaction is run under an atmosphere of 95:5 N$_2$:H$_2$ to convert the intermediate ruthenium alkylidene into an olefin-isomerization catalyst. Note that alkene migration can convert isomeric metathesis products into the same 2,3-enol ether. A single example of the formation of a 6-membered tosyl enamide was reported in this manuscript.

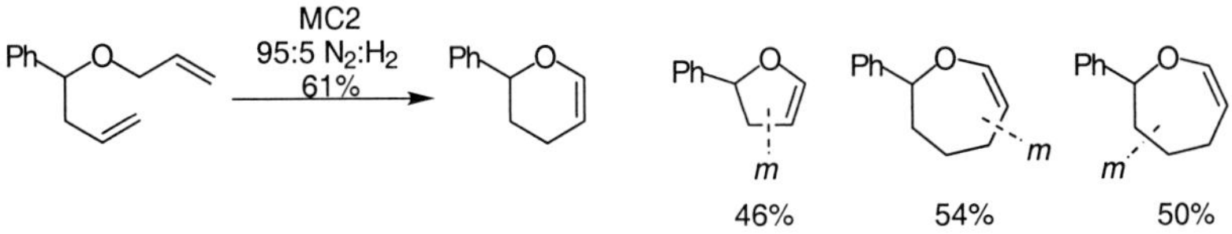

Scheme 36

Cossy and co-workers reported the synthesis of a variety of 3-oxacycloalkenes via RCM <02TL7263> (Scheme 37). Further transformations of these compounds, such as α-oxyalkylation and cuprate addition to the enone were also reported.

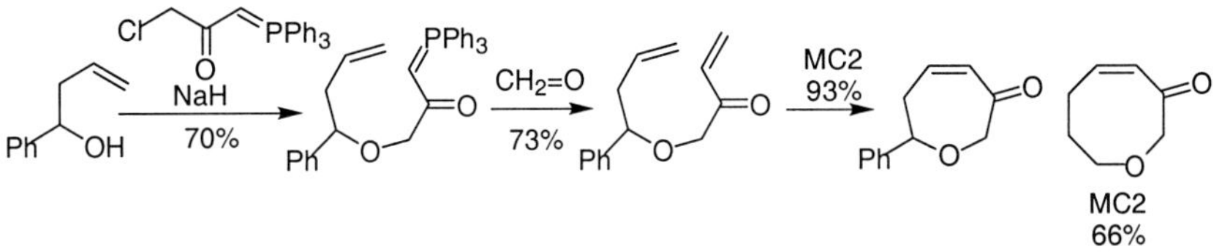

Scheme 37

The preparation of fused, polyether rings via RCM of enol ethers was reported by Rainier and co-workers <02T1997, 01TL179> (Scheme 38). The enol ether was prepared from the corresponding acetate using Takai's reduced-Ti protocol <94JOC2668>. Seven-membered, cyclic enol ethers could also be formed by this methodology.

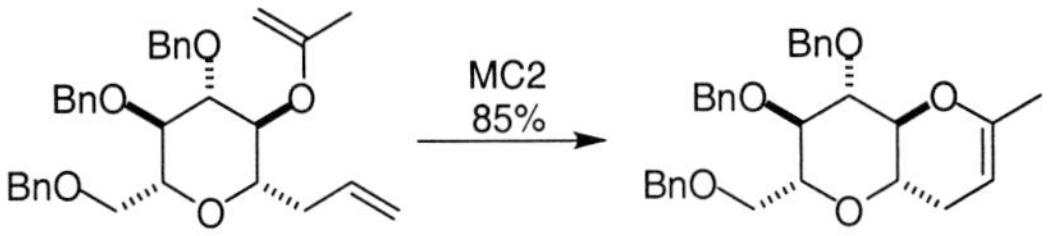

Scheme 38

A variety of spirocyclic ethers were prepared by double RCM of the appropriate polyalkene <01SL357> (Scheme 39). The yields of these processes were generally excellent and modest diastereoselection was observed in some cases.

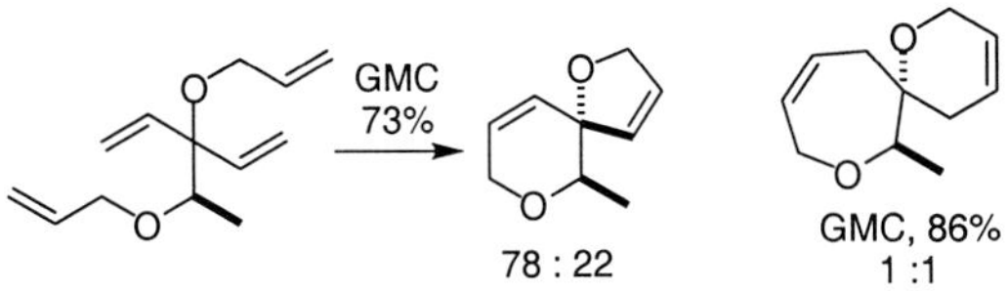

Scheme 39

1.5.2 5-Membered *O*-Containing Heterocycles

Kazmaier reported the synthesis of the furanomycin analog by RCM <02BMCL3905> (Figure 4).

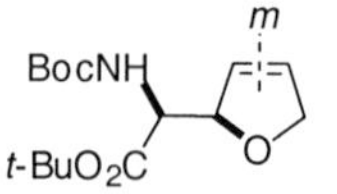

Figure 4

1.5.3 Six-Membered O-Containing Heterocycles

Walters and co-workers developed a synthesis of the library template **56** that was amenable to large-scale preparation (Scheme 40) <02JCO125>. Parallel synthesis employing this scaffold leads to a library of amino alcohols (major regioisomer shown).

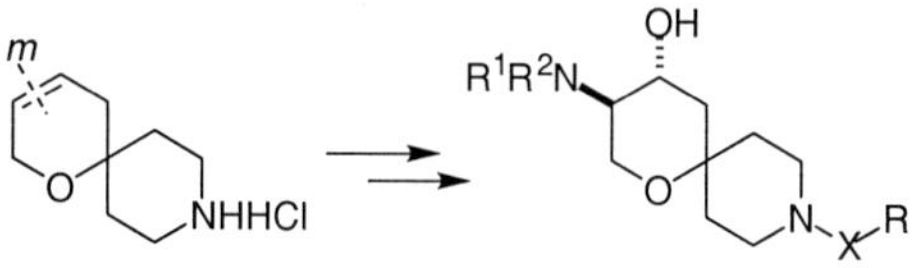

Scheme 40

Wright and co-workers developed a novel entry into spirocyclic, six-membered O-containing heterocycles as they combined oxyallyl cation chemistry profitably with a ROM-RCM process <02AG(E)4560> (Scheme 41). An analogous process, involving the Diels-Alder reaction of **57** with *N*-phenylmaleimide followed by ROM-RCM, gave **58** in excellent yield.

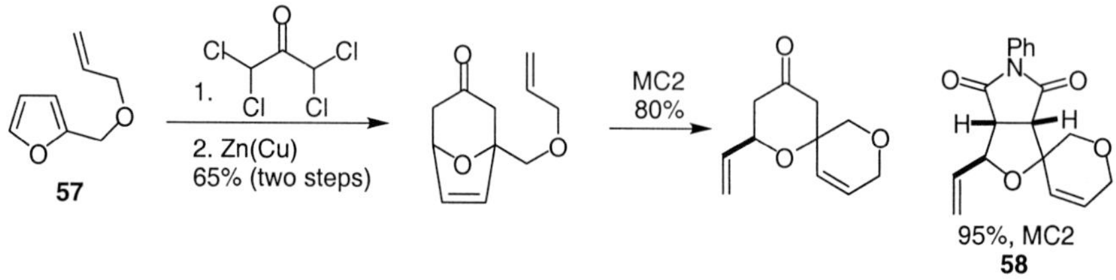

Scheme 41

The synthesis of 2-substituted chromenes was reported by Hiemstra and co-workers. Use of MC2 in this process led to similar results in shorter time <00TL5979> (Scheme 42).

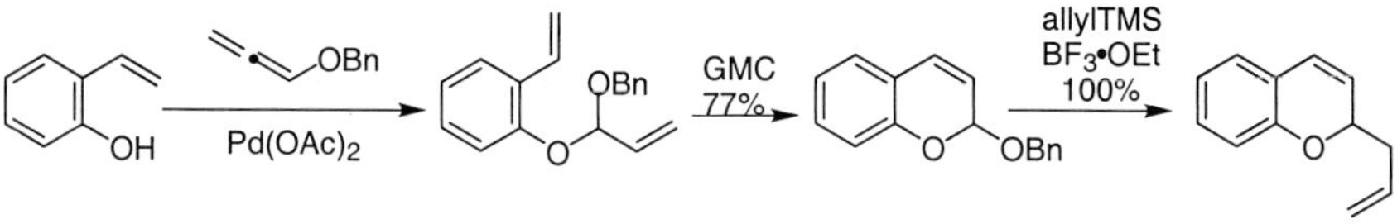

Scheme 42

Nicolaou and co-workers reported the use of *cis*-3,4-dichlorocyclobutene (**59**) as a useful building block to prepare complex polyether systems via ROM-RCM <01AG(E)4441> (Scheme 43). The GMC failed to give product in this transformation.

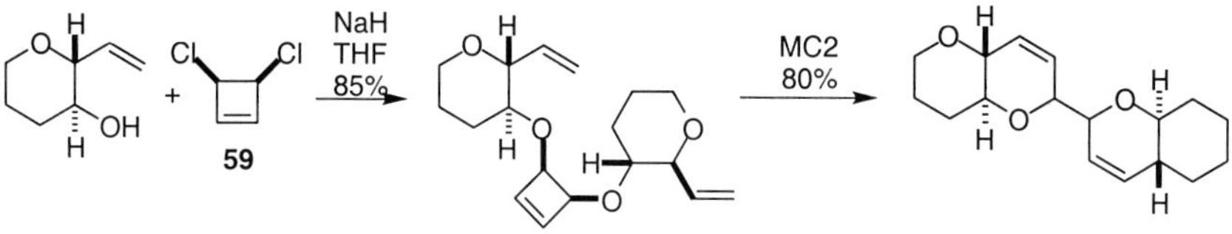

Scheme 43

Roush and Holson reported a diastereoselective preparation of the C-17–C-28 fragment of spongistatin 1 which utilized RCM to form the glycal **60** <02OL3719> (Scheme 44). Wang and co-workers reported the preparation of a wide variety of 2,6-dideoxysugars via RCM and subsequent reactions <02OL3875>. Lactone **61** is an immediate precursor to L-digitoxose.

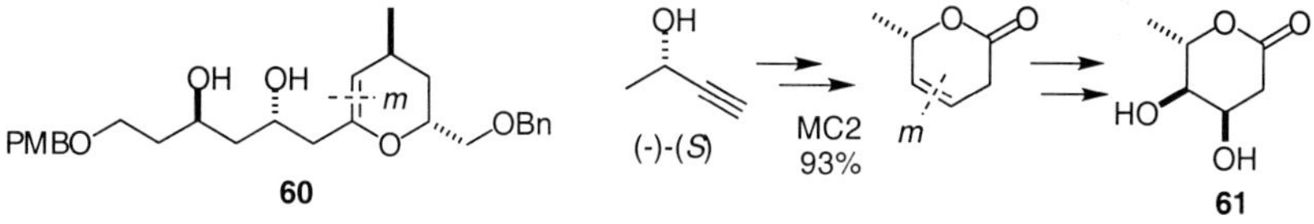

Scheme 44

The preparation of carbohydrate-derived spiroacetals by RCM was reported by van Boom and co-workers <00EJO873> (Scheme 45). The Pauson-Khand cyclization of these enynes was reported in the same manuscript.

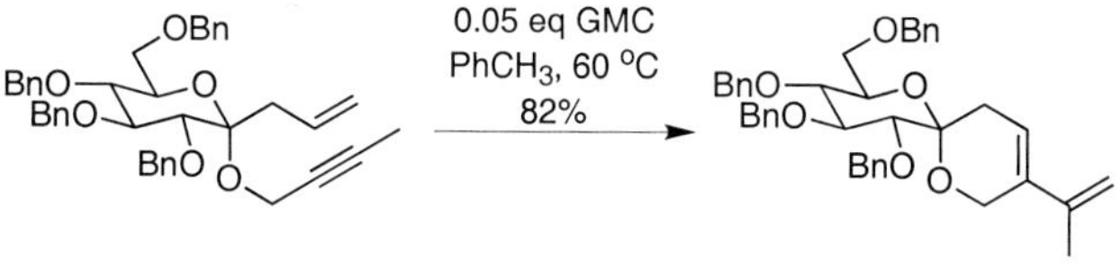

Scheme 45

1.5.4 Seven-Membered and Larger *O*-Containing Heterocycles

A general preparation of 2,5-dihydro[*b*]oxepins was reported by Wang (Scheme 46) <02H1997> in a process involving Claisen rearrangement, alkylation, and RCM. Related chemistry was employed to form the eight-membered cyclic ether **62** <02H2021>. Methodology that also paired the aromatic Claisen rearrangement with RCM was employed to give good yields of annulated coumarins **63** and **64** <02TL7781>.

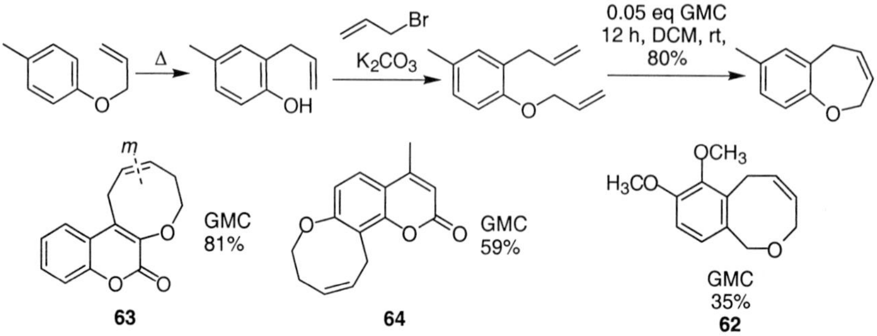

Scheme 46

The preparation of the unique spirocyclic ether **65** by double RCM was reported <02TL7851> (Scheme 47).

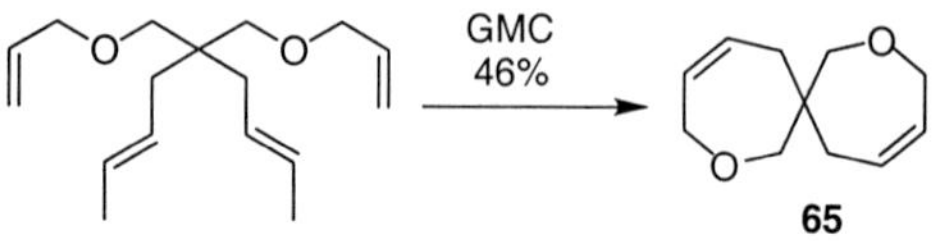

Scheme 47

Yamamoto and co-workers reported an exquisite approach to the CDEFG ring system of gambierol (a polycyclic ether natural product) via intramolecular allylation of an α-acetoxy ether followed by RCM <01JA6702> (Scheme 48).

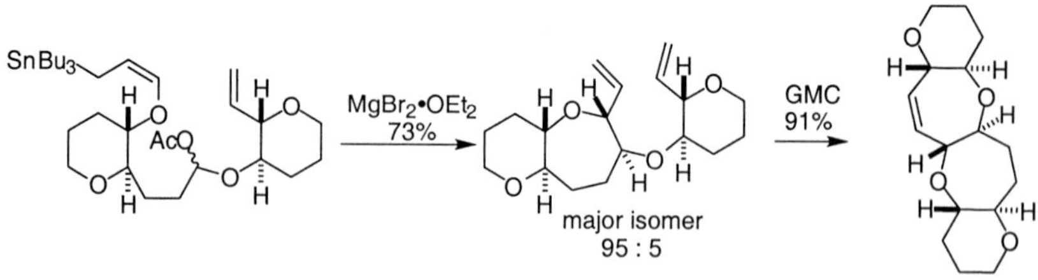

Scheme 48

Templates for ABE ring analogs of methyllycaconitine (**66**) were prepared by RCM <02TL6019> (Scheme 49). These were to be investigated as ligands for the α7-nicotinic acetylcholine receptor.

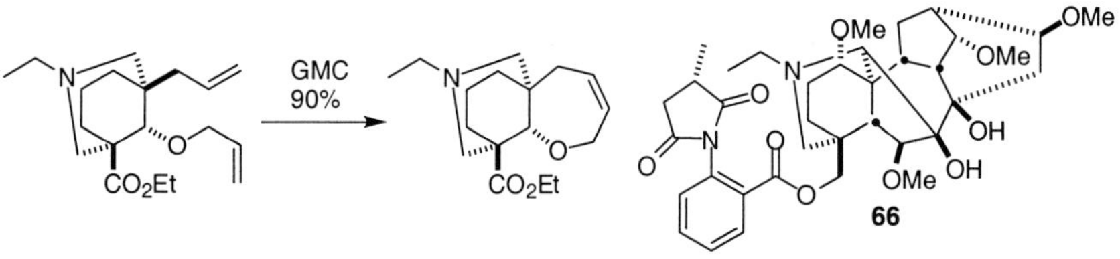

Scheme 49

The preparation of 2'-analogs of zoapatanol (**67**) from D-glucose was reported by van Boom and co-workers <01TL5749> (Scheme 50). These compounds may possess antifertility properties.

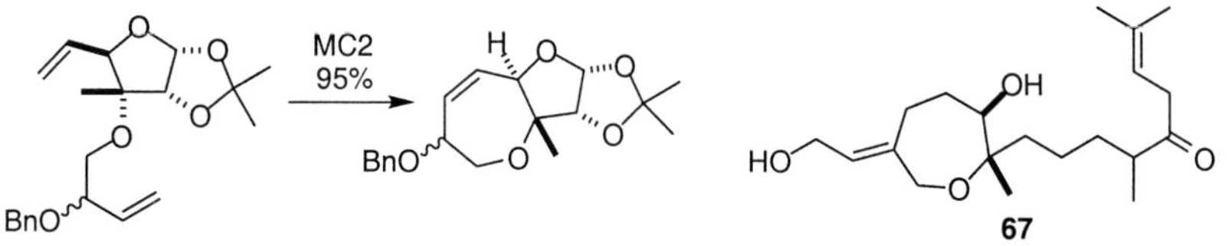

Scheme 50

1.5.5 Bridged Ring O-Heterocycles

The enantioselective preparation of both optical antipodes of oxabicyclo[4.2.1]-, -[5.2.1], and -[6.2.1]alkenes was reported by García-Tellado and co-workers <01EJO4423> (Scheme 51). The starting diene in these cases is derived in several steps from D-mannose; use of D-xylose as starting material allows the synthesis of the opposite enantiomers.

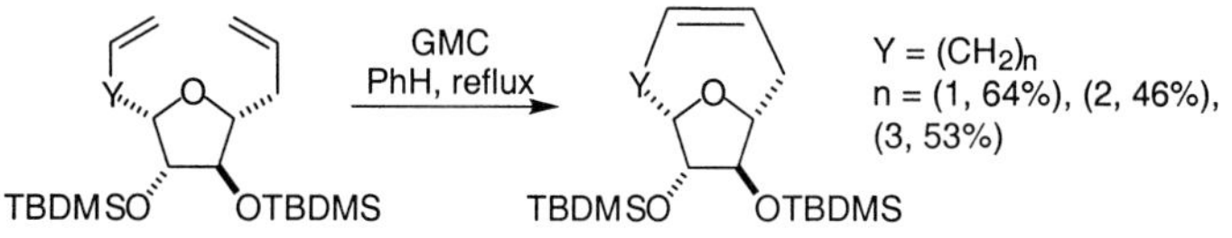

Y = (CH$_2$)$_n$
n = (1, 64%), (2, 46%), (3, 53%)

Scheme 51

The double allylation of **68** to give diene **69** was reported by Hanna and Michaut <00OL1141> (Scheme 52). In this exemplary case, RCM leads to excellent yields of the oxygen-bridged tricycle **70**.

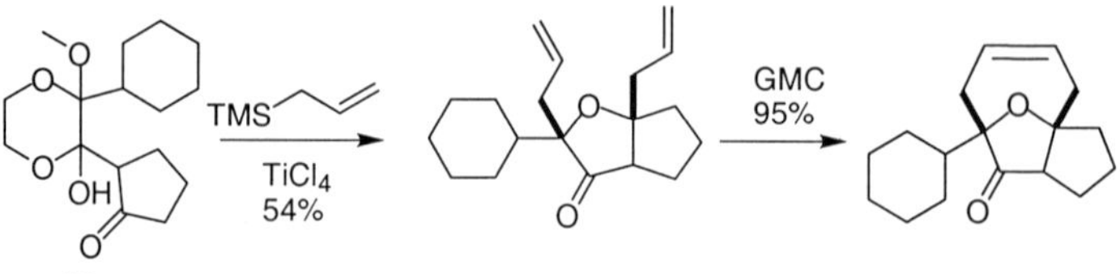

Scheme 52

1.6 BORON-, SILICON-, PHOSPHORUS- AND SULFUR-CONTAINING HETEROCYCLES

1.6.1 Boron-Containing Heterocycles

Ashe and co-workers reported the preparation of unique B-N and B-S heterocycles via RCM <00OL2089, 00OM4935> (Scheme 53).

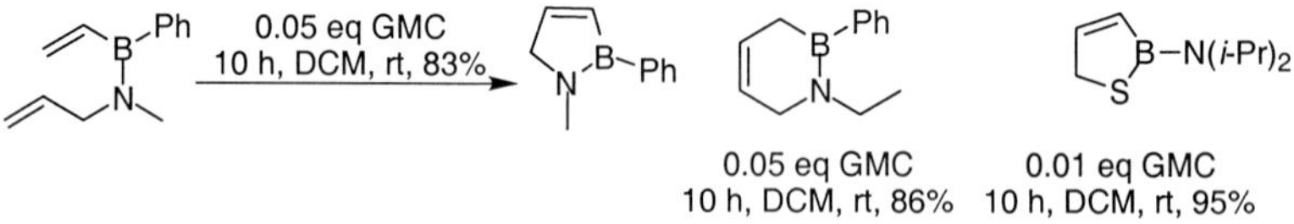

Scheme 53

1.6.2 Silicon-Containing Heterocycles

The synthesis of silicon-containing heterocycles was reported employing Si-C bond-forming <01JOM160> or aldehyde allylation <01TL581> reactions to prepare the metathesis substrates (Scheme 54).

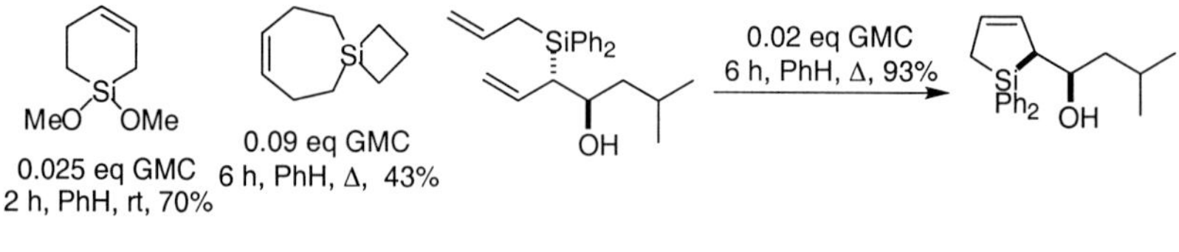

Scheme 54

Several approaches to what are formally Si-O and O-Si-O heterocycles have been reported, although these reactions invariably involve the use of the Si-O bond only as a temporary tether. The coupling and RCM of dissymmetric alcohols was reported by Eustache and co-workers <01TL239> and this method was applied in the synthesis of attenol A <02OL4105>. Denmark and Yang performed the RCM of Si-O tethered dienes in tandem with silicon-assisted cross-coupling to prepare acyclic unsaturated alcohols <01OL1749> and medium-sized ether rings <02JA2102>. It is noteworthy that efficient RCM of vinylsilyl ethers requires the use of the less sterically sensitive SMC. Postema and Piper have reported the RCM of O-Si-O tethered alkenes with olefinic monosaccharides <02TL7095> and Barrett and co-workers have shown the utility of silicon-tethered RCM in the synthesis of glycosphingolipids <00JOC6508>.

1.6.3 Sulfur-Containing Heterocycles

Sulfur-containing dienes were shown to be viable substrates for RCM by Mioskowski and co-workers <02OL1767> (Scheme 55).

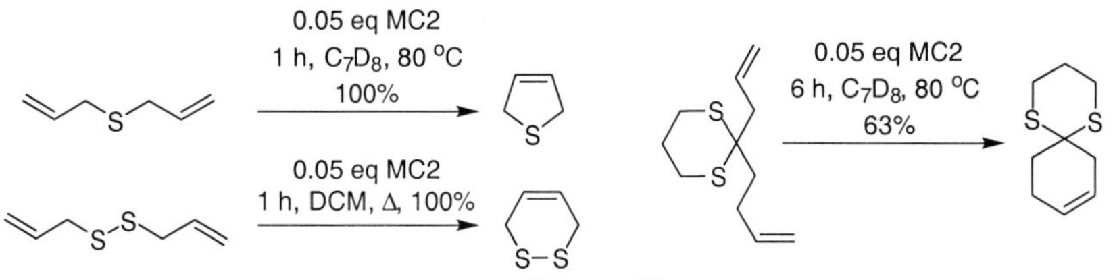

Scheme 55

Hanson and co-workers reported the preparation of cyclic α-thiophosphonates by RCM <01SL605> (Figure 5).

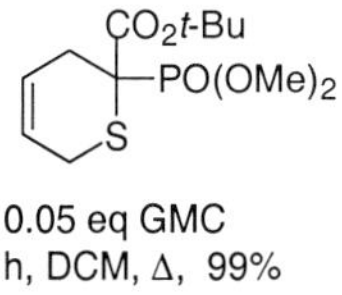

Figure 5

Cyclic sulfones have been prepared by RCM in what appears to be a general process <02OL427> (Scheme 56). These sulfones readily undergo the Ramberg-Bäcklund reaction to form cyclic dienes in excellent yields.

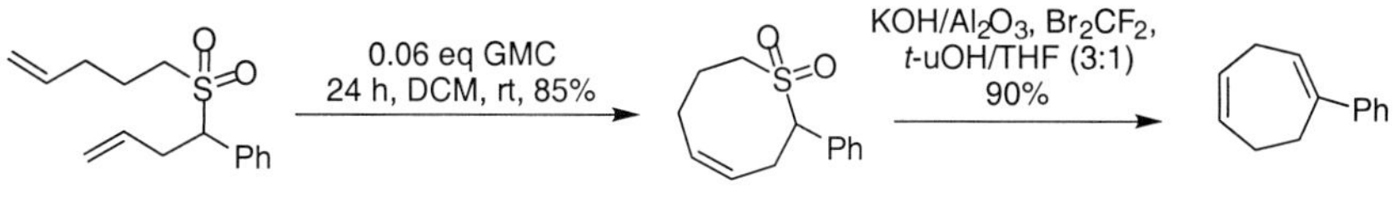

Scheme 56

1.6.4 Phosphorus-Containing Heterocycles

A wide variety of phosphorus-containing heterocycles were prepared using RCM. For example, Gouverneur and co-workers reported the synthesis of borane complexes of cyclic phosphanes <00AG(E)2491> (Figure 6) and P-O, P-N, and N-P-N heterocycles <00T2053> (Figure 7) employing this strategy.

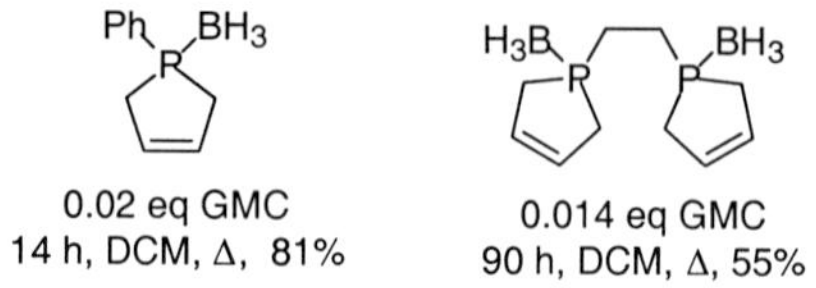

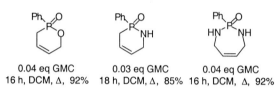

Figure 6

Figure 7

The preparation of P-O containing heterocycles from bis(diisopropylamino)vinylphosphine (**71**) was reported by van Boom and co-workers <00TL8635> (Figure 8).

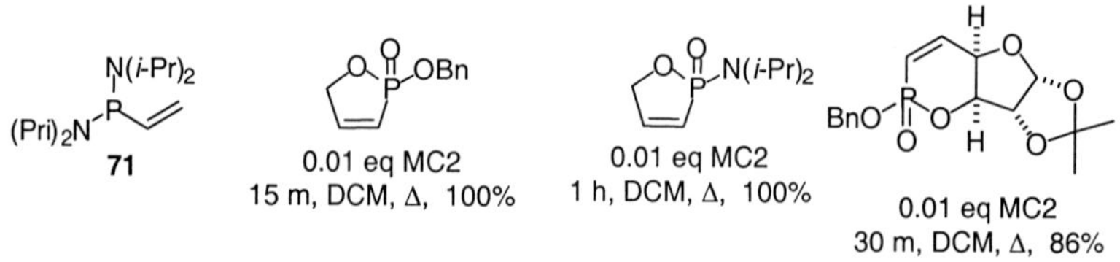

Figure 8

van Boom and co-workers also investigated the use of bis-(diisopropylamino)-ethynylphosphine as a linchpin for the preparation of metathesis substrates <01TL8231> (Scheme 57). These reactions lead to products of ene-yne (**72**, **74**) or tandem yne-diene (**73**) metathesis.

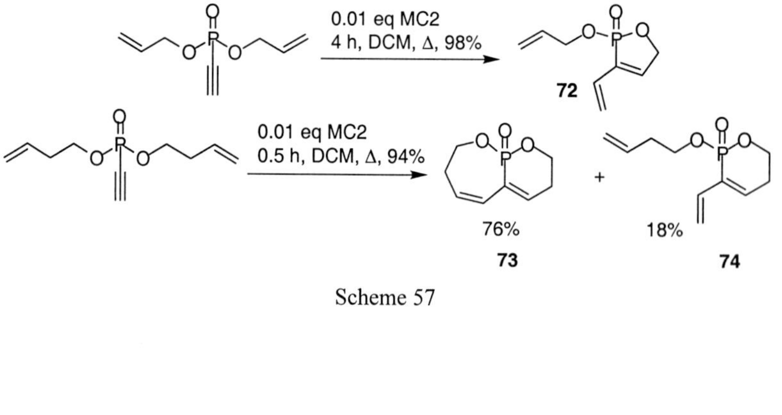

Scheme 57

Hanson and co-workers have reported extensively on the RCM of N-P-N containing dienes <01OL3939>, one example of this methodology being shown in Scheme 58. This process not only allows the preparation of *P*-heterocycles but also provides a unique entry into highly functionalized 1,4-diamines <02PS1807, 02OL4673, 00JOC7913, 00JOC4721, 00OL1769, 01S612>.

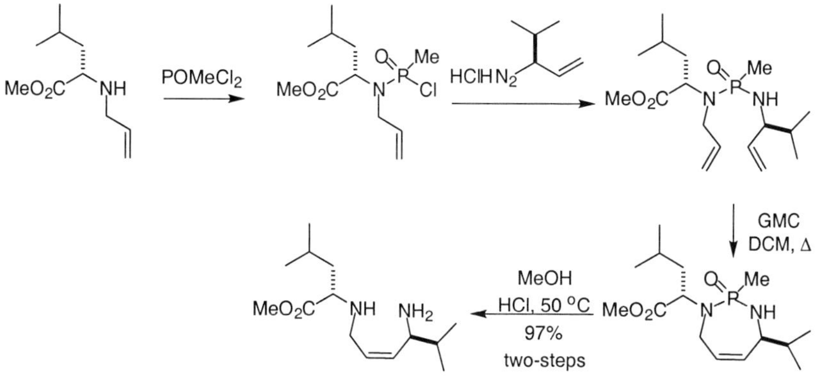

Scheme 58

Hanson and Stoianova have also reported the preparation of phosphonosugars by a strategy that incorporates RCM (Scheme 59) <01OL3285, 02PS1967>.

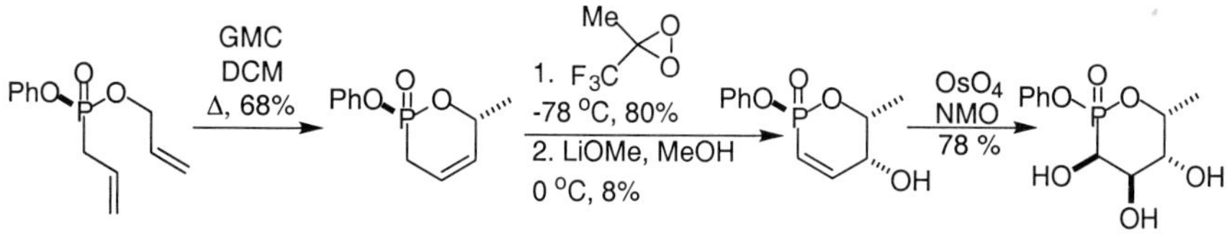

Scheme 59

1.7 A GENERAL APPROACH TO HETEROCYCLES

Snieckus reported a combination directed *ortho*-metalation (D*o*M)-RCM strategy for the synthesis of benzazepine, benzazocine, and benzannulated sulfonamide heterocycles <00SL1294> (Scheme 60). For example, the Boc-protected aniline (**75**) was sequentially allylated to give **76** which underwent RCM in excellent yield to give benzazepine **77**. Use of similar methodology led to **78** and **79** starting from *N*-methylbenzamide and *p*-tolylsulfonamide, respectively.

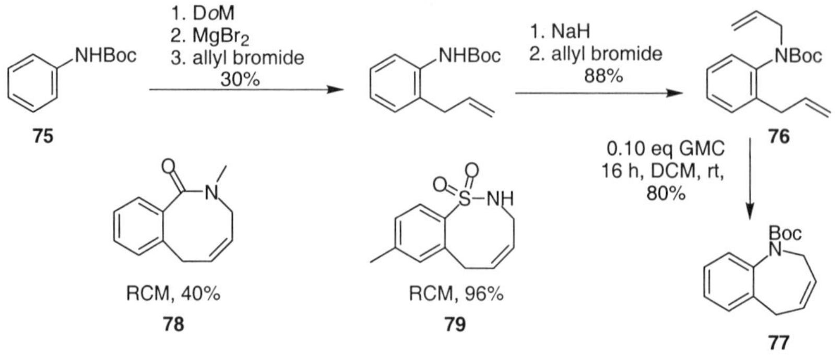

Scheme 60

1.8 NATURAL PRODUCTS

Given its ease of use and general toleration of many functional groups, RCM has had a major impact on the synthesis of heterocyclic natural products. The very scope of this impact limits the depth with which this subject can be covered, and the following review of this area is only cursory. Syntheses of several compounds that had been the object of RCM strategies prior to the period covered by this review were reported. For example, the total synthesis of *ent*-(-)-roseophilin <01JA8515>, manzamine A (the synthesis of which features a pair of sequential RCMs) <02JA8584>, epothilone 490 <02JA9825>, and woodrosin I (not shown) <02AG(E)2097> appeared in the literature (Figure 9). Syntheses of epothilone derivatives were also divulged <02OL4081, 02JOC7737> and a least one review of work in this area was published <00OS251>. Several approaches to polyether-containing natural products were reported: brevetoxin <00TL7673, 02JA3562>; ciguatoxin <02T1889, 02T10017, 02JOC3301, 01H93, 01SL952, 01TL6219, 02SL1496, 02T1835>. The total synthesis of ciguatoxin CTX3C was reported by Hirama and co-workers <01SCI1904, 02OL4551>.

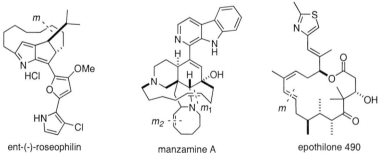

Figure 9

The microtubule stabilizing natural product laulimalide has been the target of several synthetic approaches <00TL6323, 00TL33, 00TL4705, 00TL2319, 01TL797, 01TL801, 01S2007, 01MI791, 01MI1179, 02TL213> and total syntheses <01JOC8973, 02JA13654, 02TL4841> that have featured RCM (Figure 10). Mulzer and co-workers recently reviewed this area as they reported their total synthesis of this compound <02MI573>. Several total syntheses <01SL1019, 01OL1817, 01MI5286, 02JA3245, 02T6455> and approaches to portions <00TL8569, 01CC255> of salicylihalamide A were reported.

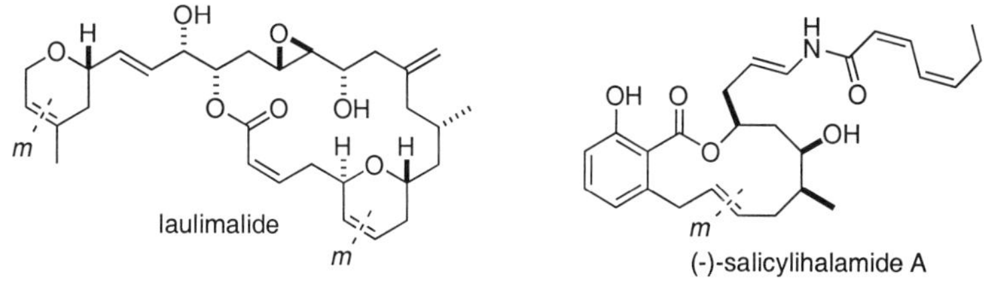

Figure 10

Total syntheses of griseoviridin <00AG(E)1664>, amphidinolide A (which features an RCM in the presence of five non-participating alkenes) <02OL2841>, and amphidinolide T4 <02AG(E)4763> were reported (Figure 11).

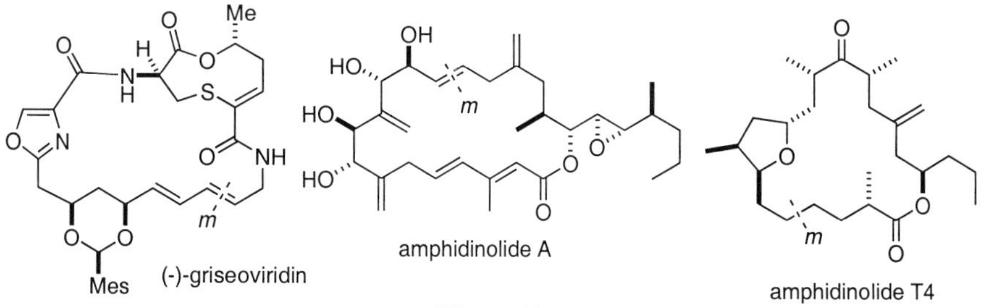

Figure 11

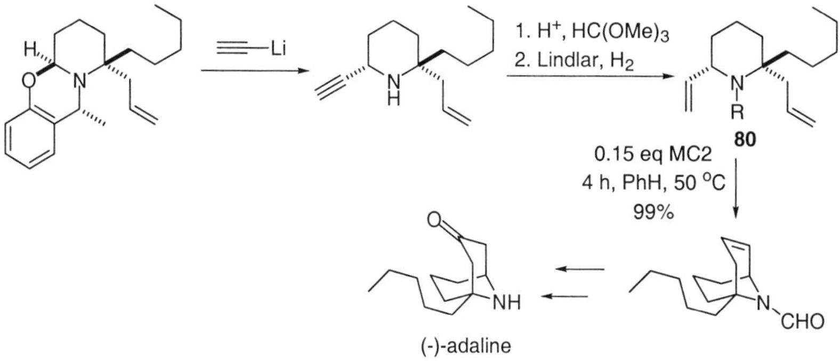

Scheme 61

The structural core of (-)-adaline was prepared by a lithium-activated S_N2-type alkynylation of an enantiomerically-enriched tricyclic *N,O*-acetal followed by reduction, *N*-formylation (**80**, R = CHO), and RCM using the MC2 <02OL2469> (Scheme 61). Yields were only slightly lower with the GMC. RCM fails with the HCl salt of **80** (R =H_2Cl), presumably because of a diequatorial arrangement of the 2,6-dialkenyl substituents in that derivative.

Wipf and co-workers reported the preparation of (-)-tuberostemonine utilizing RCM as a key step <02JA14848> (Scheme 62).

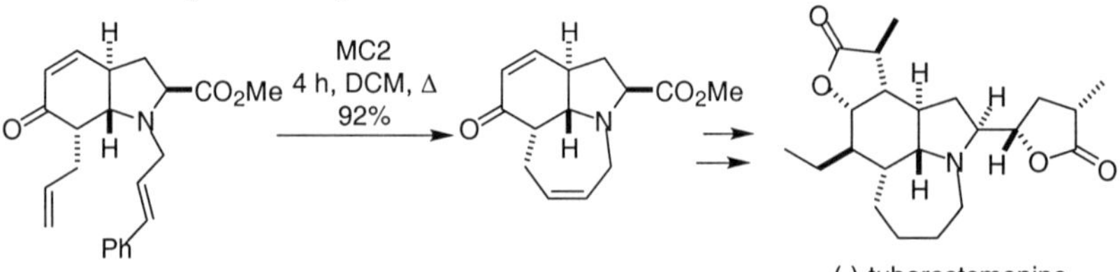

(-)-tuberostemonine

Scheme 62

The preparation of radicicol was reported by Danishefsky and co-workers (Scheme 63). It is noteworthy both that RCM takes place efficiently between one double bond of a diene and a vinyl epoxide and in the presence of a dithiane <01JA10903>.

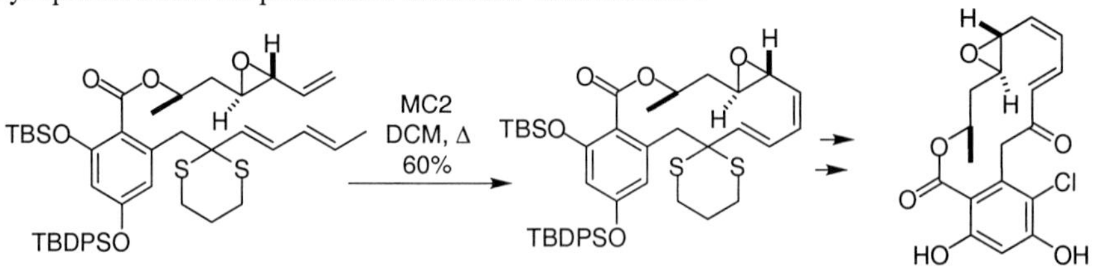

radicicol

Scheme 63

White and Hrnciar reported the total synthesis of (+)-australine via tandem RCM-transannular cyclization <00JOC9129> (Scheme 64).

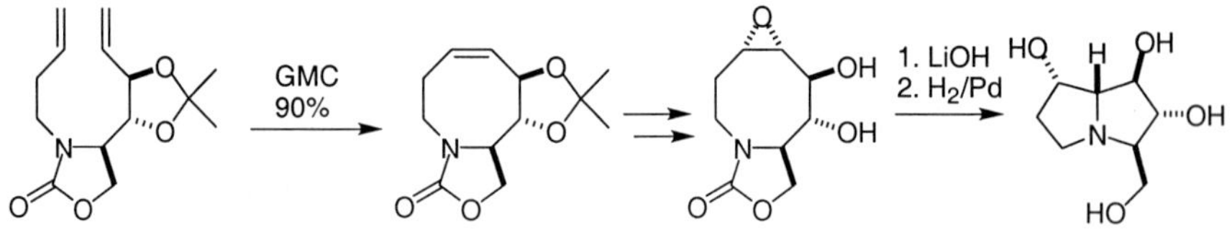

(+)-australine

Scheme 64

Burke and co-workers developed a desymmetrization RCM approach to the naturally occurring deaminated sialic acid KDN (Scheme 65) that was also applied to the preparation of *N*-

acetylneuraminic acid <02JOC8489>, after formal total syntheses of these compounds by the same process were reported <01OL237, 01TL8747>. The application of this methology to 1,7-dioxaspiro[5.5]undecanes was also published <02OL467>.

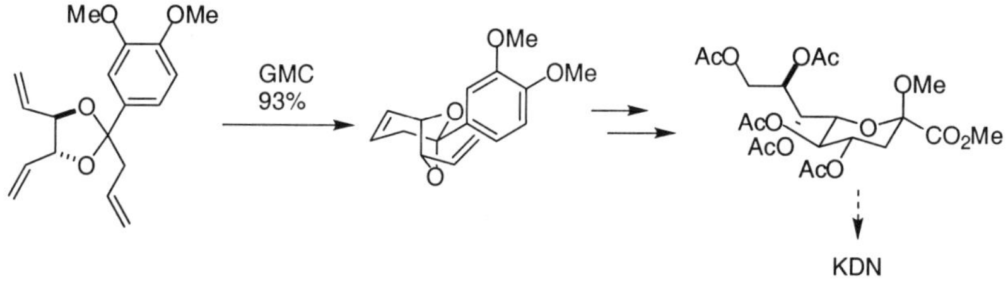

Scheme 65

Martin and co-workers reported the preparation of the indole alkaloid dihydrocorynantheol from indole-3-acetic acid in a process that featured two RCM <02OL3243> (Scheme 66).

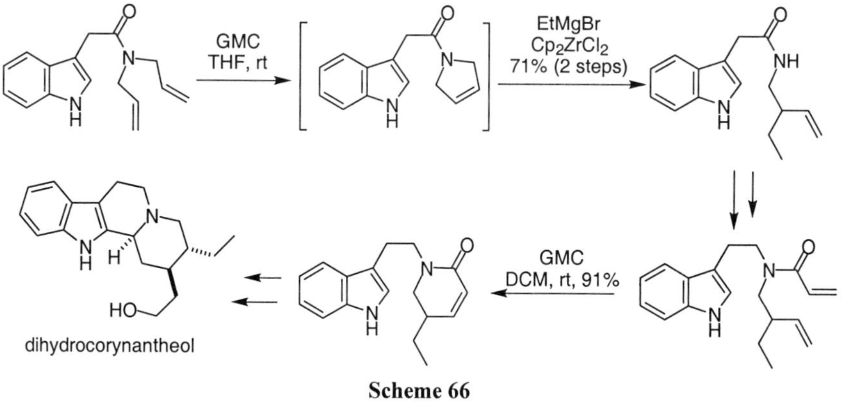

Scheme 66

The total synthesis of the novel fungal metabolite (±)-mycoepoxydiene was reported by Tadano and co-workers <02OL2941> (Scheme 67).

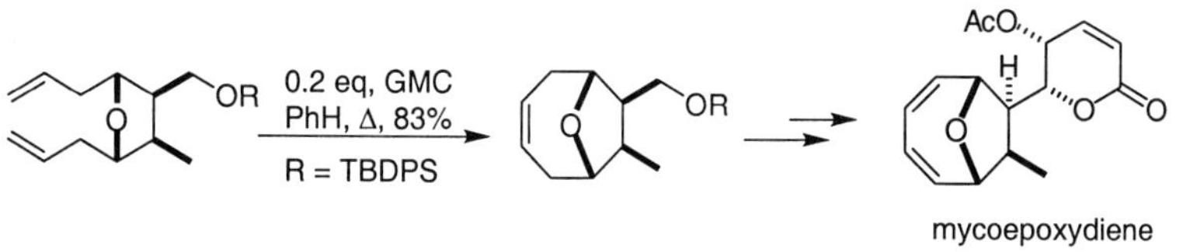

Scheme 67

Nicolaou and co-workers reported a divergent synthesis of coleophomone B and C via a common precursor <02AG(E)3276> (Scheme 68).

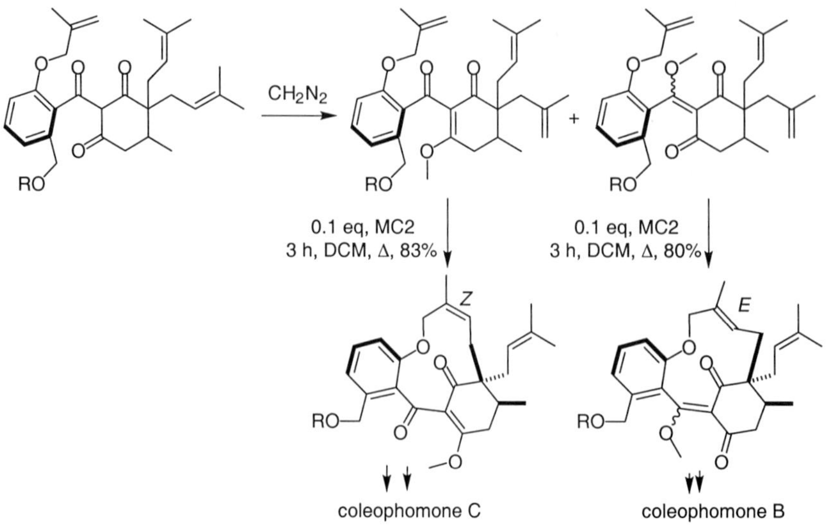

Scheme 68

In addition to the syntheses of these classes mentioned in previous sections, RCM was applied to the total synthesis of the pyrrolizidine alkaloid (-)-croalbinecine <00JOC9249> and the indolizidine alkaloid (-)-coniceine <01TA2621, 00CC1027> (Scheme 69).

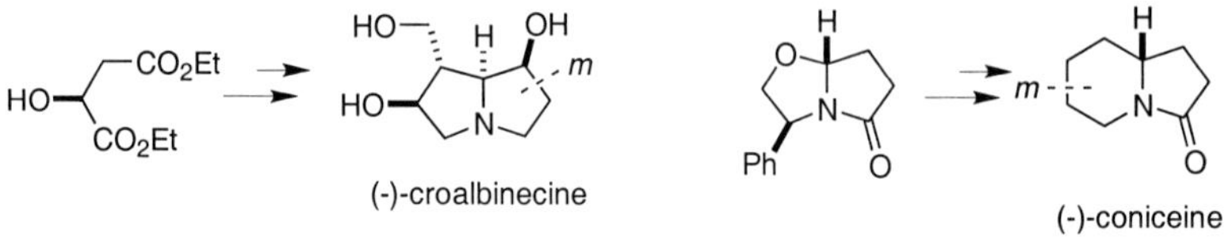

Scheme 69

Crimmins and co-workers applied their tactical approach to medium ring ethers <00S899> to the synthesis of prelaureatin <00JA5473> and (-)-isolaurallene <01JA1533, 02T1817> (Scheme 70). Prelaureatin was also prepared by Murai and co-workers using a similar strategy <02SL1493>.

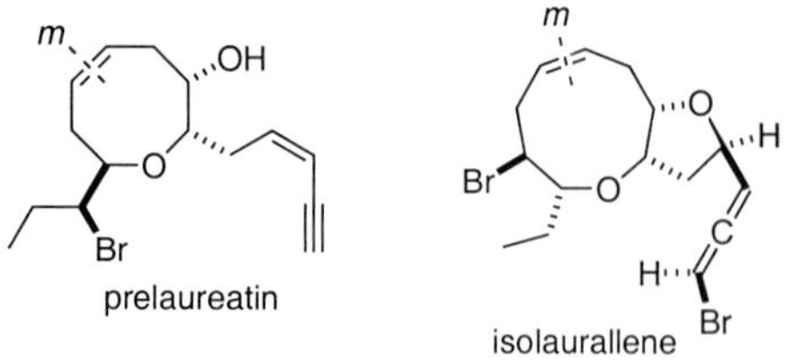

Scheme 70

The first total synthesis of microcarpalide was reported <02OL3447> as was the preparation of (-)-pinolidoxin, a potent modulator of plant pathogenesis <02OL3005> (Scheme 71).

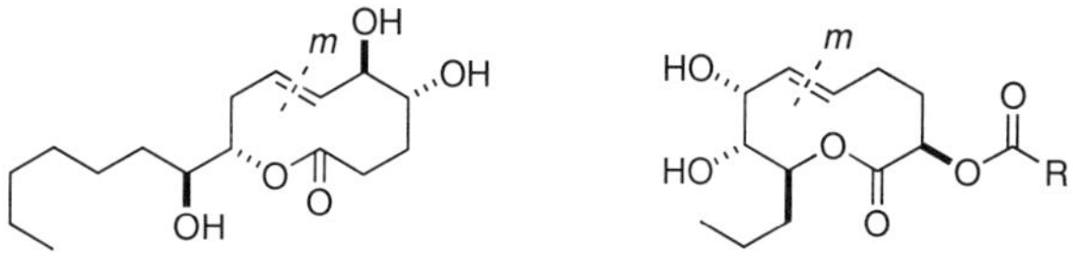

Scheme 71

microcarpalide

(-)-pinolidoxin

Ley and co-workers reported the total synthesis of (+)-aspicilin using 2,3-butane diacetal protected butane tetrols <01CJC1668> and Danishefsky and Gaul published the synthesis of the 14-membered macrolide core of migrastatin <02TL9039> (Scheme 72).

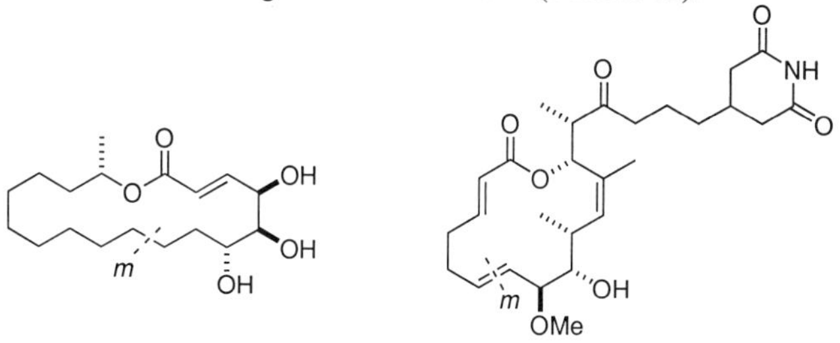

(+)-aspicilin

migrastatin

Scheme 72

1.9 CONCLUSIONS

Significant advances in the synthesis of heterocycles via ring closing metathesis have occurred in the past few years, thanks in large part to the development of new, robust catalysts, and to the concerted efforts of many research groups to exploit the synthetic utility of this reaction process. The influence of catalyst structure on the stereochemistry of cyclic alkenes formed by this reaction is certain to remain a lively area of research. A more detailed understanding of the subtle factors influencing this stereoselectivity should eventually, at least in large ring formation, enable the stereochemical control of a given RCM simply by appropriate catalyst choice. Additionally, given the diversity of products presented herein, it appears that the only limitation on this reaction with respect to the formation of heterocyclic rings may currently be the lack of convenient processes for the preparation of the desired starting materials. Overcoming this limitation will be another focus of further synthetic research in this area.

1.10 ACKNOWLEDGEMENTS

The Document Delivery Group of Pfizer Global Research and Development is thanked for their support in obtaining the articles required to write this review. The patience and forbearance of Stacy, Nicole and John is appreciated.

1.11 REFERENCES

94JA3414	R. O'Dell, D.H. McConville, G.E. Hofmeister, R.R. Schrock, *J. Am. Chem. Soc.* **1994**, *116*, 3414
94JOC2668	K. Takai, T. Kakiuchi, Y. Kataoka, K. Utimoto, *J. Org. Chem.* **1994**, *59*, 2668
95AG(E)1833	H.-G. Schmalz, *Angew. Chem., Int. Ed. Engl.* **1995**, *34*, 1833
96JA100	P. Schwab, R.H. Grubbs, J.W. Ziller, *J. Am. Chem. Soc.* **1996**, *118*, 100
97AG(E)2036	M. Schuster, S. Blechert, *Angew. Chem., Int. Ed. Engl.* **1997**, *36*, 2036
97S792	A. Fürstner, K. Langemann, *Synthesis* **1997**, 792
98JMOC29	M.L. Randall, M.L. Snapper, *J. Mol. Cat. A.* **1998**, *133*, 29
98MI275	M. Schuster, S. Blechert, *Transition Met. Org. Synth.* **1998**, *1*, 275
98T4413	R.H. Grubbs, S. Chang, *Tetrahedron* **1998**, *54*, 4413
99ACA75	A.J. Phillips, A.D. Abell, *Aldrichim. Acta* **1999**, *32*, 75
99JA2674	J. Huang, E.D. Stevens, S.P. Nolan, J.L. Petersen, *J. Am. Chem. Soc.* **1999**, *121*, 2674
99MI211	D.L. Wright, *Curr. Org. Chem.* **1999**, *3*, 211
99OL953	M. Scholl, S. Ding, C.W. Lee, R.H. Grubbs, *Org. Lett.* **1999**, *1*, 953
99TL2247	M. Scholl, T.M. Trnka, J.P. Morgan, R.H. Grubbs, *Tetrahedron Lett.* **1999**, 2247
00AG(E)1664	C.A. Dvorak, W.D. Schmitz, D.J. Poon, D.C. Pryde, J.P. Lawson, R.A. Amos, A.I. Meyers, *Angew. Chem. Int. Ed.* **2000**, *39*, 1664
00AG(E)2491	M. Schuman, M. Trevitt, A. Redd, V. Gouverneur, *Angew. Chem. Int. Ed* **2000**, *39*, 2491
00AG(E)3012	A. Fürstner, *Angew. Chem, Int. Ed.* **2000**, *39*, 3012
00AG(E)3101	J. Renaud, C.-D. Graf, L. Oberer, *Angew. Chem, Int. Ed.* **2000**, *39*, 3101
00CC699	K.C.M.F. Tjen, S.S. Kinderman, H. Hiemstra, F.P.J.T. Rutjes, H.E. Schoemaker, *Chem. Commun.*. **2000**, 699
00CC1027	M.D. Groaning, A.I. Meyers, *Chem. Commun.* **2000**, 1027
00CC1501	H. Ovaa, G.A. van der Marel, J.H. van Boom, R. Stragies, S. Blechert, *Chem. Commun.*. **2000**, 1501
00CC1771	J.C.A. Hunt, P. Laurent, C.J. Moody, *Chem. Commun.* **2000**, 1771
00CPB1593	M. Arisawa, M. Takahashi, E. Takezawa, T. Yamaguchi, Y. Torisawa, A. Nishida, M. Nakagawa, *Chem. Pharm. Bull.* **2000**, *48*, 1593
00EJO873	M.A. Leeuwenburgh, C.C.M. Appeldoorn, P.A.V. Van Hooft, H.S. Overkleeft, G.A. Van der Marel, J.H. Van Boom, *Eur. J. Org. Chem.* **2000**, 873
00JA5473	M.T. Crimmins, E.A. Tabet, *J. Am. Chem. Soc.* **2000**, *122*, 5473
00JOC4721	K.T. Sprott, P.R. Hanson, *J. Org. Chem.* **2000**, *65*, 4721
00JOC6508	A.G.M. Barrett, J.C. Beall, D.C. Braddock, K. Flack, V.C. Gibson, M.M. Salter, *J. Org. Chem.* **2000**, *65*, 6508
00JOC7913	K.T. Sprott, P.R. Hanson, *J. Org. Chem.* **2000**, *65*, 7913
00JOC9129	J.D. White, P. Hrnciar, *J. Org. Chem.* **2000**, *65*, 9129
00JOC9249	J.-B. Ahn, C.-S. Yun, K.H. Kim, D.-C. Ha, *J. Org. Chem.* **2000**, *65*, 9249
00MI58	V. Dragutan, I. Dragutan, A.T. Balaban, *Plat. Met. Rev.* **2000**, *44*, 58
00MI91	M. Karle, U. Koert, *Org. Syn. Highlights IV* **2000**, 91
00MI565	M. Jorgensen, P. Hadwiger, R. Madsen, A.E. Stutz, T.M. Wrodnigg, *Curr. Org. Chem.* **2000**, *4*, 565
00OL1141	I. Hanna, V. Michaut, *Org. Lett.* **2000**, *2*, 1141
00OL1769	D.S. Stoianova, P.R. Hanson, *Org. Lett.* **2000**, *2*, 1769
00OL1847	D.L. Wright, J.P. Schulte, II, M.A. Page, *Org. Lett.* **2000**, *2*, 1847
00OL2089	A.J. Ashe, III, X. Fang, *Org. Lett.* **2000**, *2*, 2089
00OL3971	U. Voigtmann, S. Blechert, *Org. Lett.* **2000**, *2*, 3971
00OM4935	A.J. Ashe, III, X. Fang, J.W. Kampf, *Organometallics* **2000**, *19*, 4935
00OS251	L.A. Wessjohann, G. Scheid, *Org. Syn. Highlights IV* **2000**, 251
00S893	U. Voigtmann, S. Blechert, *Synthesis* **2000**, 893
00S899	M.T. Crimmins, K.A. Emmitte, *Synthesis* **2000**, 899
00SI646	F.-X. Felpin, G. Vo-Thanh, R.J. Robins, J. Villieras, J. Lebreton, *Synlett* **2000**, 1646
00SL319	M. Lennartz, E. Steckhan, *Synlett* **2000**, 319
00SL1294	C. Lane, V. Snieckus, *Synlett* **2000**, 1294
00T2053	L. Hetherington, B. Greedy, V. Gouverneur, *Tetrahedron* **2000**, *56*, 2053
00TA753	C.M. Schuch, R.A. Pilli, *Tetrahedron: Asymmetry* **2000**, *11*, 753

00TA4639	C. Agami, F. Couty, G. Evano, *Tetrahedron: Asymmetry* **2000**, *11*, 4639
00TL33	J. Mulzer, M. Hanbauer, *Tetrahedron Lett.* **2000**, *41*, 33
00TL2319	A.K. Ghosh, Y. Wang, *Tetrahedron Lett.* **2000**, *41*, 2319
00TL3967	P. Evans, R. Grigg, M. York, *Tetrahedron Lett.* **2000**, *41*, 3967
00TL4113	C. Agami, F. Couty, N. Rabasso, *Tetrahedron Lett.* **2000**, *41*, 4113
00TL4705	A.K. Ghosh, Y. Wang, *Tetrahedron Lett.* **2000**, *41*, 4705
00TL5979	R. Doodeman, F.P.J.T. Rutjes, H. Hiemstra, *Tetrahedron Lett.* **2000**, *41*, 5979
00TL6323	E.K. Dorling, E. Ohler, J. Mulzer, *Tetrahedron Lett.* **2000**, *41*, 6323
00TL7255	R. Grigg, M. York, *Tetrahedron Lett.* **2000**, *41*, 7255
00TL7673	G. Matsuo, H. Matsukura, N. Hori, T. Nakata, *Tetrahedron Lett.* **2000**, *41*, 7673
00TL7843	A.O. Maldaner, R.A. Pilli, *Tetrahedron Lett.* **2000**, *41*, 7843
00TL8157	R. Kumareswaran, T. Balasubramanian, A. Hassner, *Tetrahedron Lett.* **2000**, *41*, 8157
00TL8569	J.T. Feutrill, G.A. Holloway, F. Hilli, H.M. Hugel, M.A. Rizzacasa, *Tetrahedron Lett.* **2000**, *41*, 8569
00TL8635	M.S.M. Timmer, H. Ovaa, D.V. Filippov, G.A. Van der Marel, J.H. Van Boom, *Tetrahedron Lett.* **2000**, *41*, 8635
00TL9553	G.S. Weatherhead, J.H. Houser, J.G. Ford, J.Y. Jamieson, R.R. Schrock, A.H. Hoveyda, *Tetrahedron Lett.* **2000**, *41*, 9553
01ACR18	T.M. Trnka, R.H. Grubbs, *Acc. Chem. Res.* **2001**, *34*, 18
01AG(E)4441	K.C. Nicolaou, J.A. Vega, G. Vassilikogiannakis, *Angew. Chem., Int. Ed.* **2001**, *40*, 4441
01CC255	G.I. Georg, Y.M. Ahn, B. Blackman, F. Farokhi, P.T. Flaherty, C.J. Mossman, S. Roy, K. Yang, *Chem. Commun* **2001**, 255
01CJC1668	D.J. Dixon, A.C. Foster, S.V. Ley, *Can. J. Chem.* **2001**, *79*, 1668
01EJO2841	J. Cossy, I. Pevet, C. Meyer, *Eur. J. Org. Chem.* **2001**, 2841
01EJO4423	P. De Armas, F. Garcia-Tellado, J.J. Marrero-Tellado, *Eur. J. Org. Chem.* **2001**, 4423
01H93	M. Maruyama, K. Maeda, T. Oishi, H. Oguri, M. Hirama, *Heterocycles* **2001**, *54*, 93
01JA1533	M.T. Crimmins, K.A. Emmitte, *J. Am. Chem. Soc.* **2001**, *123*, 1533
01JA6702	I. Kadota, A. Ohno, K. Matsuda, Y. Yamamoto, *J. Am. Chem. Soc.* **2001**, *123*, 6702
01JA6991	S.J. Dolman, E.S. Sattely, A.H. Hoveyda, R.R. Schrock, *J. Am. Chem. Soc.* **2002**, *124*, 6991
01JA8515	D.L. Boger, J. Hong, *J. Am. Chem. Soc.* **2001**, *123*, 8515
01JA10903	R.M. Garbaccio, S.J. Stachel, D.K. Baeschlin, S.J. Danishefsky, *J. Am. Chem. Soc.* **2001**, *123*, 10903
01JOC8973	A.K. Ghosh, Y. Wang, J.T. Kim, *J. Org. Chem.* **2001**, *66*, 8973
01JOC9056	S.H. Lim, S. Ma, P. Beak, *J. Org. Chem.* **2001**, *66*, 9056
01JOM160	I. Ahmad, M.L. Falck-Pedersen, K. Undheim, *J. Organometallic Chem.* **2001**, *625*, 160
01JOM359	K. Pachamuthu, Y.D. Vankar, *J. Organometallic Chem.* **2001**, *624*, 359
01MI155	V. Dragutan, I. Dragutan, A.T. Balaban, *Plat. Met. Rev.* **2001**, *45*, 155
01MI791	H.W. Lee, C.-S. Jeong, S.H. Yoon, I.-Y.C. Lee, *Bull. Korean. Chem. Soc.* **2001**, *22*, 791
01MI945	A.H. Hoveyda, R.R. Schrock, *Chemistry* **2001**, *7*, 945
01MI1179	H.W. Lee, S.H. Yoon, I.-Y.C. Lee, B.Y. Chung, *Bull. Korean. Chem. Soc.* **2001**, *22*, 1179
01MI4811	A. Fürstner, O. Guth, A. Duffels, G. Seidel, M. Liebl, B. Gabor, R. Mynott, *Eur. J. Chem.* **2001**, *7*, 4811
01MI5286	A. Fürstner, T. Dierkes, O.R. Thiel, G. Blanda, *Eur. J. Chem.* **2001**, *7*, 5286
01OL237	S.D. Burke, E.A. Voight, *Org. Lett.* **2001**, *3*, 237
01OL1161	T. Kitamura, M. Mori, *Org. Lett.* **2001**, *3*, 1161
01OL1327	W.H. Pearson, A. Aponick, *Org. Lett.* **2001**, *3*, 1327
01OL1749	S.E. Denmark, S.-M. Yang, *Org. Lett.* **2001**, *3*, 1749
01OL1817	B.B. Snider, F. Song, *Org. Lett.* **2001**, *3*, 1817
01OL2045	S.S. Kinderman, J.H. van Maarseveen, H.E. Schoemaker, H. Hiemstra, F.P.J.T. Rutjes, *Org. Lett.* **2001**, *3*, 2045
01OL3285	D.S. Stoianova, P.R. Hanson, *Org. Lett.* **2001**, *3*, 3285
01OL3939	K.T. Sprott, M.D. McReynolds, P.R. Hanson, *Org. Lett.* **2001**, *3*, 3939
01S612	K.T. Sprott, M.D. McReynolds, P.R. Hanson, *Synthesis* **2001**, 612
01S2007	A. Ahmed, E. Ohler, J. Mulzer, *Synthesis* **2001**, 2007
01SCI1904	M. Hirama, T. Oishi, H. Uehara, M. Inoue, M. Maruyama, H. Oguri, M. Satake, *Science* **2001**, *294*, 1904

01SL37	G. Vo-Thanh, V. Boucard, H. Sauriat-Dorizon, F. Guibe, *Synlett* **2001**, 37
01SL357	D.J. Wallace, P.G. Bulger, D.J. Kennedy, M.S. Ashwood, I.F. Cottrell, U.-H. Dolling, *Synlett* **2001**, 357
01SL605	J.D. Moore, K.T. Sprott, P.R. Hanson, *Synlett* **2001**, 605
01SL952	T. Oishi, S.-i. Tanaka, Y. Ogasawara, K. Maeda, H. Oguri, M. Hirama, *Synlett* **2001**, 952
01SL1019	A.B. Smith, III, J. Zheng, *Synlett* **2001**, 1019
01T5393	C. Agami, F. Couty, N. Rabasso, *Tetrahedron* **2001**, *57*, 5393
01TA817	Y. Banba, C. Abe, H. Nemoto, A. Kato, I. Adachi, H. Takahata, *Tetrahedron: Asymmetry* **2001**, *12*, 817
01TA2621	S.H. Park, H.J. Kang, S. Ko, S. Park, S. Chang, *Tetrahedron: Asymmetry* **2001**, *12*, 2621
01TL179	J.D. Rainier, J.M. Cox, S.P. Allwein, *Tetrahedron Lett.* **2001**, *42*, 179
01TL239	J.G. Boiteau, P. Van de Weghe, J. Eustache, *Tetrahedron Lett.* **2001**, *42*, 239
01TL581	Y. Landais, S.S. Surange, *Tetrahedron Lett.* **2001**, *42*, 581
01TL797	G.T. Nadolski, B.S. Davidson, *Tetrahedron Lett.* **2001**, *42*, 797
01TL801	B.T. Messenger, B.S. Davidson, *Tetrahedron Lett.* **2001**, *42*, 801
01TL1209	M. Sabat, C.R. Johnson, *Tetrahedron Lett.* **2001**, *42*, 1209
01TL4079	T. Subramanian, C.C. Lin, *Tetrahedron Lett.* **2001**, *42*, 4079
01TL4633	C. Agami, F. Couty, N. Rabasso, *Tetrahedron Lett.* **2001**, *42*, 4633
01TL5245	A. Ruckert, D. Eisele, S. Blechert, *Tetrahedron Lett.* **2001**, *42*, 5245
01TL5605	C.F. Klitzke, R.A. Pilli, *Tetrahedron Lett.* **2001**, *42*, 5605
01TL5749	H. Ovaa, G.A. van der Marel, J.H. van Boom, *Tetrahedron Lett.* **2001**, *42*, 5749
01TL6219	H. Imai, H. Uehara, M. Inoue, H. Oguri, T. Oishi, M. Hirama, *Tetrahedron Lett.* **2001**, *42*, 6219
01TL8029	M. Arisawa, C. Theeraladanon, A. Nishida, M. Nakagawa, *Tetrahedron Lett.* **2001**, *42*, 8029
01TL8231	M.S.M. Timmer, H. Ovaa, D.V. Filippov, G.A. van der Marel, J.H. van Boom, *Tetrahedron Lett.* **2001**, *42*, 8231
01TL8673	H.A. Dondas, G. Balme, B. Clique, R. Grigg, A. Hodgeson, J. Morris, V. Sridharan, *Tetrahedron Lett.* **2001**, *42*, 8673
01TL8747	E.A. Voight, C. Rein, S.D. Burke, *Tetrahedron Lett.* **2001**, *42*, 8747
02AG(E)2097	A. Fürstner, F. Jeanjean, P. Razon, *Angew. Chem. Int. Ed.* **2002**, *41*, 2097
02AG(E)3276	K.C. Nicolaou, G. Vassilikogiannakis, T. Montagnon, *Angew. Chem. Int. Ed.* **2002**, *41*, 3276
02AG(E)4560	L. Usher, M. Estrella-Jimenez, I. Ghiviriga, D.L. Wright, *Angew. Chem. Int. Ed.* **2002**, *41*, 4560
02AG(E)4732	M. Arisawa, T. Yukiyoshi, M. Nakagawa, A. Nishida, *Angew. Chem. Int. Ed.* **2002**, *41*, 4732
02AG(E)4763	A. Fürstner, C. Aissa, R. Riveiros, J. Ragot, *Angew. Chem. Int. Ed.* **2002**, *41*, 4763
02BMC3905	U. Kazmaier, S. Pahler, R. Endermann, D. Habich, H.-P. Kroll, B. Riedl, *Bioorg. Med. Chem.* **2002**, *10*, 3905
02BMCL1993	A.K. Ghosh, L.M. Swanson, C. Liu, K.A. Hussain, H. Cho, D.E. Walters, L. Holland, J. Buthod, *Bioorg. Med. Chem. Lett.* **2002**, *12*, 1993
02CC1542	A.S. Edwards, R.A.J. Wybrow, C. Johnstone, H. Adams, J.P.A. Harrity, *Chem. Commun.* **2002**, 1542
02CC2182	A. Fürstner, K. Radkowski, *Chem. Commun.* **2002**, 2182
02CEN(51)29	A.M. Rouhi, *Chem. Eng. News* **2002**, *80*, 29
02CEN(51)34	A.M. Rouhi, *Chem. Eng. News* **2002**, *80*, 34
02H1997	E.-C. Wang, M.-K. Hsu, Y.-L. Lin, K.-S. Huang, *Heterocycles* **2002**, *57*, 1997
02H2021	E.-C. Wang, C.-C. Wang, M.-K. Hsu, K.-S. Huang, *Heterocycles* **2002**, *57*, 2021
02HCA3033	L.A. Paquette, K. Basu, J.C. Eppich, J.E. Hofferberth, *Helv. Chim. Acta* **2002**, *85*, 3033
02JA2102	S.E. Denmark, S.-M. Yang, *J. Am. Chem. Soc.* **2002**, *124*, 2102
02JA2868	A.F. Kiely, J.A. Jernelius, R.R. Schrock, A.H. Hoveyda, *J. Am. Chem. Soc.* **2002**, *124*, 2868
02JA3245	Y. Wu, X. Liao, R. Wang, X.-S. Xie, J.K. De Brabander, *J. Am. Chem. Soc.* **2002**, *124*, 3245
02JA3562	I. Kadota, A. Ohno, K. Matsuda, Y. Yamamoto, *J. Am. Chem. Soc.* **2002**, *124*, 3562
02JA7061	A. Fürstner, K. Radkowski, C. Wirtz, R. Goddard, C.W. Lehmann, R. Mynott, *J. Am. Chem. Soc.* **2002**, *124*, 7061
02JA8584	J.M. Humphrey, Y. Liao, A. Ali, T. Rein, Y.-L. Wong, H.-J. Chen, A.K. Courtney, S.F. Martin, *J. Am. Chem. Soc.* **2002**, *124*, 8584
02JA9825	K. Biswas, H. Lin, J.T. Njardarson, M.D. Chappell, T.-C. Chou, Y. Guan, W.P. Tong, L. He, S.B. Horwitz, S.J. Danishefsky, *J. Am. Chem. Soc.* **2002**, *124*, 9825
02JA10779	X. Teng, D.R. Cefalo, R.R. Schrock, A.H. Hoveyda, *J. Am. Chem. Soc.* **2002**, *124*, 10779

02JA13390	A.E. Sutton, B.A. Seigal, D.F. Finnegan, M.L. Snapper, *J. Am. Chem. Soc.* **2002**, *124*, 13390
02JA13654	S.G. Nelson, W.S. Cheung, A.J. Kassick, M.A. Hilfiker, *J. Am. Chem. Soc.* **2002**, *124*, 13654
02JA14848	P. Wipf, S.R. Rector, H. Takahashi, *J. Am. Chem. Soc.* **2002**, *124*, 14848
02JCO125	M.A. Walters, F. La, P. Deshmukh, D.O. Omecinsky, *J. Comb. Chem.* **2002**, *4*, 125
02JCS(P1)2378	J.C.A. Hunt, P. Laurent, C.J. Moody, *J. Chem. Soc., Perkin Trans. 1* **2002**, 2378
02JOC1982	J. Cossy, C. Willis, V. Bellosta, S. BouzBouz, *J. Org. Chem.* **2002**, *67*, 1982
02JOC3301	M. Sasaki, T. Noguchi, K. Tachibana, *J. Org. Chem.* **2002**, *67*, 3301
02JOC4325	N. Buschmann, A. Rueckert, S. Blechert, *J. Org. Chem.* **2002**, *67*, 4325
02JOC6456	C. Stapper, S. Blechert, *J. Org. Chem.* **2002**, *67*, 6456
02JOC6896	R. Martin, M. Alcon, M.A. Pericas, A. Riera, *J. Org. Chem.* **2002**, *67*, 6896
02JOC7737	A. Rivkin, J.T. Njardarson, K. Biswas, T.-C. Chou, S.J. Danishefsky, *J. Org. Chem.* **2002**, *67*, 7737
02JOC7774	B. Lindsay Karl, G. Pyne Stephen, *J. Org. Chem.* **2002**, *67*, 7774
02JOC8489	E.A. Voight, C. Rein, S.D. Burke, *J. Org. Chem.* **2002**, *67*, 8489
02MI1	K.J. Ivin, *NATO Science Series, II: Mathematics, Physics and Chemistry* **2002**, *56*, 1
02MI573	J. Mulzer, E. Ohler, V.S. Enev, M. Hanbauer, *Adv. Synth. Catal.* **2002**, *344*, 573
02MI615	K. Basu, J.C. Eppich, L.A. Paquette, *Adv. Synth. Catal.* **2002**, *344*, 615
02MI631	S. Randl, N. Lucas, S.J. Connon, S. Blechert, *Adv. Synth. Catal.* **2002**, *344*, 631
02MI657	A. Fürstner, M. Schlede, *Adv. Synth. Catal.* **2002**, *344*, 657
02MI678	T. Kitamura, Y. Sato, M. Mori, *Adv. Synth. Catal.* **2002**, *344*, 678
02MI736	S.S. Kinderman, R. Doodeman, J.W. Van Beijma, J.C. Russcher, K.C.M.F. Tjen, T.M. Kooistra, H. Mohaselzadeh, J.H. Van Maarseveen, H. Hiemstra, H.E. Schoemaker, F.P.J.T. Rutjes, *Adv. Synth. Catal.* **2002**, *344*, 736
02OL427	Q. Yao, *Org. Lett.* **2002**, *4*, 427
02OL467	V.A. Keller, J.R. Martinelli, E.R. Streiter, S.D. Burke, *Org. Lett.* **2002**, *4*, 467
02OL639	S. Ma, B. Ni, *Org. Lett.* **2002**, *4*, 639
02OL803	N. Saito, Y. Sato, M. Mori, *Org. Lett.* **2002**, *4*, 803
02OL1767	G. Spagnol, M.-P. Heck, S.P. Nolan, C. Mioskowski, *Org. Lett.* **2002**, *4*, 1767
02OL1971	J. Barluenga, C. Mateos, F. Aznar, C. Valdes, *Org. Lett.* **2002**, *4*, 1971
02OL2029	J.M. Harris, A. Padwa, *Org. Lett.* **2002**, *4*, 2029
02OL2469	T. Itoh, N. Yamazaki, C. Kibayashi, *Org. Lett.* **2002**, *4*, 2469
02OL2841	R.E. Maleczka, Jr., L.R. Terrell, F. Geng, J.S. Ward, III, *Org. Lett.* **2002**, *4*, 2841
02OL2941	K.-i. Takao, G. Watanabe, H. Yasui, K.-i. Tadano, *Org. Lett.* **2002**, *4*, 2941
02OL3005	D. Liu, S.A. Kozmin, *Org. Lett.* **2002**, *4*, 3005
02OL3243	A. Deiters, S.F. Martin, *Org. Lett.* **2002**, *4*, 3243
02OL3359	H. Kogen, N. Toda, K. Tago, S. Marumoto, K. Takami, O. Mayuko, N. Yamada, K. Koyama, S. Naruto, K. Abe, R. Yamazaki, T. Hara, A. Aoyagi, Y. Abe, T. Kaneko, *Org. Lett.* **2002**, *4*, 3359
02OL3447	J. Murga, E. Falomir, J. Garcia-Fortanet, M. Carda, J.A. Marco, *Org. Lett.* **2002**, *4*, 3447
02OL3719	E.B. Holson, W.R. Roush, *Org. Lett.* **2002**, *4*, 3719
02OL3875	P.R. Andreana, J.S. McLellan, Y. Chen, P.G. Wang, *Org. Lett.* **2002**, *4*, 3875
02OL4081	A. Rivkin, K. Biswas, T.-C. Chou, S.J. Danishefsky, *Org. Lett.* **2002**, *4*, 4081
02OL4105	P. Van de Weghe, D. Aoun, J.G. Boiteau, J. Eustache, *Org. Lett.* **2002**, *4*, 4105
02OL4499	X.E. Hu, N.K. Kim, B. Ledoussal, *Org. Lett.* **2002**, *4*, 4499
02OL4551	M. Inoue, H. Uehara, M. Maruyama, M. Hirama, *Org. Lett.* **2002**, *4*, 4551
02OL4673	M.D. McReynolds, K.T. Sprott, P.R. Hanson, *Org. Lett.* **2002**, *4*, 4673
02OM331	A. Fürstner, O.R. Thiel, C.W. Lehmann, *Organometallics* **2002**, *21*, 331
02PS1807	K.T. Sprott, M.D. McReynolds, P.R. Hanson, *Phosphorus, Sulfur Silicon Relat. Elem.* **2002**, *177*, 1807
02PS1967	D.S. Stoianova, P.R. Hanson, *Phosphorus, Sulfur Silicon Relat. Elem.* **2002**, *177*, 1967
02SL731	K.B. Lindsay, M. Tang, S.G. Pyne, *Synlett* **2002**, 731
02SL1146	S.G. Davies, K. Iwamoto, C.A.P. Smethurst, A.D. Smith, H. Rodriguez-Solla, *Synlett* **2002**, 1146
02SL1302	T. Bach, A. Lemarchand, *Synlett* **2002**, 1302
02SL1493	K. Fujiwaraa, S.-I. Souma, H. Mishima, A. Murai, *Synlett* **2002**, 1493
02SL1496	K. Fujiwara, Y. Koyama, E. Doi, K. Shimawaki, Y. Ohtaniuchi, A. Takemura, S.-I. Souma, A. Murai, *Synlett* **2002**, 1496
02SL1724	A.W. Baltrusch, F. Bracher, *Synlett* **2002**, 1724

02SL1889	N. Ostergaard, B.T. Pedersen, N. Skjaerbaek, P. Vedso, M. Begtrup, *Synlett* **2002**, 1889
02T1185	S. Rodriguez, E. Castillo, M. Carda, J.A. Marco, *Tetrahedron* **2002**, *58*, 1185
02T1817	M.T. Crimmins, K.A. Emmitte, A.L. Choy, *Tetrahedron* **2002**, *58*, 1817
02T1835	M. Maruyama, M. Inoue, T. Oishi, H. Oguri, Y. Ogasawara, Y. Shindo, M. Hirama, *Tetrahedron* **2002**, *58*, 1835
02T1889	M. Sasaki, M. Ishikawa, H. Fuwa, K. Tachibana, *Tetrahedron* **2002**, *58*, 1889
02T1997	S.P. Allwein, J.M. Cox, B.E. Howard, H.W.B. Johnson, J.D. Rainier, *Tetrahedron* **2002**, *58*, 1997
02T5407	L. Perez-Serrano, L. Casarrubios, G. Dominguez, G. Freire, J. Perez-Castells, *Tetrahedron* **2002**, *58*, 5407
02T6455	A.B. Smith, J. Zheng, *Tetrahedron* **2002**, *58*, 6455
02T7503	H. Ovaa, C. Stapper, G.A. van der Marel, H.S. Overkleeft, J.H. van Boom, S. Blechert, *Tetrahedron* **2002**, *58*, 7503
02T10017	H. Tanaka, K. Kawai, K. Fujiwara, A. Murai, *Tetrahedron* **2002**, *58*, 10017
02T10181	L. Chacun-Lefevre, V. Beneteau, B. Joseph, J.-Y. Merour, *Tetrahedron* **2002**, *58*, 10181
02TL213	A. Sivaramakrishnan, G.T. Nadolski, I.A. McAlexander, B.S. Davidson, *Tetrahedron Lett.* **2002**, *43*, 213
02TL779	X. Ginesta, M.A. Pericas, A. Riera, *Tetrahedron Lett.* **2002**, *43*, 779
02TL1779	C.E. Neipp, S.F. Martin, *Tetrahedron Lett.* **2002**, *43*, 1779
02TL4765	P. Gonzalez-Perez, L. Perez-Serrano, L. Casarrubios, G. Dominguez, J. Perez-Castells, *Tetrahedron Lett.* **2002**, *43*, 4765
02TL4841	D.R. Williams, L. Mi, R.J. Mullins, R.E. Stites, *Tetrahedron Lett.* **2002**, *43*, 4841
02TL6019	D. Barker, M.D. McLeod, M.A. Brimble, G.P. Savage, *Tetrahedron Lett.* **2002**, *43*, 6019
02TL6739	J. Zaminer, C. Stapper, S. Blechert, *Tetrahedron Lett.* **2002**, *43*, 6739
02TL7095	M.H.D. Postema, J.L. Piper, *Tetrahedron Lett.* **2002**, *43*, 7095
02TL7263	J. Cossy, C. Taillier, V. Bellosta, *Tetrahedron Lett.* **2002**, *43*, 7263
02TL7781	S.K. Chattopadhyay, S. Maity, S. Panja, *Tetrahedron Lett.* **2002**, *43*, 7781
02TL7851	R.A.J. Wybrow, L.A. Johnson, B. Auffray, W.J. Moran, H. Adams, J.P.A. Harrity, *Tetrahedron Lett.* **2002**, *43*, 7851
02TL8177	M.G. Organ, J. Xu, B. N'Zemba, *Tetrahedron Lett.* **2002**, *43*, 8177
02TL9039	C. Gaul, S.J. Danishefsky, *Tetrahedron Lett.* **2002**, *43*, 9039
03EJO611	O. Arjona, A.G. Csaky, J. Plumet, *Eur. J. Org. Chem.* **2003**, 611
03JA2591	P.W.C. Tsang, J.A. Jernelius, G.A. Cortez, G.S. Weatherhead, R.R. Schrock, A.H. Hoveyda, *J. Am. Chem. Soc.* **2003**, *125*, 2591
03TL527	F.-X. Felpin, J. Lebreton, *Tetrahedron Lett.* **2003**, *44*, 527

Chapter 2

Photochemical Isomerizations of Some Five-Membered Heteroaromatic Azoles

James W. Pavlik
Department of Chemistry and Biochemistry, Worcester Polytechnic Institute
Worcester, MA 01609, USA
jwpavlik@wpi.edu

2.1 INTRODUCTION

The photochemistry of *N*-substituted pyrazoles and of isothiazoles has been of considerable interest <76PHC123; 80RGES501; 94PP803; 94PP1063; 970P57> since the first report that 1-methylpyrazole (**1**) undergoes photoisomerization to 1-methylimidazole (Scheme 1) <67HCA44>.

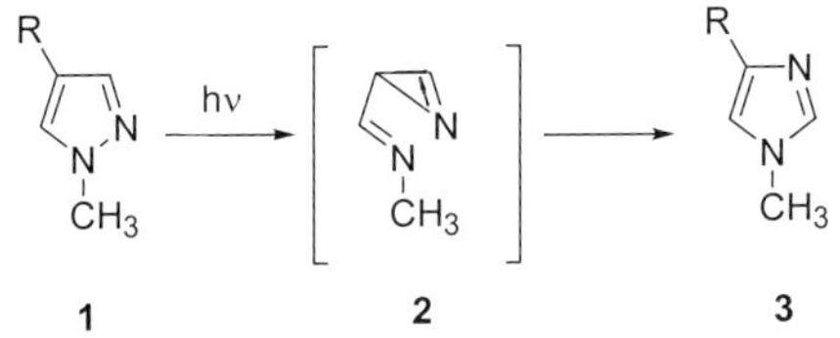

Scheme 1

At that time it was suggested that the isomerization occurred by way of an initial photo-ring contraction to an undetected 2-(*N*-methylimino)-2*H*-azirine intermediate (**2**) and subsequent ring expansion to the observed product, 1-methylimidazole (**3**). Although the intermediacy of acylazirines has been adequately demonstrated in the analogous isoxazole-to-oxazole phototransposition, such iminoazirines have not been detected in a pyrazole to imidazole isomerization, and thus, this mechanistic suggestion has never been experimentally substantiated.

2.2 PHOTOCHEMISTRY OF 1-METHYLPYRAZOLES

More recent work in this laboratory has confirmed that 1-methylpyrazoles phototranspose to 1-methylimidazoles and has also shown that they undergo two different photocleavage reactions leading to enaminonitriles and to enaminoisocyanides respectively <91JOC6313; 95JOC8138>. Although both types of photocleavage products can be spectroscopically detected and isolated, upon further irradiation these compounds also undergo photocyclization to 1-methylimidazoles.

Short-duration irradiation of 1-methylpyrazole (**1a**) (Scheme 2) led to the formation of 1-methylimidazole (**3a**), to a mixture of the *(E)* and *(Z)* isomers of enaminonitrile (**4a**), and to a mixture of the *(E)* and *(Z)* isomers of enaminoisocyanide (**5a**). More prolonged irradiation, however, led to the consumption of **4a** and **5a** and to the continued formation of **2a** indicating that **4a** and **5a** are intermediates in the phototransposition of **1a** to **3a**. 1,4-Dimethylpyrazole (**1b**) reacted similarly. Thus, short-duration irradiation led to the formation of 1,4-dimethylimidazole (**3b**), *(E)/(Z)*-enaminonitrile (**4b**), and *(E)/(Z)*-enaminoisocyanide (**5b**), whereas continued irradiation was accompanied by the consumption of **4b** and **5b** and to the formation of **3b**. The phototransposition of **1b** to **3b** via **4b** and **5b** thus appears to involve only interchange of N-2 and C-3 of the pyrazole ring.

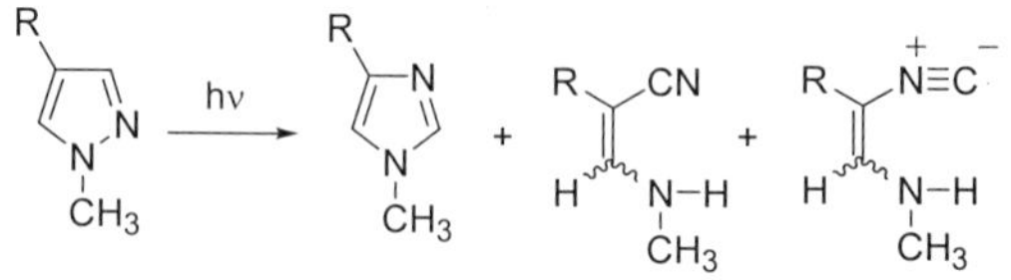

Scheme 2

1-Methyl-4-phenylpyrazole (**6**) also phototransposed regiospecifically to yield 1-methyl-4-phenylimidazole (**7**) and the two photocleavage products *(E/Z)*-3-*N*-methylamino-2-phenylpropenenitrile (**8**) and (*E,Z*)-2-(*N*-methylamino)-1-phenylethenylisocyanide (**9**) in the chemical and quantum yields shown in Scheme 3 <97JOC8325>.

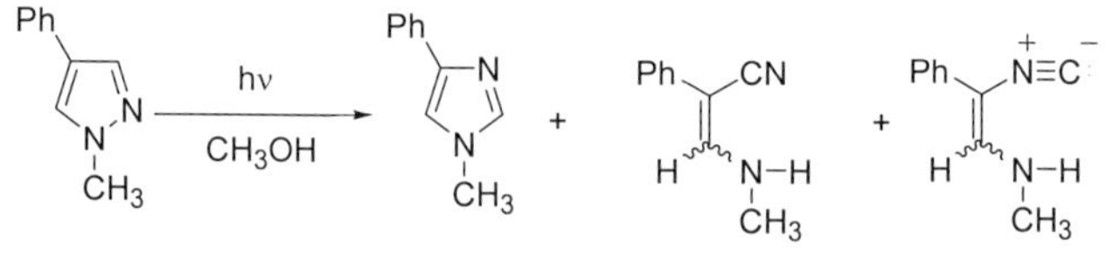

	6	7	8	9
%	- 20.5	17.2	12.0	68.1
φ	0.200+/- 0.03	0.037+/-0.002	0.008+/-0.001	0.114+/-0.001

Scheme 3

5-Deuterio-1-methyl-4-phenylpyrazole (**6-5d**) (Scheme 4) also phototransposes to 5-deuterio-1-methyl-4-phenylimidazole (**7-5d**). This proves that C-4 and C-5 of the 1-

methylpyrazole reactant transpose to C-4 and C-5 respectively of the 1-methylimidazole product and confirms that the phototransposition involves only interchange of the N-2 and C-3 ring atoms.

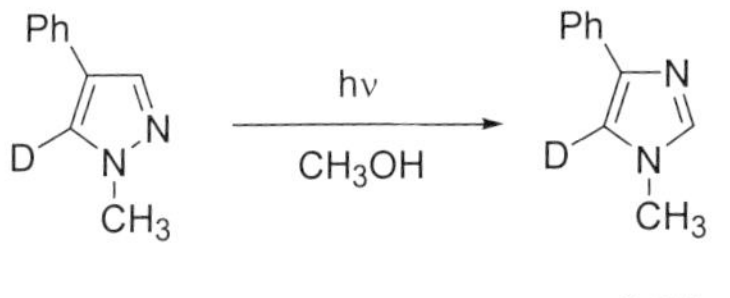

Scheme 4

Direct irradiation of Z-enaminonitrile (**8**) and Z-enaminoisocyanide (**9**) revealed (Scheme 5) that both undergo Z→E isomerization and photocyclization to **7**, the N-2–C-3 interchange product. In addition, cyclization of **Z-9** to **7** was observed when the former was heated to 80 °C. Although the quantum yields shown in Scheme 3 reveal that the photocyclization of enaminonitrile **Z-8** to imidazole **7** is a very inefficient reaction, chemical and quantum yields show that the photocleavage-photocyclization pathway *via* an isocyanide intermediate is a major route for the N-2–C-3 interchange phototransposition reaction <97JOC8325>.

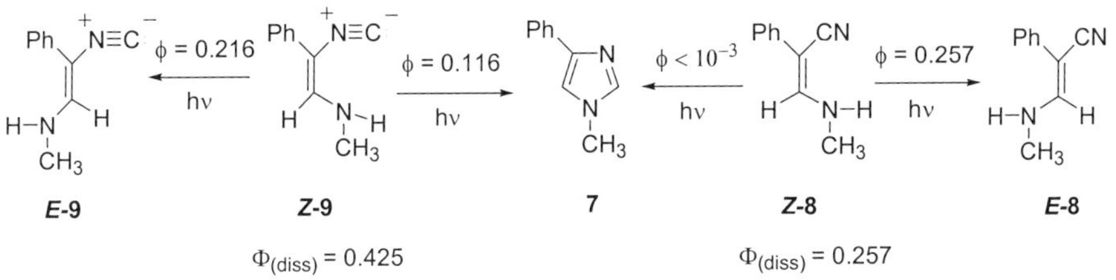

Scheme 5

From these studies a general mechanistic scheme for the N-2–C-3 interchange pathway has emerged. As shown in Scheme 6, photocleavage of the N-1–N-2 bond in **10** was suggested <97JOC8325> to result in the formation of a species that can be viewed as a β-imino vinyl nitrene (**11**). The isomerization of terminal vinyl nitrenes to nitriles is a well-documented reaction <86JOC3176>. Accordingly, nitrene (**11**) would be expected to rearrange as shown in Scheme 6 to the enaminonitrile photocleavage product (**13**) by way of ketene imine (**12**). In addition, **11** would be expected to be in photoequilibrium with the isomeric iminoazirine (**14**) <62JOC3557; 63CB399; 67TL1545; 67JA2077>, which could isomerize to the enaminoisocyanide photo-cleavage product (**18**), a known photochemical and thermal precursor of imidazole (**16**), the N-2–C-3 interchange product. This reaction requires proton transfer from carbon in **14** to nitrogen in **18**. Although in the case of acylazirines <78H1207> and thioformylazirines <98JOC5592> this proton transfer requires an added base, pyrazoles, and certainly imidazoles, are sufficiently basic to deprotonate iminoazirine (**14**) (Scheme 6).

Alternatively, electrocyclic ring opening of azirine **14** would yield nitrile ylide (**15**), which could undergo direct cyclization to imidazole (**16**) <76ACR371> or proton transfer resulting in the formation of isocyanide (**18**). If azirine (**14**) is formed in the ground state, calculations suggest that in the absence of deprotonation by a base, the most likely reaction pathway is back to the pyrazole. Calculations also reveal that excitation to the S_1 surface is accompanied by an increase in the carbon-carbon bond length from 1.49 to 1.74 Å. Accordingly, if an azirine-like structure is reached on the S_1 reaction coordinate before crossing to the S_0 surface, the excited species would be expected to undergo easy conversion to the nitrile ylide (**15**) rather than resulting in an isolable iminoazirine product

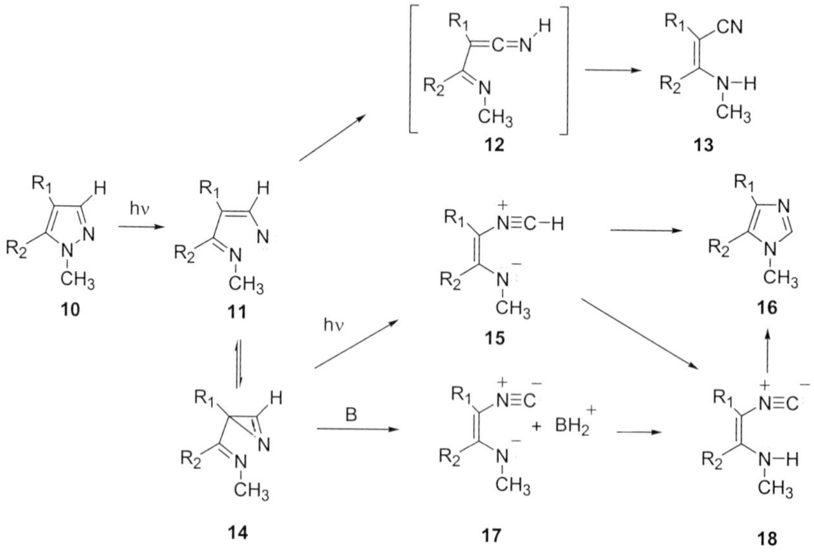

Scheme 6

1,3-Dimethylpyrazole (**19a**) phototransposes (Scheme 7) to a mixture of 1,2-dimethylimidazole (**20a**) and 1,4-dimethylimidazole (**21a**). Labelling studies showed that the deuterium at ring position 4 in **19b** transposed to ring position 4 in **20b** confirming that this is an N-2-C-3 interchange product which can be formed by the mechanistic pathway shown in Scheme 6, and to ring position 5 in **21b**, showing that the isomerization of **19** to **21** requires the interchange of the N-2-C-4 ring atoms <91JOC6313>. This product was suggested to arise *via* the transposition pathway shown in Scheme 8 that includes electrocyclic ring closure resulting in 1,5-diazabicyclo[2.1.0]pentene (**BC-19a,b**), [1,3]-sigmatropic shift of nitrogen, and rearomatization of the resulting 2,5-diazabicyclo[2.1.0]pentene (**BC-21a,b**) to provide the observed product (**21a,b**). Similarly (Scheme 9), 1,3,5-trimethylpyrazole (**22**) has also been shown to phototranspose to 1,2,5-trimethylimidazole (**23**), which can result from the N-2–C-3 interchange pathway, and to 1,2,4-trimethylimidazole (**24**), which can be formed by the electrocyclic ring closure-nitrogen migration mechanism shown in Scheme 8 <67TL5315; 69T3287>.

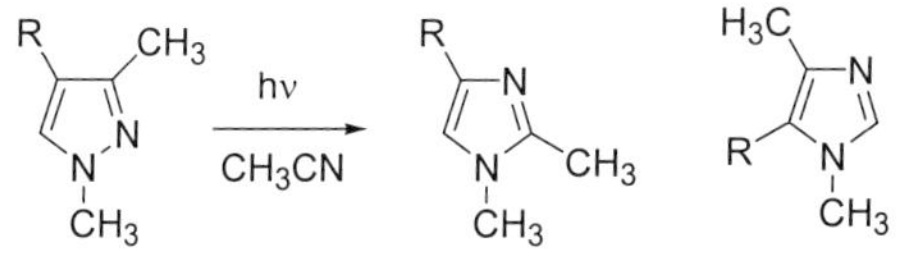

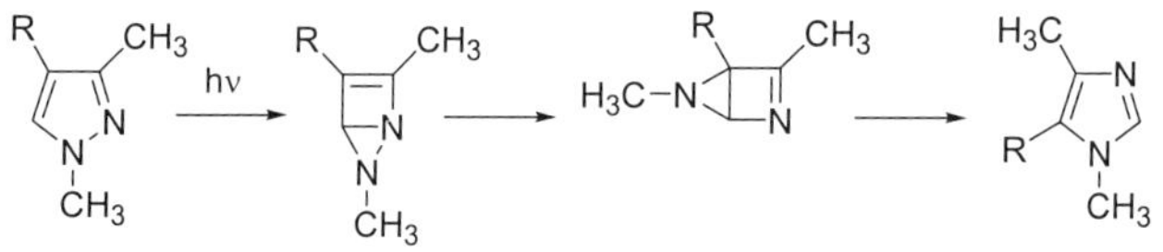

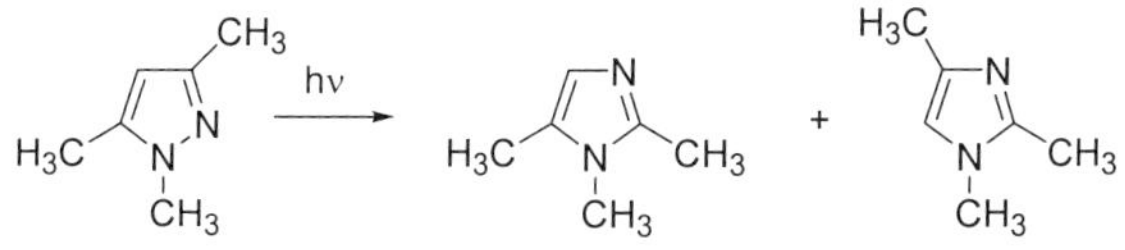

Scheme 7

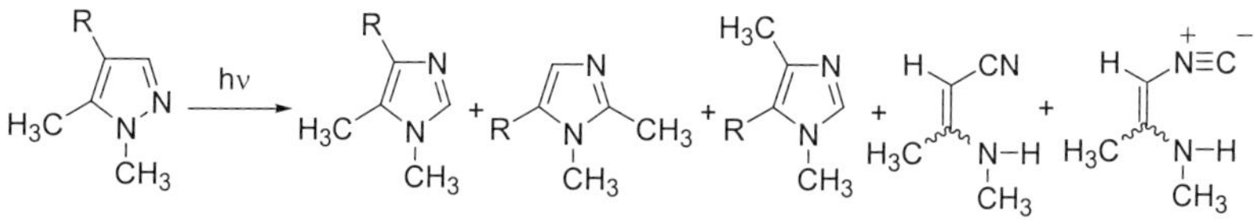

Scheme 8

Scheme 9

Scheme 10

1,5-Dimethylpyrazole (**25a**) and its 4-deuterio-derivative (**25b**) were observed (Scheme 10) to undergo photocleavage to **29** and **30**, and also phototransposition to three primary products, **26a,b** and **27a,b**, which can be rationalized by the N-2–C-3 interchange pathway (Scheme 6), and the one-step nitrogen migration mechanism (Scheme 8) respectively, and **28a,b**, which cannot arise by either of these pathways but was suggested <91JOC6313> to arise *via* a mechanism (Scheme 11) involving two successive nitrogen migrations. Such a double walk mechanism has also been implicated in the photochemistry of 3-cyano-1,5-dimethylpyrazole (**31**) <81CC604> which phototransposes (Scheme 12) to imidazole (**32**), the N2-C3 interchange product, and to imidazoles (**33**) and (**34**), products which can be formed by electrocyclic ring closure followed by one or two successive nitrogen migrations and rearomatization. This double walk mechanism also has analogy in the photochemistry of cyanothiophenes <79CC881; 79CC966> and 5-methyl-substituted 2-cyanopyrroles <75CC786; 78CC131>.

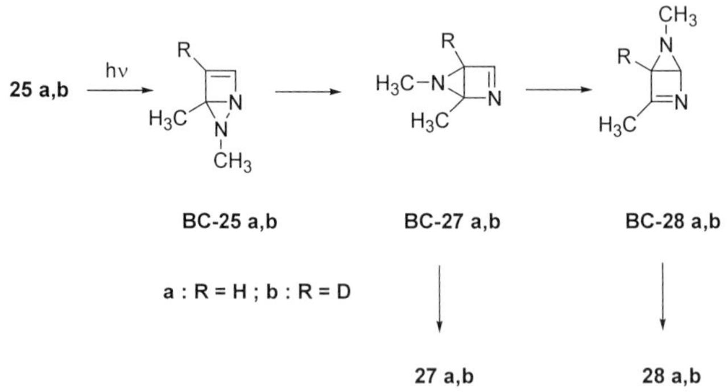

Scheme 11

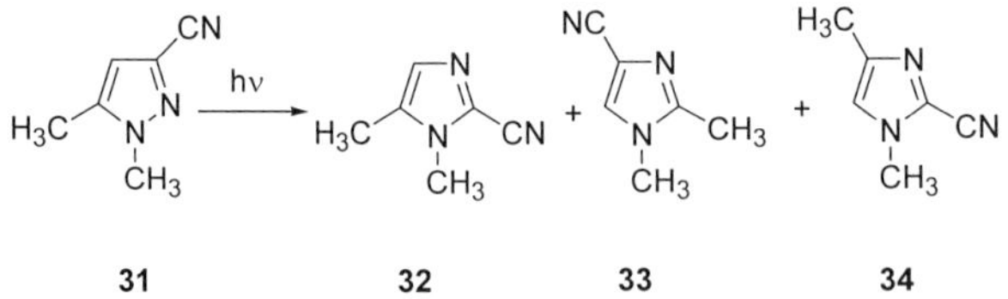

Scheme 12

Fluoro-substituted 1-methylpyrazoles also phototranspose to fluoro-substituted 1-methylimidazoles <91JOC6313>. Scheme 13 shows the primary products formed upon direct irradiation of each isomeric fluoro-1-methylpyrazole in acetonitrile. Although deuterium

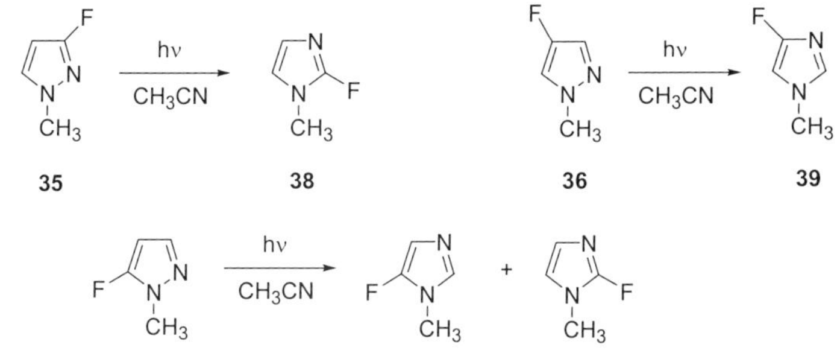

Scheme 13

labeling studies were not carried out in these cases, it was presumed that **38**, **39**, and **40** are N2-C3 interchange products of **35**, **36**, and **37**, respectively, and that **38** is formed from **37** by the electrocyclic ring closure-one step nitrogen migration mechanism. By comparison with methyl substitution, these results show that fluorine substitution significantly enhances reaction *via* the N-2–C-3 interchange pathway. In striking contrast, it has been reported <81CC604> that irradiation of 1,5-dimethyl-3-trifluoromethylpyrazole (**41**) in acetonitrile (Scheme 14) gave only 1,2-dimethyl-4-trifluoromethylimidazole (**42**) in 41% yield, the product expected from the electrocyclic ring closure-one-nitrogen migration mechanism. This pronounced effect of trifluoromethyl substitution clearly deserves more thorough study.

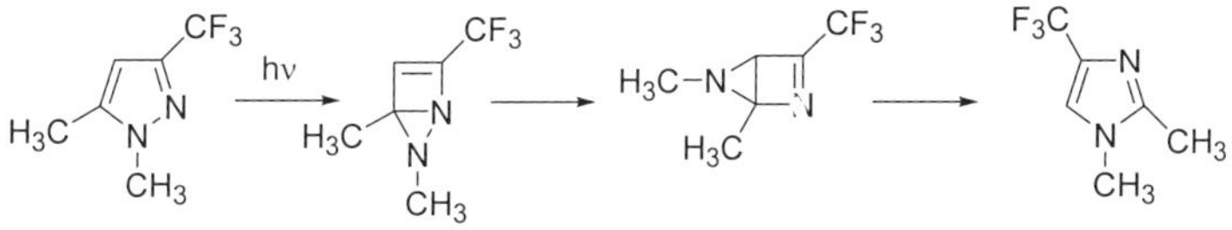

Scheme 14

Electrocyclic ring closure leading to 1,5-diazabicyclo[2.1.0]pentene species is a key step in the electrocyclic ring closure-heteroatom migration mechanism shown in Schemes 8 and 11. Interestingly, MNDO calculations revealed that as the planar Franck-Condon excited singlet state relaxes, the pyrazole ring begins to undergo disrotatory deformation, resulting in an energy minimize S_1 state in which the 1-methyl nitrogen is 12° out of the plane of the ring. From this point the molecule undergoes facile electrocyclic ring closure to yield the suggested 1,5-diazabicyclic species <91JOC6321>.

Experimental evidence also supports the intermediacy of this species. Thus, whereas irradiation of 1-methyl-5-phenylpyrazole **43** in methanol (Scheme 15) results in the formation of the anticipated imidazoles (**44-46**) and the expected photocleavage products (**47**) and (**48**), in the chemical and quantum yields shown in Scheme 15, photolysis of **43** in neat furan leads only to the pyrazole-furan [4+2] adduct (**44**) (Scheme 16). Formation of **44** is consistent with furan trapping of a photochemically generated 1,5-diazabicyclo[2.1.0]pentene species (**BC-43**) <97JOC8325>.

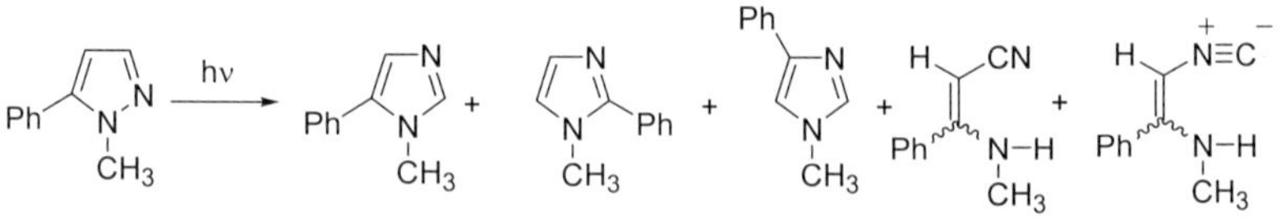

Scheme 15

	43	44	45	46	47	48
%	-23.2	16.0	30.3	45.6	trace	trace
φ	0.041	0.0080	0.140	0.0220	---	---

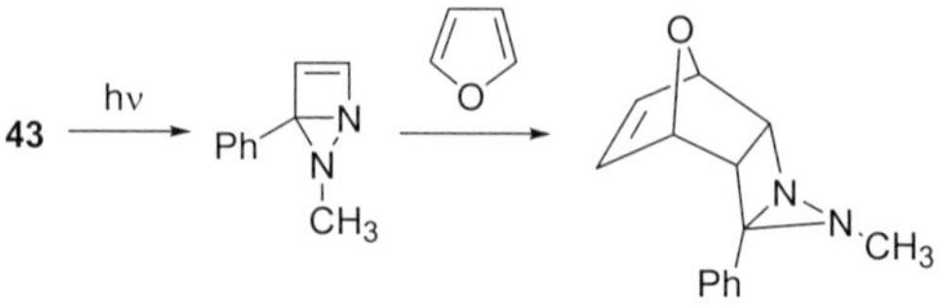

Scheme 16

It is interesting to note that the electrocyclic ring closure-heteroatom migration mechanism results only in pyrazole-to-imidazole transpositions. The [1,3] sigmatropic shift of nitrogen must therefore take place away from the azetine nitrogen to form the 2,5-diazabicyclo species but not in the opposite direction toward the azetine nitrogen to yield an isomeric 1,5-diazabicyclic species and eventually to a pyrazole-to-pyrazole transformation.

Scheme 17 summarizes the various mechanistic pathways for the 1-methylpyrazole to 1-methylimidazole transposition.

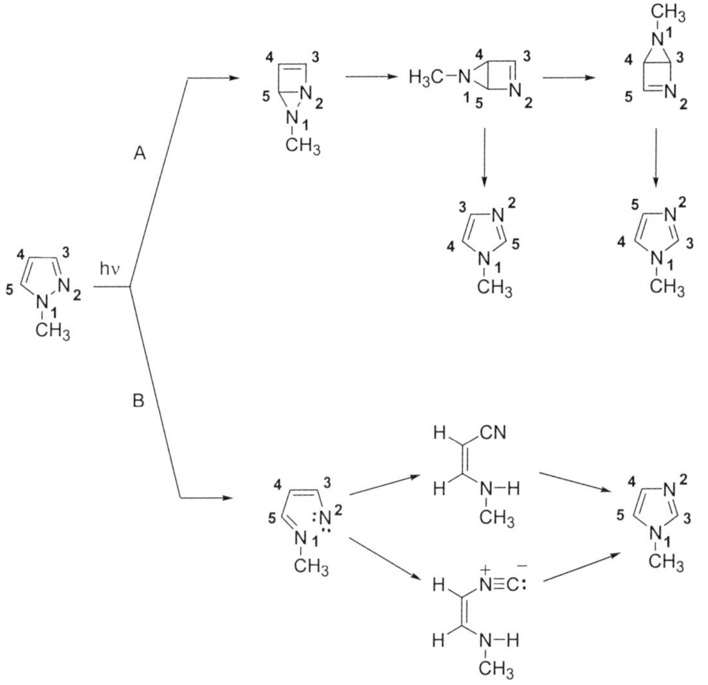

Scheme 17

According to this interpretation, the photochemistry of 1-methylpyrazoles involves a competition between photocleavage of the N–N bond to generate a vinyl nitrene, the precursor of enaminonitrile and enaminoisocyanide photocleavage products and the N-2–C-3 interchange transposition product and electrocyclic ring closure to form a 1,5-diazabicyclo[2.1.0]pentene, the precursor of the one and two nitrogen walk phototransposition products. The actual pathway followed is sensitive to the nature and position of the substituent on the pyrazole ring. Thus, whereas 4-substituted-1-methylpyrazoles have been observed to phototranspose regiospecifically via Path B, 1,3- and 1,5-disubstituted pyrazoles also transpose by way of the electrocyclic ring closure-one nitrogen walk and two nitrogen walks respectively.

In order to study the 1-methylpyrazole to 1-methylimidazole phototransposition process with minimum substituent perturbation, the phototransposition chemistry of 3,4-dideuterio-1-methylpyrazole (**1-3,4d2**) has been studied. This labelling pattern allows distinction between the three pathways since Scheme 17 shows that the conversion of **1** to **2** via the three pathways is accompanied by transposition of C-5 of the reactant to ring position 5 by the N-2–C-3 interchange pathway or to ring positions 2 or 4 if the transposition occurs by electrocyclic ring closure followed by one or two nitrogen migrations respectively.

After less than 10% conversion of **1-3,4d2**, ^{1}H NMR analysis (Scheme 18) revealed that the product consisted of a mixture of **2-2,4d2**, **2-4,5d2**, and **2-2,5d2** confirming that **1-3,4d2** undergoes phototransposition by all three pathways in a ratio of 4.8:6.5:1.0. <91JOC6313>. Thus, although the phototransposition of 1-methylpyrazole (**1**) to 1-methylimidazole (**2**) is formally a simple transformation, it is mechanistically quite complex.

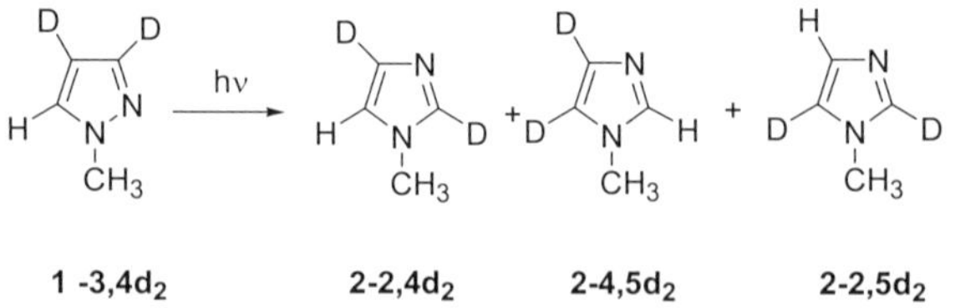

$$\underset{\text{1 -3,4d}_2}{\underset{\overset{|}{\text{CH}_3}}{\text{H}\diagdown\text{N}\diagup\text{N}}\diagup\overset{\text{D}}{\diagdown}\overset{\text{D}}{\diagup}} \xrightarrow{h\nu} \underset{\text{2-2,4d}_2}{\underset{\overset{|}{\text{CH}_3}}{\text{H}\diagdown\text{N}\diagup\text{N}}\diagup\overset{\text{D}}{\diagdown}\overset{\text{D}}{\diagup}\text{D}} + \underset{\text{2-4,5d}_2}{\underset{\overset{|}{\text{CH}_3}}{\text{D}\diagdown\text{N}\diagup\text{N}}\diagup\overset{\text{D}}{\diagdown}\overset{\text{D}}{\diagup}\text{H}} + \underset{\text{2-2,5d}_2}{\underset{\overset{|}{\text{CH}_3}}{\text{D}\diagdown\text{N}\diagup\text{N}}\diagup\overset{\text{H}}{\diagdown}\overset{\text{D}}{\diagup}\text{D}}$$

Scheme 18

2.3 PHOTOCHEMISTRY OF 1-PHENYLPYRAZOLES

In contrast to 1-methylpyrazoles, 1-phenylpyrazole (**45a**) and methyl-substituted 1-phenylpyrazoles **45b-d** phototranspose (Scheme 19) regiospecifically by the N-2–C-3 interchange pathway <93JA7645>. This suggests that the species resulting from photocleavage of the N–N bond may have diradical character (Scheme 20), which is stabilized by the *N*-phenyl group. Interestingly, unlike 1-methylpyrazole (**1**), which is non-planar in the first excited singlet state, computational studies show that the pyrazole ring in 1-phenylpyrazole (**45a**) remains planar upon excitation. This is consistent with its lack of reactivity *via* the electrocyclic ring closure pathway. Calculations do reveal that excitation is accompanied by an increase in the N–N bond length from 1.35 Å in S_0 to 1.78 Å in S_1. This corresponds to a change in the N–N bond order from 1.12 in S_0 to 0.38 in S_1 and indicates that by the time the molecule reaches the energy minimized S_1 state, the N–N bond is essentially broken and the molecule is well along the N-2–C-3 interchange reaction pathway <93HA7645>.

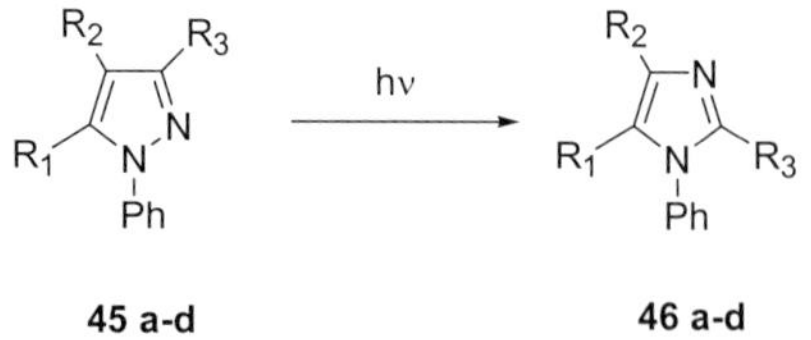

45 a-d → 46 a-d

a: $R_1 = R_2 = R_3 = H$ b: $R_1 = R_2 = H, R_3 = CH_3$
c: $R_1 = R_3 = H, R_2 = CH_3$ d: $R_2 = R_3 = H, R_1 = CH_3$

Scheme 19

45 a

Scheme 20

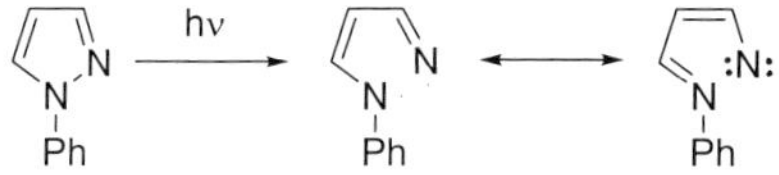

2.4 PHOTOCHEMISTRY OF 1-METHYLIMIDAZOLES

Although less studied, 1-methylimidazoles have been observed to undergo imidazole-to-imidazole transposition <91JOC6313>. Thus irradiation of 1,2-dimethylimidazole (**20a**) or 1,4-dimethylimidazole **21a** (Scheme 21) resulted in a photoequilibrium mixture of the two compounds. Although 1,5-dimethylimidazole (**28**) appeared to be photostable, irradiation of 2-deuterio-1,5-dimethylimidazole (**28-2d**) (Scheme 22) led to a photo-equilibrium mixture of **28-2d** and **28-4d**. Interchange of ring positions 2 and 4 in these reactions is consistent with a mechanism involving electrocyclic ring closure (Scheme 23) followed by regiospecific migration of the heteroatom away from the azetine ring nitrogen.

20 a **21 a**

Scheme 21

28-2d **28-4d**

Scheme 22

Scheme 23

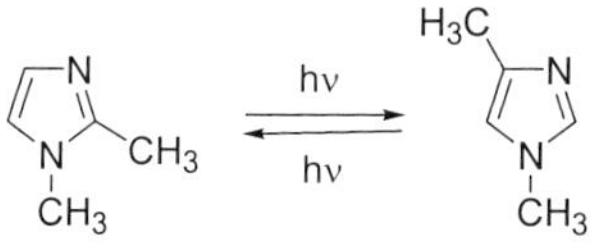

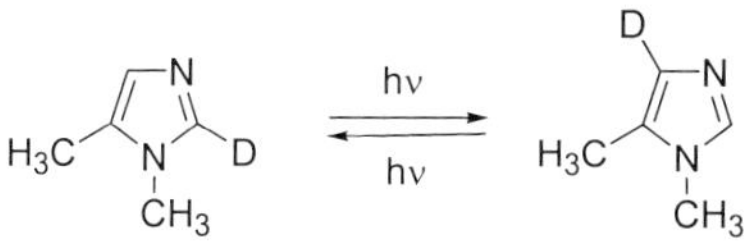

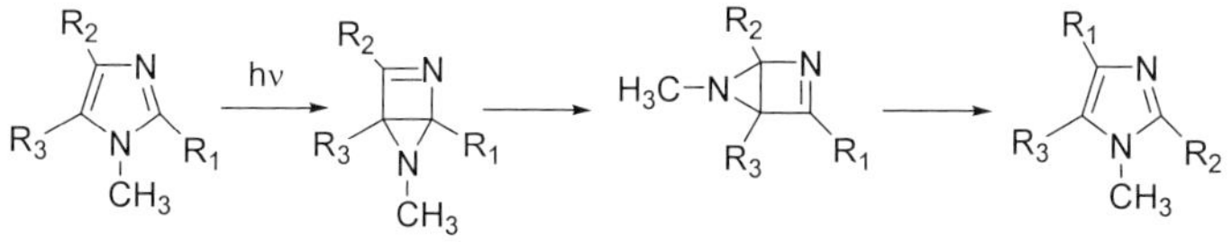

2.5 PHOTOCHEMISTRY OF METHYL-SUBSTITUTED ISOTHIAZOLES AND THIAZOLES

The photoisomerization of isothiazole (**47a**) (Scheme 24) to thiazole (**48a**) was the first reported phototransposition in the isothiazole-thiazole heterocyclic system <69CC1018>.

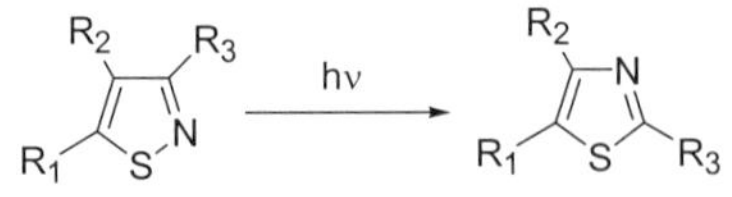

a: $R_1 = R_2 = R_3 = H$ b: $R_1 = R_2 = H$; $R_3 = CH_3$
c: $R_1 = R_3 = H$; $R_2 = CH_3$ d: $R_2 = R_3 = H$; $R_1 = CH_3$

Scheme 24

Methylisothiazoles have also been shown to undergo transposition. Although early studies appeared to implicate tricyclic zwitterionic intermediates in these reactions, <72T3141> later investigations showed that isomeric methylisothiazoles (**47b**), (**47c**), and (**47d**) each transpose to a single methylthiazole photoproduct **48b**, **48c**, and **48d** respectively. These results indicate that the transposition occurs *via* the N-2–C-3 exchange process analogous to that observed for pyrazole photochemistry <93JOC3407>.

Although methylthiazoles do not undergo phototransposition upon irradiation in a variety of neutral solvents, methylthiazolium cations (**49a-c**), formed by dissolving the corresponding neutral methylthiazoles in trifluoroacetic acid (TFA), do undergo (Scheme 25) phototransposition to methylisothiazolium ions (**50a-c**) <93JOC3407>.

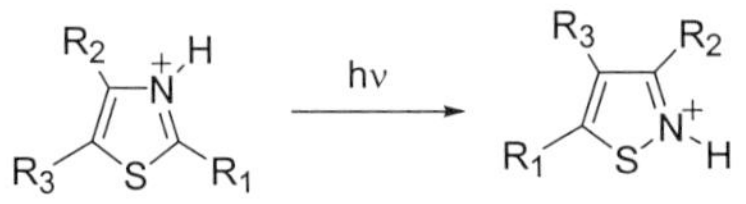

a: $R_1 = CH_3$; $R_2 = R_3 = H$ b: $R_2 = CH_3$; $R_1 = R_3 = H$
c : $R_3 = CH_3$; $R_1 = R_2 = H$

Scheme 25

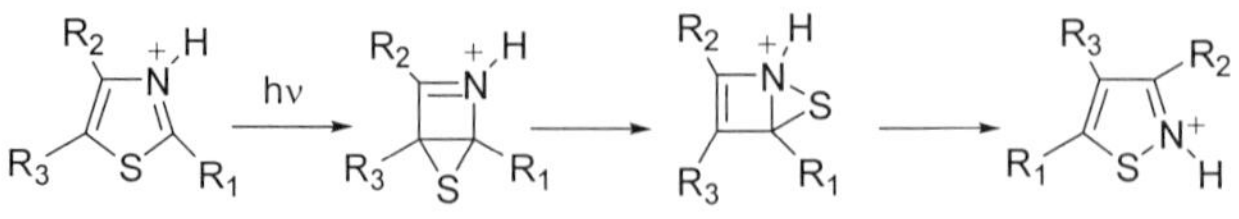

Scheme 26

Interchange of ring atoms 3 and 5 demanded by these reactions is consistent with the electrocyclic ring closure – heteroatom migration mechanism shown in Scheme 26 in which the heteroatom migrates regiospecifically toward the positively charged azetine nitrogen. This regiospecificity is in marked contrast to that observed in 1-methylimidazoles, which undergo 1-methylimidazole-to-1-methylimidazole transpositions. Thus, as shown in Scheme 23, in neutral imidazoles the heteroatom migrates only away from the azetine ring nitrogen.

2.6 PHOTOCHEMISTRY OF PHENYL SUBSTITUTED ISOTHIAZOLES AND THIAZOLES

The photochemistry of phenylthiazoles (**57-59**) and phenylisothiazoles (**60-62**) has been extensively studied in our laboratory <94JA2292> and elsewhere <70CC386; 71BCF1101; 72BCF2673; 72JCS(P2)1145; 73BCF1743; 73TL3523; 74T879; 78H389; 78JCS(P1)685>. According to the observed primary photochemical products and the results of deuterium labeling studies <94JA2292>, the six isomers can be organized (Scheme 27) into a tetrad of four interconverting isomers, **51**, **52**, **54**, and **56**, and a dyad in which 5-phenylthiazole (**53**) transposes to 4-phenylisothiazole (**55**), the only isomer that did not yield a transposition product upon irradiation in benzene solution. With one minor exception, no interconversions between the tetrad and dyad were observed. In that case, in addition to transposing to members of the tetrad, 5-phenylisothiazole (**56**) also transposed to 5-phenylthiazole (**53**), the first member of the dyad, in less that 1% yield. This conversion was assumed to occur *via* the N-2–C-3 interchange pathway.

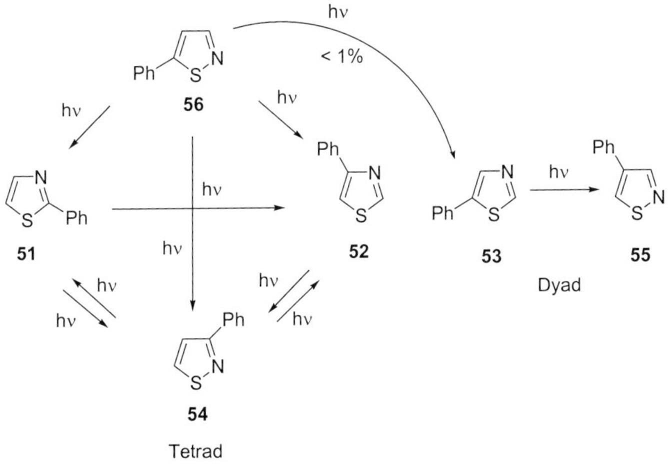

Scheme 27

The interconversions within the tetrad are consistent with the electrocyclic ring closure heteroatom migration mechanistic pathway shown in Scheme 28. Thus, photochemical excitation of any member of the tetrad results in electrocyclic ring closure, leading directly to the azathiabicyclo[2.1.0]pentene intermediates. The four bicyclic intermediates, and hence the four members of the tetrad, are interconvertible via 1,3-sigmatropic shifts of the sulfur around the four sides of the azetine ring. Thus, sulfur migration followed by rearomatization allows sulfur insertion into the four different sites in the carbon-nitrogen sequence. According to this mechanism, in principle, irradiation of any one member of the tetrad should lead to the formation of the other three. In practice, however, 4-phenylthiazole (**52**) is the dominant product from the irradiation of the other three members of the group, and shows little tendency to phototranspose. Both facts presumably reflect the greater stability of **BC-52** in which the phenyl group is in conjugation with the polar double bond of the imino group.

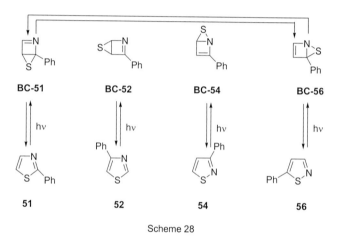

Scheme 28

In the dyad, because of the symmetry of the 3-phenylazetine ring in **BC-53** (Scheme 29), insertion of sulfur between ring positions C-1 and C-4 or CC-1 and C-2 leads to the same compound, 5-phenylthiazole (**53**). Similarly, insertion of a sulfur atom between N3 and C4 or N3 and C2 leads to 4-phenylisothiazole (**55**). Accordingly, because of this symmetry, only a dyad results. This symmetry is removed, however, in the case of 2-deuterio-5-phenylthiazole (**53-2d**). Thus, irradiation of **53-2d** resulted in the formation of three isomeric products (Scheme 30), viz., 4-deuterio-5-phenylthiazole (**53-4d**), 5-deuterio-4-phenylisothiazole (**55-5d**), and 3-deuterio-4-phenylisothiazole (**55-3d**). Deuterium labelling has thus expanded the dyad into a tetrad. These results are entirely consistent with the electrocyclic ring closure – heteroatom migration mechanism shown in Scheme 31 <94JA2292>.

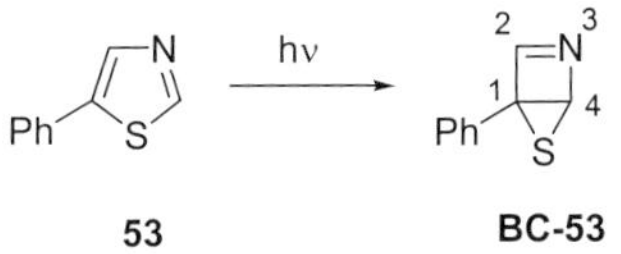

Scheme 29

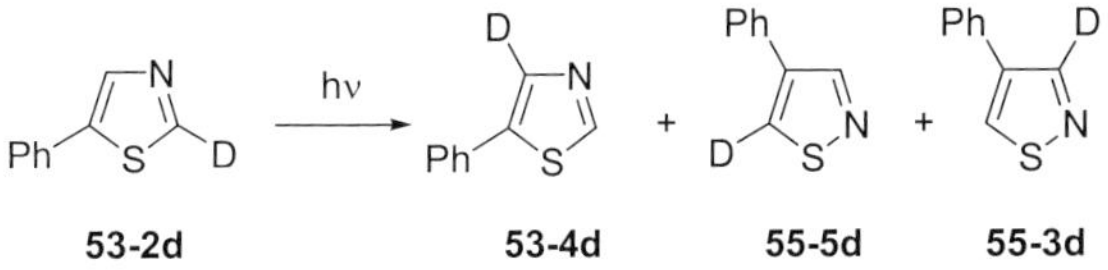

53 → BC-53 (hv)

Scheme 30

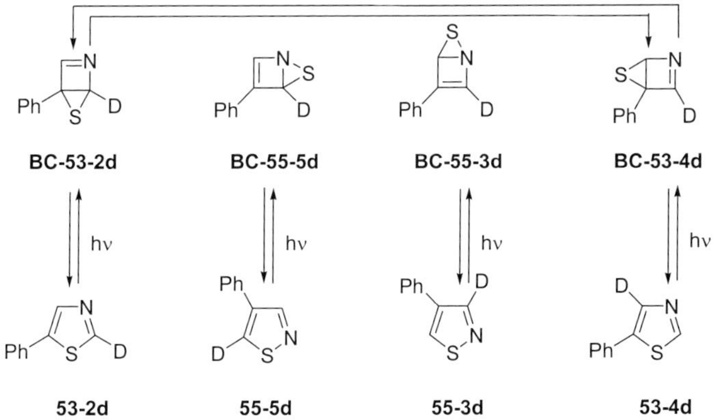

53-2d → 53-4d + 55-5d + 55-3d (hv)

Scheme 31

BC-53-2d ⇌ 53-2d (hv)
BC-55-5d ⇌ 55-5d (hv)
BC-55-3d ⇌ 55-3d (hv)
BC-53-4d ⇌ 53-4d (hv)

These results reveal that both 1-methylpyrazoles and phenylthiazoles and isothiazoles phototranspose by the electrocyclic ring closure – heteroatom migration mechanistic pathway. In the case of 1-methylpyrazoles, heteroatom migration in the initially formed 1,5-diazabicyclopentene occurs regiospecifically (Scheme 8) away from the azetine ring nitrogen resulting in the formation of a 2,5-diazabicyclopentene isomer that can aromatize to an imidazole. In the case of phenylisothiazoles, heteroatom migration in the initially formed 1-aza-5-thiabicyclopentene (Scheme 28) occurs in both directions, resulting in the formation of 2-aza-5-thiabicyclopentene or 1-aza-5-thiabicyclopentene isomers that upon aromatization result in the formation of an isomeric thiazole or isothiazole respectively. The regiospecificity of the diazabicyclopentene migration is presumably due to the greater stability of a 2,5-diazabicyclopentene relative to a 1,5-diazabicyclopentene because of the greater strength of the C–N bond in the 2,5-isomer relative to the N–N bond in the 1,5-diaza species. In the azathiabicyclopentene case, the difference between the S–N and S–C bond strengths is not expected to be as great and is therefore not anticipated to exert such a profound effect on the regiochemistry of the migration.

A more significant difference between 1-methylpyrazole and isothiazole photochemistry, however, appears to be the minor role of the N-2–C-3 interchange pathway in isothiazole chemistry. Thus, although N2-C3 interchange is a major transposition pathway in pyrazole chemistry, it is only a minor pathway upon irradiation of phenylisothiazoles in benzene solution. In fact, 4-phenylisothiazole (**55**), the compound most expected to react *via* the N-2–C-3 interchange pathway, was the only isomer that did not yield a transposition product upon irradiation in benzene solution (Scheme 27). This is not due to the photostability of the compound. Indeed, **55** is the most reactive of the six isomers. Nevertheless, even after consumption of 85% of **55**, no phototransposition product could be detected.

2.7 EFFECT OF ADDED BASE ON THE PHOTOCHEMISTRY OF ISOTHIAZOLES AND THIAZOLES.

In striking contrast, when the reaction was carried out in methanol or benzene containing a small amount of a base such as aqueous NH_3 or triethylamine (TEA) (Scheme 32), the N-2–C-3 interchange product, 4-phenylthiazole (**52**), was obtained in 90% yield along with a small quantity of the deprotonated photocleavage product (**57**), which was trapped by reaction with benzyl bromide to yield the (*E/Z*)-benzyl-thioether (**58**) <98JOC5592>. Conversion of **55-5d** to **52-5d** (Scheme 33) under these conditions confirmed that the transposition occurred *via* the

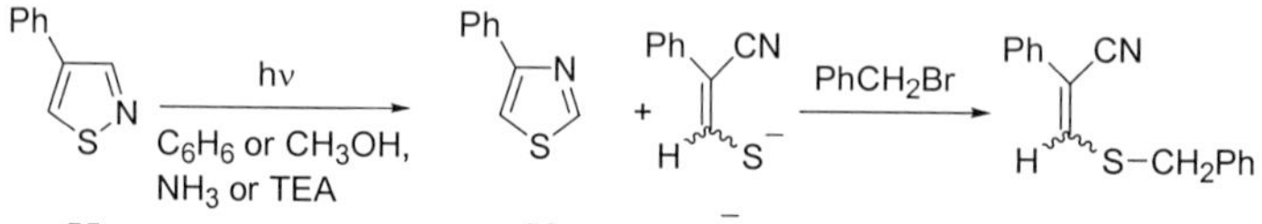

Scheme 32

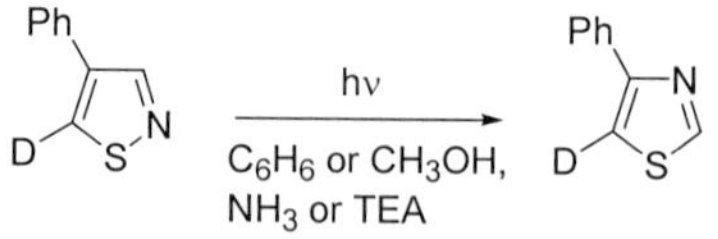

Scheme 33

N-2–C-3 interchange pathway. Although phototransposition was enhanced by the addition of base to the reaction medium, when the irradiation was carried out in ether or in methanol containing a small quantity of aqueous HCl (Scheme 34), the phototransposition was completely quenched and the only product observed was a large yield of the photocleavage product (**57H**).

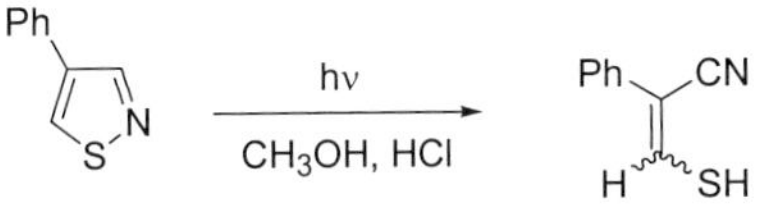

Scheme 34

55 (4-Ph-isothiazole) → hv, CH$_3$OH, HCl → **57H** (Ph(CN)C=CH(SH))

In addition to cyanosulfides, isocyanosulfides are also involved in this phototransposition process. Thus, irradiation of 4-benzylisothiazole (**59**) in methanol containing NH$_3$ (Scheme 35) led to the formation of the cyanosulfide photocleavage product (**(E/Z)-60**), which was trapped by reaction with benzyl bromide to yield benzyl-thioether (**(E/Z)-61**), to the formation of the *(E)*-isocyanosulfide photocleavage product (**(E)-62**), which was trapped by reaction with benzyl bromide followed by hydrolysis to yield *(E)*-formamide (**(E)-63**), and the *(Z)*-isocyanide photocleavage product (**(Z)-62**) which was thermally converted to the observed phototransposition product (**64**) <98JOC5592>. The N-2–C-3 transposition pathway was again confirmed by showing (Scheme 36) that irradiation of 4-benzyl-5-deuterioisothiazole (**59-5d**) resulted in the formation of 4-benzyl-5-deuteriothiazole (**64-5d**).

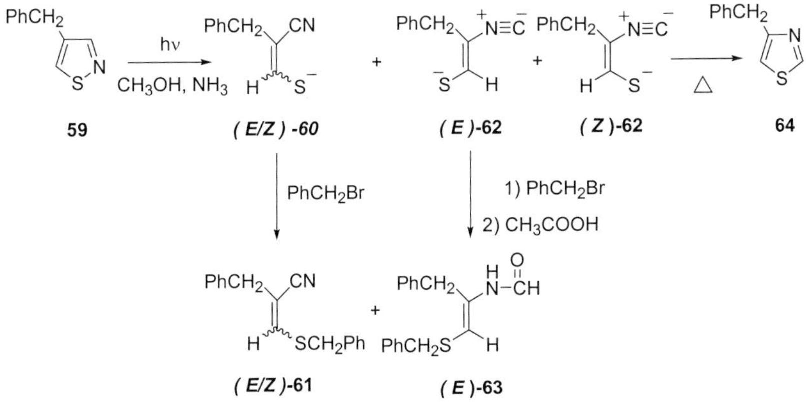

Scheme 35

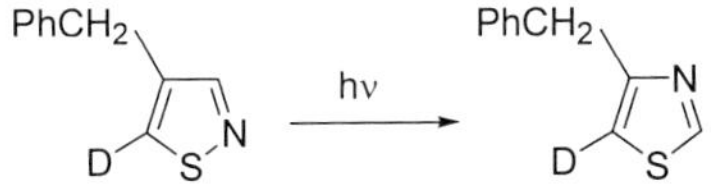

59 → hv, CH$_3$OH, NH$_3$ → (E/Z)-60 + (E)-62 + (Z)-62 → Δ → **64**

(E/Z)-60 + PhCH$_2$Br → (E/Z)-61

(E)-62 + 1) PhCH$_2$Br, 2) CH$_3$COOH → (E)-63

Scheme 36

59-5d → hv → **64-5d**

As a result of these studies it has been suggested that the N-2–C-3 interchange phototransposition of isothiazoles to thiazoles is mechanistically similar to the analogous conversion of pyrazoles to imidazoles shown in Scheme 6. Thus, photocleavage of the S-N bond in **65** (Scheme 37) was suggested to result in the formation of β-thioformylvinyl nitrene (**66**), which is expected to rearrange to the cyano-thiol photocleavage product (**67**) and also to be in equilibrium with an undetected thioformylazirine (**68**). Unlike pyrazoles and imidazoles (see Scheme 6), isothiazoles and thiazoles are not sufficiently basic to deprotonate azirine (**68**). Thus, in the absence of an external base, β-thioformylvinyl nitrene (**66**) is converted totally to cyano-thiol (**67**). As a result, upon irradiation in benzene or methanol containing HCl, 4-substituted isothiazoles undergo only photocleavage to cyano-thiols. In the presence of NH$_3$ or TEA however, the base converts cyanothiol (**67**) into cyano-sulfide (**69**), and, more importantly, the base can also deprotonate azirine (**68**) resulting in its conversion into isocyano-sulfide(**70**).

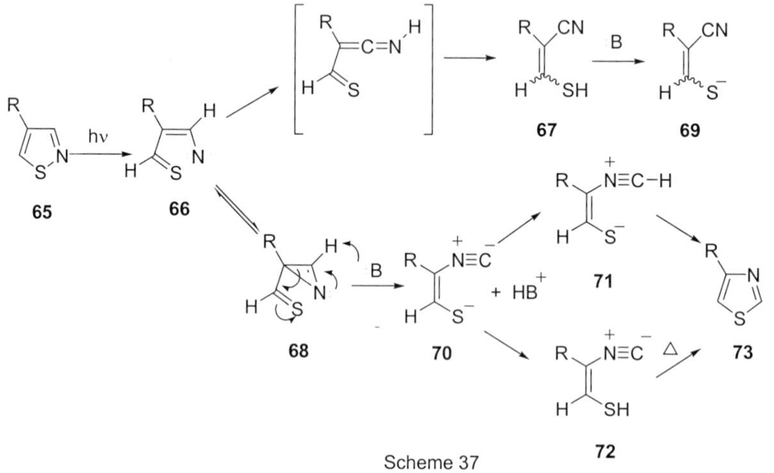

Scheme 37

The fate of isocyanosulfide (**70**) depends on the nature of the substituent originally at C-4 of the isothiazole ring. If R is aryl, as in **55**, the extended conjugation of the sulfide and the aryl group is expected to lower the basicity of the sulfide resulting in reprotonation at the isocyano carbon to yield **71**. Protonation at this site renders the carbon more susceptible to nucleophilic attack by the negative sulfur. As a result, these substituted isocyanides spontaneously cyclize to 4-arylthiazoles **73** (R=Ar).

If the C-4 substituent is alkyl or a substituted alkyl, as in **59**, protonation occurs at the more basic sulfur resulting in the formation of **72**. This reduces the nucleophilic character of the sulfur and also leaves the negatively charged carbon less susceptible to nucleophilic attack. As a result, cyclization requires a higher energy of activation, and hence, such substituted isocyanides can be detected spectroscopically and trapped by benzyl bromide.

Although the photochemistry of 4-substituted isothiazoles in the presence of NH$_3$ or TEA is dominated by reactions initiated by cleavage of the S–N bond, 5-phenyl- and 3-phenylthiazoles, (**56**) and (**54**), react under these conditions by competing photochemical pathways <00JOC3626>.

In benzene solution **56** phototransposes (Scheme 38) only by the electrocyclic ring closure-heteroatom migration pathway to yield **51**, **52** and **54** in 2%, 15%, and 5% yields respectively. In the presence of TEA, however, the major product (Scheme 39) of the photoreaction of 56 is 5-phenylthiazole (**53**) formed in 14% yields, with **52** and **54** formed in 4% and 5% yields respectively. 4-Deuterio-5-phenylisothiazole (**56-4d**) also phototransposed (Scheme 40) to 4-deuterio-5-phenylthiazole (**53-4d**), confirming that this product was formed by the N-2–C-3 interchange pathway, and **52-5d** and **54-4d** as expected for products formed by the electrocyclic ring closure-heteroatom migration mechanistic pathway.

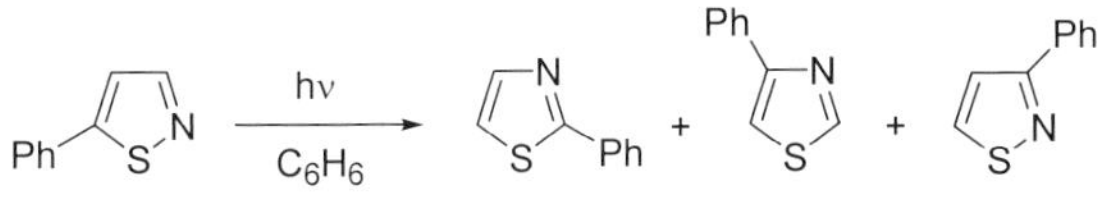

56 **51** **52** **54**

Scheme 38

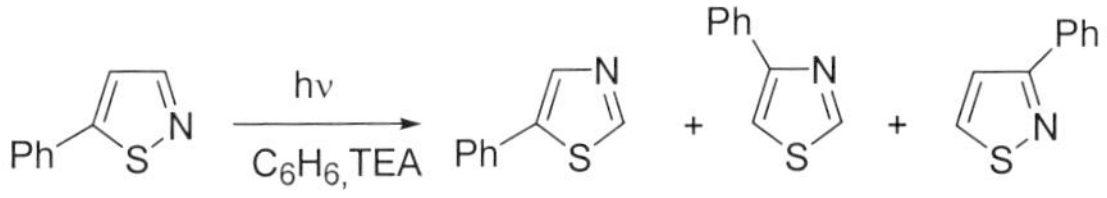

56 **53** **52** **54**

Scheme 39

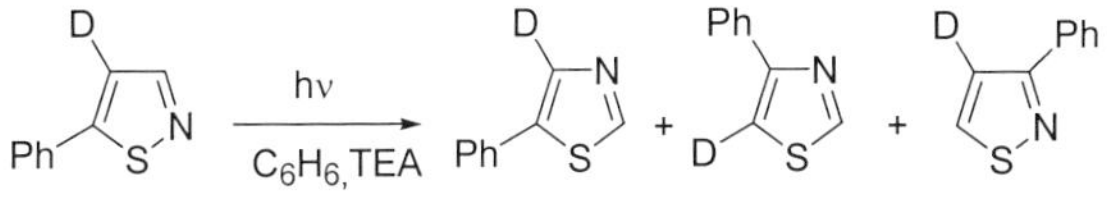

56-4d **53-4d** **52-5d** **54-4d**

Scheme 40

The yield of the N-2–C-3 interchange product **53** was 23% when **56** was irradiated in methanol solvent, 34% when the irradiation was carried out in methanol containing TEA, and 42% when the irradiation of **56** was carried out in the more polar 2,2,2-trifluoroethanol (TFE) as solvent, containing TEA. In either methanol or TFE solvents **56-4d** was converted to **53-4d** showing that the increased solvent polarity had not changed the reaction pathway.

3-Phenylisothiazole (**54**) also phototransposed (Scheme 41) to phenylthiazoles (**51**), the N-2–C-3 interchange product, and (**52**), an electrocyclic ring closure-heteroatom migration product. In the absence of a hydrogen atom at C-3, however, the yield of the N-2–C-3 interchange product was not enhanced by the addition of a base.

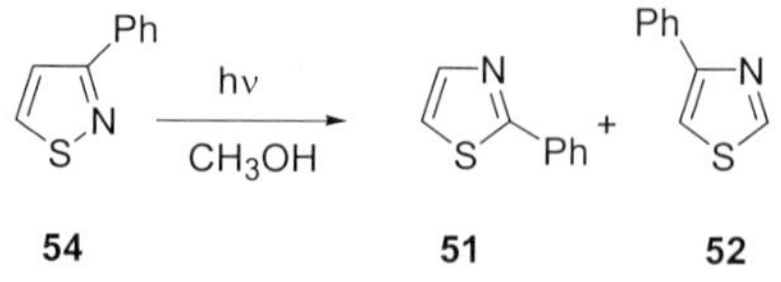

[Scheme 41: compound 54 (3-phenylisothiazole) → hv, CH₃OH → 51 (2-phenylthiazole) + 52 (4-phenylthiazole)]

Scheme 41

2.8 CONCLUSIONS

These studies show that the photochemistry of pyrazoles and isothiazoles can be rationalized by a single unifying mechanistic scheme. Thus excitation of either heterocycle is followed by a competition between electrocyclic ring closure and subsequent heteroatom migration or cleavage of the bond between the two heteroatoms. Whereas in the case of pyrazoles the former pathway leads only to pyrazole-to-imidazole transpositions, isothiazoles undergo the analogous isothiazole to thiazole isomerization and also isothiazole-to-isothiazole conversions. Imidazoles and thiazoles also react by the electrocyclic ring closure-heteroatom migration pathway resulting in imidazole to imidazole interconversions, or thiazole to either thiazole or isothiazole transpositions. Photocleavage of the bond between the two heteroatoms competes with electrocyclic ring closure and converts pyrazoles and isothiazoles to a species that can be viewed as a vinyl nitrene which is the precursor of nitrile and isocyanide photocleavage products and also leads to the N-2–C-3 interchange phototransposition of pyrazoles to imidazoles and isothiazoles to thiazoles.

Acknowledgements
I would like to thank Professor John A. Joule for his kind invitation to contribute this Review. I also gratefully acknowledge the diligent and dedicated collaboration of the graduate students whose names appear as my co-authors in the references.

2.9 REFERENCES

62JOC3557	G. Smolinsky, *J. Org. Chem.* **1962**, 27, 3557.
63CB399	L. Horner, A. Christman, A. Gross, *Chem. Ber.* **1963**, 96, 399.
67JA2077	F.W. Fowler, A. Hassner, L.A. Levy, *J. Am. Chem. Soc.* **1967**, 89, 2077.
67HCA244	H. Tiefenthaler, W. Dörscheln, H. Goth, H. Schmid, *Helv. Chim. Acta* **1967**, 50, 2244.
67TL1545	A. Hassner, F.W. Fowler, *Tetrahedron Lett.* **1967**, 1545.
67TL5315	P. Beak, J.L. Miesel, W.R. Messer, *Tetrahedron Lett.* **1967**, 5315.
69CC1018	J.P. Catteau, A. Leblache-Cambier, A. Pollet, *J. Chem. Soc., Chem. Commun.* **1969**, 1018.
69T3287	P. Beak, W.R. Messer, *Tetrahedron* **1969**, 25, 3287.
70CC386	M. Kojima, M Maeda, *J. Chem. Soc., Chem. Commun.* **1970**, 386.
71BCF1101	G. Vernin, J.C. Poite, J. Metzger, J.P. Aune, H.J.M. Dou, *Bull. Soc. Chim. Fr.* **1971**, 1101.
72BCF2673	C. Riou, G. Vernin, H.J.M. Dou, J. Metzger, *Bull. Soc. Chim. Fr.* **1972**, 3157.
72JCS(P2)1145	G. Vernin, R. Jauffred, C. Richard, H.J.M. Dou, J. Metzger, *J. Chem. Soc., Perkin Trans. 2* **1972**, 1145.
72T3141	A. Lablache-Cambier, A.Pollet, *Tetrahedron* **1972**, 28, 3141.
73BCF1743	G.Vernin, C.Riou, H.J.M. Dou, L. Bouscasse, J. Metzger, G. Loridan, *Bull. Soc. Chim. Fr.* **1973**, 1743.

73TL3523	M. Maeda, M. Kojima, *Tetrahedron Lett.* **1973**, 3523.
74T879	C. Riou, J.C. Poite, G. Vernin, J. Metzger, *Tetrahedron* **1974**, 30, 879.
75CC786	J.A. Barltrop, A.C. Day, P.D. Moxon, R.W. Ward, *J. Chem. Soc., Chem. Commun.* **1975**, 786.
76ACR371	A. Padwa., *Acc. Chem. Res.* **1976**, 9, 371.
76PHC123	A. Lablache-Combier, *Photochemistry of Heterocyclic Compounds*; O. Buchardt, Ed.; Wiley: New York, **1976**; p 123.
78CC131	J.A. Barltrop, A.C. Day, R.W. Ward, *J. Chem. Soc., Chem. Commun.* **1978**, 131.
78H389	M. Maeda, A. Kawahora, M. Kai, M. Kojima, *Heterocycles*, **1978**, 3, 389.
78H1207	K. Isomura, Y. Hirose, H. Shuyama, S. Abe, G-i. Ayabe, H. Taniguichi, *Heterocycles*, **1978**, 9, 1207.
78JCS(P1)685	M. Maeda, M. Kojima, *J. Chem. Soc., Perkin Trans. 1* **1978**, 685.
79CC881	J.A. Barltrop, A.C. Day, E. Irving, *J. Chem. Soc., Chem. Commun.* **1979**, 881.
79CC966	J.A. Barltrop, A.C. Day, E. Irving, *J. Chem. Soc., Chem. Commun.* **1976**, 966.
80RGES501	A. Padwa, *'Rearrangements in Ground and Excited States'*; P. deMayoEd., Academic Press: New York, **1980**; vol. 3, p. 501.
81CC604	J.A. Barltrop, A.C. Day, A.G. Mac, A. Shahrisa, S. Wakamatsu. *J. Chem. Soc., Chem. Commun.* **1981**, 604.
86JOC3176	A. Hassner, N.H. Wiegand, H.E. Gottlieb, *J. Org. Chem.* **1986**, 51, 3176.
91JOC6313	J.W. Pavlik, E.M. Kurzweil, *J. Org. Chem.* **1991**, 56, 6313.
91JOC6321	R.E. Connors, J.W Pavlik, D.S. Burns, E.M. Kurzweil, *J. Org. Chem.* **1991**, 56, 6321.
93JA7645	J.W. Pavlik, R.E. Connors, D.S. Burns, E.M. Kurzweil, *J. Am. Chem. Soc.* **1993**, 115, 7645.
93JOC3407	J.W. Pavlik, C.R. Pandit, C.R.Samuel, A.C. Day, *J. Org. Chem.* **1993**, 58, 3407.
94JA2292	J.W. Pavlik, P. Tongcharoensirikul, N.P.Bird, A.C. Day, J.A. Barltrop, *J. Am. Chem. Soc.* **1994**, 116, 2292.
94PP803	A. Lablache-Combier, *CRC Handbook of Photochemistry and Photobiology*; W.M. Horspool, Song, P.-S. Eds.; CRC Press: New York, **1994**; p. 803.
94PP1063	A. Lablache-Combier, *CRC Handbook of Photochemistry and Photobiology*; W.M. Horspool, P.-S. Song, Eds.; CRC Press: New York, **1994**; p. 1063.
95JOC8138	J.W. Pavlik, N. Kebede, N.P. Bird, A.C. Day, J.A. Barltrop, *J. Org. Chem.* **1995**, 60, 8138.
97JOC8325	J.W. Pavlik, N. Kebede, *J. Org. Chem.* **1997**, 62, 8325.
97OP57	J.W. Pavlik, *Organic Photochemistry*; V. Ramamuthy, K. S. Schanze, Eds.; Marcel Dekker: New York, **1997**; Vol. 1, p. 57.
98JOC5592	J.W. Pavlik, P. Tongcharoensirikul, K.M. French, *J. Org. Chem.* **1998**, 63, 5592.
00JOC3626	J.W. Pavlik, P. Tongcharoensirikul, *J. Org. Chem.* **2001**, 65, 3626.

Chapter 3

Naturally Occurring Halogenated Pyrroles and Indoles

Gordon W. Gribble
Dartmouth College, Hanover, NH, USA
ggribble@dartmouth.edu

3.1 INTRODUCTION

More than 3800 naturally occurring organohalogen compounds are now known to exist, and new examples of these natural products continue to be discovered <03ED40>. This area has been extensively reviewed both in total <92JNP1353, 93MI33, 94JCE907, 95CBR127, 95MI117, 95MI451, 95N2, 96MI1, 96MI16, 96MI155, 97MI375, 98ACR141, 02MI315, 03ED40, 03MI289> and by individual organohalogen (organochlorine <94MI310A, 96PAC1699, 00GC173>; organobromine <99CSR335, 00MI37>; organofluorine <94NPR123, 99JFC127, 02MI121>).

Given the enormous reactivity of the π-excessive heterocycles pyrrole and indole towards electrophilic halogenation <68JCS(B)392, 00MI19>, it is not surprising that the biohalogenation of these heterocycles has led to a large number of natural halogenated pyrroles and indoles. The present chapter focuses mainly on recent discoveries in this area, which have not been previously reviewed, but also summarizes some of the classical examples of these fascinating natural products, some of which have extraordinary biological activity. In addition to halogenated pyrroles and indoles, chapter coverage also includes carbazoles and carbolines. Synthetic and biosynthetic efforts in these areas are not covered.

3.2 PYRROLES

3.2.1 Simple Pyrroles

Several simple halogenated pyrroles are produced by marine organisms. Even though the oceans are richer in chloride than bromide, most marine halogenated compounds, of any type, contain bromine rather than chlorine. This is likely a consequence of the greater susceptibility of bromide to enzymatic oxidation by bromoperoxidase or other peroxidases.

The marine bacterium *Chromobacterium* sp. produces 2,3,4,5-tetrabromopyrrole (**1**) <74MB281>, and the marine acorn worm *Polyphysia crassa* has yielded the antibacterial metabolite 2,3,4-tribromopyrrole (**2**), an extremely labile compound <90JNP703>. The Red Sea sponge *Acanthella carteri* contains bromopyrrole amide (**3**) <97JNP180>, and the isomeric amide (**4**) along with **5** were isolated from the Papua New Guinea sponge *Agelas nakamurai* <98JNP1310>. The dibromo analogue (**6**) of **5** has been characterized from the sponge *Homoaxinella* sp. <98JNP1433>. Although very few simple chlorinated pyrroles are known to occur naturally, two notable exceptions are the nitro-containing pyrrolomycin A (7) isolated from an *Actinomyces* sp. <81JAN1569> and pyrrolnitrin (**8**) produced by several *Pseudomonas* microbes <90JAFC538>. A *Streptomyces* sp. has yielded the optically active neopyrrolomycin (**9**) <90JAN1192>.

1

2

3

4 R = H
5 R = CH_2OCH_3

6

7

8

9

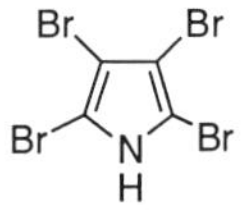

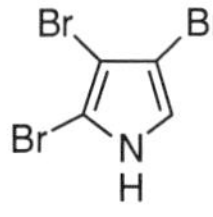

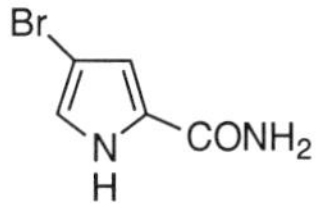

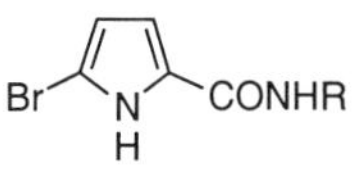

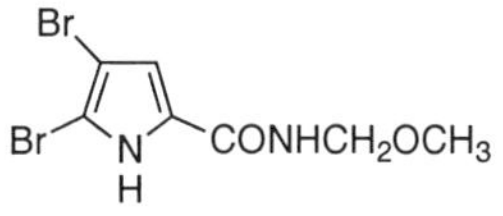

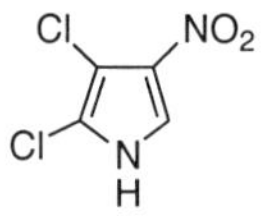

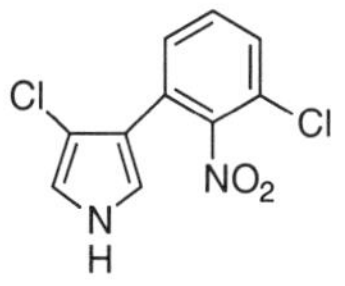

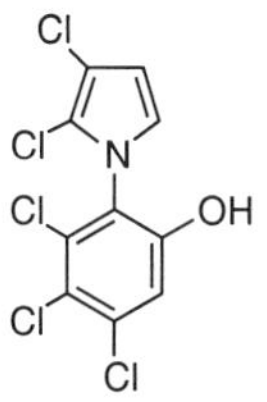

Recent years have seen the isolation from marine sponges of several brominated pyrroles with a more complex amide side chain. These include slagenin A (**10**) (*Agelas nakamurai*) <99TL5709>, **11** (*Agelas wiedenmayeri*) <99OL455>, mukanadin B (**12**) (*Agelas nakamurai*) <99JNP1581>, taurodispacamide A (**13**) <00TL9917>, sventrin (**14**) (*Agelas sventres*) <01JNP1593>, and **15** (*Axinella brevistyla*) <01JNP1576>.

10 (slagenin A)

11

12 (mukanadin B)

13 (taurodispacamide A)

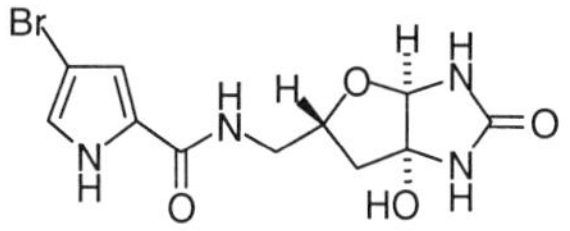

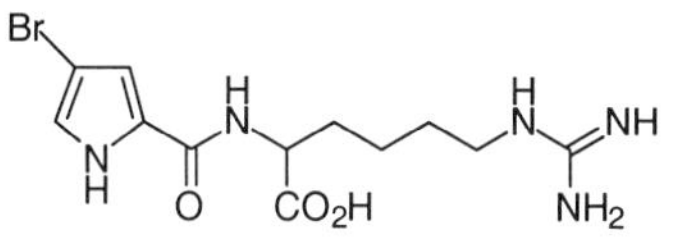

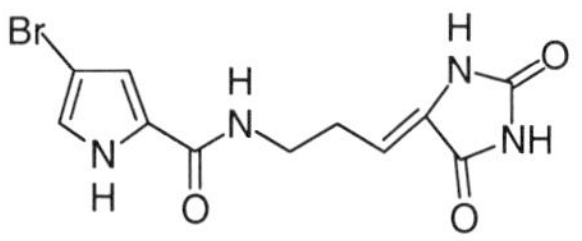

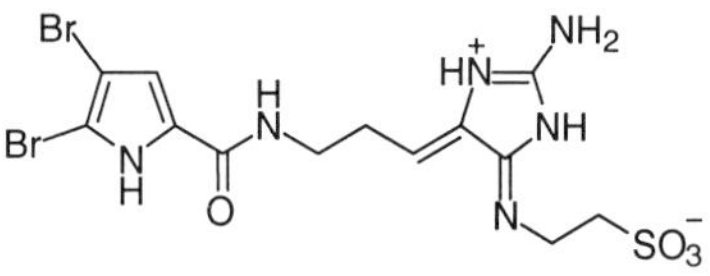

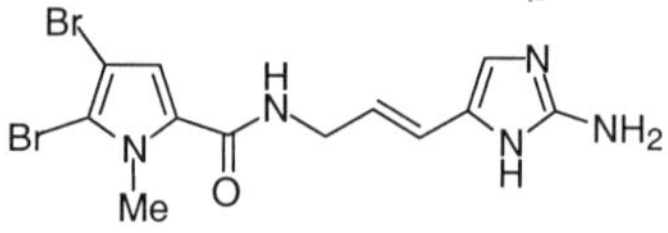

14 (sventrin) 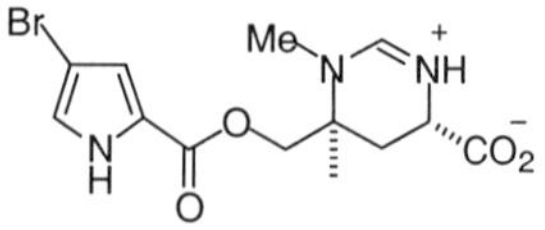   **15**

3.2.2 Complex Pyrroles

Halogenated pyrrole rings are also embodied in more complex natural products. Some recent examples of these structurally unique compounds are presented here. Marine sponges are a prolific source of halogenated pyrroles, including ugibohlin (**16**) (*Axinella carteri*) <01JNP1581>, tetracyclic **17** (*Axinella brevistyla*) <01JNP1576>, debromostevensine (**18**) (*Stylissa carteri*) <99JNP184>, and *N*-methyldibromoisophakellin (**19**) (*Stylissa caribica*) <01JNP1345>.

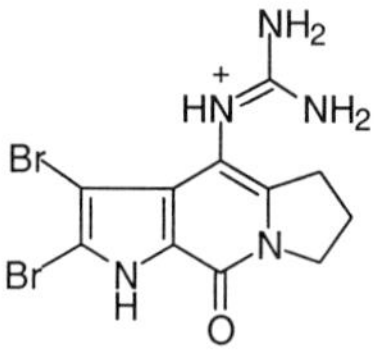

16 (ugibohlin) 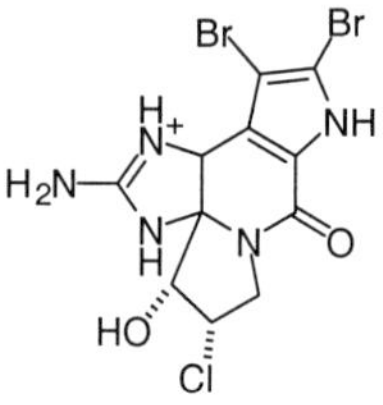   **17** 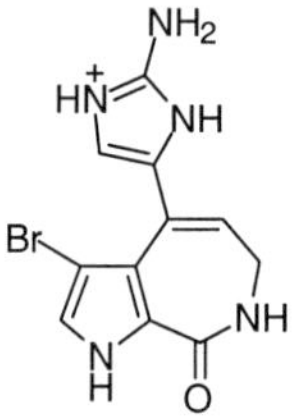   **18** (debromostevensine) 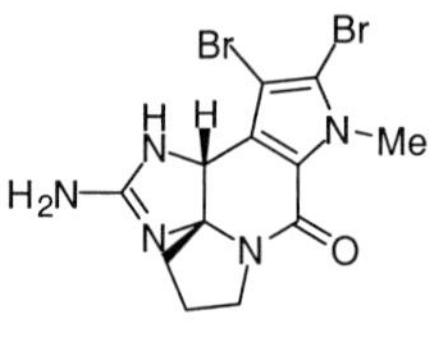   **19**

Several bis-bromopyrroles have been discovered in sponges, including the antifouling spermidine metabolite pseudoceratidine (**20**) (*Pseudoceratina purpurea*) <96TL1439>, nakamuric acid (**21**) (*Agelas nakamurai*) <99JNP1295>, and the highly complex axinellamine A (**22**) (*Axinella* sp.) <99JOC731>

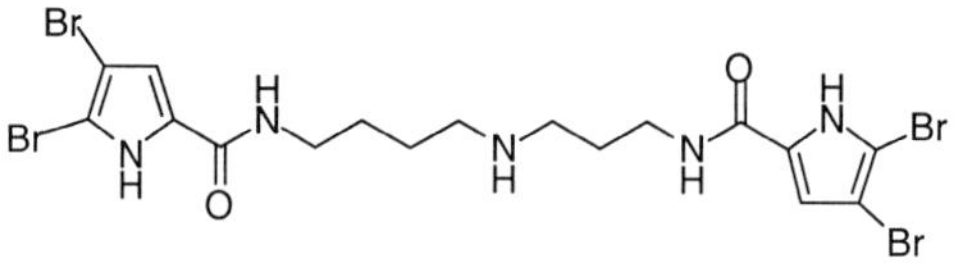

20 (pseudoceratidine)

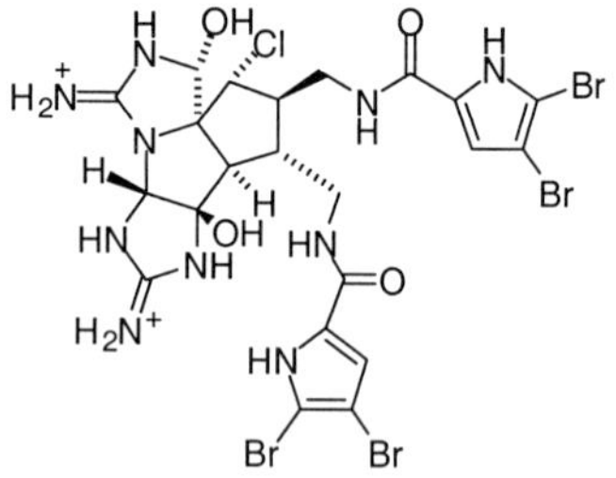

22 (axinellamine A)

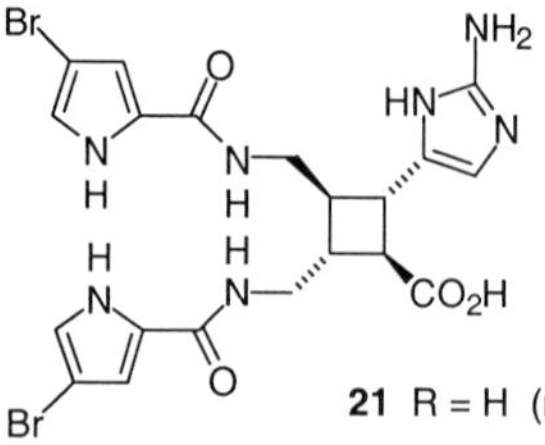

21 R = H (nakamuric acid)

Although complex chlorinated pyrroles are not generally found in marine organisms, these compounds are produced in abundance by terrestrial microorganisms and fungi. The Arizona soil fungus *Auxarthron conjugatum* has furnished the novel metabolites auxarconjugatins A (**23**) and B (**24**) <99P459>. Cultures of *Streptomyces rimosus* have yielded several new streptopyrroles, e.g., **25-30** <00JAN1>, following an earlier report on the isolation of streptopyrrole (**31**) itself from *Streptomyces armeniacus* <98ACS1040>. The new antibiotics decatromicins A (**32**) and B (**33**) are produced by *Actinomadura* sp. MK 73-NF4 <99JAN781, 99JAN787>. These compounds inhibit Gram-positive bacteria including methicillin-resistant *Staphylococcus aureus*. A *Streptomyces* sp. has yielded the macrolide colubricidin A (**34**) <99TL9219>.

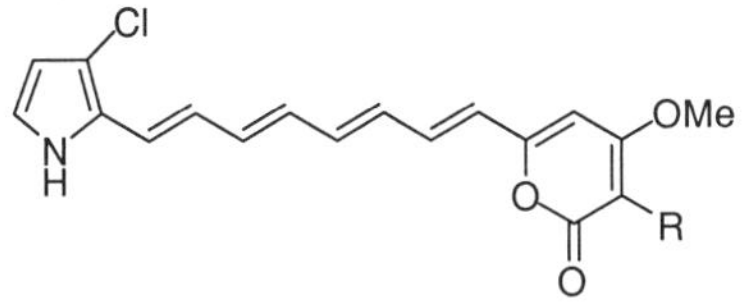

23 R = Me (auxarconjugatin A)
24 R = H (auxarconjugatin B)

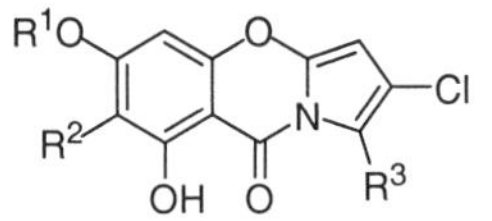

25 $R^1 = R^3 = H, R^2 = Et$
26 $R^1 = H, R^2 = Pr, R^3 = Cl$
27 $R^1 = Me, R^2 = Pr, R^3 = H$
28 $R^1 = R^3 = H, R^2 = Bu$
29 $R^1 = Me, R^2 = Et, R^3 = H$
30 $R^1 = Me, R^2 = Pr, R^3 = Cl$
31 $R^1 = R^3 = H, R^2 = Pr$

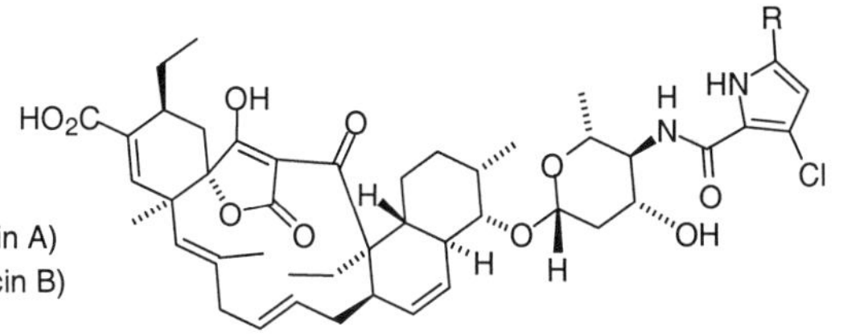

32 R = H (decatromicin A)
33 R = Cl (decatromicin B)

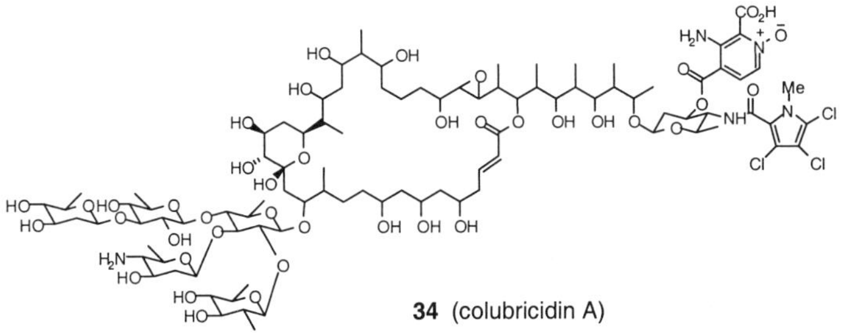

34 (colubricidin A)

3.2.3 Bipyrroles

Although few in number, halogenated bipyrroles represent one of the most interesting and potentially important classes of natural organohalogens yet to be discovered. Like DDT, PCBs, and other anthropogenic organochlorine compounds, some halogenated bipyrroles seem to bioaccumulate up the food chain and at least one is found in humans.

The first example of a halogenated bipyrrole was the marine bacterium *Chromobacterium* sp. metabolite **35** <74MB281>. Both nudibranchs (sea slugs) and bryozoans (moss animals) secrete brominated bipyrroles ("tambjamines") as chemical defensive agents. Some recent examples from the bryozoan *Bugula dentata* are tambjamines G-J (**36-39**) <94AJC1625>.

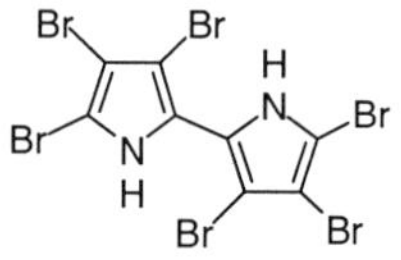

35

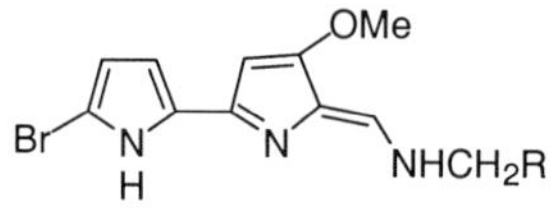

36 R = CH_3 (tambjamine G)
37 R = CH_2CH_3 (tambjamine H)
38 R = $CH(CH_3)_2$ (tambjamine I)
39 R = $CH(CH_3)CH_2CH_3$ (tambjamine J)

A search for PCBs (polychlorinated biphenyls) in the eggs of Pacific and Atlantic Ocean seabirds (albatross, puffin, gull, petrel, auklet) and in bald eagle liver samples revealed the presence of at least four halogenated bipyrroles of unknown structure <99EST26>. The structures of the two major compounds were established by synthesis as **40** and **41** <99CC2195>. These halogenated bipyrroles have also been found in the Arctic marine food web (plankton, fish, seabirds, seal) <02EP85>, thus giving further credence to the notion that these compounds have a marine bacterial origin, similar to that of bipyrrole **35**, and have a widespread environmental presence. The structures of the other two major halogenated bipyrroles in these organisms have not been confirmed by synthesis, but are tentatively thought to be **42** and **43**. A related bipyrrole heptachloro-1,2'-bipyrrole "Q1" (**44**) has been found in a myriad of marine fishes and mammals, and also in the milk of Eskimo women who eat whale blubber <01AECT221, 00EP401>. The structure of Q1 (**44**) has been established by synthesis <02AG(E)1740>. This novel compound, which has no obvious anthrogenic origin, is probably derived from marine bacteria.

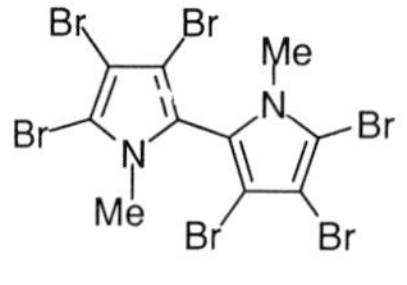

40 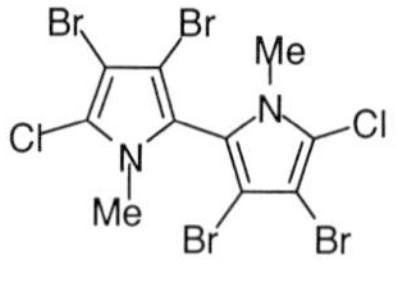   **41** 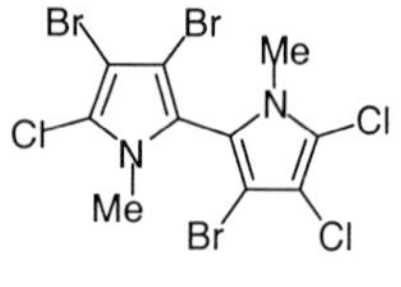   **42**

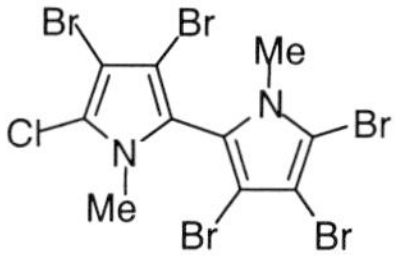

3.3 INDOLES

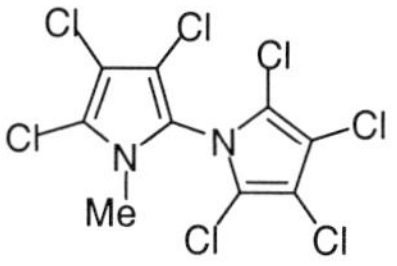

3.3.1 Simple Indoles

Given the ubiquity of the amino acid tryptophan in all living organisms, it comes as no surprise that numerous halogenated tryptophan derivatives and indoles are found to occur naturally.

Several brominated and chlorinated indoles have been isolated from marine organisms, especially acorn worms and seaweeds <96MI1>. Some notable examples are the labile 3-chloroindole (**45**), 3-bromoindole (**46**), and 3-chloro-6-bromoindole (**47**) from the acorn worm *Ptychodera flava laysanica* <75N395>. These compounds are responsible for the peculiar "iodoform-like" odor of this animal. The red alga *Laurencia brongniartii* has yielded four polybromoindoles, e.g., **48** <78TL4479>, and the New Zealand red seaweed *Rhodophyllis membranacea* contains ten tri- and tetrahaloindoles, e.g., **49** <78TL1637>. Recently, the Palauan ascidian *Distaplia regina* furnished 3,6-dibromoindole (**50**) <99NPL59>, which had been previously isolated from an acorn worm <80CBP525>.

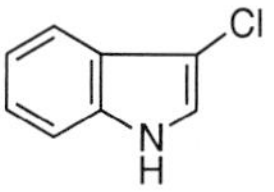

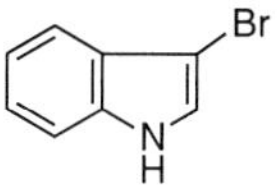

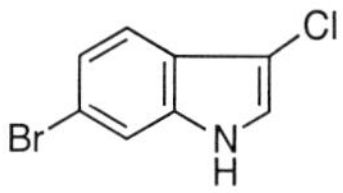

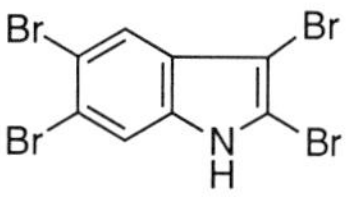

More highly substituted halogenated simple indoles are also known to occur in nature. Thus, the gastropod *Drupella fragum*, which is a predator of Madreporaria corals, contains in its mid-intestinal gland the novel brominated hydroxyindoles **51-53**, which are also potent antioxidants <98JNP1043>. The unusual sulfate-sulfamate chlorinated ancorinolates A (**54**) and C (**55**) were characterized from the sponge *Ancorina* sp. along with several non-chlorinated compounds

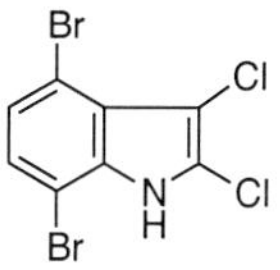

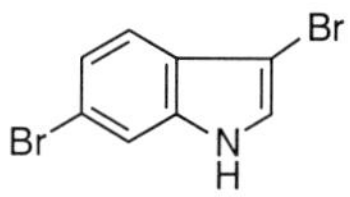

<02JOC6671>. Matemone (**56**) was found in a recent study of the Indian Ocean sponge *Iotrochota purpurea* along with the previously known 6-bromoindole-3-carbaldehyde <00JNP981>. Both compounds exhibit inhibition of the division of sea-urchin eggs. The bryozoan *Amathia convoluta* produces convolutamydine E (**57**) <99CCC1147> and convolutindole A (**58**) <02JNP938> in separate collections from Florida and Tasmania, respectively. The tribromogramine **59**, which was isolated from the bryozoan *Zoobotryon pellucidum*, is a potent inhibitor of larval settlement by the notorious encrusting barnacle *Balanus amphitrite* <94MI773, 94MI2178>.

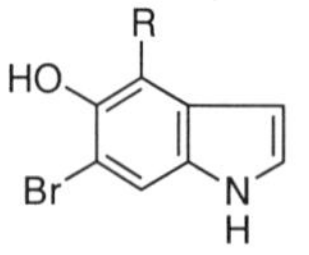

51 R = H
52 R = OH

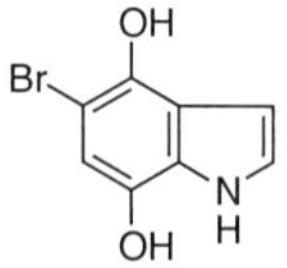

53

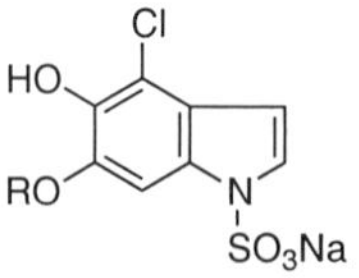

54 R = SO$_3$Na (ancorinolate A)
55 R = H (ancorinolate C)

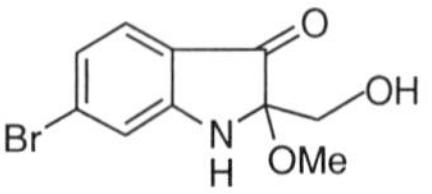

56 (matemone)

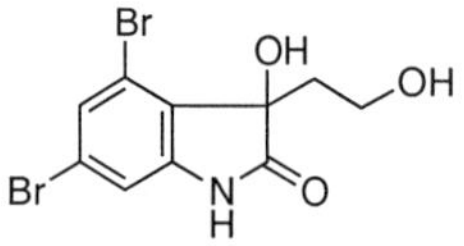

57 (convolutamydine E)

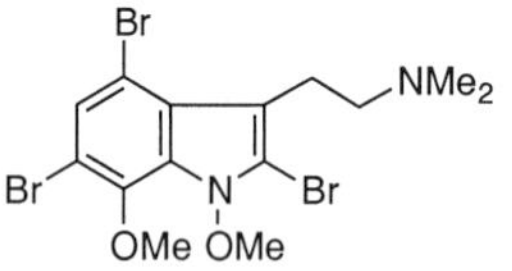

58 (convolutindole A)

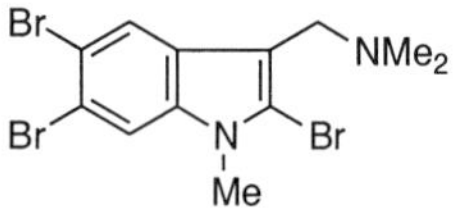

59

Although organoiodines are exceedingly rare in nature — only 100 are known — some iodinated indoles have been found in marine organisms. The Caribbean sponge *Plakortis simplex* has yielded the tryptophan betaines plakohypaphorines A-C (**60-62**) <03EJO284>. The prolific bryozoan *Flustra foliacea* from the North Sea has afforded bromoindoles **63-65**, and the physostigmine-like **66** <02JNP1633>.

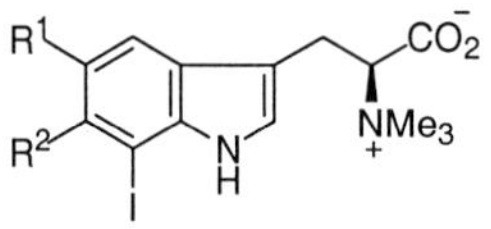

60 R^1 = R^2 = H (plakohypaphorine A)
61 R^1 = H, R^2 = I (plakohypaphorine B)
62 R^1 = I, R^2 = H (plakohypaphorine C)

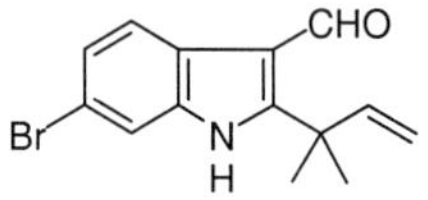

63

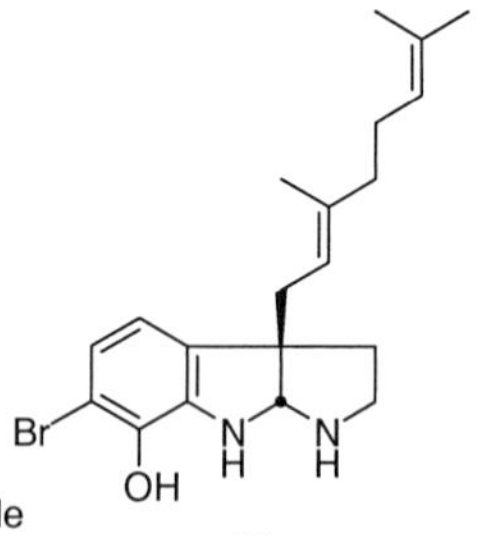

66

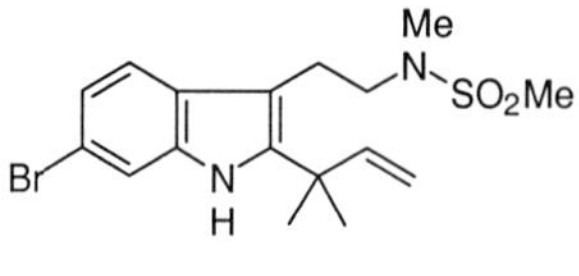

64

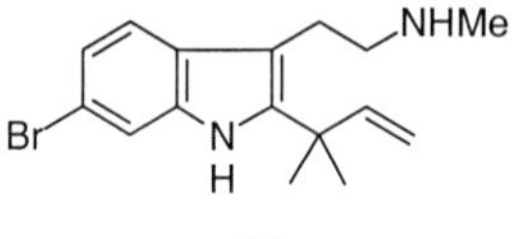

65

A relatively large collection of brominated tryptophan derivatives occurs in marine organisms <96MI1>. The New Zealand ascidian *Pycnoclavella kottae* contains four brominated kottamides, e.g., kottamide A (**67**), which display antiinflammatory, antimetabolic, and antitumor activity <02JOC5402>. The sponge *Hyrtios erecta* has yielded the new aplysinopsin **68** <01CPB1372>. The venomous cone snails *Conus imperialis*, *Conus radiatus*, and other members of the *Conus* genus, of which 500 species are known, possess peptides ("conotoxins") that invariably contain 6-bromotryptophan (**69**) <97JBC4689, 99MI239>. The ascidian *Phallusia mammillata* contains the polypeptide morulin Pm, which is rich in 6-bromotryptophan <97ABB278>.

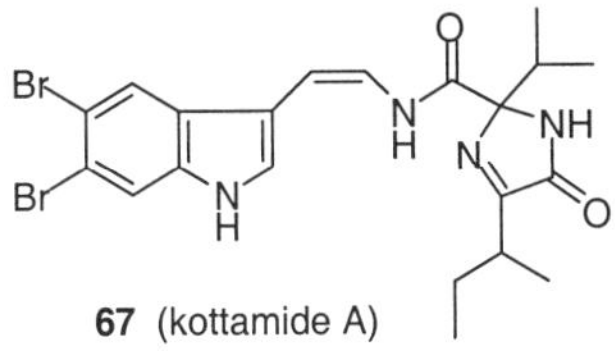

67 (kottamide A) 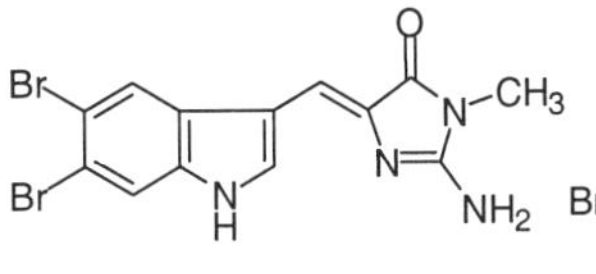   **68** 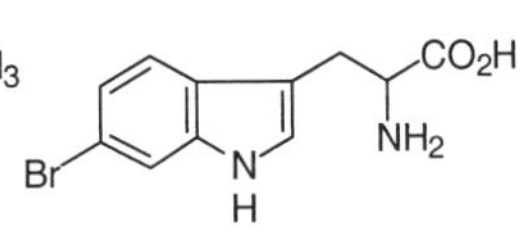   **69**

3.3.2 Complex Indoles

Like their pyrrole counterparts, halogenated indoles are found embedded in several highly complex natural products, which are probably tryptophan-derived. No other marine organisms surpass the unobtrusive and nondescript bryozoans (moss animals) in their synthetic virtuosity. Indeed, the bryozoan *Chartella papyracea* produces an array of stunningly intricate indolo-ß-lactams, e.g., chartelline A (**70**) <85JA4542, 87JOC4709> and chartellamide B (**71**) <87JOC5638>. The bryozoan *Hincksinoflustra denticulata* has yielded hinckdentine A (**72**), which embraces the novel pyrimidino[3',4':1,2]pyrrolo[2,3-*d*]azepine ring system <87TL5561>. In addition to these classic examples, the bryozoan *Securiflustra securifrons* has been found to contain a series of securamines, e.g., securamine E (**73**) <97JNP175, 96JOC887>.

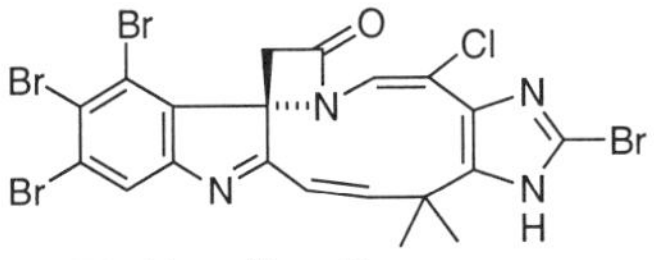

70 (chartelline A)

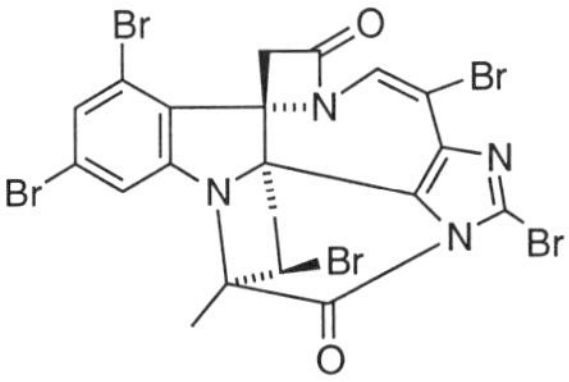

71 (chartellamide B)

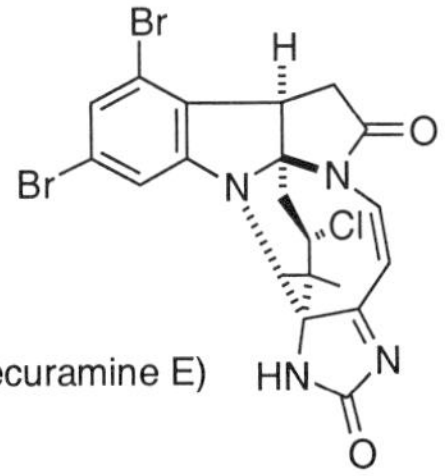

73 (securamine E)

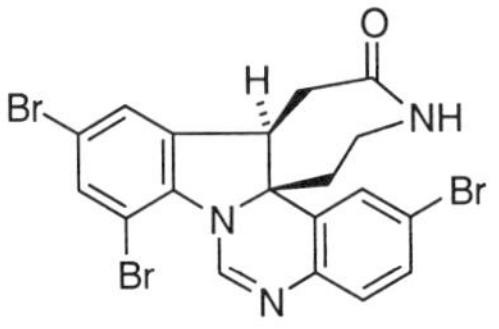

72 (hinckdentine A)

Other marine animals also possess complex bromoindoles, and two recent examples are 6-bromogranulatimide (**74**) from the Brazilian ascidian *Didemnum granulatum* <01JNP254>, and dragmacidin F (**75**) from the Mediterranean sponge *Halicortex* sp. <00T3743>. The first two ergoline marine alkaloids, pibocin A (**76**) and pibocin B (**77**), were isolated from a Far-Eastern *Eudistoma* sp. ascidian <99TL1591, 01JNP1559>.

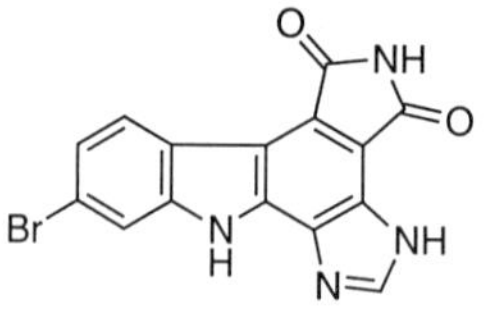

74 (6-bromogranulatimide)

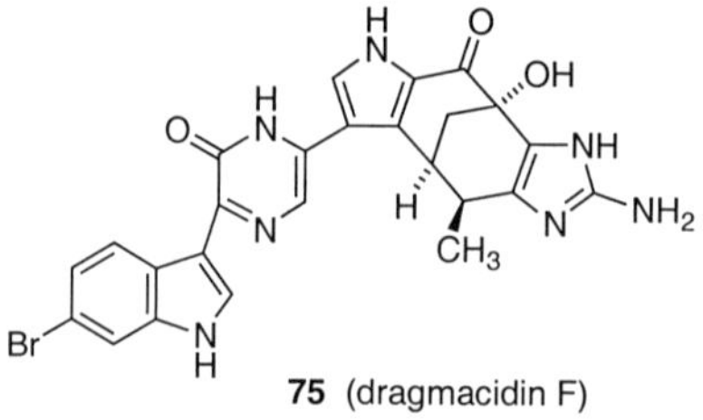

75 (dragmacidin F)

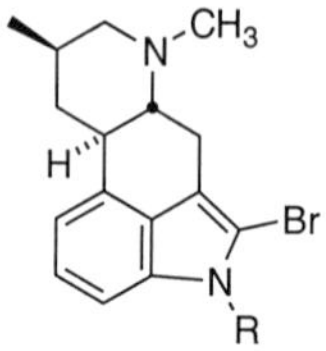

76 R = H (pibocin A)
77 R = OMe (pibocin B)

In addition to the 6-bromotryptophan-containing cone snail peptides (*vide supra*), halogenated indoles are found in a number of cyclic peptides. Several of these compounds have been summarized previously <96MI1>. Recently characterized examples are jaspamides B (**78**) and C (**79**) from the Vanuatu sponge *Jaspis splendans* <99JNP332>, keramamide E (**80**) from the Okinawan sponge *Theonella* sp. <95T2525>, and cyclocinamide A (**81**) from the sponge *Psammocinia* sp. <97JA9285>.

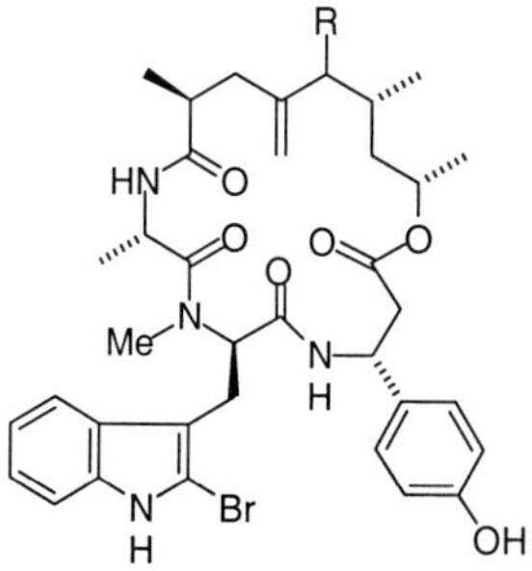

78 R = =O (jaspamide B)
79 R = OH (jaspamide C)

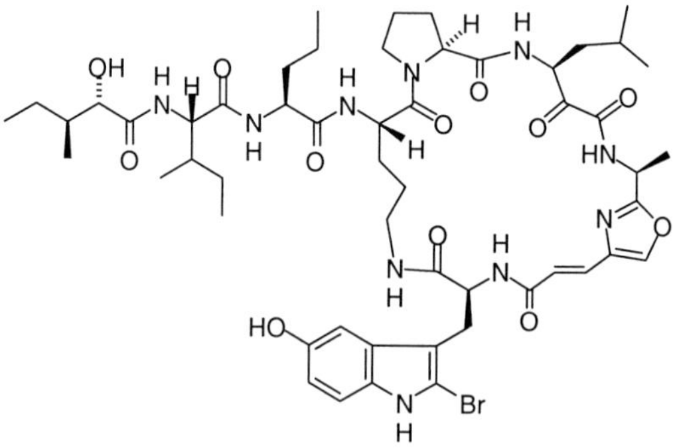

80 (keramamide E)

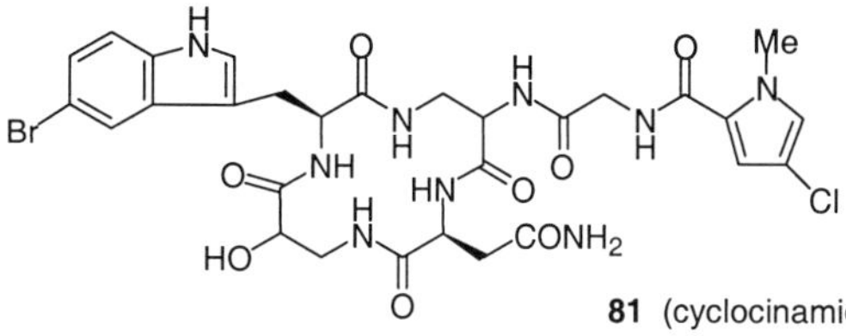

81 (cyclocinamide A)

Several cyclic peptides that contain a 6-chloroindole unit have been isolated and characterized. Microsclerodermins C (**82**) and D (**83**) are found in the sponge *Theonella* sp., and **83** is also present in the sponge *Microscleroderma* sp. <98T3043>. An Okinawan *Theonella* sp. sponge has afforded keramamide L (**84**) <98T6719>.

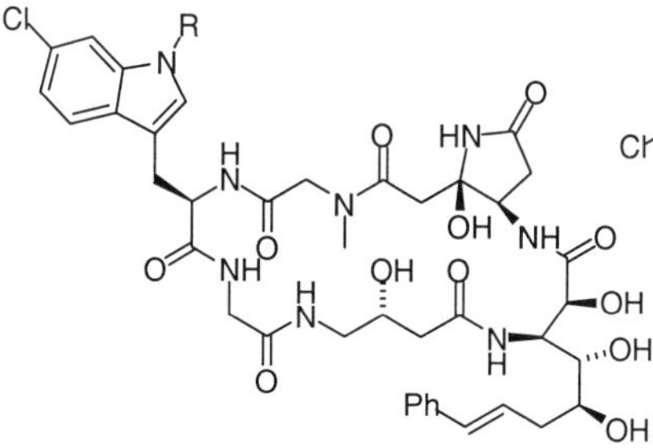

82 R = CONH$_2$ (microsclerodermin C)
83 R = H (microsclerodermin D)

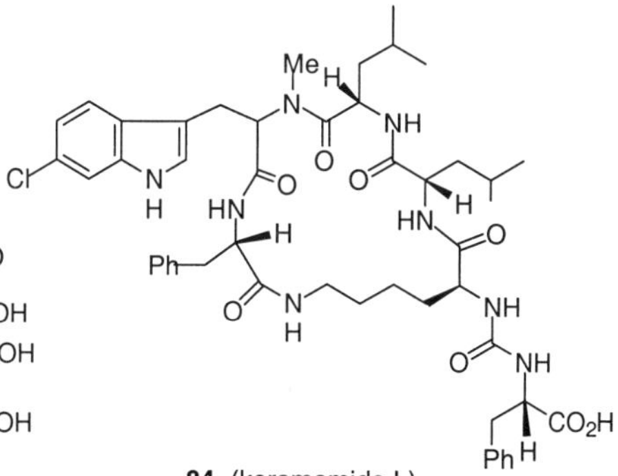

84 (keramamide L)

Terrestrial organisms also produce some chlorinated indoles. In addition to the classical example of the plant growth hormone 4-chloroindoleacetic acid (**85**) and the methyl ester (**86**), which are produced by fava beans, lentil, pea, and vetch <97P675, 96MI1>, several chlorinated indole alkaloid isoprenoids are produced by fungi. For example, *Penicillium crustosum* has afforded thomitrem A (**87**) <02P979>, which is closely related to the penitrems <96MI1>.

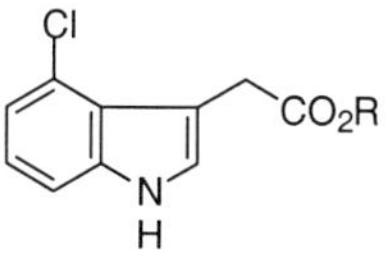

85 R = H
86 R = Me

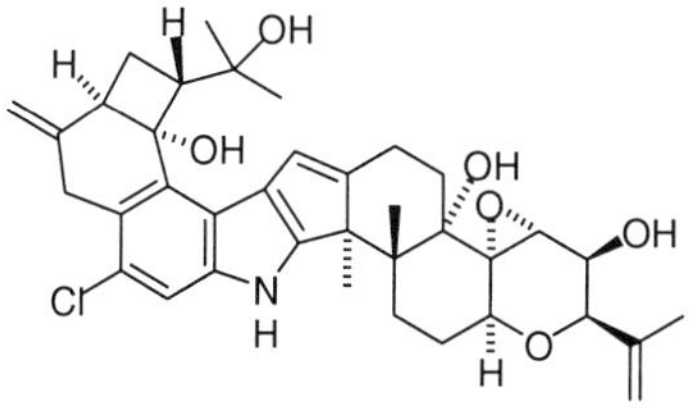

87 (thomitrem A)

3.3.3 Biindoles

The prototypical example of a halogenated biindole is Tyrian Purple (**88**) (dibromoindigo), the ancient Egyptian dye extracted from Mediterranean molluscs and the focus of a significant industry <74MI11, 90ACR152, 92MI 69, 96MI1, 01JCE1442, 01JCE1444>. This mollusc, *Murex trunculus*, has now yielded 6-bromoindigotin (**89**) <92MI145>. A *Streptomyces* sp. produces the novel chlorinated indigo glycosides akashins A-C (**90-92**) for which significant antitumor activity is found <02AG(E)597>.

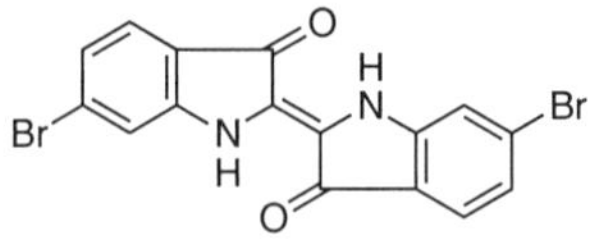

88 (Tyrian Purple)

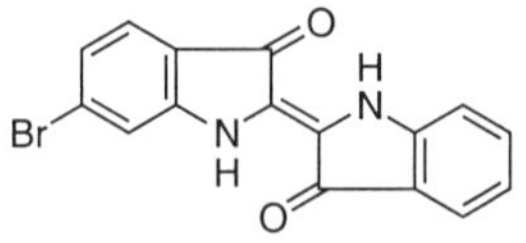

89

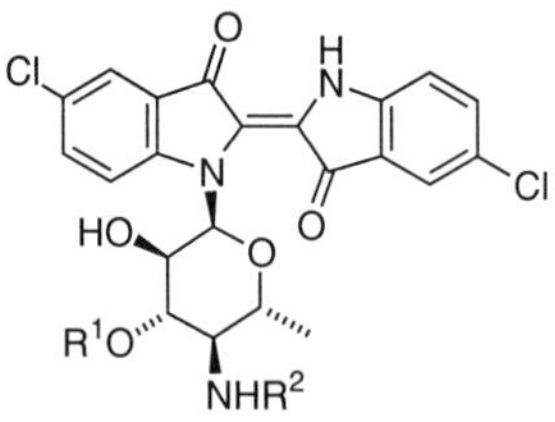

90 $R^1, R^2 = H$ (akashin A)
91 $R^1 = H, R^2 = COCH_3$ (akashin B)

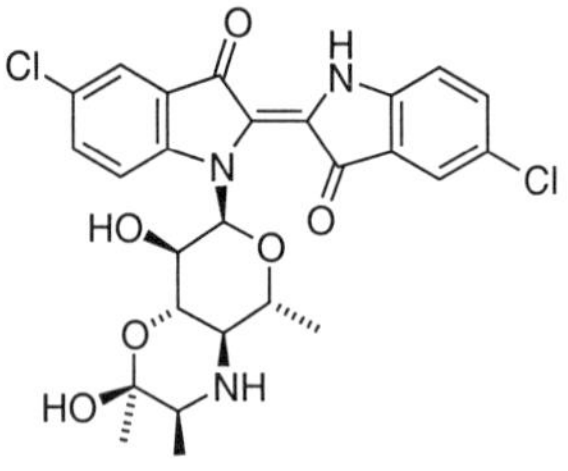

92 (akashin C)

3.3.4 Bis-Indoles

Numerous naturally occurring halogenated compounds have been identified that contain two or even three indole units. For example, the New Caledonian sponge *Orina* sp. contains the brominated indoles **93** and **94** <95JNP1254>. Several examples of these brominated bis-indoles are tethered by an imidazole ring, and two recent examples are the topsentin derivatives **95** and **96** from the sponge *Spongosorites genitrix* <99JNP647>. The sponge *Rhaphisia lacazei* contains seven brominated indoles, e.g., **97** and **98**, all related to the well-known topsentin and hamacanthin <00JNP447>. The South Australian sponge *Echinodictyum* sp. has yielded the novel echinosulfonic acids B and C (**99-100**) and echinosulfone A (**101**), the latter which is a very unusual stable *N*-carboxylic acid <99JNP1246>.

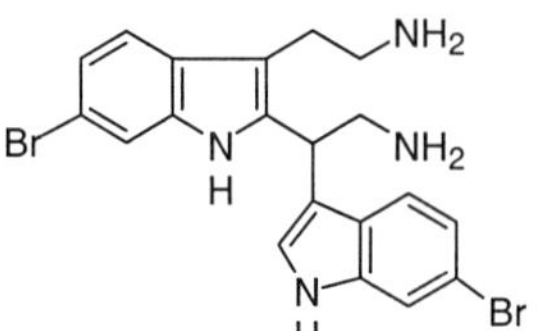

93

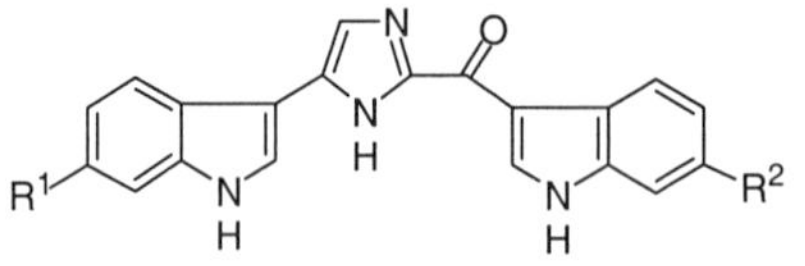

95 $R^1 = Br, R^2 = H$ (bromodeoxytopsentin)
96 $R^1 = H, R^2 = Br$ (isobromodeoxytopsentin)

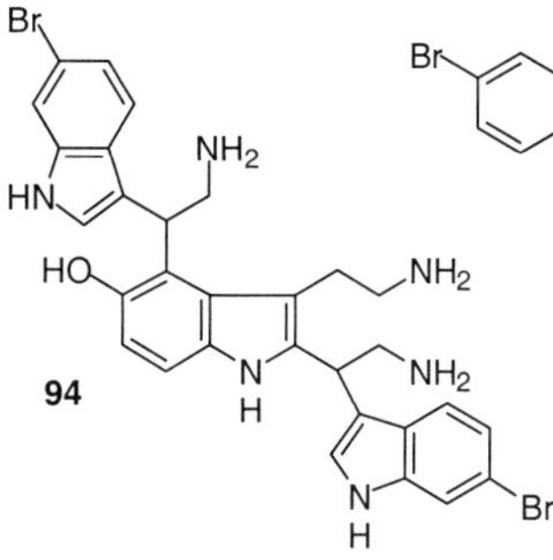

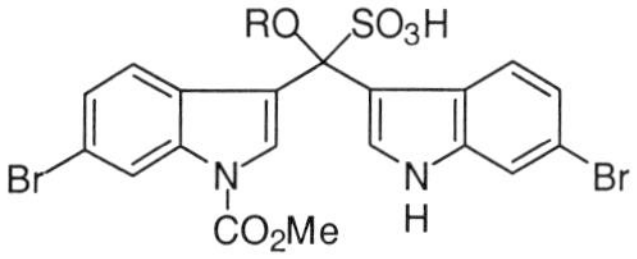

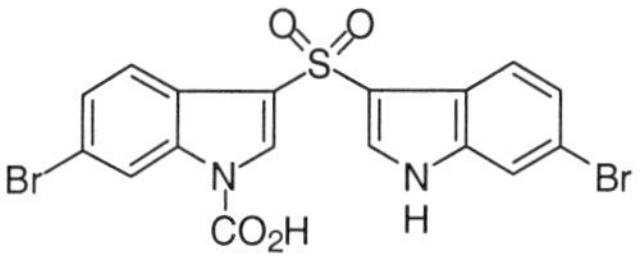

99 R = Me (echinosulfonic acid B)
100 R = H (echinosulfonic acid C)

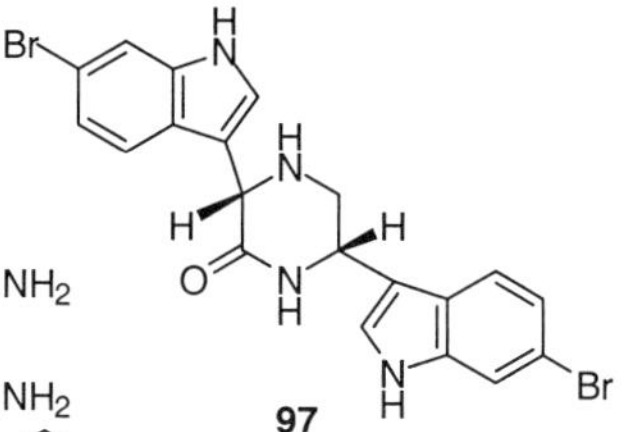

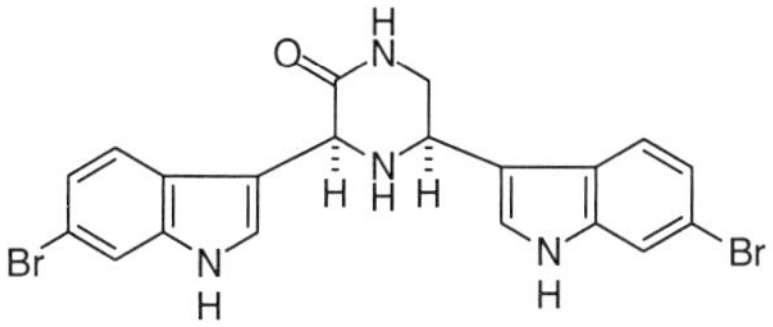

101 (echinosulfone A)

The stony coral *Tubastraea* sp. produces tubastrindole A (**102**), in addition to some nonhalogenated analogues <03TL2533>. Species of this interesting hard coral, which contains several other novel brominated compounds <96MI1>, is avoided by the highly destructive predatory Crown-of-Thorns seastar *Acanthaster planci* <86JCR(S)450>. A *Streptomyces* sp. has yielded the complex bis-indole chloptosin (**103**), which induces apoptosis and shows strong antimicrobial activity against Gram-positive bacteria including methicillin-resistant *Staphylococcus aureus* <00JOC459>.

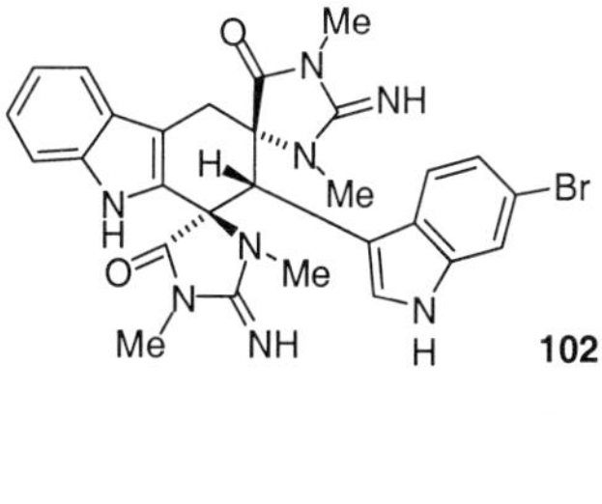

102 (tubastrindole)

103 (chloptosin)

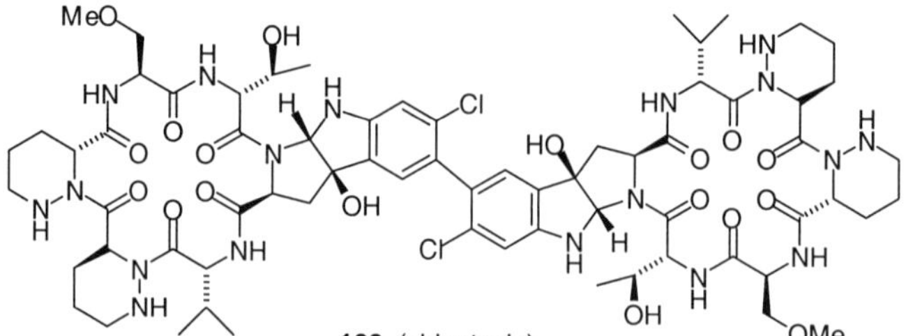

3.3.5 Carbazoles and Carbolines

Compared to the number of halogenated pyrroles and indoles, the number of known halogenated carbazoles, carbolines, indolocarbazoles, and related compounds is much smaller <96MI1>, and the number of recently discovered examples of these types is fewer still.

The encrusting bacterium *Kyrtuthrix maculans* contains the halogenated carbazoles **104-106**, the first examples of bromine- and iodine-containing carbazoles to be discovered <99P537>. The only previous examples of halogenated carbazoles are chlorohyellazole (**107**, produced by the blue-green alga *Hyella caespitosa* <79TL4915>, and 3-chlorocarbazole (**108**) which was isolated from bovine urine and is a potent monoxime oxidase inhibitor <83JNP852, 86JNT147>.

104

105

106

107 (chlorohyellazole)

108

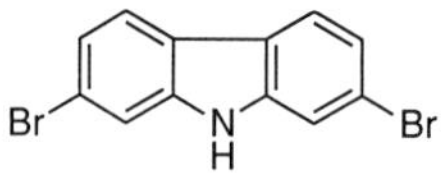

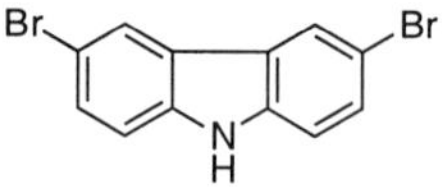

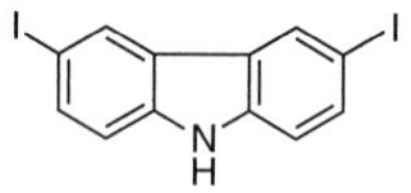

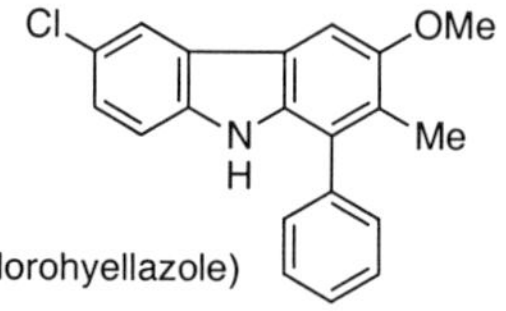

The tunicate genus *Eudistoma* has yielded most of the extant halogenated ß-carbolines, some of which have significant antiviral (Polio, Herpes) and antimicrobial activity. For example, the Caribbean tunicate *Eudistoma olivaceum* produces at least 15 brominated carbolines <96MI1>. A recent study of *Eudistoma gilboverde* has discovered the novel eudistomins **109-111** <01JNP1454>. The Australian ascidian *Pseudodistoma aureum* contains eudistomin V (**112**) <98JNP959>, and the Palauan sponge *Plakortis nigra*, which was collected at a depth of 380 feet, furnished plakortamines A-D (**113-116**) <02JNP1258>. These compounds exhibit activity against the HCT-116 human colon tumor cell line with plakortamine B (**114**) being the most active.

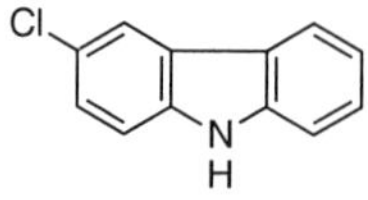

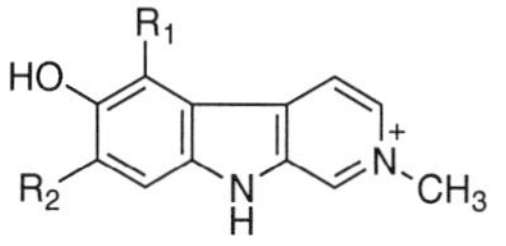

109 R₁ = Br, R₂ = H (2-methyleudistomin D)
110 R₁ = H, R₂ = Br (2-methyleudistomin J)

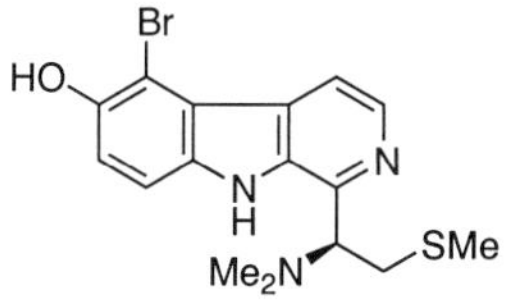

111

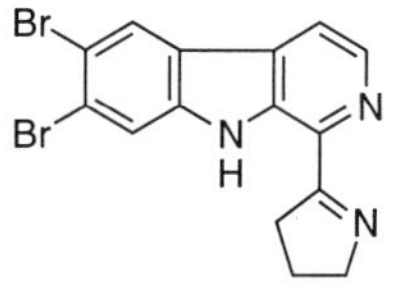

112 (eudistomin V)

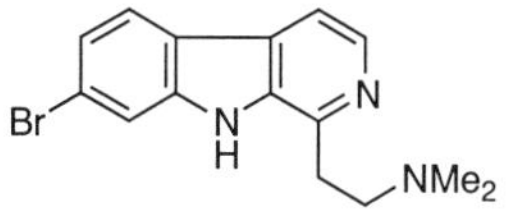

113 (plakortamine A)

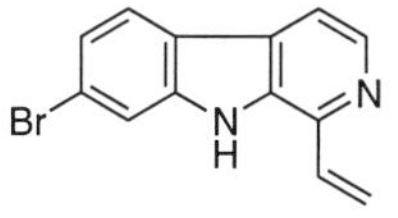

114 (plakortamine B)

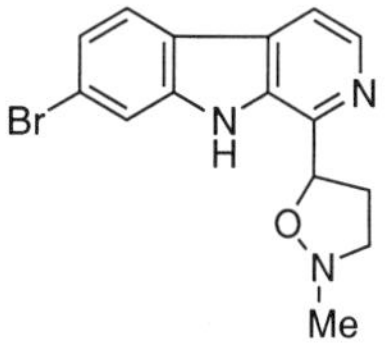

116 (plakortamine D)

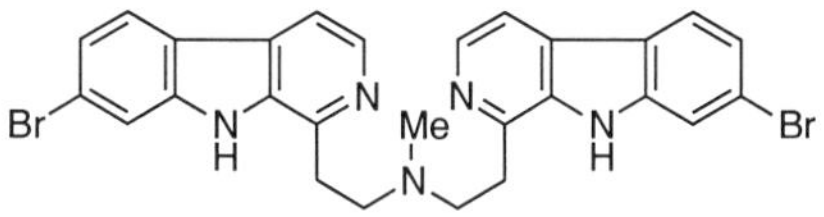

115 (plakortamine C)

3.4 CONCLUDING REMARKS

Although naturally occurring organohalogens represent a relatively new category of natural products, the discovery of such compounds has increased rapidly in recent years, and halogenated pyrroles and indoles are a major subset of this class.

This increase parallels a general revitalization of natural products research in the search for new medicinal drugs. At least in part, this renewed and increased activity in natural products is attributed to modern collection methods (e.g., SCUBA and remote submersibles for the collection of previously inaccessible marine organisms), selective bioassays for identifying biologically active compounds, powerful multidimensional NMR and mass spectral techniques for characterizing sub-milligram amounts of compounds, and new separation and purification techniques (e.g., counter-current chromatography, HPLC, LC-MS). Finally, knowledge and appreciation of folk medicine and ethnobotany will continue to guide chemists to new biologically active natural products, including organohalogen compounds.

3.5 REFERENCES

68JCS(B)392	P. Linda, G. Marino, *J. Chem. Soc. (B)* **1968**, 392.
74MB281	R.J. Andersen, M.S. Wolfe, D.J. Faulkner, *Mar. Biol.* **1974**, *27*, 281.
74MI11	J.T. Baker, *Endeavour* **1974**, *33*, 11.
75N395	T. Higa, P.J. Scheuer, *Naturwissenschaften* **1975**, *62*, 395.
78TL1637	M.R. Brennan, K.L. Erickson, *Tetrahedron Lett.* **1978**, 1637.
78TL4479	G.T. Carter, K.L. Rinehart, L.H. Li, S.L. Kuentzel, J.L. Connor, *Tetrahedron Lett.* **1978**, 4479.
79TL4915	J.H. Cardellina, M.P. Kirkup, R.E. Moore, J.S. Mynderse, K. Seff, C.J. Simmons, *Tetrahedron*

	Lett. **1979**, 4915.
80CBP525	T. Higa, T. Fujiyama, P.J. Scheuer, *Comp. Biochem. Physiol.* **1980**, *65B*, 525.
81JAN1569	M. Koyama, Y. Kodama, T. Tsuruoka, N. Ezaki, T. Niwa, S. Inouye, *J. Antibiot.* **1981**, *34*, 1569.
83JNP852	K.-C. Luk, L. Stern, M. Weigele, R.A. O'Brien, N. Spirt, *J. Nat. Prod.* **1983**, *46*, 852.
85JA4542	L. Chevolot, A.-M. Chevolot, M. Gajhede, C. Larsen, U. Anthoni, C. Christophersen, *J. Am. Chem. Soc.* **1985**, *107*, 4542.
86JCR(S)450	R. Sanduja, M.A. Alam, G.M. Wellington, *J. Chem. Res.(S)* **1986**, 450.
86JNT147	D. Dewar, V. Glover, J. Elsworth, M. Sandler, *J. Neural. Transm.* **1986**, *65*, 147.
87JOC4709	U. Anthoni, L. Chevolot, C. Larsen, P.H. Nielsen, C. Christophersen, *J. Org. Chem.* **1987**, *52*, 4709.
87JOC5638	U. Anthoni, K. Bock, L. Chevolot, C. Larsen, P.H. Nielsen, C. Christophersen, *J. Org. Chem.* **1987**, *52*, 5638.
87TL5561	A.J. Blackman, T.W. Hambley, K. Picker, W.C. Taylor, N. Thirasasana, *Tetrahedron Lett.* **1987**, *28*, 5561.
88H719	M. Alam, R. Sanduja, G.M. Wellington, *Heterocycles* **1988**, *27*, 719.
90ACR152	P.E. McGovern, R.H. Michel, *Acc. Chem. Res.* **1990**, *23*, 152.
90JAFC538	J.N. Roitman, N.E. Maloney, W.J. Janisiewicz, M. Benson, *J. Agric. Food Chem.* **1990**, *38*, 538.
90JAN1192	T. Nogami, Y. Shigihara, N. Matsuda, Y. Takahashi, H. Naganawa, H. Nakamura, M. Hamada, Y. Muraoka, T. Takita, Y. Iitaka, T. Takeuchi, *J. Antibiot.* **1990**, *43*, 1192.
90JNP703	R. Emrich, H. Weyland, K. Weber, *J. Nat. Prod.* **1990**, *53*, 703.
92JNP1353	G.W. Gribble, *J. Nat. Prod.* **1992**, *55*, 1353.
92MI69	R.H. Michel, J. Lazar, P.E. McGovern, *Archeomaterials* **1992**, *6*, 69.
92MI145	R.H. Michel, J. Lazar, P.E. McGovern, *J. Dyers Colourists* **1992**, *108*, 145.
93MI33	K. Naumann, *Chem. Zeit.* **1993**, *27*, 33.
94AJC1625	A.J. Blackman, C. Li, *Aust. J. Chem.* **1994**, *47*, 1625.
94JCE907	G.W. Gribble, *J. Chem. Ed.* **1994**, *71*, 907.
94MI310A	G.W. Gribble, *Environ. Sci. Technol.* **1994**, *28*, 310A.
94MI773	K. Kon-ya, N. Shimidzu, K. Adachi, W. Miki, *Fisheries Sci.* **1994**, 773.
94MI2178	K. Kon-ya, N. Shimidzu, W. Miki, M. Endo, *Biosci. Biotech. Biochem.* **1994**, *58*, 2178.
94NPR123	D.B. Harper, D. O'Hagan, *Nat. Prod. Rep.* **1994**, *11*, 123.
95CBR127	E.J. Hoekstra, E.W.B. DeLeer, *Chem. Br.* **1995**, 127.
95JNP1254	G. Bifulco, I. Bruno, R. Riccio, J. Lavayre, G. Bourdy, *J. Nat. Prod.* **1995**, *58*, 1254.
95MI16	M.M. Humanes, C.M. Matoso, J.A.L. da Silva, J.J.R. Fraústo da Silva, *Química* **1995**, *58*, 16.
95MI117	P. Mastalerz, *Wiad. Chem.* **1995**, *49*, 117; *Chem. Abstr.* **1995**, *124*, 14159.
95MI451	J.A. Field, F.J.M. Verhagen, E. de Jong, *TIBTECH* **1995**, *13*, 451.
95N2	K.E. Geckeler, W. Eberhardt, *Naturwissenshaften* **1995**, *82*, 2.
95T 2525	J. Kobayashi, F. Itagaki, I. Shigemori, T. Takao, Y. Shimonishi, *Tetrahedron* **1995**, *51*, 2525.
96JOC887	L. Rahbaek, U. Anthoni, C. Christophersen, P.H. Nielsen, B.O. Petersen, *J. Org. Chem.* **1996**, *61*, 887.
96MI1	G.W. Gribble, *Prog. Chem. Org. Nat. Prod.* **1996**, *68*, 1.
96MI155	J.M. Salietti Vinué, *Afinidad* **1996**, *53*, 155; *Chem. Abstr.* **1996**, *125*, 93972.
96PAC1699	G.W. Gribble, *Pure Appl. Chem.* **1996**, *68*, 1699.
96TL1439	S. Tsukamoto, H. Kato, H. Hirota, N. Fusetani, *Tetrahedron Lett.* **1996**, *37*, 1439.
97JA 9285	W.D. Clark, T. Corbett, F. Valeriote, P. Crews, *J. Am. Chem. Soc.* **1997**, *119*, 9285.
97JBC4689	A.G. Craig, E.C. Jimenez, J. Dykert, D.B. Nielsen, J. Gulyas, Fe C. Abogadie, J. Portier, J.E. Rivier, L.J. Cruz, B.M. Olivera, J.M. McIntosh, *J. Biol. Chem.* **1997**, *272*, 4689.
97JNP175	L. Rahbaek, C. Christophersen, *J. Nat. Prod.* **1997**, *60*, 175.
97JNP180	G.R. Pettit, J. McNulty, D.L. Herald, D.L. Doubek, J.-C. Chapuis, J.M. Schmidt, L.P. Tackett, M.R. Boyd, *J. Nat. Prod.* **1997**, *60*, 180.
97MI375	E. de Jong, J.A. Field, *Annu. Rev. Microbiol.* **1997**, *51*, 375.
97P 675	V. Magnus, J.A. Ozga, D.M. Reinecke, G.L. Pierson, T.A. Larue, J.D. Cohen, M.L. Brenner, *Phytochemistry* **1997**, *46*, 675.
98ACR141	G.W. Gribble, *Acc. Chem. Res.* **1998**, *31*, 141.
98ACS1040	J. Breinholt, H. Gürtler, A. Kjaer, S.E. Nielsen, C.E. Olsen, *Acta Chem. Scand.* **1998**, *52*, 1040.
98JNP959	R.A. Davis, A.R. Carroll, R.J. Quinn, *J. Nat. Prod.* **1998**, *61*, 959.
98JNP1043	M. Ochi, K. Kataoka, S. Ariki, C. Iwatsuki, M. Kodama, Y. Fukuyama, *J. Nat. Prod.* **1998**, *61*,

	1043.
98JNP1310	T. Iwagawa, M. Kaneko, H. Okamura, M. Nakatani, R.W.M. van Soest, *J. Nat. Prod.* **1998**, *61*, 1310.
98 JNP1433	A. Umeyama, S. Ito, E. Yuasa, S. Arihara, T. Yamada, *J. Nat. Prod.* **1998**, *61*, 1433.
98T 3043	E.W. Schmidt, D.J. Faulkner, *Tetrahedron* **1998**, *54*, 3043.
98T 6719	H. Uemoto, Y. Yahiro, H. Shigemori, M. Tsuda, T. Takao, Y. Shimonishi, J. Kobayashi, *Tetrahedron* **1998**, *54*, 6719.
99ABB278	S.W. Taylor, B. Kammerer, G.J. Nicholson, K. Pusecker, T. Walk, E. Bayer, S. Scippa, M. de Vincentiis, *Arch. Biochem. Biophys.* **1997**, *348*, 278.
99CC2195	G.W. Gribble, D.H. Blank, J.P. Jasinski, *Chem. Commun.* **1999**, 2195.
99CCC1147	Y. Kamano, A. Kotake, H. Hashima, I. Hayakawa, H. Hiraide, H.-P. Zhang, H. Kizu, K. Komiyama, M. Hayashi, G.R. Pettit, *Collect. Czech. Chem. Commun.* **1999**, *64*, 1147.
99CSR335	G.W. Gribble, *Chem. Soc. Rev.* **1999**, *28*, 335.
99EST26	S.A. Tittlemier, M. Simon, W.M. Jarman, J.E. Elliott, R.J. Norstrom, *Environ. Sci. Technol.* **1999**, *33*, 26.
99JAN781	I. Momose, H. Iinuma, N. Kinoshita, Y. Momose, S. Kunimoto, M. Hamada, T. Takeuchi, *J. Antibiot.* **1999**, *52*, 781.
99JAN787	I. Momose, S. Hirosawa, H. Nakamura, H. Naganawa, H. Iinuma, D. Ikeda, T. Takeuchi, *J. Antibiot.* **1999**, *52*, 787.
99JBC30664	C.S. Walker, D. Steel, R.B. Jacobsen, M.B. Lirazan, L.J. Cruz, D. Hooper, R. Shetty, R.C. DelaCruz, J.S. Nielsen, L.M. Zhou, P. Bandyopadhyay, A.G. Craig, B.M. Olivera, *J. Biol. Chem.* **1999**, *274*, 30664.
99JFC127	D. O'Hagan, D.B. Harper, *J. Fluorine Chem.* **1999**, *100*, 127.
99JNP184	C. Eder, P. Proksch, V. Wray, K. Steube, G. Bringmann, R.W.M. van Soest, Sudarsono, E. Ferdinandus, L.A. Pattisina, S. Wiryowidagdo, W. Moka, *J. Nat. Prod.* **1999**, *62*, 184.
99JNP332	A. Zampella, C. Giannini, C. Debitus, M.V. D'Auria, *J. Nat. Prod.* **1999**, *62*, 332.
99JNP647	J. Shin, Y. Seo, K.W. Cho, J.-R. Rho, C.J. Sim, *J. Nat. Prod.* **1999**, *62*, 647.
99JNP1246	S.P.B. Ovenden, R.J. Capon, *J. Nat. Prod.* **1999**, *62*, 1246.
99JNP1295	C. Eder, P. Proksch, V. Wray, R.W.M. van Soest, E. Ferdinandus, L.A. Pattisina, Sudarsono, *J. Nat. Prod.* **1999**, *62*, 1295.
99JNP1581	H. Uemoto, M. Tsuda, J. Kobayashi, *J. Nat. Prod.* **1999**, *62*, 1581.
99JOC731	S. Urban, P. de Almeida Leone, A.R. Carroll, G.A. Fechner, J. Smith, J.N.A. Hooper, R.J. Quinn, *J. Org. Chem.* **1999**, *64*, 731.
99MI239	M.B. Lirazan, A.G. Craig, R. Shetty, C.S. Walker, B.M. Olivera, L.J. Cruz, *Philippine J. Sci.* **1999**, *128*, 239.
99NPL59	A. Qureshi, D.J. Faulkner, *Nat. Prod. Lett.* **1999**, *13*, 59.
99OL455	M. Assmann, E. Lichte, R.W.M. van Soest, M. Köck, *Org. Lett.* **1999**, *1*, 455.
99P 459	T. Hosoe, K. Fukushima, K. Takizawa, M. Miyaji, K. Kawai, *Phytochemistry* **1999**, *52*, 459.
99P 537	S.-C. Lee, G.A. Williams, G.D. Brown, *Phytochemistry* **1999**, *52*, 537.
99TL1591	T.N. Makarieva, S.G. Ilyin, V.A. Stonik, K.A. Lyssenko, V.A. Denisenko, *Tetrahedron Lett.* **1999**, *40*, 1591.
99TL5709	M. Tsuda, H. Uemoto, J. Kobayashi, *Tetrahedron Lett.* **1999**, *40*, 5709.
99TL9214	F. Kong, D.Q. Liu, J. Nietsche, M. Tischler, G.T. Carter, *Tetrahedron Lett.* **1999**, *40*, 9219.
00EP401	W. Vetter, L. Alder, R. Kallenborn, M. Schlabach, *Environ. Poll.* **2000**, *110*, 401.
00GC173	N. Winterton, *Green Chem.* **2000**, *2*, 173.
00JAN1	S.J. Trew, S.K. Wrigley, L. Pairet, J. Sohal, P. Shanu-Wilson, M.A. Hayes, S.M. Martin, R.N. Manohar, M.I. Chicarelli-Robinson, D.A. Kau, C.V. Byrne, E.M.H. Wellington, J.M. Moloney, J. Howard, D. Hupe, E.R. Olson, *J. Antibiot.* **2000**, *53*, 1.
00JNP447	A. Casapullo, G. Bifulco, I. Bruno, R. Riccio, *J. Nat. Prod.* **2000**, *63*, 447.
00JNP981	I. Carletti, B. Banaigs, P. Amade, *J. Nat. Prod.* **2000**, *63*, 981.
00JOC459	K. Umezawa, Y. Ikeda, Y. Uchihata, H. Naganawa, S. Kondo, *J. Org. Chem.* **2000**, *65*, 459.
00MI19	J.A. Joule, K. Mills, *Heterocyclic Chemistry*, 4th Ed.; Blackwell Science, Oxford, 2000, pp19-20.
00MI37	G.W. Gribble, *Environ. Sci. Pollut. Res.* **2002**, *7*, 37.
00T 3743	A. Cutignano, G. Bifulco, I. Bruno, A. Casapullo, L. Gomez-Paloma, R. Riccio, *Tetrahedron* **2000**, *56*, 3743.
00TL9917	E. Fattorusso, O. Taglialatela-Scafati, *Tetrahedron Lett.* **2000**, *41*, 9917.

01AECT221	W. Vetter, E. Scholz, C. Gaus, J.F. Müller, D. Haynes, *Arch. Environ. Contam. Toxicol.* **2001**, *41*, 221.
01CPB1372	S. Aoki, Y. Ye, K. Higuchi, A. Takashima, Y. Tanaka, I. Kitagawa, M. Kobayashi, *Chem. Pharm. Bull.* **2001**, *49*, 1372.
01JCE1442	P.F. Schatz, *J. Chem. Ed.* **2001**, *78*, 1442.
01JCE1444	C.E. Steinhart, *J. Chem. Ed.* **2001**, *78*, 1444.
01JNP254	R. Britton, J.H.H.L. de Oliveira, R.J. Andersen, R.G.S. Berlinck, *J. Nat. Prod.* **2001**, *64*, 254.
01JNP1345	M. Assmann, R.W.M. van Soest, M. Köck, *J. Nat. Prod.* **2001**, *64*, 1345.
01JNP1454	M.A. Rashid, K.R. Gustafson, M.R. Boyd, *J. Nat. Prod.* **2001**, *64*, 1454.
01JNP1559	T.N. Makarieva, A.S. Dmitrenok, P.S. Dmitrenok, B.B. Grebnev, V.A. Stonik, *J. Nat. Prod.* **2001**, *64*, 1559.
01JNP1576	S. Tsukamoto, K. Tane, T. Ohta, S. Matsunaga, N. Fusetani, R.W.M. van Soest, *J. Nat. Prod.* **2001**, *64*, 1576.
01JNP1581	G.H. Goetz, G.G. Harrigan, J. Likos, *J. Nat. Prod.* **2001**, *64*, 1581.
01JNP1593	M. Assmann, S. Zea, M. Köck, *J. Nat. Prod.* **2001**, *64*, 1593.
02AG(E)597	R.P. Maskey, I. Grün-Wollny, H.H. Fiebig, H. Laatsch, *Angew. Chem., Int. Ed. Engl.* **2002**, *41*, 597.
02AG(E)1740	J. Wu, W. Vetter, G.W. Gribble, J.S. Schneekloth, Jr., D.H. Blank, H. Görls, *Angew. Chem., Int. Ed.* **2002**, *41*, 1470.
02EP85	S.A. Tittlemier, A.T. Fisk, K.A. Hobson, R.J. Norstrom, *Environ. Pollut.* **2002**, *116*, 85.
02NP938	C.K. Narkowicz, A.J. Blackman, E. Lacey, J.H. Gill, K. Heiland, *J. Nat. Prod.* **2002**, *65*, 938.
02NP1258	J.S. Sandler, P.L. Colin, J.N.A. Hooper, D.J. Faulkner, *J. Nat. Prod.* **2002**, *65*, 1258.
02JNP1633	L. Peters, G.M. König, H. Terlau, A.D. Wright, *J. Nat. Prod.* **2002**, *65*, 1633.
02JOC5402	D.R. Appleton, M.J. Page, G. Lambert, M.V. Berridge, B.R. Copp, *J. Org. Chem.* **2002**, *67*, 5402.
02JOC6671	K.M. Meragelman, L.M. West, P.T. Northcote, L.K. Pannell, T.C. McKee, M.R. Boyd, *J. Org. Chem.* **2002**, *67*, 6671.
02MI121	G.W. Gribble, in *Organofluorines*, A.H. Neilson, Ed.: Springer-Verlag, 2002; Vol. 3/N, p121.
02MI315	V.M. Dembitsky, M. Srebnik, *Prog. Lipid Res.* **2002**, *41*, 315.
02P 979	T. Rundberget, A.L. Wilkins, *Phytochemistry* **2002**, *61*, 979.
03ED40	G.W. Gribble, *Educ. Chem.* **2003**, *40*, 40.
03EJO284	C. Campagnuolo, E. Fattorusso, O. Taglialatela-Scafati, *Eur. J. Org. Chem.* **2003**, 284.
03MI289	G.W. Gribble, *Chemosphere* **2003**, *52*, 289.
03TL2533	T. Iwagawa, M. Miyazaki, H. Okamura, M. Nakatani, M. Doe, K. Takemura, *Tetrahedron Lett.* **2003**, *44*, 2533.

Chapter 4.1

Three-Membered Ring Systems

Albert Padwa
Emory University, Atlanta, GA 30322
chemap@emory.edu

Shaun Murphree
Allegheny College, Meadville, PA 16335
smurphre@allegheny.edu

4.1.1 INTRODUCTION

Representing the smallest class of heterocycles, three-membered ring systems continue to grow in synthetic utility. Occasionally of interest as target molecules in their own right, more frequently these versatile moieties are employed as reactive intermediates and building blocks for the rapid regio- and stereospecific construction of complex organic substrates. This year has seen many advances in the practicality, selectivity, and predictability of the processes involved in the preparation and subsequent reactivity of epoxides and aziridines. The scope of the following overview does not include a comprehensive review or detailed treatment of phenomena, but rather is meant to provide a snapshot of the current state of the art from a synthetic chemist's perspective. The organization of the chapter is similar to that of previous years.

4.1.2 EPOXIDES

4.1.2.1 Preparation of Epoxides

Much activity continues to be centered around the preparation of enantioenriched epoxides using chiral Co(III)-, Mn(III)- and Cr(III)-salen complexes, particularly in the area of innovative methods. A recent brief review <02CC919> focuses on the synthesis, structural features, and catalytic applications of Cr(III)-salen complexes. In an illustrative example, Jacobsen and co-workers <02JA1307> have applied a highly efficient hydrolytic kinetic resolution to a variety of terminal epoxides using the commercially available chiral salen-Co(III) complex **1**. For example, treatment of racemic *m*-chlorostyrene oxide (**2**) with 0.8 mol% of catalyst **1** in the presence of water (0.55 equiv) led to the recovery of practically enantiopure (> 99% ee) material in 40% yield (maximum theoretical yield = 50%). This method appears to be effective for a variety of terminal epoxides, and the catalyst suffered no loss of activity after six cycles.

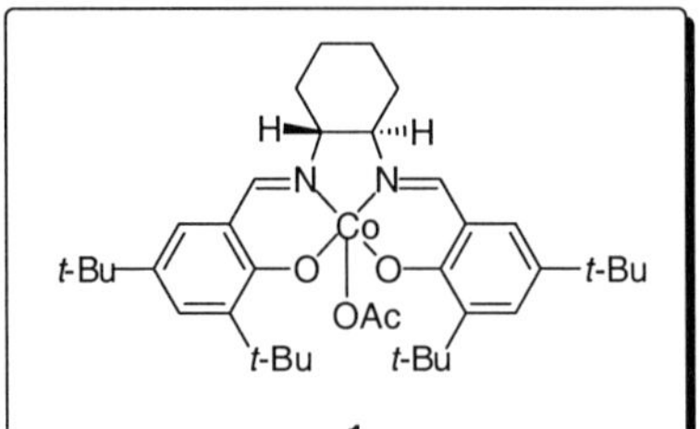

In the example above, the catalyst could be recovered by distilling off the products, suspending the residue in methanol, and isolating the resulting solid by vacuum filtration. However, since they are essentially homogeneous catalysts, salen complexes are not always so easily recovered. Therefore, novel systems have been developed with ease of catalyst isolation in mind. For example, Jacobsen's catalyst has been prepared in polymeric form (*i.e.* **3**) <02JCS(P1)870> and immobilized on silica gel by radical grafting (*e.g.* **4**) <02HCA913>. The performance of the latter approaches that of the conventional Jacobsen catalyst in terms of yield and enantioselectivity in the epoxidation of styrene (**5**); however, a drop in ee was observed for reactions using recycled catalyst.

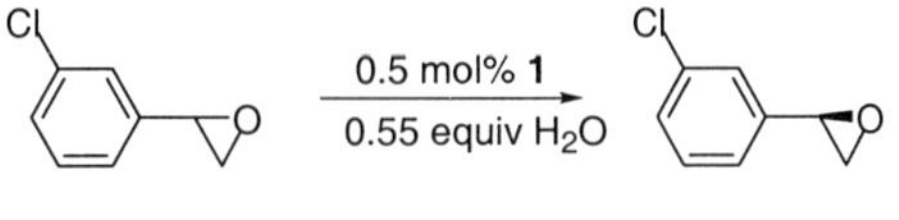

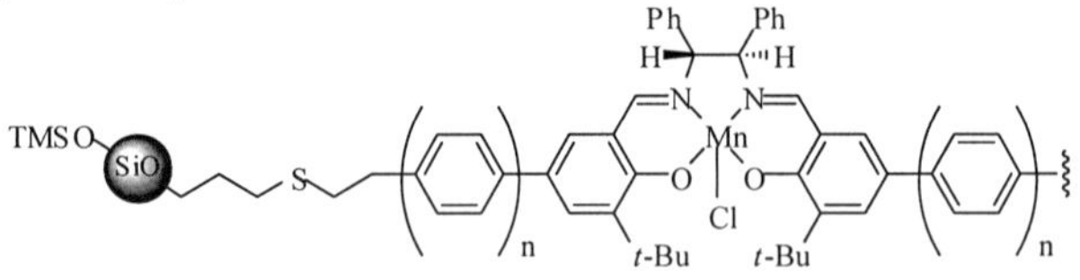

Catalyst	Yield	ee
Jacobsen	Quant.	75%
3	83%	51%
4	95%	74%

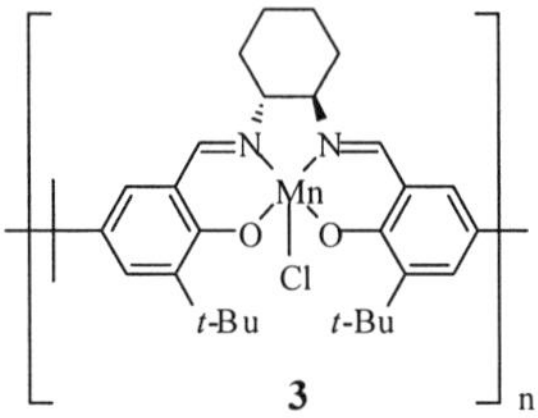

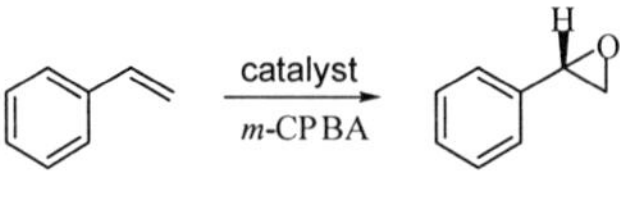

Smith and Liu <02CC886> have immobilized a Katsuki-type salen ligand by an ester linkage to Merrifield's resin to produce catalyst **7**. In a test epoxidation of 1,2-dihydronaphthalene (**8**)

using sodium hypochlorite as an oxidant and 4-phenylpyridine *N*-oxide (4-PPNO) as an activator, the immobilized (salen)Mn complex gave variable (and somewhat lower) overall yields, but exhibited sustained high enantioselectivity (> 90%), even after being recycled six times.

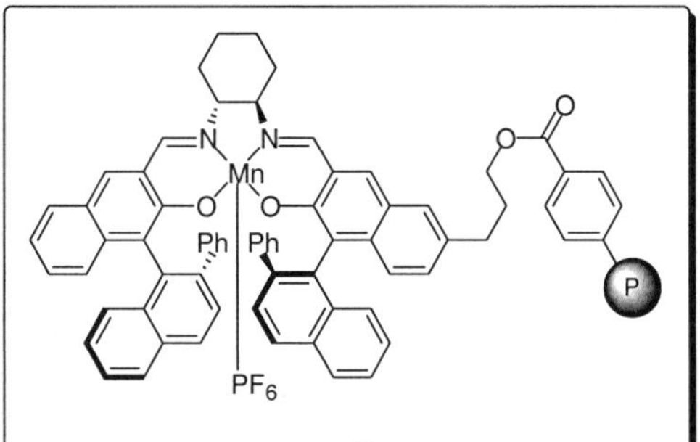

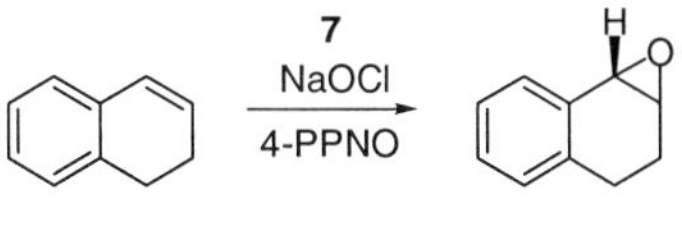

A somewhat different approach to catalyst separation has been devised by engineering the chiral salen catalyst to have built-in phase-transfer capability, as exemplified by the Mn(III) complex **10** <02TL2665>. Thus, enantioselective epoxidation of chromene derivatives (*e.g.* **11**) in the presence of 2 mol% catalyst **10** under phase transfer conditions (methylene chloride and aqueous sodium hypochlorite) proceeded in excellent yield and very good ee's. The catalyst loading could be reduced to about 0.4% with only marginal loss of efficiency.

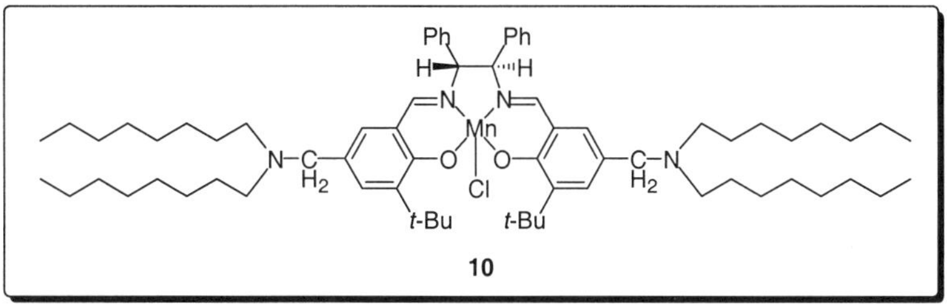

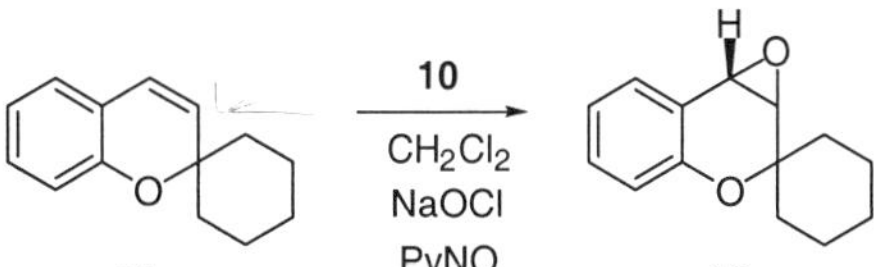

Other metal-centered catalysts that have been studied include (*bis*)strapped chiral porphyrins derived from L-proline, which can induce modest (< 30%) enantioselectivity <02EJIOC1666>. Supported amidate-bridged platinum blue complexes, which have not yet been applied to chiral epoxidation, but which show promise of utilizing molecular oxygen as the terminal oxidant, have also been used. Thus, cyclohexene (**13**) was epoxidized in 99% yield in the presence of isobutyraldehyde, atmospheric oxygen, and a carbon-supported platinum complex prepared from [$Pt_4(NH_3)_8((CH_3)_3CCONH)_4](NO_3)_5$ <02SC17>.

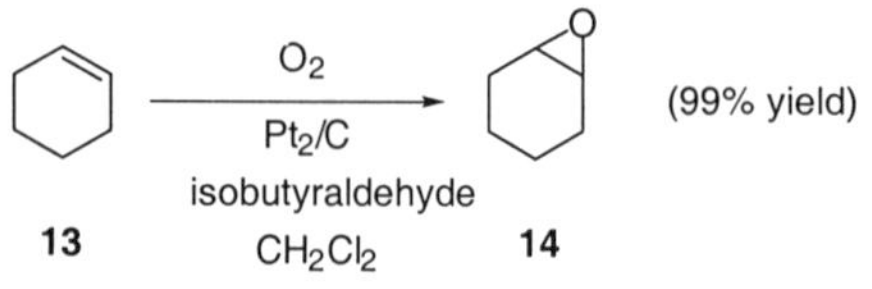

Building upon earlier studies of fructose-derived catalysts, Shi and co-workers <02JOC2435> have developed a chiral oxazolidinone ketone catalyst (**15**) for asymmetric epoxidation *via* dioxirane intermediates. This system has the advantage of exhibiting high ee's for *cis*- and terminal olefins, which can be otherwise recalcitrant. Interestingly, for olefins with aromatic substituents, it appears that the transition state shows a preference for positioning the π-system proximal to the oxazolidinone moiety (as in **19**), so that aromatic groups can be efficiently differentiated during the epoxidation. Another non-metal system that uses Oxone as a terminal oxidant is represented by the chiral iminium salt **16**, although in this case the enantiocontrol is lower (< 60%) and less predictable <02SL580>.

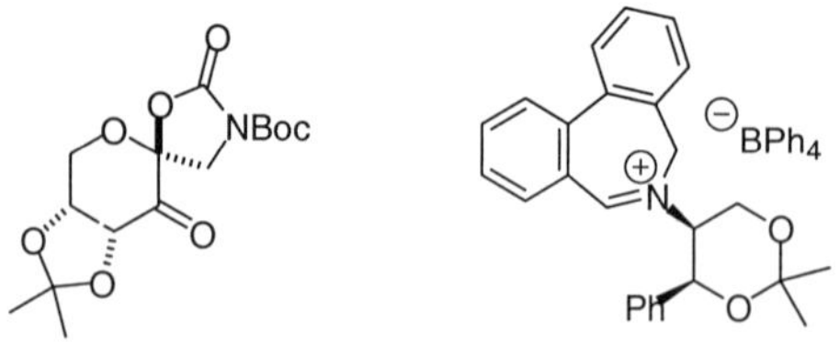

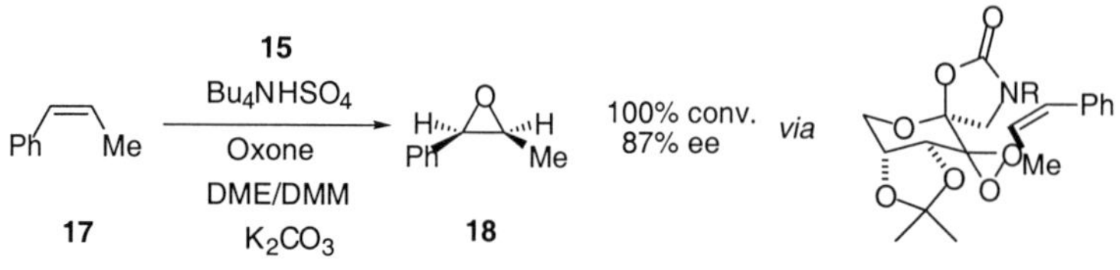

Under the rubric of functionalized olefins, the epoxidation of allylic alcohols is well-known, but continues to benefit from further study. Prompted by the observation that peracid epoxidations can be far more selective in basic aqueous medium than in organic solvents <97JOC3748>, Washington and Houk studied transition structures from the epoxidation of allylic alcohol with performate ion at the B3LYP/6-31+G(d,p) level in a CPCM continuum model for water <02OL2661>. Their findings indicate a preferential hydrogen bonding of performate with the substrate hydroxyl group rather than with water, leading to a directed epoxidation *via* transition state **20**. This reaction is similar to the corresponding cyclopropanation of allylic alcohols in the presence of aqueous sodium hydroxide. To test this model experimentally, the chiral allylic alcohol **21** was treated with monoperoxyphthalic acid (**22**) in 1M NaOH to give the *syn* epoxy alcohol **23** in nearly quantitative yield with 98% *syn/anti* selectivity.

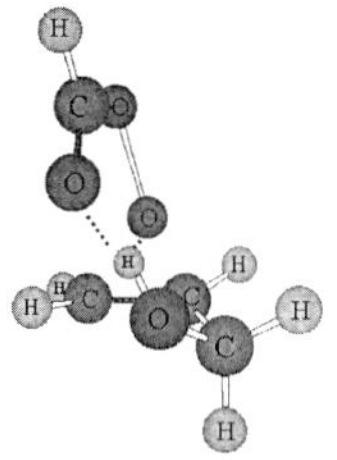

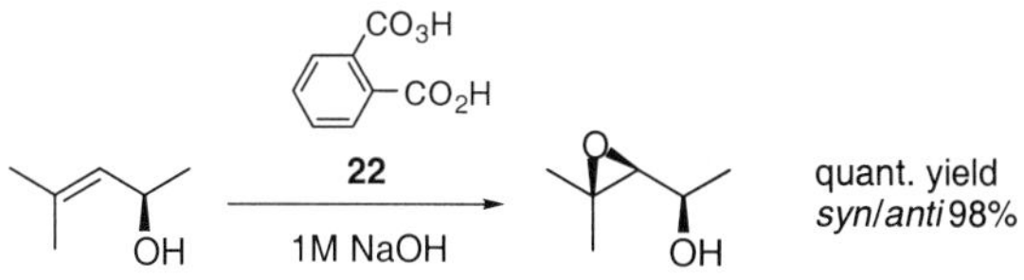

Very high asymmetric induction has been reported in a modified catalytic Sharpless epoxidation of allylic alcohols using the tertiary furyl peroxide **25** in the presence of L-diisopropyl tartrate (L-DIPT). Thus, *trans*-2-methyl-3-phenylprop-2-en-1-ol (**24**) was converted to epoxide **26** in 87% yield and with 97% ee using a catalyst loading of 20 mol% <02TL5629>. Allylic alcohols can also be converted directly to α,β-epoxyketones using the Cinchona derived chiral phase-transfer catalyst **28**. For example, the diaryl allylic alcohol **27** provided the epoxy ketone **29** in > 90% yield and 87% ee using 5 mol% of the catalyst **28** and sodium hypochlorite as the terminal oxidant <02CC2360>.

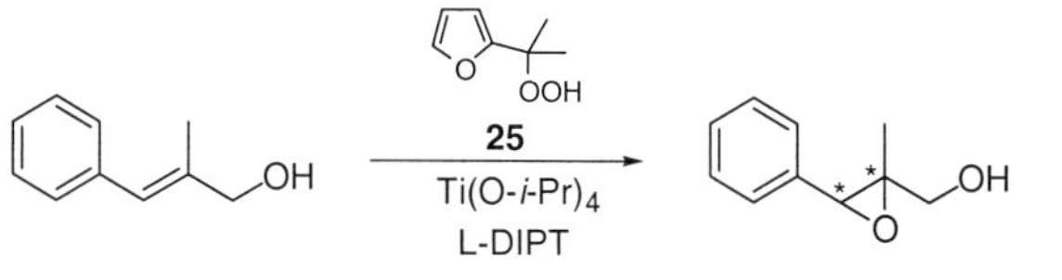

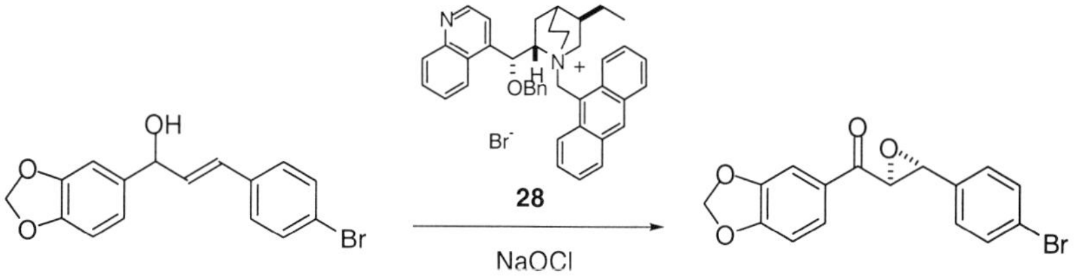

Similar conditions are also effective for the direct enantioselective epoxidation of α,β-unsaturated carbonyls, as exemplified by the conversion of chalcone (**30**) to the corresponding *R,R*-epoxide in 97% yield and 84% ee using the 4-iodophenyl cinchonine derivative **32** <02T1623>. Alternatively, the novel polyethylene glycol-supported oligo(L-leucine) catalyst

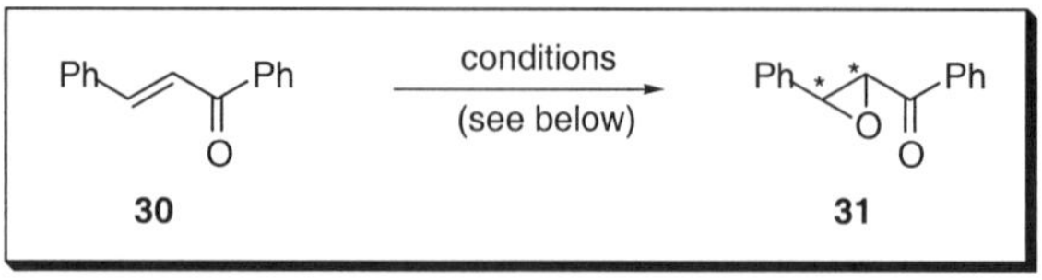

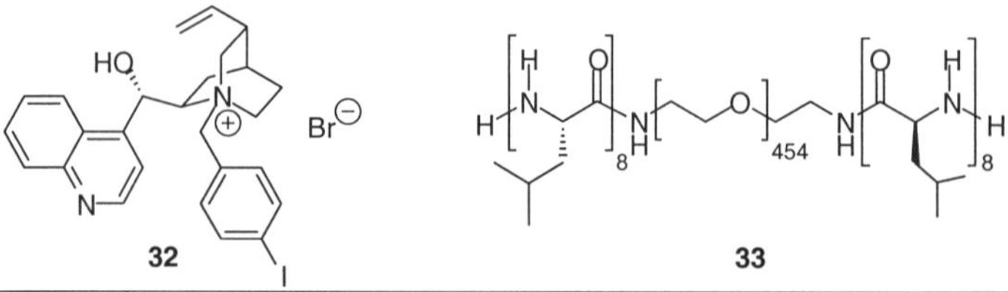

catalyst	oxidant	yield	ee	config.
32	30% H_2O_2	97%	84%	R,R
33	urea-H_2O_2	>99%	94%	S,S

33 has been used in a continuously operated chemzyme membrane reactor with urea-hydrogen peroxide as the terminal oxidant to effect a Juliá-Colonna epoxidation of 30 with the opposite sense of chirality and with even more efficient chiral induction <02SL707>. Electron-deficient olefins are generally regarded as poor substrates for the fructose-derived dioxirane precursor 34, which is otherwise effective in the epoxidation of *trans-* and trisubstituted olefins <01OL715>. In the case of enones, the sluggishness of the epoxidation reaction *per se* allows for decomposition of the dioxirane species, presumably *via* Baeyer-Villiger pathways. The structurally similar diacetate analog 35 overcomes this limitation and facilitates the formation of epoxy esters in good yields and very good ee's. For example, when ethyl cyclohexenecarboxylate (36) is treated with 25 mol% of 35 in the presence of Oxone for 24 h, the dextrorotatory epoxide (37) is formed in 77% yield and 93% ee <02JA8792>.

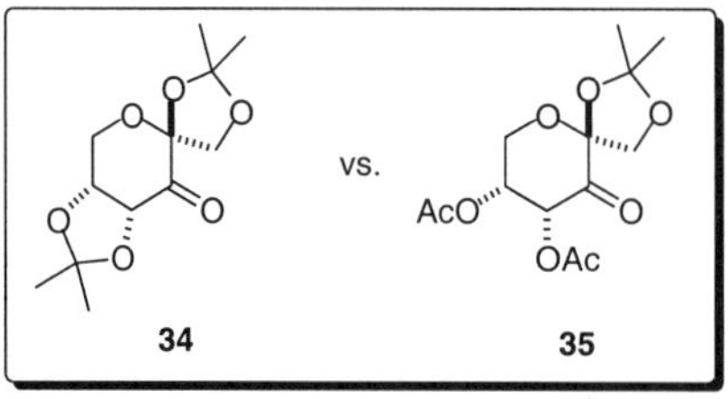

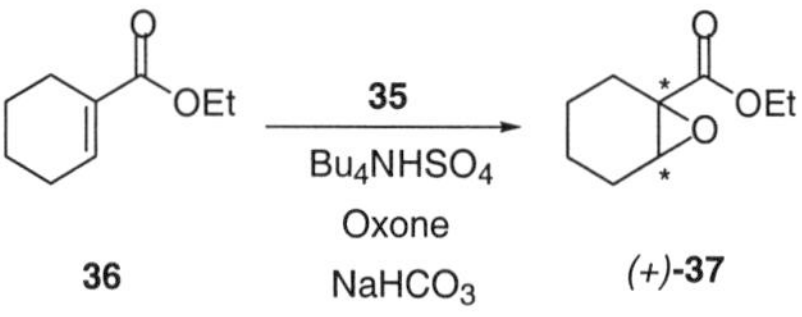

With regard to olefins bearing other functionality, the urea-hydrogen peroxide complex (UHP) with dicyclohexylcarbodiimide (DCC) in alcoholic solvent has been applied to the epoxidation of vinylsilanes (*e.g.* 38) to form the labile epoxysilanes 39 in reasonable yield <02SC2677>. The vinyl phosphonate 40 has been reported to undergo asymmetric epoxidation under the catalysis of cyclohexanone monooxygenase (CHMO) to give the practically enantiopure (> 98% ee) fosfomycin analog 41 in 40% yield <02TL1797>.

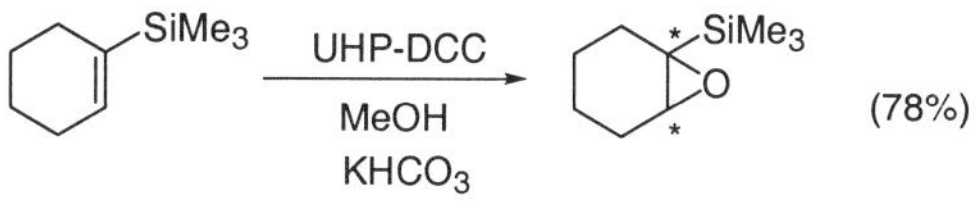

The transfer of oxygen to a double bond is not the only synthetic route to epoxides represented in the literature. Another common approach involves the addition of a carbon center, usually in the form of a sulfur ylide, to a carbonyl compound, generally an aldehyde. These reactions tend to give predominantly *trans*-epoxides, and the origin of this selectivity has recently been given an unexpected rationale. Drawing from detailed studies of a model system using density functional theory, Aggarwal and co-workers <02JA5747> have proposed a mechanism involving initial quasi [2+2]-addition of the ylide to give a cisoid betaine (*i.e.* **42**), which undergoes torsional rotation to form a transoid betaine (**43**) followed by a facile *trans*-elimination of sulfide. The two orientations of *syn* [2+2] (*i.e.* substituents on the same *vs.* opposite sides) have surprisingly little difference in transition state energies (*ca.* 0.2 kcal). However, if the [2+2] occurs with the substituents on opposite sides, the barrier to torsional rotation (*i.e.* **45** → **46**) is 1.7 kcal higher than the alternative orientation (**42** → **43**) and thus becomes the rate-limiting step.

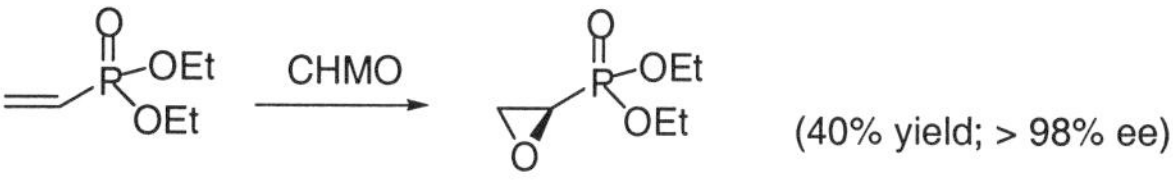

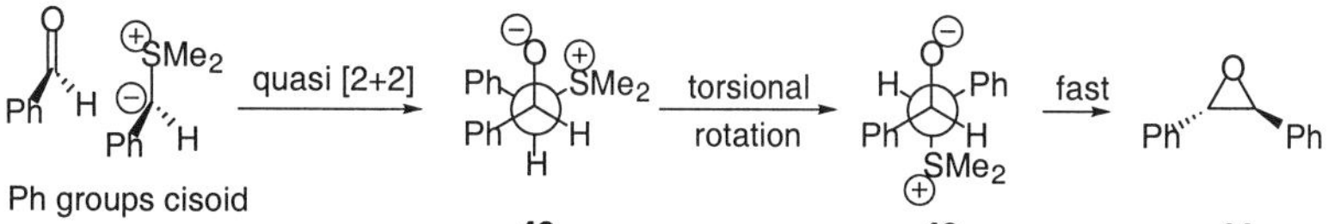

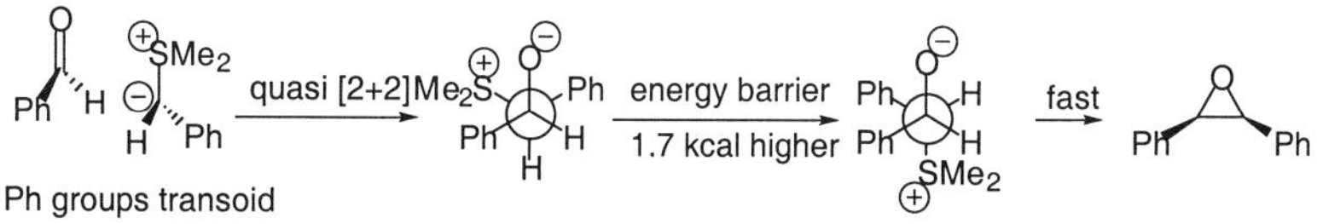

According to this model, the diastereoselectivity of the epoxidation depends upon the reversibility of the initial [2+2] addition. Thus, a more stable ylide should lead to a more reversible addition, and *trans/cis* ratios should increase. This reasoning led to the exploration of arsonium ylides as carbon-transfer reagents. Indeed, the triphenylarsine catalyzed coupling of cyclohexanecarboxaldehyde (**48**) with the tosylhydrazone salt (**49**) derived from *p*-methoxybenzaldehyde produces the *trans*-epoxide **50** in 64% yield and with greater than 98:2 *trans/cis* selectivity <02CC1514>.

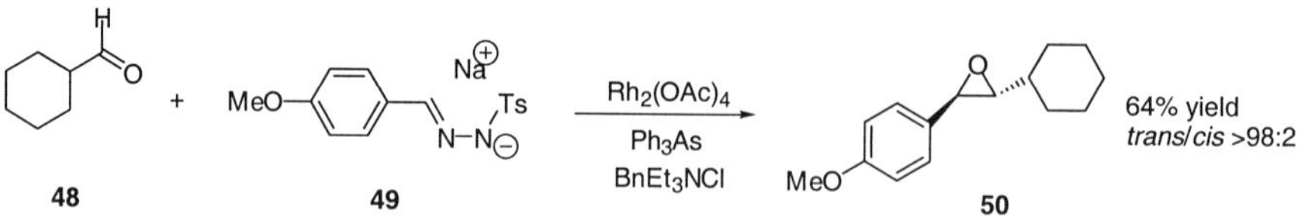

The sulfur ylide approach is amenable to asymmetric induction through the use of various chiral sulfides. One potential bottleneck, however, is the lack of ready access to these auxiliaries in optically pure form. In answer to this problem, recent reports have focused on the development of strategies that rely on asymmetric sulfides that can be obtained in a few steps and/or on large scale from the chiral pool. Two examples are the *bis*-sulfide **52**, obtained from (*R,R*)-tartaric acid <02H(8)1399>, and the tricyclic sulfide **54**, derived from from D-mannitol <02TL5427>, both of which possess C_2-symmetry and which operate in a catalytic one-pot environment.

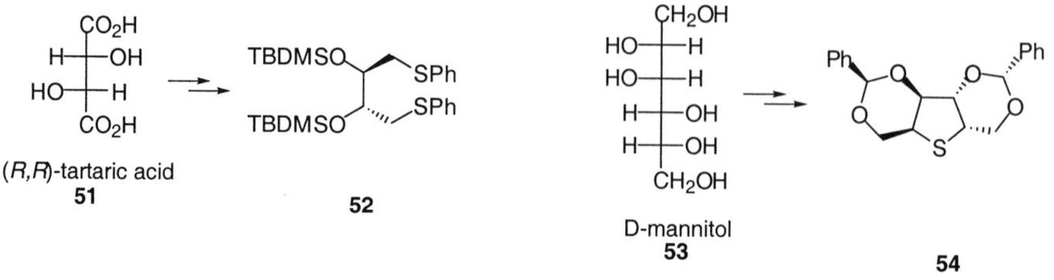

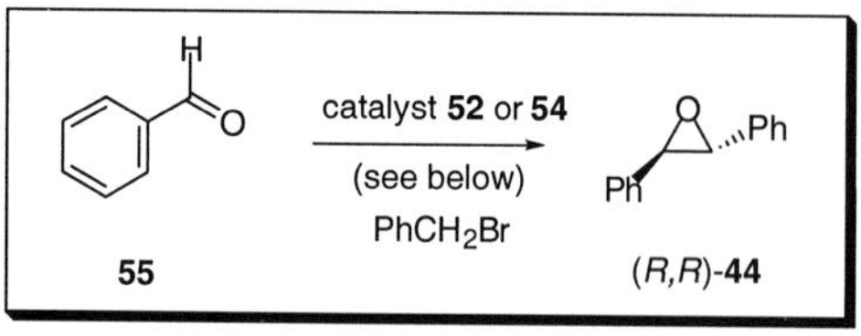

catalyst	reference	yield	ee	trans/cis
52	02H(57)1399	22%	68%	74:26
54	02TL5427	42%	94%	91:9

4.1.2.2 Reactions of Epoxides

In an unusual example of the epoxide moiety acting as a nucleophile, Hodgson and co-workers <02TL7895> have reported on a convenient method of deprotonating terminal oxiranes with lithium 2,2,6,6-tetramethylpiperidide (LTMP), followed by trapping of the anion with silyl-based electrophiles, to provide α,β-epoxysilanes in good yield. For example, chloro-epoxide **56** underwent clean conversion to epoxysilane **57** at 0 °C. This approach improves upon an earlier method, which employed sparteine derivatives at very low temperature (-90 °C) <01OL461>.

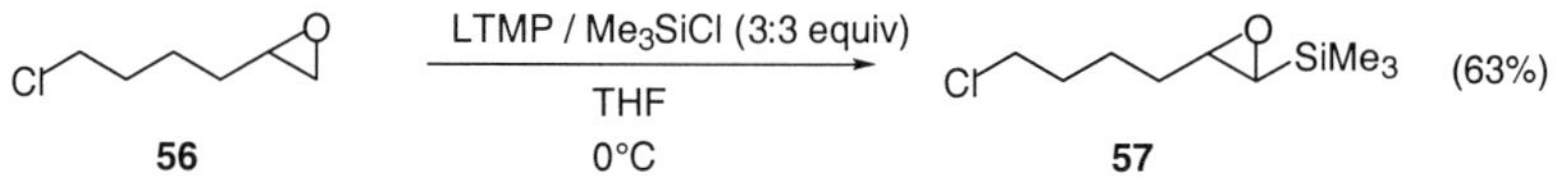

The more common role of reactivity for epoxides is that of electrophile, particularly in ring-opening reactions. During this past year reports in the literature focused on advances in selectivity, generality, and mildness of the ring-opening conditions. For example, two groups independently reported on the use of bismuth trichloride as a catalyst for the ring opening of meso-epoxides with aromatic amines <02TL7891, 02SC2307>. Thus, cyclooctene oxide (**58**) was treated with aniline in the presence of 5 mol% bismuth trichloride to afford the amino alcohol **59** in 97% isolated yield.

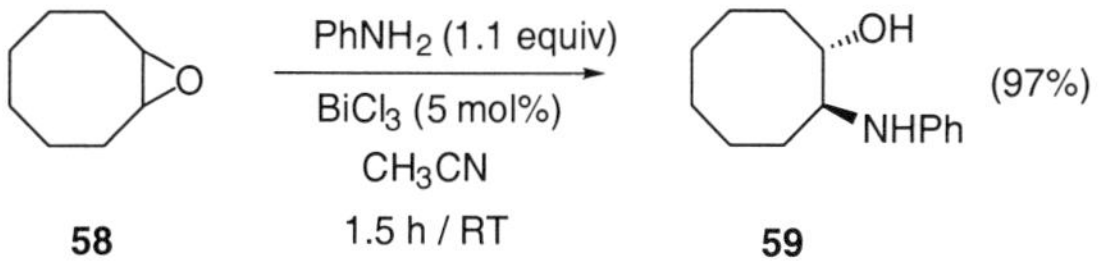

In the case of nonsymmetrical epoxides, the regioselectivity is determined by the particular reaction conditions. Thus, styrene oxide (**6**) undergoes methanolysis in the presence of the Lewis acid catalyst copper(II) tetrafluoroborate to give the hydroxy ether **60**, derived from attack of the nucleophile at the more substituted oxiranyl carbon. Similar outcomes have been observed in the solvolysis of **6** with the assistance of aminopropyl silica gel (APSG) supported iodine in catalytic quantities <02SL1251>. This selectivity appears to be much less decisive, however, in the case of monoalkyl epoxides, as illustrated in the corresponding reaction of 1-octene oxide (**61**), which yields an almost 1:1-mixture of isomers under the same conditions <02OL2817>.

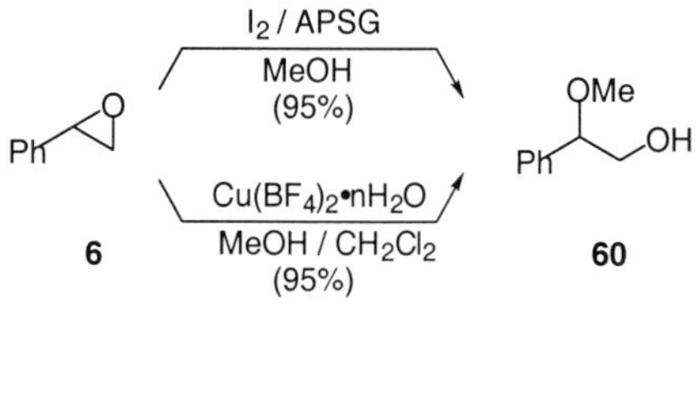

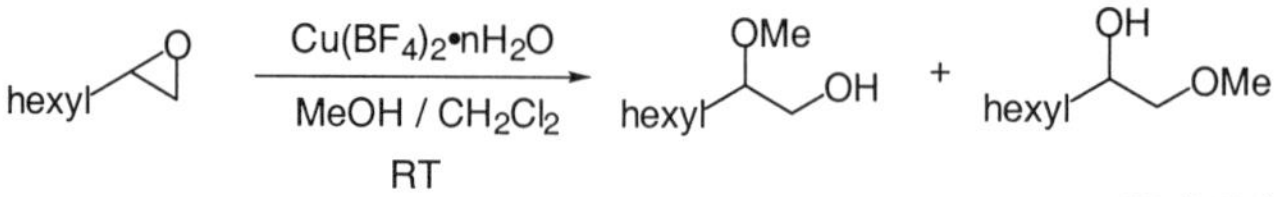

hexyl—[epoxide] → (Cu(BF$_4$)$_2$·nH$_2$O, MeOH / CH$_2$Cl$_2$, RT) → hexyl-CH(OMe)-CH$_2$-OH + hexyl-CH(OH)-CH$_2$-OMe

61   **62** (26%)   **63** (34%)

Meanwhile, it is generally easier to select for attack at the lesser substituted carbon of an unsymmetrical epoxide. For example, treatment of the epoxy ether **64** with benzylamine in the presence of 10 mol% lithium *bis*-trifluoromethanesulfonimide (LiNTf$_2$) led to the formation of amino alcohol **65** in 95% yield. This protocol was also amenable to the use of secondary amines. When the more substituted position is benzylic (*i.e.* **6**) there is still competition in the site of attack, so that as much as 25% of the total product yield was represented by ring opening at the benzylic position <02TL7083>. Rao and co-workers <02T6003> have investigated a clever

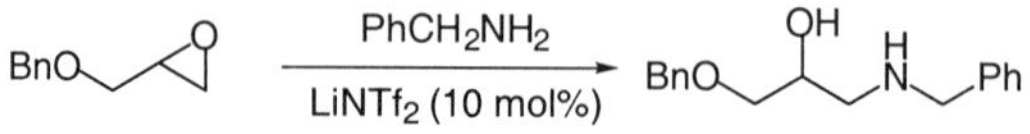

BnO—[epoxide] **64** → (PhCH$_2$NH$_2$, LiNTf$_2$ (10 mol%)) → BnO-CH$_2$-CH(OH)-CH$_2$-NH-Ph   **65** (95%)

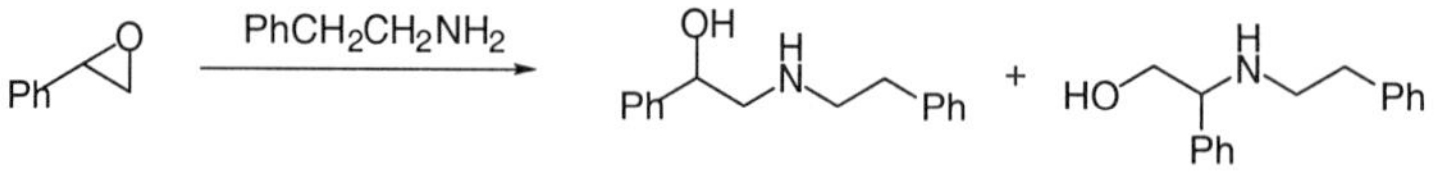

Ph—[epoxide] **6** → (PhCH$_2$CH$_2$NH$_2$) → Ph-CH(OH)-CH$_2$-NH-CH$_2$-Ph   **66** (62%)   +   HO-CH$_2$-CH(Ph)-NH-CH$_2$-Ph   **67** (15%)

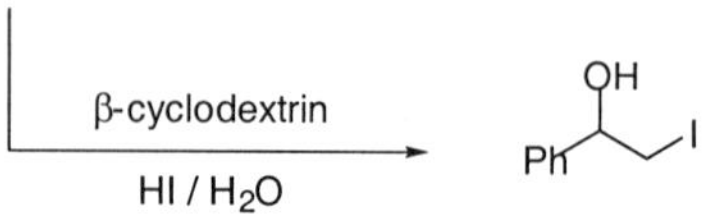

↓ (β-cyclodextrin, HI / H$_2$O)

Ph-CH(OH)-CH$_2$-I   **68** (95%)

workaround for this problem with the use of β-cyclodextrin (β-CD), which presumably operates by forming a guest-host complex in which only the terminal epoxide carbon is exposed for nucleophilic attack. With this protocol, styrene oxide (**6**) is cleanly opened to the terminal iodide **68** in the presence of β-CD and hydriodic acid.

A new epoxide inversion method has been developed which uses cesium propionate as a key reagent to promote nucleophilic ring opening, initially providing a mixture of regio-isomeric hydroxyesters (*e.g.* **70** & **71**). Subsequent mesylation of the hydroxylic groups in the crude reaction mixture, followed by treatment with potassium carbonate, provides the inverted epoxide (*R,R*-**69**) <02TL4111>.

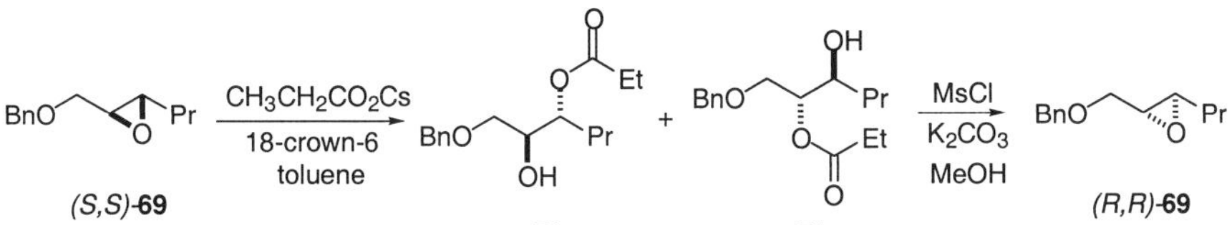

Three-Membered Ring Systems   85

An unusual *syn* addition to epoxides occurs when 1,3-diene monoepoxides are treated with organozinc reagents. Thus, the cyclic vinyl epoxide **72** was converted to the *cis*-ethyl-cyclohexenol **75** with diethyl zinc in methylene chloride and trifluoroacetic acid. The *syn* addition is believed to derive from an initial coordination of the oxiranyl oxygen to the organozinc compound, which then delivers the alkyl group to the same face. This transfer is facilitated by a relaxation of the sp³ hybridization brought about by the Lewis acidic zinc center and the allylic character of the incipient carbocation <02OL905>.

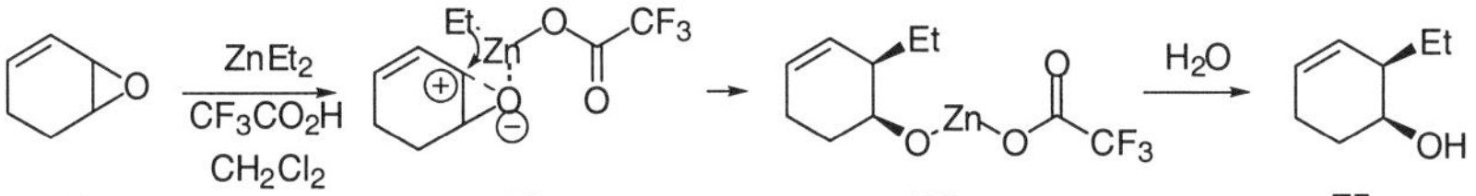

Vinyl epoxides can also be ring-opened *via* an S_N2' sense, as exemplified in the macrocyclization of the epoxy-tethered cyclopentenone **76**, which was induced to occur by treatment with lithium 2,2,6,6-tetramethylpiperidide (LTMP) followed by the mild Lewis acid diethylaluminum chloride in THF. The enolate attacked exclusively from the α-position of the cyclopentenone, presumably due to the steric hindrance at the γ-position. The mechanistic reason for the Z-geometry of the newly formed double bond has not been elucidated, but it may involve an initial diethylaluminum chloride mediated opening of the epoxide to form an intermediate Z-allylic halide, which subsequently undergoes *in situ* macrocyclization <02T6397>.

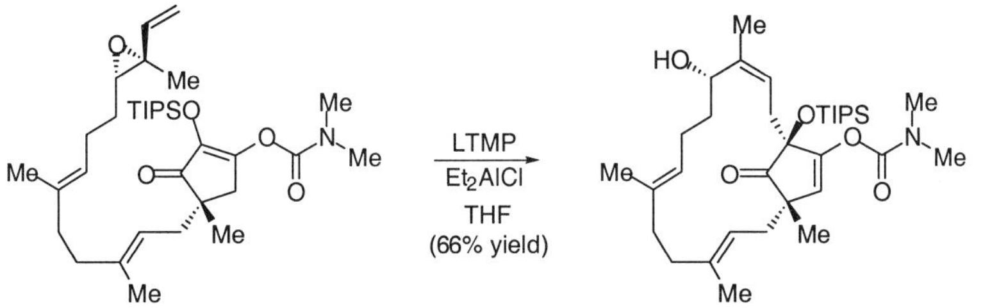

Other synthetically useful intramolecular epoxide ring openings have been reported. For example, the strained methylene epoxide **79**, derived from DMD epoxidation of the corresponding allene **78**, undergoes spontaneous isomerization to the lactone **81** *via* attack of the pendant carboxylate onto the distal epoxide carbon <02T7027>. Intramolecular arene-epoxide coupling of the highly functionalized epoxide **82** is promoted by catalytic tin tetrachloride, forming the tetracyclic species **83** in 98% yield <02JOC6954>, and the triepoxide **84** underwent *endo*-oxacyclization in the presence of boron trifluoride etherate to provide the polycyclic ether **85** in 52% yield <02JOC2515>.

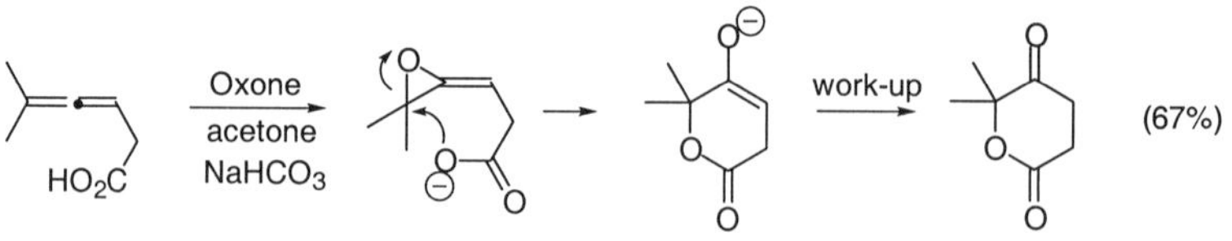

The enantioselective isomerization of meso epoxides to allylic alcohols continues to be a promising route for the preparation of these materials in high optical purity. In an extension of their ongoing work in this area with lithium amide bases <02T4665>, Andersson and co-workers have designed the optically active (*1S,3R,4R*)-[*N*-(*trans*-2,5-dialkyl)pyrrolidinyl]-methyl-2-azabicyclo[2.2.1]heptane (**87**), which was found to exhibit superior chiral induction in catalytic quantities using lithium diisopropylamide (LDA) as the stoichiometric base. Thus, the challenging substrate cyclopentene oxide (**88**) was cleanly isomerized to the chiral cyclopentenol **89** in 81% yield and with 96% ee, a significant improvement over the 49% ee obtained with higher loadings of the earlier generation catalyst **86** <02JOC1567>.

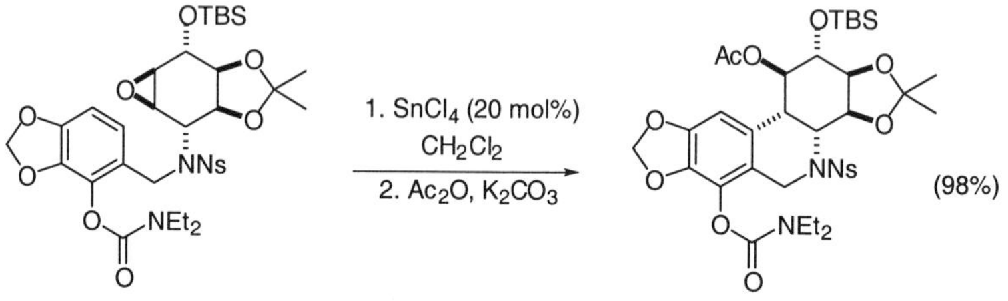

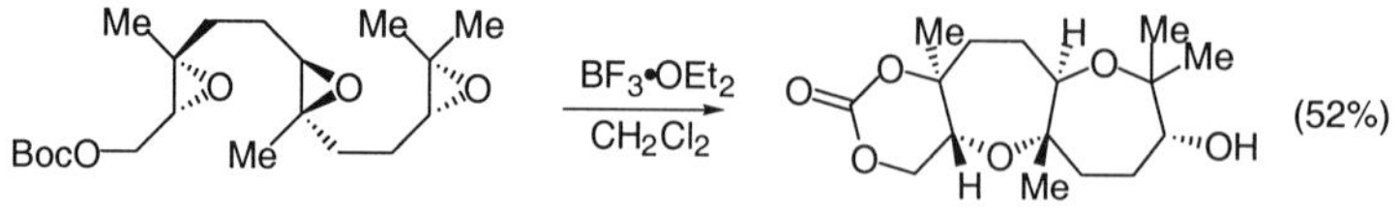

The chiral amide approach has also been applied to the catalytic kinetic resolution of racemic epoxides. For example, exposure of the tricyclic epoxide **90** with 10 mol% **86** and stoichiometric LDA at 0°C led to the recovery of the chiral spiro[4.5]decenol **91** with 90% ee and in 45% isolated yield, compared to the theoretical 50% maximum <02OL3777>. This halfway barrier

88 89

was cleverly circumvented in another protocol using a chiral Lewis acid. Thus, the *(S,S)*-enantiomer of the racemic ester enol epoxide **92** could be converted to the acyloxy ketone **93** with inversion of the acyloxy center under kinetic resolution conditions using the chiral Lewis acid catalyst [*(R)*-BINOL]$_2$-Ti(Oi-Pr)$_4$. The remaining enantiomer of the epoxide was then carefully isomerized with retention of configuration *via* acyl transfer using *p*-toluenesulfonic acid to provide **93** in 78% yield and with 93% ee <02JOC2831>.

90 91

rac-**92** **93** *(R,R)*-**92** **93**

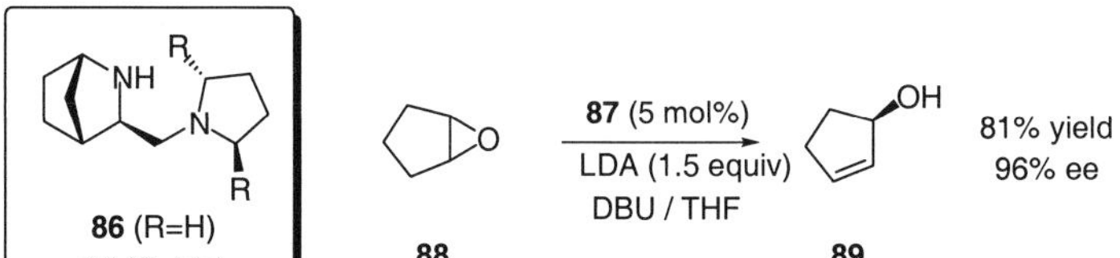

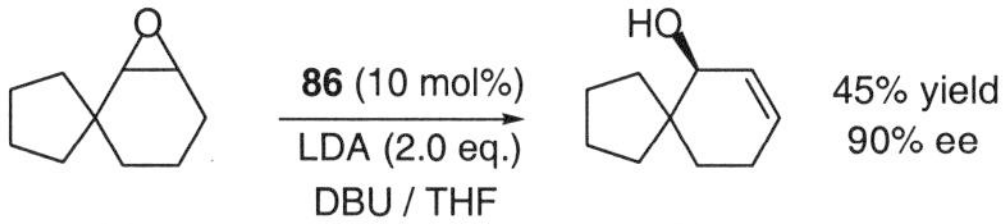

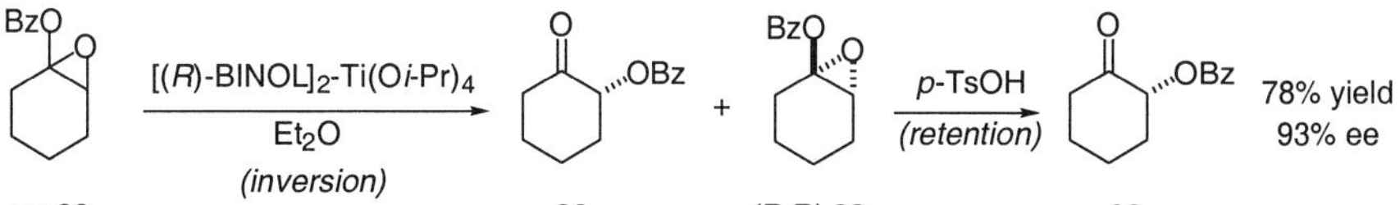

These isomerizations can also occur with ring enlargement, as illustrated by the high-yielding and stereospecific conversion of epoxy alcohol **94** to hydroxy ketone **95** in the presence of tin(IV) chloride <02TL6637>. In another set of conditions, the macrocyclic epoxy alcohol **96**

94 **95**

96 **97**

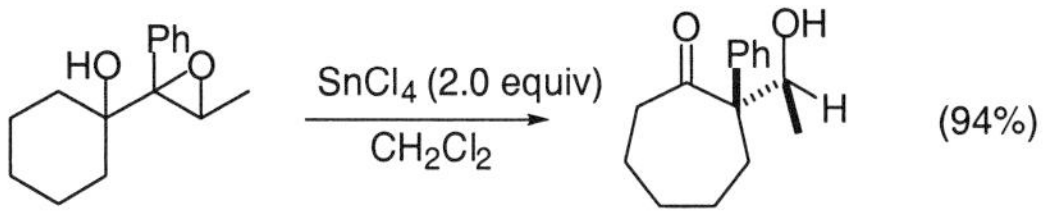

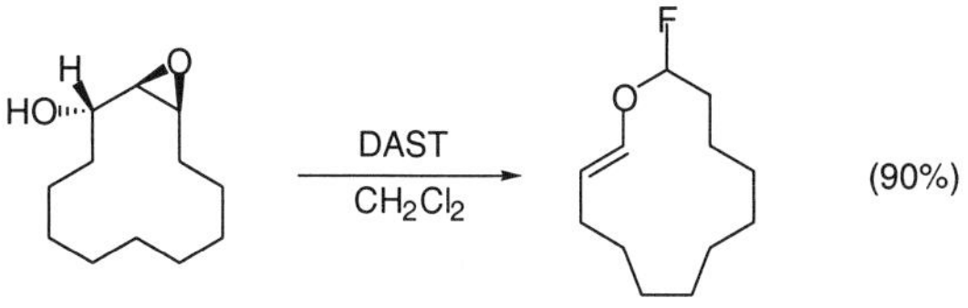

underwent ring expansion with oxygen insertion upon treatment with diethylaminosulfur trifluoride (DAST) in methylene chloride to give the novel fluoro vinyl ether **97** in 90% yield. The course of this reaction shows marked dependence upon ring size of the substrate <02OL451>.

Among the more unusual modes of epoxide reactivity is a technique for reductive deoxygenation using a system of triphenylphosphine and iodine in dimethylformamide (DMF). Under these conditions, the functionalized epoxide **98** was quantitatively converted to the diallyl ether **99**. Similar conditions were found to convert epoxides to the interestingly substituted β-bromoformate derivatives (*e.g.* **101**) <02T7037>.

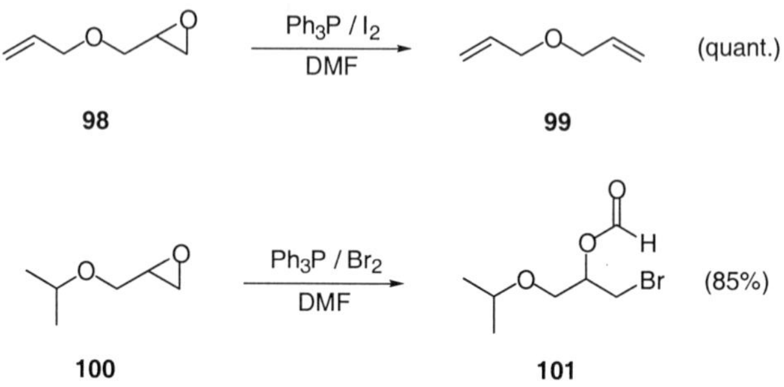

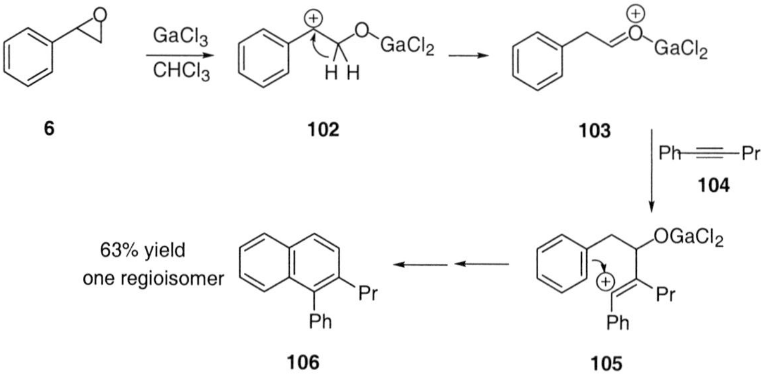

Finally, epoxides can serve as useful substrates for cycloaddition strategies. For example, alkynes react with epoxides in the presence of gallium trichloride through tandem ring opening and cyclization to give naphthalene derivatives (*e.g.* **106**). This protocol, which generally exhibits high regiocontrol, is believed to proceed through rearrangement of the gallium-induced benzylic cation **102** <02SL1553>. Similarly, Jacobsen and Movassaghi <02JA2456> have reported a direct method for the conversion of terminal epoxides (*e.g.* **107**) into γ-butenolides (*e.g.* **110**) by treatment with ynamines and a Lewis acid catalyst, a reaction which proceeds *via* an intermediate ketene aminal species (*i.e.* **109**).

(*S*)-**107** **108** **109** **110**

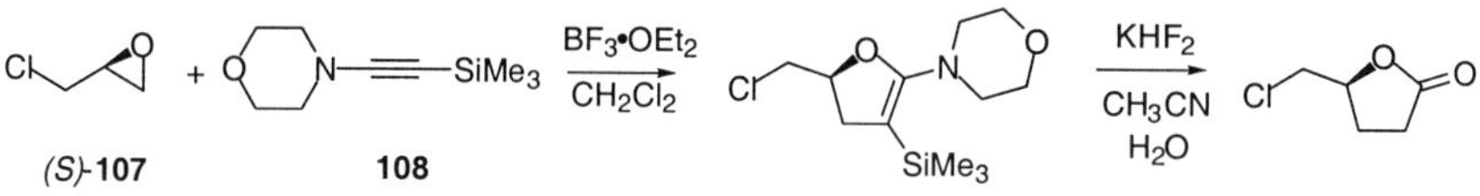

4.1.3 AZIRIDINES

4.1.3.1 Preparation of Aziridines

Aziridines are the aza-analogs of epoxides, and as such they exhibit much of the same behavior, both from a preparative and a reactive standpoint. Synthetic interest in this useful class of compounds is on the increase, and a recent review has deftly captured the current state of the art <02CSR247>.

One straightforward route to the aziridine ring system is available through the ring closure of vicinal amino alcohols, an approach which has been used to prepare vinyl N-H aziridines. Thus, 4-amino-1-phenylhex-5-en-3-ol (**111**) was treated with sulfuryl chloride to provide the sulfamidate **112**, which underwent clean thermolysis at 70 °C to form the vinyl aziridine **113** in 97% overall yield <02T5979>.

111 **112** **113** (97%)

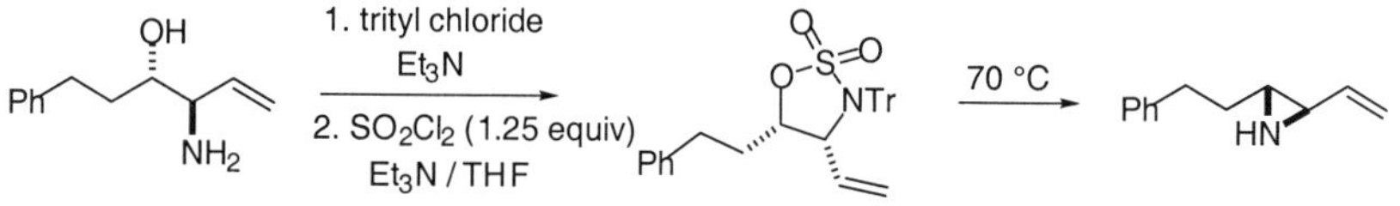

A more commonly encountered method for the preparation of aziridines is that of nitrogen transfer onto an olefin. A novel contribution in this regard was provided by DuBois and Guthikonda, who developed the broadly applicable aziridination of alkenes using a sulfamate ester (*e.g.* **115**), a rhodium carboxamide catalyst, and iodosylbenzene as a terminal oxidant <02JA13672>. An intriguing electrochemical approach has also been reported using *N*-aminophthalimide (**118**) as the nitrogen donor <02JA530>.

114 **115** **116**

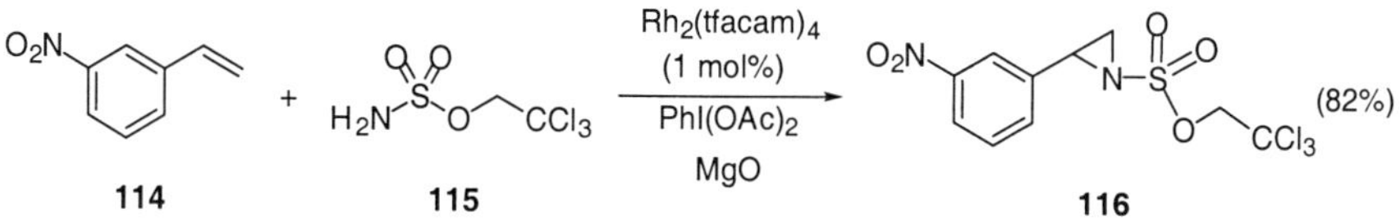

117 **118** **119**

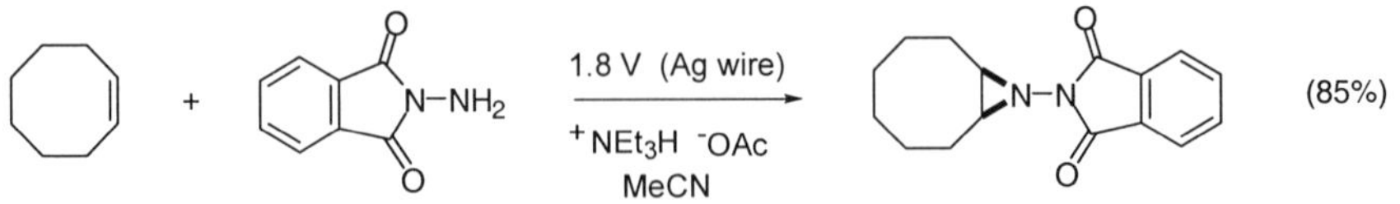

N-Aminophthalimide (**118**) can also be added to olefins in an asymmetric fashion. Thus, reaction of *N*-enoyl oxazolidinone **122** with **118** and lead tetraacetate in the presence of the camphor-derived chiral ligand **120** provides aziridine **123** in 83% yield and with 95% ee <02OL1107>. Other useful chiral ligands include imine **121**, derived from the condensation of 2,2'-diamino-6,6'-dimethylbiphenyl with 2,6-dichlorobenzaldehyde. The corresponding monometallic Cu(I) complex was found to be very efficient in chiral nitrogen transfer onto chromene derivative **124** using (*N*-(*p*-toluenesulfonyl)imino)phenyliodinane (PhI=NTs) to provide aziridine **125** in 87% yield and 99% ee <02JOC3450>.

120 (*R*)-**121**

122 **123**

124 **125**

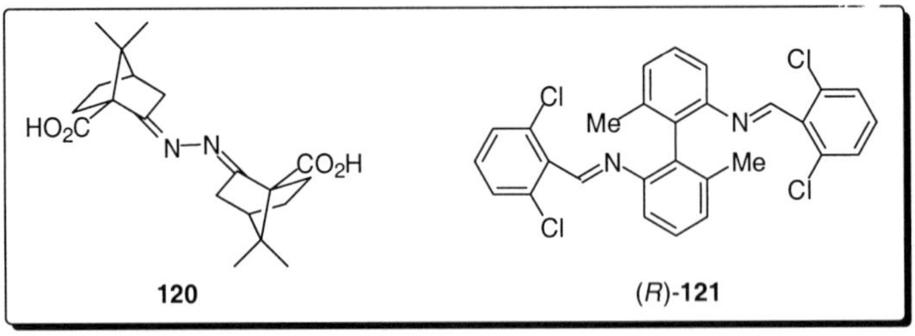

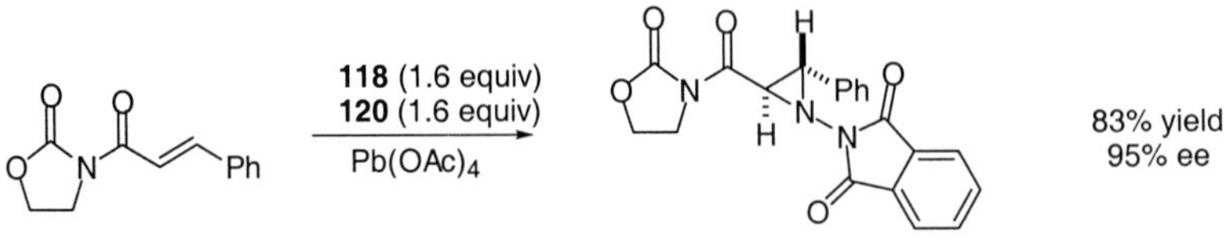

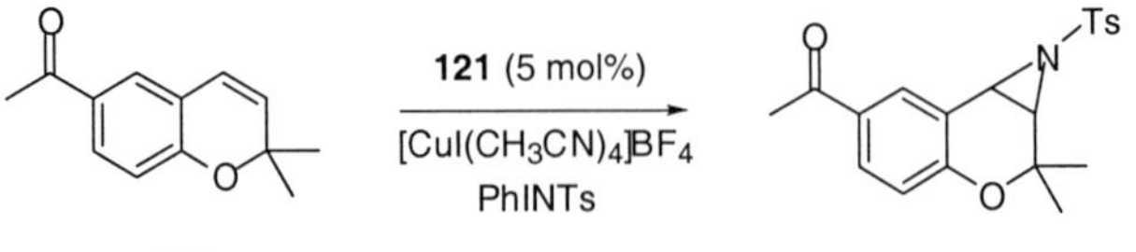

The chiral nitridomanganese complex **126** represents a novel self-contained asymmetric nitrogen transfer reagent which has been used to convert alkenes to scalemic aziridines directly, although a stoichiometric amount of transfer reagent is required. This protocol makes use of *N*-2-(trimethylsilyl)ethanesulfonyl chloride (SESCl) as an activator, providing *N*-SES-aziridines that are easily deprotected under mild conditions using *tris*(dimethylamino)-sulfonium difluorotrimethylsilicate (TASF) to give N-H aziridines in good yield. The enantioselectivity can be quite high, especially with *trans*-olefins bearing sterically bulky substituents, as is the case with the cyclohexyl stilbene derivative **127**. However, terminal and *cis*-alkenes tend to be less amenable to chiral induction <02JOC2101>.

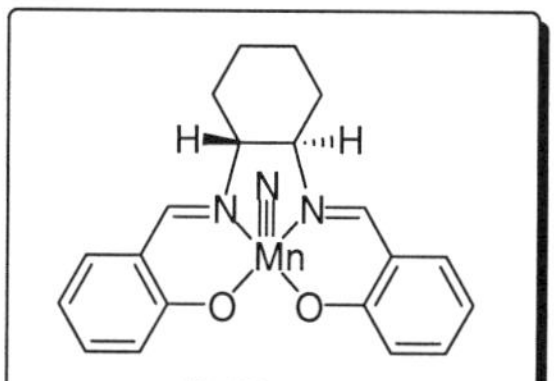

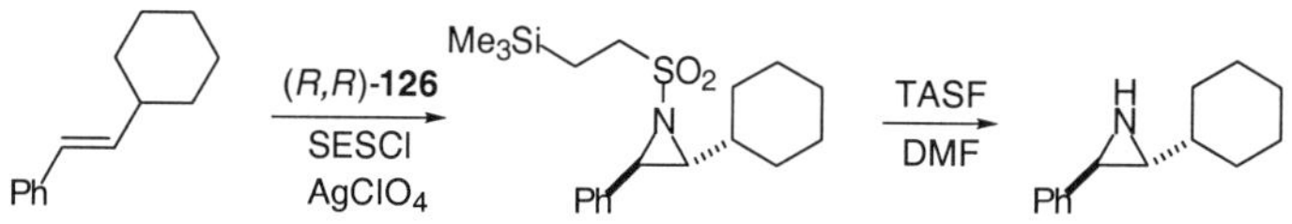

Electron-deficient alkenes can be converted to aziridines using nucleophilic nitrogen donors attached to good leaving groups. For example, *N,N'*-diazoniabicyclo[2.2.2]octane dinitrate (**130**) forms a nitrogen-nitrogen ylide in the presence of sodium hydride, which converts enones directly to the unprotected aziridines (*e.g.* **132**) by way of initial Michael addition and subsequent cyclization <02JCS(P1)1491>. Another method employs the anion of nosyloxycarbamate **134** as the aziridinating agent and a diactivated substrate which is equipped with a phenylmenthol chiral auxiliary appended to the ester moiety. Thus, enone ester **133** is converted to the azabicyclo[3.1.0]hexanone derivative **135** in 91% yield and 99% de upon treatment with **134** and calcium oxide <02JOC4972>. In the case of α-iodoenones (*e.g.* **136**), even simple primary amines can engage in aziridination, a process which is mediated by cesium carbonate <02TL4329>.

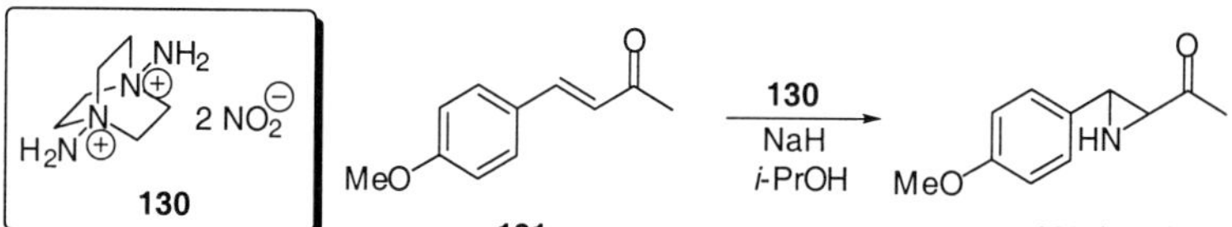

133 **134** (*S,S*)-**135**

136 **137** **138**

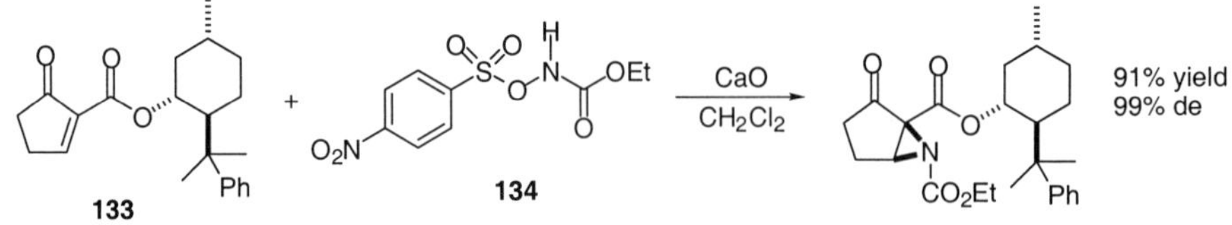

Some interesting intramolecular variants have been reported. For example, homoallylic alcohols (*e.g.* **139**) can be treated with sulfamoyl chloride to form the corresponding sulfamates (**140**), which then engage in a direct intramolecular copper-catalyzed aziridination mediated by iodosylbenzene <02OL2481>. A carbamate tether is also effective in delivering the nitrene center to the olefin, as is the case with the cyclohexenyl derivative **142**, which spontaneously cyclizes in the presence of iodosylbenzene <02OL2137>. The acetoxy-aminoquinazolinone **144** is converted to the lactone **145** *via* intramolecular aziridination upon treatment with lead tetraacetate and hexamethyldisilazane (HMDS) <02TL2083>.

139 **140** **141**

142 **143**

144 **145**

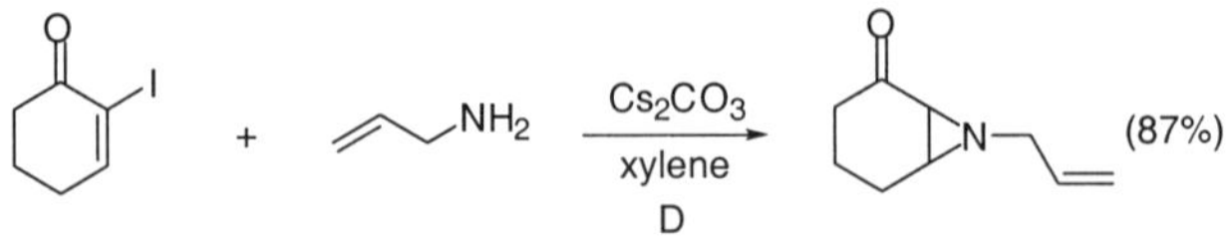

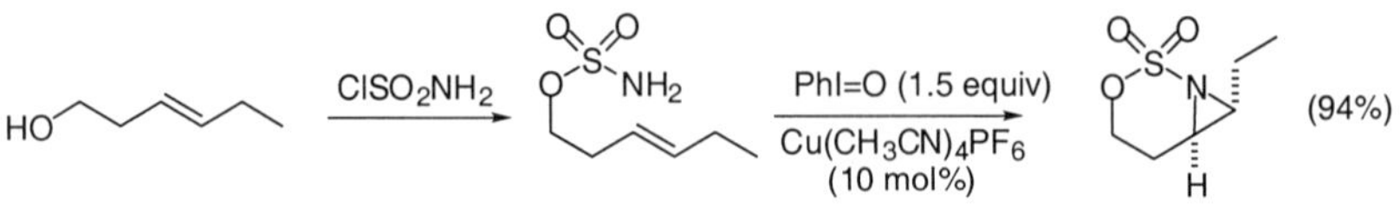

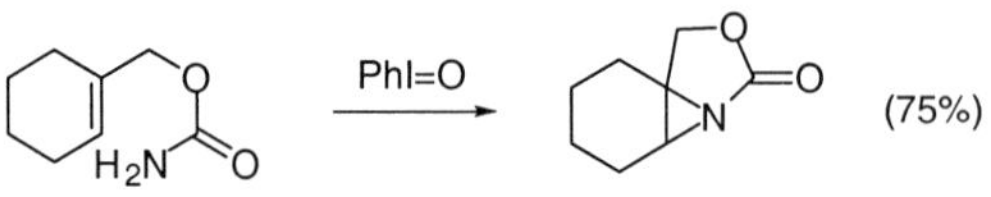

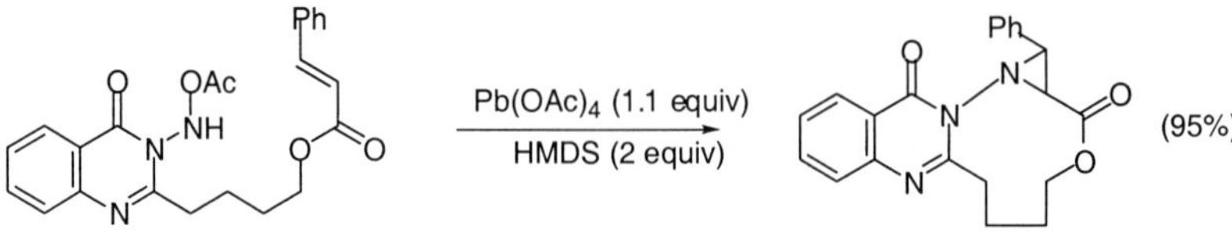

Finally, analogous to the epoxides, aziridines can also be prepared by the addition of carbenoid centers to a carbon-nitrogen double bond. In this arena, Aggarwal and co-workers have reported a highly diastereoselective aziridination of imines with trimethylsilyldiazomethane (TMSD). Thus, tosylimine **146** was converted to the *cis* aziridine **147** in 65% yield <02JOC2335>.

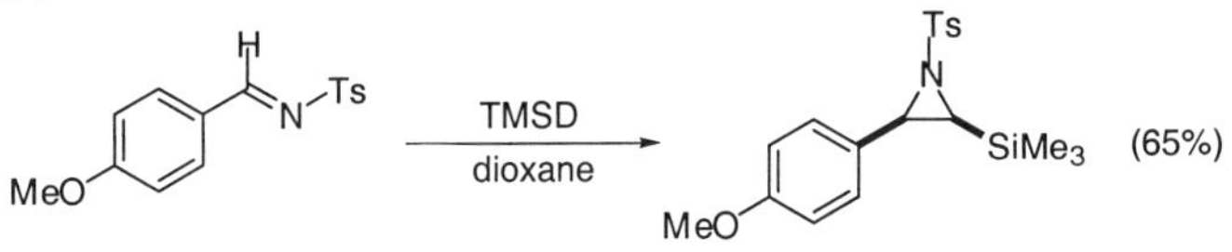

4.1.3.2 Reactions of Aziridines

Aziridines undergo a number of synthetically useful ring-opening reactions under a variety of conditions. The ring can be attacked by oxygen-centered nucleophiles, as demonstrated by the facile ring cleavage of the unsymmetrical bicyclic aziridine **148** with methanol in the presence of boron trifluoride etherate to give the product of attack at the more substituted aziridine carbon <02T7355>. Indium triflate catalyzes the opening of aziridines with carboxylic acids. For example, reaction of the cyclohexyl tosyl aziridine **150** with acetic acid and 5 mol% indium triflate resulted in the formation of amino acetate **151** in 89% yield, with a small amount of the regioisomer resulting from attack at the less substituted position <02TL2099>. Phenol-based nucleophiles are also capable of this behavior, and adding tributylphosphine to the mix has been shown to facilitate the reaction. Thus, the cyclohexene aziridine **152** provided the corresponding aryl ether **154** in 97% isolated yield in the presence of *m*-chlorophenol and tributylphosphine, but no reaction was observed in the absence of phosphine additive <02JOC5295>.

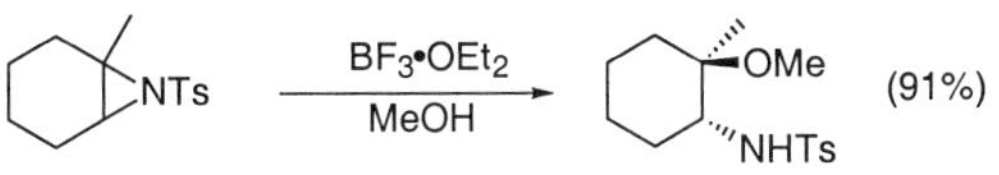

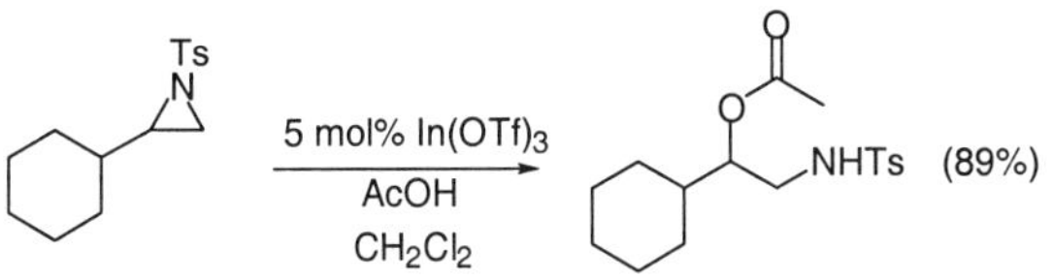

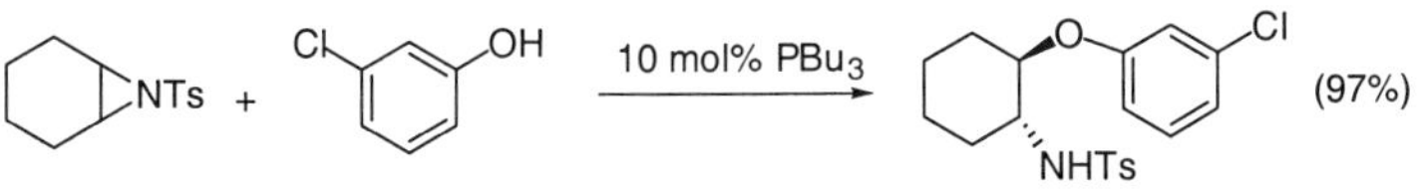

Sulfur-containing nucleophiles are equally suitable nucleophilic partners, as shown by the reaction of *p*-chlorophenol (**156**) with the functionalized sulfinylaziridine **155** to give the corresponding sulfide (**157**) in 80% yield <02JOC2902>. Even thiophene (**159**) can attack the ring in the presence of indium trichloride. The regioselectivity of this reaction prefers the 2-position of the heterocycle and the benzylic site of the aziridine, although *ca.* 10% of product formation derives from terminal attack <02TL1565>.

155 **156** **157**

158 **159** **160** (major product)

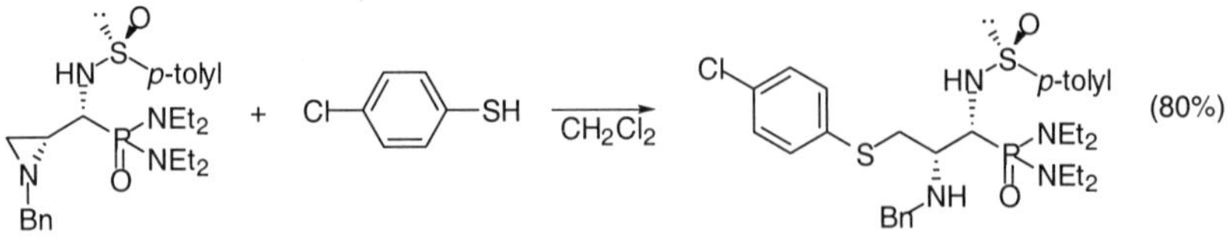

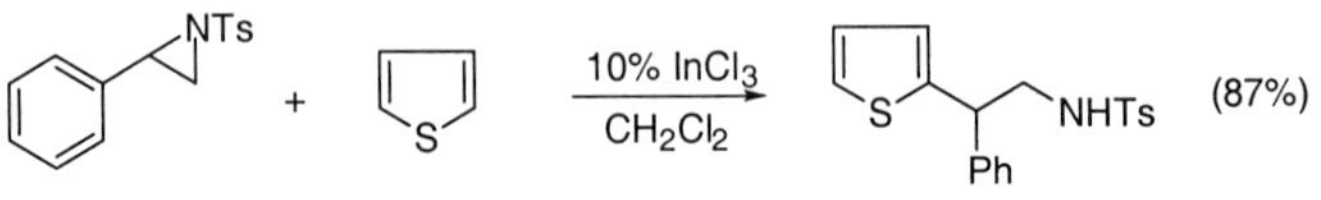

Amines are frequently encountered nucleophiles in this type of reaction, and a recent report described the ring-opening of aziridines using silica gel under solvent free conditions. Thus, aziridine **161** was combined with a slight excess of aniline in the presence of silica gel to give the diaminocyclohexane **162** in 91% yield <02TL3975>. For the simplest amine nucleophile, the azide anion is often used in preference to ammonia, since the course of the reaction is more controllable. For example, the valinol-derived nosyl aziridine **163** was converted to the primary amine **164** by treatment with lithium azide followed by triphenylphosphine <02OL949>. Sometimes azidolysis is sluggish and accompanied by unwanted isomerization reactions, so catalytic methods continue to be of interest. Along these lines, cerium(III) chloride promotes the regioselective ring opening of unsymmetrical tosyl aziridines at the terminal carbon, as exemplified by the conversion of alkyl aziridine **165** to azido amine **166** in 90% yield <02OL343>. Indium trichloride catalyzes the analogous reaction using the somewhat milder reagent trimethylsilyl azide ($TMSN_3$) <02SC1797>.

161 **162**

163 **164**

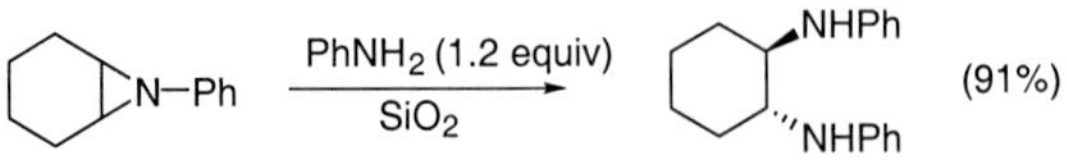

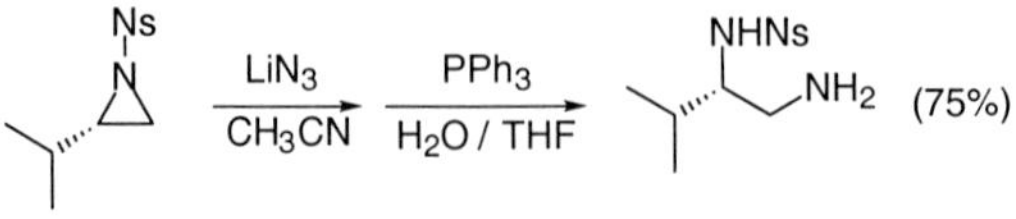

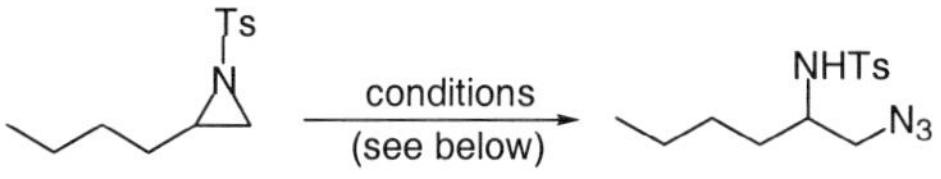

Three-Membered Ring Systems

[Scheme: aziridine **165** → **166** (NHTs, N3) under conditions (see below)]

Nu	cat	yield	ref.
NaN$_3$	CeCl$_3$	90%	02OL343
TMSN$_3$	InCl$_3$	85%	02SC1797

Other useful nucleophiles for the ring opening of aziridines include bromide, as shown in the Amberlyst-15 catalyzed reaction of lithium bromide with vinyl aziridine **167** <02TL5867> and hydride, which can be delivered by lithium triethylborohydride. This is illustrated by the conversion of tosyl azabicyclo[3.1.0]hexene **169** to the corresponding protected cyclopentenyl amine (**170**) in 79% yield <02TL723>.

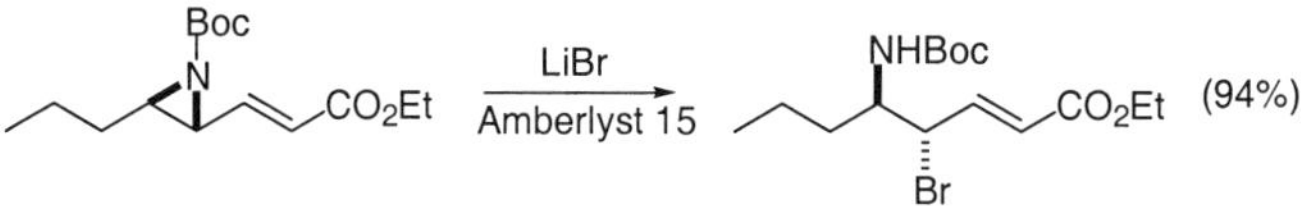

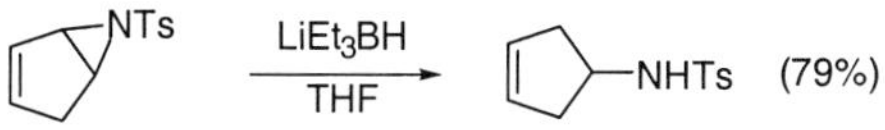

[Scheme: **167** + LiBr / Amberlyst 15 → **168** (94%)]

[Scheme: **169** + LiEt$_3$BH / THF → **170** (79%)]

The same ring strain that lends aziridines reactivity towards nucleophiles also makes them prone to ring-opening isomerizations and rearrangements. For example, the tetracyclic aziridine **171** underwent aza-pinacol rearrangement in the presence of boron trifluoride to give the tosyl imine **172**, which in turn could be trapped as the Diels-Alder adduct **173** <02CC134>. Alternatively, base-catalyzed eliminative ring-opening can be promoted with superbasic mixtures such as lithium diisopropylamide/potassium *t*-butoxide (LIDAKOR), as illustrated by the conversion of the protected aziridinyl alcohol **174** to the allyl amine derivative **175** <02T7153, 02CC778>. In the case of α-bromo aziridines (*e.g.* **176**), this elimination to allylamines can be promoted by electron transfer from magnesium metal in methanol <02T7145>.

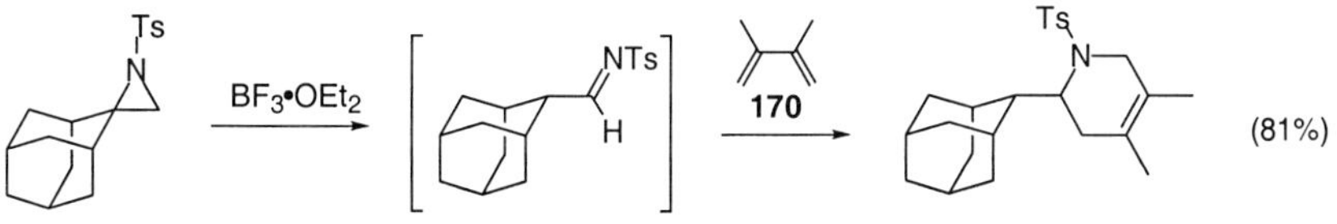

[Scheme: **171** + BF$_3$·OEt$_2$ → [**172**] + **170** → **173** (81%)]

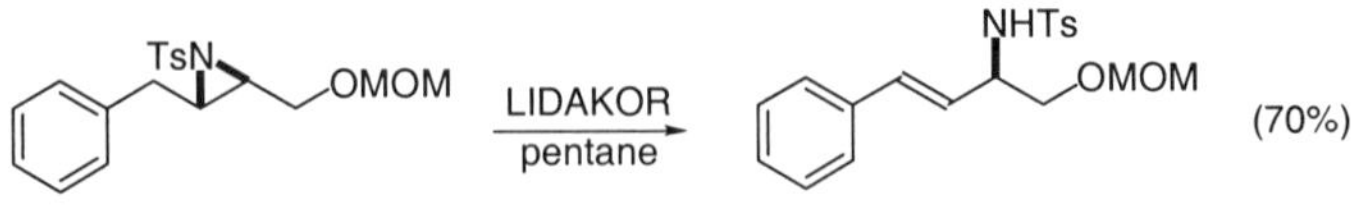

[Scheme showing conversion of 174 to 175 with LIDAKOR/pentane (70%)]

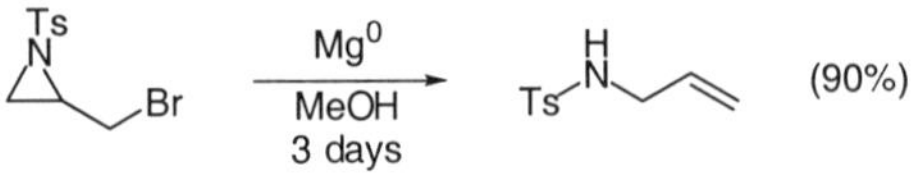

[Scheme showing 176 to 177 with Mg⁰/MeOH, 3 days (90%)]

The synthetically important subclass of alkynyl aziridines was included in a recent review <02COC539>, specifically from the standpoint of preparation and ring opening reactions with carbon nucleophiles. Such substrates (e.g. **178**) tend to suffer S_N2' attack from Grignard reagents to give aminoallenes (e.g. **179**).

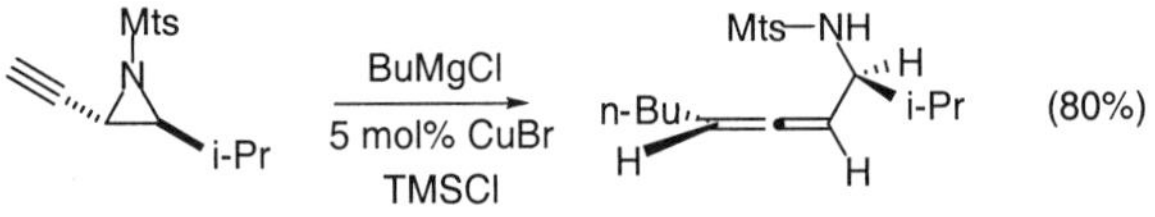

[Scheme showing 178 to 179 with BuMgCl, 5 mol% CuBr, TMSCl (80%), 179 (anti/syn 96:4)]

An organometallic reagent has also been used to ring-open an aziridinium ion (**181**), formed *in situ* by the treatment of the amino alcohol derivative **180** with lithium chloride. Subsequent addition of the aryl magnesium bromide **182** led to the formation of amine **183** in 95% overall yield <02TL6121>.

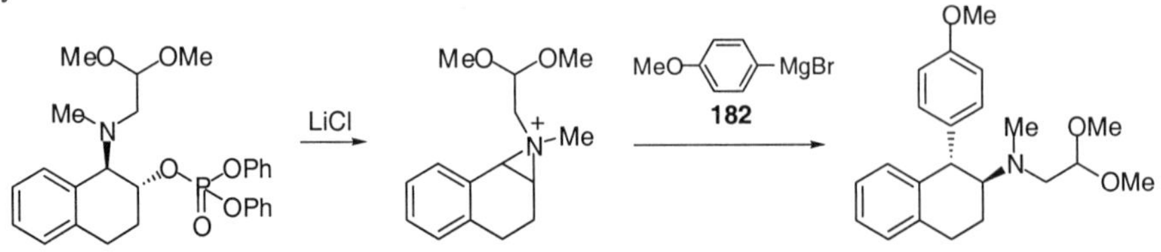

[Scheme showing 180 → 181 (LiCl) → 183 (95%) via 182]

4.1.4 AZIRINES

The preparation, properties, and synthetic applications of 2*H*-azirines have been summarized in an extensive and timely review <02OPP219>.

One significant (and indeed the first) preparative pathway to this ring system is the Neber rearrangement of oxime sulfonates. Recently it has been shown that this process can be mildly influenced by the introduction of chiral auxiliaries. Thus, rearrangement of the tosyl oxime **186** (formed *in situ* from the oxime **185**) in the presence of catalytic amounts of the chiral quaternary ammonium bromide **184** led to the formation of enantiomerically enriched amino ketone **189**, which is presumed to arise from the preferential formation of the intermediate azirine **188**. Association of the cationic chiral auxiliary with an anionic intermediate (**187**) has been invoked to rationalize the stereochemical outcome <02JA7640>.

184

185 **186** **187** **188** **189**

81% yield
51% ee

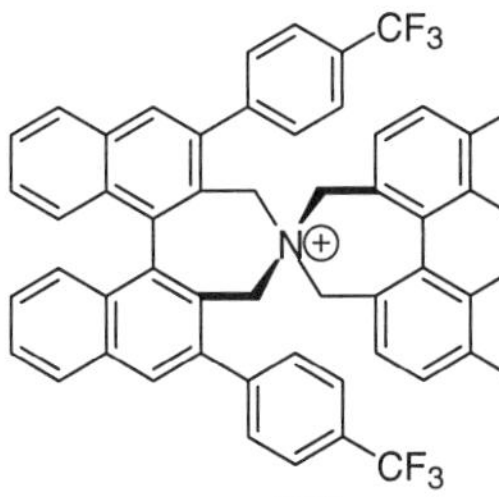

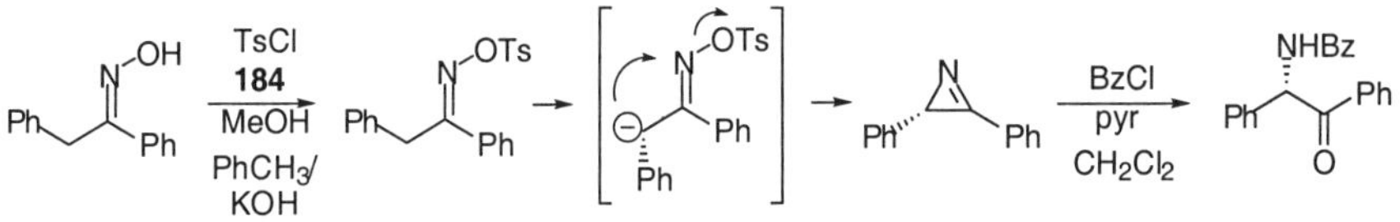

Preferential chirality can also be imposed on the sp^2 carbon of the azirine through the asymmetric transfer hydrogenation protocol. Thus, the azirine **191** was converted to the scalemic aziridine **192** in 83% yield and 72% ee in the presence of a ruthenium catalyst and the chiral auxiliary **190** in an isopropanol medium <02CC1752>.

190 **191** **192**

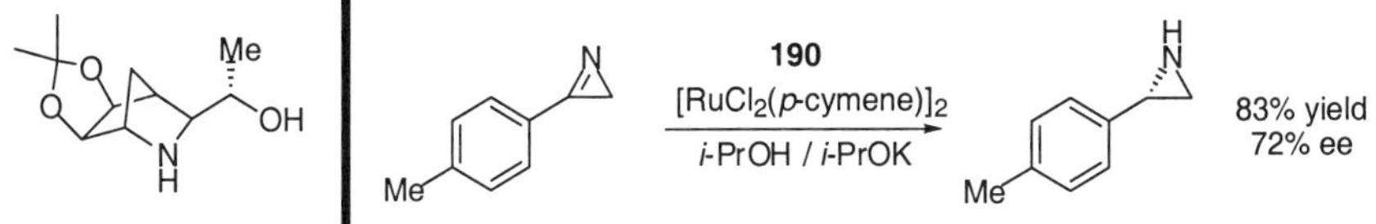

4.1.5 REFERENCES

97JOC3748 D. Ye, F. Fringuelli, O. Piermatti, F. Pizzo, *J. Org. Chem.* ***1997***, *62*, 3748.

02AGE2837 S. Matsubara, H. Yamamoto, K. Oshima, *Angew. Chem. Int. Ed.* **2002**, *41*, 2837.

02CC134 Y. Sugihara, S. Iimura, J. Nakayama, *Chem. Commun.* **2002**, 134.

02CC778 A. Mordini, L. Sharagli, M. Valasshi, F. Russo, G. Reginato, *Chem. Commun.* **2002**, 778.

02CC886	K. Smith, C.-H. Liu, *Chem. Commun.* **2002**, 886.
02CC904	A. Armstrong, R.S. Cooke, *Chem. Commun.* **2002**, 904.
02CC919	M. Bandini, P.-G. Cozzi, A. Umani-Ronchi, *Chem. Commun.* **2002**, 919.
02CC1514	V.K. Aggarwal, M. Patel, J. Studley, *Chem. Commun.* **2002**, 1514.
02CC2360	B. Lygo, D.C.M. To, *Chem. Commun.* **2002**, 2360.
02CC17522	P. Roth, P.G. Andersson, P. Somfai, *Chem. Commun.* **2002**, 1752.
02COC539	F. Chemla, F. Ferreira, *Curr. Org. Chem.* **2002**, *6*, 539.
02CSR247	J.B. Sweeney, *Chem. Sci. Rev.* **2002**, 247.
02EJIOC1666	B. Boitrel, V. Baveaux-Chambenoît, P. Richard, *Eur. J. Inorg. Chem.* **2002**, 1666.
02EJOC319	V.K. Aggarwal, M.P. Coogan, R.A. Stenson, R.V.H. Jones, R. Fieldhouse, J. Blacker, *Eur. J. Org. Chem.* **2002**, 319.
02H(57)1399	M. Ishizaki, O. Hoshino, *Heterocycles* **2002**, *57*, 1399.
02HCA913	A. Heckel, D. Seebach, *Helv. Chim. Acta* **2002**, *85*, 913.
02JA530	T. Siu, A.K. Yudin, *J. Am. Chem. Soc.* **2002**, *124*, 530.
02JA1307	S.E. Schaus, B.D. Brandes, J.F. Larrow, M. Tokunaga, K.B. Hansen, A.E. Gould, M.E. Furrow, E.N. Jacobsen, *J. Am. Chem. Soc.* **2002**, *124*, 1307.
02JA2456	M. Movassaghi, E.N. Jacobsen, *J. Am. Chem. Soc.* **2002**, *124*, 2456.
02JA2890	B.L. Lin, C.R. Clough, G.L. Hillhouse, *J. Am. Chem. Soc.* **2002**, *124*, 2890.
02JA5747	V.K. Aggarwal, J.N. Harvey, J. Richardson, *J. Am. Chem. Soc.* **2002**, *124*, 5747.
02JA7640	T. Ooi, M. Takahashi, K. Doda, K. Maruoka, *J. Am. Chem. Soc.* **2002**, *124*, 7640.
02JA8792	X.-Y. Wu, X. She, Y. Shi, *J. Am. Chem. Soc.* **2002**, *124*, 8792.
02JA13672	K. Guthikonda, J. Du Bois, *J. Am. Chem. Soc.* **2002**, *124*, 13672.
02JCS(P1)870	Y. Song, X. Yao, H. Chen, G. Pan, X. Hu, Z. Zheng, *J. Chem. Soc., Perkin Trans. 1*, **2002**, 870.
02JCS(P1)1491	J. Xu, P. Jiao, *J. Chem. Soc., Perkin Trans. 1* **2002**, 1491.
02JOC304	P. O'Brien, T.D. Towers, *J. Org. Chem.* **2002**, *67*, 304.
02JOC1567	S.K. Bertilsson, M.J. Söddergren, P.G. Andersson, *J. Org. Chem.* **2002**, *67*, 1567.
02JOC2101	M. Nishimura, S. Minakata, T. Takahashi, Y. Oderaotoshi, M. Komatsu, *J. Org. Chem.* **2002**, *67*, 2101.
02JOC2335	V.K. Aggarwal, E. Alonso, M. Ferrara, S.E. Spey, *J. Org. Chem.* **2002**, *67*, 2335.
02JOC2435	H. Tian, X. She, H. Yu, L. Shu, Y. Shi, *J. Org. Chem.* **2002**, *67*, 2435.
02JOC2515	F.E. McDonald, F. Bravo, X. Wang, X. Wei, M. Toganoh, J.R. Rodriguez, B. Do, W.A. Neiwert, K.I. Hardcastle, *J. Org. Chem.* **2002**, *67*, 2515.
02JOC2831	X. Feng, L. Shu, Y. Shi, *J. Org. Chem.* **2002**, *67*, 2831.
02JOC2902	B.-F. Li, M.-J. Zhang, X.-L. Hou, L.-X. Dai, *J. Org. Chem.* **2002**, *67*, 2902.
02JOC3450	K.M. Gillespie, C.J. Sanders, P. O-Shaughnessy, I. Westmoreland, C.P. Thickitt, P. Scott, *J. Org. Chem.* **2002**, *67*, 3450.
02JOC4972	S. Fioravanti, A. Morreale, L. Pellacani, P.A. Tardella, *J. Org. Chem.* **2002**, *67*, 4972.
02JOC5295	X.-L. Hou, R.-H. Fan, L.-X. Dai, *J. Org. Chem.* **2002**, *67*, 5295.
02JOC6954	S. Elango, T.-H. Yan, *J. Org. Chem.* **2002**, *67*, 6954.
02OL343	G. Sabitha, R.S. Babu, M. Rajkumar, J.S. Yadav, *Org. Lett.* **2002**, *4*, 343.
02OL451	P. Lakshmipathi, D. Grée, R. Grée, *Org. Lett.* **2002**, *4*, 451.
02OL905	S. Xue, Y. Li, K. Han, W. Yin, M. Wang, Q. Guo, *Org. Lett.* **2002**, *4*, 905.
02OL949	B. Moon Kim, M. So, H.J. Choi, *Org. Lett.* **2002**, *4*, 949.
02OL1107	K.-S. Yang, K. Chen, *Org. Lett.* **2002**, *4*, 1107.
02OL2137	A. Padwa, T. Stengel, *Org. Lett.* **2002**, *4*, 2137.
02OL2481	F. Duran, L. Leman, A. Ghini, G. Burton, P. Dauban, R.H. Dodd, *Org. Lett.* **2002**, *4*, 2481.
02OL2661	I. Washington, K.N. Houk, *Org. Lett.* **2002**, *4*, 2661.
02OL2817	J. Barluenga, H. Vazquez-Villa, A. Ballesteros, J.M. González, *Org. Lett.* **2002**, *4*, 2817.
02OL3777	A. Gayet, S. Bertilsson, P.G. Andersson, *Org. Lett.* **2002**, *4*, 3777.
02OPP219	F. Palacios, A.M.O. de Retana, E.M. de Marigorta, J.M. de los Santos, *Org. Prep. Proc. Int.* **2002**, *34*, 219.
02SC17	W. Chen, J. Yamada, K. Matsumoto, *Synth. Commun.* **2002**, *32*, 17.
02SC1251	B. Tamami, N. Iranpoor, H. Mahdavi, *Synth. Commun.* **2002**, *32*, 1251.
02SC1797	J.S. Yadav, B.V.S. Reddy, G.M. Kumar, Ch.V.S.R. Murthy, *Synth. Commun.* **2002**, *32*, 1797.
02SC2307	N. Raghevendra Swamy, G. Kondaji, K. Nagaiah, *Synth. Commun.* **2002**, *32*, 2307.
02SC2677	G.S. Patil, G. Nagendrappa, *Synth. Commun.* **2002**, *32*, 2677.

02SC3041	D.J. Claffey, *Synth. Commun.* **2002**, *32*, 3041.
02SL580	P.C. Bulmman Page, G.A. Rassias, D. Barros, A. Ardakani, D. Bethell, E. Merifield, *Synlett*, **2002**, 580.
02SL707	S.B. Tsogoeva, J. Wöltinger, C. Jost, D. Reichert, A. Kühnle, H.-P. Krimmer, K. Drauz, *Synlett* **2002**, 707.
02SL1553	G.S. Viswanathan, C.-J. Li, *Synlett* **2002**, 1553.
02T1623	S. Arai, H. Tsuge, M. Oku, M. Miura, T. Shioiri, *Tetrahedron*, **2002**, *58*, 1623.
02T4665	S.K. Bertilsson, P.G. Andersson, *Tetrahedron* **2002**, *58*, 4665.
02T5979	B. Olofsson, R. Wijtmmans, P. Somfai, *Tetrahedron* **2002**, *58*, 5979.
02T6003	M.A. Reddy, K. Surendra, N. Bhanumathi, K.R. Rao, *Tetrahedron* **2002**, *58*, 6003.
02T6397	A.G. Myers, M. Siu, *Tetrahedron* **2002**, *58*, 6397.
02T7001	D. Frank, S.I. Kozhushkov, T. Labahn, A. de Meijere, *Tetrahedron* **2002**, *58*, 7001.
02T7027	J.K. Crandall, E. Rambo, *Tetrahedron* **2002**, *58*, 7027.
02T7037	N. Iranpoor, H. Firouzabadi, M. Chitsaza, A.A. Jafari, *Tetrahedron Lett.* **2002**, *58*, 7037.
02T7145	K.A. Tehrani, T.N. Van, M. Karikomi, M. Rottiers, N. De Kimpe, *Tetrahedron* **2002**, *58*, 7145.
02T7153	A. Mordini, F. Russo, M. Valacchi, L. Zani, A. Degl'Innocenti, G. Reginato, *Tetrahedron* **2002**, *58*, 7153.
02T7355	B.A. Bhanu Prasad, R. Sanghi, V.K. Singh, *Tetrahedron* **2002**, *58*, 7355.
02TL723	E. Baron, P. O'Brien, T.D. Towers, *Tetrahedron Lett.* **2002**, *43*, 723.
02TL803	L. Bohé, M. Kammoun, *Tetrahedron Lett.* **2002**, *43*, 803.
02TL1565	J.S. Yadav, B.V.S. Reddy, S. Abraham, G. Sabitha, *Tetrahedron Lett.* **2002**, *43*, 1565.
02TL1797	S. Colonna, N. Gaggero, G. Carrea, G. Ottolina, P. Pasta, F. Zambianchi, *Tetrahedron Lett.* **2002**, *43*, 1797.
02TL2083	R.S. Atkinson, R.D. Draycott, D.J. Hirst, M.J. Parratt, T.M. Raynham, *Tetrahedron Lett.* **2002**, *43*, 2083.
02TL2099	J.S. Yadav, B.V.S. Reddy, K. Sadashiv, K. Harikishan, *Tetrahedron Lett.* **2002**, *43*, 2099.
02TL2665	R.I. Kureshy, N.-u. H. Kyan, S.H.R. Abdi, S.T. Patel, P.K. Iyer, R.V. Jasra, *Tetrahedron Lett.* **2002**, *43*, 2665.
02TL3975	R.V. Anand, G. Pandey, V.K. Singh, *Tetrahedron Lett.* **2002**, *43*, 3975.
02TL4111	D.O. Arbelo, J.A. Prieto, *Tetrahedron Lett.* **2002**, *43*, 4111.
02TL4329	M. Teresa Barros, C.D. Maycock, M.R. Ventura, *Tetrahedron Lett.* **2002**, *43*, 4329.
02TL5427	C.L. Winn, B.R. Bellenie, J.M. Goodman, *Tetrahedron Lett.* **2002**, *43*, 5427.
02TL5629	A. Lattanzi, P. Iannnece, A. Scettri, *Tetrahedron Lett.* **2002**, *43*, 5629.
02TL5867	G. Righi, C. Potini, P. Bovicelli, *Tetrahedron Lett.* **2002**, *43*, 5867.
02TL6121	D.R. Andrews, V.H. Dahanukar, J.M. Eckert, D. Gala, B.S. Lucas, D.P. Schumacher, I.A. Zavialov, *Tetrahedron Lett.* **2002**, *43*, 6121.
02TL6637	C.M. Maarson, A. Khan, R.A. Porter, A.J.A. Cobb, *Tetrahedron Lett.* **2002**, *43*, 6637.
02TL7083	J. Cossy, V. Bellosta, C. Hamoir, J.-R. Desmurs, *Tetrahedron Lett.* **2002**, *43*, 7083.
02TL7891	T. Ollevier, G. Lavie-Compin, *Tetrahedron Lett.* **2002**, *43*, 7891.
02TL7895	D.M. Hodgson, N.J. Reynolds, S.J. Coote, *Tetrahedron Lett.* **2002**, *43*, 7895.
02TL8257	J. Lacour, D. Monchaud, C. Marsol, *Tetrahedron Lett.* **2002**, *43*, 8257.
02TL8523	A.F. Khlebnnikov, M.S. Novikov, A.A. Amer, *Tetrahedron Lett.* **2002**, *43*, 8523.

Chapter 4.2

Four-Membered Ring Systems

Benito Alcaide
Departamento de Química Orgánica I. Facultad de Química. Universidad Complutense de Madrid, 28040-Madrid. Spain
alcaideb@quim.ucm.es

Pedro Almendros
Instituto de Química Orgánica General, CSIC, Juan de la Cierva 3, 28006-Madrid, Spain
iqoa392@iqog.csic.es

4.2.1 INTRODUCTION

The synthesis of strained four-membered heterocycles has provided a challenge, and the emergence of asymmetric approaches has increased the significance of these compounds. Oxygen- and nitrogen-containing heterocycles, in particular β-lactams, dominate the field in terms of the number of publications, but obviously, one Chapter cannot incorporate all the exciting chemistry emanating from research groups active in this area.

4.2.2 AZETINES, AZETIDINES AND 3-AZETIDINONES

A review on the chemistry of 3-azetidinones, 3-oxetanones and 3-thietanones has appeared <02CRV29>.

A three component coupling reaction of *N*-acetyl-2-azetine, aromatic imines and aromatic amines allows a rapid stereoselective entry to 2,3,4-trisubstituted tetrahydroquinolines, via fused tricyclic azetidines **1** <02CC444>. Fused heterocycles **1** were formed through an aza Diels–Alder reaction between aromatic imines and *N*-acetyl-2-azetine, which acts as an enamide substrate. This strategy has been used for the formal synthesis of luotonin A <02TL5469>.

A synthesis of 2-aryl-3,3-dichloroazetidines **2**, a relatively unexplored class of azaheterocycles, was accomplished by reduction of the corresponding 4-aryl-3,3-dichloro-2-azetidinones with monochloroalane. Reacting these 3,3-dichloroazetidines with sodium hydride in DMSO, followed by aqueous work up, afforded 1-alkyl-2-aroylaziridines, by hydrolysis of the intermediate 2-azetines <02JOC2075>.

Enantiopure 2-acylazetidines were prepared in good yields from 2-cyano-azetidines. The ketones produced were then transformed into azetidinic amino alcohols **3** by addition of phenyllithium. The latter compounds were found to be efficient ligands for enantioselective catalysis <02TA2619>.

Substituted azetidine 2-phosphonic acids **4** were prepared in enantiopure form via *N*-alkylation of the starting β-amino alcohols with a methylene phosphonate moiety, followed by sequential chlorination and stereoselective 4-*exo-tet* ring closure <02TL4633>.

A new synthesis of 1,2-diazetines via simple cycloacylation reactions has been reported <02H1257>.

Addition of several lithium ester enolates to chiral 1-aminoalkyl chloromethyl ketones gave the corresponding chlorohydrins which, when the solvent was completely evaporated to dryness at room temperature, underwent an intramolecular heterocyclization affording 3-hydroxyazetidinium salts **5** <02OL1299>. Azetidinium salts **5** were isolated in enantiomerically pure form, and with total or high diastereoselectivity. The degree of stereoselectivity was only moderately affected by the size of R^1 in the α-amino-ketone and the substituents in the ester enolate.

Bicyclic azetidines **6** have been obtained by reduction of the corresponding fused β-lactams, or alternatively by reduction of the appropriate bicyclic thiono-γ-lactams <02T7303>. These fused azetidines have been used as catalysts for the asymmetric Baylis–Hillman reaction. However, significant difficulties in the preparation of the 1-azabicyclo[3.2.0]heptanes **6** precluded further investigations and reduced the attractiveness of these azacycles as catalysts.

Regio- and stereoselective alkylations at the C-4 position of 1-alkyl-2-substituted azetidin-3-ones have been investigated. Imination of 1-benzhydryl-2-methoxymethylazetidin-3-one, followed by alkylation under kinetic conditions and final hydrolysis of the imino group, gave 4-alkyl-1-benzhydryl-2-methoxymethylazetidin-3-ones **7** in which the substituent at C-2 and C-4 had the *cis*-stereochemistry. The reduction of the carbonyl group afforded the corresponding 4-alkyl-2-methoxymethylazetidin-3-ols **8** <02T2763>.

Hydroxylation of *N*-substituted azetidines with *Sphingomonas* sp. HXN-200 gave the corresponding 3-hydroxyazetidines with excellent regioselectivity <02OL1859>. The unexpected formation of an azetidine-carborane derivative by dehydration of *N*-(1,12-dicarba-*closo*-dodecaboran-1-yl)formamide together with the first X-ray structure of a 2,3-bis(imino)azetidine **9** have been described <02IC6493>. It has been reported that the treatment of α-alkyl and α,α-dialkyl homoallylic amines with PhSeCl in CH$_3$CN containing sodium carbonate produced mixtures of azetidines and pyrrolidines <02EJO995>.

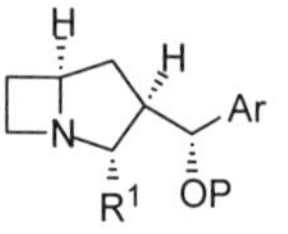

6 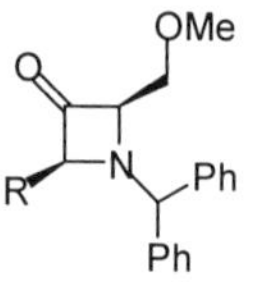 **7** 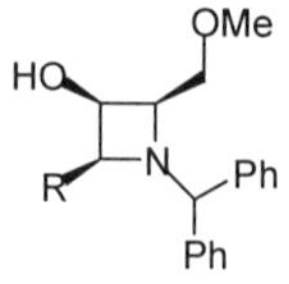 **8** 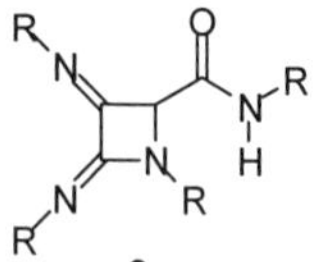 **9**
(R = 1,12-dicarba-*closo*-dodecaboran-1-yl)

Acidic treatment of 1-azabicyclo[1.1.0]butane gave the corresponding 3-monosubstituted and 1,3-disubstituted azetidine derivatives <02H433>. On the other hand, a series of aminoalkylazetidines has been identified as novel ORL1 receptor ligands. Structure-activity relationships have been investigated at the azetidine nitrogen and the alkyl side-chain sites <02BMCL3157>.

Monocyclic imino-ketenimines, in which the reactive functionalities are linked by an allylic or vinylic tether connecting the imino and ketenimino nitrogen atoms, undergo a formal intramolecular [2+2] cycloaddition to yield fused tricyclic 2-iminoazetidines, namely, azeto[1,2-*a*][1,3]thiazolo[4,5-*d*][pyrimidines **10** <02EJO4222>, and azeto[1,2-*a*]benzimidazoles **11** <02TL6259>. The observed diastereoselctivity in the newly formed C–C bond, when applicable, was excellent.

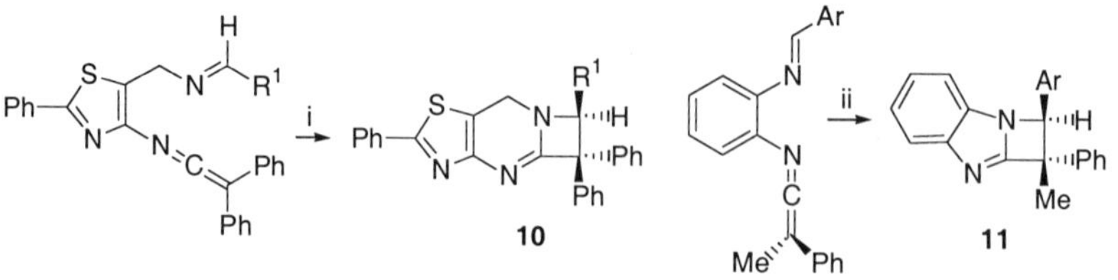

10   **11**

Key: i) Toluene, reflux. ii) Toluene, RT.

4.2.3 MONOCYCLIC 2-AZETIDINONES (β-LACTAMS)

A review on the application of the β-lactam nucleus in stereocontrolled synthesis has been published <02SL381>. The synthesis of α-amino acids, α-amino amides, and dipeptides from azetidine-2,3-diones <02CEJ3646>, a divergent β-lactam-based route to indolizidine and quinolizidine derivatives <02SL85>, the preparation of substituted glutarimides via a two-carbon ring expansion of β-lactams <02TL2491>, the use of β-lactams as building blocks in the synthesis of macrocyclic spermine and spermidine alkaloids <02T7177>, the SmI_2-mediated cyclizations of derivatized β-lactams for the highly diastereoselective construction of functionalized prolines <02JOC2411>, and a versatile synthesis of ibotenic acid analogues from oxoazetidine-carbaldehydes <02TL3951>, deserve to be mentioned as well.

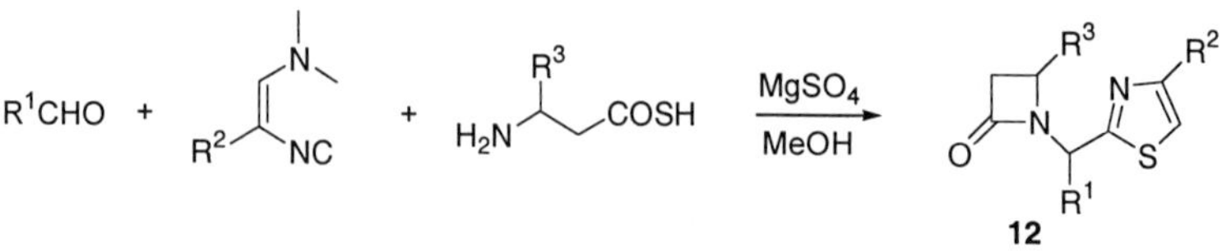

12

A novel multicomponent reaction of β-aminothiocarboxylic acids, aldehydes, and 3-dimethylamino-2-isocyanoacrylates, which involves a remarkable increase in molecular complexity, has been used for the preparation of substituted 1-thiazol-2-ylmethyl-2-azetidinones **12** <02TL6897>.

An efficient use of triphosgene, as an acid activator, for the synthesis of substituted 2-azetidinones via ketene–imine cycloaddition reaction using various acids and imines has been achieved <02T2215>. Novel routes to monocyclic β-lactams **13** and **14** through the photochemical decomposition of oxime oxalate amides <02CC2086> and α-oxoamides <02OL1443> have also been described.

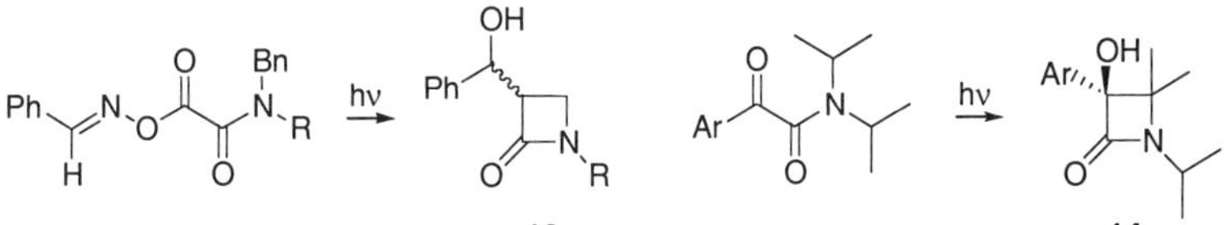

An efficient and operationally simple synthesis of tetrahydrofuran-derived spiro-β-lactams **15** and **16** using the Staudinger reaction of unsymmetrical cyclic ketones has been described <02JCS(P1)571; 02SL69>. Similarly, spiro-β-lactams **17** were synthesized via Staudinger reaction of imines derived from 7-oxanorbornenone with alkoxyacetyl chlorides <02TL6405>.

The asymmetric synthesis of 4-alkyl-4-carboxy-2-azetidinones **18** has been achieved through base-mediated intramolecular cyclization of the corresponding N-α-chloroacetyl derivatives bearing (+)- or (−)-10-(N,N-dicyclohexylsulfamoyl)isoborneol as chiral auxiliary (ee up to 82%) <02JOC3953> <02H501>.

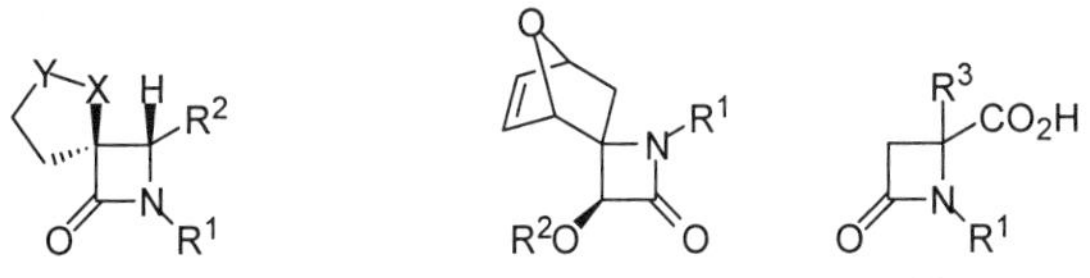

Regio- and stereoselective metal-mediated 1,3-butadien-2-ylation reactions between 1,4-dibromo-2-butyne or 1,4-bis(methanesulfonyl)-2-butyne and optically pure azetidine-2,3-diones in aqueous media offers a convenient asymmetric entry to potentially bioactive 3-substituted-3-hydroxy-β-lactams **19** <02JOC1925>. In addition, 2-azetidinone-tethered 1,3-butadienes can easily be transformed into other functionalities via Diels–Alder reaction.

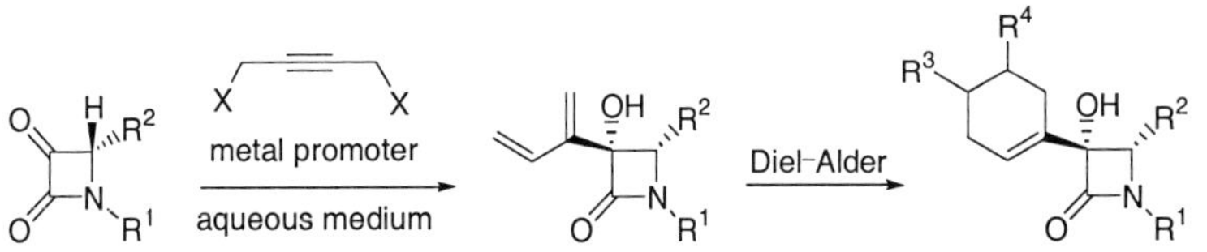

(R^1 = PMP, allyl, propargyl; R^2 = (S)-2,2-dimethyl-1,3-dioxolan-4-yl; X = Br, MsO)

The intermolecular 1,3-dipolar cycloaddition reactions of racemic as well as enantiopure 2-azetidinone-tethered nitrones with substituted alkenes and alkynes yielded isoxazolinyl-, or isoxazolidinyl-β-lactams **20**, exhibiting good regio- and facial stereoselectivity <02JOC7004>.

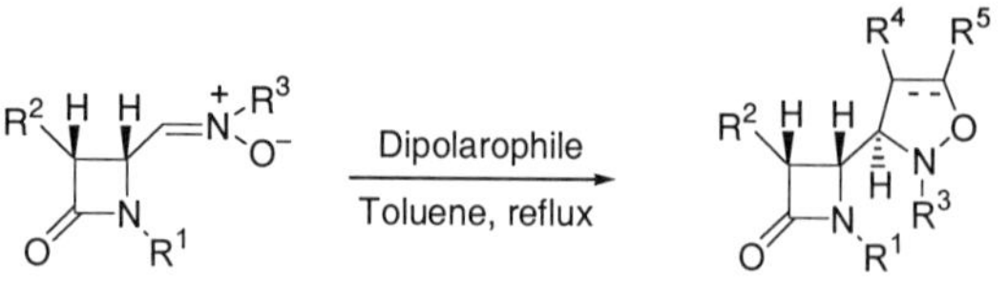

The Staudinger cycloaddition of chiral N,N-dialkylhydrazones to benzyloxyketene appears to be a new general approach to the enantioselective synthesis of 4-substituted 3-alkoxyazetidin-2-ones **21** <02AG(E)831>.

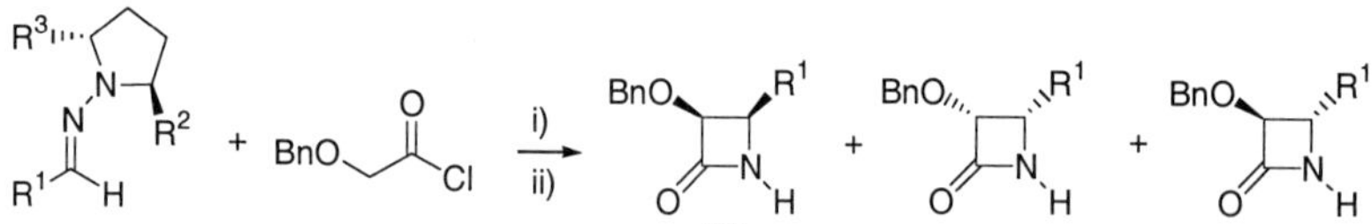

Yield: 84-98 %, d.r. = 99:1, cis:trans 99:1

Key: i) Et$_3$N, benzene, reflux. ii) MMPP, MeOH, RT.

Iridium-catalyzed reductive coupling of acrylates and imines provides *trans*-β-lactams **22** bearing aromatic, alkenyl, and alkynyl side-chains, with high diastereoselection <02OL2537>. This reaction appears to proceed through a reductive Mannich addition–cyclization mechanism.

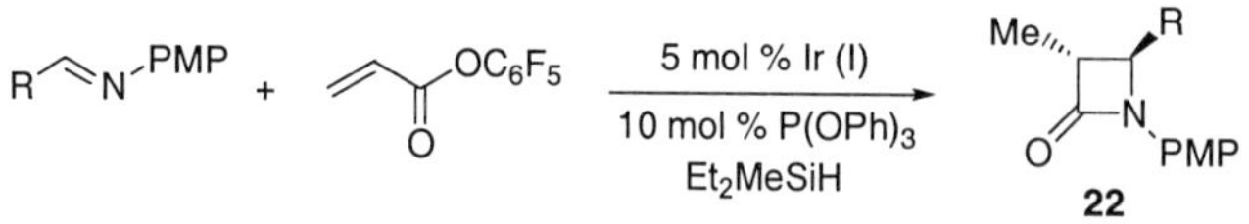

An unprecedented one-pot stereoselective synthesis of 2-azetidinone β-chlorinated allylic alcohols **23**, which can also be considered as functionalized allylsilanes, has been developed, by tin(IV) chloride-mediated reaction of propargyltrimethylsilane and 4-oxoazetidine-2-carbaldehydes <02CEJ1719>. An explanation for the formation of β-chlorovinyl alcohols involves a stepwise process with the chlorination proceeding via a silicon stabilized carbocation.

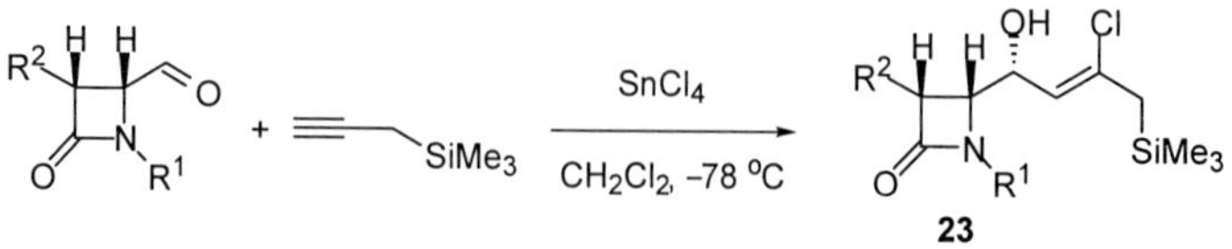

Lectka and colleagues have reacted achiral ketenes with achiral imines to achieve asymmetric induction in the synthesis of cis-β-lactams through the use of a bifunctional catalyst system consisting of a chiral base (benzoylquinine) and an achiral Lewis acid <02OL1603> <02JACS6626>.

It has also been demonstrated that a planar-chiral azaferrocene derivative of 4-(pyrrolidino)pyridine is an excellent catalyst for the enantioselective Staudinger reaction, providing β-lactams **24** with very good stereoselection and yield <02JACS1578>.

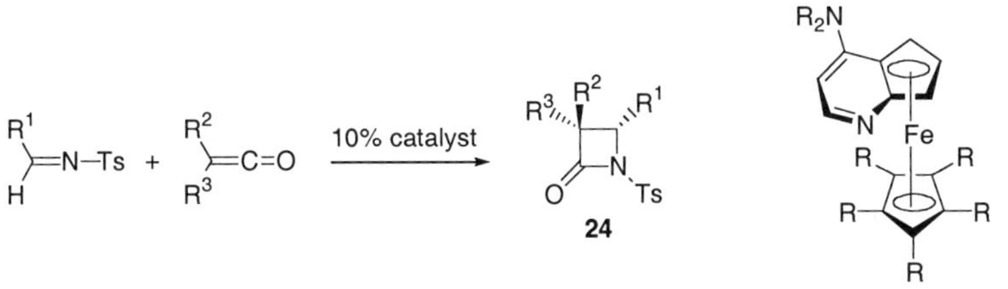

The first versatile system for the copper-catalyzed asymmetric coupling of alkynes with nitrones to form cis-β-lactams has been developed using a bis(azaferrocene) ligand <02JACS4572>.

The Kinugasa reaction has been used as well for the asymmetric synthesis of β-lactams **25** via cycloaddition between chiral oxazolidinyl propynes and nitrones, in the presence of copper(I) chloride <02TL5499>.

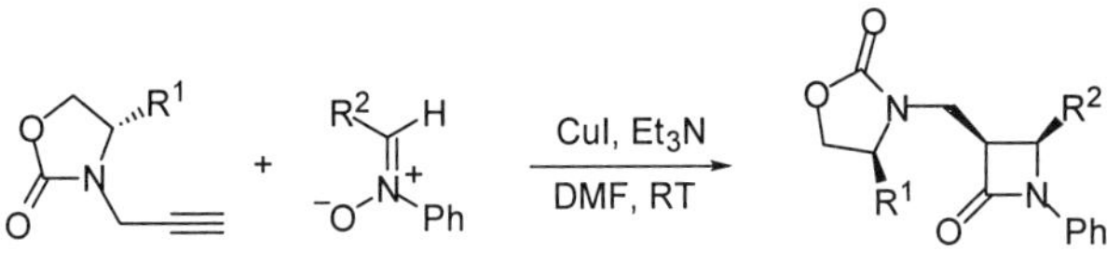

A cis-aziridine was employed in a carbonylation reaction by treatment with $Co_2(CO)_8$ and the trans-β-lactam (**26**) was obtained as a single distereo- and regioisomer in good yield <02JOC2335>. The nucleophilic ring opening of the cis starting material resulted in inversion of configuration, thus leading to the trans-β-lactam. The exclusive formation of the 3-TMS-substituted 2-azetidinone (**26**) is a consequence of the completely regioselective ring opening of the aziridine, which occurred exclusively at the silicon-bearing carbon.

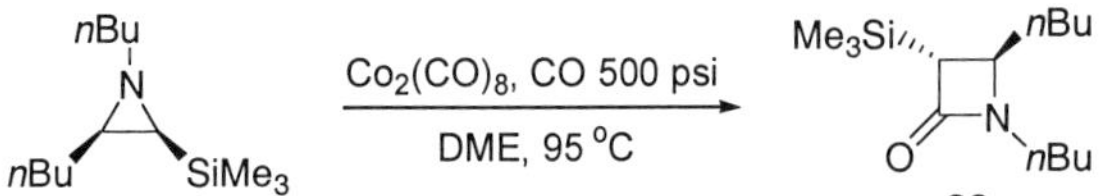

Polymer-bound β-lactams have been prepared via the ester enolate imine condensation route <02JOC8034>. On the other hand, an efficient asymmetric synthesis of 2-azetidinones was accomplished when chiral acid chlorides or chiral aldehydes were used in the polymer-supported Staudinger reaction <02TA905>.

A new solid-support strategy – the sequential column asymmetric catalysis – has been developed for the synthesis of enantiopure β-lactams <02CEJ4115>. In this strategy, reagents and catalysts are attached to a solid-phase support and loaded onto sequentially-linked colums. The substrates are present in the liquid phase that flows through the column. As a substrate encounters each successive column, it grows in complexity.

Many studies concerning the biological activities of β-lactams were published in 2002. A highly stereoselective synthesis of the novel azetidinone-based tryptase inhibitor BMS-262084 was developed <02JOC3595>. An efficient and highly stereoselective synthesis of a β-lactam inhibitor of the serine protease prostate-specific antigen was accomplished <02BMC1813>. A series of 1,4-bis(alkoxycarbonyl)azetidin-2-ones, designed as potential suicide-inhibitors of serine proteases, has been synthesized and evaluated against porcine pancreatic elastase <02T2423>. A report describes a new family of N-thiolated β-lactams that have antibacterial activity against methicillin-resistant *Staphylococcus aureus*, showing unprecedented structure-activity features and an unusual mode of action for a β-lactam antibiotic <02BMCL2229>. A new turn mimic derived from PLG (prolyl-leucyl-glycine amide) containing a β-lactam in the turn area has been prepared <02EJO2686>. A novel β-lactam antimicrobial – the penem faropenem – has been developed and shown to have excellent activity against a wide range of bacteria, including extended-spectrum β-lactamase (ESBL)-producing strains <02MI223>.

4.2.4 FUSED POLYCYCLIC β-LACTAMS

A review discusses the stereocontrolled synthesis of polycyclic-β-lactams with non-classical structure <02MI245>.

A short asymmetric synthesis of the 2-ketocarbacepham (**27**) involving as the initial step for the preparation of the starting piperidone, a hetero Diels–Alder reaction of the benzylimine derived from the enantiomer of Garner's aldehyde with Danishesky's diene, has been described <02JOC598>. The key cyclization step to form the bicyclic β-lactam system was achieved from a β-amino acid precursor using the Mukaiyama reagent, 2-chloro-N-methylpyridinium iodide.

Key: i) 2-Chloro-N-methylpyridinium iodide, Et$_3$N, MeCN, 70 °C.

Alicyclic β-lactams **28** were successfully synthesized via a parallel liquid-phase Ugi four-center three-component reaction, starting from alicyclic β-amino acids such as *cis*-2-aminocyclohexanecarboxylic acid, *cis*-2-aminocyclopentanecarboxylic acid, 2,3-*diexo*-3-aminobicyclo-[2.2.1]heptane-2-carboxylic acid and some of their partially unsaturated analogues <02OL1967>.

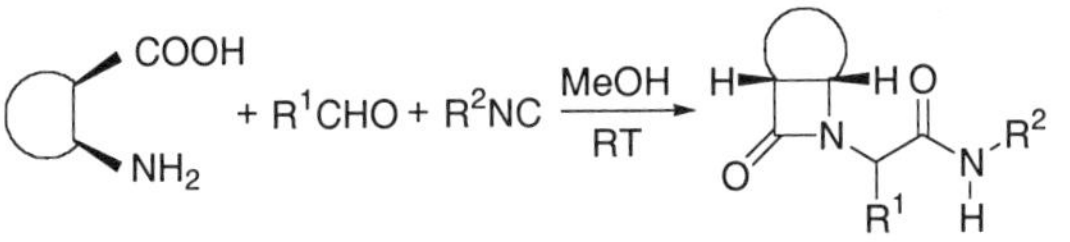

Highly enantioselective photocyclization of 1-alkyl-2-pyridones, in the solid state, to β-lactams **29** has been achieved in inclusion crystals with optically active host compounds <02OL3255>.

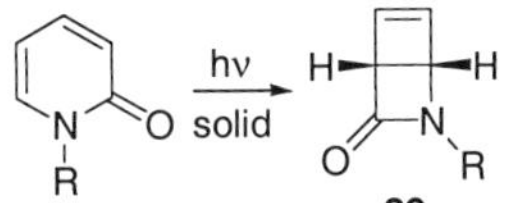

Methodology for the conversion of glucuronic acid glycosides to novel bicyclic β-lactams **30** has been reported <02OL135>. Key features of this strategy include a diastereoselective Ferrier reaction of a glucuronic acid glucal, selective β-lactam ring formation using a cyclic allylic alcohol, and a chemoselective benzylic oxidation.

A carbene insertion route to the β-lactam fused cyclic enediyne (**31**) has been developed <02TL4241>.

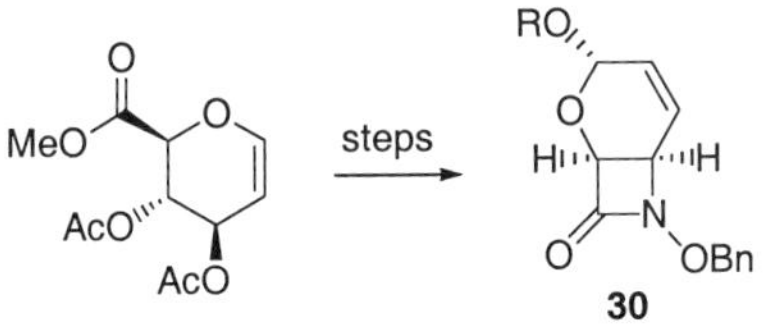

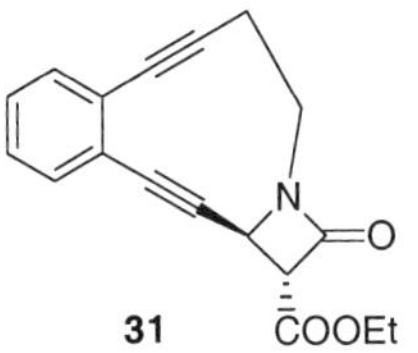

The formation of bicyclic nitrones of the 2-azetidinone *N*-oxide type, **32** and (**33**), has been achieved in a two-step route, through condensation of the corresponding 2-azetidinone tethered-alkenyl(alkynyl)aldehyde with hydroxylamine followed by phenylselenyl bromide treatment <02JOC7004>.

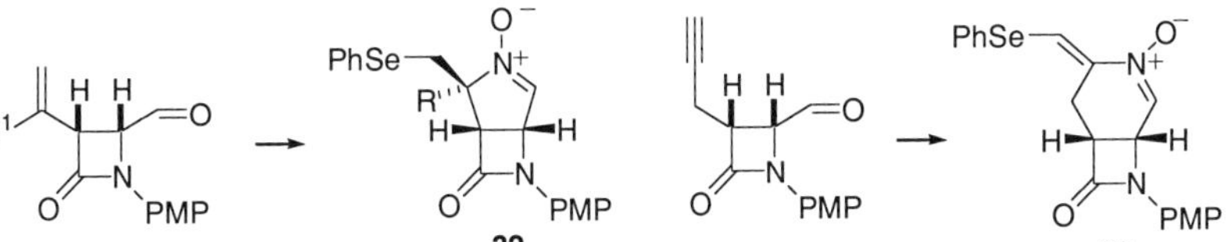

A recent paper explored the 1,2-functionalization of the allene moiety in monocyclic β-lactam allenynol derivatives under palladium-catalyzed reaction conditions <02CEJ1719>. Remarkably, a novel domino process, namely allene cyclization/intramolecular Heck reaction

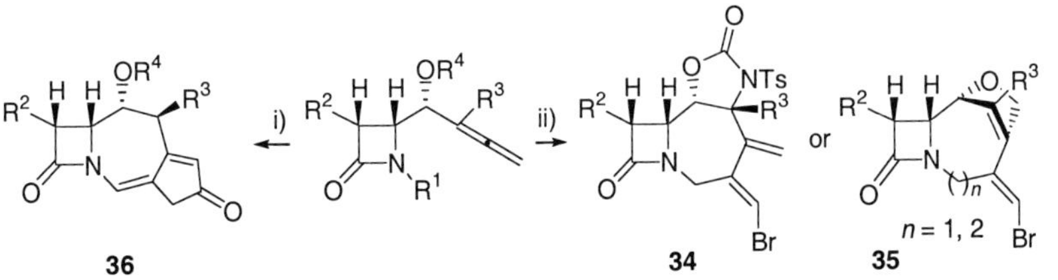

Key: i) $Co_2(CO)_8$, Me_3NO, CH_2Cl_2, RT. ii) $Pd(OAc)_2$ 10 mol%, LiBr, $Cu(OAc)_2$, K_2CO_3, O_2, MeCN, RT.

was found to give fused tricyclic β-lactams **34** or bridged medium-sized ring tricycles **35**. 2-Azetidinones **35** are remarkable since they possess an unusual pyramidalized bridgehead structure. A likely mechanism for the cascade reaction should involve an intramolecular cyclization on a (π-allyl)palladium complex and a Heck–type reaction. The allenic variant of the [2+2+1] cycloaddition reaction in the above β-lactam allenynols produced tricyclic 2-azetidinones **36** bearing a central seven-membered ring as the only isomer. Cycloadducts **36** presumably arise from the isomerization of the initially formed Pauson–Khand adducts.

The reactivity of 2H-azirines as 1,3-dipolarophiles towards β-lactam-based azomethine ylides derived from oxazolidinones has been evaluated <02JCS(P1)2014>, providing cycloadducts **37**, which incorporate the novel 2,6-diazatricyclo[4.2.0.0^{2,4}]octan-7-one ring system.

The reductive opening of epoxyimonobactams with titanocene(III) chloride gives rise to radicals that can be trapped intramolecularly by π systems to give new tricyclic 2-azetidinones **38** <02JOC8243>.

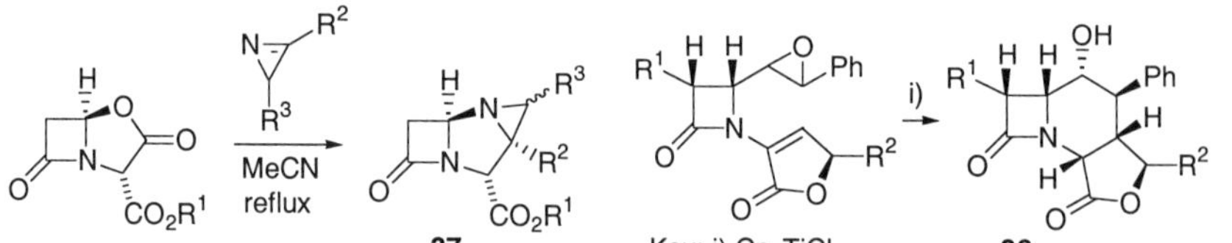

A novel and direct synthetic strategy to prepare fused tricyclic β-lactams **39** has been developed from monocyclic allenols, masked functionalized dienes, which underwent an unprecedented domino allenol transposition/intramolecular Diels-Alder reaction <02CC1472>. This transformation might be tentatively explained through a migration of the methanesulfonyl group in the initially formed α-allenic methanesulfonate to give the corresponding mesyloxy-diene counterpart, which evolves via an IMDA reaction.

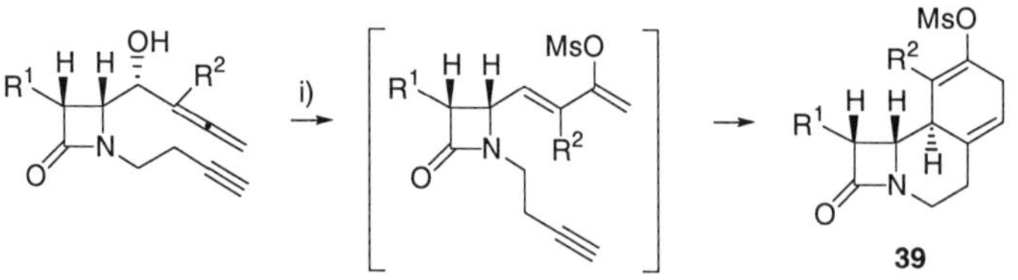

Key: i) CH$_3$SO$_2$Cl, Et$_3$N, toluene, 190 °C.

4.2.5 OXETANES AND 2-OXETANONES (β-LACTONES)

A review of the formation of four-membered heterocycles through electrophilic heteroatom cyclisation is available <02EJO3099>.

The synthesis of the oxetane intermediate (**40**) *en route* to the alkaloid gelsemine has been achieved in a straightforward way from a diol precursor <02JACS9812> <02TL545>.

N-Hydroxyphthalimide-catalyzed radical addition of oxetane to alkenes bearing electron-withdrawing substituents produced the adducts **41**. Unfortunately, this reaction gave a diastereomeric mixture of products <02TL3617>.

The 1,5-dioxaspiro[3,2]hexane (**42**) has been shown to be a useful precursor for both aminodiol and aminotriol sphingoid bases. The synthesis of oxetane (**42**) started with serine via a Mitsunobu lactone formation followed by sequential titanium-mediated methylenation and oxidation with dimethyldioxirane <02OL1719>.

Photoinduced [2+2] cycloaddition (the Paterno-Büchi reaction) of 1-acetylisatin with acyclic enol ethers afford the spiro(3H-indole-3,2'-oxetane)s **43** with moderate regio- and diastereoselectivity via the nπ* triplet state of the isatin derivative without involvement of single electron transfer <02JCS(P1)345>.

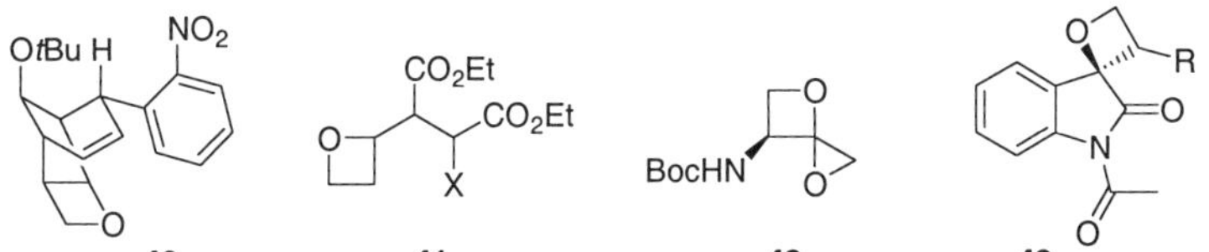

The Paterno–Büchi reaction between 2-furylmethanol derivatives and aliphatic aldehydes and ketones induced by irradiation through Vycor shows high regioselectivity but no stereoselectivity, affording a 1:1 diastereoisomeric mixture of fused oxetanes (e.g., **44**). This behavior was rationalized by assuming that the starting materials react through both single and triplet excited states <02T5045>.

Catalyst (**45**) (5 mol%) allows regioselective carbonylation of a variety of epoxides under mild conditions, yielding the corresponding β-lactones **46** <02AG(E)2781>. Thus propylene oxide was converted into β-butyrolactone in 95% yield in 4 h at 60 °C; the carbonylation is highly regioselective producing exclusively the 4-methyloxetan-2-one isomer. Retention of stereochemistry allows the synthesis of enantiomerically enriched 2-oxetanones from chiral epoxides. Catalyst (**45**) also effects carbonylation of *cis*- and *trans*-2,3-epoxybutanes regioselectively giving *trans*- and *cis*-lactones, respectively, in high yields and with inversion of configuration.

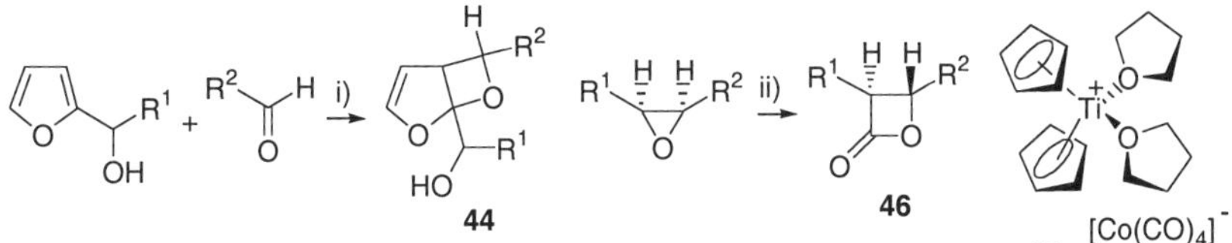

Key: i) hv, 8 h, benzene. ii) **45**, CO, DME.

A series of allenic carboxylic acids has been converted to funtionalized β-lactones **47** by oxidation-cyclization promoted by pre-prepared solutions of dimethyldioxirane <02T7027>. This transformation is rationalized by the involvement of unisolated spirodioxide intermediates, which give lactones with appropriately situated α-hydroxyketone moieties.

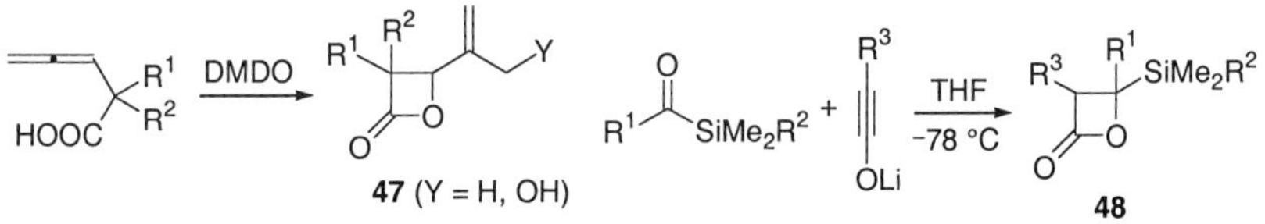

Ynolate anions react with acylsilanes at low temperature to give β-lactones **48** in good yields. When the reaction is conducted at room temperature the isolated product is a β-silyl-α,β-unsaturated ester via a β-lactone enolate intermediate <02JACS6840>.

A series of water-dispersible, surface-active poly(fluorinated oxetane)s was prepared by ring-opening polymerization of fluorinated oxetane monomers using Lewis acid catalysis <02L5993>.

A concise and efficient synthesis of racemic α,α,β-trisubstituted β-lactones **49** has been described in five steps from precursor diethyl oxalopropionate <02TL9513>. In the first step, racemic diethyl oxalopropionate was alkylated under basic conditions. The lactonization is the last step of the synthesis and was achieved according to the Mitsunobu protocol.

Serine and threonine β-lactone derivatives **50** have been prepared from the corresponding β-hydroxy acid. β-Lactones **50** were tested against HAV 3C proteinase, some of them displaying potent inhibition <02JOC1536>. Mass spectrometry and HMQC NMR studies show that inactivation of the enzyme occurs by nucleophilic attack of the cysteine thiol (Cys-172) at the β-position of the oxetanone ring. However, other related analogues, such as five-membered ring homoserine γ-lactones, failed to give significant inhibition of HAV 3C proteinase, thus demonstrating the importance of the β-lactone ring for binding. The racemic β-lactone (**51**), lacking the α-nitrogen functionality, was synthesized from a diazoester precursor by treatment with a refluxing solution of rhodium(II) acetate dimer. Tests of **51** against HAV 3C proteinase showed poor inhibition.

The ebelactones are a small group of β-lactone enzyme inhibitors, isolated from a cultured strain of soil actinomycetes. The highly stereocontrolled hydroboration of an alkene, a subsequent Suzuki–Miyaura cross-coupling reaction, a silylcupration on a non-terminal acetylene, an iododesilylation, and a β-lactonization from the corresponding β-hydroxy acid using phenylsulfonyl chloride were the key steps in a convergent total synthesis of (–)-ebelactone A (**52**) <02OL2043>.

A recent study provides comments on the mechanism and origin of stereoselectivity in the Lewis acid catalyzed [2+2] cycloaddition reaction between ketenes and aldehydes to give β-lactones. The observations made in this study highlight the broad range of factors which must be pondered in order to understand and control stereoselectivity in a multistep reaction <02AG(E)1572>.

4.2.6 THIETANES, THIETONES, THIAZETES AND β-SULTAMS

An expeditious and highly diastereoselective synthesis of (Z)-2-alkylsulfenyl (or 2-cyano) -2,4-diarylthietanes **53** has been reported by nucleophile (CN⁻, RS⁻) induced cyclization of the corresponding O,O-diethyl S-(1,3-diaryl-3-oxopropyl) phosphorodithionates under microwave irradiation in solvent free conditions <02S1502>. This reductive cyclization can be considered to be induced by the attack of a hydride ion on the carbonyl carbon of the Michael adduct precursor to give an alkoxide ion, which attacks the phosphorus atom intramolecularly.

A fused thietone, the pyridothietone (**54**), was produced as the major product by flash vacuum pyrolysis of 2-mercaptonicotinic acid at 550 °C and ca. 1×10^{-2} torr, together with a trimer <02TL5285>.

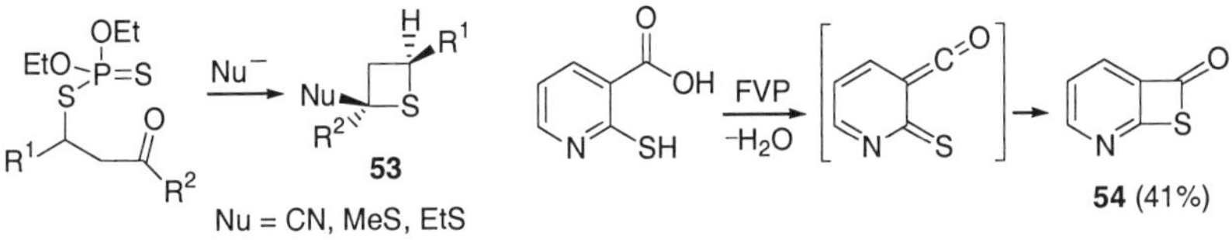

1,2-Thiazetidine 1,1-dioxides (β-sultams) are the sulfonyl analogues of β-lactams, which contain three different heteroatom bonds: N–S, S–C and N–C. An efficient four-step asymmetric synthesis of 3-substituted β-sultams **55** has been described <02TL5109>. The key step of this synthesis is the Lewis acid catalyzed aza-Michael addition of (R,R,R)-2-amino-3-methoxymethyl-2-azabicyclo[3.3.0]octane (RAMBO) to the appropriate alkenylsulfonic esters, in good yields and excellent enantiomeric excesses. Cleavage of the sulfonates followed by chlorination resulted in the corresponding sulfonyl chlorides. 3-Substituted β-sultams were obtained in moderate to good yields and high ee's over two steps, an acidic N-deprotection and an *in situ* cyclization promoted by triethylamine.

The reaction of sodium azide with 5-substituted 3-diethylamino-4-(4-methoxyphenyl)-isothiazole 1,1-dioxides afforded [1,2]thiazete S,S-dioxides **56**. It has been proposed that a nucleophilic addition of the azide ion to C-5 followed by ring closure gave a triazoline intermediate, which cyclized intramolecularly with concomitant nitrogen extrusion <02T5173>.

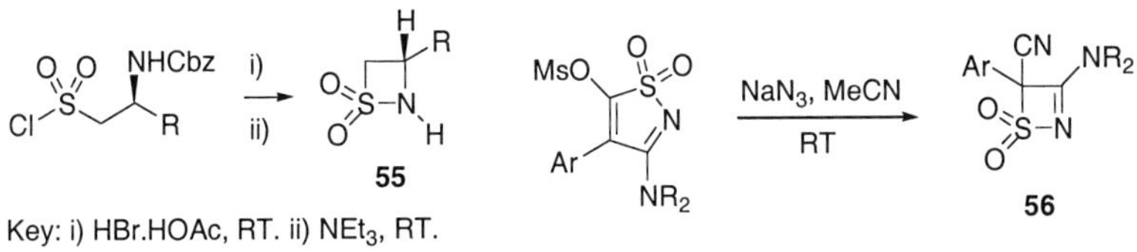

Key: i) HBr.HOAc, RT. ii) NEt₃, RT.

4.2.7 DIOXETANES AND OXATELLURETANES

Dioxetanes with an annelated five- or six-membered ring were synthesized by singlet oxygenation <02TL1523> <02TL8955>.

The first example of a chiral-auxiliary-induced [2+2] cycloaddition between $^1O_2$ and oxazolidinone-functionalized enecarbamates, which proceeds with complete diastereoselectivity as a result of steric repulsions, has been reported to afford **57** <02JACS8814>. The optically active enecarbamates bearing Evans' chiral auxiliary were photooxygenated at −35 °C with 5,10,15,20-tetrakis(pentafluorophenyl)porphine (TPFPP) as sensitizer and an 800 W sodium lamp as light source. The dioxetanes **57** were obtained exclusively, but they readily decomposed at room temperature to the expected carbonyl products because of their thermally labile nature. The absolute configuration of the dioxetanes **57** was established by reduction to the corresponding diols with L-methionine.

The first 1,5-dioxa-4λ^4-telluraspiro[3.3]heptane (**58**), was synthesized and its structure was determined by X-ray analysis <02TL6775>. The oxidative cyclization of the corresponding bis(β-hydroxyalkyl)tellurides, which was effective for the synthesis of its selenium analogue, failed. However, successive treatment of the appropriate β-hydroxyalkyl phenyl sulfide with n-BuLi/Li in the presence of catalytic amount of 4,4'-di-t-butylbiphenyl (DTBB), followed by addition of an excess amount of $TeCl_4$ gave a single isomer of the 1,2-oxatelluretane (**58**) in modest yield as a stable solid in the air.

Key: i) $^1O_2$, TPFPP, $CDCl_3$, −35 °C. ii) a) n-BuLi, Li, DTBB; b) $TeCl_4$, −78 °C.

4.2.8 SILICON AND PHOSPHORUS HETEROCYCLES

A review on the synthesis and properties of phosphetanes has appeared <02CRV201>. The first isolation and crystallographic results of a Pt-containing silacyclobutene, the stable 4-sila-3-platinacyclobutene **59**, has been described and its formation pathway was demonstrated via Pt-promoted γ-Si–H bond activation of a 3-sila-1-propenylplatinum precursor <02JACS4550>.

The first synthesis of a silacyclobutene diannelated [a,d]benzene (**60**), <02OM1101> as well as the preparation of novel 1-aza-2-silacyclobut-3-enes **61** appeared in the literature <02JCSD3253>.

It has been reported that 1,3-silyl migration on acylpolysilane leads to the formation of silene. Silene thus formed undergoes [2+2] cycloaddition with acetylene in a two-step manner, giving rise to the formation of silacyclobutene <02OM3271>.

The synthesis and crystal structure of a complex containing a 1-hafna-3-silacyclobutane ring was described <02OM1912>.

The first isolation and characterization of an anti-apicophilic spirophosphorane bearing an oxaphosphetane ring (**62**) has been achieved <02JACS7674>. 2-Furyl substituents at the phosphorus atom in Wittig ylides give rise to significant increases in the thermal stability of the oxaphosphetane intermediates, a 1,2-λ^5-oxaphosphetane bearing three 2-furyl substituents at the phosphorus atom being isolated <02EJO1143>.

The synthesis of the first 2,3-dihydro-1,2,3-azadiphosphete complex, **63**, was achieved via thermal reaction of a 2H-azaphosphirene complex in o-xylene, whereby, a 2,5-dihydro-

1,2-diaza-2,5-diphosphinine complex and a further unidentified byproduct were formed. The structure of the four-membered heterocycle (**63**) was established through an X-ray single-crystal diffraction study <02CC2204>.

4.2.9 REFERENCES

02AG(E)1572	D.A. Singleton, Y. Wang, H.W. Yang, D. Romo, *Angew. Chem. Int. Ed.* **2002**, *41*, 1572.
02AG(E)2781	V. Mahadevan, Y.D.Y.L. Getzler, G.W. Coates, *Angew. Chem. Int. Ed.* **2002**, *41*, 2781.
02AG(E)831	R. Fernández, A. Ferrete, J.M. Lassaletta, J.M. Llera, E. Martín-Zamora, *Angew. Chem. Int. Ed.* **2002**, *41*, 831.
02BMC1813	R. Annunziata, M. Benaglia, M. Cinquini, F. Cozzi, A. Puglisi, *Biorg. Med. Chem.* **2002**, *10*, 1813.
02BMCL2229	E. Turos, T.E. Long, M.I. Konaklieva, C. Coates, J.Y. Shim, S. Dickey, D.V. Lim, A. Cannons, *Biorg. Med. Chem. Lett.* **2002**, *12*, 2229.
02BMCL3157	W.L. Wu, M.A. Caplen, M.S. Domalski, H.T. Zhang, A. Fawzi, D.A. Burnett, *Biorg. Med. Chem. Lett.* **2002**, *12*, 3157.
02CC1472	B. Alcaide, P. Almendros, C. Aragoncillo, M.C. Redondo, *Chem. Commun.* **2002**, 1472.
02CC2086	E.M. Scanian, J.C. Walton, *Chem. Commun.* **2002**, 2086.
02CC2204	E. Ionescu, P.G. Jones, R. Streubel, *Chem. Commun.* **2002**, 2204.
02CC444	P.J. Stevenson, M. Nieuwenhuyzen, D. Osborne, *Chem. Comun.* **2002**, 444.
02CEJ1719	B. Alcaide, P. Almendros, C. Aragoncillo, *Chem. Eur. J.* **2002**, *8*, 1719.
02CEJ3646	B. Alcaide, P. Almendros, C. Aragoncillo, *Chem. Eur. J.* **2002**, *8*, 3646.
02CEJ4115	A.M. Hafez, A.E. Taggi, T. Lectka, *Chem. Eur. J.* **2002**, *8*, 4115.
02CRV201	A. Marinetti, D. Carmichael, *Chem. Rev.* **2002**, *102*, 201.
02CRV29	Y. Dejaegher, N.M. Kuz'menok, A.M. Zvonok, N. De Kimpe, *Chem. Rev.* **2002**, *102*, 29.
02EJO1143	M. Appel, S. Blaurock, S. Berger. *Eur. J. Org. Chem.* **2002**, 1143.
02EJO2686	T.C. Maier, W.U. Frey, J. Podlech, *Eur. J. Org. Chem.* **2002**, 2686.
02EJO3099	S. Robin, G. Rousseau. *Eur. J. Org. Chem.* **2002**, 3099.
02EJO4222	M. Alajarín, A. Vidal, R.A. Orenes, *Eur. J. Org. Chem.* **2002**, 4222.
02EJO995	X. Pannecoucke, F. Outurquin, C. Paulmier, *Eur. J. Org. Chem.* **2002**, 995.
02H1257	D. Pufky, R. Beckert, M. Doring, O. Walter, *Heterocycles* **2002**, *56*, 1257.
02H433	K. Hayashi, S. Hiki, T. Kumagai, Y. Nagao, *Heterocycles* **2002**, *56*, 433.
02H501	G. Gerona-Navarro, M.A. Bonache, N. Reyero, M.T. García-López, R. González-Muñiz, *Heterocycles* **2002**, *56*, 501.
02IC6493	P. Schaffer; P. Morel; J.F. Britten; J.F. Valliant, *Inorg. Chem.* **2002**, *41*, 6493.
02JACS1578	B.L. Hodous, G.C. Fu, *J. Am. Chem. Soc.* **2002**, *124*, 1578.
02JACS4550	M. Tanabe, K. Osakada, *J. Am. Chem. Soc.* **2002**, *124*, 4550.
02JACS4572	M.M.-C. Lo, G.C. Fu, *J. Am. Chem. Soc.* **2002**, *124*, 4572.
02JACS6626	A.E. Taggi, A.M. Hafez, H. Wack, B. Young, D. Ferraris, T. Lectka, *J. Am. Chem. Soc.* **2002**, *124*, 6626.
02JACS6840	M. Shindo, K. Matsumoto, S. Mori, K. Shishido, *J. Am. Chem. Soc.* **2002**, *124*, 6840.
02JACS7674	S. Kojima, M. Sugino, S. Matsukawa, M. Nakamoto, K. Akiba, *J. Am. Chem. Soc.* **2002**, *124*, 7674.
02JACS8814	W. Adam, S.G. Bosio, N.J. Turro, *J. Am. Chem. Soc.* **2002**, *124*, 8814.
02JACS9812	F.W. Ng, H. Lin, S.J. Danishefsky, *J. Am. Chem. Soc.* **2002**, *124*, 1578.
02JCS(D)3253	R.J. Bowen, M.A. Fernandes, P.B. Hitchcock, M.F. Lappert, M. Layh, *J. Chem. Soc., Dalton Trans.* **2002**, 3253.

02JCS(P1)2014 D. Brown, G.A. Brown, M. Andrews, J.M. Large, D. Urban, C.P. Butts, N.J. Hales, T. Gallagher, *J. Chem. Soc., Perkin Trans. 1* **2002**, 2014.
02JCS(P1)345 Y. Zhang, J. Xue, Y. Gao, H.-K- Fun, J.-H. Xu, *J. Chem. Soc., Perkin Trans. 1* **2002**, 345.
02JCS(P1)571 E. Alonso, C. del Pozo, J. González, *J. Chem. Soc., Perkin Trans. 1*, **2002**, 571.
02JOC1536 M.S. Lall, Y.K. Ramtohul, M.N.G. James, J.C. Vederas, *J. Org. Chem.* **2002**, *67*, 1536.
02JOC1925 B. Alcaide, P. Almendros, R. Rodríguez-Acebes, *J. Org. Chem.* **2002**, *67*, 1925.
02JOC2075 Y. Dejaegher, S. Mangelinckx, N. De Kimpe, *J. Org. Chem.* **2002**, *67*, 2075.
02JOC2335 V.K. Aggarwal, E. Alonso, M. Ferrara, S.E. Spey, *J. Org. Chem.* **2002**, *67*, 2335.
02JOC2411 M.F. Jacobsen, M. Turks, R. Hazell, T. Skrydstrup, *J. Org. Chem.* **2002**, *67*, 2411.
02JOC3595 X.H. Qian, B. Zheng, B. Burke-B, M.T. Saindane, D.R. Kronenthal, *J. Org. Chem.* **2002**, *67*, 3595.
02JOC3953 G. Gerona-Navarro, M.T. García-López, R. González-Muñiz, *J. Org. Chem.* **2002**, *67*, 3953.
02JOC598 A. Avenoza, J.H. Busto, C. Cativiela, F. Corzana, J.M. Peregrina, M.M. Zurbano, *J. Org. Chem.* **2002**, *67*, 598.
02JOC7004 B. Alcaide, P. Almendros, J.M. Alonso, M.F. Aly, C. Pardo, E. Sáez, M.R. Torres, *J. Org. Chem.* **2002**, *67*, 7004.
02JOC8034 S. Schunk, D. Enders, *J. Org. Chem.* **2002**, *67*, 8034.
02JOC8243 G. Ruano, M. Grande, J. Anaya, *J. Org. Chem.* **2002**, *67*, 8243.
02L5993 M. Kausch, J.E. Leising, R.E. Medsker, V.M. Russell, R.R. Thomas, *Langmuir* **2002**, *18*, 5993.
02MI223 L.A. Sorbera, M. Del Fresno, R.M. Castaner, X. Rabasseda, *Drugs of the Future*, **2002**; *27*, 223.
02MI245 B. Alcaide, P. Almendros, *Curr. Org. Chem.* **2002**, *6*, 245.
02OL1299 J.M. Concellón, E. Riego, P.L. Bernad, *Org. Lett.* **2002**, *4*, 1299.
02OL135 T.B. Durham, M.J. Miller, *Org. Lett.* **2002**, *4*, 135.
02OL1443 A. Natarajan, K.Wang, V. Ramamurthy, J.R. Scheffer, B. Patrick, *Org. Lett.* **2002**, *4*, 1443.
02OL1603 S. France, H. Wack, A.M. Hafez, A.E. Taggi, D.R. Witsil, T. Lectka, *Org. Lett.* **2002**, *4*, 1603.
02OL1719 A.J. Ndakala, M. Hashemzadeh, R. C. So, A. R. Howell, *Org. Lett.* **2002**, *4*, 1719.
02OL1859 D. Chang, H.-J. Feiten, K.-H. Engesser, J.B. van Beilen, B. Witholt, *Org. Lett.* **2002**, *4*, 1859.
02OL1967 S. Gedey, J. Van der Eycken, F. Fülöp, *Org. Lett.* **2002**, *4*, 1967.
02OL2043 A.K. Mandal, *Org. Lett.* **2002**, *4*, 2043.
02OL2537 J.A. Townes, M.A. Evans, J. Queffelec, S.J. Taylor, J.P. Morken, *Org. Lett.* **2002**, *4*, 2537.
02OL3255 K. Tanaka, T. Fujiwara, Z. Urbanczyk-Lipkowska, *Org. Lett.* **2002**, *4*, 3255.
02OM1101 L.E. Gusel'nikov, V.V. Volkova, E.N. Buravtseva, A.S. Redchin, N. Auner, B. Herrschaft, B. Solouki, G. Tsantes, Y.E. Ovxchinnikov, S.A. Pogozhikh, F.M. Dolgushin, V.V. Negrebetsky, *Organometallics* **2002**, *21*, 1101.
02OM1912 C. Visser, A. Meetsma, B. Hessen, *Organometallics* **2002**, *21*, 1912.
02OM3271 K. Yoshizawa, Y. Kondo, S.Y. Kang, A. Naka, M. Ishikawa, *Organometallics* **2002**, *21*, 3271.
02S1502 L.D.S. Yadav, R. Kapoor, *Synthesis* **2002**, 1502.
02SL381 B. Alcaide, P. Almendros, *Synlett* **2002**, 381.
02SL69 E. Alonso, C. del Pozo, J. González, *Synlett* **2002**, 69.
02SL85 B. Alcaide, C. Pardo, E. Sáez, *Synlett* **2002**, 85.
02T2215 D. Krishnaswamy, V.V. Govande, V.K. Gumaste, B.M. Bhawal, A.A.S. Deshmukh, *Tetrahedron* **2002**, *58*, 2215.
02T2423 S. Gerard, G. Dive, B. Clamot, R. Touillaux, J. Marchand-Brynaert, *Tetrahedron* **2002**, *58*, 2423.
02T2763 A. Salgado, M. Boeykens, C. Gauthier, J.-P. Declercq, N. De Kimpe, *Tetrahedron* **2002**, *58*, 2763.
02T5045 M. D'Auria, L. Emanuele, G. Poggi, R. Racioppi, G. Romaniello, *Tetrahedron* **2002**, *58*, 5045.
02T5173 F. Clerici, M.L. Gelmi, R. Soave, L. Lo Presti, *Tetrahedron* **2002**, *58*, 5173.
02T7027 J.K. Crandall, E. Rambo, *Tetrahedron* **2002**, *58*, 7027.
02T7177 H.H. Wasserman, H. Matsuyama, R.P. Robinson, *Tetrahedron* **2002**, *58*, 2491.
02T7303 A.G.M. Barrett, P. Dozzo, A.J.P. White, D.J. Williams, *Tetrahedron* **2002**, *58*, 7303.
02TA2619 F. Couty, D. Prim, *Tetrahedron: Asymmetry* **2002**, *13*, 2619.
02TA905 C.M.L. Delpiccolo, E.G. Mata, *Tetrahedron: Asymmetry* **2002**, *13*, 905.
02TL1523 M. Matsumoto, J. Murayama, M. Nishiyama, Y. Mizoguchi, T. Sakuma, N. Watanabe, *Tetrahedron Lett.* **2002**, *43*, 1523.
02TL2491 L.A. Cabell, J.S. McMurray, *Tetrahedron Lett.* **2002**, *43*, 2491.
02TL3617 K. Hirano, S. Sakaguchi, Y. Ishii, *Tetrahedron Lett.* **2002**, *43*, 3617.

02TL3951	K. Papadopoulos, D.W. Young, *Tetrahedron Lett.* **2002**, *43*, 3951.
02TL4241	A. Basak, S. Mandal, *Tetrahedron Lett.* **2002**, *43*, 4241.
02TL4633	C. Agami, F. Couty, N. Rabasso, *Tetrahedron Lett.* **2002**, *43*, 4633.
02TL5109	D. Enders, S. Wallert, *Tetrahedron Lett.* **2002**, *43*, 5109.
02TL5285	C.-H. Chou, S.-J. Chiu, W.-M. Liu, *Tetrahedron Lett.* **2002**, *43*, 5285.
02TL545	F.W. Ng, H. Lin, Q. Tan, S.J. Danishefsky, *Tetrahedron Lett.* **2002**, *43*, 545.
02TL5469	D. Osborne, P.J. Stevenson, *Tetrahedron Lett.* **2002**, *43*, 5469.
02TL5499	A. Basak, S.C. Ghosh, T. Bhowmick, A.K. Das, V. Bertolasi, *Tetrahedron Lett.* **2002**, *43*, 6259.
02TL6259	M. Alajarín, A. Vidal, F. Tovar, P. Sánchez-Andrada, *Tetrahedron Lett.* **2002**, *43*, 6259.
02TL6405	O. Arjona, A.G. Csákÿ, M.C. Murcia, J. Plumet, *Tetrahedron Lett.* **2002**, *43*, 6405.
02TL6775	N. Kano, T. Takahashi, T. Kawashima, *Tetrahedron Lett.* **2002**, *43*, 6775.
02TL6897	J. Kolb, B. Beck, A. Dömling, *Tetrahedron Lett.* **2002**, *43*, 6897.
02TL8955	M. Matsumoto, J.T. Sakuma, N. Watanabe, *Tetrahedron Lett.* **2002**, *43*, 8955.
02TL9513	C. Barbaud, M. Guerrouache, P. Guérin, *Tetrahedron Lett.* **2002**, *43*, 9513.

Chapter 5.1

Five-Membered Ring Systems: Thiophenes & Se, Te Analogs

Erin T. Pelkey
Hobart and William Smith Colleges, Geneva, NY, USA
pelkey@hws.edu

5.1.1 INTRODUCTION

The synthesis and chemistry (reactions) of thiophenes, benzo[*b*]thiophenes, and related selenium and tellurium ring systems that has been reported during the past year (Jan-Dec 2002) is the subject of this review. Aspects of thiophene chemistry have been reviewed during the past year including transition metal-catalyzed reactions <02JOMC177, 02JOMC195, 02PAC1327> and the synthesis <02CHC503> and structure <02CHC632> of thiophene 1,1-oxides and related compounds. The relative aromaticity of numerous heterocyclic compounds has been explored quantitatively <02JOC1333>, while theoretical calculations for π-stacking between thiophene rings has been investigated in order to further define the intermolecular interactions between thiophene oligomers and polymers <02JACS12200>.

5.1.2 THIOPHENE RING SYNTHESIS

Treatment of 1,4-dicarbonyl compounds with Lawesson's reagent is a common method of *de novo* thiophene synthesis and this was recently applied to the preparation of thienoazepines <02BMCL1675>. A novel preparation of 2,5-disubstituted thiophenes **2** involved the treatment of 3,5-dihydro-1,2-dioxins **1**, prepared by photo-oxidation of the corresponding dienes, with Lawesson's reagent <02TL3199>. Treatment of 1,4-dilithiated dienes with various sulfur transfer reagents including sulfur (S_8), benzenesulfonic acid thianhydride [$(PhSO_2)_2S$] <02TL1553>, and carbon disulfide (e.g. **5** → **6**) <02TL3533> provided methods for the preparation of thiophene derivatives. Specifically, bis-lithiation of dibromide **3** with *t*-butyllithium followed by quenching with sulfur gave fused thiophene **4**, the reaction of which with triplet oxygen led to a stable radical cation <02CC1192>. A sulfur transfer reaction between the bis-anion derived from phenylenediacetonitrile and thionyl chloride gave the corresponding benzo[*c*]thiophenes (and similarly thieno[3,4-*c*]thiophenes) <02JOC2453>. A novel synthesis of the interesting structure, tetrakis(2-thienyl)methane **8**, involved treating alkyne **7** with sodium sulfide in the presence of base <02TL3049>. C_2-Symmetric thiophene derivatives were prepared by conversion of the corresponding bis-alkynes to the zirconocycles with Negishi's reagent followed by treatment with sulfur monochloride (S_2Cl_2) <02TL3313>.

Thioacetamide has been utilized as a sulfur transfer reagent in the synthesis of azuleno[1,2-c]thiophenes <02H(58)405>.

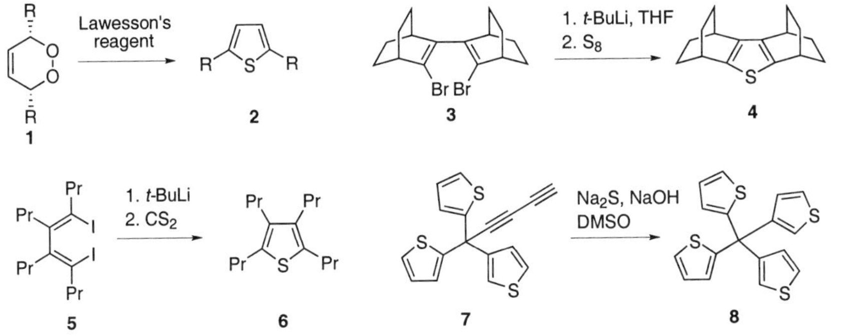

Condensation reactions of α-thioglycolate derivatives and related activated thiols onto unsaturated β-halocarbonyl compounds have been a common strategy for preparing complex thiophenes and many new examples appeared during the previous year. This approach has been utilized to prepare pyrazolo[3,4-b]thieno[2,3-d]pyridines <02BMCL9>, thieno[2,3-c]pyridines <02JOC943>, and thieno[3,2-c]pyridines <02BMCL2549>. The development of a cost-effective, scalable synthesis of the wheat fungicide silthiofam (13) involved an addition-elimination reaction of an activated thiol compound <02OPRD357>. Conversion of 2-chlorobutanone (9) into the corresponding thiol 10 followed by base-induced addition to acrylate gave thiophene 12. Alkylation of thiol 14 with α-benzotriazol-1ylalkyl chloride 15 gave 16 which underwent a base-induced cyclization reaction to give benzo[b]thiophene 17 <02CHC156>. Related intramolecular condensation reactions leading to benzo[b]thiophenes <02JHC1177, 02T4529> and dibenzothiophenes <02T1709> were also reported.

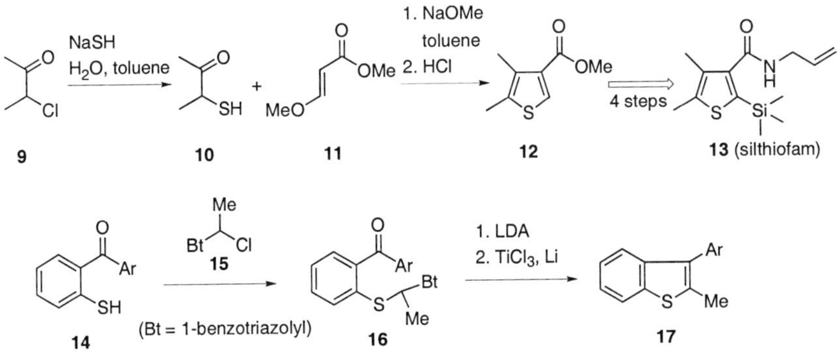

Three component condensation reactions (Gewald reaction) between an activated nitrile, ketone or aldehyde, and sulfur has been utilized to prepare a variety of α-aminothiophenes. Recent examples utilizing this condensation reaction include the synthesis of 4,5-dialkyl-2-

aminothiophenes (allosteric enhancers of agonist activity at A_1 adenosine receptor) <02JMC382>, thieno[2,3-c]pyridines <02BMCL1607, 02BMCL1897>, and 2-amino-4,5-diarylthiophenes <02BMC3113>. The latter two types of compounds were investigated as inhibitors of tumor necrosis factor-α. A new synthesis of α-aminothiophenes involved an isothiocyanate. Specifically, the base-induced addition of 2,4-pentanedione (**18**) with phenyl isothiocyanate led to intermediate **19** which underwent an alkylation reaction followed by a Dieckmann-type cyclization reaction leading to 2-aminothiophene **20** <02TL257>. A related synthetic sequence involving thioacrylamides (Hantzsch method) was utilized to prepare *N,N*-diphenyl 2-aminothiophenes, which were tested as hole transport materials <02CL896, 02T2137>. Condensation reactions involving nitriles have also been widely utilized to prepare β-aminothiophenes <02CPB1215, 02H(57)317> and a recent example involves the preparation of thieno[2,3-*d*]pyrimidines, agonists for the luteinizing hormone receptor <02CBC1053>. β-Aminothiophenes prepared via condensation reactions onto nitriles have served as useful building blocks for the synthesis of more complex fused heterocycles including thieno[*b*]azepinediones <02S1096> and pyrido[2,3-*b*]thieno[3,2-*d*]pyrimidines <02SC3493>. An alternative synthesis of β-aminothiophenes involved the condensation of 3-oxotetrahydrothiophenes with hydroxylamine <02SC2565>. The rhodium-mediated condensation between ethyl diazoacetate (**22**) and ketene *S,N*-acetal **21** gave β-aminothiophene **23** <02JCS(P1)2414, 02OL873>. The reaction of ketene *S,S*-acetals with lithiotrimethylsilyldiazomethane led to the formation of α-sulfidothiophenes <02H(57)1313>.

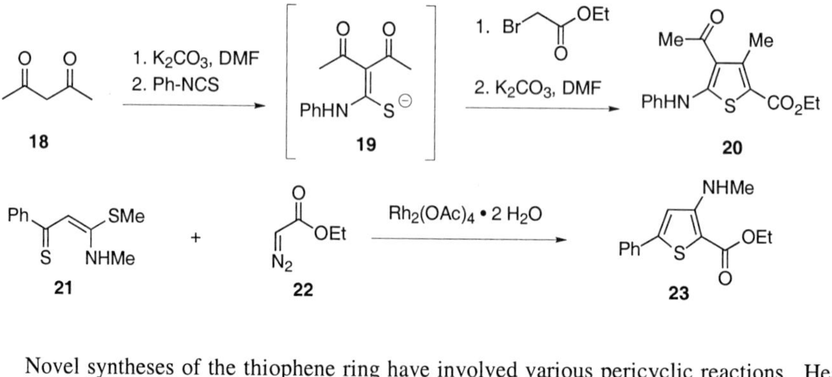

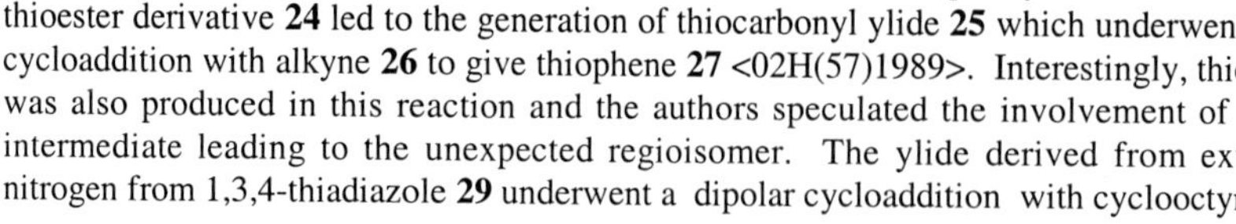

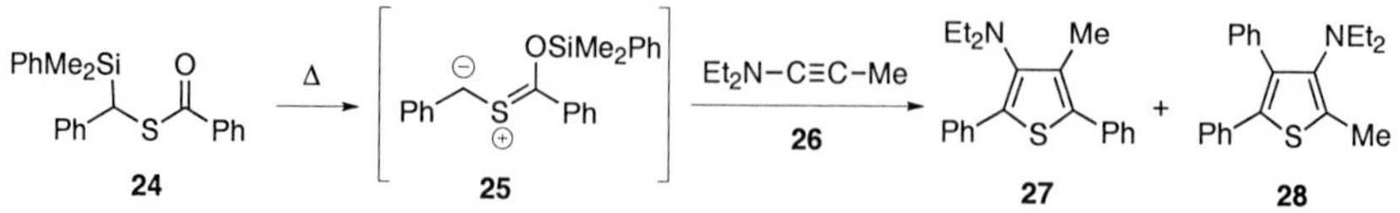

Novel syntheses of the thiophene ring have involved various pericyclic reactions. Heating thioester derivative **24** led to the generation of thiocarbonyl ylide **25** which underwent a dipolar cycloaddition with alkyne **26** to give thiophene **27** <02H(57)1989>. Interestingly, thiophene **28** was also produced in this reaction and the authors speculated the involvement of a thiirane intermediate leading to the unexpected regioisomer. The ylide derived from extrusion of nitrogen from 1,3,4-thiadiazole **29** underwent a dipolar cycloaddition with cyclooctyne (**30**) to give 2,5-dihydrothiophene **31** <02HCA451>. Treatment of **31** with acid led to the ring-opened thiophene product **32**. Oxidation of sulfide **33** with *m*-CPBA gave sulfoxide **34** which upon stirring at rt led to the formation of fused thiophene **35** <02T10047>. The authors propose a series of pericyclic reactions ([2,3] sigmatropic rearrangement followed by [3,3] sigmatropic rearrangement) to account for the formation of the thiophene product. Closely related synthetic sequences have been investigated leading to a variety of different thiophene products <02S669, 02SC1271, 02T4551, 02T10309, 02TL2123>. A thermal rearrangement of bis-allenyl thiosulfonates led to the formation of novel 3,4-fused thiophenes <02TL9615>.

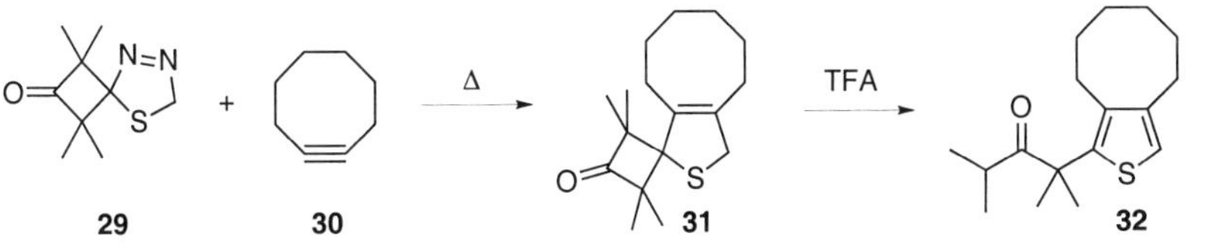

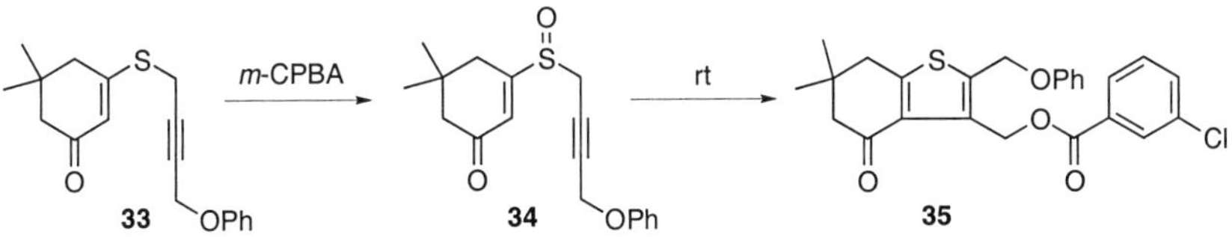

Electrophile-mediated cyclizations of sulfides onto pendant alkynes have been utilized to prepare benzo[*b*]thiophenes <02JOC1905>. Sonagashira coupling of sulfide **36** with phenylacetylene gave **37** which underwent cyclization reactions in the presence of various electrophiles to give benzo[*b*]thiophenes **38**. Generation of benzyne (**40**), by treatment of anthranilic acid (**39**) with isoamyl nitrite, in the presence of Barton ester **41** led to the formation benzo[4,5]thieno[2,3-*b*]pyridine (**42**) <02JOC3409>.

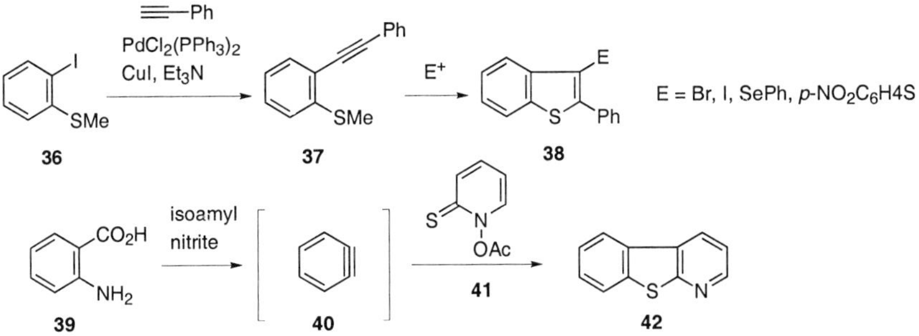

Directed metallation of *N,N*-diethylbenzamide **43** and trapping with dimethyl disulfide gave sulfide **44** <02SL325>. Side-chain deprotonation of the thiomethyl group by LDA followed by an intramolecular cyclization reaction gave thioindoxyl **45**, which upon treatment with sodium borohydride gave benzo[*b*]thiophene **46**. An isothianaphthene precursor was prepared utilizing a DIBAL reduction reaction of a 2,5-dihydro-3-oxothiophene <02TL8485>.

5.1.3 THIOPHENE RING SUBSTITUTION

Electrophilic aromatic substitution (EAS) of the thiophene ring preferentially occurs at unsubstituted α-positions and this regioselectivity is often exploited for the preparation of α-bromothiophenes via NBS bromination <02CEJ2384, 02CEJ5415, 02CM1742, 02CM4662>. Thiophene analogues of Tröger's base have been prepared and functionalized using a variety of electrophilic aromatic substitution reactions (NBS, I_2, TMS-Cl, $POCl_3$/DMF) <02JCS(P1)1963>. Preparation of 3,4-diaminothiophene (**48**) involved EAS chemistry <02JOC9073>. Bis-bromination of thiophene followed by bis-nitration under forcing conditions gave thiophene **47**. Treatment of **47** with tin and HCl led to formation of **48** via reductive removal of the α-bromo groups along with the reduction of the nitro groups to the corresponding amino groups. Condensation of **48** with diketone **49** afforded thieno[3,4-*b*]pyrazine **50**, the electrochemistry of which was studied. The regioselectivity of EAS chemistry is dependent on the functional groups present on the thiophene ring. Thieno[3,4-*b*]-1,4-oxathiazine (**53**) was prepared by acid-catalyzed transetherification of 3,4-dimethoxythiophene (**51**) with 2-mercaptoethanol (**52**) <02OL607>. NBS bromination of **53** gave a mixture of mono-brominated products **54** and **55** (85:15) favoring the product with the bromine syn to the oxygen substituent. A higher level of regioselectivity was observed with the corresponding iodination reaction (I_2, $Hg(OAc)_2$). A regioselective formylation was utilized in the preparation of thiophene-containing azacryptand Mannich bases <02JCS(P1)717>. A regioselective Friedel-Crafts acylation was utilized in the preparation of a thiophene analog of ketorolac (NSAID) <02H(56)91>. Condensation reactions of thiophenes with diphenylketones were studied <02JCS(P1)1944>. The dimerization of 2,2'-bithienyl with aldehydes led to the preferential formation of bis(2,2'-bithienyl)methanes <02SC909>. The preparation of 2-arylthiophenes was reported utilizing a novel oxidative coupling reaction of thiophene with phenylhydrazines mediated by manganese(III) acetate <02T8055>. Finally, thermolysis of 1-phenylbenzo[*b*]thiophenium triflates led to the formation of 2-phenylbenzo[*b*]thiophene via a novel phenyl migration reaction <02TL2239>.

Sulfur oxidation of the thiophene ring leading to thiophene-1-oxides or thiophene-1,1-oxides is useful for modifying the chemistry of the thiophene ring and this subject has been reviewed <02CHC632>. The structure and physical properties of quinquethiophene-1,1-dioxides have been studied in detail <02T10151>. The mechanism for the oxidation of thiophene to a mixture of thiophene-1-oxide (trapped as a Diels-Alder adduct) and thiophen-2-one was studied using deuterium incorporation experiments <02JOC7261>.

Nucleophilic aromatic substitution reactions of nitro-substituted thiophenes have been utilized to prepare biologically active thiophenes including reverse transcriptase <02H(57)97> and nitric oxide synthase <02JHC857> inhibitors. The addition-elimination reaction of 2-chloro-3-nitrothiophene (**56**) with metallated indole **57** afforded **58** which was transformed into the corresponding thiophene-fused azepino[5,4,3-*cd*]indoles **59** <02H(57)1831>.

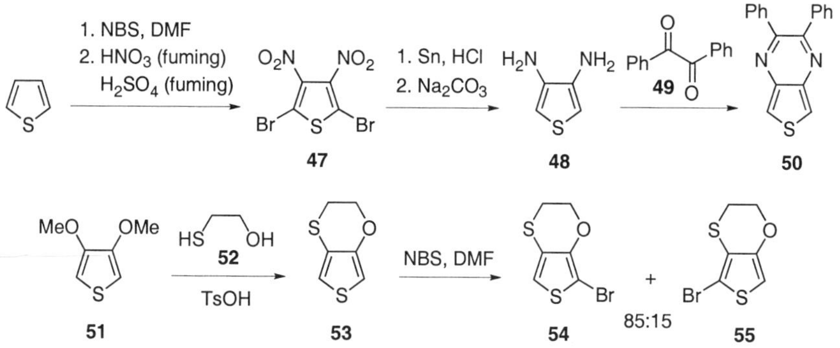

An intramolecular nucleophilic aromatic substitution reaction of thiophene **60** led to the formation of thieno[2,3-*b*][1,5]benzoxazepine **61** <02JHC163>. A vicarious nucleophilic substitution reaction (replacement of an aromatic hydrogen by a nucleophile containing a leaving group) was utilized in the synthesis of lactam-fused thiophenes, inhibitors of serine proteases <02CC1274>. The nucleophilic substitution of an oligothiophene radical cation by pyridine was reported <02JCS(P1)1135>.

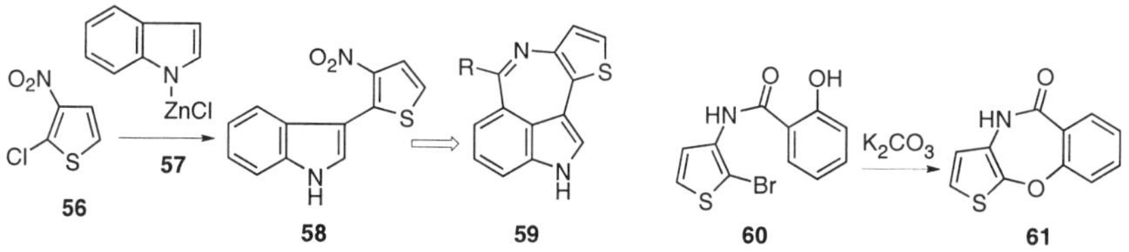

A general method for preparing α-substituted thiophenes involves α-lithiation followed by quenching with various electrophiles with recent examples involving the following electrophiles: Me₃Sn-Cl <02CEJ2384>, BuSSBu <02OL4535>, MeSSMe <02BCJ1795>, Ts-N₃ <02T3507>, DMF <02T2551, 02TL8485>, and 3-pentanone <02TL9493>. The latter example was utilized for preparing building blocks for the synthesis of bithiophene-containing calixpyrroles. α-Lithiation of thiophene derivatives followed by quenching with aromatic aldehydes provided building blocks for the preparation of thiaporphyrin analogs <02CC2642, 02JMC449>. Lithiation of benzo[*b*]thiophene (**62**) followed by quenching with excess sulfur led to the formation of the interesting sulfur rich heterocycle, pentathiepino[6,7-*b*]benzo[*d*]thiophene (**63**) <02JOC6220>. A similar sulfur-rich heterocycle, pentathiepino[2,3-*b*]thiophene, was prepared by treatment of thiophene with diisopropylethylamine and disulfur dichloride <02CC1204>. Bis-lithiation of dibenzothiophene (**64**) followed by quenching with solid CO_2 provided 4,6-dibenzothiophenedicarboxylic acid (**65**) which was utilized in the synthesis of a novel dibenzothiophene-based chiral auxiliary <02TL4907>.

Halogen-metal exchange is a powerful, regioselective method for preparing highly functionalized thiophene derivatives and selected applications (of the many reported) include the synthesis of dithieno[3,2-*b*:2',3'0-*d*]thiophene <02TL1553>, azulene-substituted thiophenes <02T7653>, and sulfide-substituted thiophenes <02JOC3961>. The preparation of a thiophene containing both a ketone and Grignard moiety was reported utilizing a halogen-metal exchange reaction with neopentylmagnesium bromide (**67**) <02SL1799>. Treatment of 2-acetyl-5-iodothiophene (**66**) with **67** followed by quenching with $SnMe_3Cl$ led to the formation of stannane **69** via intermediate Grignard **68**.

The organometallic cross-coupling of metallated thiophenes remains a prevalent method for the preparation of regioregular oligothiophenes. Selected recent examples (of the many reported) of metallated thiophenes utilized in the synthesis of oligothiophenes include thiophene-2-borates (Suzuki coupling) <02JOC4924, 02T2245>, thiophene-3-borates (Suzuki coupling) <02BMCL2011>, thiophene-2-stannanes (Stille coupling) <02CEJ3027, 02CEJ5415, 02CM1742, 02OL2067, 02T2245>, thiophene-2-zincates (Negishi coupling) <02JOC4924>, and thiophene-2-mercury derivatives <02JOC7523>. A detailed study of the preparation of thiophene oligomers using microwave-assisted Suzuki coupling reactions was reported <02JOC8877>. A novel iterative approach to the synthesis of oligothiophenes utilized a "double-coupling" strategy involving orthogonal Si/Ge protecting groups <02OL1899>. The germanium group contained a linker suitable for conjugation to a solid-phase resin. A Suzuki coupling reaction of benzothiophene-2-borates with aryl bromides produced 2-arylbenzothiophenes, compounds which were studied as agonists for estrogen receptor subtype ERβ <02JMC1399>. A Suzuki coupling of benzothiophene-2-borates with vinyl bromides provided sulfur analogs of dehydrotryptophan <02EJO2524>. A Stille coupling between thiophene-2-stannane **71** and dibromide **70** gave bis-thiophene **72** <02JACS7762>. Oxidative ring closure of **72** provided naphthodithiophene **73**, the electrochemistry of which was studied. Platinum complexes of thieno[3,2-*b*]thiophenes were prepared from the corresponding stannanes by treatment with $(COD)PtCl_2$ <02JOMC56>. Thiophene-2-stannanes were utilized in the synthesis of the 2-thiophene(aryl)iodonium triflates by treatment with PhI(CN)OTf <02JOC2798>. Finally, the preparation of novel pinacol-based thiophene-2-borates utilized an iridium-catalyzed reaction <02TL5649>.

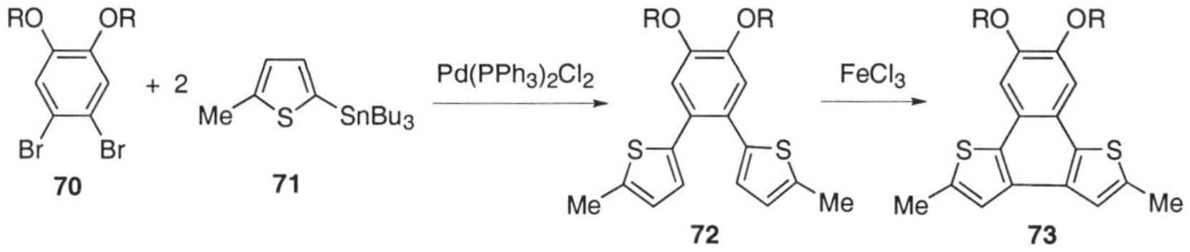

Organometallic cross-coupling reactions of halogenated thiophenes represent an important method for the preparation of highly functionalized thiophenes (for reviews see: <02JOMC177, 02JOMC195>) including regioregular oligothiophenes. Selected recent examples (of the many reported) of applications of halogenated thiophenes utilized in the synthesis of oligothiophenes include Sonogashira <02OL3043> and Suzuki <02CEJ2384, 02OL4439> coupling reactions of α-bromothiophenes. Palladium cross-coupling reactions have been utilized to prepare 3,4-dialkynylthiophenes <02JOC8812> and 2,5-dialkynylthiophenes <02CM4543, 02JACS12742> as building blocks for [14]annulenes and bent-rod liquid crystals, respectively. A novel recyclable palladium catalyst was successfully used in the Suzuki coupling of 2-iodothiophene <02JACS7642>. The preparation of sterically hindered 2,3-diarylbenzo[*b*]thiophenes were reported using a Suzuki coupling reaction <02S213>. Heterocyclic fluoren-9-one analogs were prepared utilizing a cyclocarbonylation reaction <02JOC5616>. Thus, treatment of benzo[*b*]thiophene **74** with CO in the presence of the optimized palladium catalyst and cesium pivalate gave benz[*b*]indeno[1,2-*d*]thiophene **75**. Heck reactions of bromothiophenes with olefins have been reported utilizing a novel palladium catalyst containing a tetraphosphine ligand <02TL5625>. Heck reactions with the thiophene moiety acting as the olefin component have been reported with 3-cyanobenzo[*b*]thiophene <02TL1829> and thiophene-2-carboxamides <02JACS5286>. In the latter case, multiple arylation substitutions were also investigated.

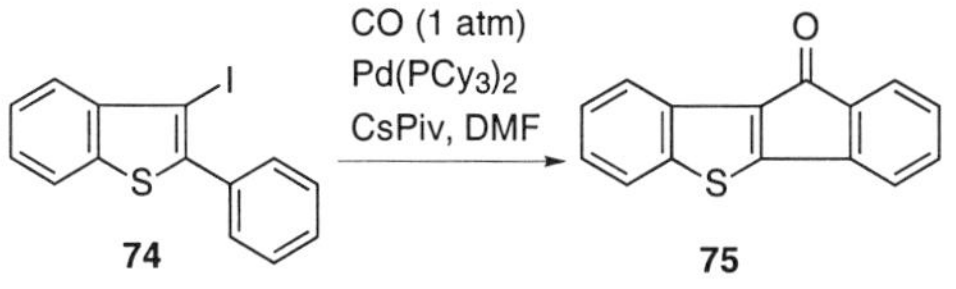

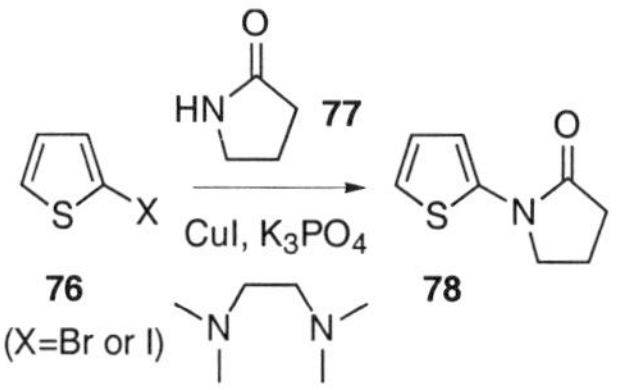

The preparation of aryl-nitrogen bonds has been the focus of intense research during the past few years. This work is now starting to be applied to the amination and amidation of halothiophenes. For example, the palladium-catalyzed amidation (Buchwald's conditions) of 2-halothiophene **76** with pyrrolidin-2-one (**77**) gave 2-amidothiophene **78** as reported by two different groups <02SL427, 02TL7365>. A palladium-catalyzed amination reaction was utilized to prepare *N,N*-diarylamino-substituted oligothiophene derivatives <02CC1530>.

Side-chain functionalization of thiophenes using Wittig and related chemistry gave vinyl-substituted thiophenes which were utilized to prepare a variety of complex thiophene derivatives <02CC932, 02CL1002, 02CM4662, 02CM4669, 02H(57)1935, 02JCS(P2)1791, 02JOC205, 02MAC4628, 02T2551, 02T7463>. Reduction of the oxime derived from thiophene-2-carboxaldehyde **79** gave 2-aminomethylthiophene **80** which was converted into the Fmoc-substituted amino acid **81** in two steps as shown <02JMC5005>. A side-chain reductive amination was utilized to prepare piperazinylmethyl-substituted thiophenes <02BMCL2145>. Asymmetric C-H insertion reactions of thiophen-3-yldiazoacetates were investigated <02TL4981>. For example, treatment of thiophene **82** with cyclohexadiene in the presence of a chiral rhodium catalyst led to the preferential formation of chiral **83**, which was elaborated into cetiedil (**84**), a potassium channel blocker. Diazonium chemistry was utilized to prepare azo-substituted thiophene derivatives <02CHC281>. Finally, the asymmetric dihydroxylation of thiophene acrylates was studied <02TL3813>.

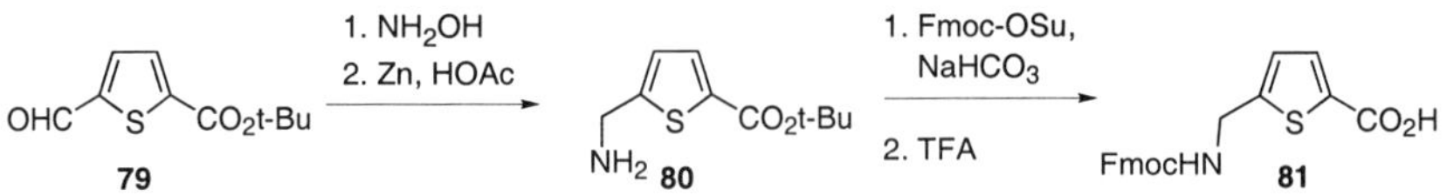

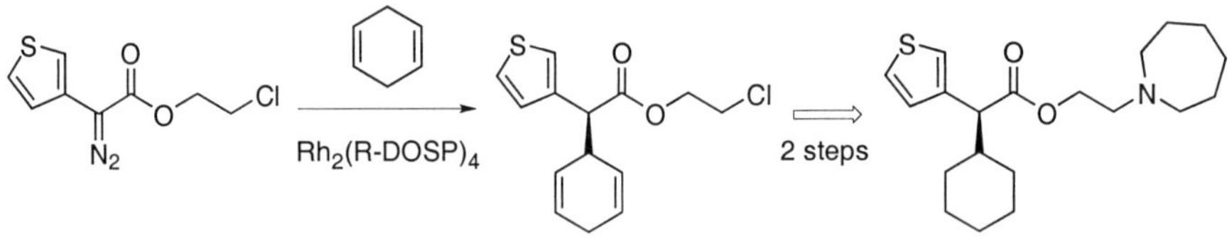

5.1.4 RING ANNELATION ON THIOPHENE

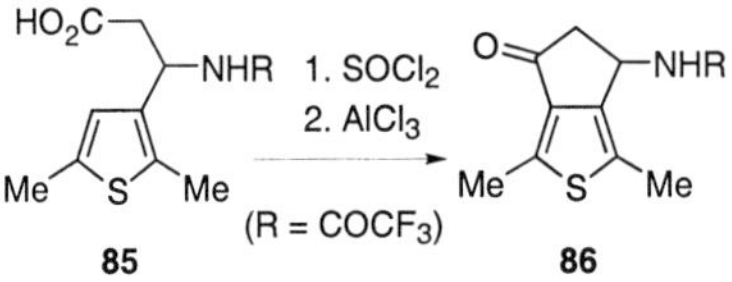

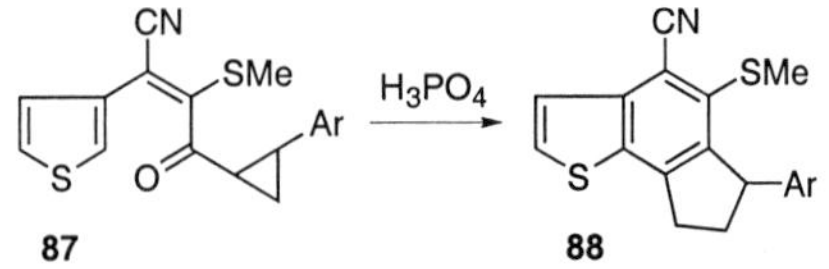

The electron-rich thiophene ring system can be converted into fused thiophenes by acid-mediated intramolecular ring annelation reactions. Intramolecular acylation reactions of carboxylic acids onto thiophenes were key steps in the formation of benzo[5,6]cyclohepta[1,2-*b*]thiophenes <02S355>, cyclopenta[*c*]thiophenes (e.g. **85** → **86**) <02BMC2185>, and thiophene-fused cyclopentadienes <02O2842>. The latter class of compounds was applied as novel ligands for zirconocene complexes. Benzoannelation reactions involving thiophene and pendant electron-rich aryl groups were mediated by the hypervalent iodine reagent PIFA <02T8581>. A "domino cationic cyclization" reaction of cyclopropyl ketone **87** led to the formation of indeno[4,5-*b*]thiophene **88** <02JOC4916>. The mechanism of this reaction seems to involve either a stepwise or concommitant acid-induced cyclopropane ring opening and benzoannelation reaction. Palladium-catalyzed oxidative cyclizations of benzothiophenes were investigated as a general approach to the preparation of thienocarbazoles <02T7943>. Finally, as previously shown (**72 → 73**), Lewis acid-catalyzed oxidative cyclizations of pendant thiophenes have been reported <02JACS7762>.

The photo-irradiation of tris(2-benzo[*b*]thienyl)methyl alcohol (**89**) gave a mixture of photocycloadducts, cyclopentanone **90** and tropone **91** <02TL8669>. The photolysis and subsequent annelation of thiophene-substituted bischromones have been described <02T3095>. The photocycloaddition of a chlorothiophene onto a pendant benzamide ring was utilized to prepare thieno[3',2':4,5]thieno[2,3-*c*]quinolines, novel antitumor agents <02CPB656>. Photocyclization of imine-substituted benzo[*b*]thiophenes (obtained via aza-Wittig reactions of the corresponding iminophosphoranes) gave benzo[*b*]thieno[2,3-*b*]pyridines <02T3507>. Isomeric naphthodithiophenes were prepared using a photocycloaddition of bis(thienyl)ethylene as a key step <02BCJ1795>.

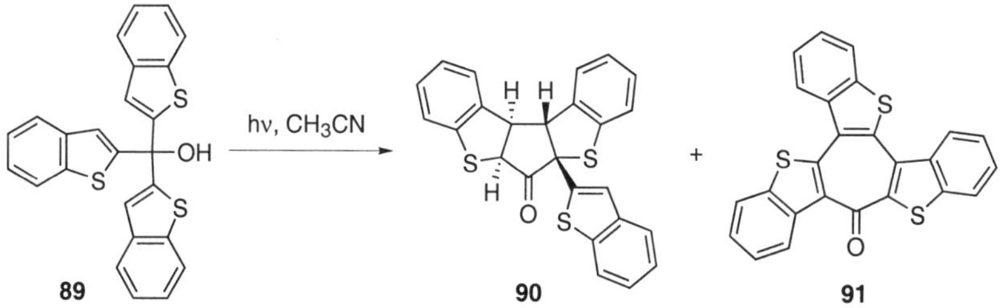

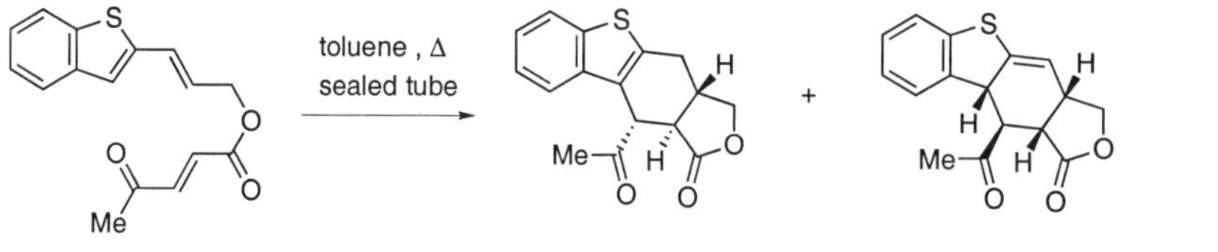

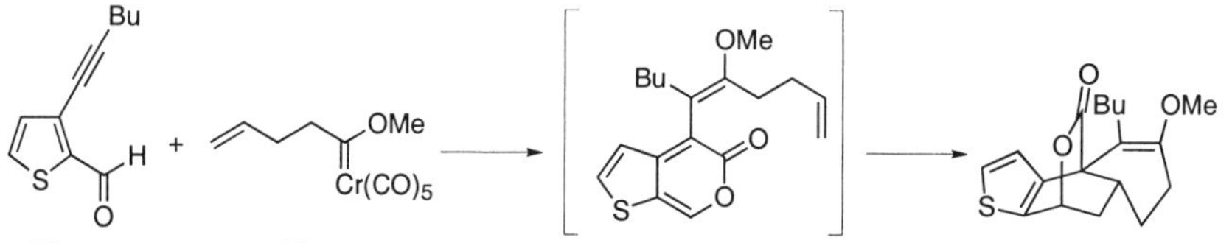

Pericyclic reactions involving thiophenes have been utilized to prepare a variety of complex heterocycles. The intramolecular Diels-Alder reaction of 2-vinylbenzo[*b*]thiophene **92** produced a pair of tetracyclic adducts **93** and **94** <02TL3963>. Coupling of Fischer carbene **96** with 3-alkynylthiophene **95** led to the formation of thieno[2,3-*c*]pyran-3-one **97** in one step <02JOC4177>. An intramolecular cycloaddition of **97** then afforded tetracyclic adduct **98**. A ruthenium-catalyzed cyclodimerization reaction involving bis-thienyl acetylene derivatives was investigated which led to the formation of tris(thienyl)-substituted benzo[*b*]thiophenes <02JHC91>. A thiophene-based *ortho*-quinodimethane was generated and subsequent cycloaddition chemistry afforded amino-substituted tetrahydro-benzo[*b*]thiophene moieties <02JMC3022>.

Perhaps the most highly investigated research area in thiophene chemistry is directed at developing novel photochromic systems containing thiophene moieties. These photochromic systems are of interest for applications such as optical data storage, optical computing, and telecommunications. One of the most well studied photochromic systems are the dithienylethenes, compounds which can undergo a photoinduced thiophene annelation reaction upon irradiation (e.g. **99** → **100**) <02JACS7481>. These types of systems have been studied by a number of research groups including Irie <02BCJ167, 02BCJ2071, 02CC2804, 02CL58, 02CL572, 02CL1224, 02EJO3796, 02JACS7481, 02JOC4574>, Feringa <02CC2734>, Krayushkin and Uzhinov <02CHC165, 02OL3879>, Lemieux <02JACS7898>, Shirota <02JMAT2612>, Gust <02JACS7668>, and Tian <02AM918, 02CC1060>. Conjugation of photochromic systems to a variety of moieties has been investigated including phthalocyanine <02CC1060>, azaporphyrin <02AM918>, fused pyrrole <02OL3879>, monoaza-crown ether (**99**) <02JACS7481>, Lucifer yellow <02JACS7481>, porphyrin-fullerene <02JACS7668>, and β-cyclodextrin <02CC2734>. Novel photochromic systems involving 4,5-dialkenylthiophenes <02JOC5208, 02OL1099> and also a bis(naphthopyran)-substituted bithiophene <02JACS1582> have also been investigated.

5.1.5 THIOPHENE INTERMEDIATES IN SYNTHESIS

Thiophenes can be utilized as synthetic scaffolds for the preparation of non-thiophene materials as the sulfur moiety can be removed by reduction (desulfurization) or extrusion (loss of SO_2). The extrusion of sulfur dioxide from 3-sulfolenes (2,5-dihydrothiophene 1,1-dioxides) gives dienes (butadienes or *o*-quinodimethanes) which can be utilized in cycloaddition chemistry. Heterocyclic *o*-quinodimethanes generated from the corresponding 3-sulfolenes were investigated during the past year <02T3655, 02TL799> as well as an aza analog <02EJO947>. The synthesis and reactions of chiral 3-phenylsulfinyl-3-sulfolenes have been reported <02TL4865>. For example, heating 3-sulfolene **101** generated diene intermediate **102** by thermal extrusion of SO_2 which then underwent an intramolecular Diels-Alder cycloaddition to give bicyclic sulfoxide **103**. The DBU-induced thermal elimination of SO_2 from 2-sulfolene **106** provided diene **107**. Compound **106** was obtained in a few steps from **104** as shown.

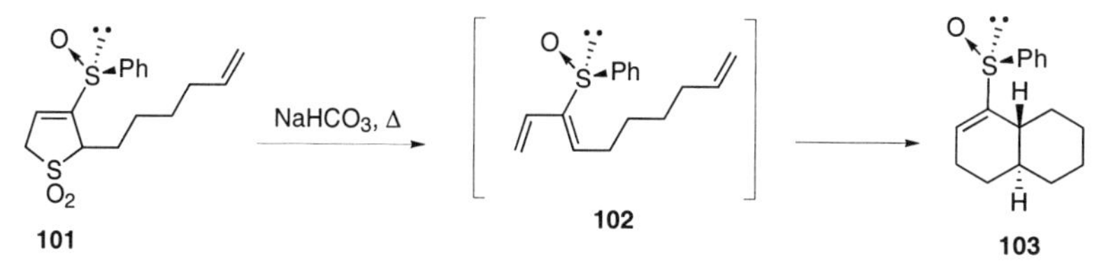

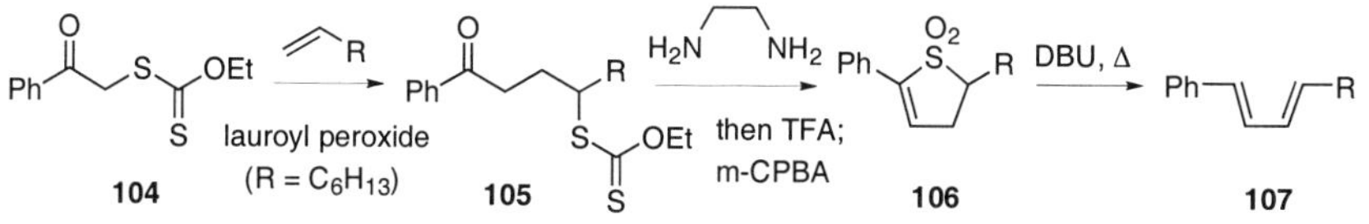

Finally, the cycloaddition of sulfur dioxide to dienes leading to either 3-sulfolenes or the corresponding sultines has been studied in detail <02CEJ1336, 02HCA712, 02HCA733, 02HCA1390>.

The cycloaddition of thiophene with maleimide was carried out under high pressure to give the exo adduct as the major product <02BMCL391>. This product is a sulfur analog of cantharidin and was tested as a protein phosphatase inhibitor.

The thiophene can be considered a four-carbon synthon via reductive desulfurization with Raney nickel. A strategy using the thiophene as a lynch-pin for the construction of complex building blocks was utilized in a total synthesis of the cytotoxic sponge metabolite haliclamine A <02JOC6474>. Thienyl cyclopropyl ketone (**108**) was elaborated into **109**. Reduction of **109** with Raney Ni gave **110** which was taken on to haliclamine A.

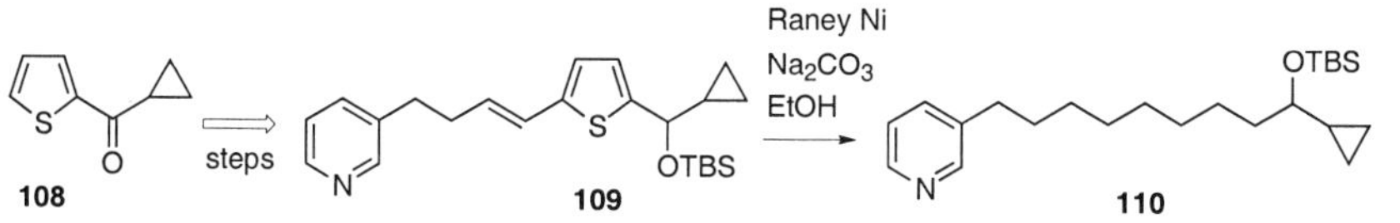

Ring opening reactions of nitrothiophenes continue to be investigated as a route to prepare other molecules including nitrocyclopropanes <02EJO1284> and functionalized isoxazoles <02T3379>. A nickel-catalyzed ring opening reaction of the thiophene ring of dinaphthothiophene provided a new route to 1,1'-binaphthyls <02JACS13396>.

5.1.6 BIOLOGICALLY IMPORTANT THIOPHENE DERIVATIVES

The isolation of a few naturally occurring thiophene derivatives from natural sources was reported during the past year. Anthrathiophene **111**, an antiangiogenic compound, was isolated from the bryozoan *Watersipora subtorquata* <02JNP1344>, bis-bithiophene **112** was isolated from *Tridax procumbens* <02NPL217>. Terthiophene derivatives were isolated from various *Tagetes* species<02JICS546>. Simple, known and previously undescribed thiophenes were isolated from *Ferula foetida* <02JNP1667>. Finally, salacinol (**113**), a potent carbohydrate-based α-glucosidase inhibitor, was isolated from *Salacia reticulata* <02BMC1547>.

A large variety of biologically important thiophene-containing compounds have been synthesized and evaluated. One of the more common scaffolds utilized in medicinal chemistry is the benzo[*b*]thiophene moiety, and recent examples include selective 5-HT reuptake inhibitors (e.g. **114**) <02JMC4128>, inhibitors of estrogen receptor subtype ERβ <02JMC1399>, acetylcholinesterase inhibitors (e.g. **115**) <02BMCL2565, 02BMCL2569>, and nitrogen mustard-conjugated antitumor agents <02BMC1611>.

The synthesis and/or biological activity of a number of complex molecules containing thiophenes fused with other heterocyclic rings have been investigated during the past year. In addition to those previously mentioned, selected examples (of the many reported) include ligands for the dopamine receptor (thianaphth[4,5-*e*][1,4]oxazine) <02JMC3022>, inverse agonists for GABA-A α5 receptor (benzothiophen-4(5*H*)-ones, e.g. **116**) <02JMC1176>, antimicrobial agents (thieno[2,3-*d*]pyrimidines) <02JHC877>, antipsychotic agents (thieno[2,3-*b*][1,5]benzoxazepines) <02JHC163>, antiherpetic agents (benzthiodiazepinones) <02BMCL2981>, anti-HTLV-1 agents (naphtho[2,3-*b*]thiophenequinones) <02CPB1215>, selective ligands for the 5-HT₇ receptor (thieno[3,2-*c*]pyridines) <02BMCL2549>, 5-HT receptor antagonists <02JMC54>, isozyme selective carbonic anhydrase inhibitors (thieno[3,2-*e*]-1,2-thiazines) <02JMC888>, anticancer agents (azeto-thienoazepinones, e.g. **117**) <02BMCL1675>, anti-inflammatory agents (pyrazolo[3,4-*b*]thieno[2,3-*d*]pyridines) <02BMCL9>, dual acting PAF/TxSI agents <02BMCL1383>, activators of ATP-regulated potassium channels (thieno[3,2-*e*]-1,2,4-thiadiazines) <02JMC4171>, and antagonists of cell adhesion molecule expressions (thieno[2,3-*c*]pyridines) <02JOC943>. Additional ring systems have also been prepared without reporting/evaluating their biological activity and these include thieno[*b*]azepinediones <02S1096>, thieno[3,2-*c*]benzazepines <02S355>, benzothieno[3,2-*b*]pyridines <02JHC783>, imidazo[1,5-*d*]thieno[2,3-*b*][1,4]thiazines (e.g. **118**) <02JHC645>, thieno[2,3-*b*]pyrazines <02H(57)97>, benzo[*b*]thiopheno[2,3-*f*]indoles and structurally related isomers <02CHC466>, thieno[3,4-b]indolizine <02H(57)17>, and 2-thia-3,5,6,7,8-pentaazabenz[*c,d*]azulene **119** <02TL695>. Finally, the conversion of a thieno[2,3-*d*]-1,3-dithiole-2-thione to a (dimerized) tetrathiocin derivative catalyzed by *Pseudomonas chlororaphis* ATCC 9447 was reported <02T2589>.

116 **117** **118** **119**

A large number of biologically important simple (non-fused) thiophenes have been synthesized and/or evaluated during the past year. In addition to those previously mentioned in this chapter, selected examples (of the many reported) include urokinase inhibitors <02BMCL491>, endothelin antagonists <02SC1615>, antitumor agents <02BMC2067>, nitric oxide synthase inhibitors <02BMCL1439>, acylhydrazide cruzain inhibitors <02BMCL1537>, thrombin inhibitors (prepared via solid-phase chemistry) <02BMCL1571>, selective α_1-adrenoreceptor antagonists <02BMCL2145>, Src kinase inhibitors <02BMCL2011>, selective ligands for human EP_3 prostanoid receptor <02BMCL2583>, antiprotozoal agents <02BMCL3475>, and protein tyrosine phosphatase inhibitors <02JMC4443>. The synthesis of the 4'-thio analog of the antitumor agent angustmycin C was reported <02TL5657>. Finally, the synthesis of various tetrahydrothiophene-based nucleosides have been reported including examples of anti-HIV agents <02BMC1499, 02BMCL2403, 02JMC4888, 02OL305> and antitumor agents <02OL529>.

5.1.7 NON-POLYMERIC THIOPHENE MATERIALS

The unique electronic and physical properties of thiophenes make them useful components of a variety of novel materials. Selected recent examples (of the many reported) include organic dyes <02CHC281, 02S1143>, fluoroionophores <02TL7419>, electroluminescent materials <02BCJ551, 02CL984, 02CM1884>, liquid crystal research <02CM4543, 02JACS12742>, photoluminescent materials <02JMAT924>, organic conductors and molecular wires <02CL1002, 02OL4535, 02TL3373>, and organic field-effect transistors <02CL958>. Thiophene derivatives functionalized with push-pull substitution have been prepared and evaluated as novel chromophores for second order non-linear optical (NLO) materials. Selected recent examples include compounds containing 2-nitrothiophene acceptors <02JCS(P2)1791>, π-extended tetrathiafulvalenes <02T7463>, rigidified π-extended systems <02CEJ1573, 02JOC205>, and tricyanovinyldihydrofuran acceptors <02CM4662, 02CM4669>. Thin films of dithienosilole **120** exhibited vapor-chromism (change in color) upon exposure to different organic solvents <02OL1891>.

The preparation and/or evaluation (e.g. electron-transfer properties) of thiophene-containing porphyrins and structurally related higher order macrocyclic materials have been reported during the past year. In addition to those previously mentioned in this chapter, selected examples of porphyrin-related macrocycles containing thiophene rings include 21,23-dithiaporphyrins <02CC2642, 02JMC449>, trithiahexaphyrins <02CEJ4542>, pentathiaheptaphyrins <02JOC6309>, azuliporphyrins <02CC1660>, oxybenziporphyrins <02CC462>, meso-furyl thiaporphyrins <02TL9453>, and phosphine-linked tetrathiophene annulenes <02T2551>. Additional types of novel thiophene-containing macrocycles and annulenes that have been synthesized and/or evaluated include tetrathiophene-fused [8]annulene **121** <02JACS12507>, a thiophene appended calix[4]arene (for ionic recognition) <02JMAT2665>, bithiophene expanded calixpyrrole analogs <02TL9493>, thiophene-based azacryptand Mannich bases <02JCS(P1)717>, and thiophene-linked terpyridine complexes <02CC3028, 02CEJ137>. Thiahelicenes, helical shaped aromatic macrocycles containing fused thiophene rings, have been synthesized and/or evaluated during the past year including [7]thiaheterohelicenes <02JOC1795> and [6]thiaheterohelicenes <02CC932>. A rational synthetic approach to dendrimers comprised entirely of thiophene rings (up to 30) has been reported <02OL2067>.

Novel, chiral thiophene- (e.g. **122**) and benzo[*b*]thiophene-based phosphines have been prepared and evaluated as ligands in an enantioselective palladium-catalyzed allylation reaction <02SL2083>.

5.1.8 THIOPHENE OLIGOMERS AND POLYMERS

The thiophene ring is an important component of many novel oligomeric and polymeric materials. The self-assembling properties of thiophene oligomers and polymers and cyclic oligothiophenes have been discussed <02AM609>, while the preparation of a library oligothiophenes utilizing solid-phase strategies has been detailed <02JCC457> and reviewed <02CC1015>. The synthesis of monodisperse thiophene oligomers continues to be an important field of study during the past year. In addition to those previously mentioned, selected recent examples include bithiopenes <02JOC6931>, trithiophenes (Ge/Si orthogonal coupling strategy) <02OL1899>, trithiophenes (isothianaphthene central heterocycle) <02CC1530>, trithiophene-based quinodimethanes (e.g. **123**) <02JACS4184, 02JACS12380, 02JOC6015>, quaterthiophenes <02CEJ3027>, sexithiophenes <02CM1742, 02JACS1269, 02T2245, 02TL6541>, octithiophenes <02AG(E)3598, 02CM4550, 02S1195>, and 12-mers (and 4-mers and 8-mers) <02OL309>. Novel radial oligothiophenes (hexakis-trithienylalkynyl-substituted benzene ring) have been synthesized and characterized <02OL3043>.

The preparation of oligothiophenes with novel side-chain substitution patterns has been reported including those containing ethylenedioxy groups <02CEJ2384>. Oligothiophene have

123 **124** **125**

been prepared with different groups located on one or both ends including those conjugated to isothiacyanates (for for tagging biomolecules) <02CEJ5072>, penta(ethylene glycol) chains <02JACS1269>, triarylimidazole groups <02CC1484>, tetrathiafulvene groups <02CEJ784>, porphyrin groups <02CEJ3027>, porphyrin and fullerene groups in the same molecule <02OL309>, fullurene groups <02CEJ5415>, and diphenylfluorene groups <02OL4439>. Heterothiophene co-oligomers have been prepared and/or studied including those containing furans <02HCA1989>, 1,2,5-oxadiazolo[3,4-c]pyridines <02H(56)421>, phenylene-vinylene groups <02H(57)1935>, ethynylene groups <02O4475, 02TL391>, thiazoles <02JACS11862>, dialkylfluorene groups <02S1136>, N-linked central pyrrole rings <02T3467>, 2,1,3-benzothiadiazole and ethyne groups <02TL3373>, a para[2.2]cyclophane central core <02OL3195>, and a 9,9'-spirobifluorene core <02JOC4924, 02JOC8104>. Substituted thiophene monomers have been prepared for use in the preparation of thiophene polymers <02CC2498, 02CC2690, 02OL607>. The preparation of oligothiophenes utilizing microwave-assisted Suzuki coupling reactions has been investigated <02JOC8877>. The synthesis of the C_2-symmetric bithieno[3,2-b][1,4]oxathiine **124**, a novel building block for thiophene-based oligomers and polymers, has been reported.

Crystallization of 2-(2-thienyl)pyrroles **125** produced gold-like and bronze-like metallic crystals <02BCJ2359, 02T10233>, while structurally related derivatives produced red-violet metallic crystals <02T10225>. A crystal structure analysis of 2,5-bis(4-biphenylyl)thiophene and the effect of structure on optical properties has been reported <02AM498>.

A review article detailing cross-coupling methods for making π-conjugated polymers including polythiophenes has appeared <02JOMC195>. Polythiophenes have been synthesized and/or evaluated as novel material applications including as organic conductors and molecular nanowires <02BCJ1795, 02JACS522, 02MAC7281>, affinitychromic materials <02JACS12463>, electroluminscent devices <02JMAT2887>, immunosensors <02CC680>, bioelectronic sensors <02CM449>, photolithography <02CM3705>, and organic bilayer photocells <02MAC4628>. Heterothiophene co-polymers have been prepared and studied including those containing pyridine groups <02JMAT2292>, aniline groups <02MAC6112>, fluorene groups and 2,1,3-benzothiadiazole groups <02JMAT2887>, and fluorene groups and phenylene/dicyanoethylene groups <02MAC1224>. Polythiophenes with novel side-chains have been synthesized including those containing polyethene groups <02CC2862>, thioether groups <02JMAT500>, chiral oxazoline groups <02CEJ4027>, and chiral alkyl groups <02MAC941>. Finally, the synthesis of hyperbranched polythiophenes has been reported <02TL6347>.

5.1.9 SELENOPHENES AND TELLUROPHENES

A small number of reports on the chemistry of selenophenes and tellurophenes appeared during the past year. A selenium transfer reaction between the bis-anion derived from phenylenediacetonitrile and selenium oxychloride ($SeOCl_2$) gave the corresponding benzo[*c*]selenophene <02JOC2453>. A novel approach to 2,5-diarylselenophenes involved the thermolysis of 1,2,3-selenadiazoles in the presence of excess arylacetylenes <02TL4817>. For example, heating **126** in the presence of pyridine **127** gave selenophene **128**. A cycloaddition-like reaction between diene **129** and selenium dioxide led to the formation of fused selenophene **130** (many additional examples reported) <02JOC6553>. The preparation of selenophene-modified porphyrin <02JMC449> and heptaphyrin <02JOC6309> macrocycles have been reported.

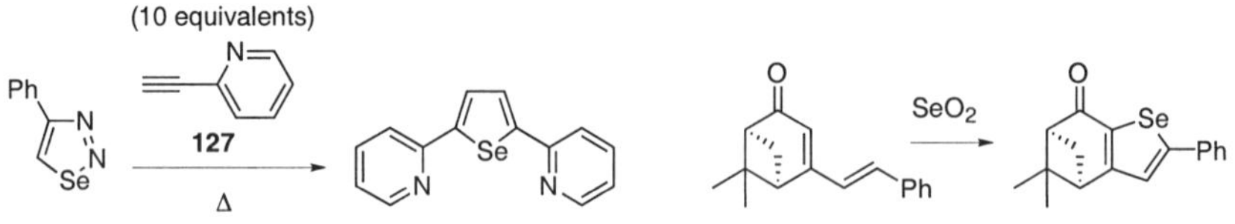

Generation of the first tellurium-containing diheteropentalene has been reported <02OL1193>. Treatment of dibromide **131** with tellurium followed by silver trifluoroacetate gave intermediate **132** which upon treatment with base gave telluro[3,4-*c*]thiophene **133**. Finally, benzo[*b*]tellurophene derivatives were synthesized and evaluated as selective dopamine D3 receptor antagonists <02JMC4594>.

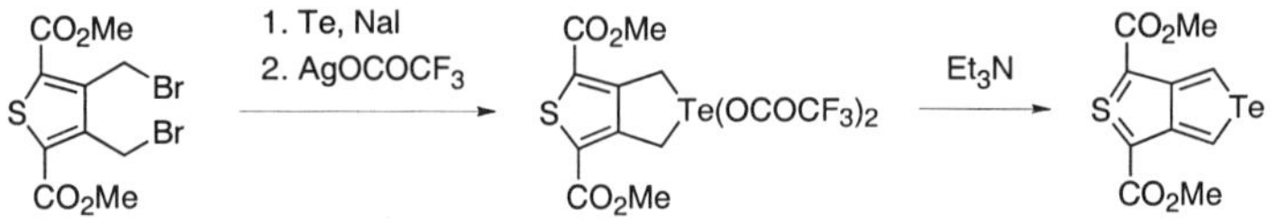

5.1.10 REFERENCES

02AG(E)3598	M.-K. Ng, L. Yu, *Angew. Chem. Int. Ed.* **2002**, *41*, 3598.
02AM498	S. Hotta, M. Goto, *Adv. Mater.* **2002**, *14*, 498.
02AM609	E. Mena-Osteritz, *Adv. Mater.* **2002**, *14*, 609.
02AM918	H. Tian, B. Chen, H. Tu, K. Müllen, *Adv. Mater.* **2002**, *14*, 918.
02BCJ167	T. Yamada, S. Kobatake, M. Irie, *Bull. Chem. Soc. Jpn.* **2002**, *75*, 167.
02BCJ551	K. Tanaka, H. Osuga, N. Tsujiuchi, M. Hisamoto, Y. Sakaki, *Bull. Chem. Soc. Jpn.* **2002**, *75*, 551.
02BCJ1795	K. Takimiya, K.-i, Kato, Y. Aso, F. Ogura, T. Otsubo, *Bull. Chem. Soc. Jpn.* **2002**, *75*, 1795.
02BCJ2071	S. Irie, M. Irie, *Bull. Chem. Soc. Jpn.* **2002**, *75*, 2071.
02BCJ2359	K. Ogura, R. Zhao, H. Yanai, K. Maeda, R. Tozawa, S. Matsumoto, M. Akazome, *Bull. Chem.Soc. Jpn.* **2002**, *75*, 2359.
02BMC1499	H.R. Moon, H.O. Kim, S.K. Lee, W.J. Choi, M.W. Chun, L.S. Jeong, *Bioorg. Med. Chem.* **2002**, *10*, 1499.
02BMC1547	M. Yoshikawa, T. Morikawa, H. Matsuda, G. Tanabe, O. Muraoka, *Bioorg. Med. Chem.* **2002**, *10*, 1547.
02BMC1611	P.G. Baraldi, R. Romagnoli, M.G. Pavani, M.D.C. Nunez, J.P. Bingham, J.A. Hartley, *Bioorg. Med. Chem.* **2002**, *10*, 1611.
02BMC2067	R.A. Forsch, J.E. Wright, A. Rosowsky, *Bioorg. Med. Chem.* **2002**, *10*, 2067.

02BMC2185	P. Dallemagne, L.P. Khanh, A. Alsaïdi, O. Renault, I. Varlet, V. Collot, R. Bureau, S. Rault, *Bioorg. Med. Chem.* **2002**, *10*, 2185.
02BMC3113	M. Fujita, T. Hirayama, N. Ikeda, *Bioorg. Med. Chem.* **2002**, *10*, 3113.
02BMCL9	C.R. Cardoso, F.C.F. de Brito, K.C.M. da Silva, A.L.P. de Miranda, C.A.M. Fraga, E.J. Barreiro, *Bioorg. Med. Chem. Lett.* **2002**, *12*, 9.
02BMCL391	A. McCluskey, M.A. Keane, C.C. Walkom, M.C. Bowyer, A.T.R. Sim, D.J. Young, J.A. Sakoff, *Bioorg. Med. Chem. Lett.* **2002**, *12*, 391.
02BMCL491	M.J. Rudolph, C.R. Illig, N.L. Subasinghe, K.J. Wilson, J.B. Hoffman, T. Randle, D. Green, C.J. Molloy, R.M. Soll, F. Lewandowski, M.-Q. Zhang, R. Bone, J.C. Spurlino, I.C. Deckman, C. Manthey, C. Sharp, D. Maguire, B.L. Grasberger, R.L. DesJarlais, Z. Zhou, *Bioorg. Med. Chem. Lett.* **2002**, *12*, 491.
02BMCL1383	M. Fujita, T. Seki, *Bioorg. Med. Chem. Lett.* **2002**, *12*, 1383.
02BMCL1439	J.J. Harnett, J. M. Auguet, I. Viossat, C. Dolo, D. Bigg, P. Chabrier, *Bioorg. Med. Chem. Lett.* **2002**, *12*, 1439.
02BMCL1537	C.R. Rodrigues, T.M. Flaherty, T. C, Springer, J.H. McKerrow, F.E. Cohen, *Bioorg. Med. Chem. Lett.* **2002**, *12*, 1537.
02BMCL1571	U.E.W. Lange, C. Zechel, *Bioorg. Med. Chem. Lett.* **2002**, *12*, 1571.
02BMCL1607	M. Fujita, T. Seki, H. Inada, N. Ikeda, *Bioorg. Med. Chem. Lett.* **2002**, *12*, 1607.
02BMCL1675	L.R. Martinez, J.G.A. Zarraga, M.E. Duran, M.T.R. Apam, R. Cañas, *Bioorg. Med. Chem. Lett.* **2002**, *12*, 1675.
02BMCL1897	M. Fujita, T. Seki, N. Ikeda, *Bioorg. Med. Chem. Lett.* **2002**, *12*, 1897.
02BMCL2011	D.H. Boschelli, D.Y. Wang, F. Ye, A. Yamashita, N. Zhang, D. Powell, J. Weber, F. Boschelli, *Bioorg. Med. Chem. Lett.* **2002**, *12*, 2011.
02BMCL2145	H. Khatuya, V.L. Pulito, L.K. Joliffe, X. Li, W.V. Murray, *Bioorg. Med. Chem. Lett.* **2002**, *12*, 2145.
02BMCL2403	H.O. Kim, Y.H. Park, H.R. Moon, L.S. Jeong, *Bioorg. Med. Chem. Lett.* **2002**, *12*, 2403.
02BMCL2549	C. Kikuchi, T. Hiranuma, M. Koyama, *Bioorg. Med. Chem. Lett.* **2002**, *12*, 2549.
02BMCL2565	J.K. Clark, P. Cowley, A.W. Muir, R. Palin, E. Pow, A.B. Prosser, R. Taylor, M.-Q. Zhang, *Bioorg. Med. Chem. Lett.* **2002**, *12*, 2565.
02BMCL2569	R. Palin, J.K. Clark, P. Cowley, A.W. Muir, E. Pow, A.B. Prosser, R. Taylor, M.-Q. Zhang, *Bioorg. Med. Chem. Lett.* **2002**, *12*, 2569.
02BMCL2583	M. Gallant, M.C. Carriére, A. Chateauneuf, D. Denis, Y. Gareau, C. Godbout, G. Greig, H. Juteau, N. Lachance, P. Lacombe, S. Lamontagne, K.M. Metters, C. Rochette, D. Sliipetz, N. Sawyer, N. Tremblay, M. Labelle, *Bioorg. Med. Chem. Lett.* **2002**, *12*, 2583.
02BMCL2981	H.W. Hamilton, G. Nishiguchi, S.E. Hagen, J.D. Domagala, P.C. Weber, S. Gracheck, S.L. Boulware, E.C. Nordby, H. Cho, T. Nakamura, T., S. Ikeda, W. Watanabe, *Bioorg. Med. Chem. Lett.* **2002**, *12*, 2981.
02BMCL3475	N. Bharti, K. Husain, M.T.G. Garza, D.E. Cruz-Vega, J. Castro-Garza, B.D. Mata-Cardenas, F. Naqvi, A. Azam, *Bioorg. Med. Chem. Lett.* **2002**, *12*, 3475.
02CBC1053	N.C.R. van Straten, G.G. Schoonus-Gerritsma, R.G. van Someren, J. Draaijer, A.E.P. Adang, C.M. Timmers, R.G.J.M. Hanssen, C.A.A. van Boeckel, *ChemBioChem* **2002**, 1053.
02CC462	S. Venkatraman, V.G. Anand, S.K. Pushpan, J. Sankar, T.K. Chandrashekar, *Chem. Commun.* **2002**, 462.
02CC680	M. Kanungo, D.N. Srivastava, A. Kumar, A.Q. Contractor, *Chem. Commun.* **2002**, 680.
02CC932	Y. Kitahara, K. Tanaka, *Chem. Commun.* **2002**, 932.
02CC1015	C.A. Briehn, P. Bäuerle, *Chem. Commun.* **2002**, 1015.
02CC1060	B. Chen, M. Wang, Y. Wu, H. Tian, *Chem. Commun.* **2002**, 1060.
02CC1192	A. Wakamiya, T. Nishinaga, K. Komatsu, *Chem. Commun.* **2002**, 1192.
02CC1204	L.S. Konstantinova, O.A. Rakitin, C.W. Rees, *Chem. Commun.* **2002**, 1204.
02CC1274	M.E. Migaud, R.C. Wilmouth, G.I. Mills, G.J. Wayne, C. Risley, C. Chambers, S.J.F. Macdonald, C. Schofield, *Chem. Commun.* **2002**, 1274.
02CC1484	C. Kikuchi, T. Iyoda, J. Abe, *Chem. Commun.* **2002**, 1484.
02CC1530	R. Kisselev, M. Thelakkat, *Chem. Commun.* **2002**, 1530.
02CC1660	S. Venkatraman, V.G. Anand, V. PrabhuRaja, H. Rath, J. Sankar, T.K. Chandrashekar, W. Teng, K.R. Senge, *Chem. Commun.* **2002**, 1660.
02CC2498	K. Zong, L. Madrigal, L. Groenendaal, J.R. Reynolds, *Chem. Commun.* **2002**, 2498.

02CC2642	N. Agarwal, S.P. Mishra, A. Kumar, C.-H. Hung, M. Ravikanth, *Chem. Commun.* **2002**, 2642.
02CC2690	D. Caras-Quintero, P. Bäuerle, *Chem. Commun.* **2002**, 2690.
02CC2734	A. Mulder, A. Jukovic, L.N. Lucas, J. van Esch, B.L. Feringa, J. Huskens, D.N. Reinhoudt, *Chem. Commun.* **2002**, 2734.
02CC2804	S. Kobatake, K. Uchida, E. Tsuchida, M. Irie, *Chem. Commun.* **2002**, 2804.
02CC2862	X. Zhang, B. Hessen, *Chem. Commun.* **2002**, 2862.
02CC3028	C. Mooriag, O. Clot, M.O. Wolf, B.O. Patrick, *Chem. Commun.* **2002**, 3028.
02CEJ137	S. Encinas, L. Flamigni, F. Barigelletti, E.C. Constable, C.E. Housecroft, E.R. Schofield, E. Figgemeier, D. Fenske, M. Neuburger, J.G. Vos, M. Zehnder, *Chem. Eur. J.* **2002**, 137.
02CEJ784	P. Frére, M. Allain, E.H. Elandaloussi, E. Levillain, F.-X. Sauvage, A. Riou, J. Roncali, *Chem. Eur. J.* **2002**, 784.
02CEJ1336	E. Roversi, R. Scopelliti, E. Solari, R. Estoppey, P. P. Braña, R. Menéndez, J.A. Sordo, *Chem. Eur. J.* **2002**, 1336.
02CEJ1573	U. Lawrentz, W. Grahn, K. Lukaszuk, C. Klein, R. Wortmann, A. Feldner, D. Scherer, *Chem. Eur. J.* **2002**, 1573.
02CEJ2384	J.J. Apperloo, L. Groenendaal, H. Verheyen, M. Jayakannan, R.A.J. Janssen, A. Dkhissi, D. Beljonne, R. Lazzaroni, J.-L. Brédas, *Chem. Eur. J.* **2002**, 2384.
02CEJ3027	F. Odobel, S. Suresh, E. Blart, Y. Nicolas, J.-P. Quintard, P. Janvier, J.-Y. Le Questel, B. Illien, D. Rondeau, P. Richomme, T. Häupl, S. Wallin, L. Hammarström, *Chem. Eur. J.* **2002**, 3027.
02CEJ4027	H. Goto, Y. Okamoto, E. Yashima, *Chem. Eur. J.* **2002**, 4027.
02CEJ4542	C.-H. Hung, J.-P. Jong, M.-Y. Ho, G.-H. Lee, G.-H., S.-M. Peng, *Chem. Eur. J.* **2002**, 4542.
02CEJ5072	G. Barbarella, *Chem. Eur. J.* **2002**, 5072.
02CEJ5415	P.A. van Hal, E.H.A. Beckers, S.C.J. Meskers, R.A.J. Janssen, B. Jousselme, P. Blanchard, J. Roncali, *Chem. Eur. J.* **2002**, 5415.
02CHC156	A.R. Katritzky, K. Kirichenko, Y. Ji, I. Prakash, *Chem. Heterocycl. Comp.* **2002**, *38*, 156.
02CHC165	M.M. Krayushkin, F.M. Stoyanovich, O.Y. Zolotarskaya, E.I. Chernoburova, N.N. Makhova, V.N. Yarovenko, I.V. Zavarzin, A.Y. Martynkin, B.M. Uzhinov, *Chem. Heterocycl. Comp.* **2002**, *38*, 165.
02CHC281	H.R. Maradiya, V.S. Patel, *Chem. Heterocycl. Comp.* **2002**, *38*, 281.
02CHC466	T.E. Khoshtariya, L.N. Kurkovskaya, N.T. Mirziashvili, N.Z. Chichinadze, L.M. Tevzadze, M.I. Sikharulidze, *Chem. Heterocycl. Comp.* **2002**, *38*, 466.
02CHC503	A.V. Mashkina, *Chem. Heterocycl. Comp.* **2002**, *38*, 503.
02CHC632	E. Lukevics, P. Arsenyan, S. Belyakov, O. Pudova, *Chem. Heterocycl. Comp.* **2002**, *38*, 632.
02CL58	T. Yamaguchi, H. Kashiyama, H. Nakazumi, T. Yamada, M. Irie, *Chem. Lett.* **2002**, 58.
02CL572	K. Morimitsu, K. Shibata, S. Kobatake, M. Irie, *Chem. Lett.* **2002**, 572.
02CL896	A. Kanitz, J. Schumann, M. Scheffel, S. Rajoelson, W. Rogler, H. Hartmann, D. Rohde, *Chem. Lett.* **2002**, 896.
02CL958	Y. Kunugi, K. Takimiya, K. Yamashita, Y. Aso, T. Otsubo, *Chem. Lett.* **2002**, 958.
02CL984	J. Yu, Y. Shirota, *Chem. Lett.* **2002**, 984.
02CL1002	K. Takahashi, H. Tanioka, H. Fueno, Y. Misaki, K. Tanaka, *Chem. Lett.* **2002**, 1002.
02CL1224	S. Kobatake, M. Morimoto, Y. Asano, A. Murakami, S. Nakamura, M. Irie, *Chem. Lett.* **2002**, 1224.
02CM449	I.F. Perepichka, M. Besbes, E. Levillain, M. Sallé, J. Roncali, *Chem. Mater.* **2002**, *14*, 449.
02CM1742	A. Afzali, T.L. Breen, C.R. Kagan, *Chem. Mater.* **2002**, *14*, 1742.
02CM1884	Y.-Z. Su, J.T. Lin, Y.-T. Tao, C.-W. Ko, S.-C. Lin, S.-S. Sun, *Chem. Mater.* **2002**, *14*, 1884.
02CM3705	J. Yu, S. Holdcroft, *Chem. Mater.* **2002**, *14*, 3705.
02CM4543	A.J. Paraskos, T.M. Swager, *Chem. Mater.* **2002**, *14*, 4543.
02CM4550	G. Zotti, A. Randi, S. Destri, W. Porzio, G. Schiavon, *Chem. Mater.* **2002**, *14*, 4550.
02CM4662	M. He, T.M. Leslie, J.A. Sinicropi, *Chem. Mater.* **2002**, *14*, 4662.
02CM4669	M. He, T.M. Leslie, J.A. Sinicropi, S.M. Garner, L.D. Reed, *Chem. Mater.* **2002**, *14*, 4669.
02CPB656	J. DoganKoruznjak, N. Slade, B. Zamola, K. Pavelic, G. Karminski-Zamola, *Chem. Pharm. Bull.* **2002**, *50*, 656.
02CPB1215	J.A. Valderrama, C. Astudillo, R.A. Tapia, E. Prina, E. Estrabaud, R. Mahieux, A. Fournet, *Chem. Pharm. Bull.* **2002**, *50*, 1215.
02EJO947	K. Wojciechowski, S. Kosinski, *Eur. J. Org. Chem.* **2002**, 947.

02EJO1284	T. Armaroli, C. Dell'Erba, A. Gabellini, F. Gasparrini, A. Mugnoli, M. Novi, G. Petrillo, C. Tavani, *Eur. J. Org. Chem.* **2002**, 1284.
02EJO2524	N.O. Silva, A.S. Abreu, AP.M.T. Ferreira, L.S. Monteiro, M.-J.R.P. Queiroz, *Eur. J. Org. Chem.* **2002**, 2524.
02EJO3796	S. Takami, T. Kawai, M. Irie, *Eur. J. Org. Chem.* **2002**, 3796.
02H(56)91	F.J. Lopez, M.-F. Jett, J.M. Muchowski, D. Nitzan, C. O'Yang, *Heterocycles* **2002**, *56*, 91.
02H(56)421	H. Gorohmaru, H., T. Thiemann, T., T. Sawada, T., T. Takahashi, T., K. Nishi-i, K., N. Ochi, N., Y. Kosugi, Y., S. Mataka, S. *Heterocycles* **2002**, *56*, 421.
02H(57)17	A. Kakehi, A., S. Ito, S., H. Suga, H., T. Miwa, T., T. Mori, T., T. Kobayashi, T. *Heterocycles* **2002**, *57*, 17.
02H(57)97	T. Erker, T., K. Trinkl, *Heterocycles* **2002**, *57*, 97.
02H(57)317	A. Genevous-Borella, M. Vuilliorgne, S. Mignani, *Heterocycles* **2002**, *57*, 317.
02H(57)1313	R. Miyabe, T. Shioiri, T. Aoyama, *Heterocycles* **2002**, *57*, 1313.
02H(57)1831	B.A. Moosa, K.A. Abu Safieh, M.M. El-Abadelah, *Heterocycles* **2002**, *57*, 1831.
02H(57)1935	C. Xue, F.-T. Luo, *Heterocycles* **2002**, *57*, 1935.
02H(57)1989	M. Komatsu, J. Choi, M. Mihara, Y. Oderaotoshi, S. Minakata, *Heterocycles* **2002**, *57*, 1989.
02H(58)405	K. Imafuku, D.-L. Wang, *Heterocycles* **2002**, *58*, 405.
02HCA451	T. Gendek, G. Mloston, A. Linden, H. Heimgartner, *Helv. Chim. Acta* **2002**, *85*, 451.
02HCA712	F. Monnat, P. Vogel, J.A. Sordo, *Helv. Chim. Acta* **2002**, *85*, 712.
02HCA733	E. Roversi, F. Monnat, P. Vogel, K. Schenk, P. Roversi, *Helv. Chim. Acta* **2002**, *85*, 733.
02HCA1390	E. Roversi, P. Vogel, K. Schenk, *Helv. Chim. Acta* **2002**, *85*, 1390.
02HCA1989	F. Garzino, A. Méou, P. Brun, *Helv. Chim. Acta* **2002**, *85*, 1989.
02JACS522	B.W. Maynor, S.F. Filocamo, M.W. Grinstaff, J. Liu, *J. Am. Chem. Soc.* **2002**, *124*, 522.
02JACS1269	A.P.H.J. Schenning, A.F.M. Kilbinger, F. Biscarini, M. Cavallini, H.J. Cooper, P.J. Derrick, W.J. Feast, R. Lazzaroni, P. Leclére, L.A. McDonell, E.W. Meijer, S.C.J. Meskers, *J. Am. Chem. Soc.* **2002**, *124*, 1269.
02JACS1582	W. Zhao, E.M. Carreira, *J. Am. Chem. Soc.* **2002**, *124*, 1582.
02JACS4184	T.M. Pappenfus, R.J. Chesterfield, C.D. Frisbie, K.R. Mann, J. Casado, J.D. Raff, L.L. Miller, *J. Am. Chem. Soc.* **2002**, *124*, 4184.
02JACS5286	T. Okazawa, T. Satoh, M. Miura, M. Nomura, *J. Am. Chem. Soc.* **2002**, *124*, 5286.
02JACS7481	L. Giordano, T.M. Jovin, M. Irie, E.A. Jares-Erijman, *J. Am. Chem. Soc.* **2002**, *124*, 7481.
02JACS7642	S.-W. Kim, M. Kim, W.Y. Lee, T. Hyeon, *J. Am. Chem. Soc.* **2002**, *124*, 7642.
02JACS7668	P.A. Liddell, G. Kodis, A.L. Moore, T.A. Moore, D. Gust, *J. Am. Chem. Soc.* **2002**, *124*, 7668.
02JACS7762	J.D. Tovar, A. Rose, T.M. Swager, *J. Am. Chem. Soc.* **2002**, *124*, 7762.
02JACS7898	K.E. Maly, M.D. Wand, R.P. Lemieux, *J. Am. Chem. Soc.* **2002**, *124*, 7898.
02JACS11862	M.-K. Ng, D.-C. Lee, L. Yu, *J. Am. Chem. Soc.* **2002**, *124*, 11862.
02JACS12200	S. Tsuzuki, K. Honda, R. Azumi, *J. Am. Chem. Soc.* **2002**, *124*, 12200.
02JACS12380	J. Casado, L.L. Miller, K.R. Mann, T.M. Pappenfus, H. Higuchi, E. Ortí, B. Milián, R. Pou-Amérigo, V. Hernández, J.T. López Navarrete, *J. Am. Chem. Soc.* **2002**, *124*, 12380.
02JACS12463	S. Bernier, S. Garreau, M. Béra-Abérem, C. Gravel, M. Leclerc, *J. Am. Chem. Soc.* **2002**, *124*, 12463.
02JACS12507	M.J. Marsella, R.J. Reid, S. Estassi, L.-S. Wang, *J. Am. Chem. Soc.* **2002**, *124*, 12507.
02JACS12742	S.H. Eichhorn, S. H., A.J. Paraskos, A. J., K. Kishikawa, K., T.M. Swager, T. M. *J. Am. Chem. Soc.* **2002**, *124*, 12742.
02JACS13396	T. Shimada, Y.-H. Cho, T. Hayashi, *J. Am. Chem. Soc.* **2002**, *124*, 13396.
02JCC457	C.A. Briehn, P. Bäuerle, *J. Comb. Chem.* **2002**, *4*, 457.
02JCS(P1)717	J.D.E. Chaffin, J.M. Marker, P.R. Huddleston, *J. Chem. Soc., Perkin Trans. 1* **2002**, 717.
02JCS(P1)1135	Y. Li, K. Kamata, T. Kawai, J. Abe, T. Iyoda, *J. Chem. Soc., Perkin Trans. 1* **2002**, 1135.
02JCS(P1)1944	B.J. Morrison, O.C. Musgrave, *J. Chem. Soc., Perkin Trans. 1* **2002**, 1944.
02JCS(P1)1963	T. Kobayashi, T. Moriwaki, M. Tsubakiyama, S. Yoshida, *J. Chem. Soc., Perkin Trans. 1* **2002**, 1963.
02JCS(P1)2414	H.M. Song, K. Kim, *J. Chem. Soc., Perkin Trans. 1* **2002**, 2414.
02JCS(P2)1791	A. Carella, A. Castaldo, A., R. Centore, A. Fort, A. SiriguJ. *Chem. Soc., Perkin Trans. 2* **2002**, 1791.
02JHC91	P. Lu, G. CaiJ. LiW.P. Weber, *J. Heterocycl. Chem.* **2002**, *39*, 91.

02JHC163	T. Kohara, H. Tanaka, K. Kimura, H. Horiuchi, K. Seio, M. Arita, T. Fujimoto, I. Yamamoto, *J. Heterocycl. Chem.* **2002**, *39*, 163.
02JHC645	T. Erker, N. Handler, *J. Heterocycl. Chem.* **2002**, *39*, 645.
02JHC783	D.H. Boschelli, F. Ye, *J. Heterocycl. Chem.* **2002**, *39*, 783.
02JHC857	T. Erker, M.E. Galanski, M. Galanski, *J. Heterocycl. Chem.* **2002**, *39*, 857.
02JHC877	F. Al-Omran, R.M. Mohareb, A.A. El-Khair, *J. Heterocycl. Chem.* **2002**, *39*, 877.
02JHC1177	F. Burkamp, S.R. Fletcher, *J. Heterocycl. Chem.* **2002**, *39*, 1177.
02JICS546	V.H.K. Verma, B. Singh, *J. Indian Chem. Soc.* **2002**, *79*, 546.
02JMAT500	C. Pozo-Gonzalo, T. Khan, J.J.W. McDouall, P.J. Skabara, D.M. Roberts, M.E. Light, S.J. Coles, M.B. Hursthouse, H. Neugebauer, A. Cravino, N.S. Sariciftci, *J. Mater. Chem.* **2002**, *12*, 500.
02JMAT924	S. Destri, M. Pasini, C. Botta, W. Porzio, F. Bertini, L. Marchio, *J. Mater. Chem.* **2002**, *12*, 924.
02JMAT2292	G.M. Chapman, S.P. Stanforth, R. Berridge, C. Pozo-Gonzalo, P.J. Skabara, *J. Mater. Chem.* **2002**, *12*, 2292.
02JMAT2612	H. Utsumi, D. Nagahama, H. Nakano, Y. Shirota, *J. Mater. Chem.* **2002**, *12*, 2612.
02JMAT2665	A. Pailleret, D.W.M. Arrigan, J.K. Browne, M.A. McKervey, *J. Mater. Chem.* **2002**, *12*, 2665.
02JMAT2887	Q. Hou, Y. Xu, W. Yang, M. Yuan, J. Peng, Y. Cao, *J. Mater. Chem.* **2002**, *12*, 2887.
02JMC54	J. Brea, J. Rodrigo, A. Carrieri, F. Sanz, M.I. Cadavid, M.J. Enguix, M. Villazón, G. Mengod, Y. Caro, C.F. Masaguer, E. Raviña, N.B. Centeno, A. Carotti, M.I. Loza, *J. Med. Chem.* **2002**, *45*, 54.
02JMC382	C.E. Tranberg, A. Zickgraf, B.N. Giunta, H. Luetjens, H. Figler, L.J. Murphree, L. R. Falke, H. Fleischer, J. Linden, P.J. Scammells, R.A. Olsson, *J. Med. Chem.* **2002**, *45*, 382.
02JMC449	D.G. Hilmey, D.M. Abe, M.I. Nelen, C.E. Stilts, G.A. Baker, S.N. Baker, F.V. Bright, S.R. Davies, S.O. Gollnick, A.R. Oseroff, S.L. Gibson, R. Hilf, M.R. Detty, *J. Med. Chem.* **2002**, *45*, 449.
02JMC888	C.-Y. Kim, D.A. Whittington, J.S. Chang, J. Liao, J.A. May, D.W. Christianson, *J. Med. Chem.* **2002**, *45*, 888.
02JMC1176	M.S. Chambers, J.R. Atack, F.A. Bromidge, H.B. Broughton, S. Cook, G.R. Dawson, S.C. Hobbs, K.A. Maubach, A.J. Reeve, G.R. Seabrook, K. Wafford, H., A.M. MacLeod, *J. Med. Chem.* **2002**, *45*, 1176.
02JMC1399	U. Schopfer, P. Schoeffter, S.F. Bischoff, J. Nozulak, D. Feuerbach, P. Floersheim, *J. Med. Chem.* **2002**, *45*, 1399.
02JMC3022	D. Dijkstra, N. Rodenhuis, E.S. Vermeulen, T.A. Pugsley, L.D. Wise, H.V. Wikström, *J. Med. Chem.* **2002**, *45*, 3022.
02JMC4128	L. Orús, S. Pérez-Silanes, A.-M. Oficialdegui, J. Martínez-Esparza, J.-C. Del Castillo, M. Mourelle, T. Langer, S. Guccione, G. Donzella, E.M. Krovat, E.K. Poptodorov, B. Lasheras, S. Ballaz, I. Hervías, R. Tordera, J. Del Río, A. Monge, *J. Med. Chem.* **2002**, *45*, 4128.
02JMC4171	F.E. Nielsen, T.B. Bodvarsdottir, A. Worsaae, P. MacKay, C.E. Stidsen, H.C.M. Boonen, L. Pridal, P.O.G. Arkhammar, P. Wahl, L. Ynddal, F. Junager, N. Dragsted, T.M. Tagmose, J.P. Mogensen, A. Koch, S.P. Treppendahl, J.B. Hansen, *J. Med. Chem.* **2002**, *45*, 4171.
02JMC4443	H.S. Andersen, O.H. Olsen, L.F. Iversen, A.L.P. Sørensen, S.B. Mortensen, M.S. Christensen, S. Branner, T.K. Hansen, J.F. Lau, L. Jeppesen, E.J. Moran, J. Su, F. Bakir, L. Judge, M. Shahbaz, T. Collins, T. Vo, M.J. Newman, W.C. Ripka, N.P.H. Møller, *J. Med. Chem.* **2002**, *45*, 4443.
02JMC4594	L. Bettinetti, K. Schlotter, H. Hübner, P. Gmeiner, *J. Med. Chem.* **2002**, *45*, 4594.
02JMC4888	Y. Chong, H. Choo, Y. Choi, J. Mathew, R.F. Schinazi, C.K. Chu, *J. Med. Chem.* **2002**, *45*, 4888.
02JMC5005	I.C. Choong, W. Lew, D. Lee, P. Pham, M.T. Burdett, J.W. Lam, C. Wisesmann, T.N. Luong, B. Fahr, W.L. DeLano, R.S. McDowell, D.A. Allen, D.A. Erlanson, E.M. Gordon, T. O'Brien, *J. Med. Chem.* **2002**, *45*, 5005.
02JNP1344	S.-J. Jeong, R. Higuchi, T. Miyamoto, M. Ono, M. Kuwano, S.F. Mawatari, *J. Nat. Prod.* **2002**, *65*, 1344.
02JNP1667	H. Duan, Y. Takaishi, M. Tori, S. Takaoka, G. Honda, M. Ito, Y. Takeda, O.K. Kodzhimatov, K. Kodzhimatov, O. Ashurmetov, *J. Nat. Prod.* **2002**, *65*, 1667.
02JOC205	J.-M. Raimundo, P. Blanchard, N. Gallego-Planas, N. Mercier, I. Ledoux-Rak, R. Hierle, J. Roncali, *J. Org. Chem.* **2002**, *67*, 205.
02JOC943	G.-D. Zhu, V. Schaefer, S.A. Boyd, G.F. Okasinski, *J. Org. Chem.* **2002**, *67*, 943.
02JOC1333	M.K. Cyrañski, T.M. Krygowski, A.R. Katritzky, P.v.R. Schleyer, *J. Org. Chem.* **2002**, *67*, 1333.

Five-Membered Ring Systems: Thiophenes & Se, Te Analogs 137

02JOC1795	K. Tanaka, H. Osuga, Y. Kitahara, *J. Org. Chem.* **2002**, *67*, 1795.
02JOC1905	D. Yue, R.C. Larock, *J. Org. Chem.* **2002**, *67*, 1905.
02JOC2453	R.R. Amaresh, M.V. Lakshmikantham, J.W. Baldwin, M.P. Cava, R.M. Metzger, R.D. Rogers, *J. Org. Chem.* **2002**, *67*, 2453.
02JOC2798	D. Bykowski, R. McDonald, R.J. Hinkle, R.R. Tykwinski, R. R. *J. Org. Chem.* **2002**, *67*, 2798.
02JOC3409	U.N. Rao, E. Biehl, *J. Org. Chem.* **2002**, *67*, 3409.
02JOC3961	P. Blanchard, B. Jousselme, P. Frére, J. Roncali, *J. Org. Chem.* **2002**, *67*, 3961.
02JOC4177	Y. Zhang, J.W. Herndon, *J. Org. Chem.* **2002**, *67*, 4177.
02JOC4574	K. Morimitsu, K. Shibata, S. Kobatake, M. Irie, *J. Org. Chem.* **2002**, *67*, 4574.
02JOC4916	S.K. Nandy, U.K.S. Kumar, H. Ila, H. Junjappa, *J. Org. Chem.* **2002**, *67*, 4916.
02JOC4924	J. Pei, J. Ni, X.-H. Zhou, X.-Y. Cao, Y.-H. Lai, *J. Org. Chem.* **2002**, *67*, 4924.
02JOC5208	S.-M. Yang, J.-J. Shie, J.-M. Fang, S.K. Nandy, H.-Y. Chang, S.-H. Lu, G. Wang, *J. Org. Chem.* **2002**, *67*, 5208.
02JOC5616	M.A. Campo, R.C. Larock, *J. Org. Chem.* **2002**, *67*, 5616.
02JOC6015	T.M. Pappenfus, J.D. Raff, E.J. Hukkanen, J.R. Burney, J. Casado, S.M. Drew, L.L. Miller, K.R. Mann, *J. Org. Chem.* **2002**, *67*, 6015.
02JOC6220	T. Janosik, B. Stensland, J. Bergman, *J. Org. Chem.* **2002**, *67*, 6220.
02JOC6309	V.G. Anand, S.K. Pushpan, S, Venkatraman, S.J. Narayanan, A. Dey, T.K. Chandrashekar, R. Roy, B.S. Joshi, S. Deepa, G.N. Sastry, *J. Org. Chem.* **2002**, *67*, 6309.
02JOC6474	S. Michelliza, A. Al-Mourabit, A. Gateau-Olesker, C. Marazano, *J. Org. Chem.* **2002**, *67*, 6474.
02JOC6553	T.M. Nguyen, I.A. Guzei, D. Lee, *J. Org. Chem.* **2002**, *67*, 6553.
02JOC6931	M. Pomerantz, A.S. Amarasekara, H.V.R. Dias, *J. Org. Chem.* **2002**, *67*, 6931.
02JOC7261	A. Treiber, *J. Org. Chem.* **2002**, *67*, 7261.
02JOC7523	L. Buzhansky, B.-A. Feit, *J. Org. Chem.* **2002**, *67*, 7523.
02JOC8104	J. Pei, J. Ni, X.-H. Zhou, X.-Y. Cao, Y.-H. Lai, *J. Org. Chem.* **2002**, *67*, 8104.
02JOC8812	A.J. Boydston, M.M. Haley, R.V. Williams, J.R. Armantrout, *J. Org. Chem.* **2002**, *67*, 8812.
02JOC8877	M. Melucci, G. Barbarella, G. Sotgui, *J. Org. Chem.* **2002**, *67*, 8877.
02JOC9073	D.D. Kenning, K.A. Mitchell, T.R. Calhoun, M.R. Funfar, D.J. Sattler, S.C. Rasmussen, *J. Org. Chem.* **2002**, *67*, 9073.
02JOMC56	M. Sato, A. Asami, G. Maruyama, M. Kosuge, J. Nakayama, S. Kumakura, T. Fujihara, K. Unoura, *J. Organomet. Chem.* **2002**, *654*, 56.
02JOMC177	A. Ricci, C. Lo Sterzo, *J. Organomet. Chem.* **2002**, *653*, 177.
02JOMC195	T. Yamamoto, *J. Organomet. Chem.* **2002**, *653*, 195.
02MAC941	Z.-B. Zhang, M. Fujiki, M. Motonaga, H. Nakashima, K. Torimitsu, H.-Z. Tang, *Macromolecules* **2002**, *35*, 941.
02MAC1224	N.S. Cho, D.-H. Hwang, J.-I. Lee, B.-J. Jung, H.-K. Shim, *Macromolecules* **2002**, *35*, 1224.
02MAC4628	L. Tan, M.D. Curtis, A.H. Francis, *Macromolecules* **2002**, *35*, 4628.
02MAC6112	B. Dufour, P. Rannou, J.P. Travers, A. Pron, M. Zagórska, M., G. Korc, I. Kulszewicz-Bajer, S. Quillard, S. Lefrant, *Macromolecules* **2002**, *35*, 6112.
02MAC7281	G.A. Sotzing, K. Lee, *Macromolecules* **2002**, *35*, 7281.
02NPL217	M.S. Ali, M. Jahangir, *Nat. Prod. Lett.* **2002**, *16*, 217.
02O2842	A.N. Ryabov, D.V. Gribkov, V.V. Izmer, A.Z. Voskoboynikov, *Organometallics* **2002**, *21*, 2842.
02O4475	W.-Y. Wong, K.-H. Choi, G.-L. Lu, Z. Lin, *Organometallics* **2002**, *21*, 4475.
02OL305	Y. Choi, H. Choo, Y. Chong, S. Lee, S. Olgen, R.F. Schinazi, C.K. Chu, *Org. Lett.* **2002**, *4*, 305.
02OL309	J. Ikemoto, K. Takimiya, Y. Aso, T. Otsubo, M. Fujitsuka, O. Ito, *Org. Lett.* **2002**, *4*, 309.
02OL529	M.H. Lim, H.O. Kim, H.R. Moon, M.W. Chun, L.S. Jeong, *Org. Lett.* **2002**, *4*, 529.
02OL607	P. Blanchard, A. Cappon, E. Levillain, Y. Nicolas, P. Frére, J. Roncali, *Org. Lett.* **2002**, *4*, 607.
02OL873	D.J. Lee, K. Kim, Y.J. Park, *Org. Lett.* **2002**, *4*, 873.
02OL1099	J.-J. Shie, S.-M. Yang, C.-T. Chen, J.-M. Fang, *Org. Lett.* **2002**, *4*, 1099.
02OL1193	D. Rajagopal, M.V. Lakshmikantham, E.H. Mørkved, M.P. Cava, *Org. Lett.* **2002**, *4*, 1193.
02OL1891	J. Ohshita, K.-H. Lee, M. Hashimoto, Y. Kunugi, Y. Harima, Y. Yamashita, A. Kunai, *Org. Lett.* **2002**, *4*, 1891.
02OL1899	A.C. Spivey, D.J. Turner, M.L. Turner, S. Yeates, *Org. Lett.* **2002**, *4*, 1899.
02OL2067	C. Xia, X. Fan, J. Locklin, R.C. Advincula, *Org. Lett.* **2002**, *4*, 2067.
02OL3043	T.M. Pappenfus, K.R. Mann, *Org. Lett.* **2002**, *4*, 3043.

02OL3195	F. Salhi, B. Lee, C. Metz, L.A. Bottomley, D.M. Collard, *Org. Lett.* **2002**, *4*, 3195.
02OL3879	M.M. Krayushkin, V.N. Yarovenko, S.L. Semenov, I.V. Zavarzin, A.V. Ignatenko, A.Y. Martynkin, B.M. Uzhinov, *Org. Lett.* **2002**, *4*, 3879.
02OL4439	K.-T. Wong, C.-F. Wang, C.-H. Chou, Y.O. Su, G.-H. Lee, S.-M. Peng, *Org. Lett.* **2002**, *4*, 4439.
02OL4535	M. Kozaki, Y. Yonezawa, K. Okada, *Org. Lett.* **2002**, *4*, 4535.
02OPRD357	G. Phillips, T.L. Fevig, P.H. Lau, G.H. Klemm, M.K. Mao, C. Ma, J.A. Gloeckner, A.S. Clark, *Org. Proc. Res. Dev.* **2002**, *6*, 357.
02PAC1327	I.P. Beletskaya, *Pure Appl. Chem.* **2002**, *74*, 1327.
02S213	A. Heynderickx, A. Samat, R. Guglielmetti, *Synthesis* **2002**, 213.
02S355	T. Kohara, H. Tanaka, K. Kimura, T. Fujimoto, I. Yamamoto, M. Arita, *Synthesis* **2002**, 355.
02S669	K.C. Majumdar, M. Ghosh, M. Jana, *Synthesis* **2002**, 669.
02S1096	E. Migianu, G. Kirsch, *Synthesis* **2002**, 1096.
02S1136	U. Asawapirom, R. Güntner, M. Forster, T. Farrell, U. Scherf, *Synthesis* **2002**, 1136.
02S1143	S. Yao, C. Hohle, P. Strohriegl, F. Würthner, *Synthesis* **2002**, 1143.
02S1195	F. von Kieseritzky, J. Hellberg, X. Wang, O. Inganäs, *Synthesis* **2002**, 1195.
02SC909	H. Halvorsen, H. Hope, J. Skramstad, *Synth. Commun.* **2002**, *32*, 909.
02SC1271	K.C. Majumdar, S.K. Ghosh, *Synth. Commun.* **2002**, *32*, 1271.
02SC1615	C. Wu, N. Blok, W. Li, G.W. Holland, *Synth. Commun.* **2002**, *32*, 1615.
02SC2565	J.M. Barker, P.R. Huddleston, P. R, M.L. Wood, *Synth. Commun.* **2002**, *32*, 2565.
02SC3493	F.A. Abu-Shanab, Y.M. Elkholy, M.H. Elnagdi, *Synth. Commun.* **2002**, *32*, 3493.
02SL325	C. Mukherjee, A. De, *Synlett* **2002**, 325.
02SL427	S.-K. Kang, D.-H. Kim, J.-N. Park, *Synlett* **2002**, 427.
02SL1799	F.F. Kneisel, P. Knochel, *Synlett* **2002**, 1799.
02SL2083	L.F. Tietze, J.K. Lohmann, *Synlett* **2002**, 2083.
02T1709	M.M. Oliveira, C. Moustrou, L.M. Carvalho, J.A.C. Silva, A. Samat, R. Guglielmetti, R. Dubest, J. Aubard, A.M.F. Oliveira-Campos, *Tetrahedron* **2002**, *58*, 1709.
02T2137	A. Noack, H. Hartmann, *Tetrahedron* **2002**, *58*, 2137.
02T2245	G. Sotgui, M. Zambianchi, G. Barbarella, C. Botta, *Tetrahedron* **2002**, *58*, 2245.
02T2551	G. Märkl, J. Amrhein, T. Stoiber, U. Striebl, P. Kreitmeier, *Tetrahedron* **2002**, *58*, 2551.
02T2589	W. Kroutil, A.A. Stämpfli, R. Dahinden, M. Jörg, U. Müller, J.P. Pachlatko, *Tetrahedron* **2002**, *58*, 2589.
02T3095	S.C.Gupta, M. Yusuf, S. Arora, S. Sharma, R.C. Kamboj, S.N. Dhawan, *Tetrahedron* **2002**, *58*, 3095.
02T3379	L. Bianchi, C. Dell'Erba, A. Gabellini, M. Novi, G. Petrillo, C. Tavani, *Tetrahedron* **2002**, *58*, 3379.
02T3467	P.E. Just, K.I. Chane-Ching, P.C. Lacaze, *Tetrahedron* **2002**, *58*, 3467.
02T3507	C. Bonini, M. D'Auria, M. Funicello, G. Romaniello, *Tetrahedron* **2002**, *58*, 3507.
02T3655	S.L. Cappelle, I.A. Vogels, T.C. Govaerts, S.M. Toppet, Compernolle, G. Hoornaert, *Tetrahedron* **2002**, *58*, 3655.
02T4529	M.G. Cabiddu, S. Cabiddu, E. Cadoni, S. Demontis, C. Fattuoni, S. Melis, *Tetrahedron* **2002**, *58*, 4529.
02T4551	K.C. Majumdar, S.K. Samanta, *Tetrahedron* **2002**, *58*, 4551.
02T7463	M. Otero, M.A. Herranz, C. Seoane, N. Martín, J. Garín, J. Orduna, R. Alcalá, B. Villacampa, *Tetrahedron* **2002**, *58*, 7463.
02T7653	K. Yamamura, N. Kusuhara, A. Kondou, M. Hashimoto, *Tetrahedron* **2002**, *58*, 7653.
02T7943	I.C.F.R. Ferreira, M.-J.R.P. Queiroz, G. Kirsch, *Tetrahedron* **2002**, *58*, 7943.
02T8055	A.S. Demir, O. Reis, M. Emrullahoglu, *Tetrahedron* **2002**, *58*, 8055.
02T8581	M.T. Herrero, I. Tellitu, E. Domínguez, S. Hernández, I. Moreno, R. SanMartín, *Tetrahedron* **2002**, *58*, 8581.
02T10047	K.C. Majumdar, M. Ghosh, *Tetrahedron* **2002**, *58*, 10047.
02T10151	A. Bongini, G. Barbarella, L. Favaretto, G. Sotgui, Zambianchi, D. Casarini, *Tetrahedron* **2002**, *58*, 10151.
02T10225	R. Zhao, M. Akazome, S. Matsumoto, M. Ogura, *Tetrahedron* **2002**, *58*, 10225.
02T10233	R. Zhao, S. Matsumoto, M. Akazome, K. Ogura, *Tetrahedron* **2002**, *58*, 10233.
02T10309	K.C. Majumdar, U.K. Kundu, S. Ghosh, *Tetrahedron* **2002**, *58*, 10309.
02TL257	G. Sommen, A. Comel, G. Kirsch, *Tetrahedron Lett.* **2002**, *43*, 257.

02TL391	J. Li, L. Liao, Y. Pang, *Tetrahedron Lett.* **2002**, *43*, 391.
02TL695	S. Tumkevicius, L.A. Agrofoglio, A. Kaminskas, G. Urbelis, T.A. Zevaco, O. Walter, *Tetrahedron Lett.* **2002**, *43*, 695.
02TL799	T.C. Govaerts, I. Vogels, F. Compernolle, G.J. Hoornaert, *Tetrahedron Lett.* **2002**, *43*, 799.
02TL1553	F. Allared, J. Hellberg, T. Remonen, *Tetrahedron Lett.* **2002**, *43*, 1553.
02TL1829	J.F.D. Chabert, C. Gozzi, M. Lemaire, *Tetrahedron Lett.* **2002**, *43*, 1829.
02TL2123	K.C. Majumdar, S.K. Ghosh, *Tetrahedron Lett.* **2002**, *43*, 2123.
02TL2239	T. Kitamura, B.-X. Zhang, Y. Fujiwara, *Tetrahedron Lett.* **2002**, *43*, 2239.
02TL3049	K. Matsumoto, H. Nakaminami, M. Sogabe, H. Kurata, M. Oda, *Tetrahedron Lett.* **2002**, *43*, 3049.
02TL3199	C.E. Hewton, M.C. Kimber, D.K. Taylor, *Tetrahedron Lett.* **2002**, *43*, 3199.
02TL3313	K.-T. Wong, R.-T. Chen, *Tetrahedron Lett.* **2002**, *43*, 3313.
02TL3373	C. Kitamura, K. Saito, M. Nakagawa, M. Ouichi, A. Yoneda, Y. Yamashita, *Tetrahedron Lett.* **2002**, *43*, 3373.
02TL3533	J. Chen, Q. Song, Z. Xi, *Tetrahedron Lett.* **2002**, *43*, 3533.
02TL3813	C. Bonini, M. D'Auria, P. Fedeli, *Tetrahedron Lett.* **2002**, *43*, 3813.
02TL3963	P. Kim, J.M. Tsuruda, M.M. Olmstead, S. Eisenberg, M.J. Kurth, *Tetrahedron Lett.* **2002**, *43*, 3963.
02TL4817	P. Arsenyan, O. Pudova, E. Lukevics, *Tetrahedron Lett.* **2002**, *43*, 4817.
02TL4865	S.-S.P. Chou, P.-W. Liang, *Tetrahedron Lett.* **2002**, *43*, 4865.
02TL4907	A. Voituriez, J.-C. Fiaud, E. Schulz, *Tetrahedron Lett.* **2002**, *43*, 4907.
02TL4981	H.M.L. Davies, A.M. Walji, R.J. Townsend, *Tetrahedron Lett.* **2002**, *43*, 4981.
02TL5625	F. Berthiol, M. Feuerstein, H. Doucet, M. Santelli, *Tetrahedron Lett.* **2002**, *43*, 5625.
02TL5649	J. Takagi, K. Sato, J.F. Hartwig, T. Ishiyama, N. Miyaura, *Tetrahedron Lett.* **2002**, *43*, 5649.
02TL5657	K. Haraguchi, H. Takahashi, H. Tanaka, *Tetrahedron Lett.* **2002**, *43*, 5657.
02TL6347	M.-H. Xu, L. Pu, *Tetrahedron Lett.* **2002**, *43*, 6347.
02TL6541	G. Loire, D. Prim, B. Andrioletti, E. Rose, A. Persoons, S. Sioneke, J. Vaissermann, *Tetrahedron Lett.* **2002**, *43*, 6541.
02TL7365	K.R. Crawford, A. Padwa, *Tetrahedron Lett.* **2002**, *43*, 7365.
02TL7419	P. Ghosh, A.D. Shukla, A. Das, *Tetrahedron Lett.* **2002**, *43*, 7419.
02TL8485	Y. Shimizu, Z. Shen, S. Ito, H. Uno, H. Daub, H.N. Ono, *Tetrahedron Lett.* **2002**, *43*, 8485.
02TL8669	N. Tanifuji, H. Huang, Y. Shinagawa, K. Kobayashi, *Tetrahedron Lett.* **2002**, *43*, 8669.
02TL9453	I. GuptaM. Ravikanth, *Tetrahedron Lett.* **2002**, *43*, 9453.
02TL9493	E.C. Lee, Y.-K. Park, J.-H. Kim, H. Hwang, Y.-R. Kim, C.-H. Lee, *Tetrahedron Lett.* **2002**, *43*, 9493.
02TL9615	M.L. Birsa, M. Cherkinsky, S. Braverman, *Tetrahedron Lett.* **2002**, *43*, 9615.

Chapter 5.2

Five-Membered Ring Systems: Pyrroles and Benzo Derivatives

Tomasz Janosik

Department of Chemistry, 6128 Burke Laboratory, Dartmouth College, Hanover, New Hampshire 03755, USA

Jan Bergman

Department of Biosciences at Novum, Karolinska Institute, Novum Research Park, SE-141 57 Huddinge, Sweden, and Södertörn University College, SE-141 04 Huddinge, Sweden e-mail: jabe@biosci.ki.se, fax: +46 8 608 1501

5.2.1 INTRODUCTION

The progress in the chemistry of pyrroles and indoles has been covered in numerous recent accounts. A review on the synthesis of heterocycles, including pyrroles and indoles, by radical cyclization, has appeared <02JCS(P1)2747>. Synthetic routes to *C*-vinylpyrroles have also been covered <02RCR563>. Cacchi and co-workers have summarized the developments regarding the synthesis of indoles by aminopalladation / reductive elimination applied to (2-aminophenyl)acetylene derivatives <02EJO2671>. Simple indole alkaloids and those possessing a non-rearranged monoterpenoid unit have been the subject of an annual review <02NPR148>. A comprehensive review on 1-hydroxy- and 1-methoxyindoles and their derivatives has been provided by Somei <02AHC(82)101>. The chemistry of carbazole alkaloids, including some aspects on indolocarbazoles, has been comprehensively reviewed <02CR4303>, while a more specialized review on indolocarbazoles appeared some time ago <01AHC(80)1>. An additional review dealing with the chemistry of isatins <01MI1> published before the reporting period should also be mentioned.

5.2.2 SYNTHESIS OF PYRROLES

3-Aryl- or 3,4-diaryl pyrroles **1** can be conveniently prepared in moderate to good yields from the readily available arylalkenes **2** and TosMIC in the presence of base in a one step operation <02OL3537>. 3,4-Disubstituted pyrroles have also been obtained by the action of TosMIC on various 2-tropanones in the presence of sodium ethoxide <02JOC5019>.

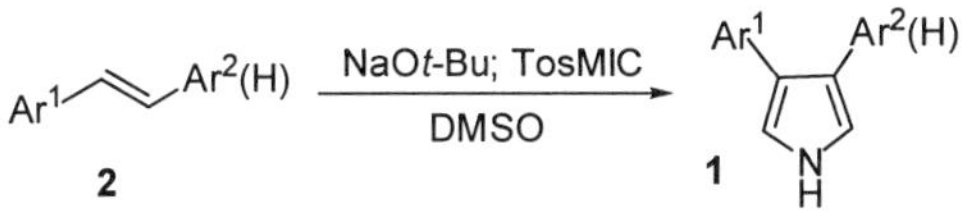

Several 2-substituted 3,4-diarylpyrroles **3** have been prepared in a regioselective process employing the base induced addition of methyl isocyanoacetate to the α,β-unsaturated nitriles **4**, and one of the pyrroles so obtained was used in a concise synthesis of the alkaloid ningalin B <02JOC9439>.

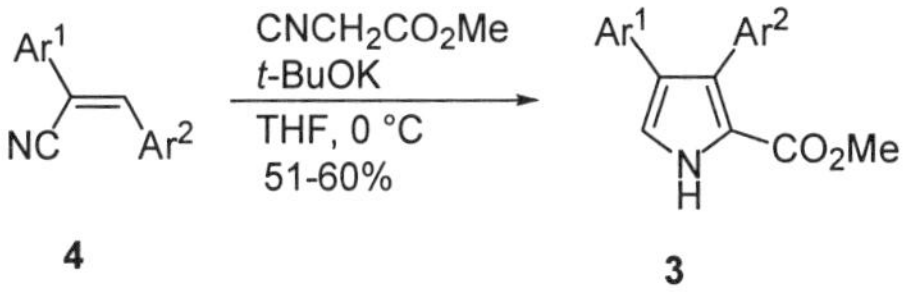

Various 3H-pyrroles have been obtained from the reaction of 1,1-diaryl-2,2-dicyanoethylenes or 1,1-diaryl-2-cyano-2-ethoxycarbonylethylenes with aromatic nitriles in the presence of 10% Sm/I_2 in refluxing THF <02SC2643>. Access has been gained to the Boc-protected pyrroles **5** via cyclodehydration of the linear precursors **6**, in turn obtained by Wittig olefination of the N-Boc-α-aminoaldehydes **7** <02EJO2565>. Precursors of type **7** have also served as substrates in Wittig reactions to produce 6-amino-3-ketosulfones, which could then undergo TFA-induced cyclization to 2,5-disubstituted pyrroles or pyrrolidines <02S331>. Formation of 2-trifluoromethylpyrroles has been observed upon treatment of an intermediate easily obtained from 1,1-bis(ethylsulfanyl)perfluorobut-1-ene and acetone with primary amines <02EJO1556>. Cyclization of boronate substituted 1,4-dichloro-2-butenes using primary amines has been reported to produce pyrrolines, which were further subjected to dehydrogenation using DDQ to give pyrrole-3-boronic esters, and/or used as partners in Suzuki-Miyaura couplings to furnish a variety of 3-substituted pyrrolines and pyrroles <02SL829>.

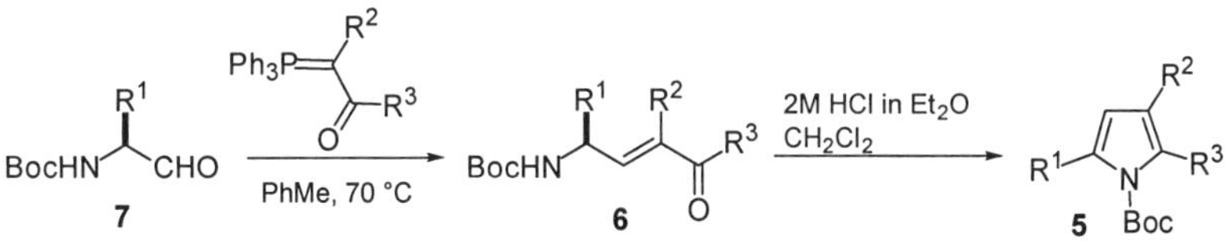

An interesting approach to 2,5-disubstituted 4-iodo-2,3-dihydropyrroles (**8**) has been reported, based on the cyclization of the sulfonamides **9** with iodine in the presence of potassium carbonate. The products could be dehydrogenated in good yields giving the corresponding pyrroles **10** with 2 equiv. of DBU at room temperature, while the use of only one equiv. of the base at elevated temperature led to partial loss of the iodine atom <02JCS(P1)622>.

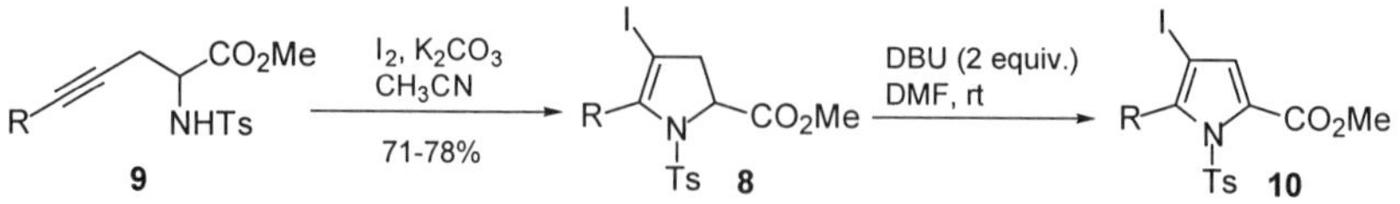

Various 3,5-disubstituted-2-pyridylpyrroles **11** were synthesized employing condensation of 2-(aminomethyl)pyridine (**12**) and suitable 1,3-dicarbonyl compounds **13** <02OL435>. The reaction of diimines with diphenylcyclopropenone has been reported to give various 4,5-diphenylpyrrol-3-ones in moderate yields <02JCS(P1)341>.

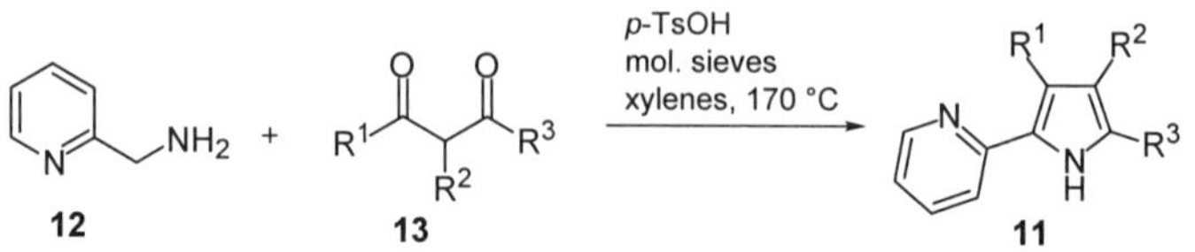

The sodium methoxide catalyzed reaction of certain diazabutadienes, e.g. **14**, with aldehydes has been used to prepare a series of 1-aminopyrrolines **15**, which could in turn be dehydrated to the corresponding pyrroles **16** as outlined below, or using TFA. Several other similar transformations have also been investigated, in all cases leading to related pyrrolines and pyrroles <02JOC8178>.

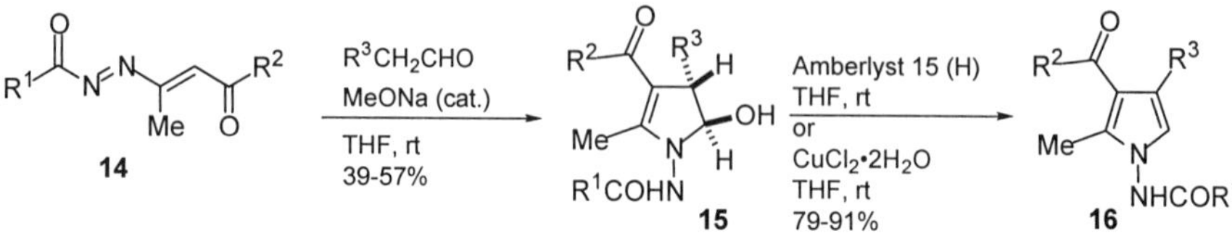

1-Acetylpyrroles such as **17** have also been prepared via cyclization of the enamines **18**, which were in turn readily prepared from the corresponding dimethylaminoalkene **19** by displacement of the dimethylamino group by amino acids. When applied to cyclic precursors, the approach also proved to be useful for the synthesis of fused pyrroles <02JCS(P1)2799>.

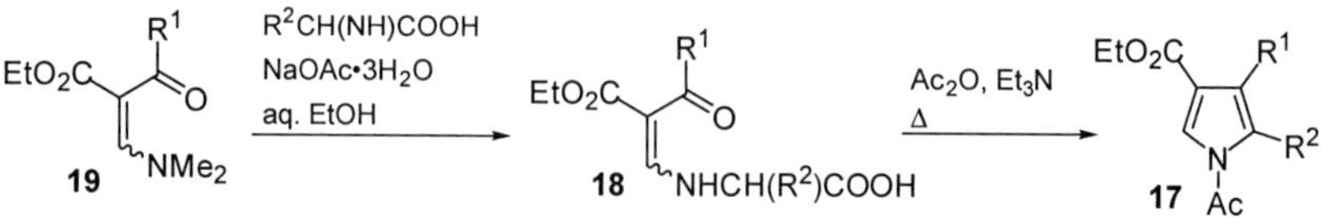

A pyrrole synthesis leading to **20** has been achieved by a CuBr catalyzed cyclization of the intermediate imine **21**, which was prepared over several steps from 1,3-hexadiyne (**22**) in a one-pot operation <02CHE748>. Aminomethyl substituted allenes have also been used for the synthesis of pyrroles by a palladium catalyzed annulation with aryl iodides <02H(57)2261>.

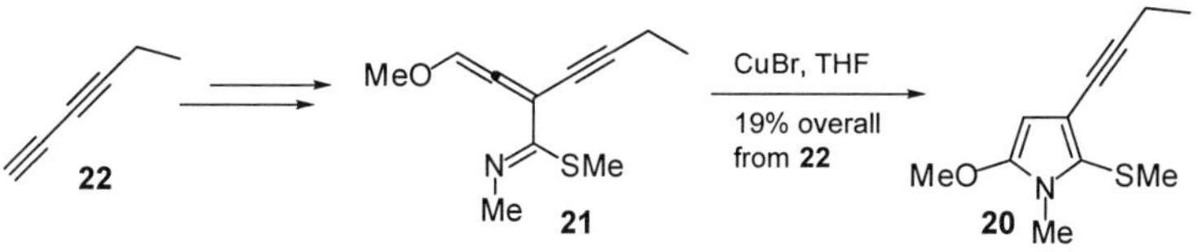

The dione **23**, obtained from acetylacetone and propargyl bromide, has been shown to undergo cyclization with ammonium chloride catalyzed by Cu_2Cl_2 in the presence of oxygen, to give the pyrrole **24** <02CHE616>.

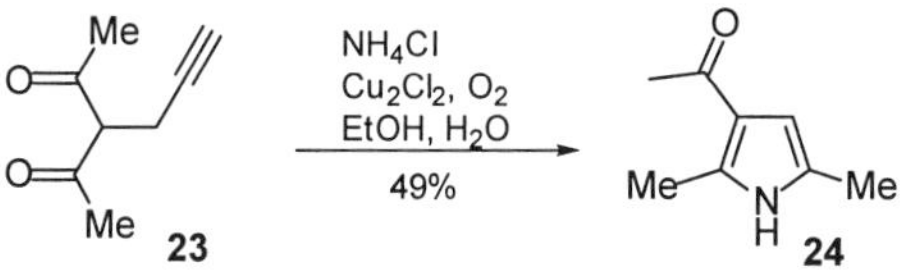

Aminocarbene complexes of chromium, such as **25** (or tungsten) have been transformed into 3-aza-1-metallahexatrienes **26**, which in turn underwent cyclization into the 2-aminopyrroles **27** in good yields upon heating <02OM1819>.

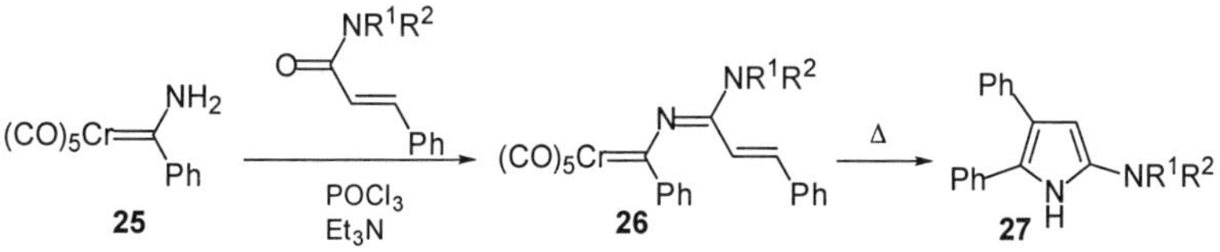

Several related carbene complexes **28** with M = Cr, Mo, or W, have been shown to undergo photocycloadditions with alkenes, in particular acrylates, to furnish a series of 1-pyrrolines **29** in moderate yields. The mechanism is believed to involve an initial cyclopropanation of the alkene, followed by a light-induced [1,3]-sigmatropic rearrangement of an intermediate *N*-cyclopropylimine <02OM4076>.

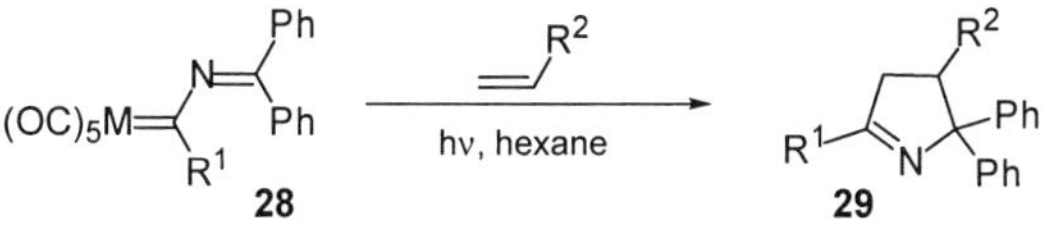

Various 1,2,4-trisubstituted pyrroles **30** have been prepared using a palladium catalyzed reaction of the enones **31** with glycine ethyl ester as outlined in the example below, or benzylamine catalyzed by $Pd(PPh_3)_4$ to give the corresponding 1-benzylpyrroles <02SL619>.

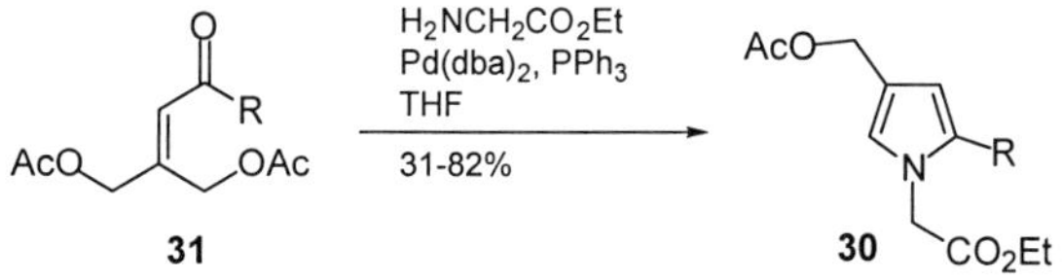

In an application of diazo decomposition, the pyrroles **32**, displaying intramolecular hydrogen bonding have been prepared in good yields from the β-keto-α-diazo carbonyl precursors **33** using a $Rh_2(OAc)_4$ catalyzed process. Decomposition induced by $Rh_2(O_2CCF_3)_4$ or $Cu(acac)_2$ in refluxing benzene also proved to be effective <02SL1913>.

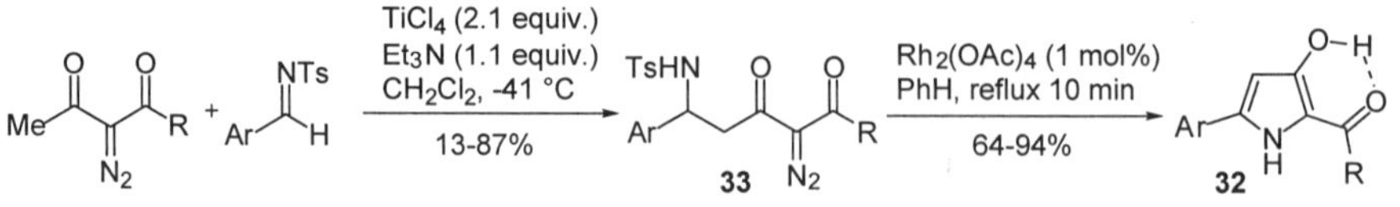

A sequential process involving a copper catalyzed cycloaddition of the vinyl sulfone **34** to N-methylpropargylamine (**35**), and a subsequent palladium meditated allylic substitution, provided a route to mixtures of the separable isomeric pyrrolines **36** and **37** <02EJO1493>.

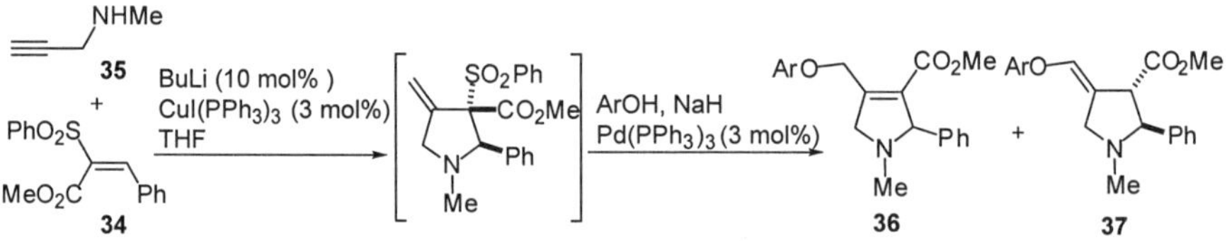

α-Xanthylketones of the general formula **38** have been demonstrated to participate in a reaction with the enesulfonamide **39** under radical conditions to give the intermediate imines **40**, which underwent spontaneous cyclization into the pyrroles **41** in moderate yields <02CC2214>.

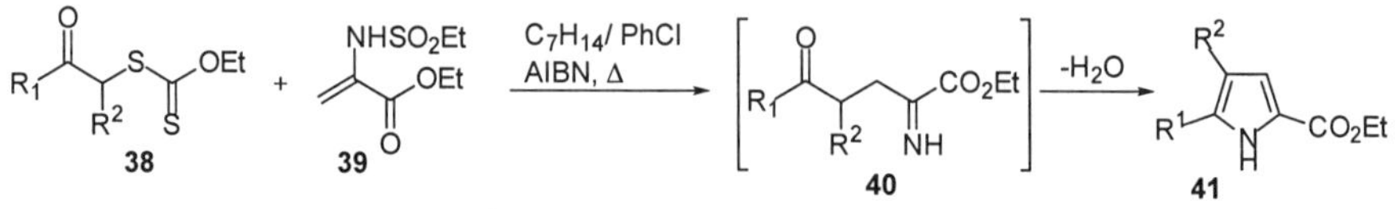

Chiral 5-methylenepyrrol-2-ones **42** were obtained as the major products from a sequence starting from chiral amines and the α,β-unsaturated ketone **43**, relying on photooxygenation of the intermediate pyrroles **44** as the key step <02TA601>.

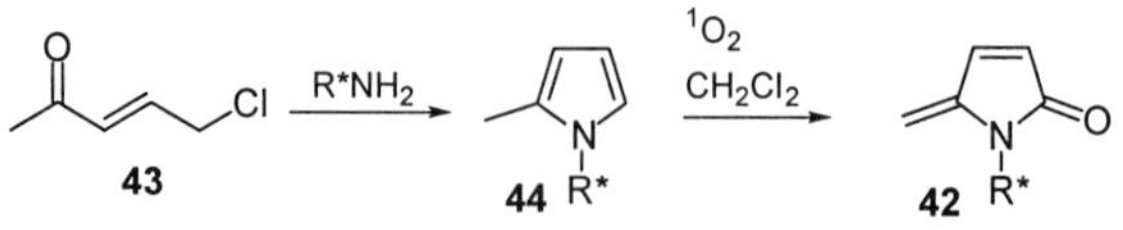

A series of fused pyrrole derivatives, such as **45**, has been synthesized in moderate to good yields by means of [4+1] cycloaddition of carbenes, generated *in situ* from the precursors **46**, and isocyanates, also formed *in situ* from the corresponding acyl azides <02OL4289>.

Several approaches to 3-fluoro-1*H*-pyrrole have been investigated, leading to its isolation and characterization, whereupon it was concluded that this material is difficult to isolate in pure form, and decomposes readily. Nevertheless, it could be utilized for the synthesis of fluorinated porphyrins <02T6713>. Other developments in pyrrole ring synthesis include the preparation of 2,4-diarylpyrroles from phenacyl azides in the presence of 2.5 equiv. of samarium iodide <02TL1863>; synthesis of 2,3,5-substituted pyrroles by acid-induced rearrangement of 2-amino-2,3-dihydrofuran derivatives <02TL4491>; synthesis of 3-(4,5-dihydroisooxazol-5-yl)pyrroles via 1,3-dipolar cycloadditions of hydroxylamines to 1-phenylsulfonyl-1,3-butadienes, followed by annulation under Barton-Zard conditions with ethyl isocyanoacetate <02TL53>; formation of a pyrrolo[2,3-*c*]quinoline from the reaction of 3-methyleneoxindole derivative with TosMIC <02T9179>; solution- and solid-phase approaches, including traceless linking to the solid support via a dioxolane, to various piperazinylmethyl substituted pyrroles <02BMCL1937>; generation of new pyrrole-containing dinuclear catalysts for olefin copolymerization via vanadium-induced aldol reaction and pinacolic coupling of 2-acetylpyrrole <02OM4390>; cyclization of 2-(2-bromoallyl)-1,3-dicarbonyl compounds to 1,2,3,5-tetrasubstituted pyrroles <02T9793>; and synthesis of methylthiomaleimides, e.g. by reaction of ketene dithioacetals with nitromethane <02JHC571>. The formation of various pyrroles has also been observed on electrocyclization of 1-azapentadienyl- or pentatrienyl cations during a detailed mechanistic and computational investigation <02EJO1523>.

5.2.3 REACTIONS OF PYRROLES

1-Methylpyrroles have been regioselectively brominated at C-3 using NBS in THF in the presence of catalytic amounts of PBr_3, while in the absence of the catalyst, bromination at C-2 was observed, providing facile routes to both isomers <02SL1152>. Anodic fluorination of 2-cyano-1-methylpyrrole has been studied, leading to various mixtures of mono-, *gem*-di-, or trifluorinated derivatives, depending on the composition of the supporting electrolyte <02S2597>.

3,4-Diiodopyrroles having *N*-blocking groups undergo Heck-reactions with amino acid derived alkenes to provide pyrroles possessing two amino acid moieties <02TL3401>. Addition of pyrroles to alkynoates catalyzed by 5% $Pd(OAc)_2$ has been used to prepare various 3-alkenylpyrroles <02CL20>. Electron deficient pyrroles have been subjected to copper(II) mediated *N*-arylation with arylboronic acids <02JOC1699>. It has been shown that 1-arylpyrroles can be regioselectively phosphorylated at C-2 with PBr_3 in the presence of pyridine. Under certain conditions, a migration of the dibromophosphino group to C-3 was observed due to the presence of pyridinium hydrobromide in the reaction medium, and the electronic nature of the *N*-aryl group <02HC223>.

The reaction of the silyloxypyrroles **47** possessing a chiral substituent at the nitrogen atom, with cyclobutanone in the presence of a Lewis acid, followed by an acid induced ring expansion of the cyclobutanol intermediate **48**, offers an asymmetric route to the 1-azaspiro[4.4]nonanes **49** in good diastereoisomeric excess <02SL1629>. In this context it might also be interesting to note that 1-*tert*-butoxycarbonyl-2-*tert*-butyldimethylsilyloxypyrrole has been reported to undergo addition to activated quinones without any catalyst <02TL4777>.

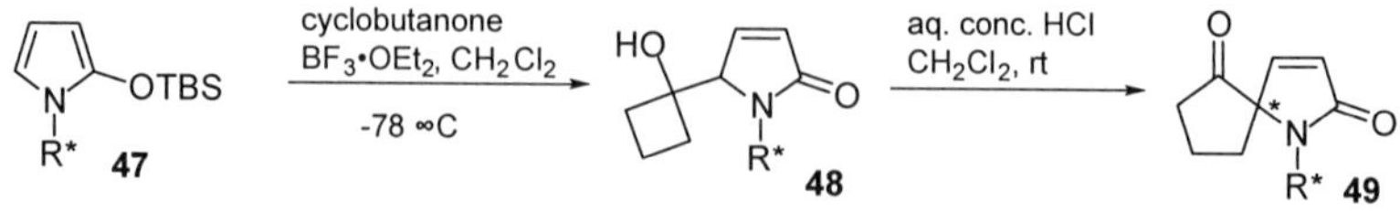

An intramolecular Michael addition has been used to effect transformation of the pyrrole **50** to the indolizidine **51**, the ester group of which was later homologated by a one-carbon unit, and the acid so obtained underwent in turn an intramolecular acylation at the pyrrole C-3 giving the tricyclic structure **52**, which eventually lead to a formal total synthesis of (±)-aspidospermidine after a few additional steps <02JCS(P1)2613>.

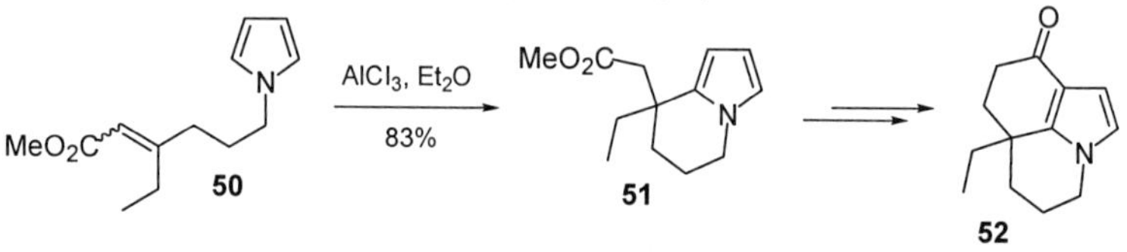

The tetracyclic systems **53** incorporating both a pyrrole and an isoindolone unit have been easily obtained utilizing a condensation of 1-(2-aminoethyl)- (**54**) or 1-(3-aminopropyl)-pyrrole (**55**) with various benzoic acid derivatives in refluxing benzene or toluene, respectively, presumably via iminum ion intermediates such as **56** <02TL2831>.

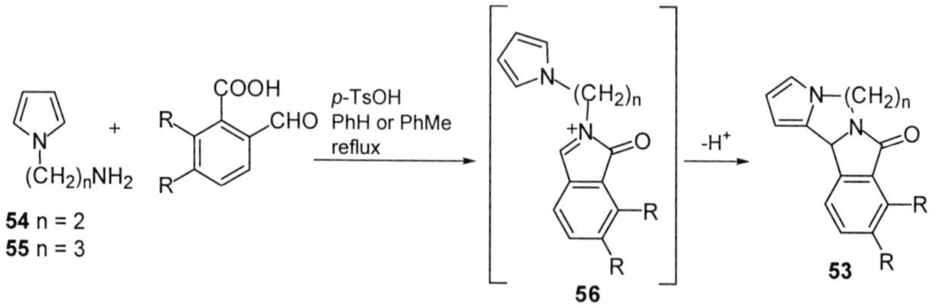

1-(2-Aminoethyl)pyrrole (**54**) has also been subjected to treatment with benzotriazole (BtH) and formaldehyde, to generate the pyrrolo[1,2-*a*]pyrazine derivative **57**, which underwent further reactions with nucleophiles (RMgX, NaCN) to provide the corresponding *N*-alkyl derivatives **58** after displacement of the benzotriazole moiety <02JOC8220>.

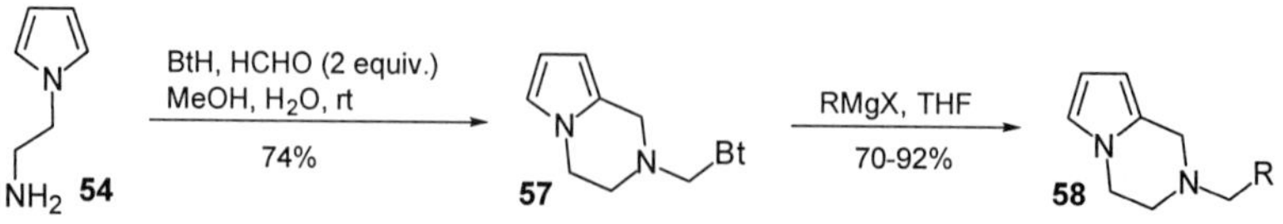

Intramolecular radical cyclizations of pyrroles such as **59** gave predominantly the pyrrolo[3,2-c]quinolines **60** via 6-*endo* cyclization, whereas systems having an electron withdrawing *N*-protecting group at the pyrrole gave instead spirooxindoles as the prevailing products, originating from a 5-*exo* process <02T1453>.

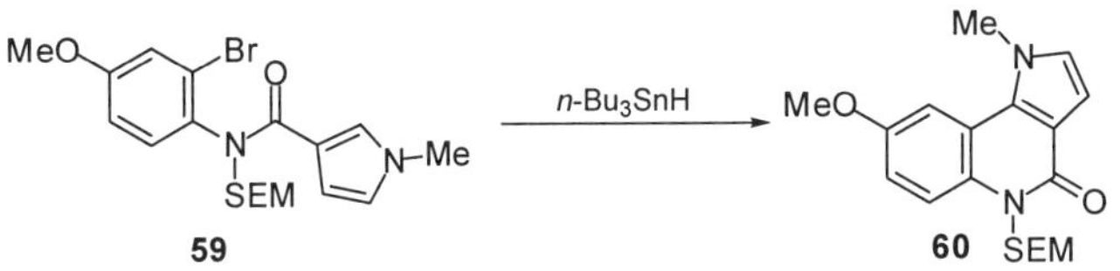

1-Methylpyrrole has been reported to undergo a [4+2] cycloaddition with 4,5-dicyanopyridazine, followed by loss of nitrogen and dehydrogenation to give a low yield (17%) of 1-methyl-5,6-dicyanoindole after chromatographic separation of the resulting complex mixture <02T8067>. The interesting furo[3,4-*b*]pyrroles **61** have been generated *in situ* over several steps from 1-phenylsulfonylpyrrole, and were found to participate readily in Diels-Alder reactions with maleic anhydride (X = O), or maleimides (X = NR) to provide the indoles **62** <02TL197>.

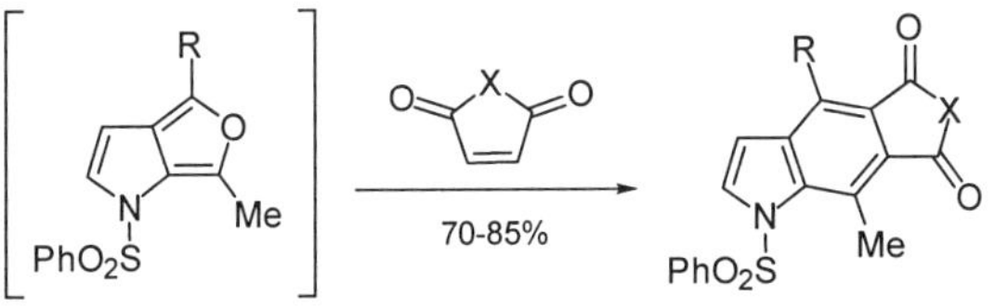

1-Methylpyrrole undergoes an interesting reaction with S_2Cl_2 in the presence of DABCO producing the pentathiepin **63**, the formation of which was also observed upon treatment of 2,5-dichloro-1-methylpyrrole (**64**) with the same combination of reagents. Related pyrrolo-fused pentathiepins were also prepared by the action of S_2Cl_2 solely or in combination with DABCO on 1-alkylpyrrolidines <02CC1204>. A study focussing on the synthesis and reactions of [1,2]dithiolo[3,4-*b*]pyrroles has also been published <02HCA4453>.

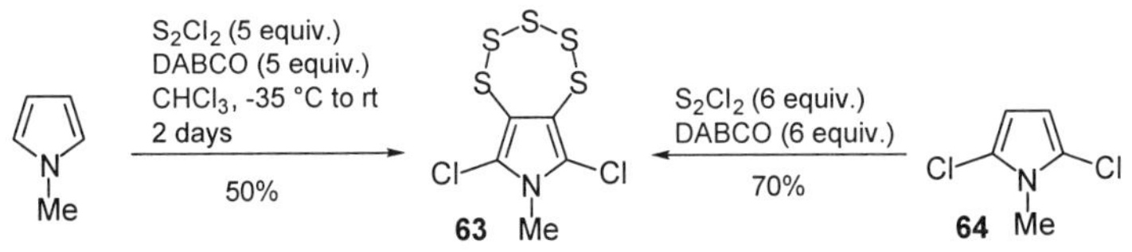

η^2-Pyrrole osmium complexes derived from amino acids have been found to display stereoselective coordination, which was also shown to lead to stereoselective electrophilic addition at the pyrrole C-3 as a consequence of the differentiation of the pyrrole enantiofaces <02OM4581>. In addition, 2-styrylpyrroles have been the subject of a study investigating their photoelectron spectra and conformational properties <02EJO551>, while 2-substituted pyrroles also having various *tert*-butyl containing bulky substituents at C-4 flanked by methyl groups at both C-3 and C-5 display atropisomerism due to restricted rotation <02TA1721>. Nucleophilic 1,2-addition of, for example Grignard reagents, to 1-acylpyrroles has been demonstrated to generate remarkably stable carbinols, which were used as intermediates for further synthetic manipulations <02AGE3188>.

Other interesting new developments in pyrrole chemistry encompass the synthesis of calix[6]pyrrole and calix[n]furan[m]pyrroles, where $m + n = 6$ <02CEJ3148>, and the formation of cyclo[8]pyrroles from 2,2′-bipyrrole fragments <02AGE1422>.

5.2.4 SYNTHESIS OF INDOLES

Syntheses of 2-substituted indoles from (2-aminobenzyl)triphenylphosphonium salts have been achieved either by cyclization with carboxylic acid anhydrides in the presence of triethylamine, or acyl chlorides in 2,6-lutidine <02TL2885>. In a related study, the acylated indoles **65** have been prepared from the phosphonium salts **66** in a process involving extrusion of a one-carbon fragment. A mechanistic rationale for these findings was also provided by the authors <02TL8893>.

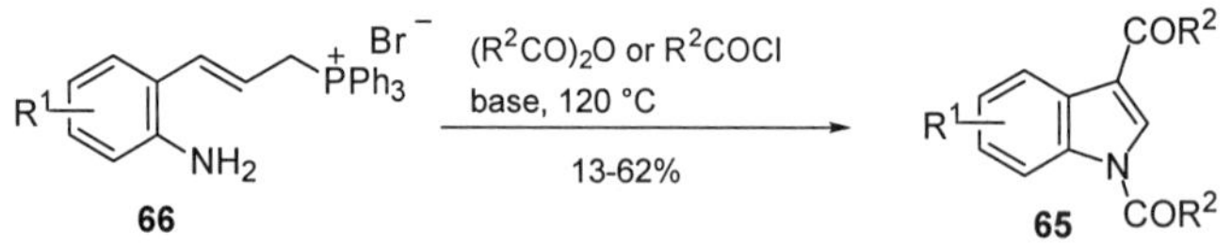

A number of 2-substituted 3-cyanoindoles were synthesized on solid phase using a modified Madelung procedure <02TL5189>. A polymer supported fluoride has proved efficient at effecting the cyclization of alkynylanilines to indoles, and was also shown to tolerate various functional groups <02TL6579>. Likewise, cyclization of alkynylanilines or alkynylanilides have also been employed in the synthesis of indoles possessing nitro or amino groups at any of the four possible positions in the six-membered ring <02TL7699>. An improved sequence for the synthesis of 7-alkyl indoles, involving generation of 2-lithionitrobenzenes from 2-bromonitrobenzenes followed by a Bartoli reaction with vinylmagnesium bromide has been developed <02SL143>. In an interesting application, densely functionalized arylmagnesium halides, e.g. **67**, have, after transmetallation, been treated with suitable allylic bromides to give intermediates which underwent a second halogen-metal exchange followed by reaction with electrophiles, and were finally cyclized to the indoles **68** under acidic conditions <02OL1819>.

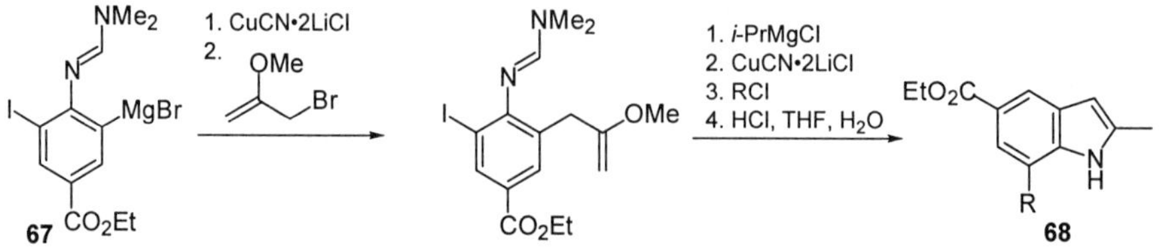

It has been demonstrated that 2,4,6-trinitrotoluene (TNT) can serve as a useful precursor for the synthesis of indoles, as well as other benzo-fused heterocyclic systems. Thus e.g. the

Five-Membered Ring Systems: Pyrroles and Benzo Derivatives 149

sulfone **69**, which can be readily obtained from TNT, underwent condensation with aldehydes at the activated methyl group, followed by displacement of the *o*-nitro group with azide, and a final thermal cyclization to the indoles **70** <02SC1465>.

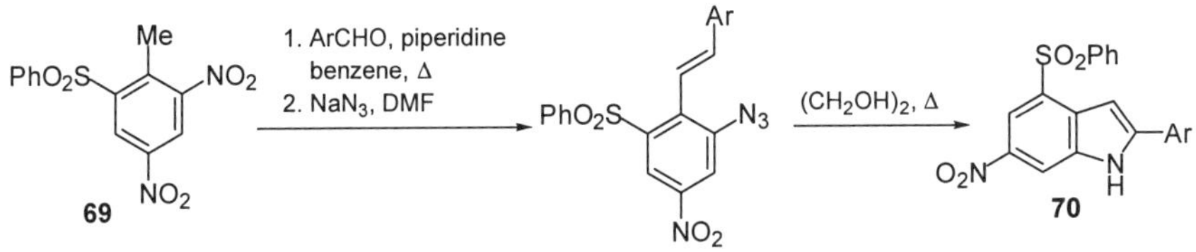

By utilizing a 5-*endo-trig* cyclization process, several β,β-difluorostyrenes (**71**) were transformed to the 2-fluoroindoles **72**. Modifications of this method also proved to be useful for the synthesis of numerous other fluorinated heterocycles <02S1917>. A series of 3-arylamino-2-chloroindoles have been prepared by thermal cyclization of 1,2-diarylamino-1,2-dichloroethenes, an event which takes place via intermediate arylaminochlorocarbenes <02S2426>.

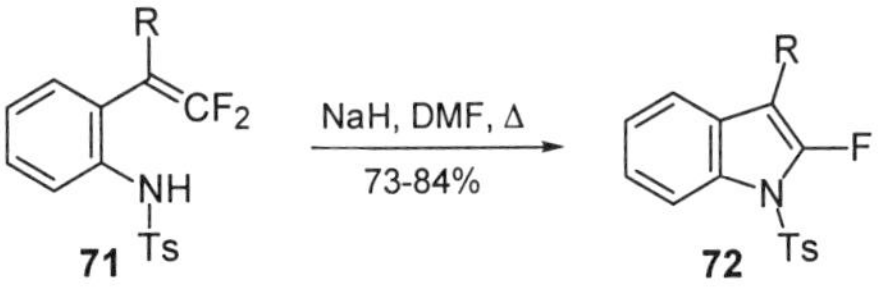

Thermally induced cyclization of the enehydrazine **73**, obtained from the sequential treatment of the corresponding arylhydrazine with acetone and TFAA, has been shown to produce the 2,4-disubstituted indole **74** along with the isomeric 2,6-disubstituted system **75**. The yields of the products were dependent on the reaction medium however, toluene or chlorobenzene were found to be the solvents of choice, giving **74** as the prevailing product <02H(57)1101>.

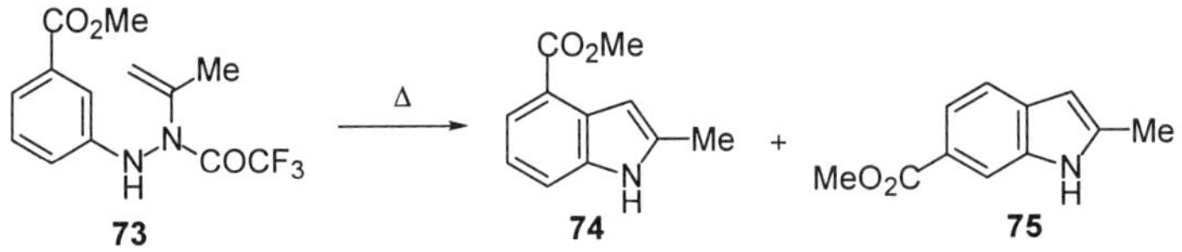

N-Alkyl-(*p*-methoxyphenyl)arylamines, such as **76**, undergo photo-induced cyclization in the presence of protic acids to produce the carbazolones **77**. The process is believed to comprise an initial cyclization, followed by two consecutive proton shifts and a final hydrolysis of the methoxy functionality <02CC270>.

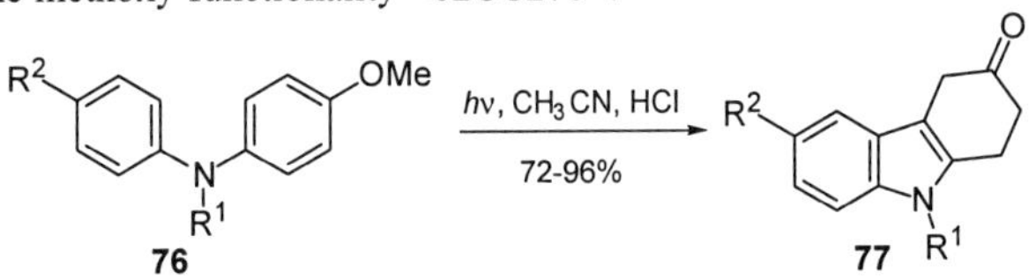

In an interesting new application, benzynes have been generated from the 2-fluoroalkylanines **78**, and underwent anionic cyclization, followed by reaction with electrophiles to furnish a variety of 4-substituted indoles **79** <02CEJ2034>.

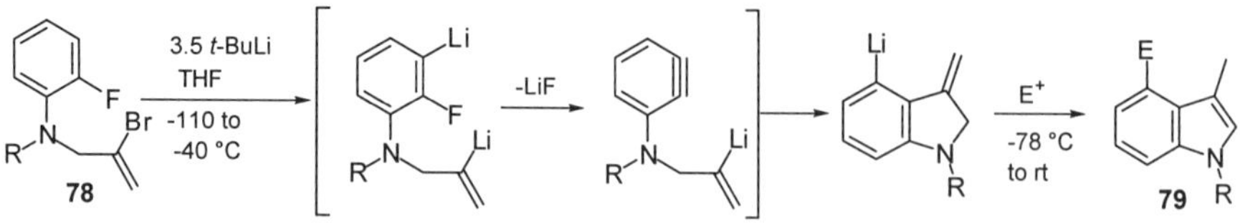

Transition metal catalyzed processes are useful and versatile tools for the preparation of nitrogen heterocycles, indoles being one of the most popular targets. Thus, for example, the [Cp*Ru(CO)$_2$]$_2$ catalyzed reaction of nitrosoaromatics with alkenes in the presence of carbon monoxide has been used to synthesize indoles. It was also demonstrated that nitrosobenzenes undergo a thermally induced reaction with alkynes to form 1-hydroxyindoles <02OL699>. Oxidative cyclization of amino alcohols **80** giving the indoles **81** has been achieved utilizing a process catalyzed by [Cp*IrCl$_2$]$_2$ <02OL2691>.

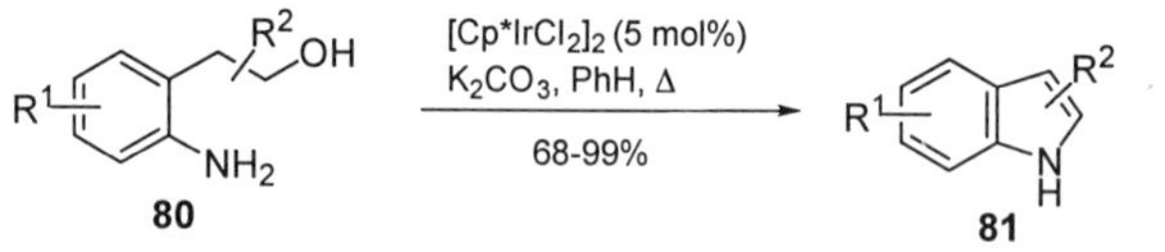

A series of indoles **82** has also been prepared using a sequence employing an N-H insertion of carbenoids generated from the α-diazophosphonates **83** in the presence of catalytic amounts of rhodium(II) into the 2-aminobenzophenones **84**, followed by cyclization of the intermediates **85** with DBU <02OL2317>.

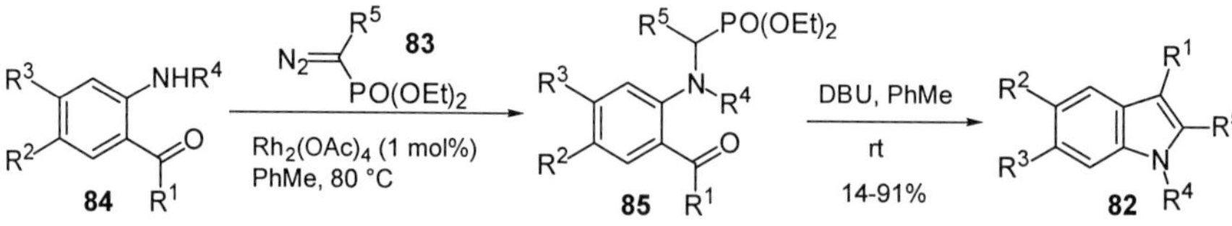

Using a three-component palladium catalyzed reaction involving the isocyanides **86**, allyl carbonates, and trimethylsilyl azide, 1-cyanoindoles **87** were prepared in moderate yields <02JA11940>.

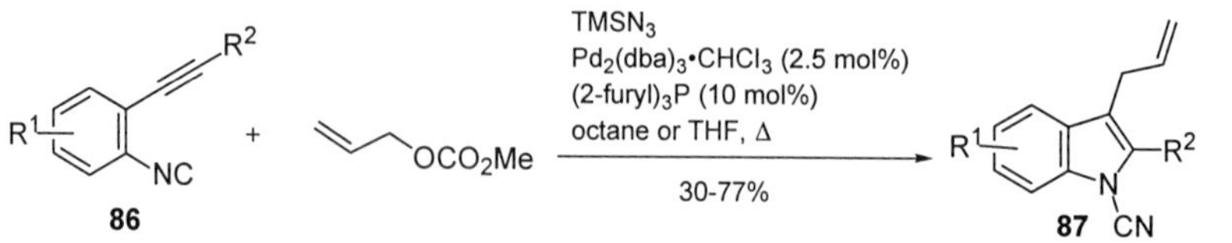

A stereoselective method for the synthesis of 2-substituted indolines **88** has been developed based on a sequence involving palladium-catalyzed coupling of zinc reagents generated from the Boc-protected secondary amines **89**, followed by an intramolecular amination of the intermediates **90** <02JCS(P1)733>.

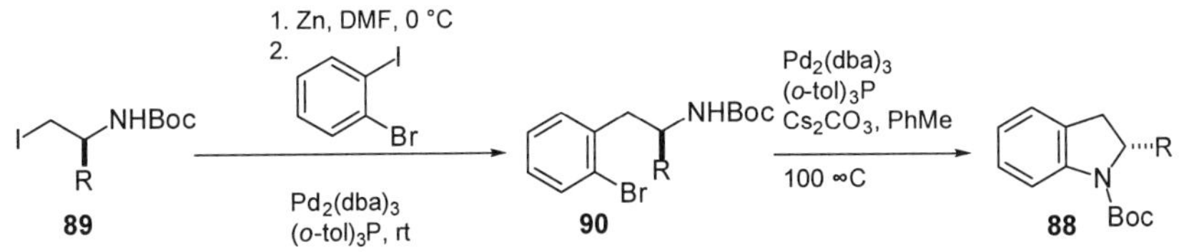

The fused indoles **91** (n = 1–3) have been synthesized by Stille coupling of the stannylbenzene **92** with 2-iodo-2-enones **93**, followed by intramolecular palladium-catalyzed reductive annulation of the intermediates **94**. The precursors **94** could be prepared in an alternative manner by coupling of aryl halides with 2-stannyl-2-enones <02TL1621>.

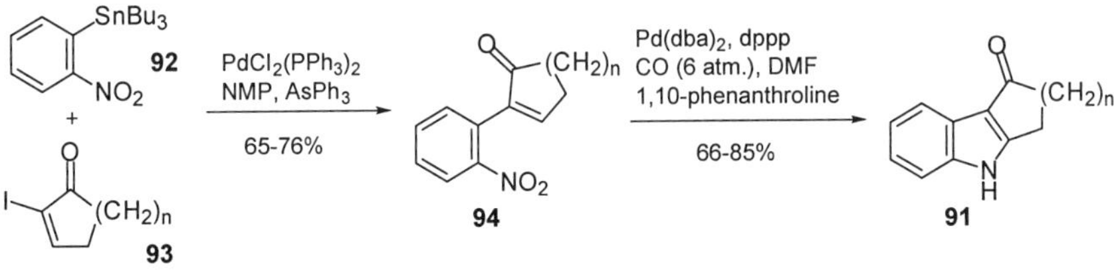

Other transition metal mediated processes leading to indoles include among others [3+2] cycloaddition of electron-rich alkenes to azomethine ylides generated from N-(o-alkynylphenyl)imines and tungsten carbonyl complexes <02JA11592>; reaction of allyl carbonates and 2-(alkynyl)phenylisocyanates mediated by a Pd(0)-Cu(I) bimetallic catalyst <02AG3230>; asymmetric carbopalladation of internal allenes derived from o-iodoaniline followed by intramolecular amination <02TA1351>; formation of 2-aryl-3-(diethylaminomethyl)indoles from a palladium catalyzed three-component reaction of aryl iodides, o-alkenylphenylisocyanide and diethylamine <02TL6197>; synthesis of the carbazole alkaloid carazostatin and its O-methyl derivative by an iron-mediated oxidative cyclization <02T8937>, preparation of 2-aryl- or 2-heteroarylindoles by a palladium catalyzed process on solid support starting from an immobilized α-diazophosphonoacetate <02CC210>; and synthesis of, in particular, 3-phenylindoles via a reductive cyclization of nitro-aromatics with alkynes catalyzed by [Cp*Ru(CO)$_2$]$_2$ in the presence of carbon monoxide <02CC484>. Various carbazoles have also been prepared using a one-pot procedure involving a palladium-catalyzed amination of aryl bromides with 2-chloroanilines, followed by an intramolecular annulation via C-H bond activation <02CC2310>.

The indolines **95** or oxindoles **96** have been reported as products originating from the base- or fluoride-mediated exo-*trig* cyclization of the intermediates **97**, derived from the precursors **98** by treatment with phenyliodine(III) diacetate (PIDA) <02JOC3425>.

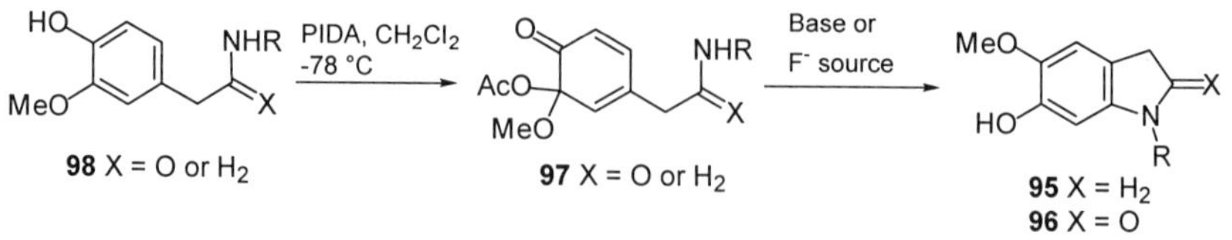

Intramolecular nucleophilic substitution of hydrogen performed on a series of anilides **99** (X = H, Cl) has provided a route to various 4-nitrooxindoles **100**. In some cases, the co-formation of minor amounts of the 6-nitro isomers was observed <02S2203>.

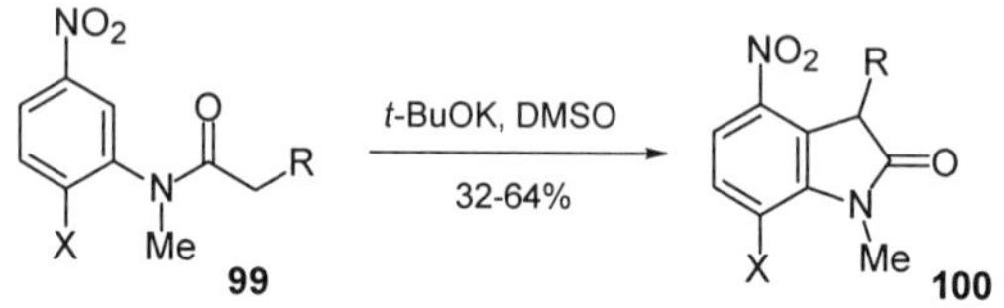

Spirooxindoles, such as, for example (±)-coerulescine (**101**) have been prepared employing a sequence starting from 2-fluoronitrobenzene, which was initially subjected to treatment with the anion of dimethyl malonate, followed by decarboxylation and concomitant installation of the methylene group using formaldehyde in the presence of potassium carbonate to produce the intermediate **102** in good yield. This material readily underwent dipolar cycloaddition with the azomethine ylide generated from sarcosine and formaldehyde, followed by a reductive cyclization of adduct **103** to furnish the natural product **101** <02TL9175>.

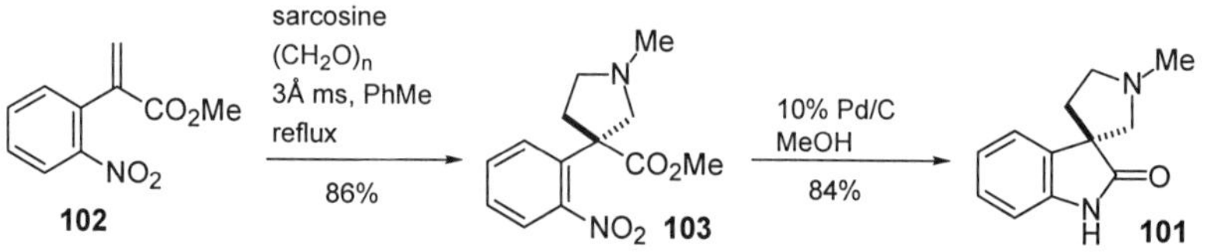

N-ω-Chloroalkylisatins have been prepared using a mild procedure employing the reaction of, for example, 1-arylpyrrolidines with oxalyl chloride and DABCO <02S34>. An efficient and regioselective synthesis of 6-iodo-4-trifluoromethylisatin, an intermeditate for the synthesis of a growth hormone secretagogue, has been achieved based on the initial preparation of 6-iodo-4-trifluoromethylindole or the corresponding oxindole, followed by transformation of these intermediates into the corresponding *gem*-3,3-dibromooxindole, and hydrolysis thereof to the desired isatin <02T3605>.

Various other synthetic approaches to indoles include the preparation of [1,4]oxazino-[2,3-*f*]indoles or -[2,3-*g*]indoles in an application of the Hemetsberger reaction <02EJO1646>; synthesis of pyrrolo[2,3-*b*]indoles by a modified Grandberg indole synthesis <02H(58)587>; a new strategy for the synthesis of 1,2-dihydro-3*H*-benzo[*e*]indoles utilizing a free radical cyclization <02JOC8958>; oxidative free radical addition of 1,3-dicarbonyl compounds to 2-amino-1,4-naphthoquinones mediated by cerium(IV) to give benzo[*f*]indole derivatives <02T7625>; preparation of indole-2-carboxaldehydes by ring transformation of quinolines via formation of intermediate 1-methoxycarbonyl-1,2-dihydroquinoline-2-phosphonates and subsequent ozonolysis and ring closure under basic conditions <02TL5295>; synthesis of spiro-derivatives of cyclopent[g]indole using the Madelung reaction <02TL4707>; and development of a palladium catalyzed process for the synthesis of indole-3-carboxylates on solid support <02JCC191>. A sequential Diels-Alder process between certain 2-amidofurans and an aminodiene for the synthesis of indoles fused to a 6-membered carbocyclic ring at C-3 and C-4 has also been developed <02OL4135>. 1,1′-Bis(2-indolylcarbonyl)ferrocene has been obtained via nitrene cyclization of a bis(o-azidostyryl) precursor <02OM2055>. A related series of 2-ferrocenylindoles **104** have been prepared by annulation of the acetylenic precursors **105** with an excess of Bu$_4$NF in THF at reflux, or 10% Pd(OAc)$_2$ in the presence of Bu$_4$NBr <02T4487>. Several interesting (η^3-indolylmethyl)palladium complexes with either phosphine or 2,2′-bipyridine ligands, having the general structure **106** have been obtained by reaction of methylpalladium complexes with o-alkenylphenyl isocyanides, and were found to undergo nucleophilic attack with amines to produce 3-(aminomethyl)indole derivatives, whereas treatment with HCl gave the corresponding indoles <02OM581>.

5.2.5 REACTIONS OF INDOLES

Reactions of metallated indoles offer some of the most valuable methods to elaborate the indole nucleus both in the pyrrole ring and the carbocyclic portion. An elegant route to 2,3-disubstituted indoles via selective metallation of 2,3-dibromo-1-methylindole (**107**) has been developed by Gribble and co-workers. Thus, the readily available 2-bromoindole (**108**) was brominated and N-methylated, followed by metallation at C-2, to give 3-bromo-2-lithio-1-methylindole, which was quenched with suitable electrophiles, and further subjected to a second halogen-metal exchange at C-3, finally leading to various 2,3-disubstituted 1-methylindoles such as **109** <02TL7135>. Moreover, bromination of **107** at C-6 with bromine in chloroform, or at both C-5 and C-6 with bromine in a mixture of formic acid and chloroform provided easy access to naturally occurring polybrominated indoles of marine origin <02JNP748>. Lithiation of 1-Boc- or 1-(phenylsulfonyl)indoles with t-BuLi, followed by reaction with dinitrogen tetraoxide at −120 °C offers a new attractive route to 2-nitroindoles <02TL4115>.

Furo[3,4-*b*]indoles (**110**) have been prepared employing an efficient route starting from the readily available acetal **111**, which was subjected to metallation at C-2, quenching with aldehydes to provide the alcohols **112**, and cyclization under acidic conditions in the presence of hydroquinone <02JOC1001>. Bis(2-indolyl)methanones obtained by reaction of 2-lithiated 1-phenylsulfonylindoles with, for example, suitable indole-2-carboxylate derivatives have been investigated as new inhibitors of platelet-derived growth factor receptor kinase <02JMC1002>. Cryogenic conditions are often required for the lithiation of indoles at C-2, however, metallation of 1-Boc-indoles with LDA at 0–5 °C followed by introduction of triisopropylborate and a final acidic hydrolysis has been demonstrated to give good yields of 2-indolylborates, useful partners for cross-coupling reactions <02JOC7551>. Metallation of 1-methoxymethylindole with BuLi, followed by generation of 1-methoxymethylindol-2-ylborates and subsequent reactions thereof with aromatic aldehydes and electrophiles, has been applied to the synthesis of some 1,2,3-trisubstituted indoles <02CC220>.

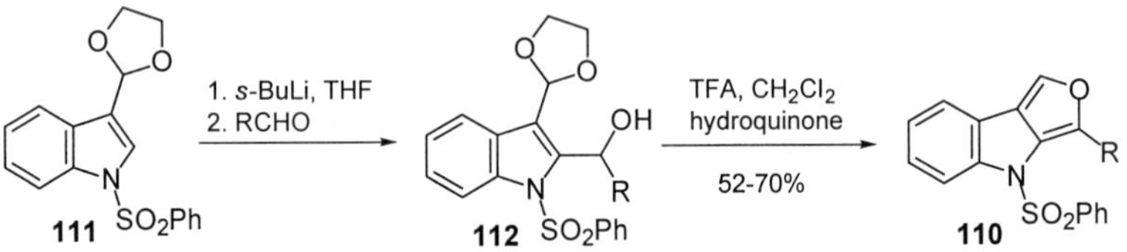

Lithiation at C-3 of the silyl protected indole **113**, followed by addition to various *N*-tosylated arylaldimines has been shown to lead to the aminomethylindoles **114**, which could be further selectively deprotected at either of the two nitrogen atoms <02JOC5850>.

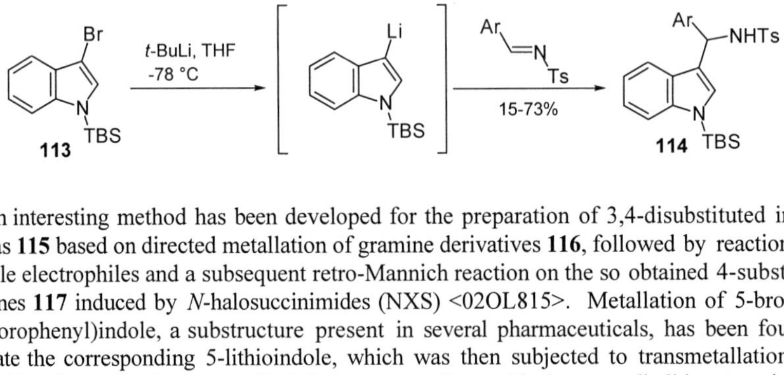

An interesting method has been developed for the preparation of 3,4-disubstituted indoles such as **115** based on directed metallation of gramine derivatives **116**, followed by reaction with suitable electrophiles and a subsequent retro-Mannich reaction on the so obtained 4-substituted gramines **117** induced by *N*-halosuccinimides (NXS) <02OL815>. Metallation of 5-bromo-1-(4-fluorophenyl)indole, a substructure present in several pharmaceuticals, has been found to generate the corresponding 5-lithioindole, which was then subjected to transmetallation with zinc chloride, and subsequent Negishi cross-coupling with heteroarylhalides to give the corresponding 5-heteroarylindoles in gram scale in a one-pot procedure <02S1509>.

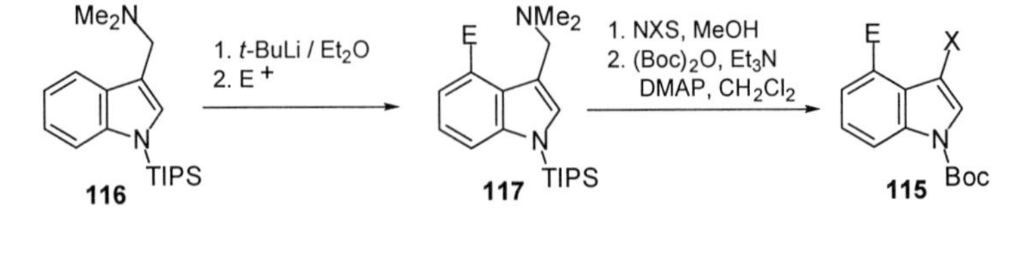

Although many efficient methods for the acylation or alkylation of indoles are known, new procedures are continually developed. 1-(Phenylsulfonyl)indoles can be acylated efficiently at C-2 in a simple procedure by lithiation with sec-BuLi, followed by quenching of the resulting lithioindoles with carboxylic acid anhydrides <02SC2035>. New synthetic routes to 3-acetylindoles encompass the use of acetic anhydride in the presence of triphenyl phosphonium perchlorate <02SC105>, indium trichloride, or indium triflate <02T1229> as catalysts. The dianion of 3-acetylindole has been generated by sequential treatment with sodium hydride and BuLi, and was alkylated at the acetyl group to furnish various extended 3-acylindoles <02H(57)1293>. A method for the enantioselective alkylation of indoles using chiral amine catalysis under acidic conditions has also been developed <02JA1172>. Yet another study on stereoselective C-3 functionalization focussed on the ring-opening of optically active styrene oxides with indoles in the presence of catalytic amounts of Lewis acids, demonstrated that the catalyst of choice in this particular case is $InBr_3$, which proved to give advantageous results in terms of ee <02JOC5386>. Aminoalkylindoles related to tryptophan have been obtained from the ytterbium triflate mediated reaction of indoles with an imine derived from ethyl glyoxylate and benzylamine <02TL4271>. Zinc triflate has been used to effect regioselective alkylation of indoles at C-3 with alkyl halides in the presence of Hünig's base and tetrabutylammonium iodide <02JOC2705>.

The mechanism of the second sulfenylation of 3-indolylsulfides yielding 2,3-indolyl bis-sulfides has been investigated in detail, and was demonstrated to take place via an intermediate indolenium 3,3-bis-sulfide, which undergoes a subsequent rearrangement via migration of one of the sulfide groups to C-2 <02JOC2854>.

Indoles and carbazoles have been N-alkylated with alcohols under Mitsunobu reaction conditions using e.g. the reagent combination $DEAD/PPh_3$, to provide a variety of 1-alkylindoles or 9-alkylcarbazoles in moderate to good yields <02TL2187>. N-Arylation of indoles has been accomplished with aryl bromides in the presence of catalysts derived from CuI and diamines such as trans-N,N'-dimethyl-1,2-cyclohexanediamine <02JA11684>, or using a copper catalyzed reaction with diaryliodonium salts <02SC903>. Bisindoles which are precursors to pyrroloindolocarbazole natural products have been successfully subjected to regio- and enantioselective allylic alkylation at one of the indole nitrogen atoms with cyclopentenyl or pyranosyl groups in a palladium catalyzed process <02OL2005>.

A protocol for the synthesis of tryptophans **118** based on the regioselective ring opening of the aziridines **119** with indoles employing scandium perchlorate as the Lewis acid has been reported, wherein it was concluded that scandium perchlorate is the Lewis acid that gives the best results in terms of regioselectivity in this particular application <02S1658>. During studies on the functionalization of the aromatic portion of tryptophan derivatives employing Friedel-Crafts methodology, it was found that an acyl group can be introduced at the indole C-6 in 1-acetyl- or 1-tosyltryptophan, while the acylation of 1'-oxotryptophan produces a C-5 substituted product <02TL7013>. Tryptophan regioisomers having the alanine moiety attached to all possible carbon atoms of indole except C-3 have been prepared using known methodology from the corresponding carboxaldehydes, followed by asymmetric reduction of the intermediate dehydrotryptophan isomers, and based on *ab initio* calculations were found to display cation binding properties similar to those of tryptophan itself <02JOC6256>.

Michael addition of electron-rich indoles to the phosphoryl acrylate **120**, followed by ion-exchange, was shown to give the interesting phosphorylated indole derivatives **121** in good overall yields. On the other hand, addition at nitrogen was observed when electron deficient indoles were used as the substrates. The drawback of this route appears to be the long reaction times, typically one week <02S1351>.

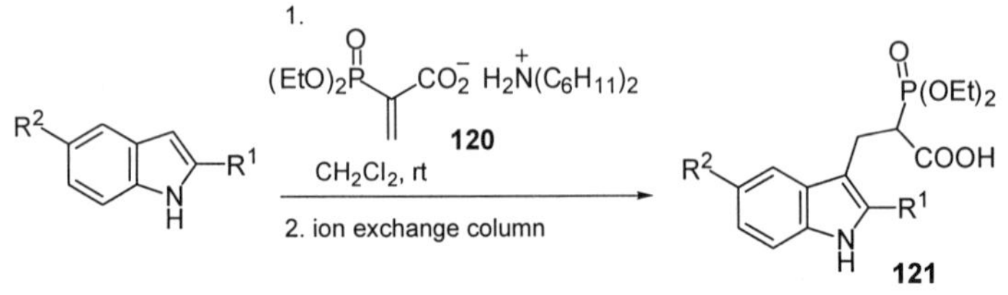

Conjugate addition of indoles to 2,5-dichlorobenzoquinone under anaerobic conditions has been investigated, leading to the conclusion that the process is efficient for a wide variety of indoles when promoted with HCl, H$_2$SO$_4$, or AcOH, followed by oxidation with DDQ. A subsequent alkaline hydrolysis step performed on the intermediate products **122** provided the corresponding 2,5-dihydroxyquinones **123**, a unit present in several naturally occurring bisindolylquinones <02JOC8374>.

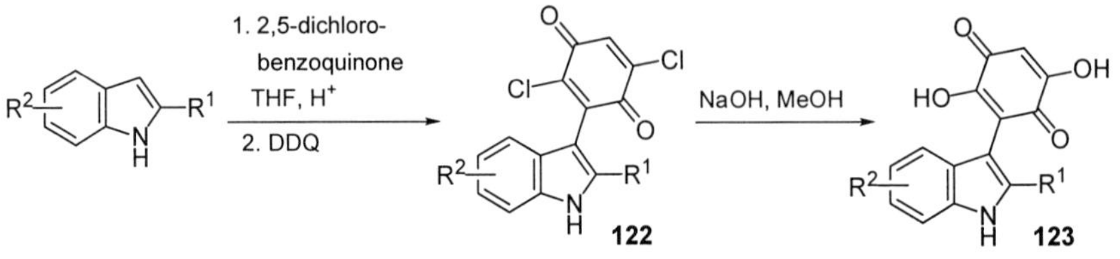

Isatins **124** have been demonstrated to undergo transformation to 3-methyleneoxindoles **125** in moderate yields employing a two-step procedure involving olefination to the β-silyl alcohols **126**, followed by elimination induced by boron trifluoride diethyl etherate <02TL4671>. Photoinduced [2+2] cycloadditions of 1-acetylisatin with e.g. furan or benzo[*b*]furan have been studied, and were found to produce indole-containing spirooxetanes with high regio- and diastereoselectivity <02JCS(P1)345>. 3-Methyleneoxindoles, generated *in situ* via intramolecular Heck-reactions of suitably substituted 2-iodoanilines, have served as substrates for 1,3-dipolar cycloadditions with azomethine-ylides leading to spiro-oxindoles <02TL2605>. Spiro-oxindoles have also been obtained via dipolar cycloadditions of methyleneoxindole derivatives to carbonyl ylides <02T7221>; asymmetric azomethine ylides generated from a 5,6-diphenylmorpholin-2-one and aldehydes <02H(58)563>; or via hydrolysis, Curtius rearrangement and intramolecular cyclization of certain 1:1:1 adducts between oxindoles, aldehydes, and Meldrum's acid <02EJO3481>.

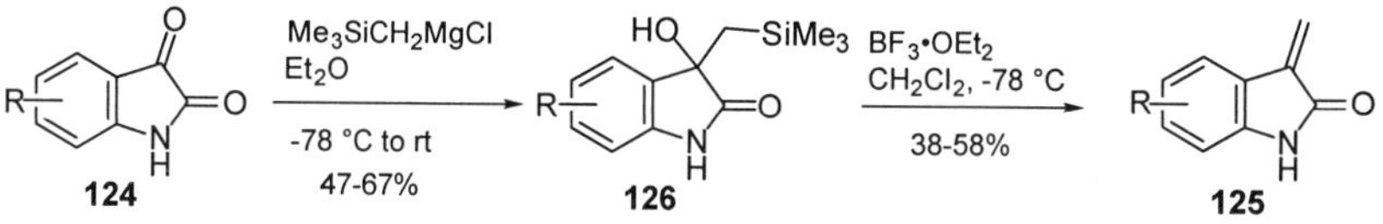

It has been shown that indolines **127** can be efficiently transformed into isatogens **128** by oxidation with *m*-CPBA in methanol. The required indolines were obtained from the corresponding indoles by reduction with tin and hydrochloric acid in ethanol <02T9187>. Various 1,3-dimethyloxindoles have been subjected to treatment with cerium(IV) ammonium nitrate, and gave the corresponding 3-alkoxy derivatives when alcohols were used as the solvents. On the other hand, when the reaction was performed in non-nucleophilic solvents, for example, THF or acetonitrile, 3-nitroxyoxindoles were produced <02T9541>. Nitration of indoline-2-carboxylic acid, followed by esterification and dehydrogenation with DDQ affords methyl 6-nitroindole-2-carboxylate, whereas a similar set of operations preformed on 1-acetylindoline-2-carboxylate produced the 5-nitro isomer <02S320>. Moreover, modified and efficient conditions for the alkylation of oxindoles at C-3 with alcohols and Raney nickel have been developed <02S595>. A mechanistic investigation of the hydrogenation of indoles to indolines with rhodium or ruthenium catalysis in the presence of protic acids has also been published <02OM1430>.

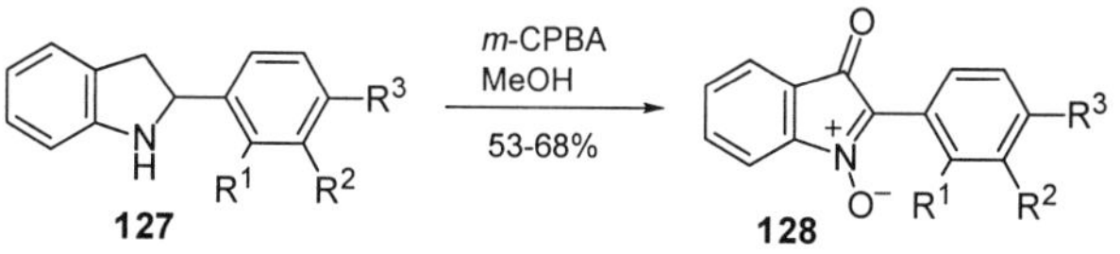

Oxidation of the protected 5-aminoindole **129** with Dess-Martin periodinane (DMP) produced the indole-quinone **130**. This methodology was employed in a new total synthesis of the *Actinomycetes* metabolite BE-10988 (**131**), which is a promising topoisomerase II inhibitor <02JA2221>.

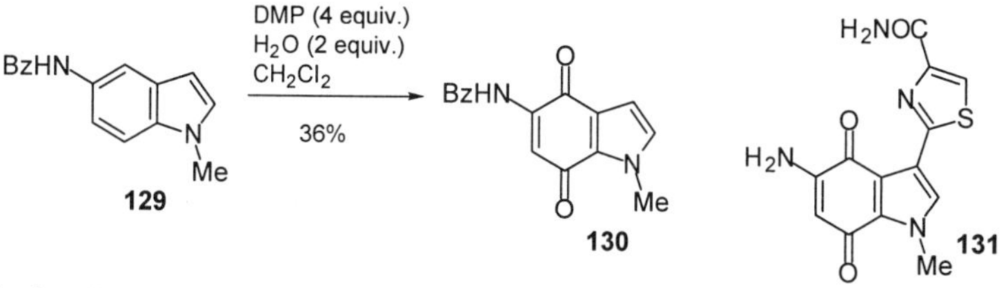

Both β- **132** and γ-carbolines have been prepared from 1-substituted 3-iodoindole-2-βcarboxaldehydes **133** or 2-bromoindole-3-carboxaldehydes respectively via a sequence involving Sonogashira coupling, followed by formation of the corresponding *tert*-butylimines, and a copper(I) catalyzed or thermal cyclization, respectively <02TL1359, 02JOC7048>. A similar intramolecular palladium catalyzed process applied on 2-bromoindole-3-carboxaldehydes possessing an alkyne containing moiety attached to the indole nitrogen atom provided tetra- or pentacyclic γ-carbolines <02OL3035>. *tert*-Butylimines generated from 1-substituted 3-iodoindole-2-carboxaldehydes or 2-bromo- or iodoindole-3-carboxaldehydes have also been annulated with acetylenes in a palladium catalyzed process leading to β- or γ-carbolines <02JOC9318>. 1,2,3,4-Tetrahydro-β-carbolines have been prepared in good yields and purity on polymer support using a microwave assisted parallel process <02SL1709>. Moreover, a strategy based on intramolecular Heck-reactions of various 3- or 2-iodoindole derivatives possessing an *N*-allylcarboxamide at either C-2 or C-3, respectively, has been employed for the synthesis of β- and γ-carbolinones <02T6673>.

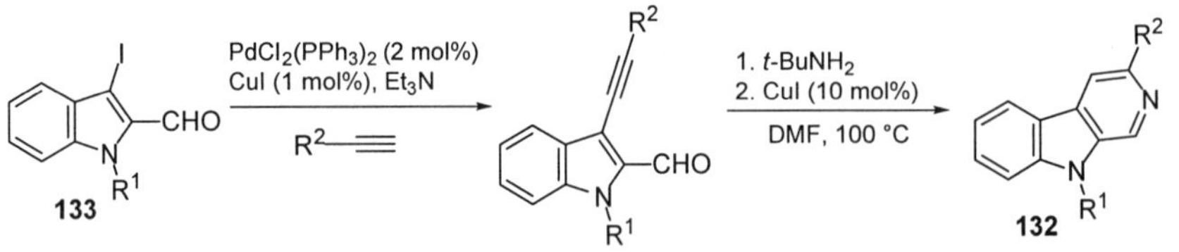

A series of functionalised 3-indolylbutadienes **134** has been subjected to cyclization induced by either HI, producing the expected carbazoles **135**, or iodine in the presence of EtSH, instead leading to the isomeric carbazoles **136** as the major products, an outcome which is the result of a migration of the tosyl group prior to annulation. This interesting result was attributed to the influence of hydrogen polyiodides which were formed from excess iodine and EtSH. A mechanism supported by a labelling experiment was also provided <02CL134>.

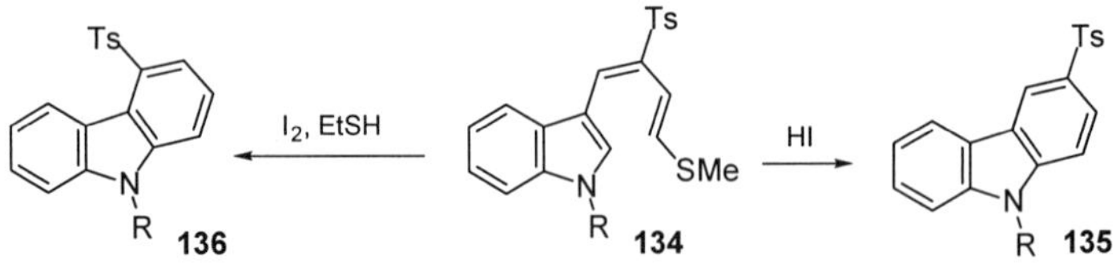

Indole-3-carboxaldehydes gave the corresponding cyanohydrin silyl ethers **137** upon heating with trimethylsilyl cyanide in acetonitrile or DME. After subsequent oxidation with DDQ, good yields of the carbonyl nitriles **138** were obtained, also providing a mild new route for the parent system **138** ($R^1 = R^2 = H$). Further elaboration provided the imidates **139**, which participated in a reaction with tryptophan esters to give moderate yields of the marine alkaloids rhopaladins A–D <02T2813>.

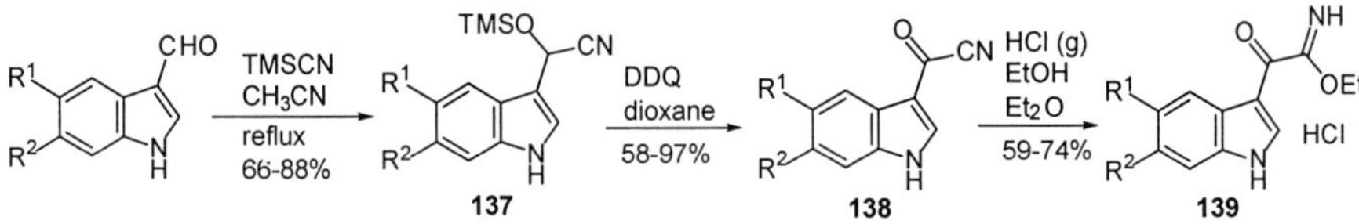

Using a modified procedure, indole-2-carboxaldehydes possessing an *N,N*-dimethylcarbamyl group at the nitrogen were converted to the corresponding 3-(oxazol-5-yl)indoles with TosMIC and DBU in hot DME <02BMCL3305>. A synthesis of a universal nucleoside derived from indole-3-carboxaldehyde has been developed, and the product was found to undergo efficient incorporation into oligonucleotides, which could in turn be further modified at the indole-3-carboxaldehyde moiety <02TL4581>. 1-Allylindole-2-carboxaldehydes have been transformed into 1-allyl-2-vinylindoles, and were demonstrated to undergo ring closing metathesis to give pyrrolo[1,2-*a*]indoles. The method was also extended to [1,2-*a*]-fused indoles containing larger rings <02TL4765>. Chiral imines derived from indole-2-carboxaldehydes and (S)-(-)-α-methylbenzylamine have been shown to participate in asymmetric aza-Diels-Alder reactions with Danishefsky's diene <02TL29>.

Based on a thermal [1,5] hydrogen shift in 3-cyanomethyl-2-vinylindoles possessing an electron-donating group at the indole nitrogen, indolo-2,3-quinodimethanes could be generated and gave substituted tetrahydrocarbazoles after reaction with suitable dienophiles, such as maleimides <02TL7925>. A new route to pyrrolo[3,4-*b*]indoles has been devised starting from 2,3-dimethyl-1-phenylsulfonylindole. Thus, radical bromination on both methyl groups, followed by cyclization by nucleophilic displacement with amines, and final dehydrogenation with DDQ provided pyrrolo[3,4-*b*]indoles in good yields <02SC2003>.

The interesting indolophane **140** has been prepared from 1,3-di(allyl)indole (**141**), and was demonstrated to undergo a transannular inverse electron demand Diels-Alder reaction leading to the formation of the pentacyclic system **142** <02OL127>. Other pentacyclic indole-containing systems have been obtained using tandem intramolecular Diels-Alder/1,3-dipolar cycloadditions of indoles possessing an 1,3,4-oxadiazole moiety <02JA11292>, or by intramolecular cyclization of 3-vinylindole motifs to *N*-attached acetylenes <02JMC1259>. Also interesting in this context is the development of indoles possessing various combinations of vinyl, allyl, ethynyl or propargyl substituents from indole-2-carboxaldehyde or indole itself <02S1810>, and the preparation of indoles fused with 7–9-membered lactams via ring closure metathesis of indole-2-carboxylic acid allylamide derivatives <02T10181>.

Intramolecular cycloaddition reactions have been performed on the furan-containing indoles **143**, to provide an efficient route to numerous derivatives of the tetracyclic ring system **144**, a structural motif of *Aspidosperma* and *Strychnos* alkaloids <02OL4643>.

An elegant approach to tetracyclic fused indole systems relying on a tandem Pummerer/Mannich cyclization cascade has been reported. For example, the enamidosulfoxide **145** underwent transformation into the tetracyclic system **146** upon treatment with TFAA. This method is also applicable for substrates containing other activated aromatics instead of indole <02JOC5928>. The formation of cyclopenta[*c*]carbazoles (**147**) has been observed upon treatment of the cyclopropane intermediates **148** with SnCl₄ in nitromethane, a process which takes place via a carbocationic cascade rearrangement involving a dimerization <02JOC9477>.

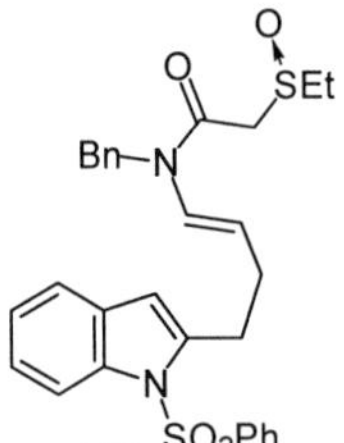

Generation of 3-indolylacyl radicals from the selenoesters **149**, using either *n*-Bu$_3$SnH or tris(trimethylsilyl)silane (TTMSS) followed by reaction with various alkenes, offers a route to 3-acylindoles **150**. On the other hand, the use of *n*-Bu$_6$Sn$_2$ under irradiation gave cyclopent[*b*]indole derivatives such as **151** via a cascade involving initial addition of the acyl radical to the alkene, and a subsequent oxidative cyclization at the indole C-2 <02JOC6268>.

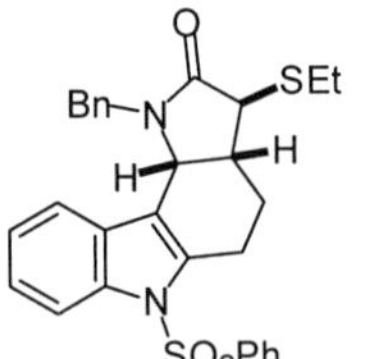

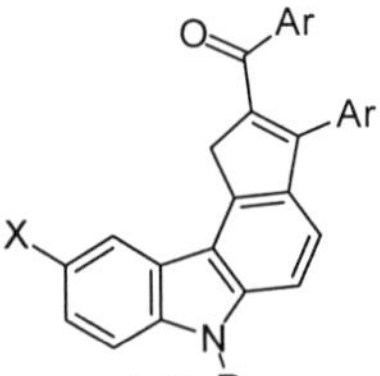

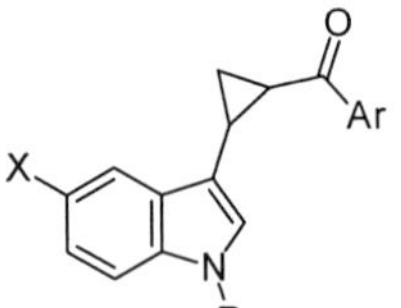

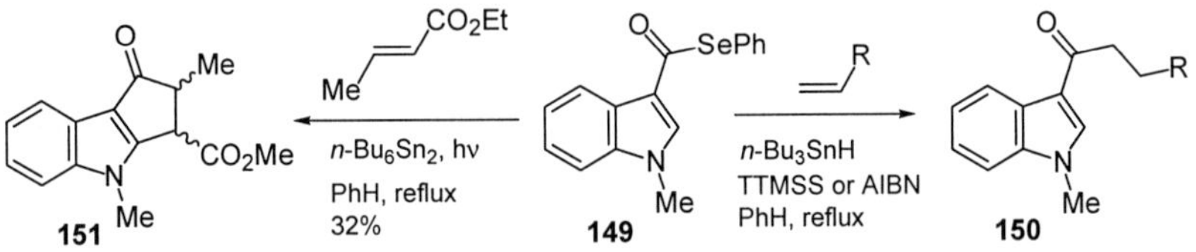

The dimerization of indole derivatives such as tetrahydrocarbazole or tryptamines with phenyliodine(III) bis(trifluoroacetate) (PIFA) has been found in each case to lead to several different dimeric products, nevertheless, this approach enabled a short total synthesis of the dimeric pyrrolidinoindole alkaloids *meso*- and *rac*-chimonanthines <02TL5637>. Oxidative dimerization of 3-methylindole using hydrogen peroxide mediated by horseradish peroxidase has been demonstrated to give the product **152**, thus correcting all structures previously suggested for this product in the literature. The identity of **152** was proved by X-ray crystallography, and a mechanism for its formation was proposed <02TL6903>. It should be added that oxidation of 2-methylindole has previously been shown to yield **153**, the same basic ring system as **152** <98TL4119>. Oxidation of the melanogenic precursor 5,6-dihydroxyindole-2-carboxylic acid under biomimetic conditions produced four new indole trimers in different atropisomeric forms <02T3681>. Mixed heterocalixarenes containing benzo[*b*]furan units and one or two indole units have been prepared from electron rich 7-benzo[*b*]furanylmethanols and activated indole-7-carboxaldehydes <02CC810>.

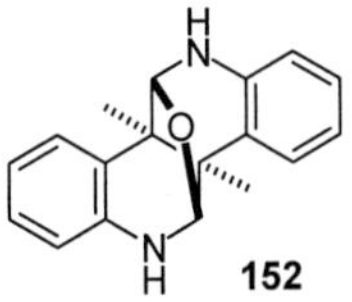

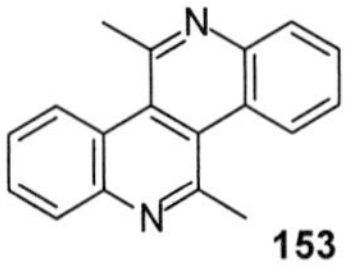

1-Methoxy-3-(2-nitrovinylindole) (**154**) has been subjected to reactions with nucleophiles, leading to attack at C-2 when DMF or HMPA were used as the reaction medium to give 1,2,3-substituted indoles **155**, whilst the use of THF as solvent changed the reaction course and a

Michael addition at the β-carbon of the nitrovinyl group was instead observed, providing the adducts **156** <02H(57)1231>. Several analogues of the Wasabi phytoalexin 1-methoxyindole-3-carboxylate possessing functional groups in the carbocyclic ring have been prepared via electrophilic substitution, taking advantage of the directing effect of the 1-methoxy group <02H(57)1627>. In a related study, the reactivity of 1-methoxy-6-nitroindole-3-carboxaldehyde has been studied, illustrating that a variety of nucleophiles can be introduced at C-2 in good yields <02H(58)53>.

5.2.6 PYRROLE AND INDOLE NATURAL PRODUCTS AND RELATED STUDIES

Indole and pyrrole alkaloids have attracted considerable interest, and numerous papers dealing with their isolation and total synthesis have appeared during the year. Some interesting efforts include the total syntheses of isoroquefortine C <02JOC620>; (+)-majvinine, (+)-10-methoxyaffinisine, (+)-N_a-methylsarpagine and macralstonidine <02OL687>; polycitones A and B <02OL3287>; alstophylline and macralstonine <02OL3339>; alstonisine <02OL4237>; (E)16-epiaffinisine, (E)16-epinormacusine B, and dehydro-16-epiaffinisine <02OL4681>; dihydroflustramine C and flustramine E <02T1479>; (+)-austamide, (+)-deoxyisoaustamide, and (+)-hydratoaustamide <02JA7904>; quadrigemine C and psycholeine <02JA9008>; (±)-tabersonine, (+)-tabersonine, (+)-16-methoxytabersonine, (+)-aspidospermidine, and (−)-quebrachamine <02JA4628>, dragmacidin D <02JA13179>; (+)-vinblastine <02JA2137>; (−)-VM55599 <02JA2556>; diazonamide A <02AGE3495>; (+)-16-epivinoxine and (−)-vinoxine <02TA95>; heptachloro-1′-methyl-1,2′-bipyrrole (Q1) <02AGE1740>; axinohydantoins <02JOC4498>; demethylasterriquinones A1 and B1 <02JOC7919>; (−)-strychnine <02JA14546>; (+)-aspidospermidine <02JA13398>, and (±)-strychnofoline <02JA14826>.

Other studies of related interest encompass the synthesis of the pyrrole part of the suggested structure of the marine natural product halitulin, the true identity of which still remains unknown <02JCS(P1)1340>; the stereoselective synthesis of the (2S,3R)-N-(1′,1′-dimethyl-2′-propenyl)-3-hydroxytryptophan fragment of the cyclic peptide cyclomarin C <02TL5291>; incorporation of enzymatically synthesized amino- and hydroxytryptophans into protein-based optical pH-sensors <02AGE4066>; preparation and screening of analogues of the pyrrole anti-inflammatory agent Ketorolac <02H(56)91>; synthesis of analogues of the alkaloid chuangxinmycin, some of which were shown to be powerful inhibitors of bacterial tryptophanyl tRNA synthetase <02BMCL3171>; total synthesis of the anti-tumor agent (−)-rhazinilam <02JA6900>; identification of a tricyclic indolic photodegradation product of the photosensitive antifungal antibiotic pyrrolnitrin <02JOC668>, and synthesis and probing of pyrrolomorphinans as δ-opioid antagonists <02JMC537>. The heterocyclic core of halitulin has together with some other related alkaloid derivatives also been prepared in connection with studies on DNA-cleaving properties of these compounds <02T6373>.

Carbazoles, indolocarbazoles, and related fused polyclyclic heterocycles also continue to capture considerable attention. New developments in indolocarbazole chemistry include preparation and screening for antiprofilerative activities of derivatives of the indolo[2,3-

a]carbazole alkaloid rebeccamycin possessing a pyranose unit attached to both indole nitrogen atoms <02JMC1330>; synthesis of indenopyrrolocarbazoles having an additional bridging oxygen-containing heterocyclic ring <02JOC3235>; construction of new indolo[2,3-*a*]carbazoles possessing alkyl, allyl, or glycosyl groups at one of the indole nitrogen atoms <02H(56)81>; synthesis and evaluation of cytotoxic properties of arcyriarubin and dechlororebeccamycin aglycon derivatives having alkoxy substituents on the indole nitrogen atoms <02JAN768>; preparation of alkyl chain bridged dimers of the indolo[2,3-*a*]pyrrolo[3,4-*c*]carbazole alkaloids (–)-(7*S*)- and (+)-(7*R*)-252a <02TL379>, and the synthesis of indolo[2,3-*b*]carbazole-6,12-diones <02T1443>. In this context it is also interesting to note that the aqueous extract of the sponge *Ancorina* sp. has been shown to contain among other products, ancorinazole (**157**), which constitutes the first example of an naturally occurring indolo[3,2-*a*]carbazole <02JOC6671>. In addition, a new approach to the alkaloid granulatimide and analogs thereof based on Stille-coupling methodology has been published <02CPB872>.

5.2.7 REFERENCES

98TL4119	J. Bergman, S. Bergman, J.-O. Lindström, *Tetrahedron Lett.* **1998**, *39*, 4119.
01AHC(80)1	J. Bergman, T. Janosik, N. Wahlström, *Adv. Heterocycl. Chem.* **2001**, *80*, 1.
01MI1	J.F.M. Da Silva, S.J. Garden, A.C. Pinto, *J. Brazilian Chem. Soc.* **2001**, *12*, 273.
02AGE1422	D. Seidel, V. Lynch, J.L. Sessler, *Angew. Chem. Int. Ed.* **2002**, *41*, 1422.
02AGE1740	J. Wu, W. Vetter, G.W. Gribble, J.S. Schneekloth, D.H. Blank, H. Görls, *Angew. Chem. Int. Ed.* **2002**, *41*, 1740.
02AGE3188	D.A. Evans, G. Borg, K.A. Scheidt, *Angew. Chem. Int. Ed.* **2002**, *41*, 3188.
02AGE3230	S. Kamijo, Y. Yamamoto, *Angew. Chem. Int. Ed.* **2002**, *41*, 3230.
02AGE3495	K.C. Nicolaou, M. Bella, D.Y.-K. Chen, X. Huang, T. Ling, S.A. Snyder, *Angew. Chem. Int. Ed.* **2002**, *41*, 3495.
02AGE4066	N. Budisa, M. Rubini, J.H. Bae, E. Weyher, W. Wenger, R. Golbik, R. Huber, L. Moroder, *Angew. Chem. Int. Ed.* **2002**, *41*, 4066.
02AHC(82)101	M. Somei, *Adv. Heterocycl. Chem.* **2002**, *82*, 101.
02BMCL1937	M. Berghauer, H. Hübner, P. Gmeiner, *Bioorg. Med. Chem. Lett.* **2002**, *12*, 1937.
02BMCL3137	M.J. Brown, P.S. Carter, A.E. Fenwick, A.P. Fosberry, D.W. Hamprecht, M.J. Hibbs, R. L. Jarvest, L. Mensah, P.H. Milner, P.J. O'Hanlon, A.J. Pope, C.M. Richardson, A. West, D.R. Witty, *Bioorg. Med. Chem. Lett.* **2002**, *12*, 3137.
02BMCL3305	T.G.M. Dhar, Z. Shen, C.A. Fleener, K.A. Rouleau, J.C. Barrish, D.L. Hollenbaugh, E.J. Iwanowicz, *Bioorg. Med. Chem. Lett.* **2002**, *12*, 3305.
02CC210	K. Yamazaki, Y. Kondo, *Chem. Commun.* **2002**, 210.
02CC220	M. Ishikura, H. Kato, N. Ohnuki, *Chem. Commun.* **2002**, 220.
02CC270	J.-H. Ho, T.-I. Ho, *Chem. Commun.* **2002**, 270.
02CC484	A. Penoni, K.M. Nicholas, *Chem. Commun.* **2002**, 484.
02CC810	D. StC Black, D.C. Craig, R. Rezaie, *Chem. Commun.* **2002**, 810.
02CC1204	L.S. Konstantinova, O.A. Rakitin, C.W. Rees, *Chem. Commun.* **2002**, 1204.
02CC2214	B. Quiclet-Sire, F. Wendeborn, S. Z. Zard, *Chem. Commun.* **2002**, 2214.
02CC2310	R.B. Bedford, C.S.J. Cazin, *Chem. Commun.* **2002**, 2310.
02CEJ2034	J. Barluenga, F.J. Fañanás, R. Sanz, Y. Fernández, *Chem. Eur. J.* **2002**, *8*, 2034.
02CEJ3148	G. Cafeo, F.H. Kohnke, G.L. La Torre, M.F. Parisi, R. Pistone Nascone, A.J.P. White, D.J. Williams, *Chem. Eur. J.* **2002**, *8*, 3148.
02CL20	J. Oyamada, W. Lu, C. Jia, T. Kitamura, Y. Fujiwara, *Chem. Lett.* **2002**, 20
02CL134	S. Matsumoto, T. Kishimoto, K. Ogura, *Chem. Lett.* **2002**, 134.

02CHE616	S.A. Vizer, E. Kh. Dedeshko, K.B. Erzhanov, *Chem. Heterocycl. Compd. (Engl. Transl.)* **2002**, *38*, 616.
02CHE748	N.A. Nedolya, L. Brandsma, S.V. Tolmachev, *Chem. Heterocycl. Compd. (Engl. Transl.)* **2002**, *38*, 748.
02CPB872	T. Yoshida, M. Nishiyachi, N. Nakashima, M. Murase, E. Kotani, *Chem. Pharm. Bull.* **2002**, *50*, 872.
02CR4303	H.-J. Knölker, K.R. Reddy, *Chem. Rev.* **2002**, *102*, 4303.
02EJO551	P. Rademacher, N. Basaric, K. Kowski, M. Sindler-Kulyk, *Eur. J. Org. Chem.* **2002**, 551.
02EJO1493	B. Clique, S. Vassiliou, N. Monteiro, G. Balme, *Eur. J. Org. Chem.* **2002**, 1493.
02EJO1523	D. Alickmann, R. Fröhlich, A.H. Maulitz, E.-U. Würthwein, *Eur. J. Org. Chem.* **2002**, 1523.
02EJO1556	J.-P. Bouillon, B. Hénin, J.-F. Huot, C. Portella, *Eur. J. Org. Chem.* **2002**, 1556.
02EJO1646	S. Mayer, J.-Y. Mérour, B. Joseph, G. Guillaumet, *Eur. J. Org. Chem.* **2002**, 1646.
02EJO2565	O. Paulus, G. Alcaraz, M. Vaultier, *Eur. J. Org. Chem.* **2002**, 2565.
02EJO2671	G. Battistuzzi, S. Cacchi, G. Fabrizi, *Eur. J. Org. Chem.* **2002**, 2671.
02EJO3481	F. Cochard, M. Laronze, É. Prost, J.-M. Nuzillard, F. Augé, C. Petermann, P. Sigaut, J. Sapi, J.-Y. Laronze, *Eur. J. Org. Chem.* **2002**, 3481.
02H(56)81	M. Somei, F. Yamada, J. Kato, Y. Suzuki, Y. Ueda, *Heterocycles* **2002**, *56*, 81.
02H(56)91	F.J. Lopez, M.-F. Jett, J.M. Muchowski, D. Nitzan, C. O'Yang, *Heterocycles* **2002**, *56*, 91.
02H(57)1101	O. Miyata, N. Takeda, T. Naito, *Heterocycles* **2002**, *57*, 1101.
02H(57)1231	K. Yamada, F. Yamada, M. Somei, *Heterocycles* **2002**, *57*, 1231.
02H(57)1293	J.H. Byers, Y. Zhang, *Heterocycles* **2002**, *57*, 1293.
02H(57)1627	K. Yamada, Y. Kanbayashi, S. Tomioka, M. Somei, *Heterocycles* **2002**, *57*, 1627.
02H(57)2261	T. Shibata, S. Kadowaki, K. Takagi, *Heterocycles* **2002**, *57*, 2261.
02H(58)53	K. Yamada, F. Yamada, T. Shiraishi, S. Tomioka, M. Somei, *Heterocycles* **2002**, *58*, 53.
02H(58)563	P.R. Sebahar, R.M. Williams, *Heterocycles* **2002**, *58*, 563.
02H(58)587	R. Tsuji, M. Nakagawa, A. Nishida, *Heterocycles* **2002**, *58*, 587.
02HC223	S.P. Ivonin, A.A. Tolmachev, A.M. Pinchuk, *Heteroatom Chem.* **2002**, *13*, 223.
02HCA4453	H.-D. Stachel, E. Eckl, E. Immerz-Winkler, C. Kreiner, W. Weigand, C. Robl, R. Wünsch, S. Dick, N. Drescher, *Helv. Chim. Acta* **2002**, *85*, 4453.
02JA1172	J.F. Austin, D.W.C. MacMillan, *J. Am. Chem. Soc.* **2002**, *124*, 1172.
02JA2137	S. Yokoshima, T. Ueda, S. Kobayashi, A. Sato, T. Kuboyama, H. Tokuyama, T. Fukuyama, *J. Am. Chem. Soc.* **2002**, *124*, 2137.
02JA2221	K.C. Nicolaou, K. Sugita, P.S. Baran, Y.-L. Zhong, *J. Am. Chem. Soc.* **2002**, *124*, 2221.
02JA2556	J.F. Sanz-Cervera, R.M. Williams, *J. Am. Chem. Soc.* **2002**, *124*, 2556.
02JA4628	S.A. Kozmin, T. Iwama, Y. Huang, V.H. Rawal, *J. Am. Chem. Soc.* **2002**, *124*, 4628.
02JA6900	J.A. Johnson, N. Li, D. Sames, *J. Am. Chem. Soc.* **2002**, *124*, 6900.
02JA7904	P.S. Baran, E.J. Corey, *J. Am. Chem. Soc.* **2002**, *124*, 7904.
02JA9008	A.D. Lebsack, J.T. Link, L.E. Overman, B.A. Stearns, *J. Am. Chem. Soc.* **2002**, *124*, 9008.
02JA11292	G.D. Wilkie, G.I. Elliott, B.S.J. Blagg, S.E. Wolkenberg, D.R. Soenen, M.M. Miller, S. Pollack, D.L. Boger, *J. Am. Chem. Soc.* **2002**, *124*, 11292.
02JA11592	H. Kusama, J. Takaya, N. Iwasawa, *J. Am. Chem. Soc.* **2002**, *124*, 11592.
02JA11684	J.C. Antilla, A. Klapars, S.L. Buchwald, *J. Am. Chem. Soc.* **2002**, *124*, 11684.
02JA11940	S. Kamijo, Y. Yamamoto, *J. Am. Chem. Soc.* **2002**, *124*, 11940.
02JA13179	N.K. Garg, R. Sarpong, B.M. Stoltz, *J. Am. Chem. Soc.* **2002**, *124*, 13179.
02JA13398	J.P. Marino, M.B. Rubio, G. Cao, A. de Dios, *J. Am. Chem. Soc.* **2002**, *124*, 13398.
02JA14546	T. Ohshima, Y. Xu, R. Takita, S. Shimizu, D. Zhong, M. Shibasaki, *J. Am. Chem. Soc.* **2002**, *124*, 14546.
02JA14826	A. Lerchner, E.M. Carreira, *J. Am. Chem. Soc.* **2002**, *124*, 14826.
02JAN768	S.A. Lakatosh, J. Balzarini, G. Andrei, R. Snoeck, E. De Clercq, M.N. Preobrazhenskaya, *J. Antibiot.* **2002**, *55*, 768.
02JCC191	K. Yamazaki, Y. Kondo, *J. Comb. Chem.* **2002**, *4*, 191.
02JCS(P1)341	M.A.-M. Gomaa, *J. Chem. Soc., Perkin Trans. 1* **2002**, 341.
02JCS(P1)345	Y. Zhang, J. Xue, Y. Gao, H.-K. Fun, J.-H. Xu, *J. Chem. Soc., Perkin Trans. 1* **2002**, 345.
02JCS(P1)622	D.W. Knight, A.L. Redfern, J. Gilmore, *J. Chem. Soc., Perkin Trans. 1* **2002**, 622.
02JCS(P1)733	H.J.C. Deboves, C. Hunter, R.F.W. Jackson, *J. Chem. Soc., Perkin Trans. 1* **2002**, 733.
02JCS(P1)1340	M.G. Banwell, A.M. Bray, A.J. Edwards, D.J. Wong, *J. Chem. Soc., Perkin Trans. 1* **2002**, 1340.
02JCS(P1)2613	M.G. Banwell, J.A. Smith, *J. Chem. Soc., Perkin Trans. 1* **2002**, 2613.
02JCS(P1)2747	W.R. Bowman, A.J. Fletcher, G.B.S. Potts, *J. Chem. Soc., Perkin Trans. 1* **2002**, 2747.
02JCS(P1)2799	C.D. Gabbutt, J.D. Hepworth, D.M. Heron, S.L. Pugh, *J. Chem. Soc., Perkin Trans. 1* **2002**, 2799.
02JHC571	Y. Tominaga, Y. Shigemitsu, K. Sasaki, *J. Heterocycl. Chem.* **2000**, *39*, 571.

02JMC537	S.K. Srivastava, S.M. Husbands, M.D. Aceto, C.N. Miller, J.R. Traynor, J.W. Lewis, *J. Med. Chem.* **2002**, *45*, 537.
02JMC1002	S. Mahboobi, S. Teller, H. Pongratz, H. Hufsky, A. Sellmer, A. Botzki, A. Uecker, T. Beckers, S. Baasner, C. Schächtele, F. Überall, M.U. Kassack, S. Dove, F.-D. Böhmer, *J. Med. Chem.* **2002**, *45*, 1002.
02JMC1259	P. Gharagozloo, S. Lazareno, M. Miyauchi, A. Popham, N. J.M. Birdsall, *J. Med. Chem.* **2002**, *45*, 1259.
02JMC1330	C. Marminon, F. Anizon, P. Moreau, S. Léonce, A. Pierré, B. Pfeiffer, P. Renard, M. Prudhomme, *J. Med. Chem.* **2002**, *45*, 1330.
02JNP748	Y. Liu, G.W. Gribble, *J. Nat. Prod.* **2002**, *65*, 748.
02JOC620	B.M. Schiavi, D.J. Richard, M.M. Joullié, *J. Org. Chem.* **2002**, *67*, 620.
02JOC668	M. Sako, T. Kihara, M. Tanisaki, Y. Maki, A. Miyamae, T. Azuma, S. Kohda, T. Masugi, *J. Org. Chem.* **2002**, *67*, 668.
02JOC1001	G.W. Gribble, J. Jiang, Y. Liu, *J. Org. Chem.* **2002**, *67*, 1001.
02JOC1699	S. Yu, J. Saenz, J.K. Srirangam, *J. Org. Chem.* **2002**, *67*, 1699.
02JOC2705	X. Zhu, A. Ganesan, *J. Org. Chem.* **2002**, *67*, 2705.
02JOC2858	P. Hamel, *J. Org. Chem.* **2002**, *67*, 2858.
02JOC3235	T.L. Underiner, J.P. Mallamo, J. Singh, *J. Org. Chem.* **2002**, *67*, 3235.
02JOC3425	L. Pouységu, A.-V. Avellan, S. Quideau, *J. Org. Chem.* **2002**, *67*, 3425.
02JOC4498	A.C. Barrios Sosa, K. Yakushijin, D.A. Horne, *J. Org. Chem.* **2002**, *67*, 4498.
02JOC5019	A.J. Airaksinen, M. Ahlgren, J. Vepsäläinen, *J. Org. Chem.* **2002**, *67*, 5019.
02JOC5386	M. Bandini, P. G. Cozzi, P. Melchiorre, A. Umani-Ronchi, *J. Org. Chem.* **2002**, *67*, 5386.
02JOC5850	J.H. Wynne, W.M. Stalick, *J. Org. Chem.* **2002**, *67*, 5850.
02JOC5928	A. Padwa, T.M. Heidelbaugh, J.T. Kuethe, M.S. McClure, Q. Wang, *J. Org. Chem.* **2002**, *67*, 5928.
02JOC6256	P.R. Carlier, P.C.-H. Lam, D.M. Wong, *J. Org. Chem.* **2002**, *67*, 6256.
02JOC6268	M.-L. Bennasar, T. Roca, R. Griera, M. Bassa, J. Bosch, *J. Org. Chem.* **2002**, *67*, 6268.
02JOC6671	K.M. Meragelman, L.M. West, P.T. Northcote, L.K. Pannell, T.C. McKee, M.R. Boyd, *J. Org. Chem.* **2002**, *67*, 6671.
02JOC7048	H. Zhang, R.C. Larock, *J. Org. Chem.* **2002**, *67*, 7048.
02JOC7551	E. Vasquez, I.W. Davies, J.F. Payack, *J. Org. Chem.* **2002**, *67*, 7551.
02JOC7919	M.C. Pirrung, Z. Li, K. Park, J. Zhu, *J. Org. Chem.* **2002**, *67*, 7919.
02JOC8178	O.A. Attanasi, L. De Crescentini, G. Favi, P. Filippone, F. Mantellini, S. Santeusanio, *J. Org. Chem.* **2002**, *67*, 8178.
02JOC8220	A.R. Katritzky, R. Jain, Y.-J. Xu, P.J. Steel, *J. Org. Chem.* **2002**, *67*, 8220.
02JOC8374	M.C. Pirrung, L. Deng, Z. Li, K. Park, *J. Org. Chem.* **2002**, *67*, 8374.
02JOC8958	S. Yang, W.A. Denny, *J. Org. Chem.* **2002**, *67*, 8958.
02JOC9318	H. Zhang, R.C. Larock, *J. Org. Chem.* **2002**, *67*, 9318.
02JOC9439	J.L. Bullington, R.R. Wolff, P.F. Jackson, *J. Org. Chem.* **2002**, *67*, 9439.
02JOC9477	C. Venkatesh, H. Ila, H. Junjappa, S. Mathur, V. Huch, *J. Org. Chem.* **2002**, *67*, 9477.
02NPR148	S. Hibino, T. Choshi, *Nat. Prod. Rep.* **2002**, *19*, 148.
02OL127	G.J. Bodwell, J. Li, *Org. Lett.* **2002**, *4*, 127.
02OL435	J.J. Klappa, A.E. Rich, K. McNeill, *Org. Lett.* **2002**, *4*, 435.
02OL687	S. Zhao, X. Liao, J.M. Cook, *Org. Lett.* **2002**, *4*, 687.
02OL699	A. Penoni, J. Volkmann, K.M. Nicholas, *Org. Lett.* **2002**, *4*, 699.
02OL815	B. Chauder, A. Larkin, V. Snieckus, *Org. Lett.* **2002**, *4*, 815.
02OL1819	D.M. Lindsay, W. Dohle, A. Eeg Jensen, F. Kopp, P. Knochel, *Org. Lett.* **2002**, *4*, 1819.
02OL2005	B.M. Trost, M.J. Krische, V. Berl, E.M. Grenzer, *Org. Lett.* **2002**, *4*, 2005.
02OL2317	Y. Nakamura, T. Ukita, *Org. Lett.* **2002**, *4*, 2317.
02OL2691	K. Fujita, K. Yamamoto, R. Yamaguchi, *Org. Lett.* **2002**, *4*, 2691.
02OL3035	H. Zhang, R.C. Larock, *Org. Lett.* **2002**, *4*, 3035.
02OL3287	A.T. Kreipl, C. Reid, W. Steglich, *Org. Lett.* **2002**, *4*, 3287.
02OL3339	X. Liu, J.R. Deschamp, J.M. Cook, *Org. Lett.* **2002**, *4*, 3339.
02OL3537	N.D. Smith, D. Huang, N.D.P. Cosford, *Org. Lett.* **2002**, *4*, 3537.
02OL4135	S.K. Bur, A. Padwa, *Org. Lett.* **2002**, *4*, 4135.
02OL4237	X.Z. Wearing, J.M. Cook, *Org. Lett.* **2002**, *4*, 4237.
02OL4289	J.H. Rigby, Z. Wang, *Org. Lett.* **2002**, *4*, 4289.
02OL4643	S.M. Lynch, S.K. Bur, A. Padwa, *Org. Lett.* **2002**, *4*, 4643.
02OL4681	J. Yu, X. Liao, J.M. Cook, *Org. Lett.* **2002**, *4*, 4681.
02OM581	K. Onitsuka, M. Yamamoto, S. Suzuki, S. Takahashi, *Organometallics* **2002**, *21*, 581.
02OM1430	P. Barbaro, C. Bianchini, A. Meli, M. Moreno, F. Vizza, *Organometallics* **2002**, *21*, 1430.
02OM1819	R. Aumann, D. Vogt, R. Fröhlich, *Organometallics*, **2002**, *21*, 1819.

02OM2055	A. Tarraga, P. Molina, J.L. López, M.D. Velasco, D. Bautista, P.G. Jones, *Organometallics* **2002**, *21*, 2055.
02OM4076	P.J. Campos, D. Sampedro, M.A. Rodríguez, *Organometallics* **2002**, *21*, 4076.
02OM4390	D. Reardon, J. Guan, S. Gambarotta, G.P.A. Yap, D.R. Wilson, *Organometallics* **2002**, *21*, 4390.
02OM4581	M.T. Valahovic, W.H. Myers, W.D. Harman, *Organometallics* **2002**, *21*, 4581.
02RCR563	L.N. Sobenina, A.P. Demenev, A.I. Mikhaleva, B.A. Trofimov, *Russ. Chem. Rev. (Engl. Transl.)* **2002**, *71*, 563.
02S34	Y. Cheng, Y.-H. Zhan, O. Meth-Cohn, *Synthesis* **2002**, 34.
02S320	S.N. Lavrenov, S.A. Lakatosh, L.N. Lysenkova, A.M. Korolev, M.N. Preobrazhenskaya, *Synthesis* **2002**, 320.
02S331	S. Banetti, C. De Risi, P. Marchetti, G.P. Pollini, V. Zanirato, *Synthesis* **2002**, 331.
02S595	B. Volk, T. Mezei, G. Simig, *Synthesis* **2002**, 595.
02S1351	H. Krawczyk, M. Sliwinski, *Synthesis* **2002**, 1351.
02S1509	T. Balle, K. Andersen, P. Vadsø, *Synthesis* **2002**, 1509.
02S1658	T. Nishikawa, S. Kajii, K. Wada, M. Ishikawa, M. Isobe, *Synthesis* **2002**, 1658.
02S1810	L. Pérez-Serrano, L. Casarrubios, G. Domínguez, P. González-Pérez, J. Pérez-Castells, *Synthesis* **2002**, 1810.
02S1917	J. Ichikawa, Y. Wada, M. Fujiwara, K. Sakoda, *Synthesis* **2002**, 1917.
02S2203	M.Makosza, M. Paszcwski, *Synthesis* **2002**, 2203.
02S2426	Y. Cheng, Y.-H. Zhan, H.-X. Guan, H. Yang, O. Meth-Cohn, *Synthesis* **2002**, 2426.
02S2597	T. Tajima, T. Fuchigami, *Synthesis* **2002**, 2597.
02SC105	R. Nagarajan, P.T. Perumal, *Synth. Commun.* **2002**, *32*, 105.
02SC903	T. Zhou, Z.-C. Chen, *Synth. Commun.* **2002**, *32*, 903.
02SC1465	V.V. Rozhkov, A.M. Kuvshinov, S.A. Shevelev, *Synth. Commun.* **2002**, *32*, 1465.
02SC2003	T.L.S. Kishbaugh, G.W. Gribble, *Synth. Commun.* **2002**, *32*, 2003.
02SC2035	J. Jiang, G.W. Gribble, *Synth. Commun.* **2002**, *32*, 2035.
02SC2643	X. Xu, Y. Zhang, *Synth. Commun.* **2002**, *32*, 2643.
02SL143	M.C. Pirrung, M. Wedel, Y. Zhao, *Synlett* **2002**, 143.
02SL619	M. Friedrich, A. Wächtler, A. de Meijere, *Synlett* **2002**, 619.
02SL829	A. Hercouet, A. Neu, J.-F. Peyronel, B. Carboni, *Synlett* **2002**, 829.
02SL1152	E. Dvornikova, K. Kamienska-Trela, *Synlett* **2002**, 1152.
02SL1629	L. Planas, J. Pérard-Viret, J. Royer, M. Selkti, A. Thomas, *Synlett* **2002**, 1629.
02SL1709	C.-Y. Wu, C.-M. Sun, *Synlett* **2002**, 1709.
02SL1913	G. Deng, N. Jiang, Z. Ma, J. Wang, *Synlett* **2002**, 1913.
02T1229	R. Nagarajan, P.T. Perumal, *Tetrahedron* **2002**, *58*, 1229.
02T1443	J. Bergman, N. Wahlström, L.N. Yudina, J. Tholander, G. Lidgren, *Tetrahedron* **2002**, *58*, 1443.
02T1453	C. Escolano, K. Jones, *Tetrahedron* **2002**, *58*, 1453.
02T1479	M.S. Morales-Ríos, R.S. Suárez-Castillo, P. Joseph-Nathan, *Tetrahedron* **2002**, *58*, 1479.
02T2813	T. Janosik, A.-L. Johnson, J. Bergman, *Tetrahedron* **2002**, *58*, 2813.
02T3605	W.E. Hume, T. Tokunaga, R. Nagata, *Tetrahedron* **2002**, *58*, 3605.
02T3681	A. Pezzella, D. Vogna, G. Prota, *Tetrahedron* **2002**, *58*, 3681.
02T4487	J.C. Torres, R.A. Pilli, M.D. Vargas, F.A. Violante, S.J. Garden, A.C. Pinto, *Tetrahedron* **2002**, *58*, 4487.
02T6373	A. Fürstner, H. Krause, O. R. Thiel, *Tetrahedron* **2002**, *58*, 6373.
02T6673	E.M. Beccalli, G. Broggini, A. Marchesini, E. Rossi, *Tetrahedron* **2002**, *58*, 6673.
02T6713	J. Leroy, E. Porhiel, A. Bondon, *Tetrahedron* **2002**, *58*, 6713.
02T7221	V. Nair, P.M. Treesa, N.P. Rath, A.C. Kunwar, K.S. KiranKumar, A. RaviSankar, M. Vairamani, S. Prabhakar, *Tetrahedron* **2002**, *58*, 7221.
02T7625	C.-C. Tseng, Y.-L. Wu, C.-P. Chuang, *Tetrahedron* **2002**, *58*, 7625.
02T8067	D. Giomi, M. Cecchi, *Tetrahedron* **2002**, *58*, 8067.
02T8937	H.-J. Knölker, T. Hopfmann, *Tetrahedron* **2002**, *58*, 8937.
02T9179	J. Bergman, S. Rehn, *Tetrahedron* **2002**, *58*, 9179.
02T9187	J. Slätt, J. Bergman, *Tetrahedron* **2002**, *58*, 9187.
02T9541	C. Escolano, L. Hallverdú, K. Jones, *Tetrahedron* **2002**, *58*, 9541.
02T9793	A.S. Demir, I.M. Akhmedov, Ö. Sesenoglu, *Tetrahedron* **2002**, *58*, 9793.
02T10181	L. Chacun-Lefèvre, V. Bénéteau, B. Joseph, J.-Y. Mérour, *Tetrahedron* **2002**, *58*, 10181.
02TA95	M.-L. Bennasar, E. Zulaica, Y. Alonso, B. Vidal, J.T. Vázquez, J. Bosch, *Tetrahedron: Asymmetry* **2002**, *13*, 95.
02TA601	A.S. Demir, F. Aydogan, I.M. Akhmedov, *Tetrahedron: Asymmetry* **2002**, *13*, 601.
02TA1351	K. Hiroi, Y. Hiratsuka, K. Watanabe, I. Abe, F. Kato, M. Hiroi, *Tetrahedron: Asymmetry* **2002**, *13*, 1351.

02TA1721	S.E. Boiadjiev, D.A. Lightner, *Tetrahedron: Asymmetry* **2002**, *13*, 1721.
02TL29	J.T. Kuethe, I.W. Davies, P.G. Dormer, R.A. Reamer, D.J. Mathre, P.J. Reider, *Tetrahedron Lett.* **2002**, *43*, 29.
02TL53	S.H. Hwang, M.J. Kurth, *Tetrahedron Lett.* **2002**, *43*, 53.
02TL197	N.V. Moskalev, G.W. Gribble, *Tetrahedron Lett.* **2002**, *43*, 197.
02TL379	K. Tamaki, E.W.D. Huntsman, D.T. Petsch, J.L. Wood, *Tetrahedron Lett.* **2002**, *43*, 379.
02TL1359	H. Zhang, R.C. Larock, *Tetrahedron Lett.* **2002**, *43*, 1359.
02TL1621	T.L. Scott, B.C.G. Söderberg, *Tetrahedron Lett.* **2002**, *43*, 1621.
02TL1863	X. Fan, Y. Zhang, *Tetrahedron Lett.* **2002**, *43*, 1863.
02TL2187	A. Bombrun, G. Casi, *Tetrahedron Lett.* **2002**, *43*, 2187.
02TL2605	R. Grigg, E.L. Millington, M. Thornton-Pett, *Tetrahedron Lett.* **2002**, *43*, 2605.
02TL2831	A.R. Katritzky, H.-Y. He, R. Jiang, *Tetrahedron Lett.* **2002**, *43*, 2831.
02TL2885	S. Taira, H. Danjo, T. Imamoto, *Tetrahedron Lett.* **2002**, *43*, 2885.
02TL3401	P.N. Collier, I. Patel, R.J.K. Taylor, *Tetrahedron Lett.* **2002**, *43*, 3401.
02TL4115	J. Jiang, G.W. Gribble, *Tetrahedron Lett.* **2002**, *43*, 4115.
02TL4271	A. Janczuk, W. Zhang, W. Xie, S. Lou, J. Cheng, P.G. Wang, *Tetrahedron Lett.* **2002**, *43*, 4271.
02TL4491	P.M.T. Ferreira, H.L.S. Maia, L.S. Monteiro, *Tetrahedron Lett.* **2002**, *43*, 4491.
02TL4581	A. Okamoto, K. Tainaka, I. Saito, *Tetrahedron Lett.* **2002**, *43*, 4581.
02TL4671	S. Rossiter, *Tetrahedron Lett.* **2002**, *43*, 4671.
02TL4707	V. Kouznetsov, F. Zubkov, A. Palma, G. Restrepo, *Tetrahedron Lett.* **2002**, *43*, 4707.
02TL4765	P. González-Pérez, L. Pérez-Serrano, L. Casarrubios, G. Domínguez, J. Pérez-Castells, *Tetrahedron Lett.* **2002**, *43*, 4765.
02TL4777	M.A. Brimble, R. Halim, M. Petersson, *Tetrahedron Lett.* **2002**, *43*, 4777.
02TL5189	D.A. Wacker, P. Kasireddy, *Tetrahedron Lett.* **2002**, *43*, 5189.
02TL5291	S.-J. Wen, H.-W. Zhang, Z.-J. Yao, *Tetrahedron Lett.* **2002**, *43*, 5291.
02TL5295	M. Sugiura, N. Yamaguchi, T. Saya, M. Ito, K. Asai, I. Maeba, *Tetrahedron Lett.* **2002**, *43*, 5295.
02TL5637	H. Ishikawa, H. Takayama, N. Aimi, *Tetrahedron Lett.* **2002**, *43*, 5637.
02TL6197	K. Onitsuka, S. Suzuki, S. Takahashi, *Tetrahedron Lett.* **2002**, *43*, 6197.
02TL6579	A. Yasuhara, N. Suzuki, T. Yoshino, Y. Takeda, T. Sakamoto, *Tetrahedron Lett.* **2002**, *43*, 6579.
02TL6903	K.-Q. Ling, T. Ren, J.D. Protasiewicz, L.M. Sayre, *Tetrahedron Lett.* **2002**, *43*, 6903.
02TL7013	Y. Jiang, D. Ma, *Tetrahedron Lett.* **2002**, *43*, 7013.
02TL7135	Y. Liu, G.W. Gribble, *Tetrahedron Lett.* **2002**, *43*, 7135.
02TL7699	W.-M. Dai, L.-P. Sun, D.-S. Guo, *Tetrahedron Lett.* **2002**, *43*, 7699.
02TL7925	M. Laronze, J. Sapi, *Tetrahedron Lett.* **2002**, *43*, 7925.
02TL8893	S. Taira, H. Danjo, T. Imamoto, *Tetrahedron Lett.* **2002**, *43*, 8893.
02TL9175	N. Selvakumar, A.M. Azhagan. D. Srinivas, G.G. Krishna, *Tetrahedron Lett.* **2002**, *43*, 9175.

Chapter 5.3

Five–Membered Ring Systems : Furans and Benzofurans

Xue-Long Hou
Shanghai–Hong Kong Joint Laboratory in Chemical Synthesis and State Key Laboratory of Organometallic Chemistry, Shanghai Institute of Organic Chemistry, The Chinese Academy of Sciences, 354 Feng Lin Road, Shanghai 200032, China.
xlhou@pub.sioc.ac.cn

Zhen Yang
Key Laboratory of Bioorganic Chemistry and Molecular Engineering of Ministry of Education, Department of Chemical Biology, College of Chemistry, Peking University, Beijing 100871, China.
zyang@chem.pku.edu.cn

Henry N.C. Wong
*Department of Chemistry, Institute of Chinese Medicine and
Central Laboratory of the Institute of Molecular Technology for Drug Discovery and Synthesis,[†]
The Chinese University of Hong Kong, Shatin, New Territories, Hong Kong SAR, China.*
hncwong@cuhk.edu.hk
*and
Shanghai–Hong Kong Joint Laboratory in Chemical Synthesis, Shanghai Institute of Organic Chemistry, The Chinese Academy of Sciences, 354 Feng Lin Road, Shanghai 200032, China.*
hncwong@pub.sioc.ac.cn

[†] An Area of Excellence of the University Grants Committee (Hong Kong).

5.3.1 INTRODUCTION

The authors of this chapter aim to review articles that were published in 2002 concerning the applications and syntheses of furans, benzofurans and their derivatives. Several reviews concerning the chemistry of these families were published in 2002. The construction of furans and benzo[*b*]furans from alkynes and organic halides or triflates via palladium–catalyzed reactions has been reviewed <02H(56)613>. A review on anodic oxidation procedures for the conversion of phenols into benzo[*b*]furans has also appeared <02SL533>. An article was published summarizing the use of polysubstituted furans as dienes in Diels-Alder cycloaddition reactions, leading to a number of complex natural products <02SL851>. The synthesis of dihydrofurans from functionalized allenes is the theme of another review <02S1759>. Total syntheses of highly symmetric squalene–derived cytotoxic polyethers have also been summarized <02JSO1112>. Reviews have also appeared focusing on the synthesis of tetrahydrofurans <02JCS(P1)2301> <02S2778>.

An overview on naturally occurring tetrahydrofurans isolated from gorgonian octocorals of the genus *Briareum* was also published <02H(57)1705>. A total synthesis of a new squalene–derived epoxy tri–THF diol isolated from *Spathelia glabrescens* led to a complete assignment of its stereostructure <02TL5849>. On the other hand, a complete reassessment of the structural assignments of the sclerophytin diterpenes has also been recorded <02JNP126>.

Pirkle's reagent, a chiral solvating agent, was used to determine the absolute configuration of the annonaceous butenolides by the NMR method <02CEJ5662>.

A large number of novel natural products containing tetrahydrofuran and dihydrofuran rings have been identified from natural sources in 2002. Those compounds whose biological activities have not been mentioned are: botryolins A and B from green microalga *Botryococcus braunii* <02P(59)839>, clavidol, 3-*epi*-dehydrothyrsiferol and lactodehydrothyrsiferol from red seaweed *Laurencia viridis* <02T8119>, renealtins A and B from seeds of *Renealmia exaltata* <02JNP375>, caloflavans A and B from leaves of *Ochna calodendron* <02P(59)435>, yesanchinosides A–F from underground part of *Panax japonicus* <02JNP346>, (22*R*,24*S*)-22,25-epoxy-9,19-cyclolanostane-3β,16β,24-triol 3-[α-L-rhamnopyranosyl-(1→4)-β-D-glucopyranoside and (22*R*,24*S*)-22,25-epoxy-9,19-cyclolanostane-3β,16β,24-triol 3-[α-D-gluocopyranosyl-(1→3)-β-D-glucopyranoside from *Corchorus depressus* L. <02HCA689>, beesiosides A–F from whole plants of *Beesia caltahefolia* <02JNP42>, algacins E–H from stem barks of *Aglaia cordata* <02TL5783>, concentriols B–D from fruiting bodies of the xylariaceous ascomycete *Daldinia concentrica* <02P(61)345>, 2α-guaicyl-4-oxo-catechyl-3,7-dioxabicyclo[3.3.0]octane and 1α-hydroxy-2a,4a-guaicyl-3,7-dioxabicyclo[3.3.0]octane from *Saussurea medusa* <02P(59)85>, laxiflorins F and G from leaves of *Isodon eriocalyx* var. *laxiflora* <02TL661>, 6-methylcryptoacetalide, 6-methyl-epicryptoacetalide and 6-methylcryptotanshinone from the whole plant of *Salvia aegyptiaca* <02P(61)361>, rel-5-hydroxy-7,4'-dimethoxy-2"*S*-(2,4,5-trimethoxy-*E*-styryl)tetrahydrofuro[4"*R*,5"*R*:2,3]flavanonol and rel-5-Hydroxy-7,4'-dimethoxy-3"*S*-(2,4,5-trimethoxy-*E*-styryl)tetrahydrofuro-[4"*R*,5"*R*:2,3]-flavanonol from leaves of *Alpinia flabellate* <02JNP389>, pachyclavulariaenone B from the soft coral *Pachyclavularia violacea* <02JNP1357>, conidione from *Aplidium conicum* <02JNP1328>, ibhayinol from the sea hare *Aplysia dactylomela* <02JNP580>, itomanallenes A and B from red alga *Laurencia ntricate* <02P(60)861>, ileabethin from *Pseudopterogorgia elisabethae* (Bayer) <02TL5601>, compounds and from *Plocamium cartilagineum* (L.) Dixon (Plocamiaceae) <02T8539>, furanovibsanin B–F and 7-*epi*-furanovibsanin B from leaves of *Viburnum awabuki* <02T10033>, and yaretol from the whole plant of *Azorella madreporica* <02JNP1678>.

Those naturally occurring compounds containing tetrahydrofuran or dihydrofuran skeletons, whose biological activities have been assessed are (biological activities shown in parentheses): muricins H and I, *cis*-annomontacin, *cis*-corossolone and annocatalin from seeds of *Annoa muricata* (muricins H and I, *cis*-annomontacin, *cis*-corossolone and annocatalin showed significant activity in *in vitro* cytotoxic assays against two human hepatoma cell lines, Hep G_2 and 2,2,15; annocatalin showed high selectivity towards the Hep 2,2,15 cell line) <02JNP470>, agastinol and agastenol from whole plant of *Agastacje rugosa* (inhibited etoposide–induced apoptosis in U937 cells with IC_{50} =15.2 and 11.4 μg/mL, respectively) <02JNP414>, varitriol from a marine–derived strain of the fungus *Emericella variecolor* (increased potency towards selected renal, CNS, and breast cancer cell lines) <02JNP364>, pachastrissamine from *Pachastrissa* sp. (cytotoxic) <02JNP1505>, plakortethers A–G from Caribbean sponge *Plakortis simplex* (plakortethers A, B, D and E showed selective cytotoxicity against the RAW 264–7 cell line of murine macrophage) <02EJO61>, 21α,25-dimethylmelianodiol[(21*R*,23*R*)-epoxy-24-hydroxy-21α,25-dimethoxy]tirucalla-7-en-3-one and 21β,25-dimethylmelianodiol [(21*S*,23*R*)-epoxy-24-hydroxy-21β,25-dimethoxy]tirucalla-7-en-3-one from stems of *Raulinoa echinata* (weak *in vitro* inhibitory activity against trypomastigote forms of *Trypanosoma cruzi*) <02JNP562>, ajugasaliciosides A–E from aerial parts of *Ajuga salicifolia* (L.) Schreber (specifically inhibited the viability and growth of Jurkat T–leukaemia cells at concentration below 10 μM) <02HCA1930>, squamocin–O_1 and squamocin–O_2 from seeds of *Annona squamosa* L. (much lower activity on human K562 leukemia and HLE hepatoma cells as compared with squamocin) <02P(61)999>, (17*S*,20*R*,24*R*)-17,25-dihydroxy-20,24-epoxy-14(18)-malabaricen-3-one and (17*R*,20*S*,24*R*)-17,25-dihydroxy-20,24-epoxy-14(18)-malabaricen-3-one from leaves of *Caloncoba echinata* (*in vitro* inhibition of growth of *Plasmodium falciparum* parasites) <02JNP1764>, (+)-5'-demethoxyepiexcelsin from *Litsea verticillata* Hance (moderate anti–HIV activity with IC_{50} = 42.7 μM) <02P(59)325>, 4'-*O*-β-D-glucosyl-9-*O*-(6"-deoxysaccharosyl)olivil and berchemol-4'-*O*-β-D-glucoside from roots of *Valeriana officinalis* (4'-*O*-β-D-glucosyl-9-*O*-(6"-deoxy-saccharosyl)olivil is a partial agonist at rat and human A_1 adenosine receptors) <02JNP1479>, terpiodiene from Okinawan sponge *Terpios hoshinota* (moderate cytotoxicity against P388 cells) <02CL38>, 1β,8β-diacetoxyl-

6α,9α-difuroyloxydihydro-β-agarofuran and 1β-acetoxyl-2β,6α,9α-trifuroyloxydihydro-β-agarofuran from roots of *Celastrus orbiculatus* (moderate inhibition in both NF–κB activation and nitric oxide production) <02JNP89>, vernoguinosterol and vernoguinoside from stem bark of *Vernonia guineensis* (trypanocidal activity) <02P(59)371>, pachyclavulariaenones D–G from soft coral *Pachyclavularia violacea* (significant cytotoxicity toward P–388 tumor cells with ED_{50} = 0.2 μg/mL and HT–29 tumor cells with ED_{50} = 3.2 μg/mL) <02JNP1475>, hyperibones A–I from aerial parts of *Hypericum scabrum* (hyperibones A, B and D showed mild *in vitro* anti–bacterial activity against methicillin–resistance *Staphylococus aureus* (MRSA) and methicillin–sensitive *Staphylococus aureus* (MSSA)) <02JNP290>, and 22,23-dihydronimocinol from fresh leaves of *Azadirachta indica* (neem) (mortality for fourth instar larvae of the mosquito (Anopheles stephensi) with LC_{50} = 60 ppm) <02JNP1216>.

Several oligostilbenes containing benzo[*b*]furan frameworks were isolated from vitaceaeous plants. According to the biogenetic pathway of oligostilbenes, their regiospecific and stereospecific transformation was mechanistically discussed <02T9265>. The chemistry and biological activity of a dibenzofuran derivative called usnic acid found in lichens have been reviewed <02P(61)729>.

Those furan–containing compounds whose biological activities were not mentioned are: crotozambefurans A, B and C from stem barks of *Croton zambesicus* <02P(60)345>, 2,19:15,16-diepoxy-*neo*-clerodan-3,13(16),14-trien-18-oic acid, 15,16-epoxy-5,10-*seco*-clerodan-1(10),2,4,13(16),14-pentaen-18,19-olide and 15,16-epoxy-*neo*-clerodan-1,3,13(16),14-tetraen-18,19-olide from aerial parts of *Baccharis flabellata* Hook. & Arn. var. flabellate (Asteraceae) <02P(61)389>, teucrolins F and G from aerial parts of *Teucrium oliveriamum* <02P(59)409>, 3β-hydroxyteubutilin A, 12-*epi*-montanin G and 20-*epi*-3,20-di-*O*-deacetylteupyreinidin from aerial parts of *Teucrium montbretii* subsp. *Libanoticum* <02JNP142>, 1α,2α-epoxy-17β-hydroxyazadiradione, 1α,2α-epoxynimolicinol and 7-deacetylnimolicinol from the neem seed oil of *Azadirachta indica* <02JNP1177>, kinalborins A–C from the whole plant of *Kinostemon alborubrum* (HEMSL.) <02HCA2547>, caesalmin H from seeds of *Caesalpinia minax* Hance <02BMC2161>, 5-(2-hydroxyphenoxymethyl)furfural from seeds of *Cassia fistula* <02JNP1165>, 4aαH-3,5α,8aβ-trimethyl-4,4a,8a,9-tetrahydronaphtho[2,3-*b*]furan-8-one and 2-hydroxy-4aαH-3,5α,8aβ-trimethyl-4,4a,8a,9-tetrahydronaphtho[2,3-*b*]furan-8-one from *Siphonochilus aethiopicus* <02P(59)405>, furanovibsanin A, 3-*O*-methylfuranovibsanin A and furanovibsanin G from leaves of *Viburnum awabuki* <02T10033>, and methyl 5-styrylfuran-2-carboxylate from roots of *Renealmia nicolaioides* <02JNP1616>.

Those naturally occurring compounds containing furan skeletons, whose biological activities have been assessed are (biological activities shown in parentheses): *epi*-sarcotin A and sarcotin F–M from marine sponge *Sarcotragus* sp. (moderate to significant cytotoxicity against five human tumor cell lines) <02JNP1307>, khayanolides D, E and khayanoside from stem bark of Egyptian *Khaya senegalensis* (Khayanolide E showed antifeedant activity (100 ppm) against the third–instar larvae of *Spodoptera littoralis* (Boisduval)) <02JNP1219>, dysoxylumic acids A, C, D and dysoxylumolides A–C from bark of *Dysoxylum hainanense* Merr (dysoxylumic acids A and C showed significant antifeedant activity against Pieris rapae L) <02T7797>, macrocaesalmin from seeds of *Caesalpinia minax* Hance (inhibitory activity against RSV with IC_{50} = 24.2 μg/mL) <02TL2415>, salmahyrtisols A and B from Red Sea sponge *Hyrtios erecta* (cytotoxic against murine leukaemia (P-388), human lung carcinoma (A-549) and human colon carcinoma (HT-29)) <02JNP2>, saurufurans A and B from root of *Saururus chinensis* (saurufuran A showed activation effect on peroxisome proliferator–activated receptor γ (PPARγ) with EC_{50} = 16.7 μM) <02JNP616>, isovouacapenol A–D from leaves of *Caesalpinia pulcherrima* (active against bacteria (*S. aureus*, *E. coli*, *P. aeruginosa*, *B. subtilis*) and fungi (*C. albicans*, *T. mentagrophytes*)) <02JNP1107>, cespitularins A–D from Formosan soft coral *Cespitularia hypotentaculata* (cespitularin C exhibited potent cytotoxicity against P–388 and A549 cells), and cespitularins A, B and D showed moderate cytotoxicity against P–388 cells) <02JNP1429>.

Those benzo[*b*]furan- or dihydrobenzo[*b*]furan–containing compounds whose biological activities were not mentioned are: skimmianine and dictamnine from *Teclea trichocarpa* Enge. (Rutaceae) <02JNP956>, millettocalyxin C and pongol methyl ether from stem barks of *Millettia erythrocalyx* <02JNP589>, latifolol from the stem of *Gnetum latifolium* Blume <02P(61)959>, 6-methoxy-[2",3":7,8]-furanoflavanone and 2,5-dimethoxy-4-hydroxy-[2",3":7,8]-

furanoflavan from roots of *Millettia erythrocalyx* <02P(61)943>, (+)-viniferol B and (+)-viniferol C from the stem of *Vitis vinifera* 'Kyohou' <02T6931>, licoagrosides D and E from *Glycyrrhiza pallidiflora* hairy root cultures <02P(60)351>, crotafurans A–C from the bark of *Crotalaria pallida* and crotafuran D from the seeds of *Crotalaria assamica* <02HCA847>, gnemonosides F, G, H from *Gnetum gnemonoides* and gnemonosides I and J from *Gnetum africanum* <02HCA2394>, gneafricanins A and B from stem lianas of *Gnetum africanum* <02H(57)1057>, 3β-(3',5'-dihydroxyphenyl)-2α-(4"-hydroxyphenyl)-dihydrobenzofuran-5-carbaldehyde from rhizomes of *Smilax bracteata* <02JNP262>, and 5,4'-dihydroxy-[2"-(1-hydroxy-1-methylethyl)dihydrofurano]-(7,8:5",4")flavanone from leaves of *Macaranga conifera* <02P(61)867>.

Those naturally occurring compounds containing benzo[*b*]furan or dihydrobenzo[*b*]furan skeletons, whose biological activities have been assessed are (biological activities shown in parentheses): ulexins C and D from *Ulex europaeus* ssp. *Europaeus* (no inhibition of the growth of *Cladosporium cucumerinum*) <02JNP175>, paradisin C from grapefruit juice (inhibition of cytochrome P_{450} (CYP) 3A4 (IC_{50} = 1.0 µM) <02T6631>, kynapcin–24 from fruiting bodies of *Polyozellus multiflex* (noncompetitively inhibited prolyl endopeptidase (PEP) with IC_{50} = 1.14 µM) <02JNP76>, stemofurans A–K from roots of *Stemona collinsae* (antifungal activity against *Cladosporium herbarum*) <02JNP820>, tournefolal and tournefolic acid A from stems of *Tournefortia sarmentosa* (anti–LDL–peroxidative activity) <02JNP745>, smiranicin from *Smirnowia iranica* (toxicity against Plasmodium falciparum 3D7 strain with IC_{50} = 78.0±18.2 µM) <02JNP1754>, 3,4:8,9-dimethylene-dioxypterocarpan from roots of *Tephrosia aequilata* Baker (low activity against gram–positive bacteria, *Bacillus subtilis* and *Micrococcus lutea*) <02P(60)375>, gneafricanins C–E from stem lianas of *Gnetum africanum* Welw (antioxidant activity with 13, 50 and 32 µM inhibition in lipid peroxide and 10, 34 and 30 µM scavenging activity of super oxide) <02H(57)1507>, ochrocarpins A–G from Bark of *Ochrocaros punctatus* (cytotoxicity against the A2780 ovarian cancer cell line) <02JNP965>, and hopeaphenols A and B from stem bark of *Vatica oblongifolia* (hopeaphenol A showed moderate activity against methicillin–resistant *Staphylococcus aureus* with MIC = 100 µg/mL and *Mycobacterium smegmatis* with MIC = 50 µg/mL).

5.3.2 REACTIONS

5.3.2.1 Furans

The cycloaddition of furan was studied theoretically using quantum mechanical calculations at the B3LYP/6–31G* theory level <02JOC959> <02OL473>. Furans were found to be highly activated when they were allowed to coordinate with an electron–rich rhenium complex. Thus, either [2+2] or [3+2] cycloaddition reactions proceeded under mild reaction conditions, depending on the position of the ring that is coordinated <02JA7395>. The behavior of 2-vinyl furan in the Diels-Alder reaction was studied <02EJO3589>. Although there are two pathways for such substrates, methyl 2-methyl-5-vinyl-3-furoate functioned as a diene involving the furan C-4–C-5 double bond and the exocyclic double bond and the adducts were prone to aromatization when maleimide and maleic anhydride were used <02EJO3589>. Diels-Alder reactions of furan were also efficiently promoted by a catalytic amount of an active Lewis acid $HfCl_4$. Even in the presence of an excess of furan, the reaction proceeded at low temperature without the usual *endo*/*exo* isomerization and gave high *endo* selectivity <02AG(E)4079>. It was also found that *endo*–selectivity (2:1 *endo* vs. *exo*) was observed when the Diels-Alder reaction of furan with acrylate was performed in ionic liquids [bmim]BF_4 and [bmim]PF_6 <02SL1815>. A Diels-Alder reaction – ring opening reaction of 2-methoxyfuran with a benzopyranylidene–metal complex was found to deliver naphthalene derivatives <02CL124>. Diels-Alder reaction of 2,4-dimethyl furan was employed in the synthesis of the C,D–subunit of taxol <02H(56)479>. Lipase–catalyzed domino kinetic resolution reaction – intramolecular Diels-Alder reaction of furfuryl alcohol and fumarate derivatives provided the corresponding chiral oxabicyclo[2.2.1]heptenes <02CEJ4264>. The Diels-Alder reaction of cyclic sulfilimine with furan has also been investigated, and a high *exo*–selectivity was observed <02JOC2919>. The theoretical study of this reaction by using B3LYP/6-31G(d)//AM1 and B3LYP/6-31G(d)//B3LYP/6-31G(d) basis sets was also carried out <02JOC2926>. Regio– and diastereoselectivity of photo–induced [2+2] cycloaddition of furan and benzofuran with 1-acetylizatin was also studied <02JCS(P1)345>. The [4+3] cycloaddition of furan with α-tosyl ketones was also reported <02SL489>. Asymmetric [4+3] cycloaddition reaction of furans having a chiral auxiliary at the C-2 position and 2,4-dihalo-pentan-3-one under sonochemical and/or thermal conditions was reported and the influence of the chiral auxiliary was studied. A cis–diastereospecificity and a high endo–diastereoselectivity were observed in almost all cases <02TL2017> <02T4769>. As shown below, reaction of pyrroles with trifluoroacetic acid followed by boron trifluoride etherate in the presence of maleic acid derivatives provided indole derivatives, presumably through the Diels-Alder reaction of the previously unknown furo[3,4-b]pyrrole ring system. These fused bicyclic heterocycles can be viewed as analogs of hetero-o-quinodimethanes and are of intrinsic theoretical interest <02TL197>.

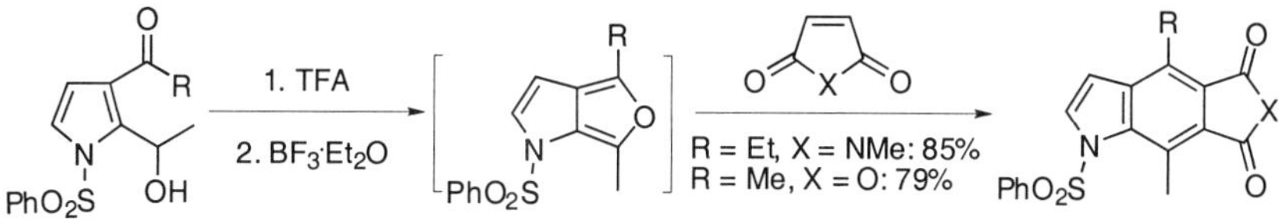

Inter– and intramolecular cycloaddition reactions of furans followed by transformation of adducts were employed in the synthesis of complex natural products <02T9903>. As shown below, an intramolecular Diels-Alder reaction served as a key step to set up the D-homo-10-epi-adrenosterone with non–natural configuration at C-10 <02CEJ1051>.

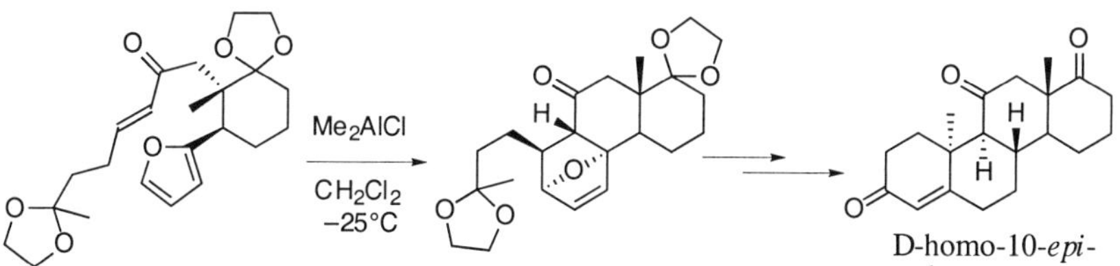

A sequence of aminodiene Diels-Alder reactions of amidofurans provided tricyclic ketones, analogues of Kornfeld's ketone, which will serve as the intermediates in the synthesis of ergot alkaloids <02OL4135>.

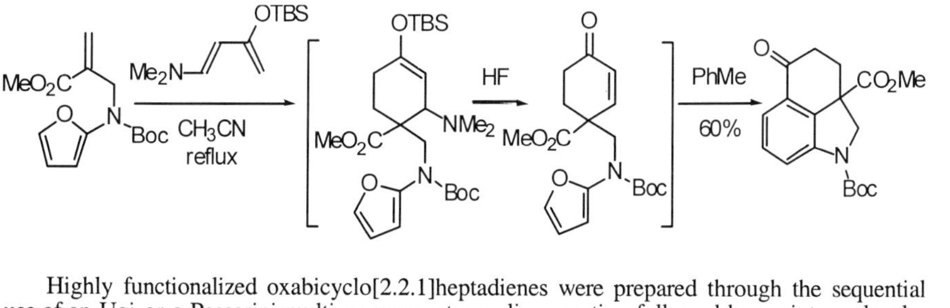

Highly functionalized oxabicyclo[2.2.1]heptadienes were prepared through the sequential use of an Ugi or a Passerini multi–component coupling reaction followed by an intramolecular Diels-Alder reaction of furan with a substituent containing acetylene functionality at C-2 position <02TL943>. Intramolecular Diels-Alder reaction of furan–containing vinylsulfonamides with a chiral auxiliary attached to the nitrogen atom provided chiral δ- and γ-sultams with variable diastereoselectivity. The two diastereomers could be separated by chromatography <02TL4753>.

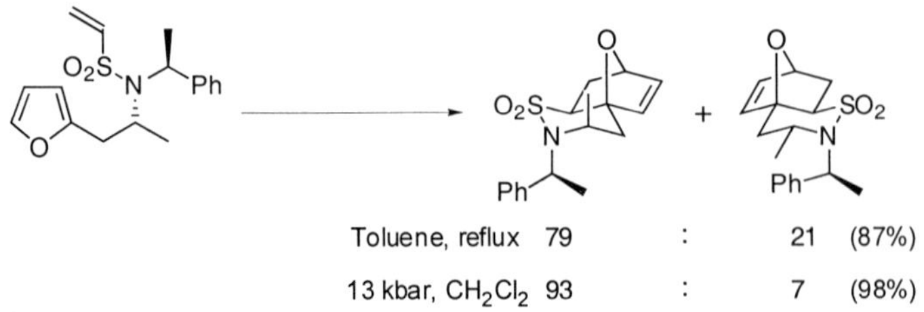

| Toluene, reflux | 79 | : | 21 | (87%) |
| 13 kbar, CH$_2$Cl$_2$ | 93 | : | 7 | (98%) |

Diels-Alder reaction of furfural derivatives with 1,4-phthalazinedione followed by rearrangement and oxidation with Pb(OAc)$_4$ provided [5,6]benza-3a,7a-diazaindane derivatives in moderate to good yields <02OL773>.

As a nucleophile, 2-trimethylsiloxyfuran attacked *N*-gulosyl-*C*-alkoxymethylnitrones to give bicyclic products. The stereoselectivity of the reaction was highly dependent on the bulkiness of the *C*-substituent of the nitrone. The major adducts were elaborated into the key intermediate for polyoxin C <02OL1111>.

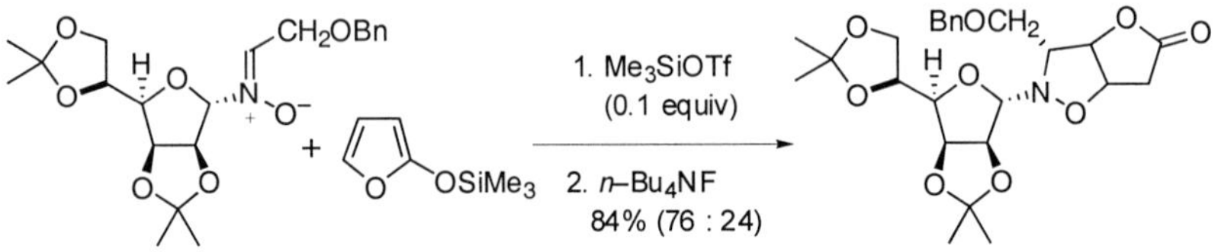

Heterocyclic cage compounds were provided when 2-trimethylsiloxyfuran was allowed to react with 4-amino or 4-hydroxy-4-[(*p*-tolylsulfinyl)-methyl]cyclohexa-2,5-dienone in a one–pot procedure through domino conjugate additions. High stereoselectivity was realized when there is methyl substituent at the 3-position of the cyclohexadienone <02CEJ208>.

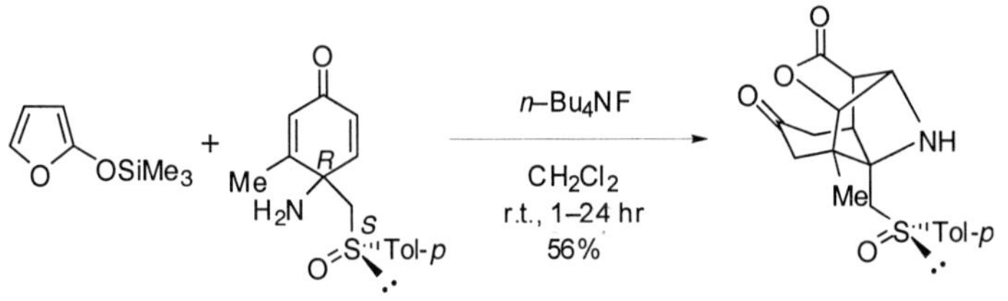

Addition of 2-trimethylsiloxyfuran to an activated 2-acetyl-1,4-naphthoquinone followed by oxidative rearrangement of the resulting furonaphthofuran ring system to a furonaphthopyran ring system was utilized as a key step in the synthesis of the azido analog of the antibiotic medermycin <02SL1318>.

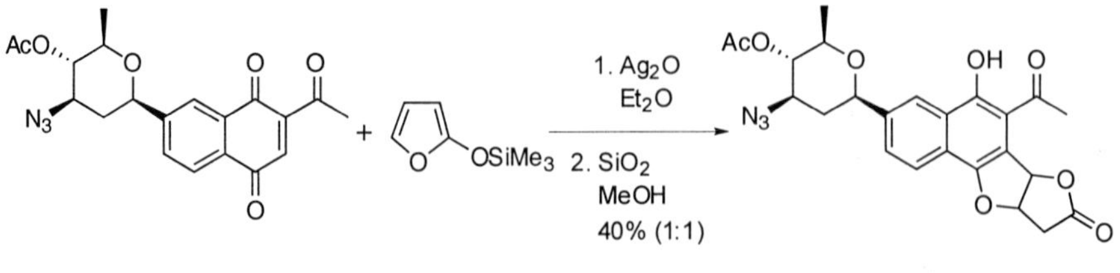

Lewis acid catalyzed intramolecular vinylogous aldol reaction of 5-substituted furoic acids afforded spiro compounds with high diastereoselectivity, which were converted to the zaragozic acids <02JOC4200>.

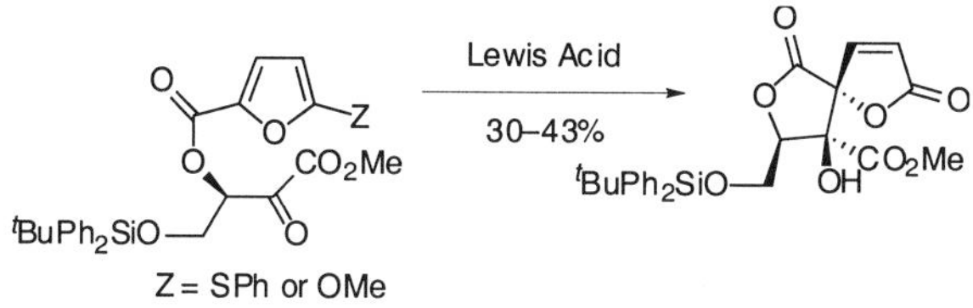

The *meso*-decamethylcalix[5]pyrrole, a long–sought target molecule, was synthesized in low but efficient yield (*ca.* 1%) by using a furan analog <02OL2695>.

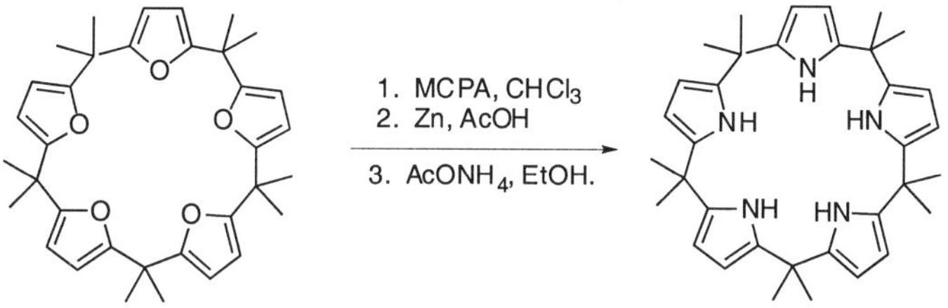

The stereoselective reductive alkylation of 3-methyl-2-furoic acids using a readily available chiral auxiliary provided chiral 2,5-dihydrofurans, which were conveniently transformed into highly functionalized and enantiopure dihydropyranones in high yield <02OL3059> <02JCS(P1)1369>. A C-3-trimethylsilyl group was found to give rise to high levels of stereoselectivity. <02JCS(P1)1748>.

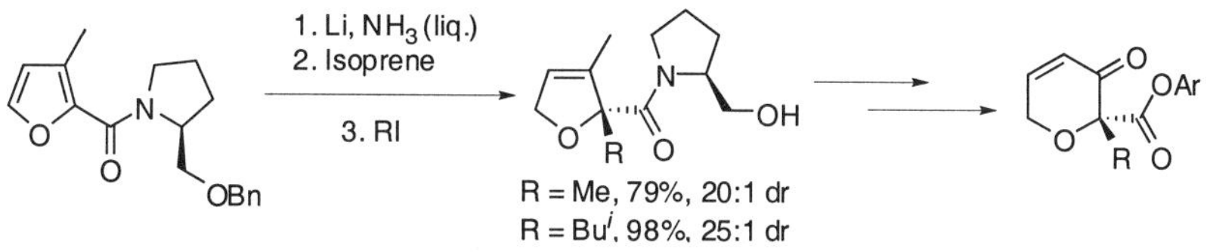

Furfuryl alcohols rearrange to 4-hydroxycyclopentenones by acid treatment, and the latter isomerize to 4-hydroxycyclopentenones (02SL1451).

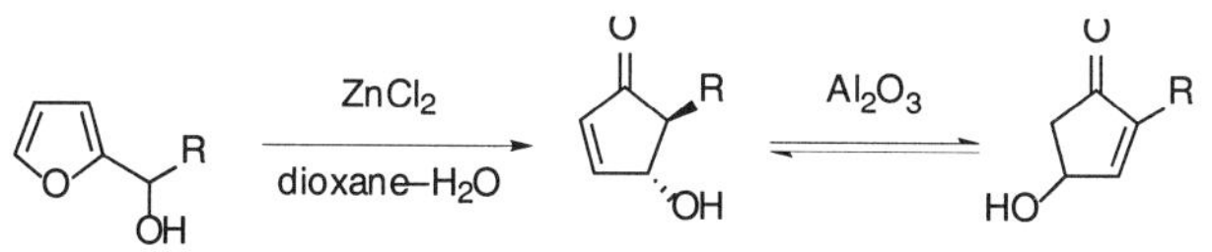

Sharpless catalytic asymmetric dihydroxylation of 2-vinylfuran followed by treatment of the diol product with NBS, and subsequent Jones oxidation and Luche reduction provided a highly functionalized α,β-unsaturated γ-lactone in 70% yield overall giving a key intermediate in the synthesis of phomopsolide C <02TL8195>. Similar procedures were adopted by the same authors to synthesize 2-deoxy and 2,3-dideoxyhexoses <02OL1771>. As depicted below, Sharpless kinetic resolution was also used to prepare a pyranone from a furfuryl alcohol in 38% yield, which served as the intermediate in the synthesis of (+)-isoaltholactone <02T6799>.

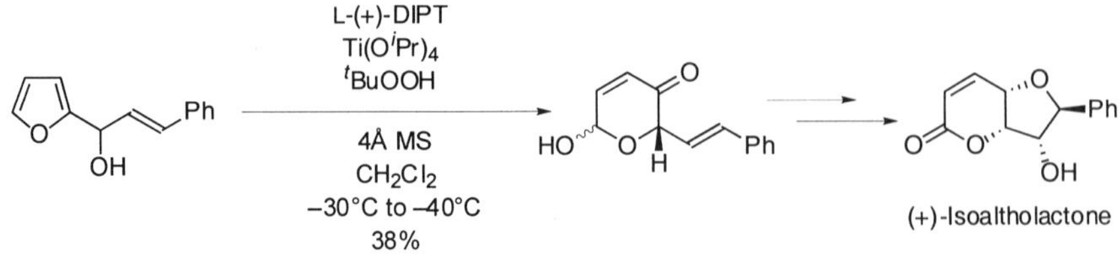

Photo–oxidation of furan was used as the final step to realize the natural product (+)-cacospongionolide B possessing antimicrobial and cytotoxic activities <02JA11584>.

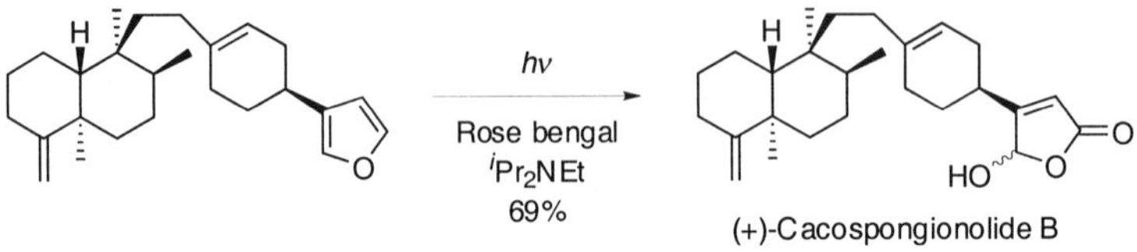

Alkylation of furans with 1-iodoperfluorohexane used sodium dithionite as radical initiator. Thus, perfluorohexylation in the 2-position accompanied with dimerization was observed <02TL443>.

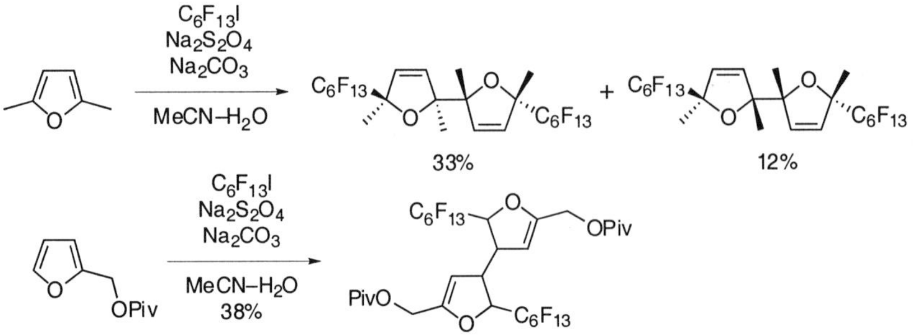

Organometallic 2-furylate–complexes decomposed through deprotonation and ring–opening of the 2-furyl group to give an ynylenolate complex <02OM1759>. Photo–reaction of arenecarbothioamides with 3-methylfuran in benzene–acetone provided dihydrothiophene derivatives. This reaction was affected by the properties of solvents and the furan used. If the reaction of benzenecarbothioamide with furan was performed in benzene, 3-benzoylfuran was produced <02H(57)1587>.

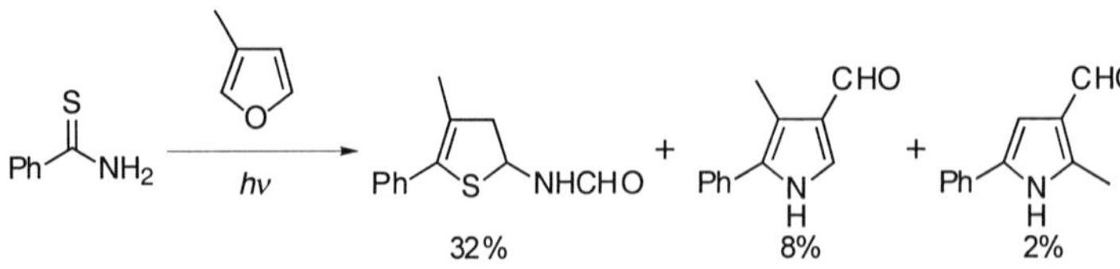

5.3.2.2 Di- and Tetrahydrofurans

The tricyclic compound shown in the following scheme underwent a 1,2–acyl shift (oxa–di–π–methane rearrangement) to form the tetracyclic compound upon irradiation in a Pyrex immersion well in degassed dry acetone as both a solvent as well as a sensitizer. On the other hand, a 1,3–acyl shift mechanism was responsible for the formation of the cyclobutanone when the same tricyclic molecule was irradiated in benzene at 10°C <02T9729>.

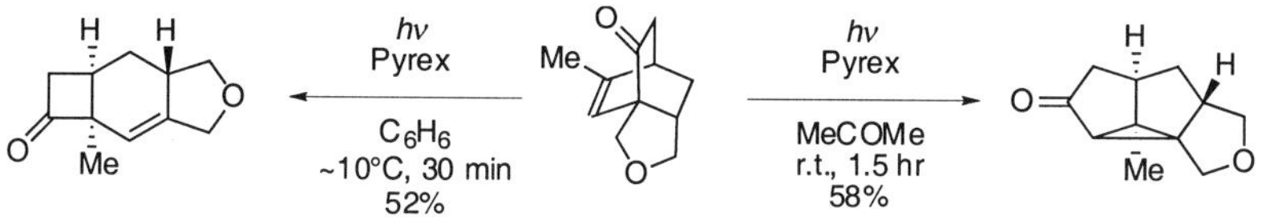

Tetrahydrofuran underwent addition to imines in the presence of dimethylzinc and air via a presumed radical process <02OL3509>. An unusual cleavage of tetrahydrofuran by thiophenol on the surface of silica gel impregnated with $InCl_3$ under microwave irradiation was also reported <02SL987>. A stereoselective 1,4–rearrangement–ring expansion of the tetrahydrofurans to oxocanes as exemplified below via a bicyclo[3.3.0]oxonium ion intermediate was recorded <02OL675>.

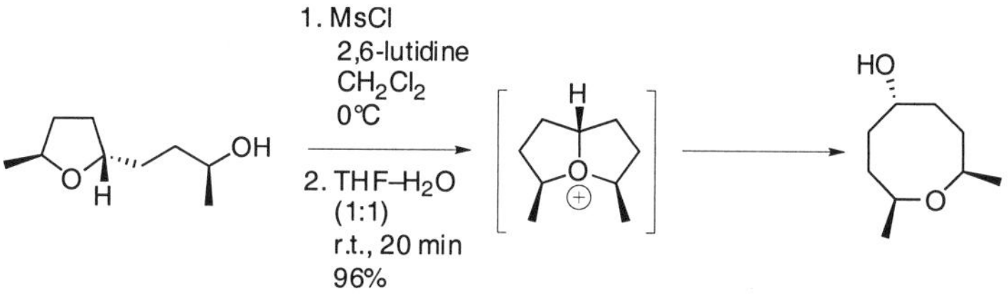

The 2-methylenetetrahydrofuran depicted in the following scheme underwent a Claisen rearrangement catalyzed by t-Bu_3Al to provide an inseparable mixture of cycloheptenes as well as a ring–opened product <02T10189>.

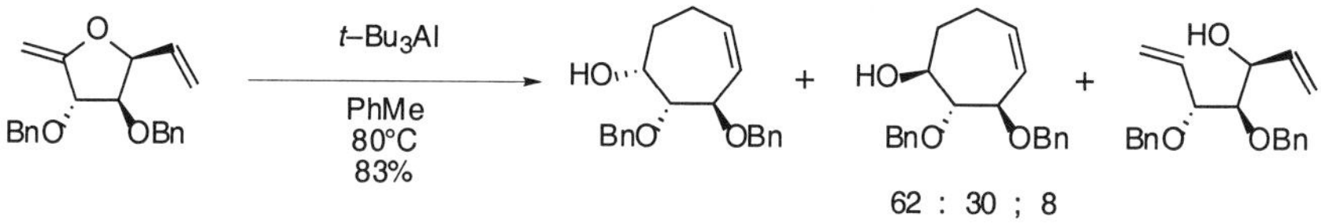

Asymmetric Heck reactions using 2,3-dihydrofuran as substrate were achieved with microwave irradiation in rapid reactions and with preparative convenience <02S1611>. Fluorous chiral BINAP was also employed for asymmetric Heck reactions of 2,3-dihydrofuran <02TL3053>. On the other hand, a stereocontrolled hydroxy group protection involving asymmetric tetrahydrofurylation in good enantiomeric excess was recorded <02CL782>.

Li <02JOC3969> and Yadav <02S2537> both reported similar procedures involving the use of $InCl_3$– or $Sc(OTf)_3$–catalyzed domino reactions of aromatic amines with 2,3-dihydrofuran. The indium trichloride example is illustrated below. However, $Sc(OTf)_3$ was found to give the same products in a similar yield but with a better diastereoselectivity (92:8). $InCl_3$ and triphenylphosphonium perchlorate were also found to be effective for the cyclization reaction between o-hydroxyaldimines and 2,3-dihydrofuran to furnish furanobenzopyran derivatives <02T10301>.

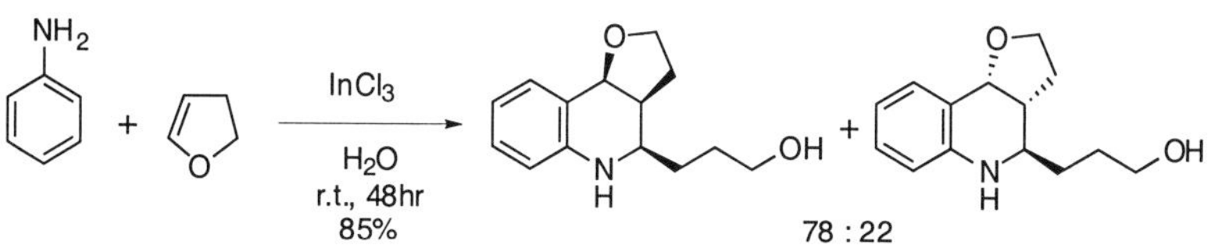

A facile synthesis of enantiopure tricyclic furyl derivatives employing 4-vinyl-2,3-dihydrofuran via Diels-Alder cycloaddition reaction was reported <02TL7983>. A new capture–ROMP–release procedure for chromatography–free purification of *N*-hydroxysuccinimide Mitsunobu reactions was reported by Hanson, who used a Mitsunobu reaction to capture a variety of alcohols onto a norbornenyl *N*-hydroxysuccinimide monomer. Treatment of this monomer under ROM–polymerization then generated a water-soluble polymer that was readily separable from other by–products. Subsequent reaction with hydrazine was utilized to release the *O*-alkylhydroxylamines in good purity from the water–soluble polymer <02OL1007>.

Rhodium–catalyzed addition of arylboronic acids to oxabenzonorbornadienes led to cleavage of the oxygen bridge. An example is shown in the following scheme <02CC390>. Copper phosphoramidite–catalyzed enantioselective alkylative ring–opening reactions of the same systems with dialkylzinc is also documented <02OL2703>. The latter procedure gave a high level of *anti*–stereoselectivity.

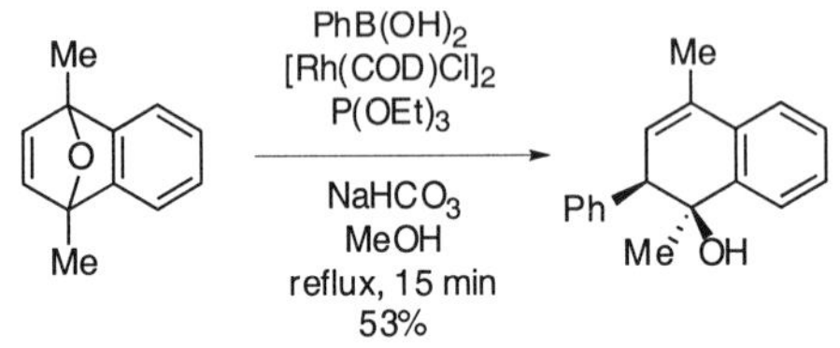

A direct ring opening of the oxabicyclo[3.2.1]octenes shown below with DIBAL-H provided an entry to the chiral C-19–C-26 and C-27–C-32 fragments of scytophycin C <02OL245>.

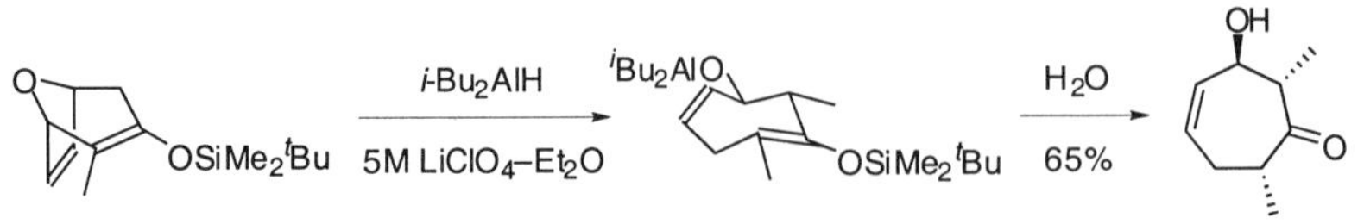

Exposure of the 2,5-dihydrofuran–embedded diene illustrated below in a dilute solution of Grubbs' catalyst led to its rearrangement to spirocycles <02AG(E)4560>.

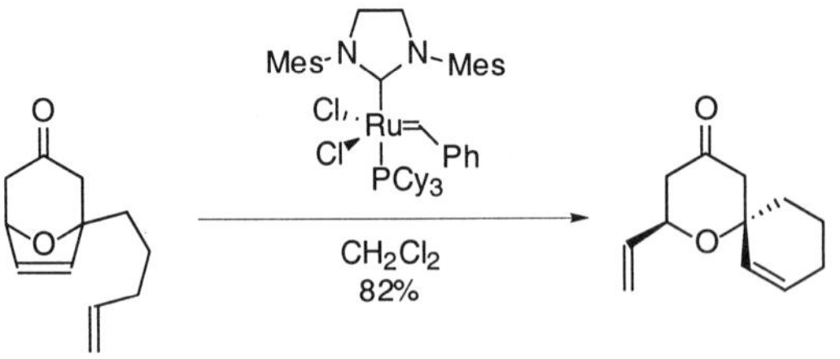

5.3.2.3 Benzo[*b*]furans and Related Compounds

The four prandiol–related stereoisomers were synthesized in five steps starting from (*R*)-(+)-2,3-dihydro-2-(1'-methylethenyl)-6-methoxybenzo[*b*]furan and its (*S*)-(-)-enantiomer. The synthesis involved dihydroxylation and condensation to afford a furocoumarin. The absolute configuration of the newly formed stereogenic center at C–1' was established by circular dichroism spectroscopy in combination with X-ray crystallographic analysis <02TA1147>. Kinetic resolution of (±)-2,3-disubstituted-2,3-dihydrobenzo[*b*]furan depicted in the following scheme was achieved through *Candida cylindracea* lipase and *Rhizopus arrhizus* lipase. The influence of enzymes and solvents on the enantioselectivity was discussed <02TA1219>.

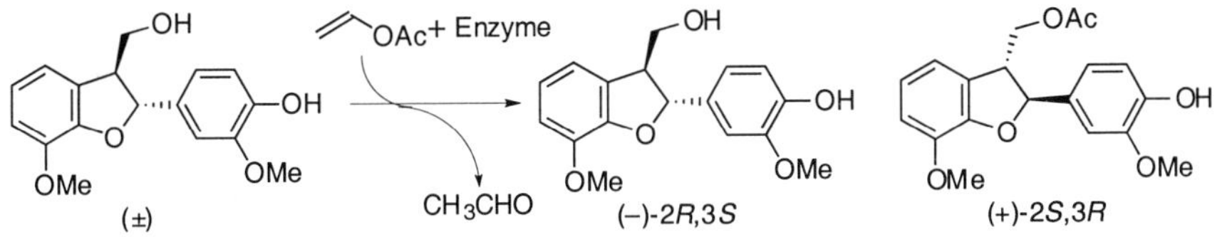

Different allyl 2-lithioaryl ethers underwent a tandem carbolithiathion/γ–elimination in Et₂O/TMEDA, affording *o*-cyclopropylphenol derivatives in a diastereoselective manner. The use of (–)-sparteine as a chiral ligand instead of TMEDA allowed the synthesis of cyclopropane derivatives with up to 81% ee. The following scheme illustrates this stereoselective process for the formation of chiral *o*-cyclopropylphenols <02OL2225>.

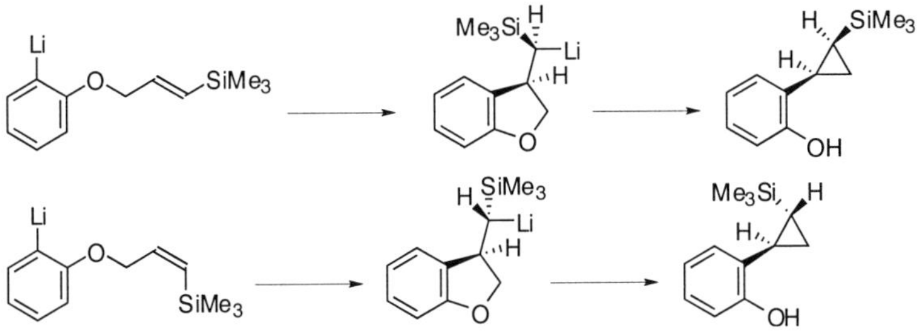

The key feature in the total synthesis of (–)-frondosin was the intramolecular Heck reaction to build up the benzo[*b*]furan based tetracyclic ring, which eventually led to the final target <02AG(E)1569>.

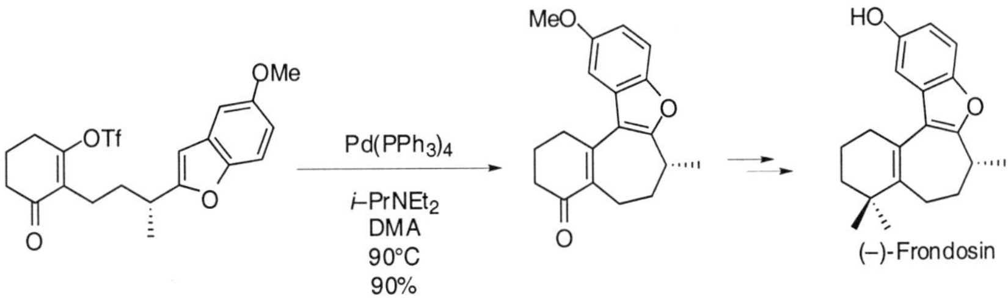

An efficient synthetic approach was developed based on regioselective transition metal–catalyzed cross–coupling reactions by using 2,3,5-tribromobenzofuran as a starting material. Thus, a palladium–catalyzed cross–coupling reaction between the tribromide and an arylzinc reagent gave the dibromide, which underwent sequential Kumada coupling with a Grignard reagent and a Negishi coupling with methyl zinc chloride to afford eventually the trisubstituted benzo[*b*]furan <02TL9125>. Benzo[*b*]furan-3-triflates also underwent standard Stille, Heck, Suzuki, and Sonogashira reactions to provide 3-substituted benzo[*b*]furans in excellent yields <02SL501>.

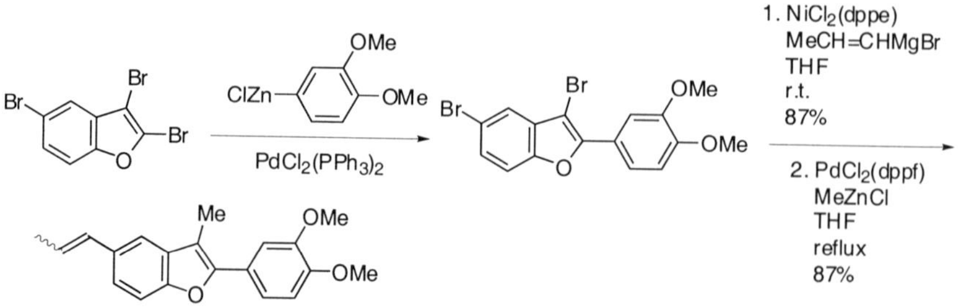

Mixed heterocalix[3]arenes and heterocalix[4]arenes containing indole and benzo[*b*]furan rings were synthesized via acid–catalyzed reactions of benzo[*b*]furylmethanols and activated indoles. These new calixarenes are of interest as potential molecular receptors. Synthetically, the benzo[*b*]furan part of these calixarenes was realized from benzo[*b*]furancarbaldehyde as illustrated in the following scheme <02CC810>.

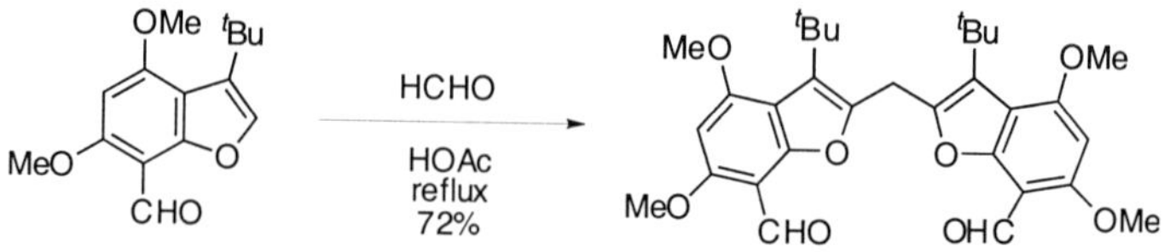

Several 4-anilinofuro[2,3-*b*]quinolines were synthesized in good yields starting from dictamine a naturally occurring alkaloid. Their cytotoxicity was evaluated in the NCI's full panel of 60 human cancer cell lines <02HCA2214>. A regioselective Friedel–Crafts reaction was utilized to directly synthesize 3-benzoyl-2-methylbenzo[*b*]furan from 2-methylbenzo[*b*]furan. The product is a useful intermediate to generate a novel type of antiarrhythmic agent <02JMC623>.

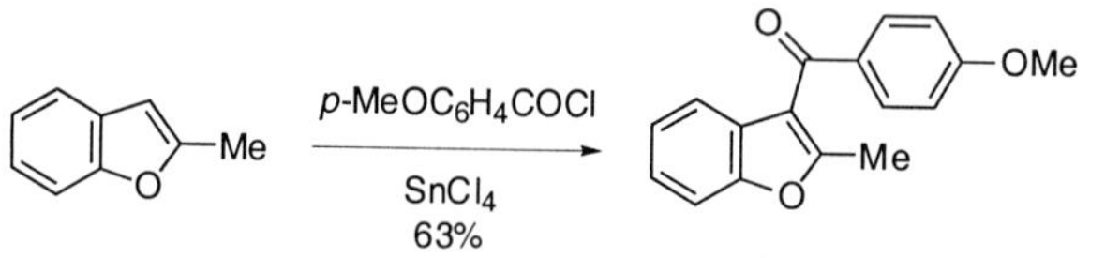

Furopyridines have a fused structure containing both a π–excessive furan ring and a π–deficient pyridine ring. Interesting results were obtained during the Birch reduction of the four different furopyridines, and a mechanistic interpretation for the results was presented. An example is shown in the following scheme <02JHC335>.

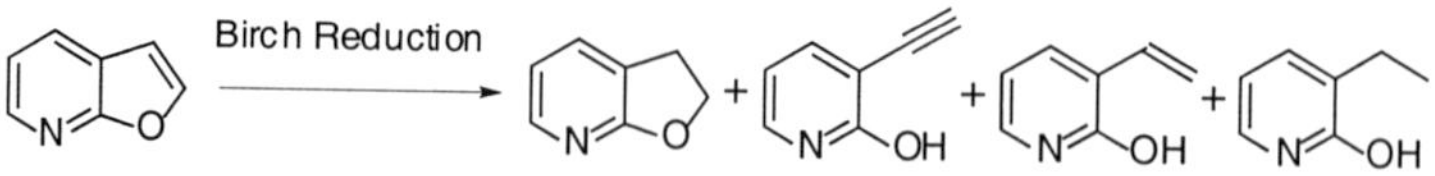

5.3.3 SYNTHESIS

5.3.3.1 Furans

Reductive annulation of 1,1,1-trichloroethyl propargyl ethers using a catalytic amount of Cr(II) generated by Mn–Me$_3$SiCl provided 3-substituted furans in high yields. Some natural products, such as perillene and dendrolasin, were prepared utilizing this procedure <02OL1387>.

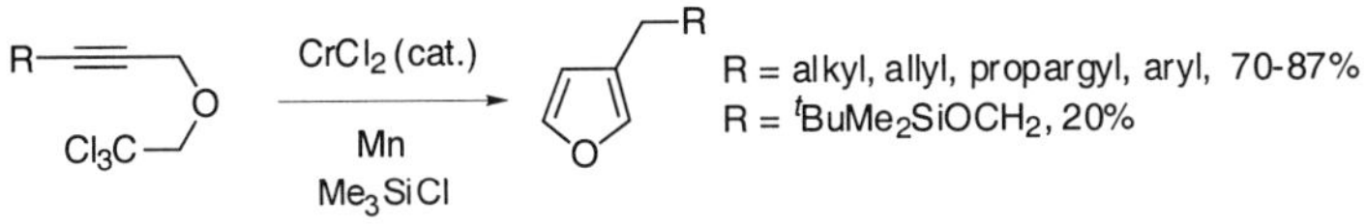

Five–Membered Ring Systems : Furans and Benzofurans

Methylfurolabdane containing a 2,3-disubstituted furan substructure, isolated from the cuticular wax of the leaves of *Nicotiana tabacum*, was synthesized from (+)-sclareolide. The furan ring was formed from a β,γ-unsaturated ketone via an oxidation–cyclization procedure <02EJO4169>.

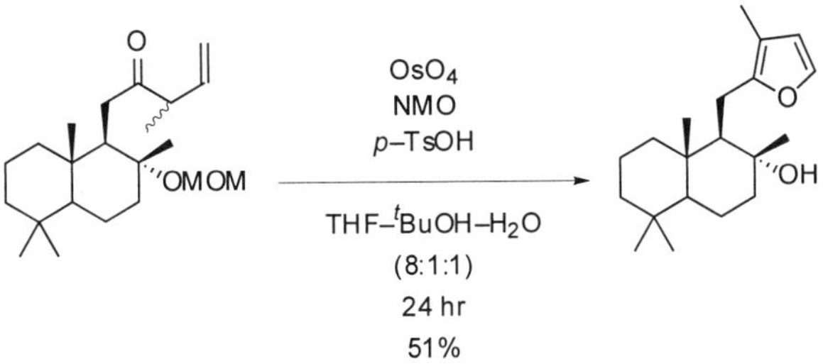

As illustrated in the following scheme, the oxa–Pictet–Spengler reaction of (3-furyl)alkanols, prepared by the Yb(fod)$_3$ catalyzed carbonyl–ene reaction of 3-methylene-2,3-dihydrofuran and aldehydes, with aldehydes in the presence of *p*-toluenesulfonic acid gave the furano[2,3-*c*]pyrans in good yields <02S1541>. A two–step electrochemical annulation directed to polycyclic systems containing annulated furans has been developed. This procedure involved an initial conjugate addition of a furyethyl cuprate and trapping of the enolate as the corresponding silyl enol ether. The second step of the annulation involved the anodic coupling of the furan and the silyl enol ether to form the *cis*–fused six–membered ring with high stereoselectivity <02OL3763>.

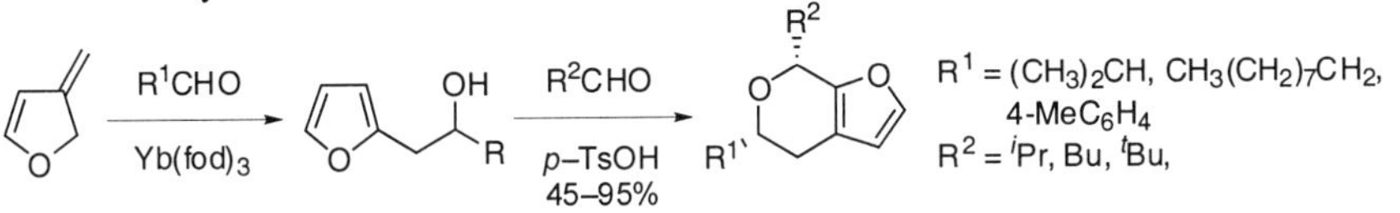

2,4-Disubstituted furans were provided in high yield through a novel oxidative cyclisation-dimerisation reaction between two different allenes as shown below <02AG(E)1775>.

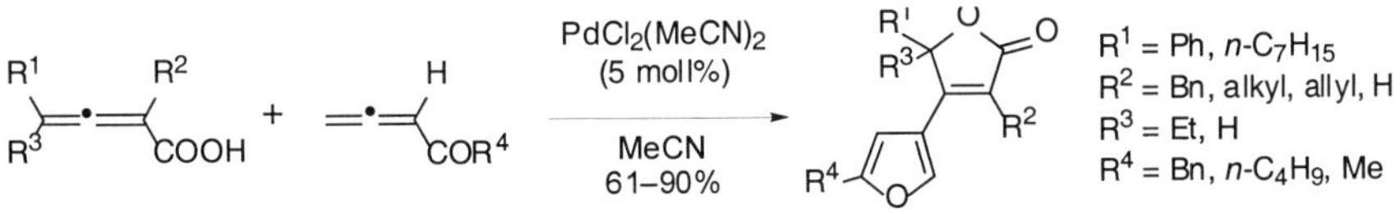

2,5-Disubstituted furans with a cyclopropane subunit at the 5–position were synthesized through metal–catalyzed cyclization of 1-benzoyl-*cis*-1-buten-3-yne derivatives in high yield. Many types of metal complexes, such as Mo, W, Ru, Rh, Pd and Pt complexes were suitable catalysts <02JA5260>.

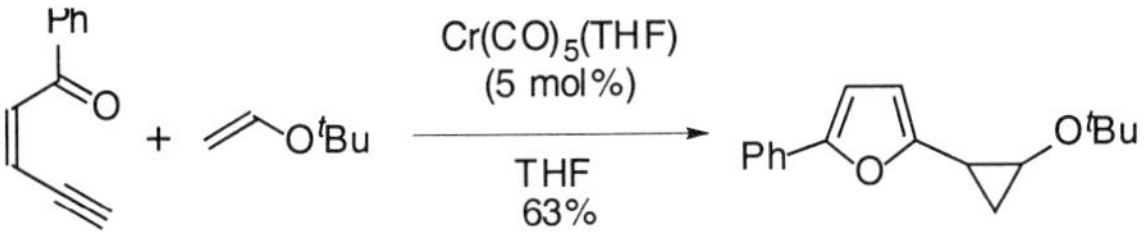

Reaction of 2-methyl-5-lithiofuran with chiral aminoaldehydes in the presence of Lewis acids gave the corresponding furyl *anti*-amino alcohols with high diastereoselectivity <02TA2133>. A library of 2,5-disubstituted furans was set up by using a 5-hydroxymethylfurfural–based scaffold on solid support <02T10469>. A general and efficient procedure for the preparation of 2,5-disubstituted furans containing acid– and base–labile groups via CuI–catalyzed cycloisomerization of alkynyl ketones has appeared, for which a plausible mechanism was proposed <02JOC95>. 2,5-Disubstituted furans can also be prepared in high yields from the reaction of Me$_3$Si-thymine with 2-substituted-2,5-dimethoxy-2,5-dihydrofurans in the presence of Me$_3$SiOTf <02OL3251>. Synthesis of an 18–membered cyclic trimer of 5-(aminomethyl)-2-furancarboxylic acid from the corresponding furan was reported, which served as a novel synthetic receptor for carboxylate recognition <02TL1317>.

Deoxygenation of a 1,2,4-trioxane with Zn powder in acetic acid–H$_2$O resulted in an unexpected formation of a furan in high yield <02S1711>.

α,β-Unsaturated carbonyl compounds with an appropriate leaving group underwent 1,5-electrocyclization reactions to yield 2,5-disubstituted furans upon heating in the presence of acid, presumably through the intermediate formed from 1-oxapentadienyl cations, the conformational and energy properties of which were also studied by DFT calculations (B3LYP/6-31+G*) <02EJO1523>. An oxa–analog of 5,10,15-triarylcorrole, where two pyrrole rings are replaced by furan moieties, was produced by condensation of 2,5-bis(arylhydroxymethyl)furan, 2-phenylhydroxymethylfuran, and pyrrole. 2-Phenylhydroxymethylfuran served as a suitable synthon to introduce a furan ring with the ability to create a direct pyrrole–furan α–α bond <02JOC5644>.

A simple procedure to prepare 5-aryl- and 5-pyridyl-2-furaldehydes from inexpensive, commercially available 2-furaldehyde diethyl acetal was reported. The reaction proceeded in a four–step, one–pot procedure and the yield of coupling step was usually between 58–91% <02OL375>. A facile route to 3,4-furandicarboxylic acids was developed. DDQ–oxidation of 2,5-dihydrofuran derivatives, which were produced from dimethyl maleic anhydride, furnished the desired esters of furan-3,4-dicarboxylic acid <02S1010>. The furan-fused tetracyclic core of halenaquinol and halenaquinone possessing antibiotic, cardiotonic, and protein tyrosine kinase inhibitory activities was synthesized. Intramolecular cycloaddition of an *o*-quinodimethane with furan gave the adduct as a single isomer via an *endo*–transition state, which was converted to trisubstituted furan by oxidation–elimination reactions <02T6097>.

Reaction of functionalized bicyclic compound with DDQ provided the tricyclic core of phomactin A. This product with a dihydrofuran subunit, was extremely unstable and dehydrated readily to form the corresponding furan <02OL2413>.

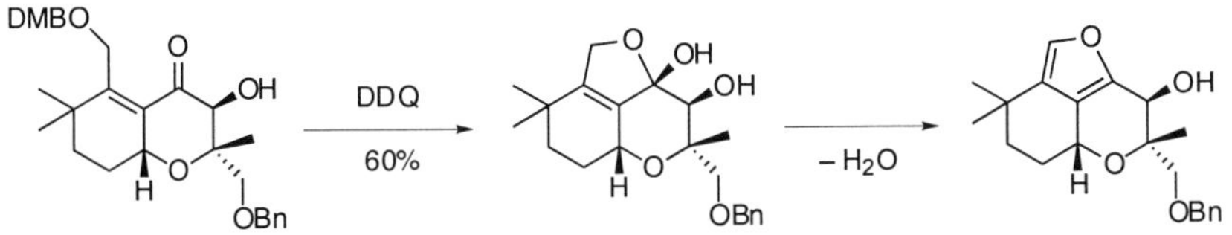

An efficient procedure for the preparation of 2,3,4-trisubstituted furans in three steps was reported. Thus, the reaction of hydroxyacrylonitrile sodium salts with chloromalonate gave rise to malonate ester intermediates, ring closure of which afforded 3-amino-4-arylfuran-2-carboxylates in 15–40% yields <02S753>.

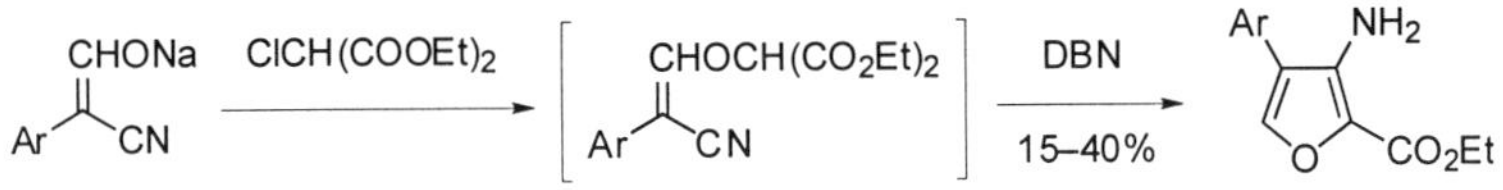

2,3,4-Trisubstituted furans fused with a heterocycle, the equivalents of 2,3-dimethylidenechroman-4-ones, were synthesized by employing a hydrolytically induced cycloreversion procedure. Acid–catalyzed unmasking of the acetal group proceeded to provide furobenzopyran derivatives, with concomitant cyclization and elimination of propanediol. The cycloaddition reactions of these with dienophiles was also studied <02TL4507>. A furo-[3,4-b]indole, an indole-2,3-quinodimethane synthon was also prepared in a good yield utilizing a similar procedure through cyclization of a hydroxymethyl substituted indole acetal in the presence of a Lewis acid <02JOC1001>.

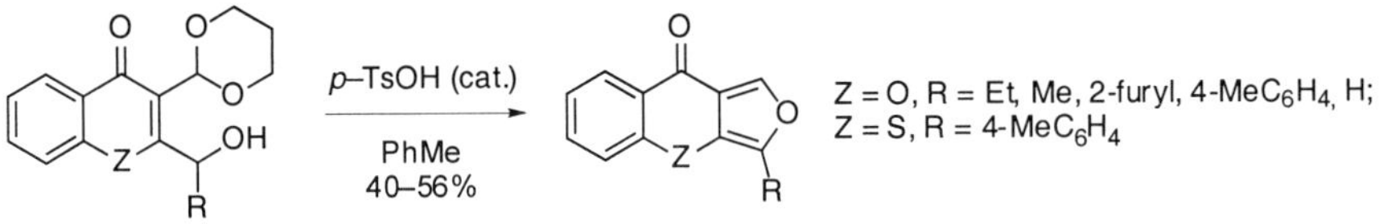

In decalin at 190 °C or in acetic acid with 1,4-benzoquinone and sodium acetate trifluoromethylbutadienyl phenyl sulfides underwent intramolecular cyclization to afford 4-trifluoromethyl-2,3-disubstituted furans in high yields <02TL665>.

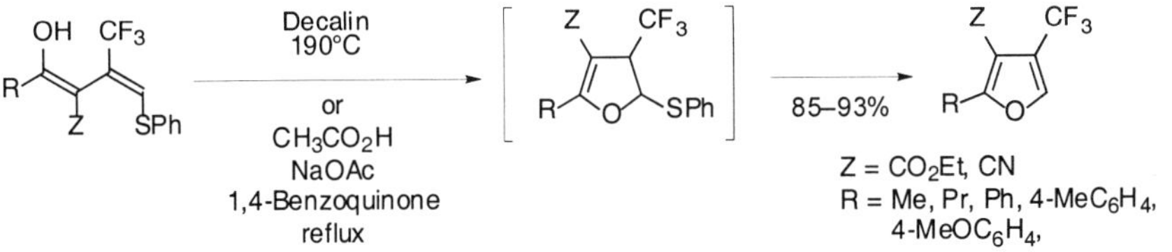

Reaction of ketones and α,α-dimethoxyketones in the presence of $TiCl_4$-Bu_3N led to the formation of α,β-unsaturated lactones, which were converted to 2,3,4-trisubstituted furans. One of the examples is the preparation of menthofuran, the natural mint perfume <02CC2542>.

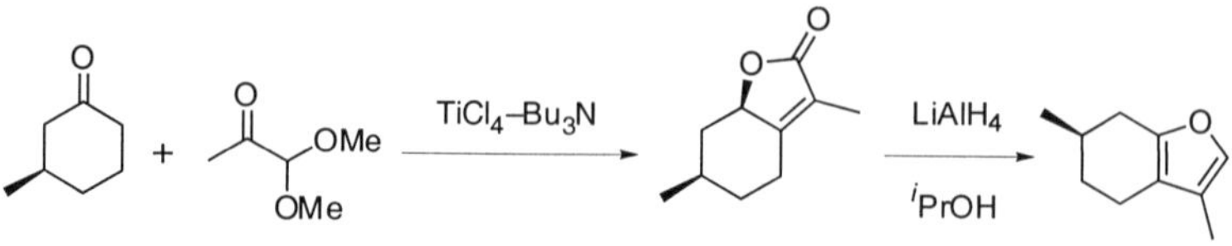

2-Alkenylfurans were synthesized from a Pd–catalyzed cyclization reaction of an α-propargyl β-keto ester. High E/Z-selectivity was realized when the Me$_3$Si– group was introduced at the α'-position of the triple bond. The geometry of the double bond was almost completely inverted by reaction with a catalytic amount of diphenyl diselenide <02OL1787>.

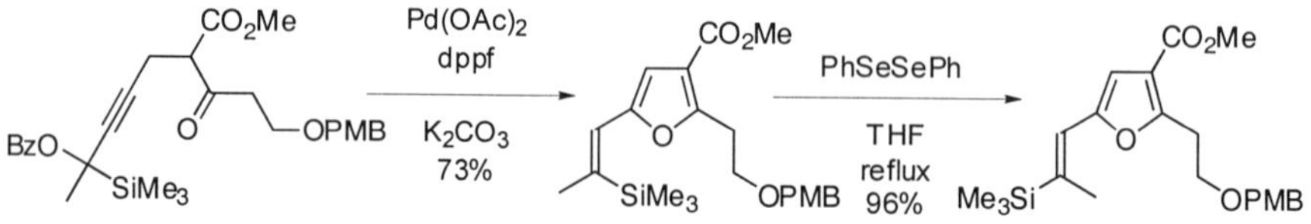

Oxidation of the 2H-pyran-2-ones as illustrated below under basic condition provided 2,3,5-trisubstituted furans albeit in low yield <02EJO1830>.

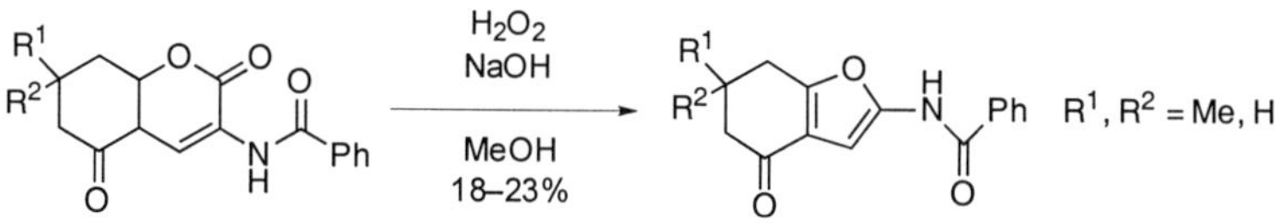

Several related methods for the preparation of differentially substituted 5-thio-2,3-trisubstituted furans were developed, which involved the formation of a thionium ion and the cyclization of this reactive intermediate into the tethered carbonyl group <02JOC1595>.

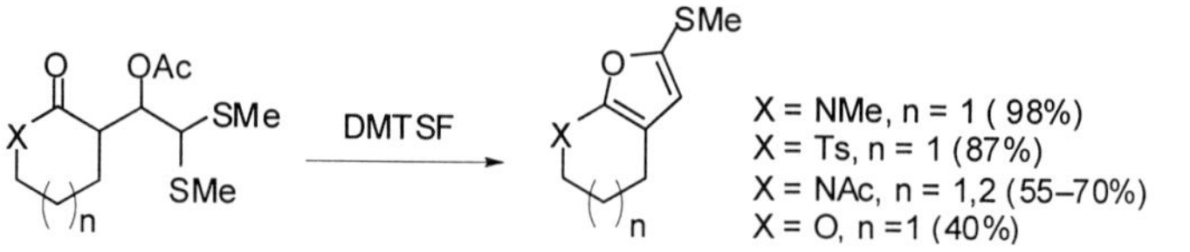

A facile three–component reaction leading to furan-annulated heterocycles appeared. The *in situ* generated quinone methides from hydroxycoumarin and hydroxy-6-methylpyrone with various aldehydes underwent facile reaction with cyclohexyl isocyanide to produce furocoumarins in good yields. Quinone methides from hydroxy-1-methylquinolinone afforded furoquinolones. The reaction presumably occurs via a [4+1] cycloaddition followed by a [1,3] H shift <02TL2293>. A similar procedure was reported by using N,N'-dimethylbarbituric acid, 4-nitrobenzaldehyde and alkyl or aryl isocyanides. The reaction proceeded in water and gave rise to the corresponding furo[2,3-d]pyrimidinediones in high yields <02TL9151>.

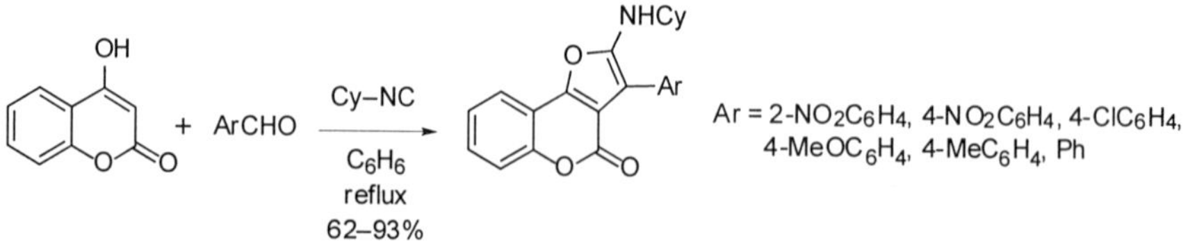

Yavari reported a procedure to prepare tetrasubstituted furans through the reaction of dibenzoylacetylene and enol systems in the presence of triphenylphosphine <02TL4503>.

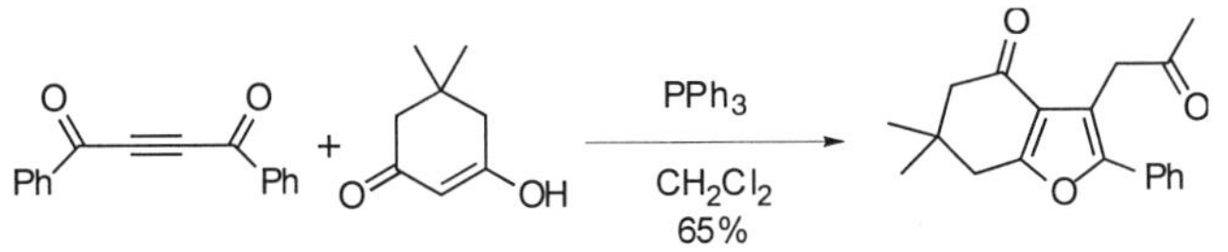

5.3.3.2 Di- and Tetrahydrofurans

The synthesis of di- and tetrahydrofurans continues to be one of the most active areas in synthetic organic chemistry. In the construction of tetrahydrofurans, the Williamson cycloetherization has always been a practical approach. Lautens employed a three–step procedure to convert the alcohol shown below to a bis-tetrahydrofuran <02OL1879>. A similar cyclization route was also utilized in the total synthesis of muconin <02TL8661>.

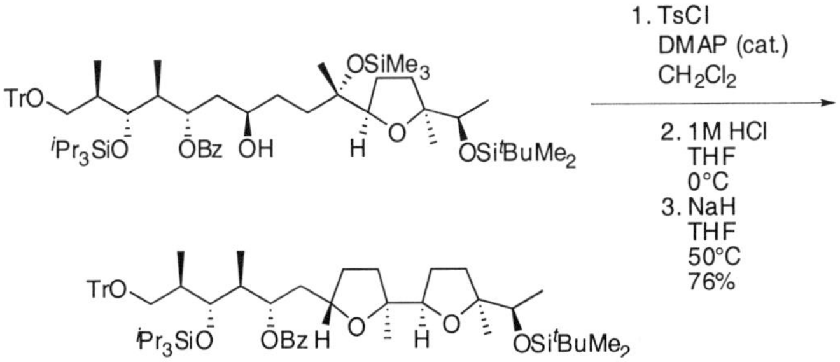

In the synthesis of new 18-substituted analogues of calcitriol, the iodoketone was stereoselectively reduced by sodium borohydride, and the product was treated with silver acetate to afford the tetrahydrofuran ring <02JOC4707>. A silver(I) oxide mediated highly selective monotosylation of symmetrical diols has been developed and applied to the synthesis of polysubstituted tetrahydrofurans <02OL2329>. An aminotetrahydrofuran was also prepared via a Williamson–type cycloetherization <02JOC1692>.

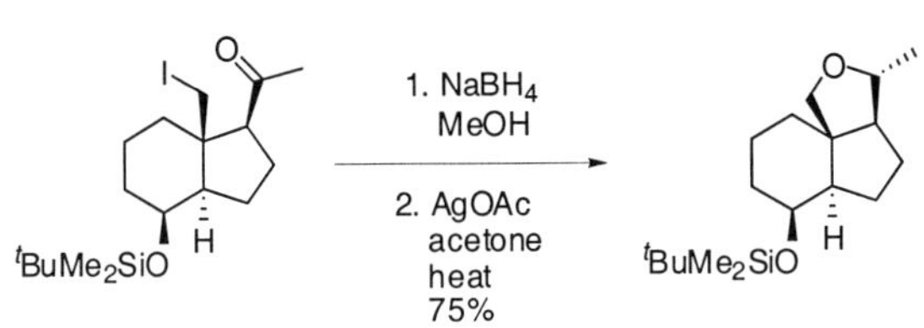

Makosza devised a procedure for the synthesis of substituted tetrahydrofurans by using propyl chloride containing an electron–withdrawing group at C-3. Thus, treatment of these alkyl chlorides with a strong base led to the formation of a carbanion, which upon reaction with an aldehyde generated a tetrahydrofuran. A Williamson–type intermediate appeared to be pivotal <02CEJ4234>. Asymmetric synthesis of tetrahydrofurans by [1,2]–phenylsulfonyl migrations under thermodynamic control conditions was dealt with by Warren <02JCS(P1)2634> <02JCS(P1)2646>, who also assessed the scope and limitations of these novel reactions <02JCS(P1)2652>. Exposure of the acetate depicted below to potassium carbonate in MeOH resulted in a saponification reaction, and was followed by attack of the alkoxide on the epoxide ring and the formation of the tetrahydrofuran framework <02T1817>. Similar cyclic ether formation pathways have been employed for the total synthesis of *cis*-solamin <02OL1083>, in the stereoselective synthesis of tetrahydrofurans <02T1865>

<02TL1495>, and in the synthesis of four types of tetrahydrofuran cores in acetogenins <02OL2977>. A similar strategy was also utilized in the formation of tetrahydrofurans containing a 3-methylene group (*vide infra*) <02JOC6690>, as well as in the synthesis of central building blocks of biogenetically intriguing oxasqualenoids <02TA2641>.

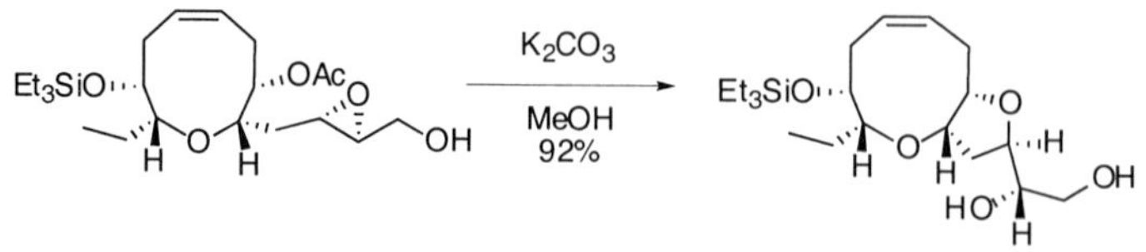

Another route by which tetrahydrofurans can be made is using a 5–*endo*–trig iodocyclization. For example, the (Z)-ene-diol shown below gave almost exclusively the polysubstituted tetrahydrofuran under excellent stereochemical control <02TL6771>. In a similar manner, the essential tetrahydrofuran skeleta in the substituted 1,5-dioxaspiro[2.4]heptanes <02HCA3262>, (+)-pamamycin–607 <02AG(E)1392> <02TL3613>, several oligo-tetrahydrofurans <02T2077> and a dihydroagarofuran sesquiterpene <02TL627> were all realized applying halocyclizations as a key step. An oxymercuration–demercuration procedure was also employed to construct the key tetrahydrofuran ring in the total synthesis of (+)-nonactic acid <02JCS(P1)2896>.

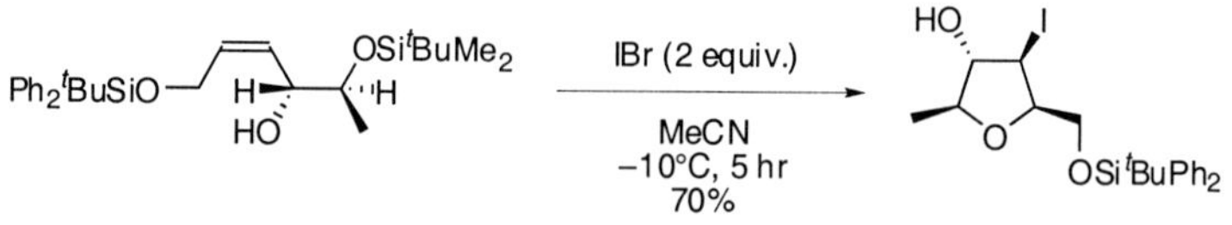

A formal synthesis of pamamycin–607 was reported by Nagumo, who constructed the tetrahydrofuran ring by making use of novel ether–ring transformation via a phenonium ion. The reaction mechanism is depicted below <02TL5333>.

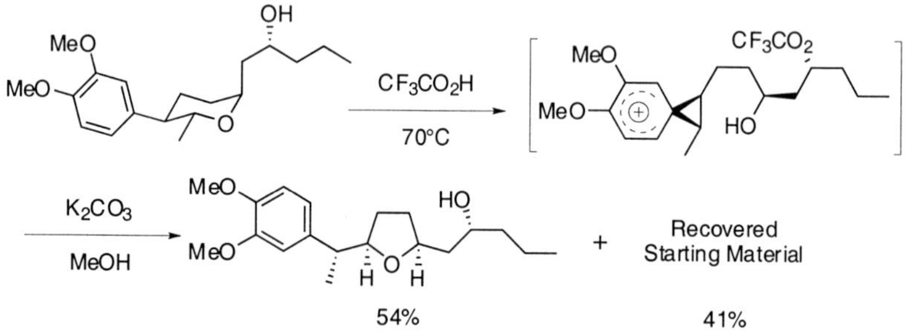

Brown reported a synthesis of *cis*-solamin using a permanganate–oxidative cyclization of a 1,5-diene under phase transfer conditions as illustrated below <02OL3715>. Permanganate oxidation <02JOC8079> as well as RuO_4–promoted oxidation <02TL9265> of farnesoate esters also furnished tetrahydrofuran–containing fragments with control of relative stereochemistry at four new stereocenters.

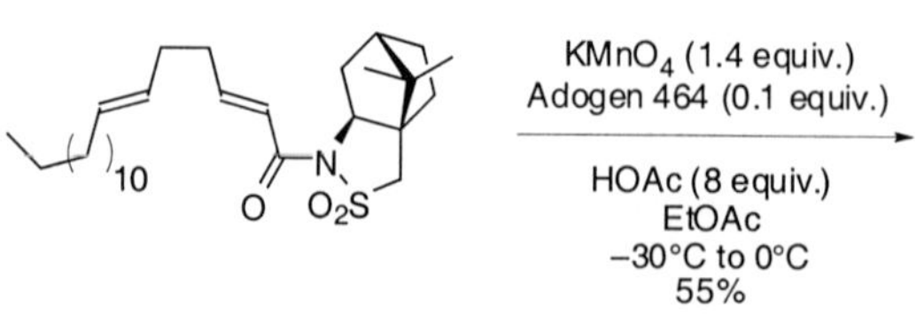

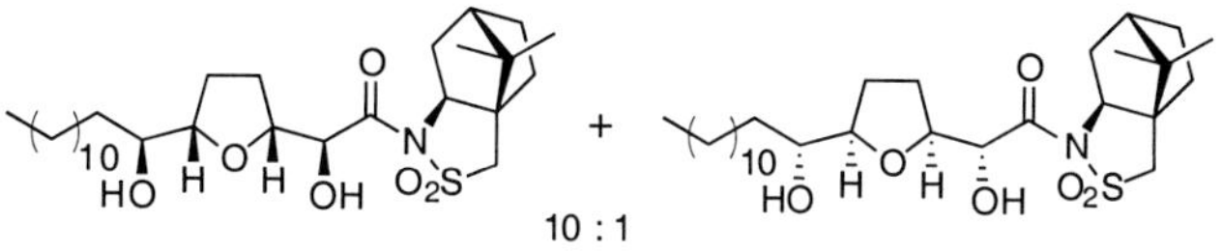

10 : 1

An oxidative spiroannulation reaction was carried out for simple phenols and as a result good yields of spiro–compounds containing tetrahydrofuran rings were obtained <02TL3597>. In the stereospecific and enantiospecific total synthesis of the sarpagine indole alkaloid dehydro-16-epinormacusine B, an oxidative cyclization of the alcohol shown below was the key and final step <02OL4681>.

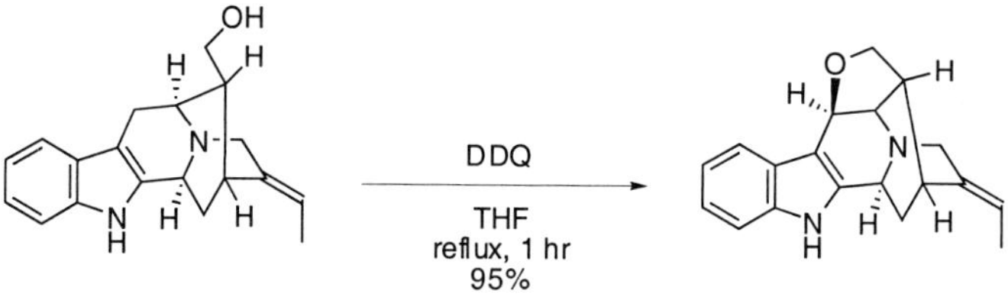

In a study towards the total synthesis of diterpene antibiotic guanacastepene A, an epoxy conjugate ketone was rearranged to the tricyclic ether on exposure to $BF_3 \cdot Et_2O$ in moist dichloromethane <02OL1063>. As depicted in the following example, treatment of allenyl–aldehyde dimethyl acetals with a Lewis acid such as Me_3SiI, $TiCl_4$ or $InCl_3$ provided an entry to a mixture of *cis-* and *trans-*haloalkenyl substituted tetrahydrofurans via an intramolecular nucleophilic attack by the allene group on the oxonium ion generated from the dimethyl acetals and the Lewis acid <02TL9105>.

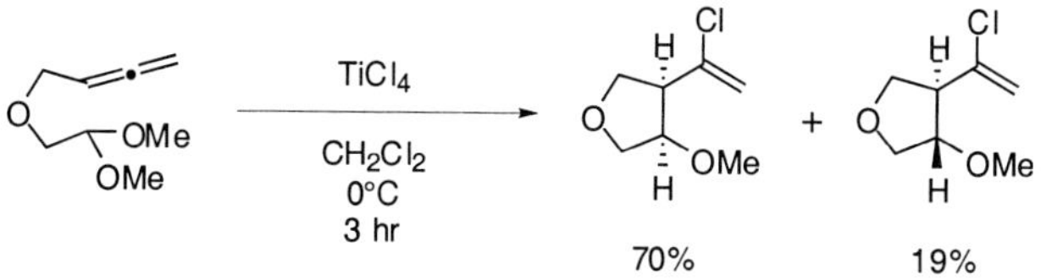

The synthesis of polysubstituted tetrahydrofurans was also achieved in a stereoselective manner by a formal [3+2] cycloaddition of an allylsilane with α-triethylsiloxy aldehydes. An example showing the mechanism is illustrated <02JA3608>. In another approach, allylsilane was also allowed to react with α-keto esters in a [3+2] annulation reaction, providing highly substituted tetrahydrofurans in good yields as single diastereomers <02OL2945>.

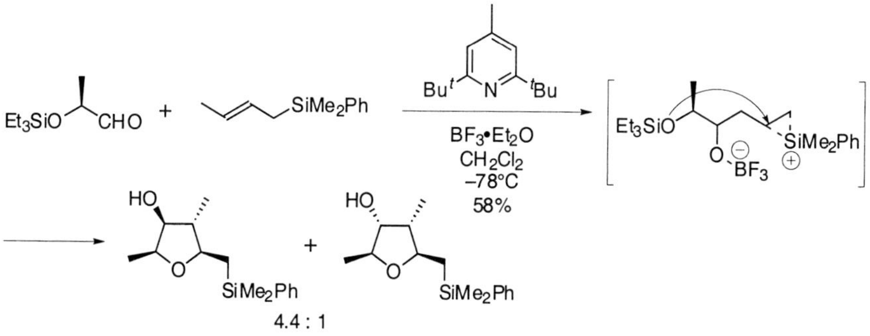

Radical cyclization is also an efficient method for the formation of tetrahydrofuran rings. Thus, treatment of the seleno radical precursor with n-Bu$_3$SnH in hot toluene gave the multi-substituted tetrahydrofuran in high yield <02T2605>.

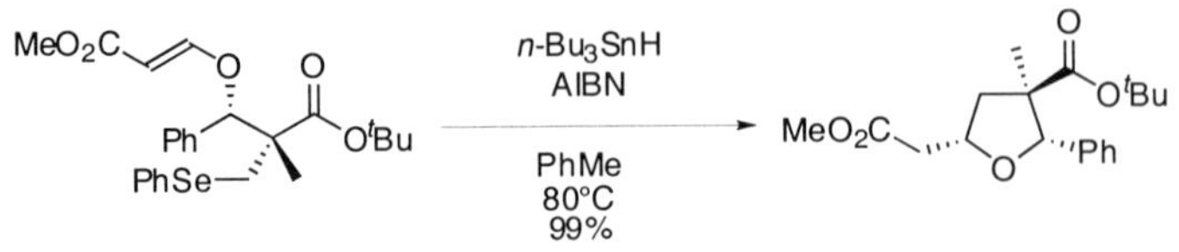

Indolines and indoles were prepared by a direct electrochemical reduction of arenediazonium salts. As a result, radical intermediates were generated from which 3,4-disubstituted tetrahydrofuran skeleta were constructed <02OL2735>. A short and stereoselective total synthesis of furano lignans was realized by radical cyclization of epoxides using a transition–metal radical source <02JOC3242>. Other preparations of tetrahydrofurans using radical cyclization include the synthesis of novel amino acids L-bis-tetrahydrofurylglycines <02TL2931> as well as the stereoselective synthesis of pamamycin–607 <02JA14655> <02TL7295>. An oxidative cyclization involving radical cations in the formation of a tetrahydrofuran ring was utilized by Moeller to realize the synthesis of (+)-nemorensic acid <02JA10101>. A radical cyclization reaction concerning a zirconocene–olefin complex as an efficient single electron transfer reagent was developed by Oshima, who converted the β-haloalkyl allyl acetals to a bicyclic oxygenated product <02SL337>.

A straightforward construction of a tetrahydrofuran ring starting from the α-diazolactone was reported, involving a rhodium–catalyzed C–H insertion <02CC2042>.

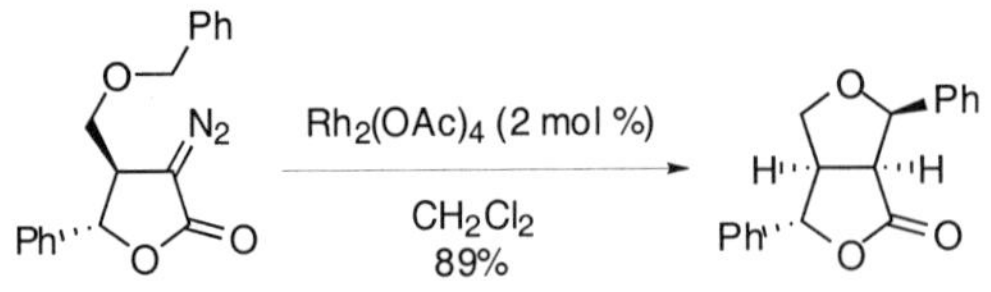

Rhodium–catalyzed tandem hydroformylation–acetalization of α,ω-alkenediols is a facile entry to perhydrofuro[2,3-b]furans. A benzoannulated tetrahydrofuro[2,3-b]furan was obtained in this manner <02OL289>. A mixture of trisubstituted tetrahydrofurans was obtained from a π-allylmolybdenum complex and this is depicted in the following scheme <02OL2001>.

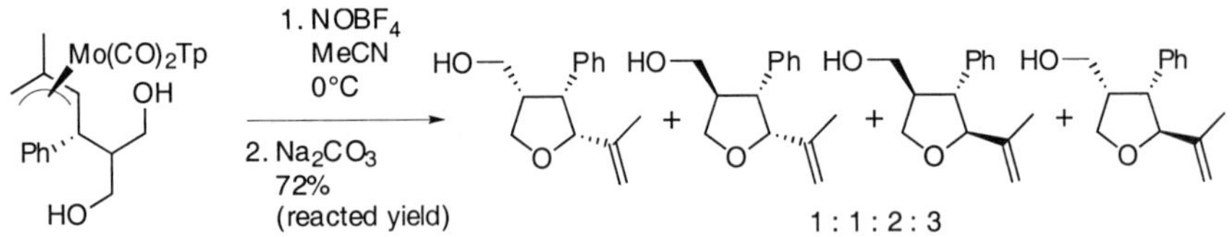

A palladium–mediated, ligand–controlled C_2 diol desymmetrization gave the desired tetrahydrofuran, which was precursor for the F–ring of halichondrin B <02OL3411>.

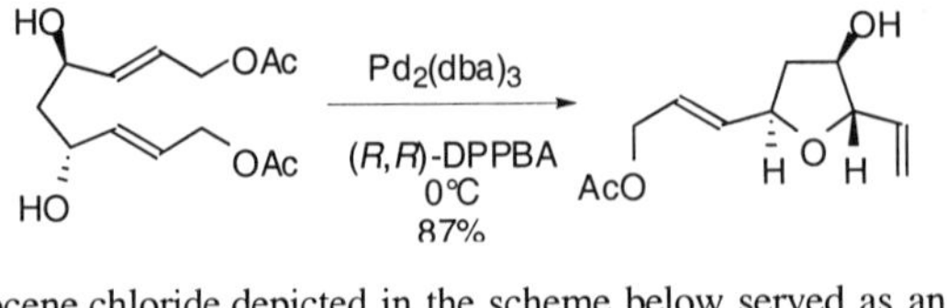

The acylzirconocene chloride depicted in the scheme below served as an acyl group donor, which reacted with ω-unsaturated α,β-enones under Pd–Me$_2$Zn–catalyzed conditions to lead to bicyclo[3.3.0] compounds in which tetrahydrofuran rings were embedded <02OL4061>.

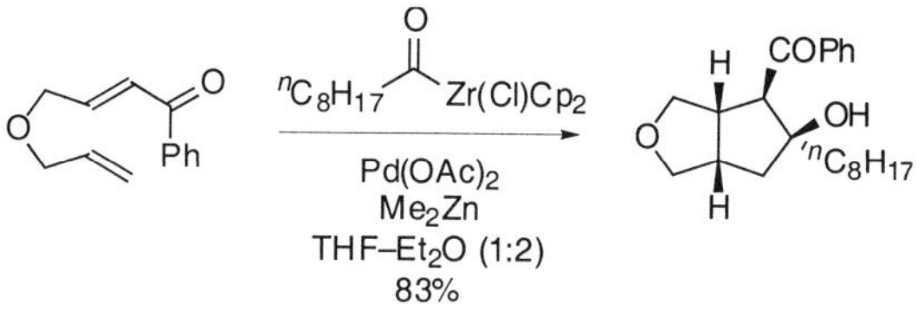

Treatment of alkynyltungsten complexes with tethered aziridines in the presence of $BF_3 \cdot Et_2O$ was known to proceed through a [3+2] cycloaddition route, providing bicyclic tungsten–enamine intermediates in a stereoselective manner. Subsequent decomplexation of these organotungsten species with iodine followed by hydrolysis gave lactams fused with a tetrahydrofuran ring <02OL4151>. A tandem intramolecular inverse electron demand Diels-Alder/1,3-dipolar cycloaddition reactions between a diene and a 1,3,4-oxadiazole led to the formation of tetrahydrofuran ring embedded in a polycyclic framework <02JA11292>.

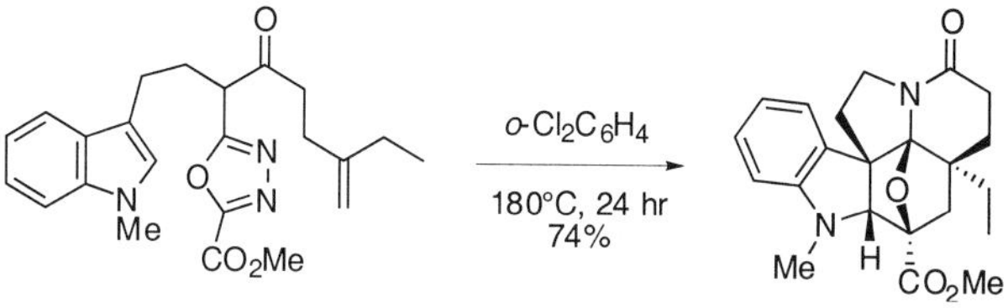

Tetrahydrofuran skeletons were obtained through a highly diastereoselective thermal [5+2] intramolecular pyrone–alkene cycloaddition procedure by appending a homochiral *p*-toluenesulfinyl group on the alkene <02CEJ884>. It was also reported in the same article that use of a sulfonimidoyl group leads to oxa-bridged carbocycles that are enantiomeric to those obtained from the sulfinyl-attached alkenes. The lactol shown below underwent consecutive Wittig reaction and cycloetherization to form the tetrahydrofuran, which was converted to (+)-oxybiotin <02TL2281>.

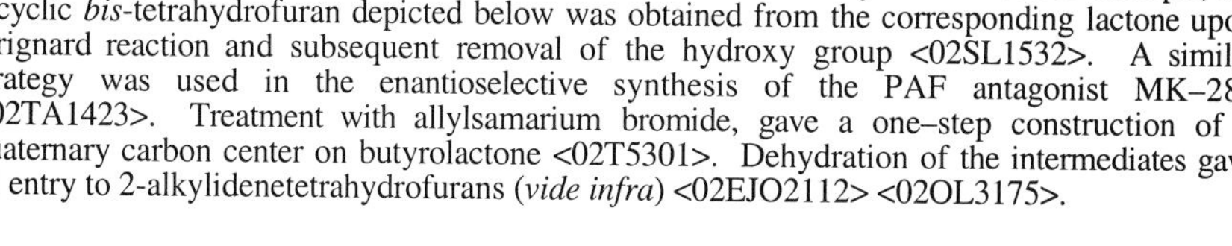

Tetrahydrofuran scaffolds could also be constructed from butyrolactones. For example, the bicyclic *bis*-tetrahydrofuran depicted below was obtained from the corresponding lactone upon Grignard reaction and subsequent removal of the hydroxy group <02SL1532>. A similar strategy was used in the enantioselective synthesis of the PAF antagonist MK-287 <02TA1423>. Treatment with allylsamarium bromide, gave a one-step construction of a quaternary carbon center on butyrolactone <02T5301>. Dehydration of the intermediates gave an entry to 2-alkylidenetetrahydrofurans (*vide infra*) <02EJO2112> <02OL3175>.

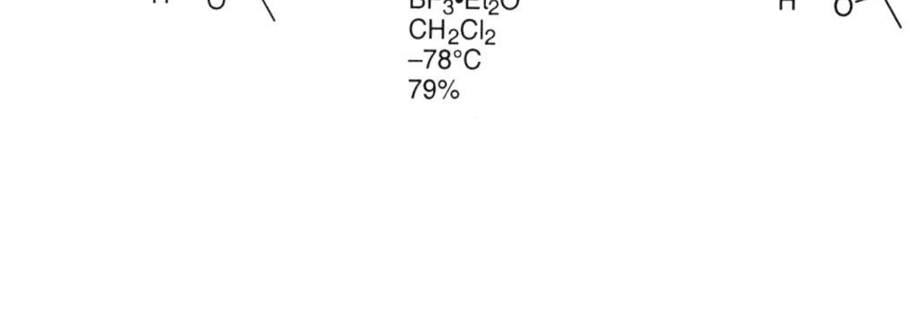

In the synthesis of the core ring systems of the sclerophytin diterpenes, a Lewis acid–promoted [4+3] annulation strategy was employed. Thus, a cyclic acetal was converted to the bicyclic keto ester upon treatment with *bis*(triethylsilyl)dienyl ether <02TL359>. Access to 2,5-disubstituted tetrahydrofurans was also made possible by reacting acetal derivatives with Grignard reagents <02CC160>. Tetrahydrofurans containing a 2-alkylidene or a 3-alkylidene group have frequently been identified in naturally occurring molecules. For this reason alone, the construction of these molecular frameworks has been rather active in the fields of organic and organometallic synthesis. To illustrate, the dicarbonyl compound shown in the following iodocyclization reaction provided the exocyclic ether <02JHC639>.

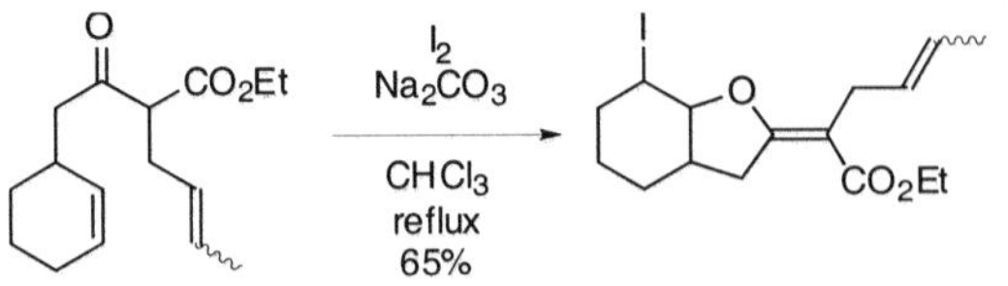

Langer reported a procedure in which dicarbonyl compounds reacted with a dibromide, leading to 2-alkylidenetetrahydrofurans <02CEJ917>. The mechanism is shown below.

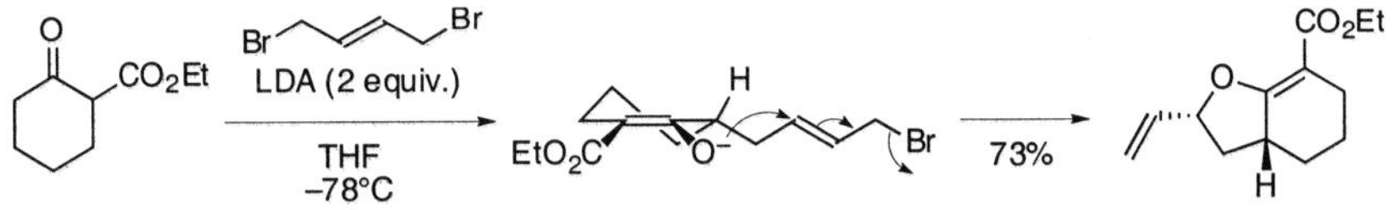

Langer also prepared other functionalized 2-alkylidenetetrahydrofurans by cyclization of 1,3-*bis*(trimethylsiloxy)-1,3-butadienes with epoxides <02CEJ1443>. Mercury(II)–induced cyclization of the hydroxyalkyne below led to the formation of the enol ether <02TL3011>.

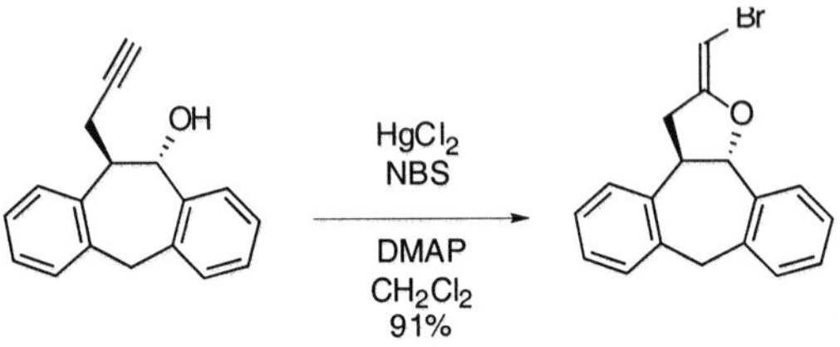

In his synthetic studies towards the total synthesis of garsubellin A, Shibasaki employed a Wacker oxidation in which the acetal shown below first underwent an acetonide deprotection, which was then followed by an intramolecular Wacker–type reaction, affording the tricyclic product in 69% yield <02OL859> <02TL3621>.

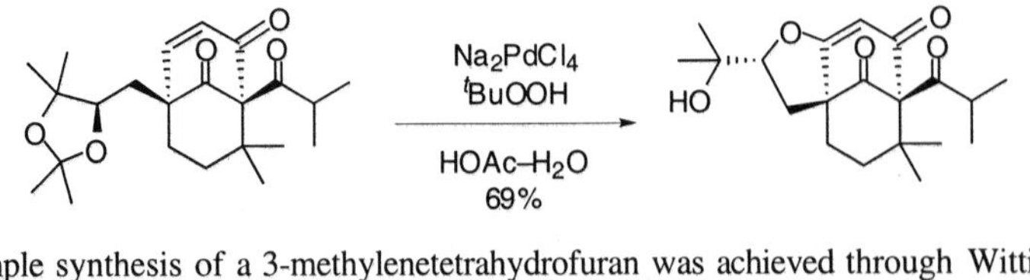

A simple synthesis of a 3-methylenetetrahydrofuran was achieved through Wittig reaction <02T4865>. 3-Methylenetetrahydrofurans were also obtained from a Wittig reagent and 1,1-disubstituted epoxides in the presence of paraformaldehyde, Me_3SiOTf and the Hünig base EtN^iPr_2 <02CL438>.

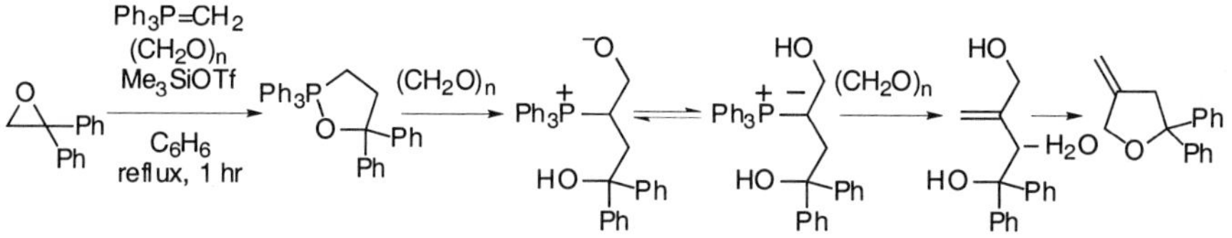

Promoted by copper iodide, a large array of 3-methylenetetrahydrofurans were synthesized from propargylic alkoxides as nucleophiles and activated alkenes as Michael acceptors. One of these reactions is shown below <02TL2609>.

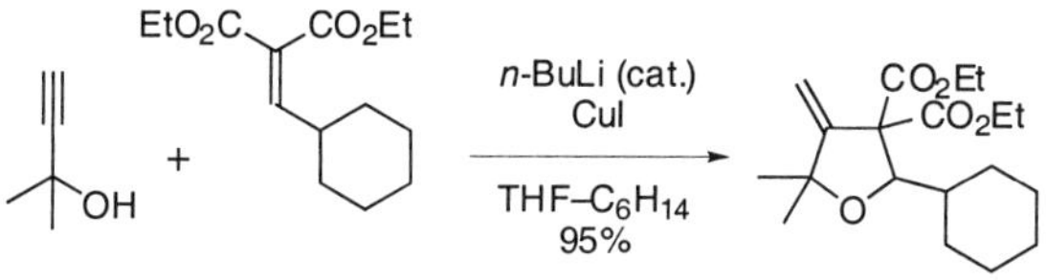

In the synthesis of the C-14–C-26 segment of halichondrins, the chromium complex (10 mol%) was shown to catalyze the reaction between the alkenyl iodide and the aldehyde in the presence of $NiCl_2$ (40 mol%), Mn, Me_3SiCl, $Et_3N \cdot HCl$ and LiCl in THF, affording the 3-methylenetetrahydrofuran in 70–80% yield <02OL4435>.

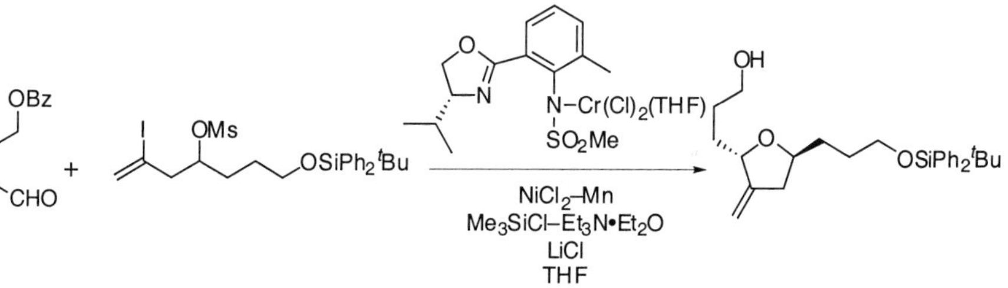

The oxazoline depicted in the following scheme gave an organolithium intermediate on treatment with MeLi in TMEDA, which led to cyclization at −78°C. The cyclized product was successfully trapped with an alkylating agent to provide the 3-alkylidenetetrahydrofuran in good yield <02OL787>.

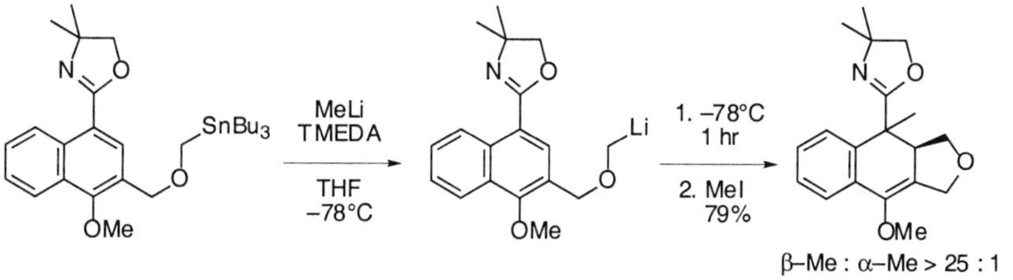

Due to their low Lewis acidities and their high reduction tendencies towards epoxides, titanocene reagents were found to be superior in converting epoxides to pivotal β-metal oxy radical intermediates that reacted intramolecularly with an alkyne to form 3-methylenetetrahydrofurans. An example is shown in the following scheme <02T7017>. Other radical induced formation of 3-methylenetetrahydrofurans includes intramolecular radical cyclization of bromoalkynes using a hypophosphite salt and this strategy was applied to the total synthesis of a naturally occurring furanolignan called dihydrosesamin <02T2435>. The stereoselective semi–synthesis of GM–237354, a potent inhibitor of fungal elongation factor 2 (EF–2), was also achieved making use of a radical cyclization step to construct a 3-methylenetetrahydrofuran framework <02TL6705>.

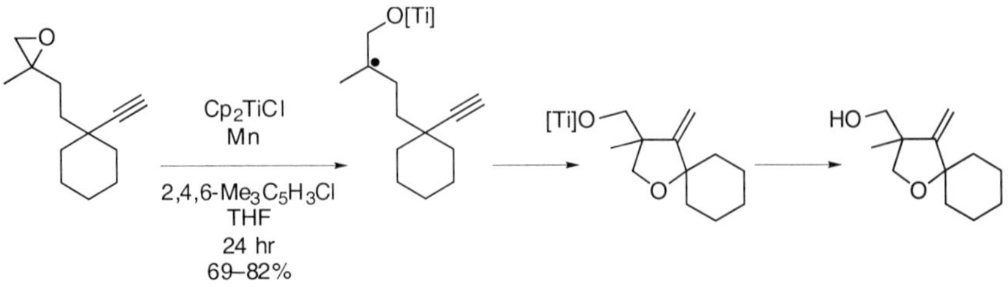

As shown below, a novel indium–mediated reaction of an iodoalkyne led to a 5-*exo*-cyclization involving radical species and as a result a 3-methylenetetrahydrofuran was obtained <02TL4585>.

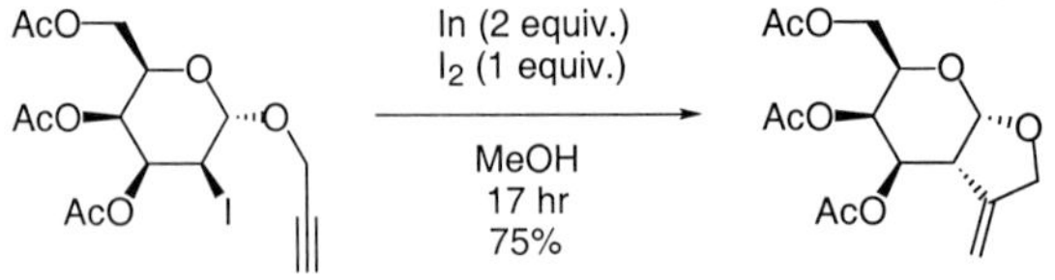

A Heck–type cobalt–catalyzed cyclization of 6-halo-1-hexenes into 3-methylenetetrahydrofurans was reported by Oshima <02OL2257>. Lu reported a palladium(II)–catalyzed cyclization of alkynes with carbonyls through the acetoxypalladation of alkynes, which was followed by the insertion of the carbonyl multiple bonds into the carbon–palladium bond. Subsequent protonolysis provided 3-alkylidenetetrahydrofurans in good yields, as shown below <02AG(E)4343>.

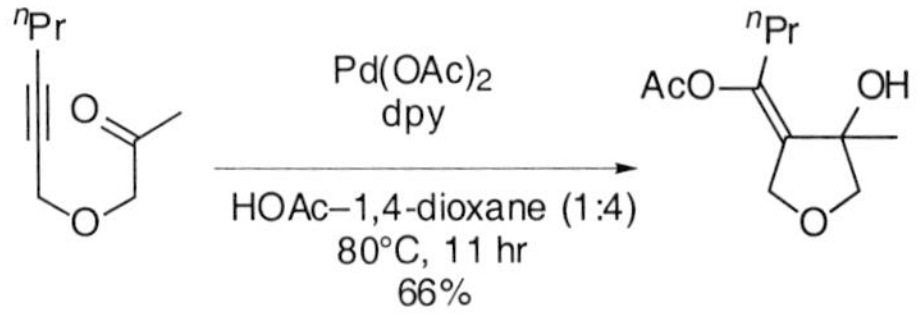

Highly enantioselective Rh–catalyzed intramolecular Alder–ene reactions for the synthesis of chiral 3-alkylidene-4-vinyltetrahydrofurans were reported by Zhang, as illustrated below <02AG(E)3457>. Metallic indium was also shown to mediate the intramolecular cyclization of tethered allyl bromides onto terminal alkynes to afford 3-methylene-4-vinyltetrahydrofurans in 50–62% yield <02SL2068>.

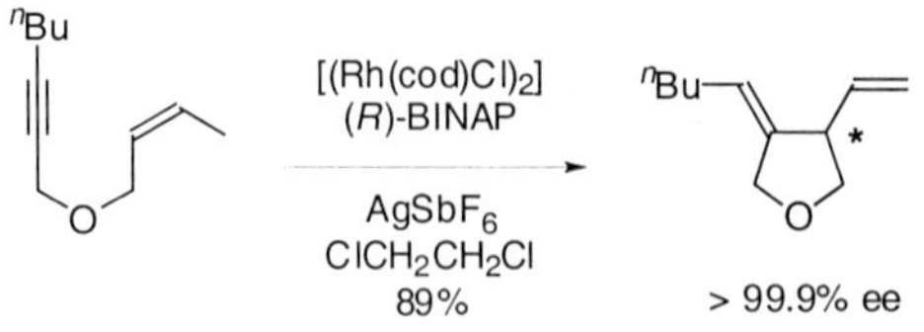

A tetrahydrofuran fused with a seven–membered ring was obtained from an enyne through a [5+2] cycloaddition reaction catalyzed by $[(C_{10}H_8)Rh(COD)]^+$ SbF_6^- complex <02AG(E)4550>. Rhodium–catalyzed carbonylative alkene–alkyne coupling reactions employing aldehydes as a CO source were reported by Shibata, who discovered that cinnamaldehyde was the best CO donor. Several cyclopentenones fused with a tetrahydrofuran ring were realized under solvent–free conditions <02OL1619>. A conventional method making use of CO and stabilized cobalt nanoparticles also led to the formation of similar products <02OL277>. Several articles published in 2002 dealt with syntheses of 2,3-dihydrofurans. As can be seen in the example shown below, the base–promoted Feist–Bénary condensation reaction between β-dicarbonyl compounds (β-ketoesters, β-diketones and β–dialdehydes) with α-haloketones led to the formation of highly substituted 2,3-dihydrofurans <02OL205>. The same strategy was employed in a rapid synthesis of the 7-deoxyzaragozic acid core <02OL209>.

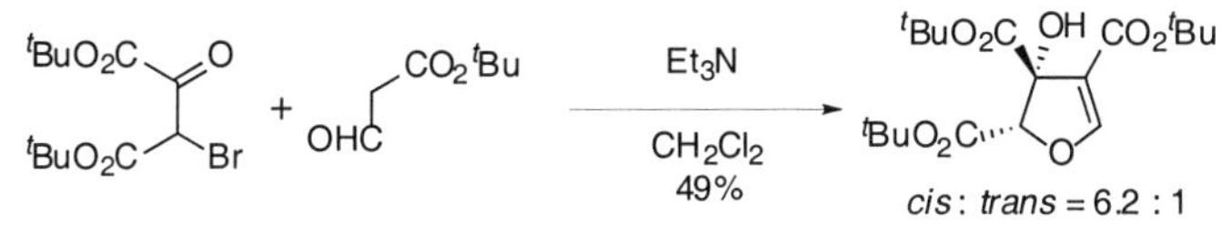

The construction of the 2,3-dihydrofuran moieties of breviones was reported in which the epoxide and the α-pyrone depicted in the following scheme underwent both C–alkylation and consecutive O-alkylation presumably via π-allylpalladium complexes <02TL1713>.

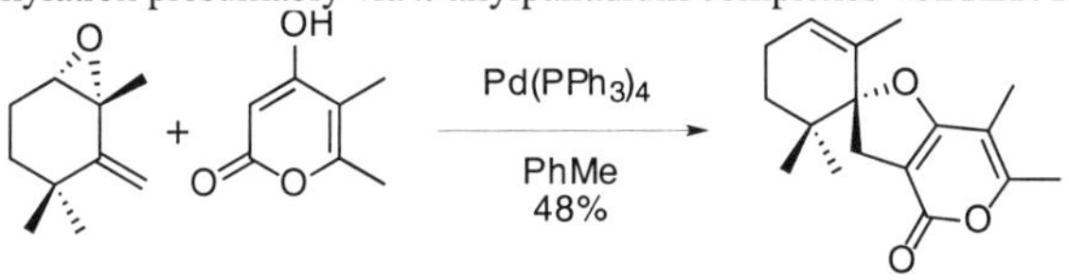

Regiospecific synthesis of similar spirodihydrofurans was also achieved with cerium(IV) ammonium nitrate mediated oxidative [3+2] cycloaddition involving addition of radical species derived from 1,3-dicarbonyl compounds to exocyclic alkenes <02SL787>. Iodoenolcyclization of 2-alkenyl-1,3-dicarbonyl compounds in the formation of 2,3-dihydrofurans was reported to be strictly dependent on the dicarbonyl structure as well as the substituents on the allyl group <02T8825>. A short synthesis of (±)-epiasarinin was reported by Steel, who used an alkenyl epoxide–dihydrofuran rearrangement route as the pivotal step, involving a conrotatory ring opening of the epoxide and a disrotatory cyclization of the intermediate <02OL1159>.

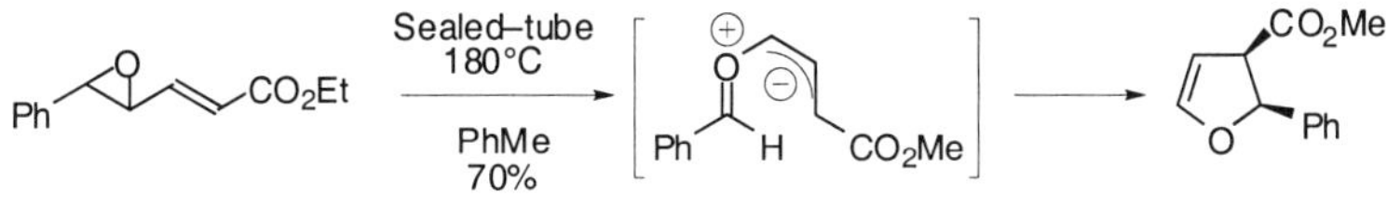

An efficient synthesis of 2,3-dihydrofurans was accomplished by rhodium–catalyzed reactions of cyclic diazodicarbonyls with allyl halides. This approach was utilized to provide a rapid entry towards furocoumarins and furophenalenones <02T2359>. 2,3-Dihydrofurans were also prepared by an unusual diversion of the well–documented di–π–methane rearrangement <02OL1155>. 2,5-Dihydrofuran systems are also common and important structural features found in naturally occurring molecules. A straightforward preparation of 2,5-dihydrofurans was achieved by reacting (Z)-1,4-dihydroxy-2-alkenes with DCC in the presence of a catalytic amount of CuCl <02JCS(P1)1555>. Treatment of 1,7-dihydroxyhepta-2,5-diyn-4-ones with hydrogen halide in acetic acid led to the formation of halogenated spiroketals containing 2,5-dihydrofuran moieties <02JCS(P1)713>. Another way from which spiro-bis-2,5-dihydrofurans can be prepared was by treating ninhydrin with dialkyl acetylenedicarboxylates in the presence of triphenylphosphine <02TL2927>. 2,5-Dihydrofurans were conveniently realized via insertion reactions of alkylidene carbenes. The best results were obtained on substrates containing electron–withdrawing substituents that are less prone to competing rearrangement reactions <02JCS(P1)965>. When the amide shown below was heated in a solution of acetic anhydride, a mesoionic species was generated, which eventually underwent an intramolecular 1,3–dipolar cycloaddition and subsequent CO_2 expulsion to form the thiazole derivative with a 2,5-dihydrofuran ring <02JOC4045>.

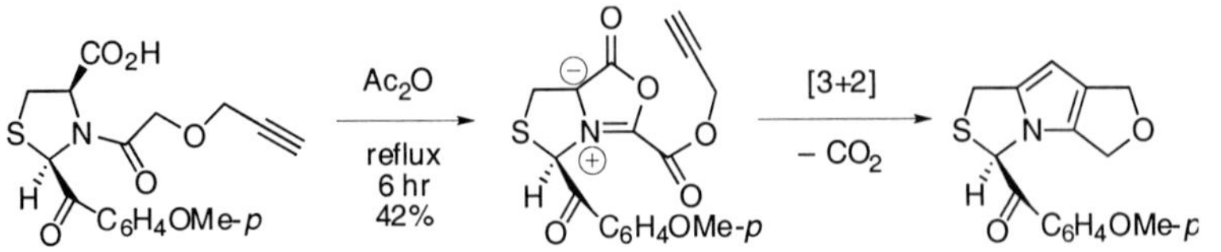

1,4-Dilithio-1,3-dienes were found to react with aldehydes to form polysubstituted 2,5-dihydrofurans in a regio– and stereoselective manner <02OL2269>. Like other metal–carbene reactions, rhodium–catalyzed tandem carbonyl ylide formation – cycloaddition with propargyl bromide gave the 2,5-dihydrofuran in good yield <02OL1809>.

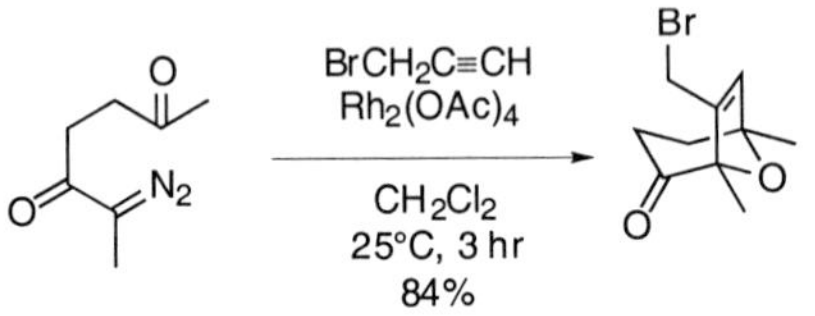

Microwave irradiation was shown to be an efficient approach to accelerate ring–closing metathesis reactions for the synthesis of 2,5-dihydrofurans using ruthenium–based catalysts in ionic liquids or microwave transparent solvents such as dichloromethane <02OL1567>. The ruthenium catalyst generated *in situ* from [RuCl$_2$(*p*-cymene)]$_2$, 1,3-*bis*(mesityl)imidazolinium chloride and cesium carbonate was also useful in converting terpenoids with geminal ethynyl and allyloxy groups to vinyl–substituted 2,5-dihydrofurans <02EJO3816>. Ruthenium–catalyzed reactions between the ether illustrated below and diethyl ketomalonate provided the 3,4-disubstituted 2,5-dihydrofuran in 42% yield <02JA6844>.

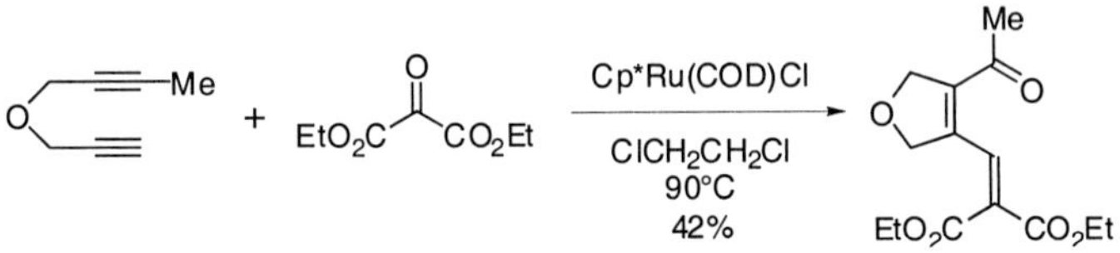

5.3.3.3 Benzo[*b*]furans and Related Compounds

Reduction of 2,6-diacetoxy-2'-bromoacetophenone with NaBH$_4$ led to 3,4-diacetoxydihydrobenzo[*b*]furan in a process involving acyl migration and cyclization. Subsequent hydrogenolysis gave 4-acetoxydihydrobenzo[*b*]furan which upon saponification led to 4-hydroxydihydrobenzo[*b*]furan in good yield. This approach was shown to be a general method for preparation of substituted dihydrobenzo[*b*]furans <02TL1923>. Iodine–magnesium exchange provided a mild preparative method to generate functionalized aryl magnesium chlorides bearing a leaving group in the molecule, which underwent a transition metal–catalyzed (a copper or cobalt salt) cyclization reactions to afford tricyclic compounds <02TL4875>.

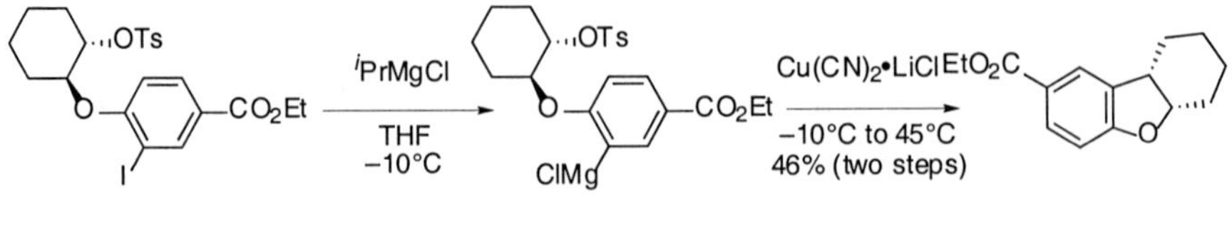

A stereoselective Michael addition was utilized in a synthesis of the macrocyclic polyamine alkaloid lunarine. The key intermediate was generated as shown below <02JCS(P1)1115>.

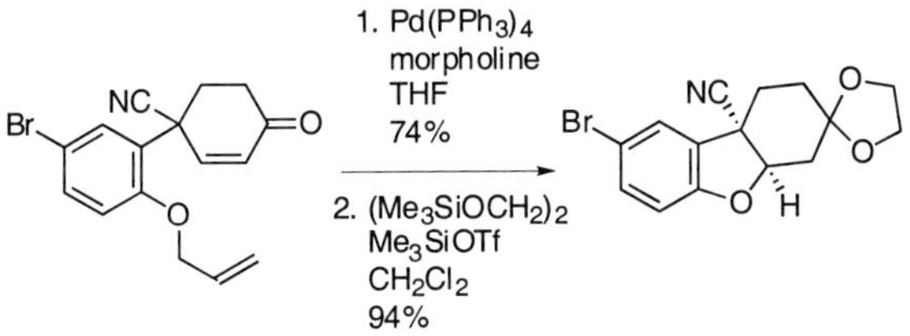

In the total synthesis of (−)-morphine, the pivotal tetracyclic intermediate was generated as depicted below. Thus, this core structure was synthesized by reduction of the enone to afford an allylic alcohol, which was then briefly exposed to BBr$_3$ to give the desired skeleton in 64% yield <02JA12416>.

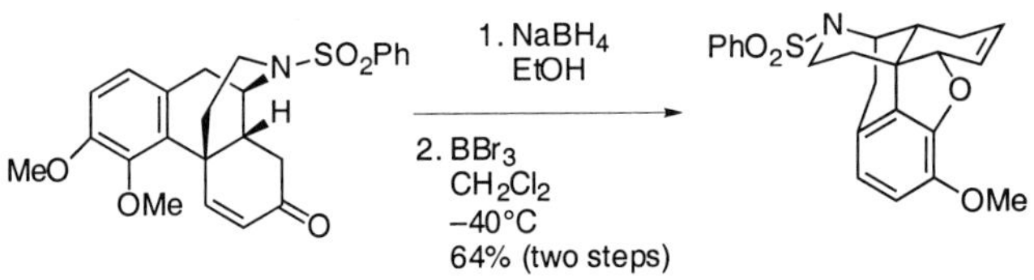

In the total synthesis of diazonamide, the later stage to form the benzo[*b*]furan ring was realized by a DIBAL-H initiated ring closure with the participation of the adjacent phenolic hydroxy group to afford the desired cyclic aminal in 55% yield <02AG(E)3495>. As shown below, the ferric ion–catalyzed cycloaddition of the styrene derivative with quinone gave the 2,3-dihydrobenzo[*b*]furan in an excellent yield with *trans*-selectivity. Remarkably, the reaction was carried out in an ionic liquid solvent system <02TL3041>. Novel formation of dihydrobenzo[*b*]furans between fulvenes and alkenes was promoted by the use of microwave irradiation <02OL663>.

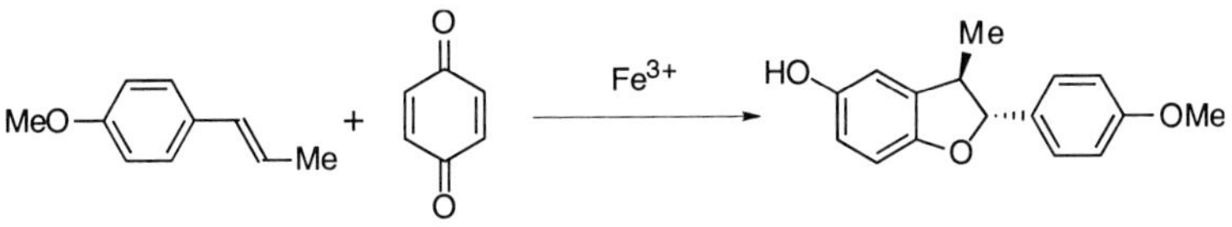

An interesting method to synthesize dihydrobenzo[*b*]furans was developed by employing a [3+2] dipolar cycloaddition between allylsilanes and *o*-benzoquinones <02TL5349>.

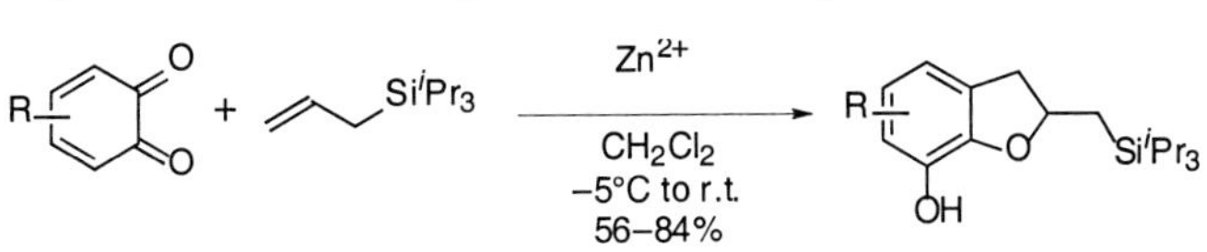

A new synthesis of galanthamine has been developed which addresses many of the shortcomings of the previous syntheses. The key step involved in this synthesis is the intramolecular Heck reaction illustrated below <02JA2795>.

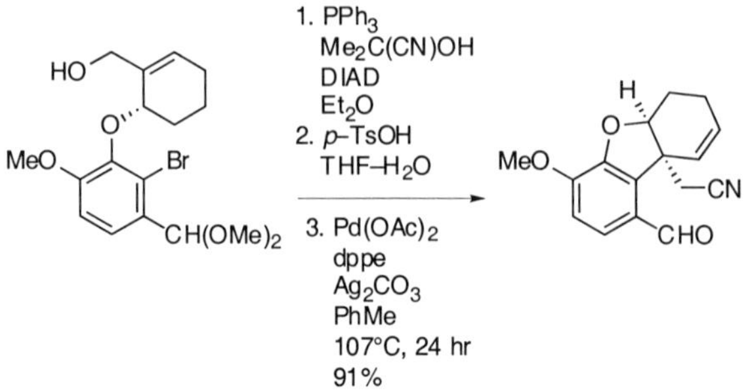

The enantioselective intramolecular C–H insertion reaction of aryldiazoacetates was explored with the use of dirhodium(II) carboxylate catalysts incorporating N-phthaloyl- or N-benzene–fused-phthaloyl-(S)-amino acids as chiral bridging ligands. Dirhodium tetrakis[N-phthaloyl-(S)-*tert*-leucinate] and $Rh_2(S$-PTTL$)_4$ were proven to be the best catalysts for this process, providing exclusively methyl *cis*-2-aryl-2,3-dihydrobenzo[*b*]furan-3-carboxylate in up to 94% ee <02OL3887>. Flavanones as shown below on oxidation with iodobenzene diacetate in the presence of sulfuric acid in trimethyl orthoformate underwent a stereospecific ring contraction to give methyl *trans*-2-aryl-2,3-dihydrobenzo[*b*]furan-3-carboxylates as major products <02T4261>.

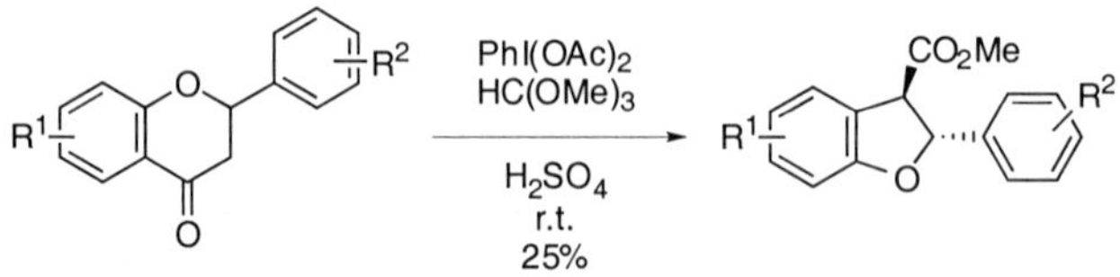

The VO(acac)$_2$ based oxidative system was employed to epoxidize various double bonds under mild conditions. Thus, a one–pot synthesis of 2,3-dihydrobenzo[*b*]furanols was achieved in high regio– and diastereoselective manner in the presence of a catalytic amount of TFA (20 mol%). This metal–catalyzed methodology was shown to be more practical and superior to the previously employed *m*-CPBA based method <02SL942>.

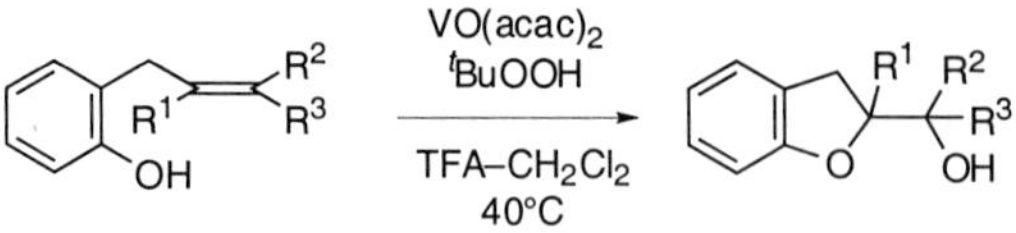

Aerobic oxidative cyclization of 2,2-dihydroxystilbenes via oxygen cation radical leading to the formation of *cis*-4b,9b-dihydrobenzofuro[3,2-*b*]benzofurans was carried out in an enantioselective manner by using (nitrosyl)Ru(salen) as a catalyst under irradiation conditions <02CL36>.

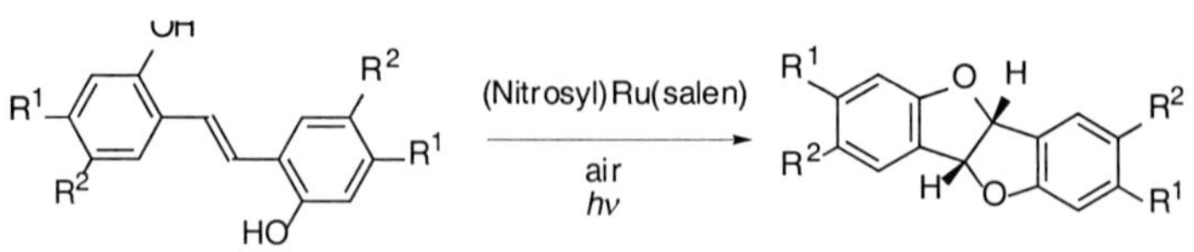

3-Substituted-2,3-dihydrobenzo[*b*]furans were obtained in very good yields by $S_{RN}1$ photo–stimulated reactions in liquid ammonia from appropriate halo–aromatic compounds *ortho*-substituted with a suitable double bond and Me$_3$Sn$^-$, Ph$_2$P$^-$, I$_3^-$ and $^-$CH$_2$NO$_2$ anions.

The novelty of this work involves the versatile application of 5-*exo* ring closure processes during the propagation cycle of the $S_{RN}1$ reaction; the alkyl radical intermediates formed then reacted with the nucleophiles to afford the ring closure–substituted heterocycles. The factors governing the observed product distribution are discussed and one of the examples is illustrated below <02JOC8500>.

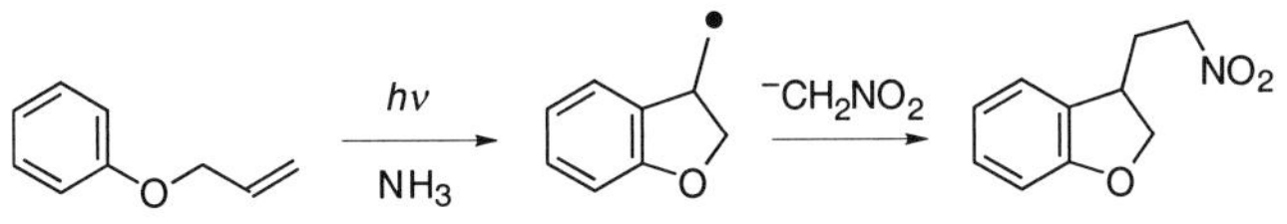

When coumarins shown below having an electron–withdrawing group at the C–3 position were treated with 2.4 equiv. of dimethylsulfoxonium methylide at room temperature in DMF or DMSO, novel tricyclic 2-substituted-cyclopenta[*b*]benzofuran-3-ols were obtained in moderate to good yields. A mechanism was proposed, and the key intermediate related to the proposed approach was provided <02T1497>.

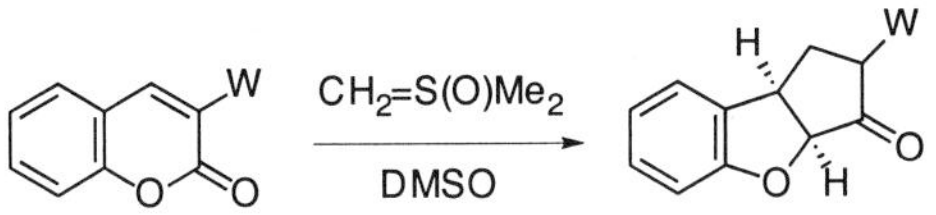

The key intermediate in the total synthesis of furaquinocin was obtained in good yield by a reductive Heck reaction that proceeded with a sterically hindered base pentamethylpiperidine (PMP) <02JA11616>. A new hypothesis for the major skeletal rearrangement (anthraquinone → xanthone → coumarin) that occurs in the complex biosynthesis of aflatoxin B$_1$ was proposed. To test this hypothesis, an intermediate 11-hydroxy-*O*-methylstergmatocystin (HOMST) was synthesized as shown below. The key transformation in this synthesis involved the treatment of an ester–aldehyde with ^{i}Pr$_3$SiOTf, which smoothly produced a mixed acetal. Direct reduction with DIBAL-H led to the aldehyde. The desired product was eventually obtained via several steps as shown <02JA5294>.

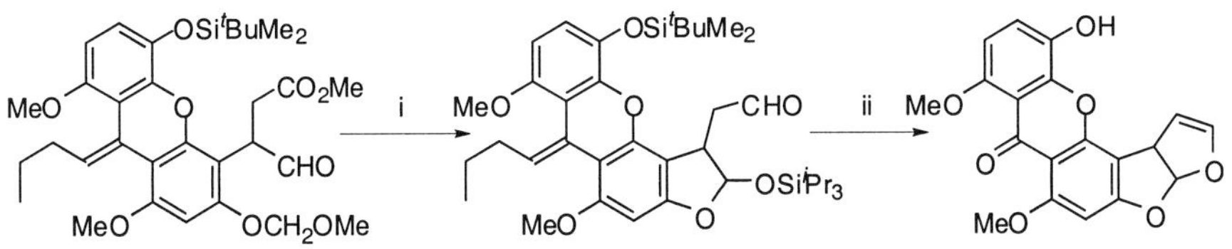

i (1) CF$_3$SO$_3$Si^{i}Pr$_3$, (2) DIBAL-H (90% two steps). ii (1) *m*-CPBA, CH$_2$Cl$_2$ (85%), (2) Et$_3$N•HF, MeCN (73%), (3) PhSH, 4Å MS, Amberlyst, MeCN (93%), (4) *m*-CPBA, CH$_2$Cl$_2$ (62%)

A new furoquinolinone derivative was synthesized in which the key steps were bromination, cyclization and elimination <02JMC1146>.

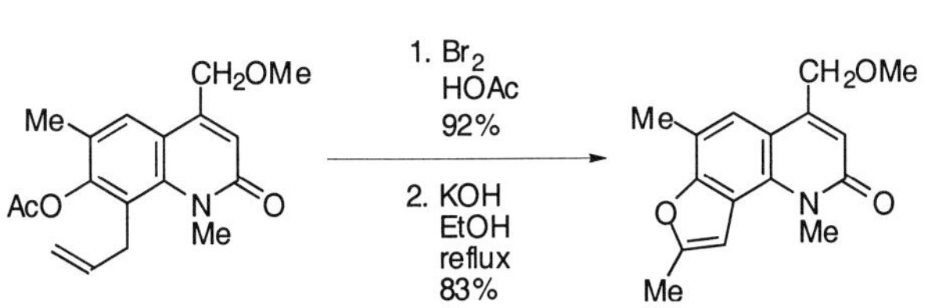

An improved procedure to synthesize the scaffold of psoralen was developed by addition of an aldehyde to a NaOH solution at reflux. A mechanistic interpretation was provided, though the exact mechanism of the reaction is still not clear and no intermediate was isolated

<02T4859>. Pyranoanthocyanidin was found to undergo rearrangement to form a new type of furoanthocyanidin with a core structure of furo[2,3-c]-1-benzopyrylium. A mechanistic interpretation was provided <02TL715>. Reactive nitrovinylquinones were shown to react with an acyclic enol ether to form dihydrobenzo[b]furans. This reaction involved a conjugated Michael addition of the enol ether to the quinone, and was followed by ring closure. The dihydrobenzo[b]furans thus generated eliminated an alcohol to give benzo[b]furans <02JOC8366>. As depicted in the following scheme, furobenzoxazin-3-one, a new tricycle nucleus, was synthesized in a straightforward route involving the condensation of a furan ring onto a pre–constituted 1,4-benzoxazinone nucleus <02EJOC1937>.

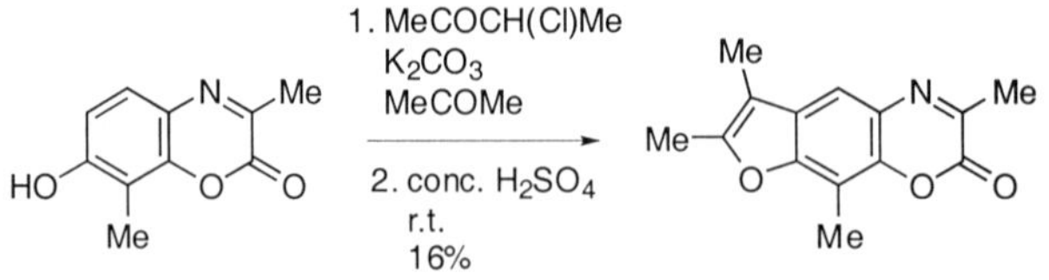

A series of nitrile–containing benzo[b]furans were synthesized and evaluated as melatonin receptor ligands <02JMC2788>. Substituted 1-(bromoacetyl)azulenes reacted with substituted salicylaldehyde to furnish 1-(2-benzofurancarbonyl)azulenes, but no yields were recorded <02JHC671>.

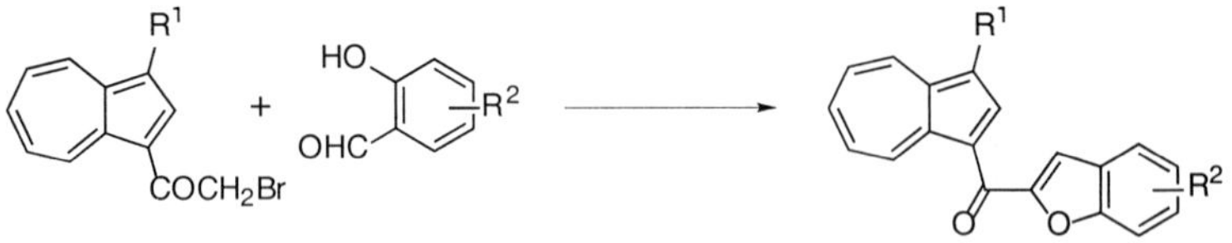

A simple route to 3-vinylbenzo[b]furans and 3-vinylfuropyridines from readily accessible acetylenic precursors was recorded. As can be seen in the following scheme, a standard halogen–lithium exchange triggered an irreversible addition on the triple bond of the precursor according to a 5-exo-*dig* heterocyclization process, and was followed by a lithium ethoxide elimination to afford a diene after isomerization of the exocyclic allene to 1,3-diene, which on reaction with an acrylate under a thermal conditions to provide the [4+2] cycloadduct <02OL2791>. On the other hand, the reaction of benzil with phenol at 180 °C in the presence of $SnCl_4 \cdot 5H_2O$ produced a mixture of benzo[b]furan based compounds. A mechanistic interpretation for the formation of these compounds was discussed <02T4255>.

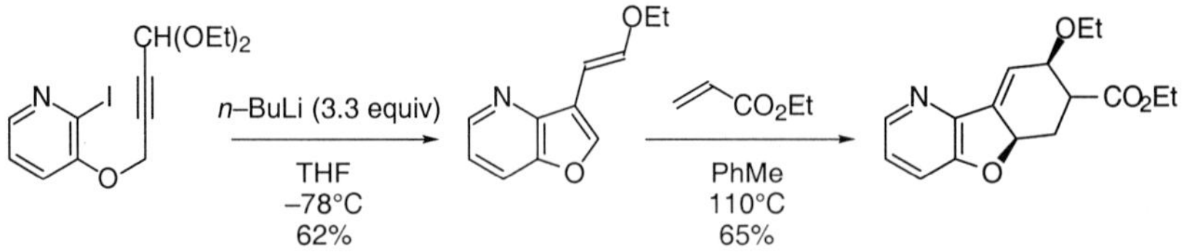

In a new total synthesis of XH-14, the key step included the bromine–promoted cyclization as illustrated below <02JOC6772>.

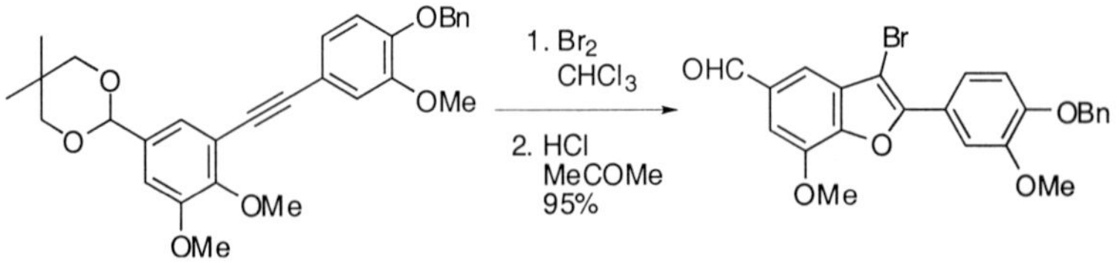

Irradiation of macrocycles having two 1,5-dibenzoyl-2,4-dialkoxybenzene moieties with a 350 nm mercury lamp followed by dehydration afforded benzo[1,2-b:5,4-b']difuran–containing cyclophanes via a quadruple photocyclization in a one–pot operation as illustrated in the following scheme <02H(57)657>. Facile routes for the synthesis of novel cyclophanes containing both benzo[1,2-b:5,4-b']difuran and naphthalene rings were developed in a similar manner <02JCS(P1)310>.

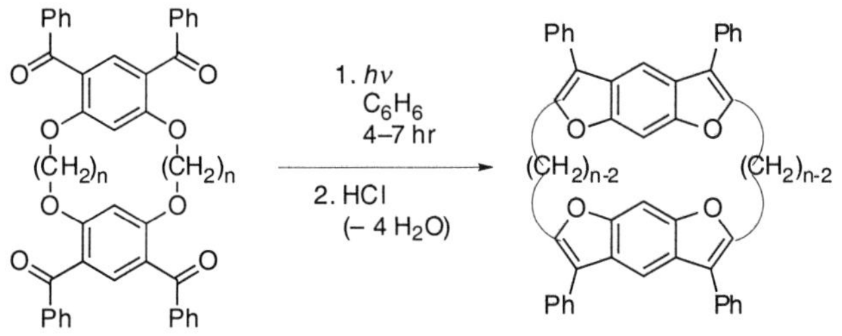

A general synthetic methodology and mechanistic details for the palladium–catalyzed carbonylative annulation of the o-alkynylphenol to construct 2-substituted-3-arylcarbonyl-benzo[b]furans was reported <02JOC2365>.

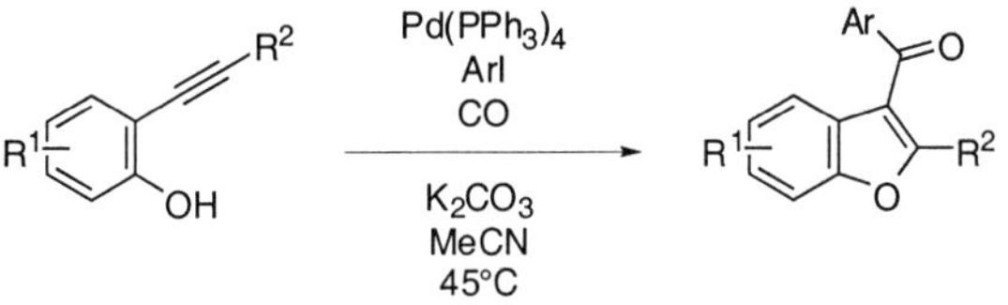

In a similar manner, the first synthesis of 4-, 5-, and 6-nitrobenzo[b]furans was achieved via an initial Sonogashira cross–coupling reaction of 2-iodonitrophenol acetates prepared from commercially available and inexpensive 2-aminonitrophenols. The 2-alkynylnitrophenol acetates obtained were subjected to a base–promoted reaction to form nitrobenzo[b]furans as illustrated in the scheme below <02TL9377>. Other benzo[b]furans were also prepared by similar procedures, which involved Sonogashira reactions and palladium–catalyzed cyclization with and without CO <02JMC2670>. A palladium–mediated cascade carbonylative annulation of o-alkynylphenols was achieved successfully on silyl linker–based macrobeads, which led to an efficient combinatorial synthesis of a 2,3-disubstituted benzo[b]furan library <02OL2607>. In a similar manner, 2-ethynyl-3-pyridinols and 3-ethynyl-2-pyridinols were converted into 2-substituted furo[3,2-b]pyridines and 2-substituted furo[2,3-b]pyridines through a coupling–cyclization process with aryl/heteroaryl halides or vinyl triflates. Two catalytic systems were utilized in this reaction, i.e., PdCl$_2$(PPh$_3$)$_2$–CuI and Pd$_2$(dba)$_3$–P(t-Bu)$_3$ <02SL453>. On the other hand, 2,3-disubstituted-furo[3,2-b]pyridines, 2,3-disubstituted-furo[2,3-b]pyridines, and 2,3-disubstituted-furo[2,3-c]pyridines were prepared in good yields under mild conditions by I$_2$ promoted cyclization or by PdCl$_2$ catalyzed carbonylative annulations under an atmosphere of CO <02OL2409>. Cu(I)–catalyzed cross–coupling reactions were also utilized to synthesize a variety of 2-arylbenzo[b]furans in good yields <02OL4727>.

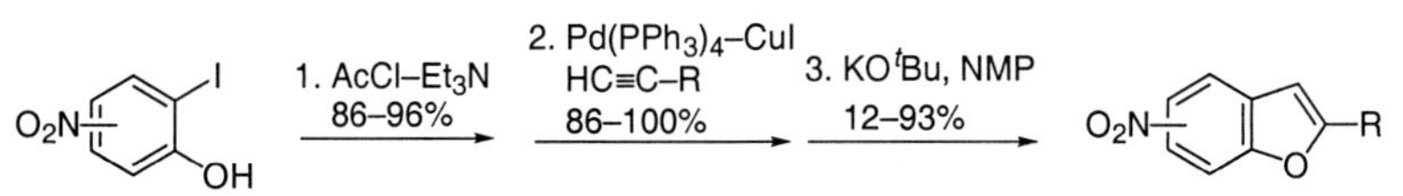

An approach for the synthesis of dibenzo[*b*]furan was developed by an intramolecular Heck reaction <02T5927>.

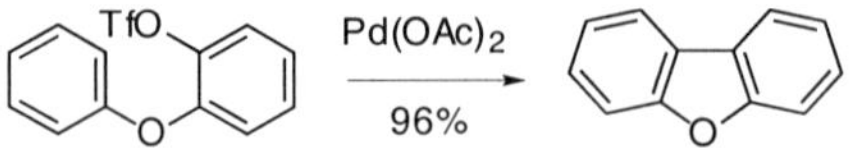

By taking advantage of the α- and β-activation of chloropyridines as well as palladium–mediated reactions, all four possible benzo[4,5]furopyridine tricyclic heterocycles, namely benzo[4,5]furo[2,3-*b*]pyridine, benzo[4,5]furo[2,3-*c*]pyridine, benzo[4,5]furo[3,2-*c*]pyridine, and benzo[4,5]-furo[3,2-*b*]pyridine were efficiently synthesized from 2-chloro-3-iodopyridine, a 3-chloro-4-stannylpyridine, 4-chloro-3-iodopyridine, and 2-chloro-3-hydroxypyridine, respectively <02OL2201>. The synthesis of benzo[3,2-*c*]isoquinoline using thermal ring transformation of a benzisoxazolo[2,3-*a*]isoquinoline salt was also recorded <02TL6035>. A general regioselective route to pyridazine analogs of benzofurocoumarins (such as the angular compound shown below) required a Diels-Alder reaction between dihydrofuro-3-ones and 3,6-*bis*(trifluoromethyl)-1,2,4,5-tetrazine. Synthetically, the desired products were generated by a regioselective Fries rearrangement, which was followed by the Diels-Alder reaction <02S43>.

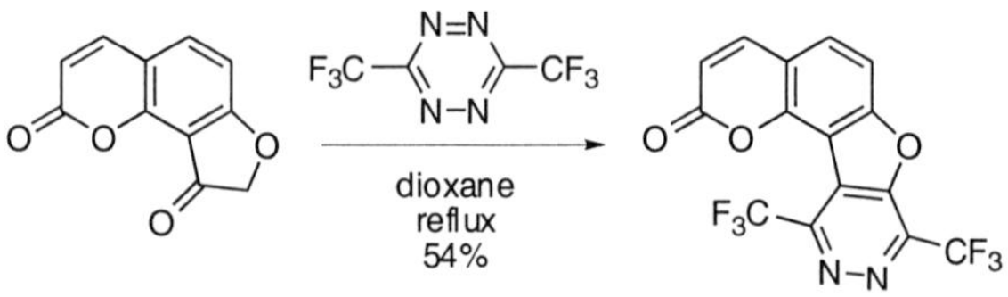

5.3.3.4 Benzo[*c*]furans and Related Compounds

A group of Taiwan chemists reported the synthesis of 1,3-dihydrobenzo[*c*]furans by a highly regio– and chemoselective nickel–catalyzed [2+2+2] cycloaddition pathway between electron–deficient diynes with allenes <02JOC7724>. An example is shown below.

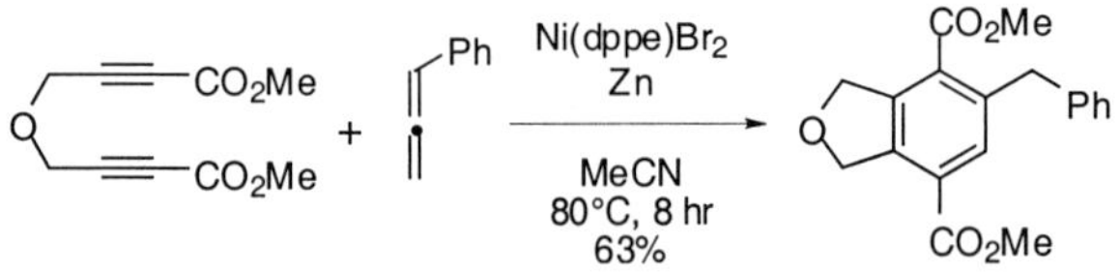

Ether–linked *p*-dialkynylarenes underwent alkyne cyclotrimerization to form monodisperse oligo-*p*-phenylenes containing dihydrobenzo[*c*]furans <02OL745>.

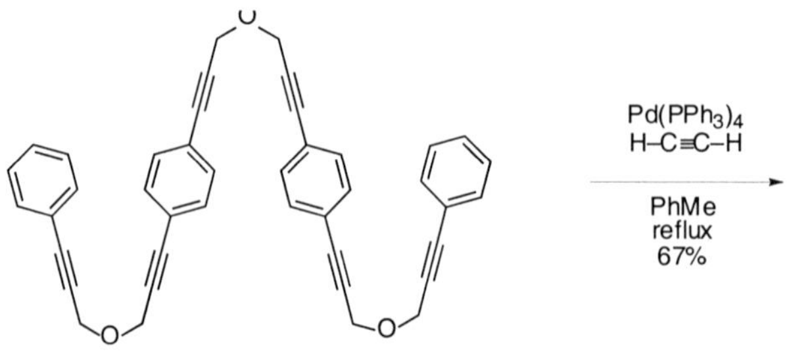

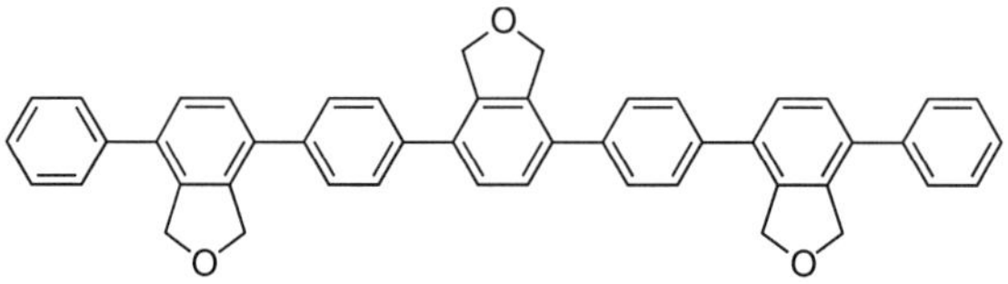

Functionalized dihydrobenzo[c]furans were synthesized using the n-BuLi–Me$_2$N(CH$_2$)$_2$OLi superbase system that led to a direct and regioselective α–lithiation and alkylation <02TL4045>. On the other hand, silylated dihydrobenzo[c]furan–Cr(CO)$_3$ complex also underwent highly regioselective lithiation and stereoselective alkylation to give substituted dihydrobenzo[c]furans <02AG(E)2525>. Dihydrobenzo[c]furans were also obtained by a direct reduction of phthalides with sodium borohydride <02HCA2458>. As illustrated in the scheme below, polysubstituted dihydrobenzo[c]furans were synthesized from the diols, which in turn were obtained from thianthrenes <02TL7205>.

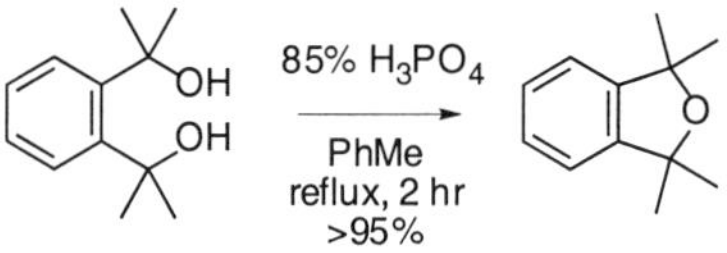

Another way in which dihydrobenzo[c]furans can be made is by reacting allylic halides having an aldehyde in the same molecule with allyltributylstannane in the presence of a catalytic amount of Pd$_2$(dba)$_3$•CHCl$_3$ <02CL158>.

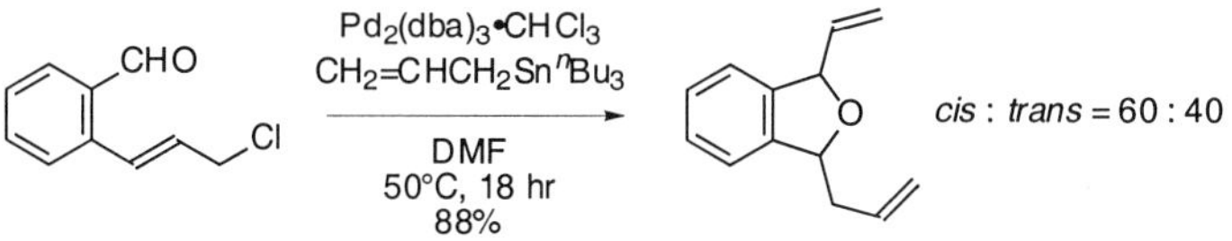

A palladium–catalyzed method for the generation of benzo[c]furan (isobenzofuran) under neutral condition has been devised by Mikami <02OL3355>. Thus, on reaction with Pd$_2$(dba)$_3$•CHCl$_3$ which was the best catalyst, the methyl ether depicted below was converted to isobenzofuran and was trapped by dimethyl acetylenedicarboxylate to give the Diels-Alder adduct in 45% yield. Mikami also reported that silyl lactols upon treatment with metal fluorides also led to the formation of silylisobenzofurans <02CL1868>. In a short synthesis of the CDEF ring of lactonamycin, an isobenzofuran intermediate played a pivotal role <02OL1527>. Wong also reported a convenient synthesis of 5,6-bis(trimethylsilyl)benzo[c]furan as an isolable molecule <02T9413>.

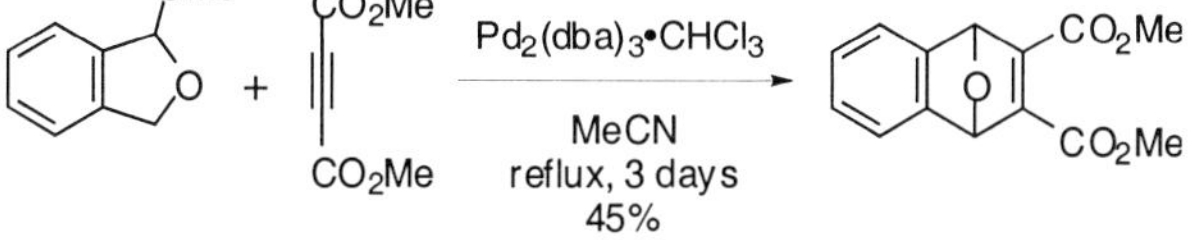

Acknowledgements: HNCW wishes to thank the Areas of Excellence Scheme established under the University Grants Committee of the Hong Kong Special Administrative Region, China (Project No. AoE/P–10/01) and the Croucher Foundation (Hong Kong) for financial support. XLH acknowledges with thanks supports from the National Natural Science Foundation of China, National Outstanding Youth Fund, the Chinese Academy of Sciences, and Shanghai Committee of Science and Technology.

5.3.5 REFERENCES

02AG(E)1392	S.H. Kang, J.W. Jeong, Y.S. Hwang, S.B. Lee, *Angew. Chem. Int. Ed.* **2002**, *41*, 1392.
02AG(E)1775	S. Ma, Z. Yu, *Angew. Chem. Int. Ed.* **2002**, *41*, 1775.
02AG(E)2525	S. Zemolka, J. Lex, H.-G. Schmalz, *Angew. Chem. Int. Ed.* **2002**, *41*, 2525.
02AG(E)3457	A.-W. Lei, M.-S. He, S.-L. Wu, X.-M. Zhang, *Angew. Chem. Int. Ed.* **2002**, *41*, 3457.
02AG(E)4079	Y. Hayashi, M. Nakamura, S. Nakao, T. Inoue, M. Shoji, *Angew. Chem. Int. Eg.* **2002**, *41*, 4079.
02AG(E)4343	L.-G. Zhao, X.-Y. Lu, *Angew. Chem. Int. Ed.* **2002**, *41*, 4343.
02AG(E)4550	P.A. Wender, T.J. Williams, *Angew. Chem. Int. Ed.* **2002**, *41*, 4550.
02AG(E)4560	L.C. Usher, M. Estrella-Jimenez, I. Ghiviriga, D.L. Wright, *Angew. Chem. Int. Ed.* **2002**, *41*, 4560.
02BMC2161	R.-W. Jiang, S.-C. Ma, Z.-D. He, X.-S. Huang, P. P.-H. But, H. Wang, S.-P. Chan, V. e.-C. Ooi, H.-X. Xu, T.C.W. Mak, *Bioorg. Med. Chem.* **2002**, *10*, 2161.
02CC160	X. Franck, R. Hocquemiller, B. Figadère, *Chem. Commun.* **2002**, 160.
02CC390	M. Murakami, H. Igawa, *Chem. Commun.* **2002**, 390.
02CC2042	N.A. Swain, R.C.D. Brown, G. Bruton, *Chem. Commun.* **2002**, 2042.
02CC2542	Y. Tanabe, K. Mitarai, T. Higahi, T. Misaki, Y. Nishii, *Chem. Commun.* **2002**, 2542.
02CEJ208	M.C. Carreño, C.G. Luzoón, M. Ribagorda, *Chem. Eur. J.* **2002**, *8*, 208.
02CEJ884	F. López, L. Castedo, J.L. Mascareñas, *Chem. Eur. J.* **2002**, *8*, 884.
02CEJ917	P. Langer, E. Holtz, N.N.R. Saleh, *Chem. Eur. J.* **2002**, *8*, 917.
02CEJ1051	S. Claeys, D. Van Haver, P.J. De Clercq, M. Milanesio, D. Viterbo, *Chem. Eur. J.* **2002**, *8*, 1051.
02CEJ1443	P. Langer, H. Armbrust, T. Eckardt, J. Magull, *Chem. Eur. J.* **2002**, *8*, 1443.
02CEJ4234	M. Makosza, M. Judka, *Chem. Eur. J.* **2002**, *8*, 4234.
02CEJ4264	S. Akai, T. Naka, S. Omura, K. Tanimoto, M. Imanishi, Y. Takebe, M. Matsugi, Y. Kita, *Chem. Eur. J.* **2002**, *8*, 4264.
02CL38	T. Teruya, S. Nakagawa, T. Koyama, K. Suenaga, D. Uemura, *Chem. Lett.* **2002**, 38.
02CL124	H. Kusama, F.Shiozawa, M. Shido, N. Iwasawa, *Chem. Lett.* **2002**, 124.
02CL158	M. Bao, H. Nakamura, A. Inoue, Y. Yamamoto, *Chem. Lett.* **2002**, 158.
02CL438	K. Okuma, Y. Hirose, K. Shioji, *Chem. Lett.* **2002**, 438.
02CL782	H. Nagano, T. Katsuki, *Chem. Lett.* **2002**, 782.
02EJO61	C. Campagnuolo, E. Fattorusso, O. Taglialatela-Scafati, A. Ianaro, B. Pisano, *Eur. J. Org. Chem.* **2002**, 61.
02EJO1523	D. Alickmann, R. Fröhlich, A.H. Maulitz, E.-U. Würthwein, *Eur. J. Org. Chem.* **2002**, 1523.
02EJO1830	A. Lévai, M. Kocevar, G. Tóth, A. Simon, L. Vranicar, W. Adam, *Eur. J. Org. Chem.* **2002**, 1830.
02EJO2112	G. Hanquet, X.J. Salom-Roig, S. Lemeitour, G. Solladié, *Eur. J. Org. Chem.* **2002**, 2112.
02EJO3589	M.G.B. Drew, A. Jahans, L.M. Harwood, S.A.B.H. Apoux, *Eur. J. Org. Chem.* **2002**, 3589.
02EJO3816	J. Le Nôtre, C. Bruneau, P.H. Dixneuf, *Eur. J. Org. Chem.* **2002**, 3816.
02EJO4169	S. Rosselli, M. Bruno, I. Pibiri, F. Piozzi, *Eur. J. Org. Chem.* **2002**, 4169.
02H(56)479	O. Arjona, M.L. León, R. Menchaca, J. Plumet, *Heterocycles* **2002**, *56*, 479.
02H(56)613	S. Cacchi, G. Fabrizi, A. Goggiomani, *Heterocycles* **2002**, *56*, 613.
02H(57)1057	I. Iliya, T. Tanaka, M. Iinuma, Z. Ali, M. Furasawa, K.-i. Nakaya, *Heterocycles* **2002**, *57*, 1057.
02H(57)1507	I. Iliya, T. Tanaka, M. Iinuma, Z. Ali, M. Furasawa, K.-i. Nakaya, N. Matsuura, M. Ubukata, *Heterocycles* **2002**, *57*, 1507.
02H(57)1587	K. Oda, H. Tsujita, M. Machida, *Heterocycles* **2002**, *57*, 1587.
02H(57)1705	P.-J. Sung, M.-C. Chen, *Heterocycles* **2002**, *57*, 1705.
02HCA689	M. Zahid, A. Ali, O. Ishurd, A. Ahmed, Z. Ali, V.U. Ahmad, Y. Pan, *Helv. Chim. Acta* **2002**, *85*, 689.
02HCA847	J.-R. Weng, M.-H. Yen, C.-N. Lin, *Helv. Chim. Acta* **2002**, *85*, 847.
02HCA1930	P. Akbay, J. Gertsch, I. Çahs, J. Heilmann, O. Zerbe, O. Sticher, *Helv. Chim. Acta* **2002**, *85*, 1930.
02HCA2394	I. Iliya, T. Tanaka, M. Iinuma, M. Furasawa, Z. Ali, K.-I Nakaya, J. Murata, D. Darnaedi, *Helv. Chim. Acta* **2002**, *85*, 2394.
02HCA2458	S. Aggarwal, N. Ghosh, R. Aneja, H. Joshi, R. Chandra, *Helv. Chim. Acta* **2002**, *85*, 2458.
02HCA2547	Y. Deng, S.-L. Peng, Q. Zhang, X. Liao, L.-S. Ding, *Helv. Chim. Acta* **2002**, *85*, 2547.

02HCA3262	F. Alonso, J. Meléndez, M. Yus, *Helv. Chim. Acta* **2002**, *85*, 3262.
02JA3608	S.R. Angle, N.A. El-Said, *J. Am. Chem. Soc.* **2002**, *124*, 3608.
02JA5260	K. Miki, F. Nishino, K. Ohe, S. Uemura, *J. Am. Chem. Soc.* **2002**, *124*, 5260.
02JA6845	Y. Yamamoto, H. Takagishi, K. Itoh, *J. Am. Chem. Soc.* **2002**, *124*, 6845.
02JA7395	L.A. Friedman, M. Sabat, W.D. Harman, *J. Am. Chem. Soc.* **2002**, *124*, 7395.
02JA10101	B. Liu, S.-Q. Duan, A.C. Sutterer, K.D. Moeller, *J. Am. Chem. Soc.* **2002**, *124*, 10101.
02JA11292	G.D. Wilkie, G.I. Elliott, B.S.J. Blagg, S.E. Wolkenberg, D.R. Soenen, M.M. Miller, S. Pollack, D.L. Boger, *J. Am. Chem. Soc.* **2002**, *124*, 11292.
02JA11584	A.K. Cheung, M.L. Snapper, *J. Am. Chem. Soc.* **2002**, *124*, 11584.
02JA14655	E.J. Jeong, E.J. Kang, L.T. Sung, S.K. Hong, E. Lee, *J. Am. Chem. Soc.* **2002**, *124*, 14655.
02JCS(P1)345	Y. Zhang, J. Xue, Y. Gao, H.-K. Fun, J.-H. Xu, *J. Chem. Soc., Perkin Trans. 1* **2002**, 345.
02JCS(P1)713	K. Tanaka, N. Harada, *J. Chem. Soc., Perkin Trans. 1* **2002**, 713.
02JCS(P1)965	L.F. Walker, A. Bourghida, S. Connolly, M. Wills, *J. Chem. Soc., Perkin Trans. 1* **2002**, 965.
02JCS(P1)1369	T.J. Donohoe, J.-B. Guillermin, D.S. Walter, *J. Chem. Soc., Perkin Trans. 1* **2002**, 1369.
02JCS(P1)1555	M.G. Duffy, D. H. Grayson, *J. Chem. Soc., Perkin Trans. 1* **2002**, 1555.
02JSC(P1)1748	T.J. Donohoe, A.A. Calabrese, J.-B. Guillermin, C.S. Frampton, D.S. Walter, *J. Chem. Soc., Perkin Trans. 1* **2002**, 1748.
02JCS(P1)2301	M.C. Elliott, *J. Chem. Soc., Perkin Trans. 1* **2002**, 2301.
02JCS(P1)2634	L. Caggiano, D.J. Fox, D. House, Z.A. Jones, F. Kerr, S. Warren, *J. Chem. Soc., Perkin Trans. 1* **2002**, 2634.
02JCS(P1)2646	L. Caggiano, D.J. Fox, D. House, Z.A. Jones, F. Kerr, S. Warren, *J. Chem. Soc., Perkin Trans. 1* **2002**, 2646.
02JCS(P1)2652	D. House, F. Kerr, S. Warren, *J. Chem. Soc., Perkin Trans. 1* **2002**, 2652.
02JHC639	C.J. Valduga, H.A. Stefani, N. Petragnani, *J. Heterocycl. Chem.* **2002**, *39*, 639.
02JNP2	D.T.A. Youssef, R.K. Yamaki, M. Kelly, P.J. Scheuer, *J. Nat. Prod.* **2002**, *65*, 2.
02JNP42	J.-H. Ju, D. Liu, G. Lin, X.-D. Xu, B. Han, J.-S. Yang, G.-Z. Tu, L.-B. Ma *J. Nat. Prod.* **2002**, *65*, 42.
02JNP76	K.-S. Song, I. Raskin, *J. Nat. Prod.* **2002**, *65*, 76.
02JNP89	H.Z. Jin, B.Y. Hwang, H.S. Kim, J.H. Lee, Y.H. Kim, J.J. Lee, *J. Nat. Prod.* **2002**, *65*, 89.
02JNP126	D. Friedrich, L.A. Paquette, *J. Nat. Prod.* **2002**, *65*, 126.
02JNP142	M. Bruno, M.L. Bondì, S. Rosselli, A. Maggio, F. Piozzi, N.A. Arnold, *J. Nat. Prod.* **2002**, *65*, 142.
02JNP175	P. Máximo, A. Lourenço, S.S. Feio, J.C. Roseiro, *J. Nat. Prod.* **2002**, *65*, 175.
02JNP262	S.Y. Li, H. Fuchino, N. Kawahara, S. Sekita, M. Satake, *J. Nat. Prod.* **2002**, *65*, 262.
02JNP290	M. Matsuhisa, Y. Shikishima, Y. Takaishi, G. Honda, M. Ito, Y. Takeda, H. Shibata, T. Higuti, O.K. Kodzhimatov, O. Ashurmetov, *J. Nat. Prod.* **2002**, *65*, 290.
02JNP346	K. Zou, S. Zhu, C. Tohda, S. Cai, K. Komatsu, *J. Nat. Prod.* **2002**, *65*, 346.
02JNP364	J. Malmstrøm, C. Christophersen, A. F. Barrero, J.E. Oltra, J. Justicia, A. Rosales, *J. Nat. Prod.* **2002**, *65*, 364.
02JNP375	M. Sekiguchi, H. Shigemori, A. Ohsaki, J. Kobayashi, *J. Nat. Prod.* **2002**, *65*, 375.
02JNP389	H. Kikuzaki, S. Tesaki, *J. Nat. Prod.* **2002**, *65*, 389.
02JNP470	C.-C. Liaw, F.-R. Chang, C.-Y. Lin, C.-J. Chou, H.-F. Chiu, M.-J. Wu, Y.-C. Wu, *J. Nat. Prod.* **2002**, *65*, 470.
02JNP562	M.W. Biavatti, P.C. Vieira, M.F.G.F. da Silva, J.B. Fernandes, S. Albuquerque, *J. Nat. Prod.* **2002**, *65*, 562.
02JNP580	R.C.B. Copley, M.T. Davies-Coleman, D.R. Edmonds, D.J. Faulkner, K.L. McPhail, *J. Nat. Prod.* **2002**, *65*, 580.
02JNP589	B. Sritularak, K. Likhitwitayawuid, J. Conrad, B. Vogler, S. Reeb, I. Klaiber, W. Kraus, *J. Nat. Prod.* **2002**, *65*, 589.
02JNP616	B.Y. Hwang, J.-H. Lee, J.B. Nam, H.S. Kim, Y.S. Hong, J.J. Lee, *J. Nat. Prod.* **2002**, *65*, 616.
02JNP745	Y.-L. Lin, Y.-Y. Chang, Y.-H. Kuo, M.-S. Shiao, *J. Nat. Prod.* **2002**, *65*, 745.
02JNP820	T. Pacher, C. Seger, D. Engelmeier, S. Vajrodaya, O. Hofer, H. Greger, *J. Nat. Prod.* **2002**, *65*, 820.
02JNP956	M.W. Muriithi, W.-R. Abraham, J. Addae-Kyereme, I. Scowen, S.L. Croft, P.M. Gitu, H. Kendrick, E.N.M. Njagi, C.W. Wright, *J. Nat. Prod.* **2002**, *65*, 956.

02JNP965	V.S.P. Chaturvedula, J.K. Schilling, D.G.I. Kingston, *J. Nat. Prod.* **2002**, *65*, 965.
02JNP1107	C.Y. Ragasa, J.G. Hofileña, J.A. Rideout, *J. Nat. Prod.* **2002**, *65*, 1107.
02JNP1165	Y.-H. Kuo, P.-H. Lee, Y.-S. Wein, *J. Nat. Prod.* **2002**, *65*, 1165.
02JNP1177	G. Hallur, A. Sivramakrishnan, S.V. Bhat, *J. Nat. Prod.* **2002**, *65*, 1177.
02JNP1216	B.S. Siddiqui, F. Afshan, S. Faizi, S.N. Naqvi, R.M. Tariq, *J. Nat. Prod.* **2002**, *65*, 1216.
02JNP1219	M. Nakatani, S. A. M. Abdelgaleil, S. M. I. Kassem, K. Takezaki, H. Okamura, T. Iwagawa, M. Doe, *J. Nat. Prod.* **2002**, *65*, 1219.
02JNP1307	Y. Liu, J. Hong, C.-O. Lee, K. S. Im, N. D. Kim, J. S. Choi, J. H. Jung, *J. Nat. Prod.* **2002**, *65*, 1307.
02JNP1328	L. Garrido, E. Zabía, M.J. Ortega, J. Salvá, *J. Nat. Prod.* **2002**, *65*, 1328.
02JNP1357	C. Anta, N. González, J. Rodríguez, C. Jiménez, *J. Nat. Prod.* **2002**, *65*, 1357.
02JNP1429	C.-Y. Duh, A.A.H. El-Gamal, S.-K. Wang, C.-F. Dai, *J. Nat. Prod.* **2002**, *65*, 1429.
02JNP1475	G.-H. Wang, J.-H. Sheu, C.-Y. Duh, M.Y. Chiang, *J. Nat. Prod.* **2002**, *65*, 1475.
02JNP1479	B. Schumacher, S. Scholle, J. Hötzl, N. Khudeir, S. Hess, C.E. Müller, *J. Nat. Prod.* **2002**, *65*, 1479.
02JNP1505	I. Kuroda, M. Musman, I.I. Ohtani, T. Ichiba, J. Tanaka, D.G. Gravalos, T. Higa, *J. Nat. Prod.* **2002**, *65*, 1505.
02JOC95	A.V. Kel'in, V. Gevorgyan, *J. Org. Chem.* **2002**, *67*, 95.
02JOC959	L.R. Domingo, M.J. Aurell, *J. Org. Chem.* **2002**, *67*, 959.
02JOC1001	G.W. Gribble, J. Jiang, Y. Liu, *J. Org. Chem.* **2002**, *67*, 1001.
02JOC1595	A. Padwa, C.K. Eidell, J.D. Ginn, M. S. McClure, *J. Org. Chem.* **2002**, *67*, 1595.
02JOC1692	S.H. Lee, H. Kohn, *J. Org. Chem.* **2002**, *67*, 1692.
02JOC2919	J.L.G. Ruano, C. Alemparte, F.R. Clemente, L.G. Gutiérrez, R. Gordillo, A.M.M. Castro, J.H.R. Ramos, *J. Org. Chem.* **2002**, *67*, 2919.
02JOC2926	J.L.G. Ruano, C. Alemparte, F.R. Clemente, L.G. Gutiérrez, R. Gordillo, A.M.M. Castro, J.H.R. Ramos, *J. Org. Chem.* **2002**, *67*, 2926.
02JOC3242	S.C. Roy, K.K. Rana, C. Guin, *J. Org. Chem.* **2002**, *67*, 3242.
02JOC3969	J.-H. Zhang, C.-J. Li, *J. Org. Chem.* **2002**, *67*, 3969.
02JOC4045	T.M.V.D. Pinho e Melo, M.I.L. Soares, A.M. d'A. R. Gonsalves, J.A. Paixão, A.M. Beja, M.R. Silva, L.A. da Veiga, J.C. Pessoa, *J. Org. Chem.* **2002**, *67*, 4045.
02JOC4200	S. Naito, M. Escobar, P.R. Kym, S. Liras, S.F. Martin, *J. Org. Chem.* **2002**, *67*, 4200.
02JOC4707	I. Cornella, J.P. Sestelo, A. Mouriño, L.A. Sarandeses, *J. Org. Chem.* **2002**, *67*, 4707.
02JOC5644	M. Pawlicki, L. Latos-Grazynski, L. Szterenberg, *J. Org. Chem.* **2002**, *67*, 5644.
02JOC6690	K.-i. Takao, T. Tsujita, M. Hara, K.-i. Tadano, *J. Org. Chem.* **2002**, *67*, 6690.
02JOC7724	M. Shanmugasundaram, M.-S. Wu, M. Jeganmohan, C.-W. Huang, C.-H. Cheng, *J. Org. Chem.* **2002**, *67*, 7724.
02JOC8079	R.C.D. Brown, C.J. Bataille, R.M. Hughes, A. Kenney, T.J. Luker, *J. Org. Chem.* **2002**, *67*, 8079.
02JSO1112	Y. Morimoto, T. Iwai, T. Kinoshita, *J. Syn. Org. Chem. Jpn.* **2002**, *60*, 1112.
02OL205	M.A. Calter, C. Zhu, *Org. Lett.* **2002**, *4*, 205.
02OL209	M.A. Calter, C. Zhu, R.J. Lachicotte, *Org. Lett.* **2002**, *4*, 209.
02OL245	K.W. Hunt, P.A. Grieco, *Org. Lett.* **2002**, *4*, 245.
02OL277	S.U. Son, S.I. Lee, Y.K. Chung, S.-W. Kim, T. Hyeon, *Org. Lett.* **2002**, *4*, 277.
02OL289	R. Roggenbuck, A. Schmidt, P. Eilbracht, *Org. Lett.* **2002**, *4*, 289.
02OL473	S.K. Bur, S.M. Lynch, A. Padwa, *Org. Lett.* **2002**, *4*, 473.
02OL593	F.E. McDonald, X.-D. Wei, *Org. Lett.* **2002**, *4*, 593.
02OL675	Y. Sakamoto, K. Tamegai, T. Nakata, *Org. Lett.* **2002**, *4*, 675.
02OL745	F.E. McDonald, V. Smolentsev, *Org. Lett.* **2002**, *4*, 745.
02OL773	A.S. Amarasekara, S. Chandrasekaran, *Org. Lett.* **2002**, *4*, 773.
02OL787	J. Clayden, M.N. Kenworthy, *Org. Lett.* **2002**, *4*, 787.
02OL859	H. Usuda, M. Kanai, M. Shibasaki, *Org. Lett.* **2002**, *4*, 859.
02OL1007	A.M. Harned, P.R. Hanson, *Org. Lett.* **2002**, *4*, 1007.
02OL1063	G. Mehta, J.D. Umarye, *Org. Lett.* **2002**, *4*, 1063.
02OL1083	H. Makabe, Y. Hattori, A. Tanaka, T. Oritani, *Org. Lett.* **2002**, *4*, 1083.
02OL1111	N. Mita, O. Tamura, H. Ishibashi, M. Sakamoto, *Org. Lett.* **2002**, *4*, 1111.
02OL1155	H.E. Zimmerman, W.-S. Chen, *Org. Lett.* **2002**, *4*, 1155.
02OL1159	D.J. Aldous, A.J. Dalençon, P.G. Steel, *Org. Lett.* **2002**, *4*, 1159.
02OL1387	D.K. Barma, A. Kundu, R. Baati, C. Mioskowski, J.R. Falck, *Org. Lett.* **2002**, *4*, 1387.

02OL1527	T.R. Kelly, D.-C. Xu, G. Martínez, H.-X. Wang, *Org. Lett.* **2002**, *4*, 1527.
02OL1567	K.G. Mayo, E.H. Nearhoof, J.J. Kiddle, *Org. Lett.* **2002**, *4*, 1567.
02OL1619	T. Shibata, N. Toshida, K. Takagi, *Org. Lett.* **2002**, *4*, 1619.
02OL1771	M.H. Haukaas, G.A. O'Doherty, *Org. Lett.* **2002**, *4*, 1771
02OL1787	P. Wipf, M.J. Soth, *Org. Lett.* **2002**, *4*, 1787.
02OL1809	D.M. Hodgson, T.D. Avery, A.C. Donohue, *Org. Lett.* **2002**, *4*, 1809.
02OL1879	M. Lautens, J.T. Colucci, S. Hiebert, N.D. Smith, G. Bouchain, *Org. Lett.* **2002**, *4*, 1879.
02OL2001	A.J. Pearson, E.F. Mesaros, *Org. Lett.* **2002**, *4*, 2001.
02OL2257	T. Fujioka, T. Nakamura, H. Yorimitsu, K. Oshima, *Org. Lett.* **2002**, *4*, 2257.
02OL2269	J.-L. Chen, Q.-L. Song, P.-X. Li, H.-R. Guan, X.-L. Jin, Z.-F. Xi, *Org. Lett.* **2002**, *4*, 2269.
02OL2329	A. Bouzider, G. Sauvé, *Org. Lett.* **2002**, *4*, 2329.
02OL2413	P.J. Mohr, R.L. Halcomb, *Org. Lett.* **2002**, *4*, 2413
02OL2695	G. Cafeo, F.H. Kohnke, M.F. Parisi, R.P. Nascone, G.L. La Torre, D.J. Williams, *Org. Lett.* **2002**, *4*, 2695
02OL2703	F. Bertozzi, M. Pineschi, F. Macchia, L.A. Arnold, A.J. Minnaard, B.L. Feringa, *Org. Lett.* **2002**, *4*, 2703.
02OL2735	F. LeStrat, J.A. Murphy, M. Hughes, *Org. Lett.* **2002**, *4*, 2735.
02OL2945	Z.-H. Peng, K.A. Woerpel, *Org. Lett.* **2002**, *4*, 2945.
02OL2977	N. Maezaki, N. Kojima, M. Asai, H. Tominaga, T. Tanaka, *Org. Lett.* **2002**, *4*, 2977.
02OL3059	T.J. Donohoe, A. Raoof, G.C. Freestone, I.D. Linney, A. Cowley, M. Helliwell, *Org. Lett.* **2002**, *4*, 3059.
02OL3175	F. Velázquez, H.F. Olivo, *Org. Lett.* **2002**, *4*, 3175.
02OL3251	M. Albert, D. De Souza, P. Feiertag, H. Hönig, *Org. Lett.* **2002**, *4*, 3251.
02OL3355	K. Mikami, H. Ohmura, *Org. Lett.* **2002**, *4*, 3355.
02OL3411	L. Jiang, S.D. Burke, *Org. Lett.* **2002**, *4*, 3411.
02OL3509	K.-i. Yamada, H. Fujihara, Y. Yamamoto, Y. Miwa, T. Taga, K. Tomioka, *Org. Lett.* **2002**, *4*, 3509.
02OL3715	A.R.L. Cecil, R.C.D. Brown, *Org. Lett.* **2002**, *4*, 3715.
02OL3763	C.R. Whitehead, E.H. Sessions, I. Ghiviriga, D.L. Wright, *Org. Lett.* **2002**, *4*, 3763.
02OL4061	Y. Hanzawa, M. Yabe, Y. Oka, T. Taguchi, *Org. Lett.* **2002**, *4*, 4061.
02OL4135	S.K. Bur, A. Padwa, *Org. Lett.* **2002**, *4*, 4135.
02OL4151	R.J. Madhushaw, C.-C. Hu, R.-S. Liu, *Org. Lett.* **2002**, *4*, 4151.
02OL4435	H.-W. Choi, K. Nakajima, D. Demeke, F.-A. Kang, H.-S. Jun, Z.-K. Wan, Y. Kishi, *Org. Lett.* **2002**, *4*, 4435.
02OL4681	J.-M. Yu, X.-B. Liao, J.M. Cook, *Org. Lett.* **2002**, *4*, 4681.
02OM1759	S.N. Ringelberg, A. Meetsma, S.I. Troyanov, B. Hessen, J.H. Teuben, *Organometallics* **2002**, *21*, 1759.
02P(59)85	H.-Q Duan, Y. Takaishi, H. Momota, Y. Ohmoto, T. Taki, *Phytochemistry* **2002**, *59*, 85.
02P(59)325	D.H. Vu, G.T. Tan, H.-J. Zhang, P.A. Tamez, V.H. Nguyen, M.C. Nguyen, D.D. Soejarto, H.H.S. Fong, J.M. Pezzuto, *Phytochemistry* **2002**, *59*, 325.
02P(59)371	A.T. Tchinda, A. Psopmo, P. Tane, J.F. Ayafor, J.D. Connolly, O. Sterner, *Phytochemistry* **2002**, *59*, 371.
02P(59)405	C.W. Holzapfel, W. Marais, P.L. Wessels, B.-E. Van Wyk, *Phytochemistry* **2002**, *59*, 405.
02P(59)409	M.A. Al-Yahya, F.S. el-Feraly, D.C. Dunbar, I. Muhammad, *Phytochemistry* **2002**, *59*, 409.
02P(59)435	B.B. Messanga, S.F. Kimbu, B.L. Sondengam, B. Bodo, *Phytochemistry* **2002**, *59*, 435.
02P(59)839	P. Metzger, M.-N. Rager, C. Largeau, *Phytochemistry* **2002**, *59*, 839.
02P(60)345	B.T. Ngadjui, B.M. Abegaz, F. Keumedjio, G.N. Folefoc, G.W.F. Kapche, *Phytochemistry* **2002**, *60*, 345.
02P(60)351	W. Li, K. Koike, Y. Asada, M. Hirotani, H. Rui, T. Yoshikawa, T. Nikaido, *Phytochemistry* **2002**, *60*, 351.
02P(60)375	P.K. Tarus, A.K. Machocho, C.C. Lang'at-Thoruwa, S.C. Chhabra, *Phytochemistry* **2002**, *60*, 375.
02P(60)861	M. Suzuki, Y. Takahashi, Y. Mitome, T. Itoh, T. Abe, M. Masuda, *Phytochemistry* **2002**, *60*, 861.
02P(61)345	N.Q. Dang, T. Hashimoto, M. Tanaka, M. Baumgartner, M. Stadler, Y. Asakawa, *Phytochemistry* **2002**, *61*, 345.
02P(61)361	M.H. Al Yousuf, A.K. Bashir, G. Blunden, T.A. Crabb, A.V. Patel, *Phytochemistry* **2002**, *61*, 361.

02P(61)389	V.E.J. Hikawczuk, P.C. Rossomando, O.S. Giordano, J.R. Saad, *Phytochemistry* **2002**, *61*, 389.
02S753	V. Lisowski, D.N. Vu, X. Feng, S. Rault, *Synthesis* **2002**, 753.
02S1010	A.M. Deshpande, A.A. Natu, N.P. Argade, *Synthesis* **2002**, 1010.
02S1541	W.H. Miles, S.K. Heinsohn, M.K. Brennan, D.T. Swarr, P.M. Eidam, K.A. Gelato, *Synthesis* **2002**, 1541.
02S1611	P. Nilsson, H. Gold, M. Larhed, A. Hallberg, *Synthesis* **2002**, 1611.
02S1711	M. Mischne, *Synthesis* **2002**, 1711.
02S1759	N. Krause, A. Hoffmann-Röder, J. Canisius, *Synthesis* **2002**, 1759.
02S2537	J.S. Yadav, B.V.S. Reddy, K.U. Gayathri, A.R. Prasad, *Synthesis* **2002**, 2537.
02SL337	K. Fujita, H. Yorimitsu, K. Oshima, *Synlett* **2002**, 337.
02SL489	S.T. Handy, M. Okello, *Synlett* **2002**, 489.
02SL533	S. Yamamura, S. Nishiyama, *Synlett* **2002**, 533.
02SL787	S. Muthusamy, C. Gunanathan, S.A. Babu, *Synlett* **2002**, 787.
02SL851	A. Padwa, S.K. Bur, D.M. Danca, J.D. Ginn, S.M. Lynch, *Synlett* **2002**, 851.
02SL987	B.C. Ranu, S. Samanta, A. Hajra, *Synlett* **2002**, 987.
02SL1318	M.A. Brimble, R.M. Davey, M.D. McLeod, *Synlett* **2002**, 1318.
02SL1451	A.G. Csákÿ, C. Contreras, M. Mba, J. Plumet, *Synlett* **2002**, 1451
02SL1532	H. Yoda, Y. Nakaseko, K. Takabe, *Synlett* **2002**, 1532.
02SL1815	I. Hemeon, C. DeAmicis, H. Jenkins, P. Scammells, R.D. Singer, *Synlett* **2002**, 1815.
02SL1868	H. Ohmura, K. Mikami, *Synlett* **2002**, 1868.
02SL2068	M.M. Salter, S. Sardo-Infffiri, *Synlett* **2002**, 2068.
02T1817	M.T. Crimmins, K.A. Emmitte, A.L. Choy, *Tetrahedron* **2002**, *58*, 1817.
02T1865	A.B. Dounay, G.J. Florence, A. Saito, C.J. Forsyth, *Tetrahedron* **2002**, *58*, 1865.
02T2077	D. Dabideen, Z.-M. Ruan, D.R. Mootoo, *Tetrahedron* **2002**, *58*, 2077.
02T2359	Y.R. Lee, J.Y. Suk, *Tetrahedron* **2002**, *58*, 2359.
02T2435	S.C. Roy, C. Guin, K.K. Rana, G. Maiti, *Tetrahedron* **2002**, *58*, 2435.
02T2605	A. Kamimura, H. Mitsudera, K. Matsuura, Y. Omata, M. Shirai, S. Yokoyama, A. Kakehi, *Tetrahedron* **2002**, *58*, 2605.
02T4769	A.M. Montaña, P.M. Grima, *Tetrahedron* **2002**, *58*, 4769.
02T4865	S. Bera, V. Nair, *Tetrahedron* **2002**, *58*, 4865.
02T5301	Z.-F. Li, Y.-M. Zhang, *Tetrahedron* **2002**, *58*, 5301.
02T6097	N. Toyooka, M. Nagaoka, E. Sasaki, H. Qin, H. Kakuda, H. Nemoto, *Tetrahedron* **2002**, *58*, 6097.
02T6631	T. Ohta, T. Maruyama, M. Nagahashi, Y. Miyamoto, S. Hosoi, F. Kiuchi, Y. Yamazoe, S. Tsukamoto, *Tetrahedron* **2002**, *58*, 6631.
02T6799	X. Peng, A. Li, J. Lu, Q. Wang, X. Pan, A.S.C. Chan, *Tetrahedron* **2002**, *58*, 6799.
02T6931	K.-X. Yan, K. Terashima, Y. Takaya, M. Niwa, *Tetrahedron* **2002**, *58*, 6931.
02T7017	A. Gansäuer, B. Rinker, *Tetrahedron* **2002**, *58*, 7017.
02T7797	X.-D. Luo, S.-H. Wu, D.-G. Wu, Y.-B. Ma, S.-H. Qi, *Tetrahedron* **2002**, *58*, 7797.
02T8119	M.L. Souto, C.P. Manríquez, M. Norte, J.J. Fernández, *Tetrahedron* **2002**, *58*, 8119.
02T8539	A.R. Díaz-Marrero, M. Cueto, E. Dorta, J. Rovirosa, A. San-Martín, J. Darias, *Tetrahedron* **2002**, *58*, 8539.
02T8825	R. Antonioletti, S. Malancona, P. Bovicelli, *Tetrahedron* **2002**, *58*, 8825.
02T9413	S.-H. Chan, C.-Y. Yick, H. N. C. Wong, *Tetrahedron* **2002**, *58*, 9413.
02T9729	V. Singh, S.Q. Alam, G.D. Praveena, *Tetrahedron* **2002**, *58*, 9729.
02T9903	M. Takadoi, T. Katoh, A. Ishiwata, S. Terashima, *Tetrahedron* **2002**, *58*, 9903.
02T10033	Y. Fukuyama, M. Kubo, T. Fujii, A. Matsuo, Y. Minoshima, H. Minami, M. Morisaki, *Tetrahedron* **2002**, *58*, 10033.
02T10189	E. Sisu, M. Sollogoub, J.-M. Mallet, P. Sinäy, *Tetrahedron* **2002**, *58*, 10189.
02T10301	M. Anniyappan, D. Muralidharan, P.T. Perumal, *Tetrahedron* **2002**, *58*, 10301.
02T10469	P. Gupta, S.K. Singh, A. Pathak, B. Kundu, *Tetrahedron* **2002**, *58*, 10469.
02TA1423	H.-X. Shi, H.-Z. Liu, R. Bloch, G. Mandville, *Tetrahedron: Asymmetry* **2002**, *13*, 1423.
02TA2133	E. Kobrzycka, D. Grykoa, J. Jurczak, *Tetrahedron: Asymmetry* **2002**, *13*, 2133.
02TL197	N.V. Moskalev, G.W. Gribble, *Tetrahedron Lett.* **2002**, *43*, 197.
02TL359	G.A. Molander, S.C. Jeffrey, *Tetrahedron Lett.* **2002**, *43*, 359.
02TL443	S. Tews, M. Hein, R. Miethchen, *Tetrahedron Lett.* **2002**, *43*, 443
02TL627	W.J. Xia, D.R. Li, L. Shi, Y.Q. Tu, *Tetrahedron Lett.* **2002**, *43*, 627.

02TL661	X.-M. Niu, S.-H. Li, Q.-S. Zhao, Z.-W. Lin, H.-D. Sun, Y. Lu, C. Wang, Q.-T. Zheng, *Tetrahedron Lett.* **2002**, *43*, 661.
02TL665	B. Jiang, F. Zhang, W. Xiong, *Tetrahedron Lett.* **2002**, *43*, 665.
02TL943	D.L. Wright, C.V. Robotham, K. Aboud, *Tetrahedron Let.* **2002**, *43*, 943.
02TL1317	T.K. Chakraborty, S. Tapadar, S.K. Kumar, *Tetrahedron Lett.* **2002**, *43*, 1317.
02TL1495	M. Karikomi, S. Watanabe, Y. Kimura, T. Uyehara, *Tetrahedron Lett.* **2002**, *43*, 1495.
02TL1713	H. Takikawa, M. Hirooka, M. Sasaki, *Tetrahedron Lett.* **2002**, *43*, 1713.
02TL2017	A.M. Montaña, P.M. Grima, *Tetrahedron Lett.* **2002**, *43*, 2017.
02TL2281	V. Popsavin, G. Benedekovic, M. Popsavin, *Tetrahedron Lett.* **2002**, *43*, 2281.
02TL2293	V. Nair, R.S. Menon, A.U. Vinod, S. Viji, *Tetrahedron Lett.* **2002**, *43*, 2293.
02TL2415	R.-W. Jiang, P.P.-H. But, S.-C. Ma, W.-C. Ye, S.-P. Chan, T.C.W. Mak, *Tetrahedron Lett.* **2002**, *43*, 2415.
02TL2609	M. Cavicchioli, X. Marat, N. Monteiro, B. Hartmann, G. Balme, *Tetrahedron Lett.* **2002**, *43*, 2609.
02TL2927	I. Yavari, M. Adib, M.H. Sayahi, *Tetrahedron Lett.* **2002**, *43*, 2927.
02TL2931	E. Ami, S. Rajesh, J. Wang, T. Kimura, Y. Hayashi, Y. Kiso, T. Ishida, *Tetrahedron Lett.* **2002**, *43*, 2931.
02TL3053	Y. Nakamura, S. Takeuchi, S.-L. Zhang, K. Okumura, Y. Ohgo, *Tetrahedron Lett.* **2002**, *43*, 3053.
02TL3011	F. Compernolle, H. Mao, A. Tahri, T. Kozlecki, E. Van der Eycken, B. Medaer, G.J. Hoornaert, *Tetrahedron Lett.* **2002**, *43*, 3011.
02TL3597	G.L. Plourde, *Tetrahedron Lett.* **2002**, *43*, 3597.
02TL3613	S.H. Kang, J.W. Jeong, *Tetrahedron Lett.* **2002**, *43*, 3613.
02TL3621	H. Usuda, M. Kanai, M. Shibasaki, *Tetrahedron Lett.* **2002**, *43*, 3621.
02TL4045	Y. Fort, P. Gros, A.L. Rodriguez, *Tetrahedron Lett.* **2002**, *43*, 4045.
02TL4503	I. Yavari, M. Anary-Abbasinejad, A. Alizadeh, *Tetrahedron Lett.* **2002**, *43*, 4503.
02TL4507	G.E. Daia, C.D. Gabbutt,a, J.D. Hepworth, B.M. Heron, D.E. Hibbs, M.B. Hursthouse, *Tetrahedron Lett.* **2002**, *43*, 4507.
02TL4585	R. Yanada, N. Nishimori, A. Matsumura, N. Fujii, Y. Takemoto, *Tetrahedron Lett.* **2002**, *43*, 4585.
02TL4753	V.O. Rogatchov, H. Bernsmann, P. Schwab, R. Fröhlich, B. Wibbeling, P. Metz, *Tetrahedron Lett.* **2002**, *43*, 4753.
02TL4865	S. Bera, V. Nair, *Tetrahedron* **2002**, *58*, 4865.
02TL5333	S. Nagumo, Y. Ishii, Y.-i. Kakimoto, N. Kawahara, *Tetraahedron Lett.* **2002**, *43*, 5333.
02TL5601	A.D. Rodríguez, I.I. Rodríguez, *Tetrahedron Lett.* **2002**, *43*, 5601.
02TL5783	B.-G. Wang, R. Ebel, C.-Y. Wang, V. Wray, P. Proksch, *Tetrahedron Lett.* **2002**, *43*, 5783.
02TL5849	Y. Morimoto, M. Takaishi, T. Iwai, T. Kinoshita, H. Jacobs, *Tetrahedron Lett.* **2002**, *43*, 5849.
02TL6705	M. Arai, S. Kaneko, T. Konosu, *Tetrahedron Lett.* **2002**, *43*, 6705.
02TL6771	D.W. Knight, E.R. Staples, *Tetrahedron Lett.* **2002**, *43*, 6771.
02TL7205	M. Yus, F. Foubelo, J.V. Ferrández, *Tetrahedron Lett.* **2002**, *43*, 7205.
02TL7295	E. Lee, H.O. Han, *Tetrahedron Lett.* **2002**, *43*, 7295.
02TL7983	H.-L. Huang, H.-C. Huang, R.-S. Liu, *Tetrahedron Lett.* **2002**, *43*, 7983.
02TI 195	J.M. Harris, G.A. O'Doherty, *Tetrahedron Lett.* **2002**, *43*, 8195.
02TI 8661	S. Takahashi, A. Kubota, T. Nakata, *Tetrahedron Lett.* **2002**, *43*, 8661.
02TL9105	S.-K. Kang, Y.-M. Kim, Y.-H. Ha, C.-M. Yu, H. Yang, Y. Lim, *Tetrahedron Lett.* **2002**, *43*, 9105.
02TL9151	A. Shaabani, M.B. Teimouri, H.R. Bijanzadeh, *Tetrahedron Lett.* **2002**, *43*, 9151.
02TL9265	G. Bifulco, T. Caserta, L. Gomez-Paloma, V. Piccialli, *Tetrahedron Lett.* **2002**, *43*, 9265.

Chapter 5.4

Five Membered Ring Systems: With More than One N Atom

Larry Yet
Albany Molecular Research, Inc., Albany, NY, USA
larryy@albmolecular.com

5.4.1 INTRODUCTION

Advances in the chemistry of pyrazoles, imidazoles, triazoles, tetrazoles, and related fused heterocyclic derivatives continued in 2002. Solid-phase combinatorial chemistry of benzimidazoles, imidazoles, and triazoles has been particularly active. Synthetic routes to all areas continue to be pursued vigorously with improvements and applications. Applications of imidazole- and 1,2,3-benzotriazole-containing reagents to a wide array of synthetic applications remained a constant theme.

5.4.2 PYRAZOLES AND RING-FUSED DERIVATIVES

Many physical studies of pyrazoles were published in 2002. Two desmotropes, 3-phenyl-1H-pyrazole and 5-phenyl-1H-pyrazole have been isolated and the conditions for their interconversion established <02HCA2763>. A series of pyrazolino[60]fullerenes has been prepared in one pot by 1,3-dipolar cycloaddition of nitrile imines to C_{60} <02T5821>. The synthesis and ^{1}H and ^{13}C NMR study of pyrazoles derived from chiral cyclohexanones were investigated <02H(57)307>. Mixed-anion nickel(II) and cobalt(II) complexes with a tetramethyl-substituted 4,4'-bipyrazolyl ligand were synthesized to provide a new entry to a realm of three-dimensional five-connected coordination topologies <02CC436>. A stepwise ring-closure synthesis and characterization of a homoleptic palladium(II)-pyrazolato cyclic trimer has been published <02CC1012>. Copper(II)-induced formation of cage-like compounds containing pyrazole macrocycles was reported <02CC936>. 3(5)-*tert*-Butylpyrazole has been found to be a ditopic receptor for zinc(II) halides <02CC704>. New pyrazole-related *C*-nucleosides with modified sugar moieties were synthesized and analyzed for biological activity <02T569>. Density functional calculation studies were performed on cycloaddition reactions between 1-aza-2-azoniaallene cation and olefins in pyrazole formation <02JOC7432>. Flash vacuum pyrolysis of pyrazoles was investigated in the presence of anionic clays having a hydrotalcite structure <02JOC8147>. β-Nitro-*meso*-tetraphenylporphyrin reacted with diazomethane to give pyrazoline-fused chlorin as the main product <02SL1155>. Enantioselective Michael additions of nitromethane to an α,β-unsaturated pyrazole auxiliary acceptor were effective under catalytic double activation in the presence of catalytic chiral Lewis acids and amines <02JA13394>. Several new 1*H*-pyrazolo[3,4-*b*]quinoxaline derivatives with *N*,*N*-dialkylamino electron-donating groups were prepared, and their photoluminescence in solution and electroluminescence have been investigated <02CC1404>. A series of novel α-amino acids containing the 1-hydroxypyrazole ring has been prepared by addition of organomagnesium or organolithium intermediates to diethyl *N*-Boc-iminomalonate as an electrophilic glycine equivalent <02T1595>. 2-Acyl-3-phenyl-*L*-menthopyrazoles were effective enantioselective agents employed in the resolution of secondary alcohols <02TA1713>.

Many methods for preparation of pyrazoles have appeared in the literature in 2002. A versatile synthesis of pyrazoles from benzophenone hydrazones was demonstrated with a variety of 1,3-bifunctional substrates under acidic conditions <02TL2171>. Ethyl 3-methoxy-1-methyl-1*H*-pyrazol-4-carboxylate (**1**) underwent efficient lithiation and quenching with various reagents to give intermediate halides, zincates, or boronic acids, which were cross-coupled under palladium catalysis with various partners to give tetrasubstituted pyrazoles **2** <02SL769>. 1,3,5-Trisubstituted pyrazolines **3** were converted to the corresponding pyrazoles **4** by treatment with a catalytic amount of palladium on carbon or by oxygen in acetic acid <02OL3955>. *N*-(1-Cycloalkenyl)pyrazoles were synthesized via elimination of benzotriazole from the corresponding 1-[1-(pyrazol-1-yl)cycloalkyl]benzotriazoles, which were obtained by cyclizations of dihaloalkanes with 1-(benzotriazol-1-ylmethyl)pyrazole <02JOC8230>. Aqueous one-pot synthesis of pyrazoles from enaminoketones promoted by microwave irradiation has been reported <02S1669>. A variety of vinylogous iminium salt derivatives have been examined as useful precursors for the regiocontrolled synthesis of heterocyclic-appended pyrazoles <02T5467>. Microwave-promoted synthesis of 5-acyloxypyrazoles from 1-phenyl-3-methylpyrazole-5-one and acid chlorides under solvent-free conditions has been achieved in moderate to good yields <02SC2549>. Reactions of trichloromethyl-substituted 1,3-dielectrophiles **5** with hydrazine afforded hydroxy pyrazoles **6** <02TL5005>. 4-Alkoxypyrazoles **8** were synthesized from 3,5-heptadiones **7** and hydrazines <02SL1170>. Haloacetylated enol ethers were shown to be useful precursors for the regiospecific synthesis of 5-trichloromethylpyrazoles in the presence of hydrazines <02SC419, 02SC1585>.

N-Alkyl-substituted phthalimides **9** were easily transformed into mono-, di- or trisubstituted pyrazoles **10** via a one-pot addition/decyclization/cyclocondensation sequence <02JCS(P1)207>. 5-Silylpyrazoles can be prepared from condensation of silylalkynones with hydrazines <02T4975>. Reactions of acylated diethyl malonates with hydrazine monohydrochloride in ethanol afforded 3,4-disubstituted pyrazolin-5-ones <02T3639>.

Substituted pyrazoline derivatives **13** were synthesized in high yields through cycloaddition reactions of azides **11** with acrylates **12** under Baylis-Hillman reaction conditions <02SL513>. Unsymmetrical azines **14** thermally cyclized to fused pyrazoles **15** <02TL6431>. Indazoles **17** were obtained from thermal cyclizations of (2-alkynylphenyl)triazenes **16** <02JOC6395>. 1-Aryl-4,6-dinitro-1H-indazoyl-3-methylcarboxylates **19** were prepared from methyl 2,4,6-trinitrophenylacetate (**18**) and arenediazonium chlorides <02SC467>. Site-selective functionalization of trifluoromethyl-substituted pyrazoles was explored with bromine additions or with lithiation followed by carboxylation additions <02EJO2913>. Densely functionalized pyrazolidines, prepared from a diastereoselective 1,3-dipolar cycloaddition of a chiral non-racemic azomethine imine ylide, were useful precursors to polysubstituted 1,3-diamines by chemoselective electroreduction <02S1885>. Nitrilimine cycloadditions with polymer-supported acrylates afforded 5-carboxy- or 5-aminocarbonyl-4,5-dihydropyrazoles <02JCS(P1)2504>. Refluxing the sodium bisulfite adduct of heteroaromatic and aromatic aldehydes and phenylhydrazine afforded a general synthesis of 3-heteroaryl and 3-aryl substituted-1H-indazoles <02SC3399>. A new flexible synthesis of pyrazoles with different, functionalized substituents at C-3 and C-5 has been reported <02JOC9200>.

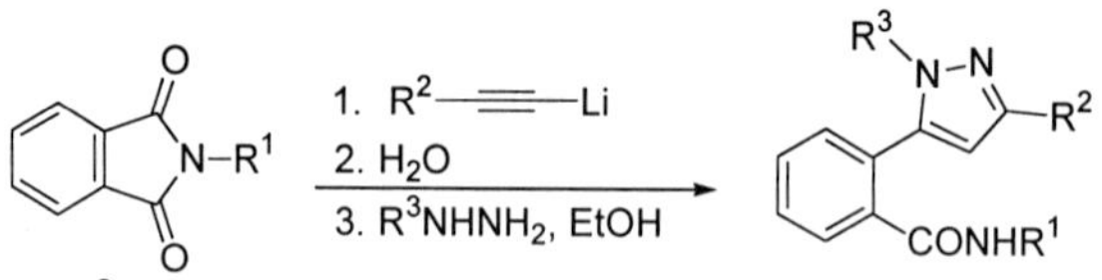

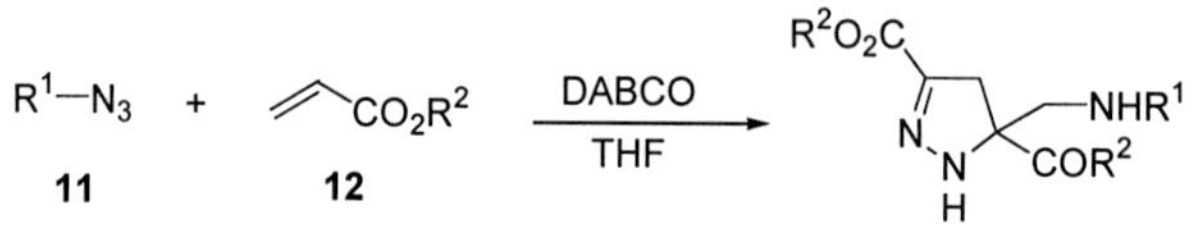

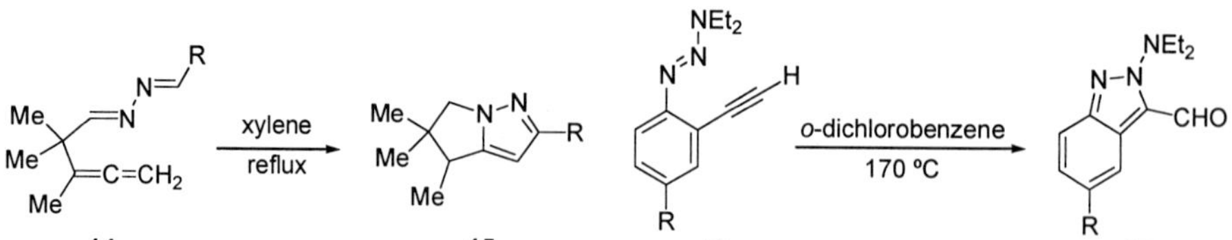

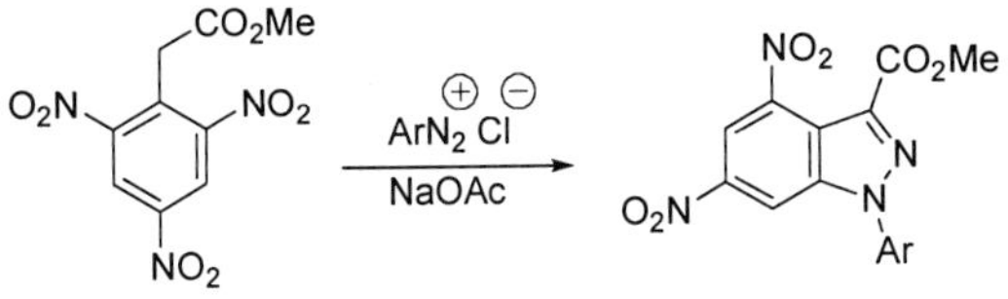

Intramolecular radical arylation of N-azinylpyridinium-N-aminides led to the synthesis of the pyrazolo[1,5-a]pyridine nucleus <02SL1093>. Regioselective bromine-magnesium exchange of 2-benzylated 3,4,5-tribromopyrazole-1-oxides and subsequent halogen dance of the resulting monometalated intermediates were used to synthesize pyrazolo[3,4-a]quinoline-1-oxides <02T7635>. 2-Thioxohydantoin ketene dithioacetals were versatile intermediates for synthesis of methylsulfanylimidazo[4,5-c]pyrazoles <02SC2245>. New condensed pyrazolo[1,5-e][1,3,5]benzoxadiazocines were prepared by cyclizations of 4,5-dihydro-3-methyl-5-(2-hydroxyphenyl)-1H-pyrazole-1-carboximidamides with triethyl orthoformate <02JCS(P1)1260>. b-Amino-b-(pyrid-4-yl)acrylonitriles reacted with hydrazine to yield fused pyrazolopyridine derivatives <02T9423>. Tröger's base analogs were obtained from reactions of 3-alkyl-5-amino-1-arylpyrazoles with formaldehyde <02JCS(P1)1588> and from 5-N-(benzotriazol-1-ylmethyl)amino-3-$tert$-butyl-1-phenylpyrazole with electron-rich alkenes in the presence of protic or Lewis acid catalysts <02TL5617>. 1-Aryl-1,2-diazabuta-1,3-dienes were useful substrates in the syntheses of 1-arylpyrazoles and 4-(2-oxopyrrol-3-yl)pyrazol-3-ones <02S1546>. Synthesis of pyrazolo[1,5-a]pyrimidine derivatives was achieved by reaction of 5-aminopyrazoles with suitable unsaturated keto compounds <02SC253>. Palladium-catalyzed intramolecular C-N bond formation of N-acetamino-2-(2-bromo)arylindolines **20**, followed by hydrolysis and air oxidation with aluminum oxide, allowed the preparation of indolo[1,2-b]indazoles **21** <02TL3577>. N-Hydroxypyrazoles were used for directed lithiation leading to a new synthesis of azaxanthones <02T2397>. Sequential functionalization of pyrazole-1-oxides via regioselective metalation led to the synthesis of 3,4,5-trisubstituted 1-hydroxypyrazoles

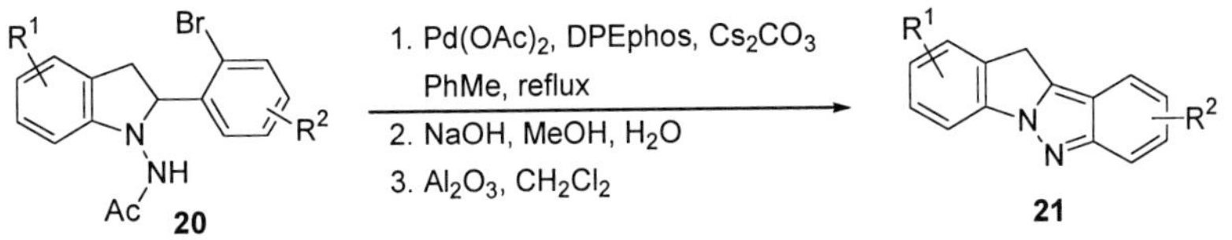

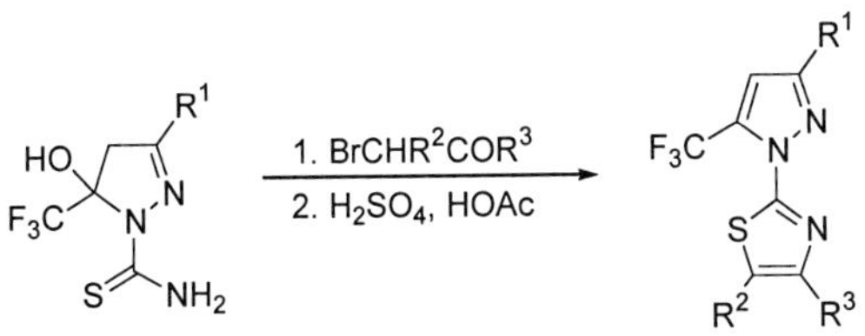

<02JOC3904>. Pyrazole thioamides **22** were transformed into pyrazolothiazoles **23** via a cyclization/dehydration sequence <02S1079>. 1-Acetyl-2-methoxyazulene (**24**) reacted with arylhydrazines in refluxing ethanol to give 1-aryl-3-methylazuleno[1,2-*d*]pyrazoles **25** in moderate to high yields <02H(56)497>.

5.4.3 IMIDAZOLES AND RING-FUSED DERIVATIVES

Several physical and biological studies of imidazoles were published in 2002. The basicity of 1,3-di-*tert*-butylimidazol-2-ylidene was measured in tetrahydrofuran against three hydrocarbon indicators <02JA5757>. A positively charged tripodal receptor with nitro groups in the imidazolium rings was designed, synthesized, and characterized for its anion binding strength <02OL2897>. Six 2,2'-biimidazoles with various amide groups at the 4- and 4'-positions were synthesized and were investigated for anion binding <02JOC5963>. A quantitative evaluation of the chloride template effect in the formation of dicationic [1₄]imidazoliophanes has been published <02JOC8463>. Pyrrole-imidazole hairpin polyamides were found to be effective DNA alkylating agents <02CEJ4781>.

Many general procedures for the synthesis of the imidazole core have been published in 2002. 2-Aminoimidazole (**27**) was prepared in an efficient and environmentally friendly manner from *O*-methylisourea (**26**) and 2,2-dimethoxyethylamine followed by acidic cyclization <02TL593>. Novel 4-substituted-5-nitroimidazoles were prepared from an $S_{RN}1$ reaction <02TL4127>. Enantiopure 1,4-disubstituted 2-imidazolines were synthesized from enantiopure β-amino alcohols <02JOC3919>. A variety of aryl- and alkyl-substituted imidazoles have been prepared by the ring opening of *N*-(2-oxoalkyl)oxazolinium salts with ammonia <02TL6997>. A variety of novel *N*-alkoxy aromatic-fused imidazoles have been prepared in a simple two-step process from 2-fluoronitroaromatics or 2-chloro-3-nitropyridines <02TL7707>. Iridinium-catalyzed reaction of 1-methylimidazole and various aldehydes with dimethyl acetylenedicarboxylate (DMAD) as the hydrogen acceptor afforded a general synthesis of 2-substituted imidazoles <02AG(E)2779>. Various substituted imidazoles can be prepared efficiently from cyclic or acyclic 1,2-aminoalcohols via a four-step procedure involving acylation of the amine, oxidation of the alcohol, imine formation and cyclization <02TL7687>. Rhodium-catalyzed addition of dichloro-*p*-toluenesulfonamide with acetonitrile to α,β-unsaturated esters **28**, using a ferric chloride-triphenylphosphine complex, afforded imidazolidine derivatives **29** with high *anti:syn* ratios <02JOC4777>. Lithium aluminum hydride was employed as a cyclization agent of urea methyl esters **30** to 3,5-disubstituted tetrahydro-1*H*-imidazole-2,4-diones **31** <02SC1015>. A highly diastereoselective multicomponent synthesis was reported of unsymmetrical imidazolines **33**, which contain a four-point diversity applicable to alkyl, aryl, acyl, and heterocyclic substitutions, from oxazolones **32** <02OL3533>. Reaction between 3-(*N,N*-dimethylamino)-2-isocyanoacrylate **34** and primary amines regioselectively afforded 1-alkyl-4-imidazolecarboxylates **35** in good yields <02OL4133>. 2,4-Disubstituted imidazoles **38** were prepared in an optimized process from condensation of amidines **36** and α-haloketones **37** <02OPRD682>. Addition of imidazoles to arynes provided a novel approach to the synthesis of *N*-alkyl-*N*-arylimidazolium salts <02OL2767>. A series of 1,2-dimethyl-3-arylsulfonyl-4,5-dihydroimidazolium iodides, precursors to trisubstituted ethylenediamine derivatives, was prepared by convenient sulfonylation and methylation from 2-methylimidazoline <02SC1447>. 4,5-Disubstituted imidazolidin-2-ones **41** were prepared from low-valent titanium-induced cyclization of arylimines **39** and triphosgene **40** <02SC2613>. The facile synthesis of 2,2-dichloro-1-[3-(2,2-dichloroacetyl)-2-(4-methoxylphenyl)imidazolidin-1-yl]ethanone from *N,N*'-bis(4-methoxybenzylidene)ethane-1,2-diamine and dichloroacetyl chloride has been described <02SC3255>. Three-component syntheses of 4(5)-sulfanyl-1*H*-imidazoles **43** from aldehydes, 2-oxo-2-arylthioacetamides **42**, and ammonium acetate with alkyl bromides under microwave conditions has been reported <02JCO87>.

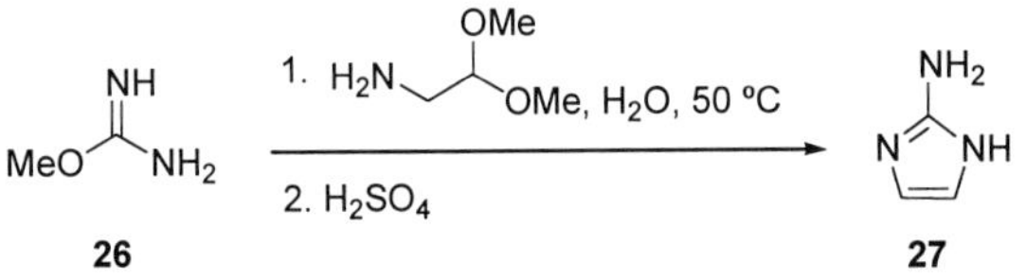

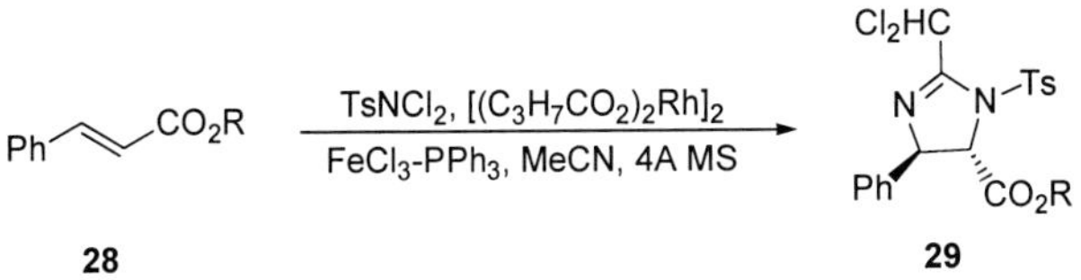

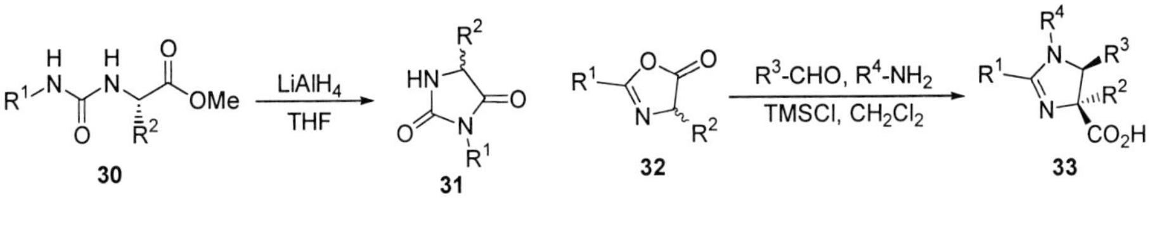

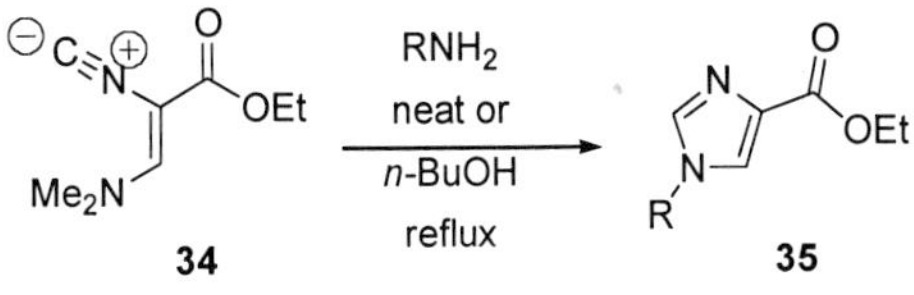

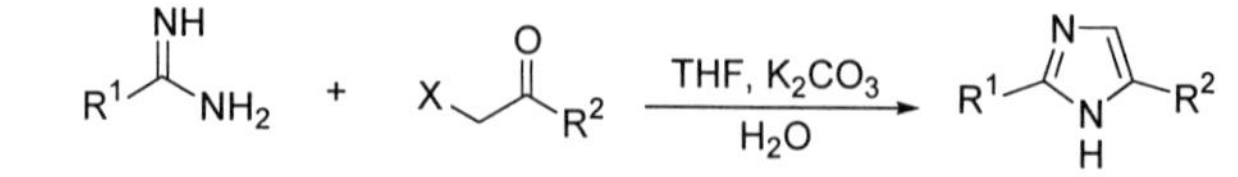

Syntheses of benzimidazoles have been accomplished by various routes. Chiral imines **44** were deprotected with hydroxylamine and then cyclized under acid conditions to give chiral benzimidazoles **45** <02JOC1708>. Intramolecular palladium-catalyzed aryl amination of aryl amidines **46** afforded benzimidazoles **47** <02TL1893>. Reductive intermolecular coupling/heterocyclization of 2-nitroaniline and aromatic aldehydes in the presence of 2-bromo-2-nitropropane and zinc afforded 2-substituted and 1,2-disubstituted benzimidazoles <02H(57)5>. Ammonium sulfate-magnesium selective reduction of N-2-nitrophenylimidates led to the synthesis of 2-substituted benzimidazoles <02SC387>. N-[(2-Benzylideneamino)phenyl]-C-methylketenimines underwent intramolecular cyclization via two different reactions, a [2 + 2] cycloaddition and a rare imino-ene reaction, to yield azeto[1,2-a]benzimidazoles and 2-(α-styryl)benzimidazoles, respectively <02TL6259>. Condensation of trifluoromethyl-o-phenylenediamines and aromatic aldehydes under microwave irradiation with zinc chloride supported on alumina resulted exclusively in the formation of trifluoromethyl-2-arylbenzimidazoles <02SC2467>. 2-(2-Diphenylphosphinyl-naphthalen-1-yl)-1-isopropyl-1H-benzimidazole has been synthesized as a new atropisomeric P,N-chelating ligand for asymmetric catalysis <02TA137>. A solution-phase parallel synthesis of substituted benzimidazoles was achieved using polymer-bound scavengers and reagents <02JCO320>. A rapid microwave-assisted liquid-phase combinatorial synthesis of 2-(arylamino)benzimidazoles has been reported <02JCO359>. A small library of benzimidazoles was prepared using polymer-bound reagents and scavengers <02SL739>.

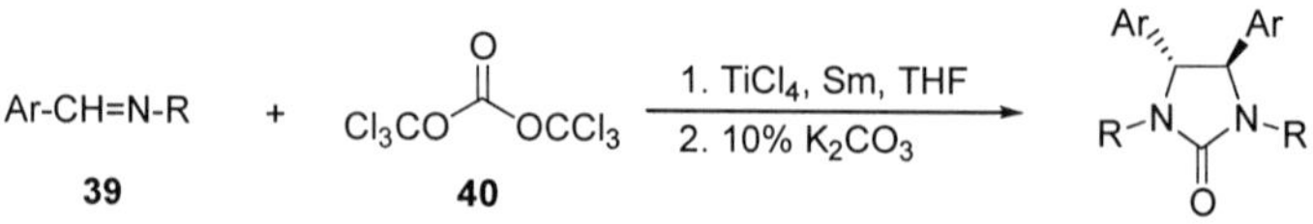

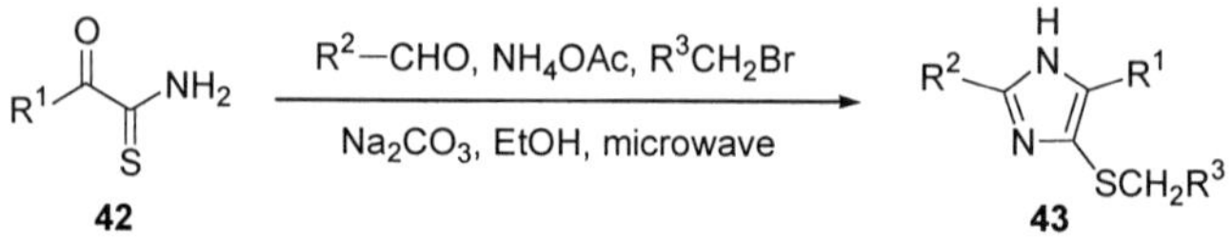

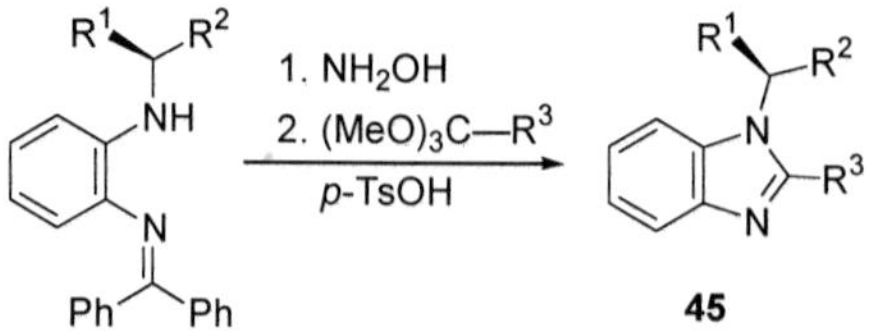

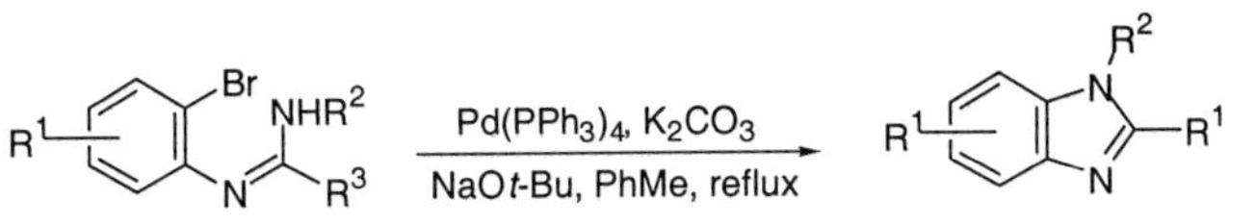

The imidazole core structure itself has been utilized as a template for several synthetic operations. 1-(3-Furyl)-1H-imidazole derivatives were obtained from the reaction of substituted imidazoles, dibenzoylacetylene and triphenylphosphine <02TL9449>. 3-Fluoro-3-imidazolylpropenoic acids were prepared from 1-trityl-3-propenoic acid <02JOC3468>. Imidazole-4,5-dicarboxylic acid was a precursor to symmetrically and dissymmetrically disubstituted imidazole-4,5-dicarboxamides <02JOC7151>. Green chemistry desulfurization of 2-mercaptoimidazoles **48** with hydrogen peroxide in the presence of catalytic tungstic acid afforded 1-substituted-5-hydroxymethylimidazoles **49** <02ORPD674>. Various isocyanates **50** reacted with *N*-heterocyclic carbenes, thermally generated from imidazoline **51**, to give various [4 + 1] cycloaddition products **52** <02OL4289>. Intermolecular coupling of isomerizable alkenes with benzimidazoles via rhodium-catalyzed C-H bond activation to heterocycles has been reported <02JA3202, 02JA13964>. Treatment of 1-methyl-4-nitroimidazole (**53**) with 1,1,1-trimethylhydrazinium iodide in the presence of sodium methoxide in dimethyl sulfoxide afforded 5-amino-1-methyl-4-nitroimidazole (**54**) in moderate yield <02TL6613>. *N*-Protected 2-substituted 4,5-diiodoimidazoles **55** were treated with excess *iso*-propylmagnesium chloride and electrophiles to give a convenient synthesis of 4,5-disubstituted imidazoles **56** <02JOC2699>. Reactions of 2,3-diaryl-1-methyl-4,5-dihydroimidazolium iodides with various nucleophilic reagents were examined <02SC1457>. Reactions of arylsulfonylimidazolium salts with aromatic amines produced *N'*-aryl-*N*-methyl-*N*-(2-arylsulfonylamino)ethylacetamidine monohydroiodides in good to excellent yields <02SC1129>. The synthesis of several 1H-imidazol-1-ylmethylpiperidines has been reported <02TL8917>.

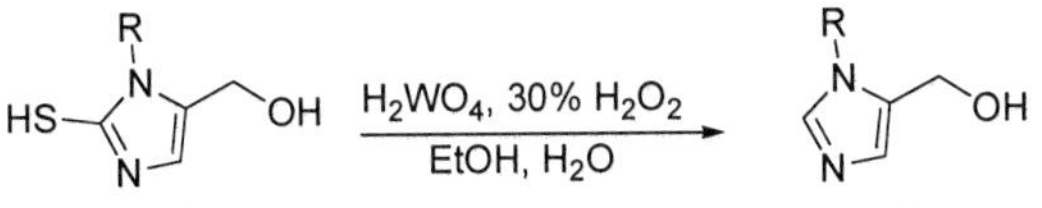

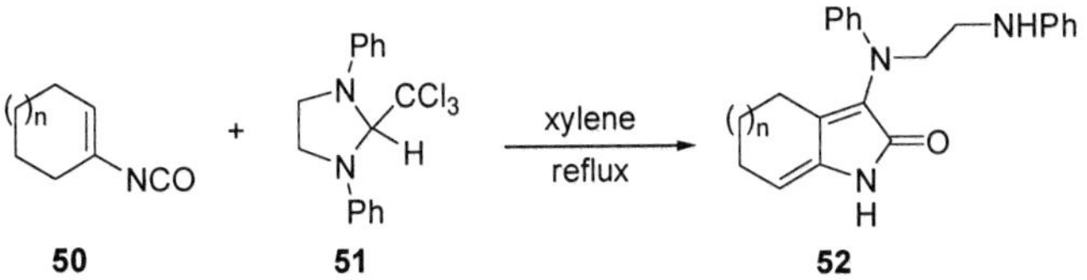

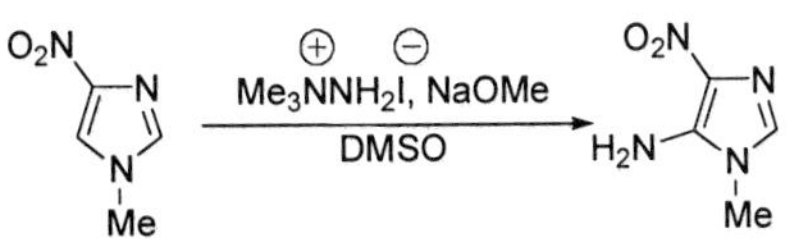

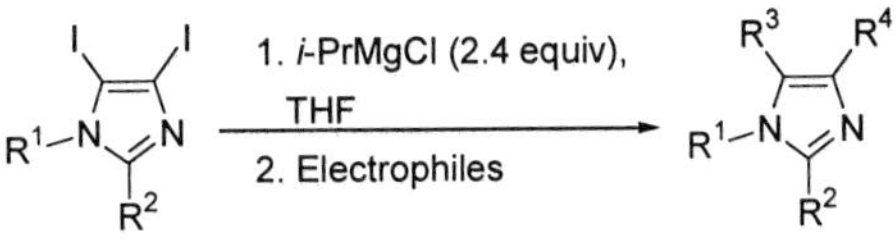

Imidazole-containing reagents have found useful applications in a variety of organic transformations. A second generation ruthenium-based olefin metathesis catalyst coordinated with 1,3-dimesityl-4,5-dihydroimidazol-2-ylidene ligand **57** has been utilized in the synthesis of symmetrical trisubstituted olefins by cross-metathesis <02OL1939>, ring-opening and ring-closing metathesis of cycloalkene-yne to give bicyclic compounds and/or dimeric compounds in good yields <02OL3855>, in the synthesis of carbocyclic methyl and silyl enol ethers <02CC2490>, and in ring-opening metathesis polymerization in donor solvents <02CC2572>. Imidazolidone **58** has been used in enantioselective organocatalytic 1,4-addition of electron-rich benzenes to α,β-unsaturated aldehydes <02JA7894>. Several novel imidazoline catalysts were screened for the organocatalytic asymmetric conjugate addition of nitroalkanes to α,β-unsaturated enones <02JOC8331>.

Bulky imidazolium salt **59** was found to greatly accelerate the intramolecular Heck reactions of aromatic chlorides <02TL9347>, to be useful ligands in the palladium-catalyzed carbonylative coupling of aryl diazonium ions and aryl boronic acids <02TL9137>, in the microwave-assisted palladium-catalyzed synthesis of pinacol boronates from aryl chlorides <02OL541>, in the nickel-catalyzed coupling of aryl chlorides and amines <02JOC3029>, and in the palladium-catalyzed Heck-coupling of aryl diazonium salts with olefins <02OL2079>. Bulky imidazolium salt **60** was a novel ligand utilized in the palladium-catalyzed synthesis of oxyindoles from o-haloanilides <02TL193>. N-Carbamoyl-substituted heterocyclic carbene palladium(II) complex **60** promoted Sonogashira cross-coupling reactions under mild conditions <02OL1411>. Other nucleophilic N-heterocyclic carbenes have been developed for transesterification reactions <02OL3583, 02OL3587> and for aryl aminations <02OL2229>.

Ionic liquids have been a popular topic of interest in 2002 and a review of the applications of these solvents in organic synthesis has been published (02ACA75>. New, densely functionalized fluoroalkyl-substituted imidazolium ionic liquids have been reported <02TL9497>. An ultrasound-assisted preparation of a series of ambient-temperature ionic liquids, 1-alkyl-3-methylimidazolium halides, which proceeds via efficient reactions of 1-methyl imidazole with alkyl halides/terminal dihalides under solvent-free conditions, has been described <02OL3161>. New hydrophilic poly(ethyleneglycol)-ionic liquids have been synthesized from

1,3-disubstituted imidazolium cations and fluorinated anions <02OPRD374>. An improved preparation of 1,3-dialkylimidazolium tetrafluoroborate ionic liquids using microwave irradiation has been reported <02TL5381>. The ionic liquid 1-ethyl-3-methylimidazolium chloride, in conjuction with aluminum chloride, has been used in the Friedel-Crafts acylation of indoles <02TL5793>. The ionic liquids 1-butyl-3-methylimidazolium halides (**62**, X = Cl, Br) have been utilized in the catalytic Rosenmund-von Braun reaction of aryl halides and sodium cyanide <02TL387> and in the base-catalyzed Baylis-Hillman reaction <02CC1612>. Ionic liquid **62** (X = NTf$_2$) has been used biocatalytically in the acylation of aliphatic alcohols with vinyl acetate <02CC992>. 1,3-Di-*n*-butylimidazolium tetrafluoroborate has been employed in ultrasound-promoted Suzuki cross-coupling reactions <02CC616>. The ionic liquid 1-butyl-3-methylimidazolium tetrafluoroborate (**63**, [bmim][BF$_4$]) has been utilized in the esterification of sodium salicylate with benzyl chloride <02TL9381>, in the study of ionic liquid mediated microwave heating of organic solvents <02JOC3145>, in nickel-catalyzed Michael reactions <02CC434>, in the Suzuki-Miyaura cross-coupling reactions of 4-iodophenol immobolized on polystyrene-Wang resin <02OL3071>, in the preparation of a C$_9$-aldehyde via aldol condensation <02CC1610>, in the fluorination of alkyl halides and mesylates with potassium fluoride <02JA10278>, in direct asymmetric aldol reactions with L-proline <02TL8741>, and in the syntheses of carbamates from reactions of primary and secondary aliphatic amines with dimethyl carbonate <02TL8145>. The ionic liquid 1-butyl-3-methylimidazolium hexafluorophosphorate (**64**, [bmim][PF$_6$]) has been utilized in the copper-free Sonogashira coupling <02OL1691>, in combination with DMAP for dihydroxylation of olefins <02OL2197>, and in the first Bischler-Napieralski room-temperature cyclization <02TL5089>.

Me–N⊕N–*n*-Bu X⊖
62 X = Cl, Br, NTf$_2$

Me–N⊕N–*n*-Bu BF$_4$⊖
63

Me–N⊕N–*n*-Bu PF$_6$⊖
64

Vicarious nucleophilic substitution of 1-benzyl-5-nitroimidazole led to the synthesis of 6,7-dihydroimidazo[4,5-*d*][1,3]diazepin-8(3*H*)-one <02TL1595>. An efficient method for the synthesis of 4,7-dihydro-1*H*-imidazo[4,5-*e*]-1,2,4-triazepin-8-one derivatives was accomplished by the use of microwave technology <02SL519>. A preparation of a new cyclic quaterbenzimidazole has been reported <02S723>. Intramolecular nitrone cycloadditions have been used in a synthetic approach to imidazo[1,2-*a*]pyridine derivatives <02T4445>. Multi-component reaction of aldehydes, isonitriles and 2-aminoazines afforded 2- and 3-aminoimidazo[1,2-*a*]pyrimidines <02TL4267>. 6-Amino-1,4,6,7-tetrahydroimidazo[4,5-*b*]pyridin-5-ones represent new heterocyclic scaffolds which can be used as conformationally restricted histidine analogs and as modified purines <02TL4343>. 1,4-Disubstituted 5-thioxoperhydroimidazo[4,5-*d*]imidazol-2-ones were prepared by one-pot criss-cross cycloaddition reactions of 1,4-disubstituted 1,4-diazabuta-1,3-dienes with HNCS and HNCO generated *in situ* from potassium salts and acetic acid <02TL4833>. Efficient syntheses of new polyfunctionalized thiadiazacenaphthylenes from imidazol[1,2-*a*]pyridines have been published <02H(57)21>. A new simple and efficient synthesis of 2-trifluoromethylimidazo[1,2-*a*]pyridines and 2-trifluoromethyl[1,2-*a*]quinolines from 1,1,1-trifluoro-4-(phenylsulfonyl)but-3-ene-2,2-diols with various 2-aminopyridines has been described <02S1379>. A versatile route to imidazo[4,5-*h*]isoquinolin-9-ones has been prepared by S$_N$Ar reaction <02TL7553>.

Rearrangement of alkoxycarbonyl imidazole acryl azides in diphenyl ether at high temperatures afforded imidazo[1,5-c]pyrimidinone or imidazo[4,5-c]pyridinone derivatives <02TL5879>. Efficient synthesis of imidazopyridodiazepines from *peri* annulation in imidazo[1,2-a]pyridine has been described <02TL9119>. A convenient synthesis of 3,6-disubstituted-2-aminoimidazo[1,2-a]pyridines has been published <02TL9051>. Novel 2,3-dihydroimidazo[2,1-b][1,3]oxazoles were prepared from intramolecular nucleophilic *ipso*-substitution of 2-alkylsulfonylimidazoles <02S2691>. 4,4'-Bi-1H-imidazol-2-ones were efficiently synthesized from 5-amino-α-imino-1H-imidazole-4-acetonitriles and isocyanates <02JOC5546>.

Several solid-phase combinatorial approaches to the benzimidazole core have been developed in 2002. Solid-supported 1,2-benzenediamines have been useful precursors for many of these reactions. For example, polymer-bound aryldiamines **65** were cyclized using the Vilsmeier reagent to yield imidazolines **66** following acidic cleavage <02JCO496>. Cyclization of polymer-supported benzenediamines **67** with 1,1-dicarbonyldiimidazole, followed by alkylation and acid cleavage afforded substituted dihydroimidazolyl dihydrobenzimidazol-2-ones **68** <02T2095>. Resin-bound benzenediamines **69** were cyclized with isothiocyanate and cleaved with acid to give 2-(arylamino)benzimidazoles **70** <02TL1529>. An efficient method for the solid-phase synthesis of trisubstituted [1,3,5]triazino[1,2-a]benzimidazole-2,4-(3H,10H)-diones from resin-bound amino acids was described <02JCO345>. A general route for the synthesis of 1-substituted-2-arylbenzimidazoles using solid-phase support has been reported <02JCO475>. The solid-phase syntheses of dihydroimidazolyl 2-alkylthiobenzimidazoles, dihydroimidazolyl 2-alkylsulfonylbenzimidazoles, dihydroimidazolyl dihydroquinoxaline-2,3-diones, and dihydroimidazolyl dihydrobenzimidazol-2-imines have been reported <02JCO214>.

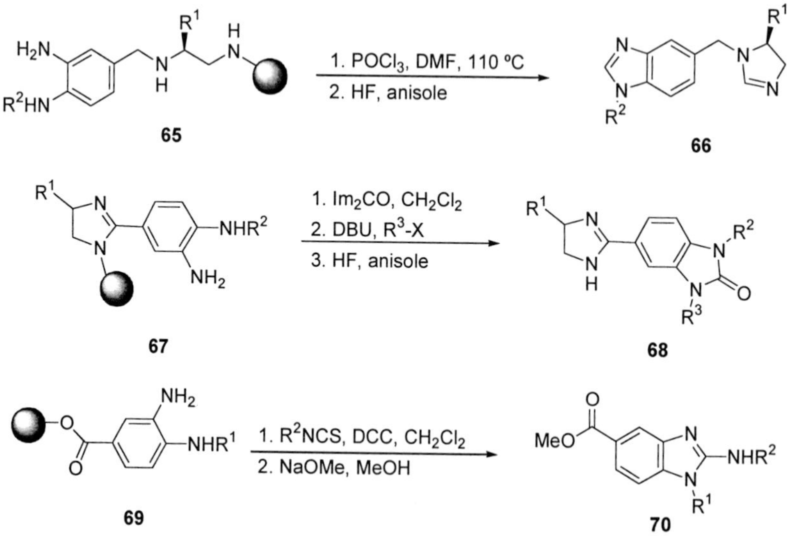

Other methods of preparing imidazoles on solid-supports have been reported. Solid-supported amino acid **71** underwent sequential reductive amination with aldehydes, benzotriazole-mediated cyclizations and acidic cleavage to give 1,2,5-trisubstituted-4-

imidazolidinones **72** <02JCO209>. A solid-phase synthesis of imidazo[4,5-*b*]pyridin-2-ones and related urea derivatives by cyclative cleavage of a carbamate linkage with concomitant cyclization has been published <02JCO352>. The cyclization reaction of resin-bound diamines **73** with arylisothiocyanates in the presence of mercuric chloride followed by acidic cleavage afforded 1,5-disubstituted 2-aryliminoimidazolidines **74** <02JOC3138>. A new approach to the preparation of 2-substituted imidazole libraries using a polystyrene-carbamyl chloride resin in a traceless fashion has been reported <02OL4017>. Resin-bound thioureas **75** were treated with Mukaiyama's reagent followed by acidic cleavage to yield 2-aminoimidazolidin-4-ones **76** <02T3349>. Resin-bound Boc-protected amidines **77** underwent deprotection, cyclization, and simultaneous cleavage promoted by a polyamine resin to give 2-aminoimidazolones **78** <02TL4463>. Parallel synthesis of 1,2,4-trisubstituted imidazoles was accomplished via a carbazate linker of *N*-alkyl-*N*-(β-keto)amides <02TL7557>. Polymer-bound glycerine resin **79** was cyclized with isocyanates followed by acidic cleavage to yield 2-imidazolones **80** <02TL4571>.

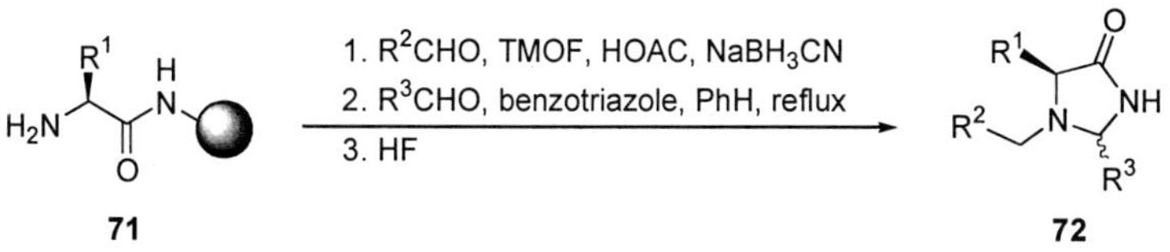

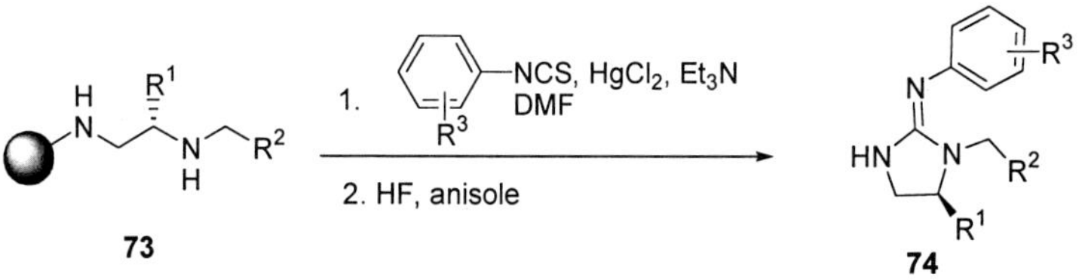

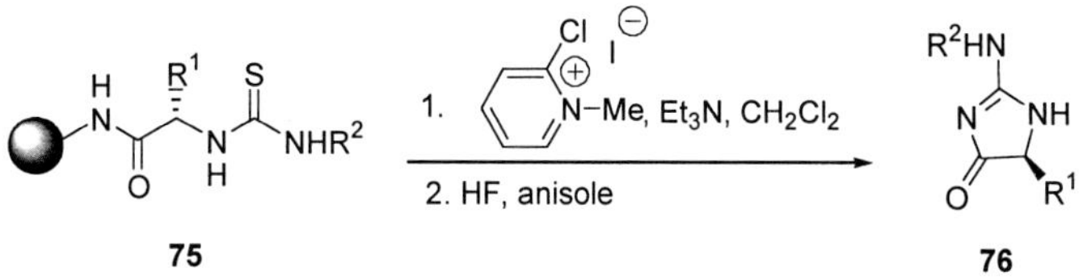

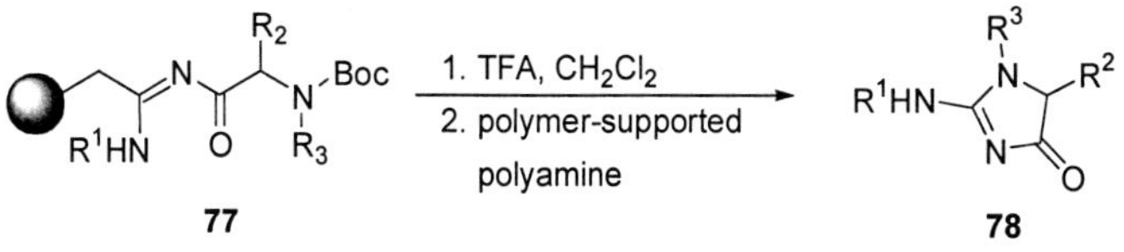

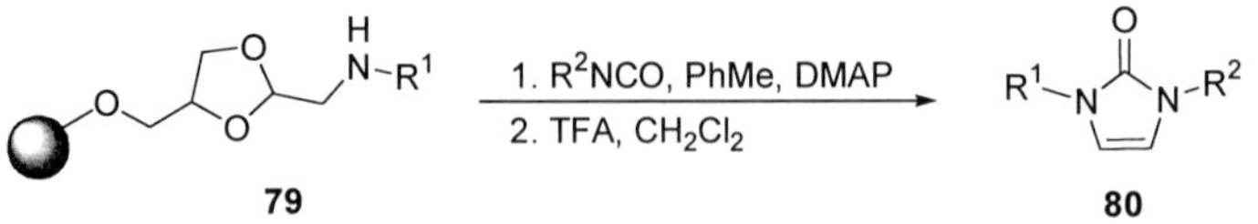

5.4.4 1,2,3-TRIAZOLES AND RING-FUSED DERIVATIVES

Irradiation of benzotriazole with a variety of maleimide derivatives led to the stereo- and regioselective formation of aryl [2 + 2] photocycloaddition products <02OL1487>. The catalytic activity of cucurbituril in 1,3-dipolar cycloadditions has been applied to the synthesis of oligotriazoles <02CC22>. The ^{15}N NMR spectra, tautomerism and diastereomerism of 4,5-dihydro-1H-1,2,3-triazoles were investigated <02JCS(P1)126>. An experimental NMR and theoretical (GIAO) study of the tautomerism of benzotriazole in solution has been reported <02T9089>. N-Hydroxybenzotriazole resin was used in optimization experiments for amide synthesis <02JCO576>.

2-Allyl-1,2,3-triazoles **81** were prepared by the palladium-catalyzed three component coupling reaction of alkynes, allyl methyl carbonate and trimethylsilyl azide <02TL9707>. Thiocarbamoylimidazolium salts were useful precursors in the presence of sodium azide in the synthesis of aminothiatriazoles <02TL7601>. 1,3-Dipolar cycloaddition of organic azides **82** with acetylenic amides **83** under solvent-free microwave irradiation produced N-substituted C-carbamoyl-1,2,3-triazoles **84** <02JOC9077>. New synthesis and alkylation reactions of 1-arylmethyleneamino- and 1-arylsulfonyl-5-hydroxy-1H-1,2,3-triazoles has been published <02JCS(P1)211>. Reduction of benzo-1,2,3,4-tetrazine 1,3-dioxides with $Na_2S_2O_4$ or $SnCl_2$ was suggested to proceed via an intermediate N-nitrosobenzotriazole, which upon release of the nitroso group afforded benzotriazoles <02OL3227>. Nitro-1H-benzotriazole derivatives containing electron-withdrawing substituents in the *ortho* and *meta* positions to the nitro group have been synthesized by the simple and direct nitration of the parent 1H-benzotriazoles <02H(57)1461>. The selective protection of the triazole function of 5-aminobenzotriazole has been reported <02H(57)1227>.

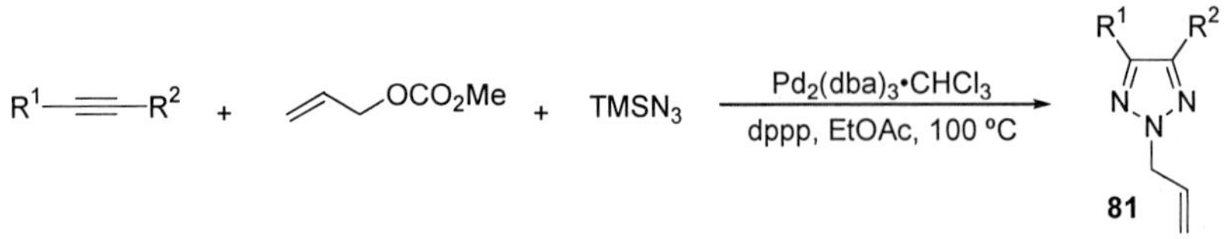

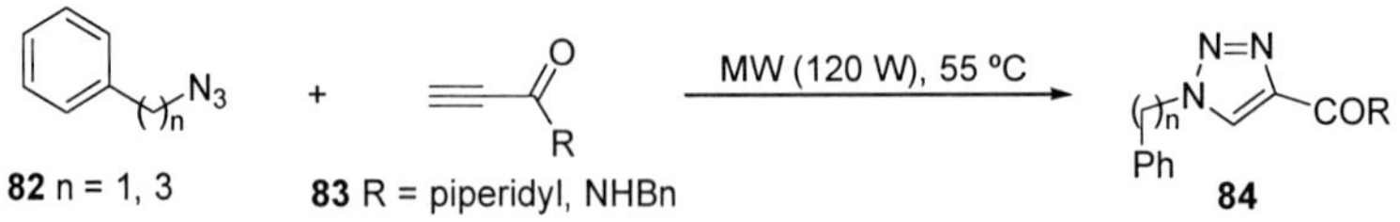

Benzotriazole-based methodologies continued to be dominant in 2002. Reaction of diaryl ketones with aminoalkylbenzotriazoles in the presence of ytterbium metal at room temperature gave 2-amino alcohols in good yields under mild and neutral conditions <02TL2251>.

Lithiation of aliphatic 1-acylbenzotriazoles followed by reaction with α,β-unsaturated ketones and aldehydes afforded either 3,4,6-trisubstituted 3,4-dihydropyran-2-ones or 1,3-dienes depending on the carbonyl reagent <02JOC3104>. Unsymmetrical and optically active imidazolidines and hexahydropyrimidines were synthesized in good yields by Mannich reactions of 1,2-ethanediamines with benzotriazole and formaldehyde, followed by nucleophilic substitution of the benzotriazolyl group with carbon nucleophiles <02JOC3109, 02JOC3115>. Convenient syntheses of 2,3,4,5-tetrahydro-1,4-benzothiazepines, -1,4-benzoxazepines and -1,4-benzodiazepines using benzotriazole methodology has been reported <02JCS(P1)592>. A new synthesis of 2-benzazepines by benzotriazole-mediated cyclization has been published <02S601>. Reactions of polymer-supported chalcones with 2-(benzotriazol-1-yl)acetamides under basic conditions followed by acidic cleavage afforded 3,4,6-trisubstituted pyridin-2-ones <02JCO249>. Condensation reactions of benzotriazole and 2-(pyrrol-1-yl)-1-ethylamine with either formaldehyde or glutaric dialdehyde afforded intermediates, which underwent further nucleophilic substitution of the benzotriazole group to give 1,2,3,4-tetrahydropyrrolo[1,2-a]pyrazines and 5,6,9,11,11a-hexahydro-8H-pyrido[1,2-a]pyrrolo[2,1-c]pyrazines <02JOC8220>. Reactions of organometallic reagents with 1-(substituted ethynyl)-1H-1,2,3-benzotriazoles yielded disubstituted acetylenes in synthetically useful yields <02JOC7526>. 2,4-Benzodiazepin-1-ones were prepared in moderate to good yields by reaction of bis(benzotriazolylmethyl)amines with *ortho*-metalated *N*-substituted benzamides <02JOC8237>. Dichloroacetamides, on heating with benzotriazole, morpholine, and triethylamine, produced, in a one-pot reaction, α-benzotriazolyl-α-morpholineacetamides, which reacted with nucleophiles to afford diverse α-morpholinoamides <02JOC8239>. Reaction of 1-(triphenylphosphoroylideneaminoalkyl)benzotriazoles with double Grignard additions of allylmagnesium bromide afforded easy access to *N,N*-bis(but-3-enyl)amines <02JOC7530>. χ-(Benzotriazol-1-yl)allylic sulfoxides underwent Pummerer-type reactions to give 2,3-benzo-1,3a,6a-triazapentalenes <02EJO493>. Novel syntheses of enantiopure hexahydroimidazo[1,5-b]isoquinolines and tetrahydroimidazo[1,5-b]isoquinolin-1(5H)-ones were accomplished via benzotriazole-mediated iminium cationic cyclizations <02JOC8224>.

Syntheses of fused mesoionic heterocycles such as [1,2,3]triazolo[1,5-a]-quinoline, -quinazoline, -quinoxaline, and -benzotriazine derivatives have been described <02T3185>. Cyclizations of alkyl 2-benzoylamino-(4,5-dicyano-1H-1,2,3-triazol-1-yl)propenoates gave [1,2,3]triazolo[1,5-a]pyrazines <02H(56)353>. Reaction of triethyl *N*-(1-ethyl-2-methyl-4-nitro-1H-imidazol-5-yl)phosphorimidate with aryl isocyanates provided a route to 2-aryl-2H,4H-imidazo[4,5-d][1,2,3]triazoles <02JCS(P1)1968>. 2-(*N,N*-Diphenylamino)-4-hydrazino-6-trifluoromethylpyrimidine reacted with aldoses to give the corresponding hydrazones, which on bromine oxidation cyclized to yield 1-(*s*-triazolo[4,3-c]pyrimidin-3-yl)substituted polyols <02SC1791>. New annelated 1,2,3-triazolo[1,5-a]pyrimidines were obtained through domino reactions of 3-azidopyrroles and methylene active nitriles <02T9723>. A rapid analog synthetic strategy has been developed to allow easy variation at the C-5 position of 1,2,3-triazolo[1,5-a]quinazolines <02T9973>.

Functionalized 1,2,3-triazoles **86** and **87** were prepared by [2 + 3] cycloadditions of resin-bound α-azido esters **85** with substituted alkynes <02TL4059>. Regiospecific copper(I)-catalyzed 1,3-dipolar cycloadditions of resin-bound alkynes **88** with azides afforded solid-supported 1,2,3-triazoles **89**, which were ligated further to give 1,4-substituted-1,2,3-triazole peptide compounds <02JOC3057>.

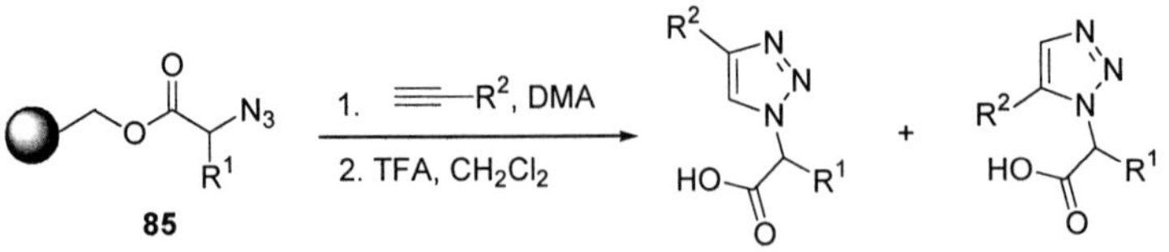

5.4.5 1,2,4-TRIAZOLES AND RING-FUSED DERIVATIVES

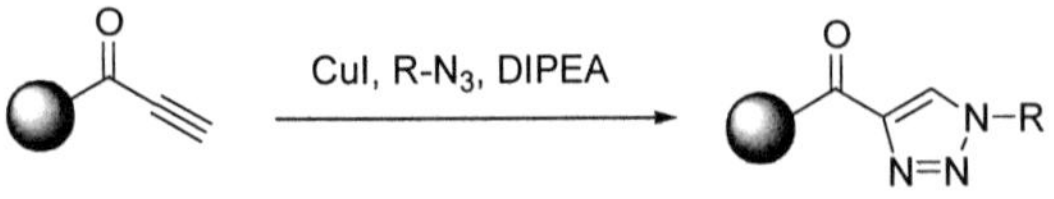

A series of selective 1,2,4-triazole non-nucleoside reverse transcriptase inhibitors were found to have activity against the HIV Type 1 *in vitro* <02HCA1883>. A facile route to poly[1-(2,4,6-trichlorophenyl)-1*H*-1,2,4-triazol-5-yl]alkane derivatives has been reported <02JCS(P1)991>. A novel *cis* amide bond surrogate incorporating 1,2,4-triazole was designed and synthesized by the reaction of a thionotripeptide, formic hydrazide, and mercury(II) acetate <02JOC3266>. The syntheses and radiostability of novel biologically active 1,2,4-triazoles, prepared from amino acids, containing the sulfonamide moiety have been reported <02HC316>.

Addition of amines to α-nitrohydrazones **90** followed by addition of sodium nitrite afforded 1,2,4-triazole **91** <02TL8925>. Iodobenzene diacetate cyclization of hydrazones **92** afforded 1,2,4-triazolo[4,3-*a*][1,8]naphthyridines **93** <02SC2377>. Hydrazonyl chlorides **94** reacted with cycloalkanone oximes **95** to give 1,2,4-triazolospiro compounds **96** <02SC2017>. Efficient synthesis of several new and novel 4-substituted triazolidinediones (urazoles) from isocyanates, amines, anilines and carboxylic acids was reported <02SC1741>. Trichloroisocyanuric acid <02SL1633> or potassium dichromate in the presence of aluminum chloride <02SC3445> were effective reagents for the oxidation of urazoles **97** to triazolinediones **98**. Reactions of isothiocyanates **99** and acyl hydrazides **100** afforded 1,2,4-triazole-3-thiones **101**, which were intercepted by alkyl halides to give substituted 3-thio-1,2,4-triazoles **102** <02JCO315>. 2-(6-Methoxy-2-naphthyl)propanoic acid (Naproxen) was utilized in the syntheses of several fused 1,2,4-triazoles <02HC199>. A facile one-pot regioselective synthesis of 1,2,4-triazolo[4,3-*a*]5(1*H*)-pyrimidinones was accomplished via tandem Japp-Klingemann, Smiles rearrangement, and cyclization reactions <02HC136>. Syntheses of 1-alkyl-1,2,4-triazoles and the formation of quaternary 1-alkyl-4-polyfluoroalkyl-1,2,4-triazolium salts led to the synthesis of ionic liquids <02JOC9340>.

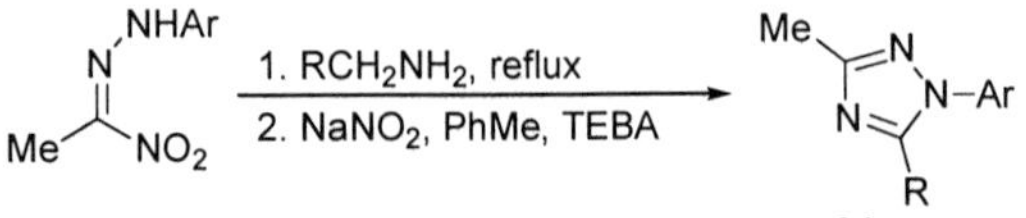

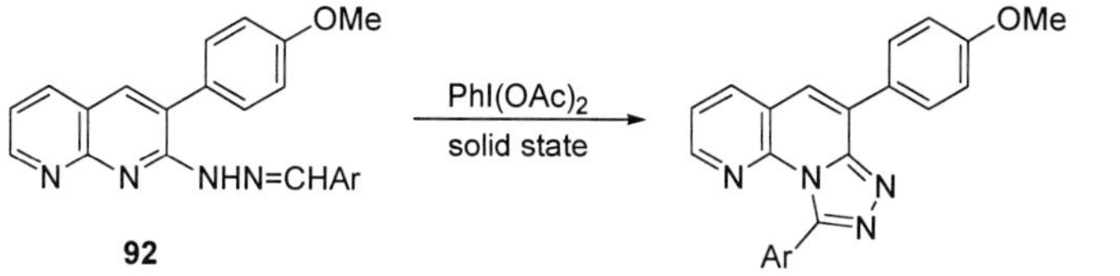

Regioselective 1,5-electrocyclization of *N*-[*as*-triazin-3-yl]nitrilimines with bromine in acetic acid or by ferric chloride in ethanol afforded *s*-triazolo[4,3-*b*]-*as*-triazin-7(8*H*)-ones <02T8559>. Various xanthates added efficiently to olefins bearing 1,2,4-triazoles in the presence of camphorsulfonic acid via a radical chain reaction initiated by a small amount of lauroyl peroxide <02OL4345>. The photocyclization of substituted 1,2,4-triazole-3-thiones, under base-mediated conditions, afforded 1,2,4-triazolo[3,4-*b*]-1,3-(4*H*)-benzothiazines along with the desulfurization product <02TL5119>. The reaction of 1,2,4-triazole with functionalized carbodiimides led to the synthesis of 1-(2-imidazolonyl)-1,2,4-triazoles and 1-(2-quinazolonyl)-1,2,4-triazoles <02SC3057>. A one-step synthesis of 1,2,4-triazolo[5,4-*b*][1,3,4]benzotriazepin-6-ones by 1,3-dipolar cycloaddition of nitrilimines with 1,3,4-benzotriazepin-5-ones was reported <02SC1815>. Cycloadditions of heterocumulene cations generated new routes to [1,2,4]triazolo[3,2-*d*][1,5]benzoxazepines <02SC1327>. An efficient procedure for the regiospecific preparation of arenesulfonamide derivatives of 3,5-diamino-1,2,4-triazole <02S185> and 9-trifluoromethylated [1,2,4]triazolo[1,5-*a*]azepine derivatives <02S349> has been described. Reactions of benzyl bromide or benzyl cinnamate with *N*-(benzotriazol-1-

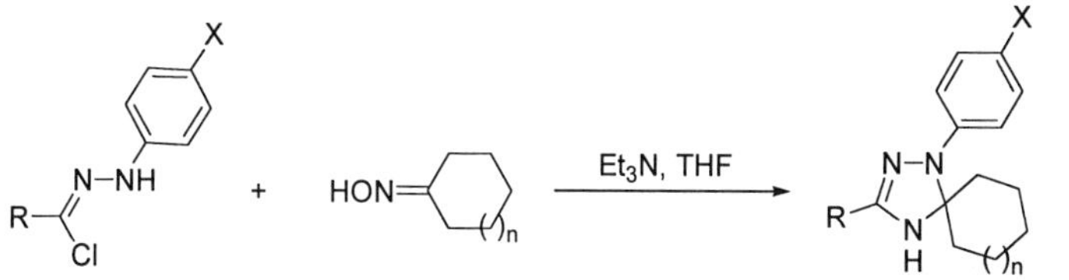

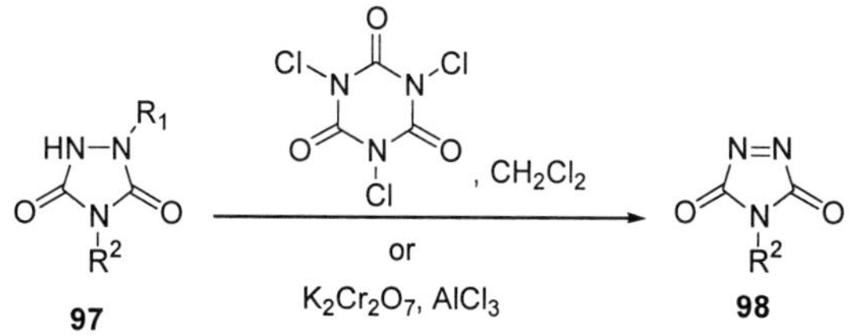

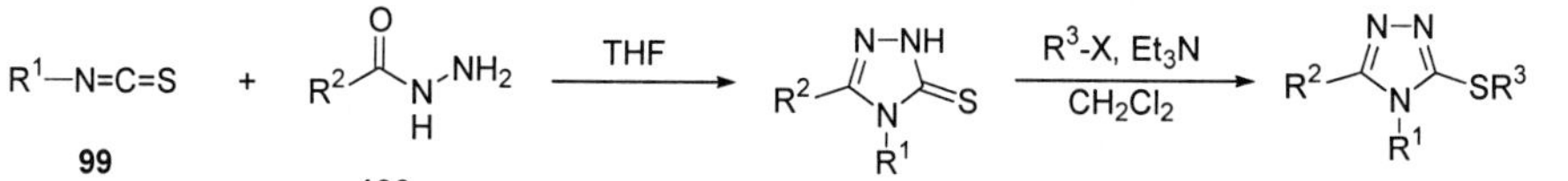

ylmethyl)arylimidoyl chlorides in the presence of potassium *tert*-butoxide occurred with opening of the benzotriazole ring to afford 1,2,4-triazolo[1,5-*a*]quinoxalines <02JOC3118>. Novel spiro-[3*H*-indole-3,3'-[1,2,4]triazolidine]-2-ones were prepared via a one-pot 1,3-dipolar cycloaddition reaction of azomethine imines with isatin imines under thermal conditions <02TL9721>. *C*-Phosphorylation of 1,2,4-triazoles with phosphorus(III) halides led to the synthesis of 4,5-dihydrobenzo[*e*][1,2,4]triazolo[5,1-*c*][1,4,2]diazaphosphinine derivatives <02HC146>. 4-Phenyl-1,2,4-triazoline-3,5-dione and its perfluoro analog were efficient reagents for trapping arene oxides <02CC1956>. 6-Aryl-3-(D-glucopentitol-1-yl)-7*H*-1,2,4-triazolo[3,4-*b*][1,3,4]thiadiazines were readily accessible in high yields by reaction of 4-amino-5-mercapto-3-(D-glucopentitol-1-yl)-1,2,4-triazole with substituted ω-bromoacetophenones <02SC3455>.

A robust 'catch, cyclize, and release' preparation of 3-thioalkyl-1,2,4-triazoles mediated by the polymer-bound base P-BEMP has been described <02TL5305>. Reaction of solid-supported hydrazides **103** with isocyanates or isothiocyanates followed by base-induced cyclization/cleavage afforded 1,2,4-trisubstituted urazoles and thiourazoles **104** <02JCO491, 02TL3899>. Polymer-supported *N*-acyl-1*H*-benzotriazole-1-carboximidamides **105** reacted with hydrazines followed by acidic cyclizative release to give 3-alkylamino-1,2,4-triazoles **106** <02OL1751>.

5.4.6 TETRAZOLES AND RING-FUSED DERIVATIVES

A review on 5-substituted-1*H*-tetrazoles as carboxylic acid isosteres in medicinal chemistry has been published <02BMC3379>. The mechanisms of tetrazole formation by addition of azides to nitriles have been investigated by density functional theory calculations <02JA12210>.

Thymidine dimers in which the natural phosphodiester linkage has been replaced by a 2,5-disubstituted tetrazole ring have been synthesized and have been incorporated into oligodeoxynucleotides <02HCA2847>. The synthesis of mono- and bis-3-substituted thymidine derivatives with a polycyclic tetrazole linker (1,5-bis(tetrazol-5-yl)-3-oxapentane) has been reported <02TL1901>. α-Methylene tetrazole-based peptidomimetics were synthesized for inhibition studies of HIV protease <02JCS(P1)172>. A catalytic amount of tetrazole was found to be useful in the syntheses of symmetrical *P,P'*-dialkyl partial esters of methylenebisphosphonic acid from the corresponding acid chloride via a facile two-step, one-pot process <02SC2683>. The relative strength of the 1*H*-tetrazol-5-yl- and the 2-(triphenylmethyl)-2*H*-tetrazol-5-yl-group in directed *ortho*-lithiation was studied <02TL3137>. New fluorine-containing tetrazolopyrimidines have been synthesized <02S901>.

5-Aryltetrazoles can be obtained either by arylation of *N*-tributylstannyl 5-aryltetrazoles **107** with diphenyliodonium chloride (**108**) in the presence of copper(II) acetate to give *N*-phenyltetrazoles **109** <02TL6217> or by palladium-catalyzed arylation of the sodium salt of 5-aryltetrazoles **110** with diaryliodonium tetrafluoroborate (**111**) in the presence of copper salts regioselectively at the N2 position to give diaryltetrazoles **112** <02TL6221>. The palladium-catalyzed three-component coupling reaction of nitriles **113**, allyl methyl carbonate (**114**), and trimethylsilyl azide (**115**) in the presence of catalytic tri(2-furyl)phosphine provided 2-allyltetrazoles **116** <02JOC7413>. A $TMSN_3$-modified Passerini three-component reaction of α-amino aldehydes **117**, isocyanides **118**, and trimethylsilyl azide (**115**) afforded tetrazoles **119**, which were transformed into *cis*-constrained norstatine mimetics **120** <02TL6833>. Huisgen 1,3-dipolar cycloaddition of azides with acyl cyanides and sulfonyl cyanides afforded 5-acyltetrazoles <02AG(E)2113> and 5-sulfonyltetrazoles <02AG(E)2110>, respectively. A short solution-phase preparation of fused azepine-tetrazoles via an Ugi/de-Boc/cyclize strategy has been reported <02TL3681>. Reactions of α-aminonitriles **121** with sodium azide and catalytic zinc bromide proceeded readily in refluxing water/2-propanol to provide tetrazole analogs of α-amino acids **122** in good yields <02OL2525>.

107 **108** **109**

110 **111** **112**

113 **114** **115** **116**

117 **118** **115** **119** **120** (X= CO or SO_2)

121 **122**

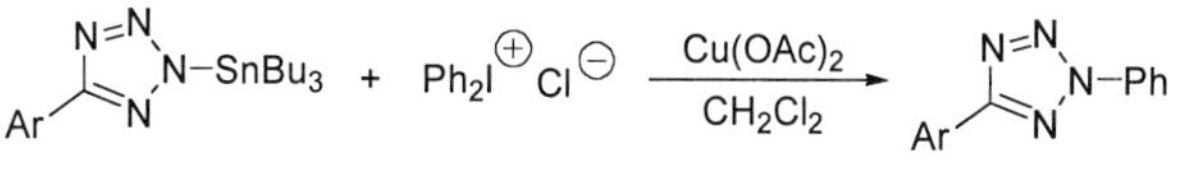

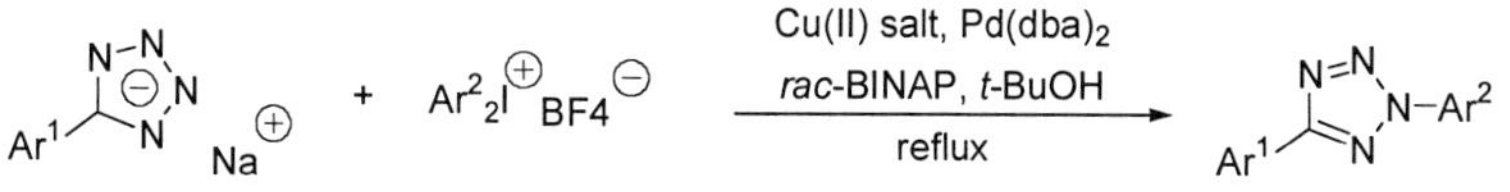

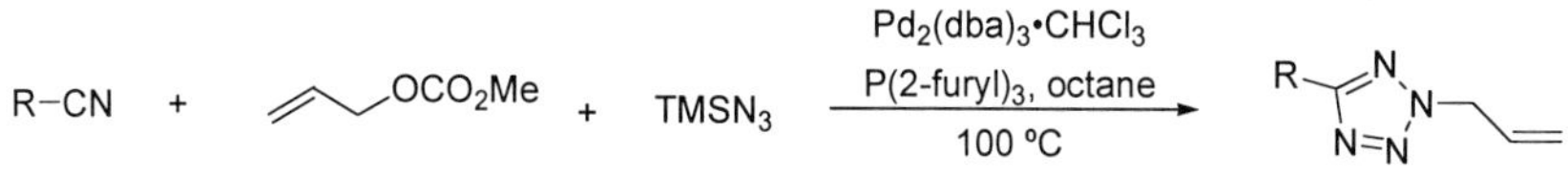

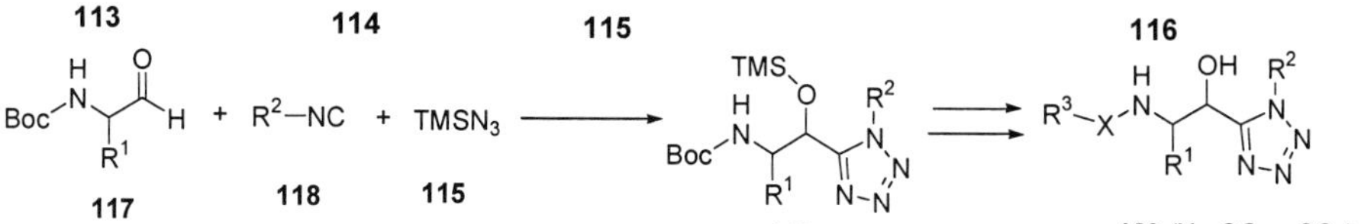

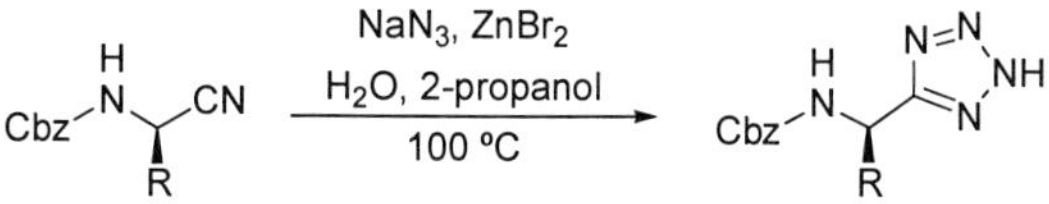

5.4.7 REFERENCES

02ACA75	H. Zhao, S.V. Malhotra, *Aldrichim. Acta* **2002**, *35*, 75.
02AG(E)2110	Z.P. Demko, K.B. Sharpless, *Angew. Chem., Int. Ed.* **2002**, *41*, 2110.
02AG(E)2113	Z.P. Demko, K.B. Sharpless, *Angew. Chem., Int. Ed.* **2002**, *41*, 2113.
02AG(E)2779	Y. Fukumoto, K. Sawada, M. Hagihara, N. Chatani, S. Murai, *Angew. Chem. Int. Ed.* **2002**, *41*, 2779.
02BMC3379	R.J. Herr, *Bioorg. Med. Chem.* **2002**, *10*, 3379.
02CC22	T.C. Krasia, J.H.G. Steinke, *Chem. Commun.* **2002**, 22.
02CC434	M.M. Dell'Anna, V. Gallo, P. Mastrorilli, C.F. Nobile, G. Romanazzi, G.P. Suranna, *Chem. Commun.* **2002**, 434.
02CC436	V.V. Ponomarova, V.V. Komarchuk, I. Boldog, A.N. Chernega, J. Sieler, K.V. Domasevitch, *Chem. Commun.* **2002**, 436.
02CC616	R. Rajagopal, D.V. Jarikote, K.V. Sriniasan, *Chem. Commun.* **2002**, 616.
02CC704	X. Liu, C.A. Kilner, M.A. Halcrow, *Chem. Commun.* **2002**, 704.
02CC936	F. Escarti, C. Miranda, L. Lamarque, J. Latorre, E. Garcia-Espana, M. Kumar, V.J. Aran, P. Navarro, *Chem. Commun.* **2002**, 936.
02CC992	M.T. Reetz, W. Wiesenhofer, G. Francio, W. Leitner, *Chem. Commun.* **2002**, 992.
02CC1012	P. Baran, C.M. Marrero, S. Perez, R.G. Raptis, *Chem. Commun.* **2002**, 1012.
02CC1404	P. Wang, Z. Xie, O. Wong, C-S. Lee, N. Wong, L. Hung, S. Lee, *Chem. Commun.* **2002**, 1404.
02CC1610	C.P. Mehnert, N.C. Dispenziere, R.A. Cook, *Chem. Commun.* **2002**, 1610.
02CC1612	V.K. Aggarwal, I. Emme, A. Mereu, *Chem. Commun.* **2002**, 1612.
02CC1956	A.P. Henderson, E. Mutlu, A. Leclercq, C. Bleasdale, W. Clegg, R.A. Henderson, B.T. Golding, *Chem. Commun.* **2002**, 1956.
02CC2490	V.K. Aggarwal, A.M. Daly, *Chem. Commun.* **2002**, 2490.
02CC2572	C. Slugove, S. Demel, F. Stelzer, *Chem. Commun.* **2002**, 2572.
02CEJ4781	T. Bando, A. Narita, I. Saito, H. Sugiyama, *Chem. Eur. J.* **2002**, *8*, 4781.
02EJO493	T. Kim, K. Kim, Y.J. Park, *Eur. J. Org. Chem.* **2002**, 493.
02EJO2913	M. Schlosser, J.-N. Volle, F. Leroux, K. Schenk, *Eur. J. Org. Chem.* **2002**, 2913.
02H(56)353	T. Jug, M. Polak, T. Trcek, B. Vercek, *Heterocycles* **2002**, *56*, 353.
02H(56)497	D.-L. Wang, K. Imafuku, *Heterocycles* **2002**, *56*, 497.
02H(57)5	B.H. Kim, R. Han, T.H. Han, Y.M. Jun, W. Baik, B.M. Lee, *Heterocycles* **2002**, *57*, 5.
02H(57)21	E. Moreau, J.-M. Chezal, C. Dechambre, D. Canitrot, Y. Blache, C. Lartigue, O. Chavignon, J.-C. Teulade, *Heterocycles* **2002**, *57*, 21.
02H(57)307	R. Faure, A. Frideling, J.-P. Galy, I. Alkorta, J. Elguero, *Heterocycles* **2002**, *57*, 307.
02H(57)1227	D. McKeown, C.J. McHugh, A. McCabe, W.E. Smith, D. Graham, *Heterocycles* **2002**, *57*, 1227.
02H(57)1461	C.J. McHugh, D.R. Tackley, D. Graham, *Heterocycles* **2002**, *57*, 1461.
02HC136	A.S. Shawalik, M.A. Abdallah, M.A.N. Mosselhi, T.A. Farghaly, *Heteroatom Chem.* **2002**, *13*, 136.
02HC146	E.V. Zarudnitskii, V.V. Ivanov, A.A. Yurchenko, A.M. Pinchuk, A.A. Tolmachev, *Heteroatom Chem.* **2002**, *13*, 146.
02HC199	Y.A. Ammar, M.M. Ghorab, A.M. El-Sharief, S.I. Mohamed, *Heteroatom Chem.* **2002**, *13*, 199.
02HC316	A.M. El-Sharief, M.M. Ghorab, M.S.A. El-Gaby, S.I. Mohamed, Y.A. Ammar, *Heteroatom Chem.* **2002**, *13*, 316.
02HCA1883	I.M. Lagoja, C. Pannecouque, L. Musumei, M. Froeyen, A.V. Aerschot, J. Balzarini, P. Herdewijn, E. De Clercq, *Helv. Chim. Acta* **2002**, *85*, 1883.
02HCA2763	M.A. Garcia, C. Lopez, R.M. Claramunt, A. Kenz, M. Pierrot, J. Elguero, *Helv. Chim. Acta* **2002**, *85*, 2763.
02HCA2847	V.V. Filichev, A.A. Malin, V.A. Ostrovskii, E.B. Pedersen, *Helv. Chim. Acta* **2002**, *85*, 2847.
02JA3202	K.L. Tan, R.G. Bergman, J.A. Ellman, *J. Am. Chem. Soc.* **2002**, *124*, 3202.
02JA5757	Y.-J. Kim, A. Streitwieser, *J. Am. Chem. Soc.* **2002**, *124*, 5757.
02JA7894	N.A. Paras, D.W.C. MacMillan, *J. Am. Chem. Soc.* **2002**, *124*, 7894.
02JA10278	D.W. Kim, C.E. Song, D.Y. Chi, *J. Am. Chem. Soc.* **2002**, *124*, 10278.
02JA12210	F. Himo, Z.P. Demko, L. Noodleman, K.B. Sharpless, *J. Am. Chem. Soc.* **2002**, *124*, 12210.
02JA13964	K.L. Tan, R.G. Bergman, J.A. Ellman, *J. Am. Chem. Soc.* **2002**, *124*, 13964.

02JA13394	K. Itoh, S. Kanemasa, *J. Am. Chem. Soc.* **2002**, *124*, 13394.
02JCO87	C.M. Coleman, J.M.D. MacElroy, J.F. Gallagher, D.F. O'Shea, *J. Comb. Chem.* **2002**, *4*, 87.
02JCO209	M. Rinnova, A. Vidal, A. Nefzi, R.A. Houghten, *J. Comb. Chem.* **2002**, *4*, 209.
02JCO214	A.N. Acharya, J.M. Ostresh, R.A. Houghten, *J. Comb. Chem.* **2002**, *4*, 214.
02JCO249	A.R. Katritzky, C. Chassaing, S.J. Barrow, Z. Zhang, V. Vvedensky, B. Forood, *J. Comb. Chem.* **2002**, *4*, 249.
02JCO315	M-E. Theoclitou, N.G.J. Delaet, L.A. Robinson, *J. Comb. Chem.* **2002**, *4*, 315.
02JCO320	B.Raju, N. Nguyen, G.W. Holland, *J. Comb. Chem.* **2002**, *4*, 320.
02JCO345	G. Klein, A.N. Acharya, J.M. Ostresh, R.A. Houghten, *J. Comb. Chem.* **2002**, *4*, 345.
02JCO352	M. Ermann, N.M. Simkovsky, S.M. Roberts, D.M. Parry, A.D. Baxter, *J. Comb. Chem.* **2002**, *4*, 352.
02JCO359	P.M. Bendale, C.-M. Sun, *J. Comb. Chem.* **2002**, *4*, 359.
02JCO475	H. Akamatsu, K. Fukase, S. Kusumoto, *J. Comb. Chem.* **2002**, *4*, 475.
02JCO491	C.W. Phoon, M.M. Sim, *J. Comb. Chem.* **2002**, *4*, 491.
02JCO496	A.N. Acharya, C. Thai, J.M. Ostresh, R.A. Houghten, *J. Comb. Chem.* **2002**, *4*, 496.
02JCO576	O.W. Gooding, L. Vo, S. Bhattacharyya, J.W. Labadie, *J. Comb. Chem.* **2002**, *4*, 576.
02JCS(P1)126	K. Banert, J. Lehmann, H. Quast, G. Meichsner, D. Regnat, B. Seiferling, *J. Chem. Soc., Perkin Trans. 1* **2002**, 126.
02JCS(P1)172	B.C.H. May, A.D. Abell, *J. Chem. Soc., Perkin Trans. 1* **2002**, 172.
02JCS(P1)207	K.-T. Chang, Y.H. Choi, S.-H. Kim, Y.-J. Yoon, W.S. Lee, *J. Chem. Soc., Perkin Trans. 1* **2002**, 207.
02JCS(P1)211	Y.A. Rozin, E.A. Savel'eva, Y.Y. Morzherin, W. Dehaen, S. Toppet, L.V. Meerelt, V.A. Bakulev, *J. Chem. Soc., Perkin Trans. 1* **2002**, 211.
02JCS(P1)592	A.R. Katritzky, Y.-J. Xu, H.-Y. He, *J. Chem. Soc., Perkin Trans. 1* **2002**, 592.
02JCS(P1)991	Q. Wang, H. Yang, Y. Liu, Z. Ding, F. Tao, *J. Chem. Soc., Perkin Trans. 1* **2002**, 991.
02JCS(P1)1260	J. Svetlik, T. Liptaj, *J. Chem. Soc., Perkin Trans. 1* **2002**, 1260.
02JCS(P1)1588	R. Abonia, A. Albornoz, H. Larrahondo, J. Quiroga, B. Insuasty, H. Hormaza, A. Sanchez, M. Nogueras, *J. Chem. Soc., Perkin Trans. 1* **2002**, 1588.
02JCS(P1)1968	A. Taher, S. Eichenseher, A.M.Z. Slawin, G. Tennant, G.W. Weaver, *J. Chem. Soc., Perkin Trans. 1* **2002**, 1968.
02JCS(P1)2504	L. Garanti, G. Molteni, P. Casati, *J. Chem. Soc., Perkin Trans. 1* **2002**, 2504.
02JOC1708	F.M. Rivas, A.J. Giessert, S.T. Diver, *J. Org. Chem.* **2002**, *67*, 1708..
02JOC2699	R.H.-J. Butz, S.D. Lindell, *J. Org. Chem.* **2002**, *67*, 2699.
02JOC3029	C. Desmarets, R. Schneider, Y. Fort, *J. Org. Chem.* **2002**, *67*, 3029.
02JOC3057	C.W. Tornoe, C. Christensen, M. Meldal, *J. Org. Chem.* **2002**, *67*, 3057.
02JOC3104	A.R. Katritzky, O.V. Denisko, *J. Org. Chem.* **2002**, *67*, 3104.
02JOC3109	A.R. Katritzky, K. Suzuki, H.-Y. He, *J. Org. Chem.* **2002**, *67*, 3109.
02JOC3115	A.R. Katritzky, S.K. Singh, H.-Y. He, *J. Org. Chem.* **2002**, *67*, 3115.
02JOC3118	A.R. Katritzky, T.-B. Huang, O.V. Denisko, *J. Org. Chem.* **2002**, *67*, 3118.
02JOC3138	Y. Yu, J.M. Ostresh, R.A. Houghten, *J. Org. Chem.* **2002**, *67*, 3138.
02JOC3145	N.E. Leadbeater, H.M. Torenius, *J. Org. Chem.* **2002**, *67*, 3145.
02JOC3266	Y. Hitotsuyanagi, S. Motegi, H. Fukaya, K. Takeya, *J. Org. Chem.* **2002**, *67*, 3266.
02JOC3468	B. Dolensky, K.L. Kirk, *J. Org. Chem.* **2002**, *67*, 3468.
02JOC3904	A.S. Paulson, J. Eskildsen, P. Vedsø, M. Begtrup, *J. Org. Chem.* **2002**, *67*, 3904.
02JOC3919	N.A. Boland, M. Casey, S.J. Hynes, J.W. Matthews, M.P. Smyth, *J. Org. Chem.* **2002**, *67*, 3919.
02JOC4777	H.-X. Wei, S.H. Kim, G. Li, *J. Org. Chem.* **2002**, *67*, 4777.
02JOC5546	A.M. Dias, I. Cabral, M.F. Proenc, B.L. Booth, *J. Org. Chem.* **2002**, *67*, 5546.
02JOC5963	C.P. Causey, W.E. Allen, *J. Org. Chem.* **2002**, *67*, 5963.
02JOC6395	D.B. Kimball, T.J.R. Weakley, M.M. Haley, *J. Org. Chem.* **2002**, *67*, 6395.
02JOC7151	A.V. Wiznycia, P.W. Baures, *J. Org. Chem.* **2002**, *67*, 7151.
02JOC7413	S. Kamijo, T. Jin, Y. Yamamoto, *J. Org. Chem.* **2002**, *67*, 7413.
02JOC7432	M.-J. Wei, D.-C. Fang, R.-Z. Liu, *J. Org. Chem.* **2002**, *67*, 7432.
02JOC7526	A.R. Katritzky, A.A.A. Abdel-Fattah, M. Wang, *J. Org. Chem.* **2002**, *67*, 7526.
02JOC7530	A.R. Katritzky, S.K. Nair, A. Silina, *J. Org. Chem.* **2002**, *67*, 7530.
02JOC8147	E.L. Moyano, M. del Arco, V. Rives, G.I. Yranzo, *J. Org. Chem.* **2002**, *67*, 8147.
02JOC8220	A.R. Katritzky, R. Jain, Y.-J. Xu, P.J. Steel, *J. Org. Chem.* **2002**, *67*, 8220.

02JOC8224	A.R. Katritzky, K. Suzuki, H.-Y. He, *J. Org. Chem.* **2002**, *67*, 8224.
02JOC8230	A.R. Katritzky, R. Maimait, Y.-J. Xu, Y.S. Gyoung, *J. Org. Chem.* **2002**, *67*, 8230.
02JOC8237	A.R. Katritzky, S.K. Nair, V. Rodriguez-Garcia, Y.-Y. Xu, *J. Org. Chem.* **2002**, *67*, 8237.
02JOC8239	A.R. Katritzky, Y. Zhang, S.K. Singh, *J. Org. Chem.* **2002**, *67*, 8239.
02JOC8331	N. Halland, R.G. Hazell, K.A. Jorgensen, *J. Org. Chem.* **2002**, *67*, 8331.
02JOC8463	S. Ramos, E. Alcalde, G. Doddi, P. Mencarelli, L. Perez-Garcia, *J. Org. Chem.* **2002**, *67*, 8463.
02JOC9077	A.R. Katritzky, S.K. Singh, *J. Org. Chem.* **2002**, *67*, 9077.
02JOC9200	D.B. Grotjahn, S. Van, D. Combs, D.A. Lev, C. Schneider, M. Rideout, C. Meyer, G. Hernandez, L. Mejorado, *J. Org. Chem.* **2002**, *67*, 9200.
02JOC9340	Y.R. Mirzaei, B. Twamley, J.M. Shreeve, *J. Org. Chem.* **2002**, *67*, 9340.
02OL541	A. Fürstner, G. Seidel, *Org. Lett.* **2002**, *4*, 541.
02OL1411	R.A. Batey, M. Shen, A.J. Lough, *Org. Lett.* **2002**, *4*, 1411.
02OL1487	K. Booker-Milburn, P.M. Wood, R.F. Dainty, M.W. Urquhart, A.J. White, H.J. Lyon, J.P.H. Charmant, *Org. Lett.* **2002**, *4*, 1487.
02OL1691	T. Fukuyama, M. Shinmen, S. Nishitani, M. Sato, I. Ryu, *Org. Lett.* **2002**, *4*, 1691.
02OL1751	G.M. Makara, Y. Ma, L. Margarida, *Org. Lett.* **2002**, *4*, 1751.
02OL1939	A.K. Chatterjee, D.P. Sanders, R.H. Grubbs, *Org. Lett.* **2002**, *4*, 1939.
02OL2079	M.D. Andrus, C. Song, J. Zhang, *Org. Lett.* **2002**, *4*, 2079.
02OL2197	Q. Yao, *Org. Lett.* **2002**, *4*, 2197.
02OL2229	M.S. Viciu, R.M. Kissling, E.D. Stevens, S.P. Nolan, *Org. Lett.* **2002**, *4*, 2229.
02OL2525	Z.P. Demko, K.B. Sharpless, *Org. Lett.* **2002**, *4*, 2525.
02OL2767	H. Yoshida, S. Sugiura, A. Kunai, *Org. Lett.* **2002**, *4*, 2767.
02OL2897	H. Ihm, S. Yun, H.G. Kim, J.K. Kim, K.S. Kim, *Org. Lett.* **2002**, *4*, 2897.
02OL3071	J.D. Revell, A. Ganesan, *Org. Lett.* **2002**, *4*, 3071.
02OL3161	V.V. Namboodiri, R.S. Varma, *Org. Lett.* **2002**, *4*, 3161.
02OL3227	M.O. Ratnikov, D.L. Lipilin, A.M. Churakov, Y.A. Strelenko, V.A. Tartakovsky, *Org. Lett.* **2002**, *4*, 3227.
02OL3533	S. Peddibhotla, S. Jayakumar, J.J. Tepe, *Org. Lett.* **2002**, *4*, 3533.
02OL3583	G.A. Grasa, R.M. Kissling, S.P. Nolan, *Org. Lett.* **2002**, *4*, 3583.
02OL3587	G.W. Nyce, J.A. Lamboy, E.F. Connor, R.M. Waymouth, J.L. Hedrick, *Org. Lett.* **2002**, *4*, 3587.
02OL3855	M. Mori, Y. Kuzuba, T. Kitamura, Y. Sato, *Org. Lett.* **2002**, *4*, 3855.
02OL3955	N. Nakamichi, Y. Kawashita, M. Hayashi, *Org. Lett.* **2002**, *4*, 3955.
02OL4017	Y. Deng, D.J. Hlasta, *Org. Lett.* **2002**, *4*, 4017.
02OL4133	C.J. Helal, J.C. Lucas, *Org. Lett.* **2002**, *4*, 4133.
02OL4289	J.H. Rigby, Z. Wang, *Org. Lett.* **2002**, *4*, 4289.
02OL4345	F. Gagosz, S.Z. Zard, *Org. Lett.* **2002**, *4*, 4345.
02OPRD374	J. Graga-Dubreuil, M.-H. Famelart, J.P. Bazureau, *Org. Proc. Res. Dev.* **2002**, *6*, 374.
02OPRD674	H.H. Chang, K.W. Lee, D.H. Nam, W.S. Kim, H. Shin, *Org. Proc. Res. Dev.* **2002**, *6*, 674.
02OPRD682	B. Li, C.K.-F. Chiu, R.F. Hank, J. Murry, J. Roth, H. Tobiassen, *Org. Proc. Res. Dev.* **2002**, *6*, 682.
02S185	K. Chibale, J. Dauvergne, P.G. Wyatt, *Synthesis* **2002**, 185.
02S349	Z. Ding, S. Xia, X. Ji, H. Yang, F. Tao, Q. Wang, *Synthesis* **2002**, 349.
02S519	P. Raboisson, B. Norberg, J.R. Casimir, J.-J. Bourguignon, *Synthesis* **2002**, 519.
02S601	A.R. Katritzky, R. Maimait, Y.-J. Xu, R.G. Akhmedova, *Synthesis* **2002**, 601.
02S723	E. Tauer, *Synthesis* **2002**, 723.
02S901	A.L. Krasovsky, A.M. Moiseev, V.G. Nenajdenko, E.S. Balenkova, *Synthesis* **2002**, 901.
02S1079	H.G. Bonacorso, A.D. Wastowski, M.N. Muniz, N. Zanatta, M.A.P. Martins, *Synthesis* **2002**, 1079.
02S1379	A.L. Krasovsky, V.G. Nenajdenko, E.S. Balenkova, *Synthesis* **2002**, 1379.
02S1546	O.A. Attanasi, L. De Crescentini, G. Favi, P. Filippone, F. Mantellini, S. Santeusanio, *Synthesis* **2002**, 1546.
02S1669	V. Molteni, M.M. Hamilton, L. Mao, C.M. Crane, A.P. Termin, D.M. Wilson, *Synthesis* **2002**, 1669.
02S1885	A. Chauveau, T. Martens, M. Bonin, L. Micouin, H.-P. Husson, *Synthesis* **2002**, 1885.
02S2691	P. Moreno, M. Heras, M. Maestro, J.M. Villalgordo, *Synthesis* **2002**, 2691.
02SC253	G.H. Elgemeie, H.A. Ali, *Synth. Commun.* **2002**, *32*, 253.

02SC387	A. Harizi, H. Zantour, *Synth. Commun.* **2002**, *32*, 387.
02SC419	M.A.P. Martins, C.M.P. Pereira, A.P. Sinhorin, G.P. Bastos, N.E.K. Zimmermann, A. Rosa, H.G. Bonacorso, N. Zanatta, *Synth. Commun.* **2002**, *32*, 419.
02SC467	V.V. Rozhkov, S.S. Vorob'ov, A.V. Lobatch, A.M. Kuvshinov, S.A. Shevelev, *Synth. Commun.* **2002**, *32*, 467.
02SC1015	M.L. Sanders, I.O. Donkor, *Synth. Commun.* **2002**, *32*, 1015.
02SC1129	C. Xia, J. Chen, B. Zhao, H. Wang, C. Kang, Y. Ni, P. Zhou, *Synth. Commun.* **2002**, *32*, 1129.
02SC1327	Q. Wang, X. Liu, F. Li, Z. Ding, F. Tao, *Synth. Commun.* **2002**, *32*, 1327.
02SC1447	C. Xia, H. Wang, B. Zhao, J. Chen, C. Kang, Y. Ni, P. Zhou, *Synth. Commun.* **2002**, *32*, 1447.
02SC1457	C. Xia, J. Hao, Y. Tang, Y. Ni, P. Zhou, *Synth. Commun.* **2002**, *32*, 1457.
02SC1585	A.F.C. Flores, M.A.P. Martins, A. Rosa, D.C. Flores, N. Zanatta, H.G. Bonacorsso, *Synth. Commun.* **2002**, *32*, 1585.
02SC1741	T. Little, J. Meara, F. Ruan, M. Nguyen, M. Qabar, *Synth. Commun.* **2002**, *32*, 1741.
02SC1791	A.H. Ismail, *Synth. Commun.* **2002**, *32*, 1791.
02SC1815	T. El Messaoudi, M. El Messaoudi, T. Zair, A. Hasnaooui, M. Esseffar, J.-P. Lavergne, *Synth. Commun.* **2002**, *32*, 1815.
02SC2017	A.-R.S. Ferwanah, N.G. Kandile, A.M. Awadallah, O.A. Miqdad, *Synth. Commun.* **2002**, *32*, 2017.
02SC2245	G.H. Elgemeie, A.H. Elghandour, H.A. Ali, A.M. Hussein, *Synth. Commun.* **2002**, *32*, 2245.
02SC2377	K. Mogilaiah, H.R. Babu, N.V. Reddy, *Synth. Commun.* **2002**, *32*, 2377.
02SC2467	G.V. Reddy, V.V.V.N.S.R. Rao, B. Narsaiah, P.S. Rao, *Synth. Commun.* **2002**, *32*, 2467.
02SC2533	Y. Bai, J. Lu, H. Gan, Z. Wang, *Synth. Commun.* **2002**, *32*, 2533.
02SC2549	Y. Bai, J. Lu, H. Gan, Z. Wang, , *Synth. Commun.* **2002**, *32*, 2549.
02SC2613	Z. Li, Y. Zhang, *Synth. Commun.* **2002**, *32*, 2613.
02SC2683	D.C. Stepinski, A.W. Herlinger, *Synth. Commun.* **2002**, *32*, 2683.
02SC3057	M.-W. Ding, G.-P. Zeng, Z.-J. Liu, *Synth. Commun.* **2002**, *32*, 3057.
02SC3255	C. Unaleroglu, B. Temelli, T. Hokelek, *Synth. Commun.* **2002**, *32*, 3255.
02SC3399	S. Servi, Z.R. Akgün, *Synth. Commun.* **2002**, *32*, 3399.
02SC3445	I. Mohammadpoor-Baltork, M.M. Sadeghi, S.E. Mallakpour, A.R. Hajipour, A.-H. Adibi, *Synth. Commun.* **2002**, *32*, 3445.
02SC3455	Y. Xiong, L. Zhang, A. Zhang, D. Xu, *Synth. Commun.* **2002**, *32*, 3455.
02SC3509	G.H. Elgemeie, A.M. Elzanate, A.H. Elghandour, S.A. Ahmed, *Synth. Commun.* **2002**, *32*, 3509.
02SL513	J.S. Yadav, B.V.S. Reddy, V. Geetha, *Synlett* **2002**, 513.
02SL519	P. Raboisson, B. Norberg, J. R. Casimir, J.-J. Bourguignon, *Synlett* **2002**, 519.
02SL739	Y.K. Yun, J.A. Porco, Jr., J. Labadie, *Synlett* **2002**, 739.
02SL769	B. Cottineau, J. Chenault, *Synlett* **2002**, 769.
02SL1093	A. Nunez, A. Viedma, V. Martinez-Barrasa, C. Burgos, J. Alvarez-Builla, *Synlett* **2002**, 1093.
02SL1155	A.M.G. Silva, A.C. Tomé, M.G.P.M.S. Neves, J.A.S. Cavaleiro, *Synlett* **2002**, 1155.
02SL1170	D.A. Price, S. Gayton, P.A. Stupple, *Synlett* **2002**, 1170.
02SL1633	M.A. Zolfigol, E. Madrakian, E. Ghaemi, S. Mallakpour, *Synlett* **2002**, 1633.
02T569	M. Popsavin, L. Torovic, S. Spaic, S. Stankov, A. Kapor, Z. Tomic, V. Popsavin, *Tetrahedron* **2002**, *58*, 569.
02T1595	P. Cali, M. Begtrup, *Tetrahedron* **2002**, *58*, 1595.
02T2095	A.N. Acharya, J.M. Ostresh, R.A. Houghten, *Tetrahedron* **2002**, *58*, 2095.
02T2397	J.L. Kristensen, P. Vedsø, M. Begtrup, *Tetrahedron* **2002**, *58*, 2397.
02T3185	P.A. Abbott, R.V. Bonnert, M.V. Caffrey, P.A. Cage, A.J. Cooke, D.K. Donald, M. Furber, S. Hill, J. Withnall, *Tetrahedron* **2002**, *58*, 3185.
02T3349	Y. Yu, J.M. Ostresh, R.A. Houghten, *Tetrahedron* **2002**, *58*, 3349.
02T3639	J.-C. Jung, E.B. Watkins, M.A. Avery, *Tetrahedron* **2002**, *58*, 3639.
02T4445	D. Basso, G. Broggini, D. Passarella, T. Pilati, A. Terraneo, G. Zecchi, *Tetrahedron* **2002**, *58*, 4445.
02T4975	J.P. Cuadrado, A.M. Gonzalez-Nogal, R. Valero, *Tetrahedron* **2002**, *58*, 4975.
02T5467	J.T. Gupton, S.C. Clough, R.B. Miller, B.K. Norwood, C.R. Hickenboth, I.B. Chertudi, S.R. Cutro, S.A. Petrich, F.A. Hicks, D.R. Wilkinson, J.A. Sikorski, *Tetrahedron* **2002**, *58*, 5467.
02T5821	E. Espildora, J.L. Delgado, O. de la Cruz, A. de la Hoz, V. Lopez-Arza, F. Langa, *Tetrahedron* **2002**, *58*, 5821.

02T7635	J. Eskildsen, N. Østergaard, P. Vedsø, M. Begtrup, *Tetrahedron* **2002**, *58*, 7635.
02T8559	A.S. Shawali, S.M. Gomha, *Tetrahedron* **2002**, *58*, 8559.
02T9089	N. Jagerovic, M.L. Jimeno, I. Alkorta, J. Elguero, R.M. Claramunt, *Tetrahedron* **2002**, *58*, 9089.
02T9423	S.A.S. Ghozlan, A.Z.A. Hassanien, *Tetrahedron* **2002**, *58*, 9423.
02T9723	A. Lauria, P. Diana, P. Barraja, A. Montalbano, G. Cirrincione, G. Dattolo, A.M. Almerico, *Tetrahedron* **2002**, *58*, 9723.
02T9973	P. Jones, M. Chambers, *Tetrahedron* **2002**, *58*, 9973.
02TA137	A. Figge, H.J. Altenbach, D.J. Brauer, P. Tielmann, *Tetrahedron: Asymmetry* **2002**, *13*, 137.
02TA1713	C. Kashima, S. Mizuhara, Y. Miwa, Y. Yokoyama, *Tetrahedron: Asymmetry* **2002**, *13*, 1713.
02TL193	T.Y. Zhang, H. Zhang, *Tetrahedron Lett.* **2002**, *43*, 193.
02TL387	J.X. Wu, B. Beck, R.X. Ren, *Tetrahedron Lett.* **2002**, *43*, 387.
02TL593	H. Weinmann, M. Harre, K. Koenig, E. Merten, U. Tilstam, *Tetrahedron Lett.* **2002**, *43*, 593.
02TL1529	C.-Y. Wu, C.-M. Sun, *Tetrahedron Lett.* **2002**, *43*, 1529.
02TL1595	B.-C. Chen, S.T. Chao, J.E. Sundeen, J. Tellew, S. Ahmad, *Tetrahedron Lett.* **2002**, *43*, 1595.
02TL1708	F.M. Rivas, A.J. Giessert, S.T. Diver, *Tetrahedron Lett.* **2002**, *43*, 1708.
02TL1893	C.T. Brain, S.A. Brunton, *Tetrahedron Lett.* **2002**, *43*, 1893.
02TL1901	V.V. Filichev, M.V. Jasko, A.A. Malin, V.Y. Zubarev, V.A. Ostrovskii, *Tetrahedron Lett.* **2002**, *43*, 1901.
02TL2171	N. Haddad, J. Baron, *Tetrahedron Lett.* **2002**, *43*, 2171.
02TL2251	W. Su, B. Yang, Y. Zhang, *Tetrahedron Lett.* **2002**, *43*, 2251.
02TL3137	P. Rhonnstad, D. Wensbo, *Tetrahedron Lett.* **2002**, *43*, 3137.
02TL3577	Y. Zhu, Y. Kiryu, H. Katayama, *Tetrahedron Lett.* **2002**, *43*, 3577.
02TL3681	T. Nixey, M. Kelly, D. Semin, C. Hulme, *Tetrahedron Lett.* **2002**, *43*, 3681.
02TL3899	K.-H. Park, L.J. Cox, *Tetrahedron Lett.* **2002**, *43*, 3899.
02TL4059	B.E. Blass, K.R. Coburn, A.L. Faulkner, C.L. Hunn, M.G. Natchus, M.S. Parker, D.E. Portlock, J.S. Tullis, R. Wood, *Tetrahedron Lett.* **2002**, *43*, 4059.
02TL4127	M.D. Crozet, P. Perfetti, M. Kaafarani, P. Vanelle, M.P. Crozet, *Tetrahedron Lett.* **2002**, *43*, 4127.
02TL4267	G.S. Mandair, M. Light, A. Russell, M. Hursthouse, M. Bradley, *Tetrahedron Lett.* **2002**, *43*, 4267.
02TL4343	C. Escolano, M. Rubiralta, A. Diez, *Tetrahedron Lett.* **2002**, *43*, 4343.
02TL4463	K. Yang, B. Lou, H. Saneii, *Tetrahedron Lett.* **2002**, *43*, 4463.
02TL4571	J.-F. Cheng, C. Kaiho, M. Chen, T. Arrhenius, A. Nadzan, *Tetrahedron Lett.* **2002**, *43*, 4571.
02TL4833	J. Verner, J. Taraba, M. Potacek, *Tetrahedron Lett.* **2002**, *43*, 4833.
02TL5005	A.F.C. Flores, N. Zanatta, A. Rosa, S. Brondani, M.A.P. Martins, *Tetrahedron Lett.* **2002**, *43*, 5005.
02TL5089	Z.M.A. Judeh, C.B. Ching, J. Bu, A. McCluskey, *Tetrahedron Lett.* **2002**, *43*, 5089.
02TL5119	A. Senthilvelan, V.T. Ramakrishnan, *Tetrahedron Lett.* **2002**, *43*, 5119.
02TL5305	T.L. Graybill, S. Thomas, M.A. Wang, *Tetrahedron Lett.* **2002**, *43*, 5305.
02TL5381	V.V. Namboodiri, R.S. Varma, *Tetrahedron Lett.* **2002**, *43*, 5381.
02TL5617	R. Abonia, E. Rengifo, J. Quiroga, B. Insuasty, A. Sanchez, J. Cobo, J. Low, M. Nogueras, *Tetrahedron Lett.* **2002**, *43*, 5617.
02TL5793	K.-S. Yeung, M.E. Farkas, Z. Qiu, Z. Yang, *Tetrahedron Lett.* **2002**, *43*, 5793.
02TL5879	Y. Jiao, E. Valente, S.T. Garner, X. Wang, H. Yu, *Tetrahedron Lett.* **2002**, *43*, 5879.
02TL6217	D.V. Davydov, I.P. Beletskaya, B.B. Semenov, Y.I. Smushkevich, *Tetrahedron Lett.* **2002**, *43*, 6217.
02TL6221	I.P. Beletskaya, D.V. Davydov, M.S. Gorovoy, *Tetrahedron Lett.* **2002**, *43*, 6221.
02TL6259	M. Alajarin, A. Vidal, F. Tovar, P. Sanchez-Andrada, *Tetrahedron Lett.* **2002**, *43*, 6259.
02TL6431	S. Man, P. Kulhánek, M. Potácek, M. Necas, *Tetrahedron Lett.* **2002**, *43*, 6431.
02TL6613	O.V. Donskaya, V.N. Elokhina, A.S. Nakhmanovich, T.I. Vakul'skaya, L.I. Larina, A.I. Vokin, A.I. Albanov, V.A. Lopyrev, *Tetrahedron Lett.* **2002**, *43*, 6613.
02TL6833	T. Nixey, C. Hulme, *Tetrahedron Lett.* **2002**, *43*, 6833.
02TL6997	M.P. John, S.A. Hermitage, J.R. Titchmarsh, *Tetrahedron Lett.* **2002**, *43*, 6997.
02TL7553	R.J. Snow, T. Butz, A. Hammach, S. Kapadia, T.M. Morwick, A.S. Prokopowicz, III, H. Takahashi, J.D. Tan, M.A. Tschantz, X.-J. Wang, *Tetrahedron Lett.* **2002**, *43*, 7553.
02TL7557	J.M. Cobb, N. Grimster, N. Khan, J.Y.Q. Lai, H.J. Payne, L.J. Payne, T. Raynham, J. Taylor, *Tetrahedron Lett.* **2002**, *43*, 7757.

02TL7601	M.G. Ponzo, G. Evindar, R.A. Batey, *Tetrahedron Lett.* **2002**, *43*, 7601.
02TL7687	K.H. Bleicher, F. Gerber, Y. Wuthrich, A. Alanine, A. Capretta, *Tetrahedron Lett.* **2002**, *43*, 7687.
02TL7707	J.M. Gardiner, A.D. Goss, T. Majid, A.D. Morley, R.G. Pritchard, J.E. Warren, *Tetrahedron Lett.* **2002**, *43*, 7707.
02TL8145	T. Sima, S. Guo, F. Shi, Y. Deng, *Tetrahedron Lett.* **2002**, *43*, 8145.
02TL8741	T.-P. Loh, L.-C. Feng, H.-Y. Yang, J.-Y. Yang, *Tetrahedron Lett.* **2002**, *43*, 8741.
02TL8917	J. Rivera, N. Jayasuriya, D. Rane, K. Keertikar, J.A. Ferreira, J. Chao, K. Minor, T. Guzi, *Tetrahedron Lett.* **2002**, *43*, 8917.
02TL8925	L. El Kaim, L. Grimaud, N.K. Jana, F. Mettetal, C. Tirla, *Tetrahedron Lett.* **2002**, *43*, 8925.
02TL9051	C. Jaramillo, J.C. Carretero, J. E. de Diego, M. del Prado, C. Hamdouchi, J.L. Roldan, C. Sanchez-Martinez, *Tetrahedron Lett.* **2002**, *43*, 9051.
02TL9119	C. Dechambre, J.M. Chezal, E. Moreau, F. Estour, B. Combourieu, G. Grassy, A. Gueiffier, C. Enguehard, V. Gaumet, O. Chavignon, J.C. Teulade, *Tetrahedron Lett.* **2002**, *43*, 9119.
02TL9137	M.B. Andrus, Y. Ma, Y. Zang, C. Song, *Tetrahedron Lett.* **2002**, *43*, 9137.
02TL9347	S. Caddick, W. Kofie, *Tetrahedron Lett.* **2002**, *43*, 9347.
02TL9381	Z.M.A. Judeh, H.-Y. Shen, B.C. Chi, L-C. Feng, S. Selvasothi, *Tetrahedron Lett.* **2002**, *43*, 9381.
02TL9449	I. Yavari, A. Alizadeh, M. Anary-Abbasinejad, *Tetrahedron Lett.* **2002**, *43*, 9449.
02TL9497	R.P. Singh, S. Manandhar, J.M. Shreeve, *Tetrahedron Lett.* **2002**, *43*, 9497.
02TL9707	S. Kamijo, T. Jin, Z. Huo, Y. Yamamoto, *Tetrahedron Lett.* **2002**, *43*, 9707.
02TL9721	J. Azizian, A.V. Morady, S. Soozangarzadeh, A. Asadi, *Tetrahedron Lett.* **2002**, *43*, 9721.

Chapter 5.5

Five-Membered Ring Systems: With N & S (Se) Atoms

David J. Wilkins
Key Organics Ltd, Highfield Industrial Estate, Camelford, Cornwall PL32 9QZ, UK.
davidw@keyorganics.ltd.uk

Paul A. Bradley
Pfizer Global Research & Development, Sandwich Laboratories, Ramsgate Road, Sandwich, Kent CT13 9NJ, UK.
Paul_A_Bradley@sandwich.pfizer.com

5.5.1 ISOTHIAZOLES

The first reported regioselective lithiation of the 5-position of 3-(benzyloxy)isothiazole (**1**) using LDA in diethyl ether has been described. Subsequent quench with a variety of electrophiles gave the desired products **2** in 54-68% yield. Only benzoylation proved to be unsuccessful, resulting in a complex reaction mixture <02JOC2375>.

This methodology was used to prepare thioibotenic acid (**3**), a potential glutamic receptor ligand <02JOC2375>.

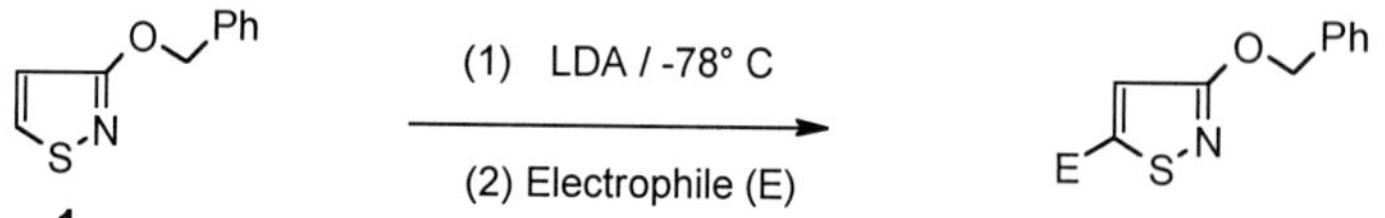

E = PhCHO, MeOCOCN, DMF, Cyclohexanone; Yields: 54-68%

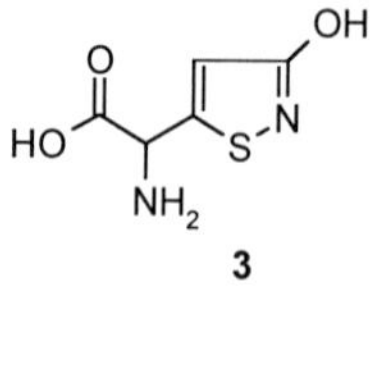

O-Lithiation of *N-tert*-butylbenzenesulfonamide (**4**) followed by reaction with a variety of ketones gave carbinol derivatives **5**. Subsequent cyclisation using TMSCl-NaI-MeCN afforded 3,3-disubstituted and spiro-2,3-dihydrobenzo[*d*]isothiazole-1,1-diones **6** in excellent yields <02JCS(P1)302>.

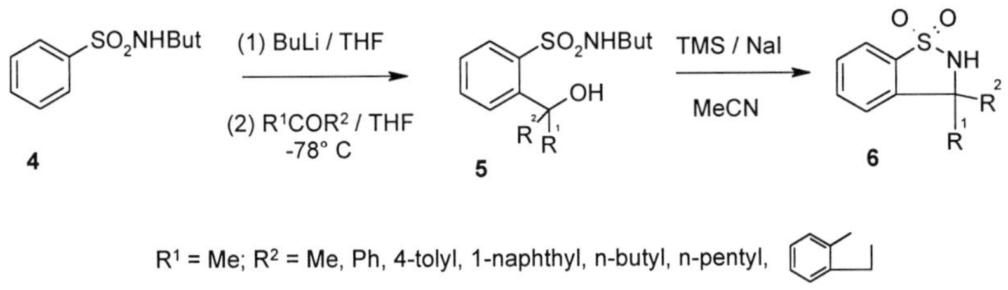

R^1 = Me; R^2 = Me, Ph, 4-tolyl, 1-naphthyl, n-butyl, n-pentyl,

The 2-(3-oxo-1,2-benzisothiazolinyl) group in 2-mercaptobenzoates **7** could easily be replaced with a variety of amines to give *N*-substituted 2-sulfenamoylbenzoates **8** or *N*-substituted 1,2-benzisothiazolin-3-ones **9** in good yield <02T3779>.

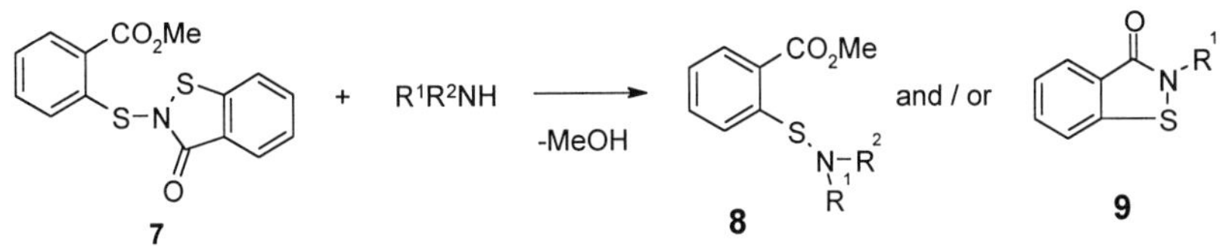

Treatment of a variety of amides **10** with excess $SOCl_2$ led to the formation of *N*-substituted 5-benzoylisothiazol-3(2*H*)-ones **11** which could easily be debenzoylated with NaOH to the corresponding *N*-substituted isothiazol-3(2*H*)-ones **12**. This procedure was high yielding affording isothiazolones *N*-substituted with a bulky alkyl group, such as *t*-butyl and a phenyl group bearing both electron-withdrawing and electron-donating substituents <02JHC149>.

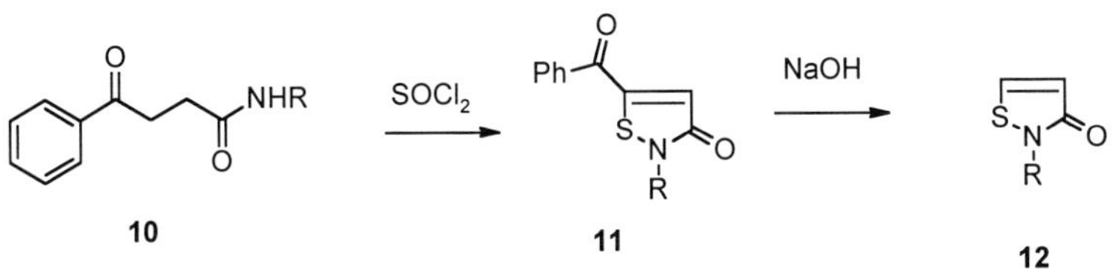

R = ^{t}Bu, 3-$NO_2C_6H_4$, 4-$NO_2C_6H_4$, 4-MeC_6H_4, 4-$MeOC_6H_4$.

The Diels-Alder reactions of the sulfilimine **13** with both furan and acyclic dienes has been investigated by Ruano *et al.* Reaction of **13** with acyclic dienes led to high *endo* selectivity whereas *exo* selectivity was observed in the reaction with furan <02JOC2919>.

13

endo-**14A** (R = OMe; 89%)
endo-**15A** (R = Me; 84%)

endo **16B** (R = Me; 11%)

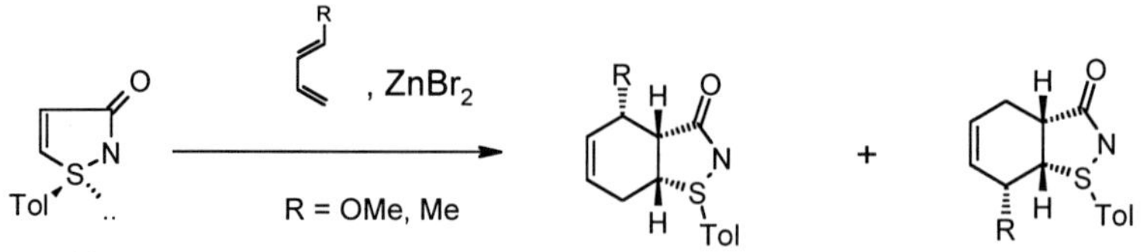

The synthesis of a series of monocyclic hydroperoxy- and hydroxy-substituted sulfin- and sulfonamides by oxidation of 4,5-dimethylisothiazolium salts has been reported <02HCA183>. These hydroperoxy derivatives have been used as oxidising agents for organic compounds containing heteroatoms such as sulfur and phosphorous <02HCA183>.

Cyclisation of 5-benzylsulfonyl-3-chloro-4-methoxycarbonylaminoisothiazole (**18**) using an excess of NaOMe in DMF gave 3-methoxy-5-oxo-6-phenyl-5,6-dihydro-4*H*-isothiazolo[5,4-*b*]-1,4-thiazine 7,7-dioxide **19**, which was the first representative of a new heterocyclic ring system. Compound **18** was prepared in good yield (81%) by the reaction of **17** with $PhI(OAc)_2$ in methanol <02RCB187>.

17 **18** **19**

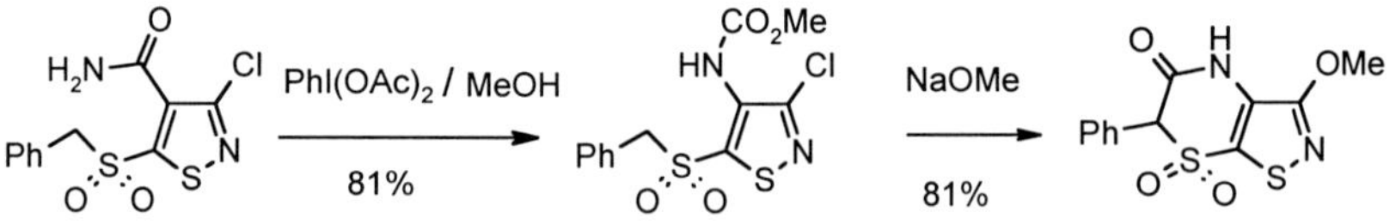

3,3-Dichloro-3*H*-benz[*c*][1,2]oxathiazol-1,1-dioxide (**20**) was found to be a very versatile compound for the preparation of a range of peptidic saccharin derivatives. Treatment of **20** with a range of amino acid esters **21** in DMF at 0 °C furnished peptides **22** in yields of 35-80%. [Reaction of **20** with amino acids did not give any of the peptides **22**]. Attempted base hydrolysis (0.1M NaOH) of the ester was unsuccessful, resulting in opening of the isothiazole ring system. Acidic hydrolysis was more successful giving the required carboxylic acids <02PHA384>.

20 **21** **22**

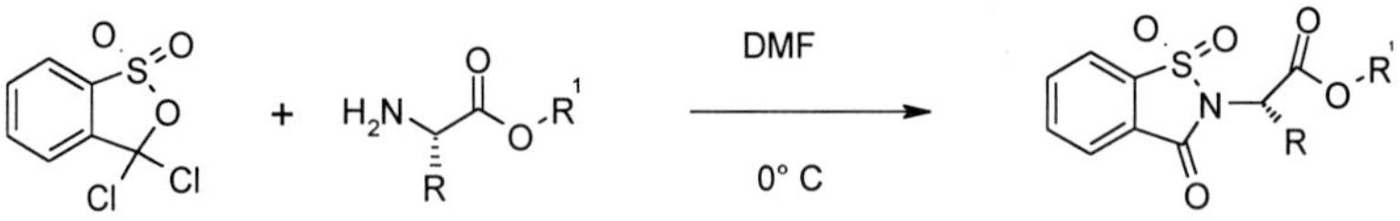

R = Me, H, CH_2, (CO_2Bn), CH(Me)Et, $CH_2CH(Me_2)$, $(CH_2)_2SMe$, CH_2OH, Bn, $CH(Me)_2$
R1 = Me, Et, Bn

The hemiacetals of (+)- and (-)-*N*-glyoxyloylborane-10,2-sultam **23a** and **23b** and their imines **23c** and **23d** reacted with allyl iodide in the presence of indium in DMF at 0 °C to give the corresponding α-hydroxy and α-amino camphor sultam derivatives **24a** and **24b** and **24c** and **24d** with high diastereoselectivites (86-90% de). Cleavage of the auxillary then furnished α-hydroxy and α-amino acids <02JCS(P1)1314>.

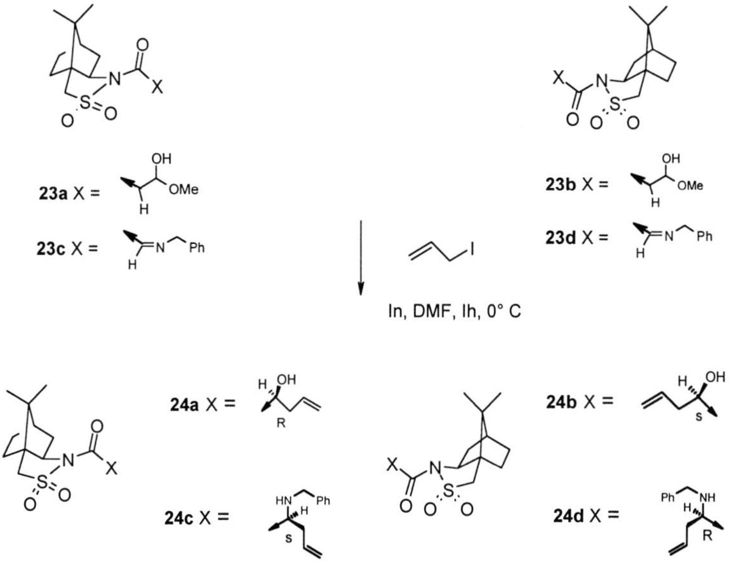

A subsequent similar synthesis of α-amino acids was then reported by Naito *et al.* which used the indium-mediated alkylation and allyation reaction of the Oppolzer camphorsultam derivative of glyoxylic oxime ether <02CC1454>.

Metalation of the phenyl-substituted *cis*-aziridine **25** followed by methyl iodide quench afforded the tricyclic isothiazole 1,1-dioxide **26** as a single diastereoisomer in 75% yield. The reaction was proposed to proceed *via* metalation of the benzylic position of the aziridine, rather than the position α to the silicon. Subsequent intramolecular attack of the benzylic anion **27** at the tosyl group *ortho* to the sulfonyl group gave **28** which was trapped with MeI to give the single diastereoisomer **26**, whose structure was confirmed by X-ray analysis <02JOC2335>.

Christiano *et al.* showed that the migration of the allyl group in pseudosaccharyl ethers **29** proceeded through both [1,3]- and [3,3]-mechanisms affording **30** and **31**. The proportion of **30** and **31** was found to depend on temperature, reaction time and the polarity of the reaction medium. It was also demonstrated that **31** could be converted into **30** by heating <02JCS(P1)1213>.

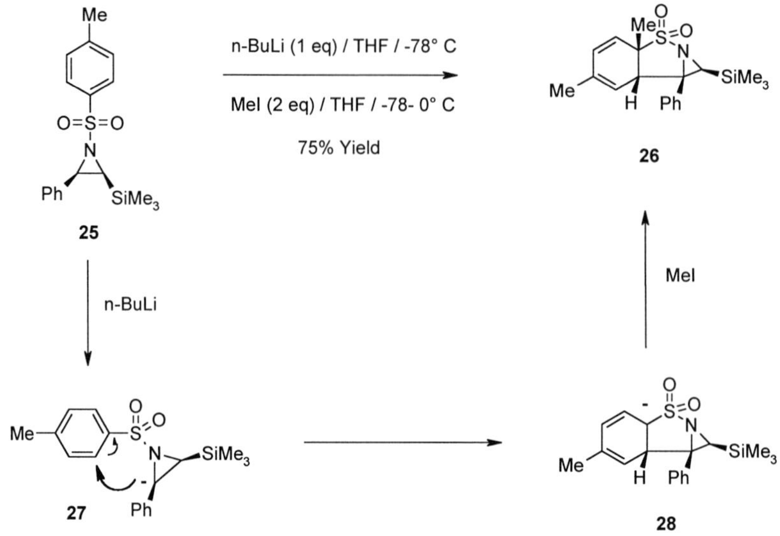

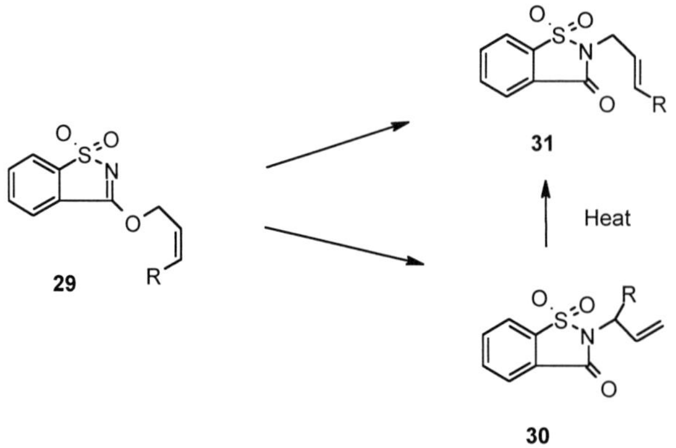

Treatment of (*E*)-3-alkylamino-3-alkylthiol-1-(thioaryl)propenes **32** with Ni(OAc)$_2$.4H$_2$O in EtOH at room temperature afforded the Ni(II)-propenethiolates **33** in excellent yields. Subsequent reaction of **33** with alkyl- or arylthiols at reflux in 1,2-dichloroethane gave novel *S,N*-acetals **34**. Compound **34** was then converted into 5-aryl-3-(arylthio)isothiazoles **35** by cyclisation and *N*-dealkylation <02JOC5375>.

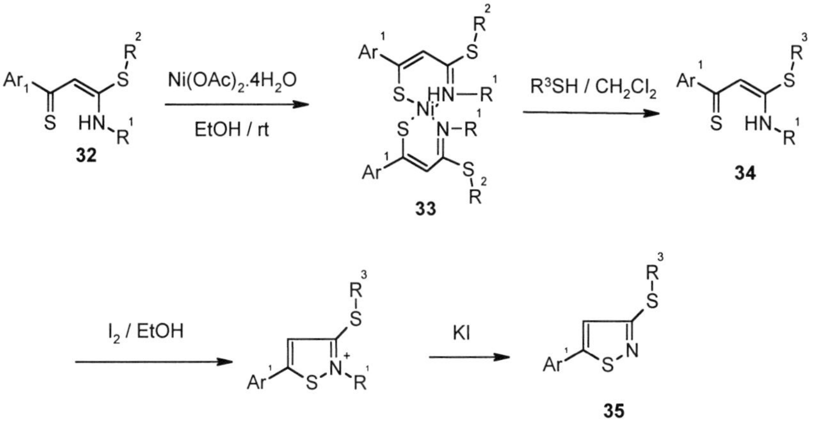

5.5.2 THIAZOLES

The synthesis of novel 2-cyanothiazolecarbazole analogues of ellipticine has been reported. The 3-aminocarbazole **36** was treated with 4,5-dichloro-1,2,3-dithiazolium chloride (Appel's salt; **37**) to give the imino-1,2,3-dithiazole **38**. Rearrangement of **38** under thermal conditions gave **39** in moderate yield <02TL2482>.

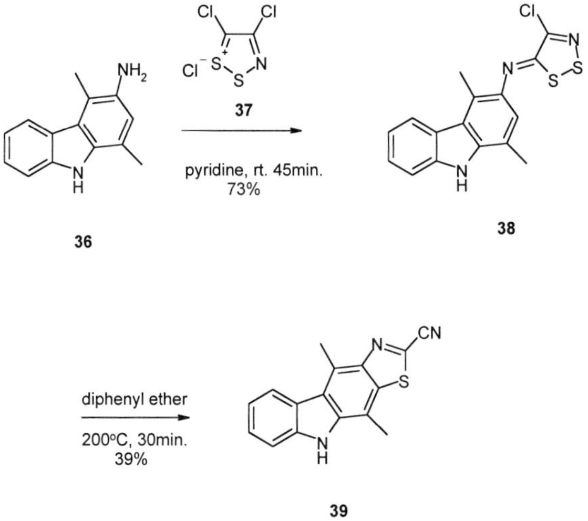

The synthesis of 2',4-disubstituted 2,4'-bithiazoles by a series of two regioselective cross coupling reactions has been reported. This bithiazole unit has been found in a number of natural products such as the bleomycins and macrocyclic antibiotics such as cyclothiazomycin.

The synthesis starts with 2,4-dibromothiazole (**40**), a regioselective Pd(0)-catalysed cross coupling step introduces a substituent at the 2-position. Alkyl or aryl zinc halides were employed as the nucleophiles to give **41**. The 4-bromothiazole derivative **41** was then converted into a carbon nucleophile either as a zinc derivative (Negishi conditions) or as a tin derivative (Stille conditions) which then underwent a second cross coupling reaction with 2,4-dibromothiazole (**40**) to give exclusively 2',4-disubstituted 2,4'-bithiazoles **42**.

In general the Negishi cross coupling conditions gave higher yields than the corresponding Stille conditions; however, the 2'-phenyl and 2'-alkynyl-4-bromo-2,4'-bithiazoles could not be synthesised under the Negishi conditions <02JOC5789>.

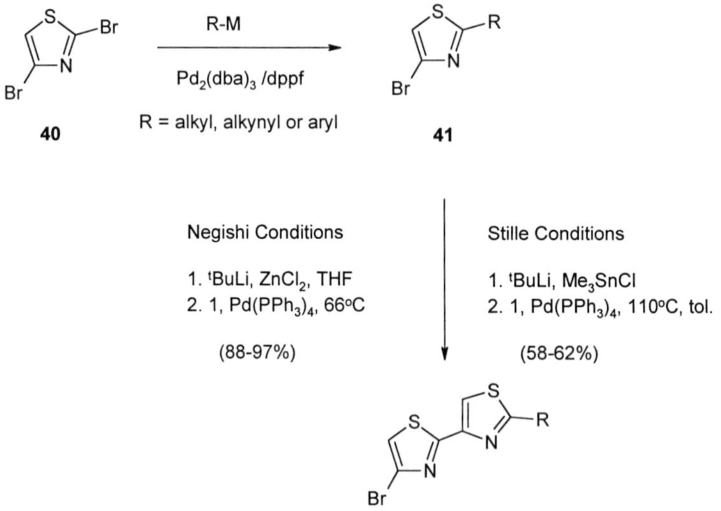

A one-pot synthesis of 2-aminothiazoles using supported reagents has been described. The method reacts α-haloketones **43** with silica gel supported potassium thiocyanate and alumina supported tertiaryammonium acetates to give 2-aminothiazoles **44** in high yields (up to 95%). If a similar reaction is carried out in one pot using unsupported reagents only a trace of product is observed because of the interaction of the amine with potassium thiocyanate <02TL1717>.

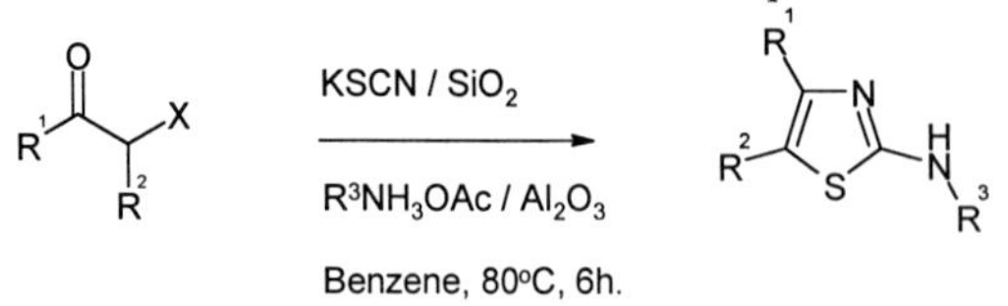

R^1, R^2 = methyl or aryl
R^3 = alkyl, cycloalkyl or aryl
X = Br or Cl

The cyclodehydration of 2-substituted-*N*-acylthiazolidine-4-carboxylic acids yields bicyclic munchnones. This mesoionic ring system acts as a cyclic azomethine ylid and can undergo 1,3-dipolar cycloaddition reactions with dipolarophiles. A range of chiral pyrrolo[1,2-c]thiazoles have been prepared by this method both intermolecularly and intramolecularly.

The chiral 2-phenacyl thiazolidine **45** derived from L-cysteine was acylated with the acid chloride **46** to give the adduct **47** as a single diastereoisomer. When **47** was converted into **48** and then heated in acetic anhydride, the mesoionic species **49** is generated *in situ* which undergoes an intramolecular 1,3-dipolar cycloaddition reaction to give the chiral pyrrolo[1,2-c]thiazole with a fused dihydrofuran ring **50** in 42% yield. A range of furan and pyran fused pyrrolo[1,2-c]thiazoles have been described along with some examples of intermolecular 1,3-dipolar cycloaddition reactions using DMAD, methyl propiolate and methyl vinyl ketone <02JOC4045>.

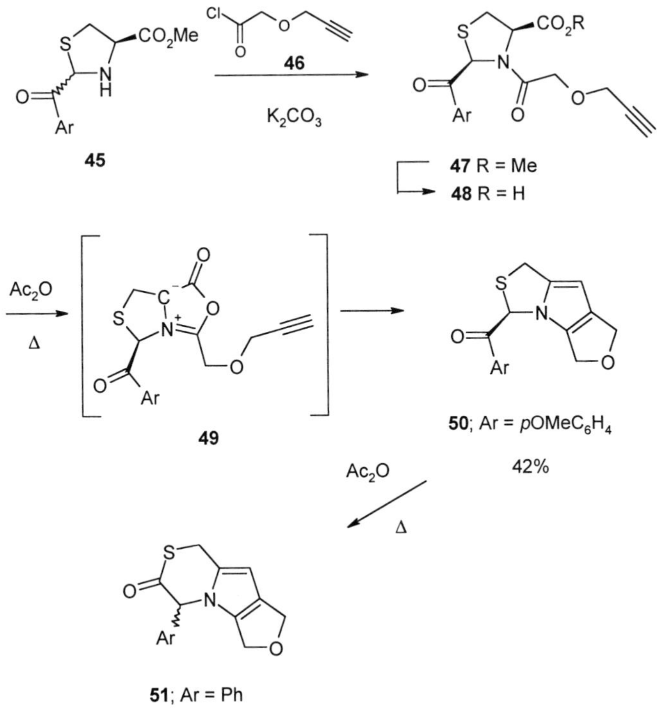

If **50** Ar = Ph is subjected to prolonged heating in acetic anhydride an interesting rearrangement to a thiazine derivative **51** occurs, a mechanism for this rearrangement is proposed <02JOC4045>.

The preparation of thiazoles on solid supports continues to attract attention. A variation on this approach, whereby instead of tethering a thioamide or thiourea to a solid support the α-bromoketone is tethered, has been reported.

An aromatic halide linked to a polystyrene resin **52** was treated with tributyl(1-ethoxyvinyl)tin under Stille conditions to give a vinyl ether which was treated with NBS to afford the α-bromoketone **53**. Subsequent reaction with a thioamide or thiourea gives 2,4-disubstituted or 2-aminothiazoles **54**, after cleavage of the resin, in good to excellent yields <02TL3193>.

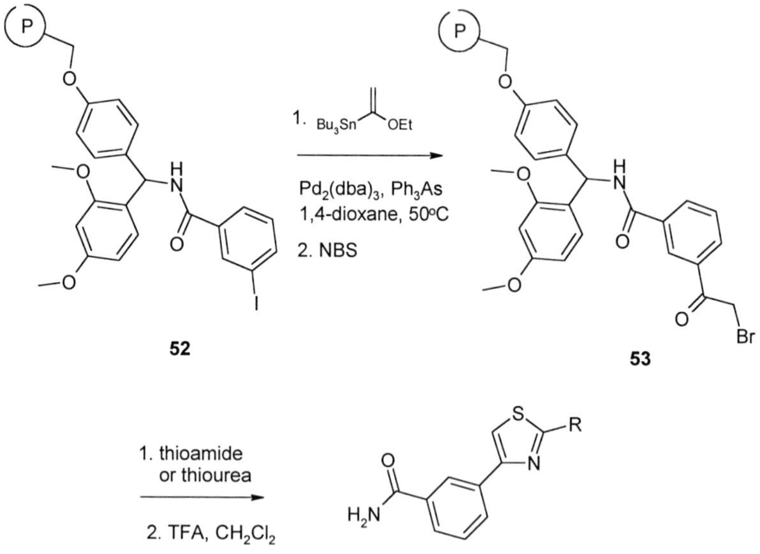

Another approach to the solid phase synthesis of thiazoles involves an interesting C-sulfanylation step. The starting material for this synthesis is a resin bound piperazine **55** which is converted into a thiourea and then treated with an α-bromoketone to give the thiazole **56**. Treatment of **56** with either thiols or disulfides and iodine or sulfonyl chlorides with iodine and triphenylphosphine afforded 5-sulfanylthiazoles **57**, which could be obtained in high yields and purity after cleavage from the resin <02EJOC2953>.

A new synthesis of thiazolines has been described. The annulation of thioamides with 2-alkynoates or 2,3-dienoates in the presence of a phosphine catalyst yields thiazolines in a selective manner <02JOC4595>.

Phosphines are known to impart bielectrophilic character to electron deficient alkynes such as **58** and promote γ-addition of nucleophiles to alkynes. Binucleophiles such as thioamides could potentially give two thiazoline isomers **59** and **60**. It is proposed that an initial vinyl phosphonium intermediate is formed which prefers to react with hard nucleophiles such as nitrogen rather than sulfur; consequently, only thiazole **59** was observed.

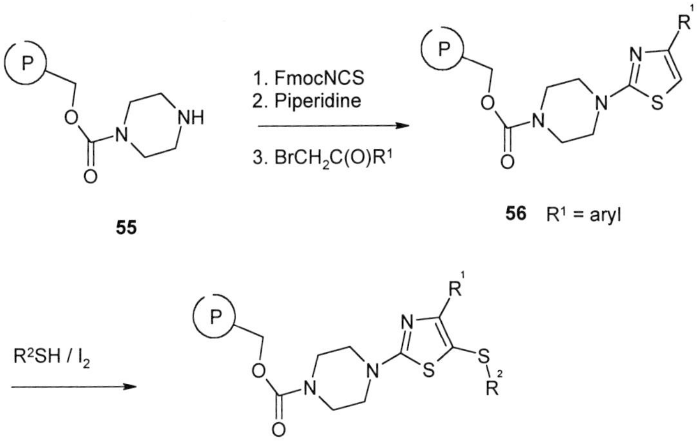

The annulation reaction is applicable to a variety of aromatic thioamides with both electron withdrawing and donating groups and heteroaryl thioamides such as 2-thienyl and 3-pyridyl. Aliphatic thioamides such as thioacetamide did not undergo clean cyclisation though. When triphenylphosphine was used as the catalyst the reaction required refluxing in toluene; however, if more nucleophilic phosphines were used such as tributylphosphine the reaction could be carried out at room temperature <02JOC4595>.

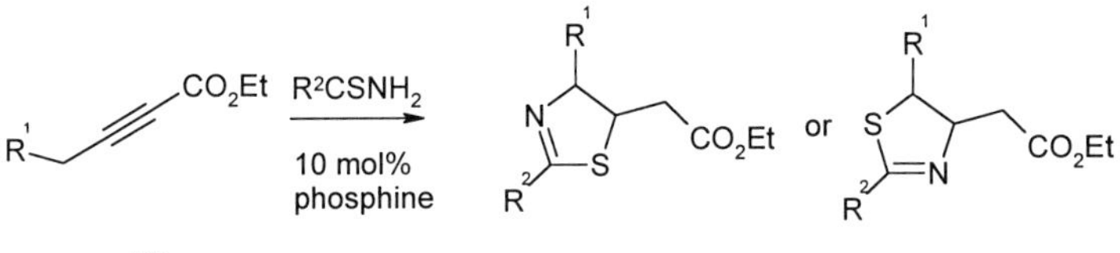

The amidine **61** when treated with alkylsulfonyl chlorides forms the intermediate thiazoline diazadiene **62** which cyclises to give the thiazolo[3,2-*b*]-1,2,4-thiadiazine-1,1-dioxide **63** in moderate yield. This is the first synthesis of this class of compound <02TL4099>

The synthesis of enantiopure thiazolines has been reported. The mixture of enantiomerically pure diastereoisomeric amido selenides **64**, obtained from the reaction of camphorselenyl sulfate with *trans* alkenes, when treated with Lawesson's reagent affords a mixture of two thioamide derivatives **65a** and **65b**. The two diastereoisomeric thioamide selenides were easily separated by chromatography, each was then treated with phenyl selenyl chloride which caused deselenylation to occur in a stereospecific manner furnishing the thiazolines **66a** and **66b** in an enantiopure form <02TA429>.

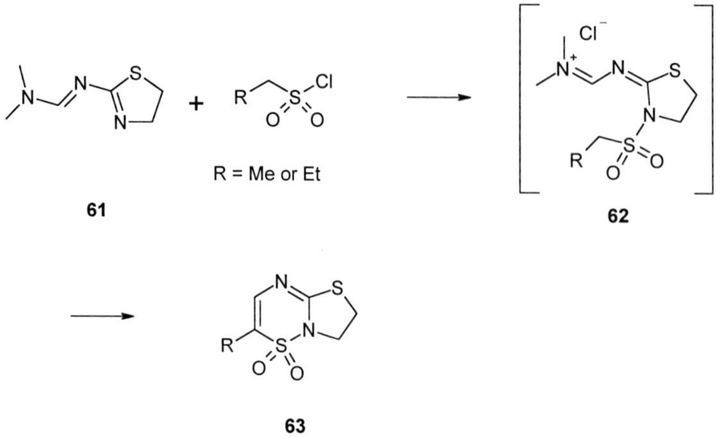

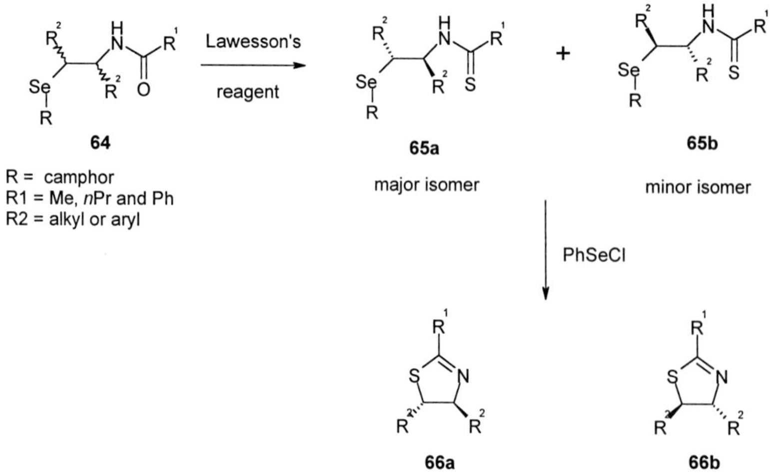

An interesting spirocyclic thiazolidine ring system has been reported. Starting from the thiazolethione **67** and the aziridine **68**, which when heated to 100 °C forms the azomethine ylid **69**, undergoes a 1,3-dipolar cycloaddition reaction to afford the cycloadduct **70** as the sole regioisomer <02H393>.

The synthesis of *N,N*-disubstituted 2-amino-5-acylthiazoles has been reported. Starting from the thiourea **71** reaction with α-halomethyl ketones **72** furnished the thiazoles **73** in good yields. This method is useful for preparing *N,N*-bisaryl derivatives <02T2137>.

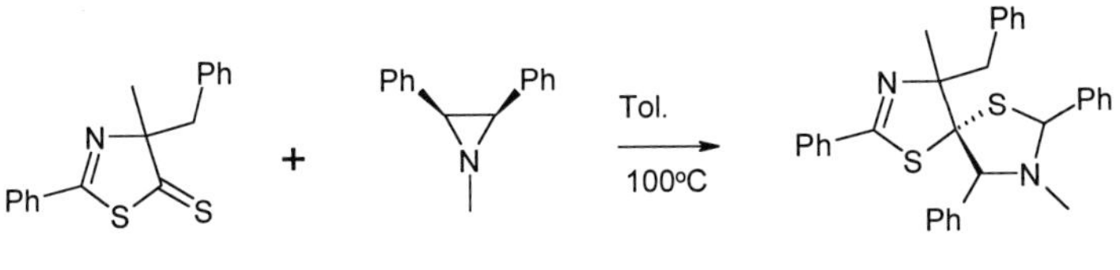

The preparation of highly functionalised thiazoles using the classical Hantzsch synthesis has been reported. Starting from diethyl 3-oxoglutarate, reaction with NBS gave the 2,4-dibromo derivative **74**, which when treated with thiourea gave the thiazole acetic ester **75** in good yield. The α-bromo substituent was then transformed into a ketone and an amino group *via* an azido intermediate <02JCS(P1)652>.

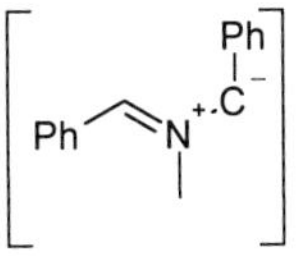

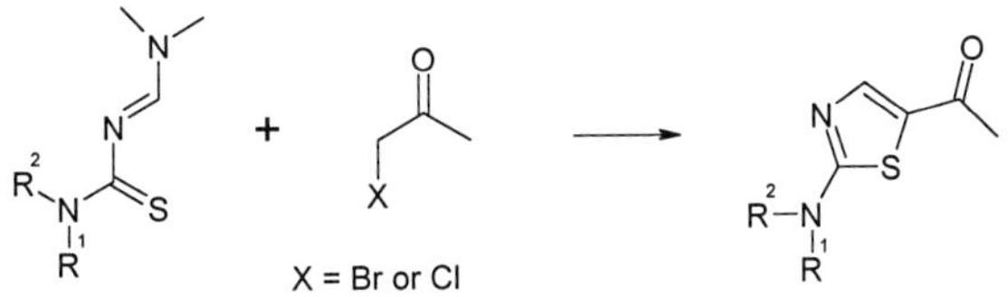

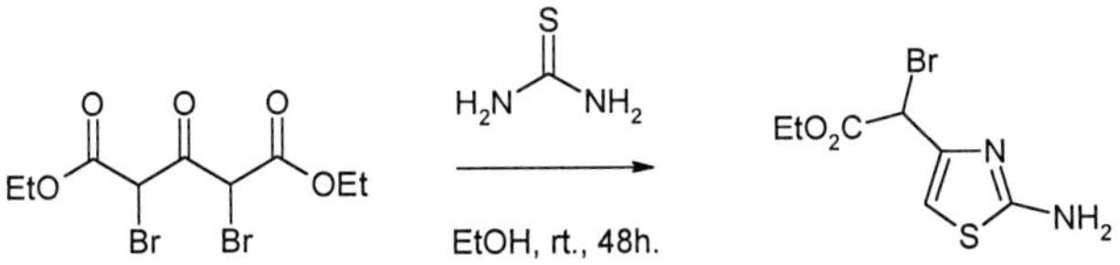

The Hantzsch synthesis of thiazoles is an excellent method for the synthesis of simple thiazoles, however for some substituted examples low yields have been reported as a result of dehalogenation of the α-haloketone. An alternative method for the synthesis of highly substituted thiazoles has been reported, thus starting from the 2-bromo-5-chlorothiazole **76** it was possible to introduce substituents selectively at the 2-position by a palladium-catalysed cross coupling reaction to give **77** (74-92%). In order to introduce a substituent into the 5-position, longer reaction times and an excess of organometallic reagent to give **78** (56-87%) were required <02OL1363>.

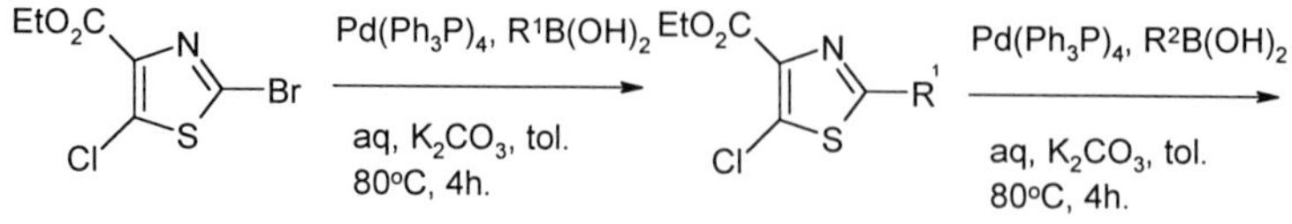

76 → **77** → **78**

R¹ = aryl, vinyl, pyridyl

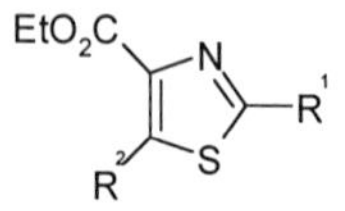

R² = aryl, vinyl, pyridyl and phenylalkynyl

Chiral 3-acyl-4-alkylthiazolidine-2-thiones have been prepared from *R*-(-)-2-amino-1-butanol and *S*-(+)-leucinol, these have been used to enantioselectively transfer an acyl group to racemic amines or amino acids to give acylated products in high yield (78-94%) and moderate optical purity (48-68% ee) <02IJC(B)593>.

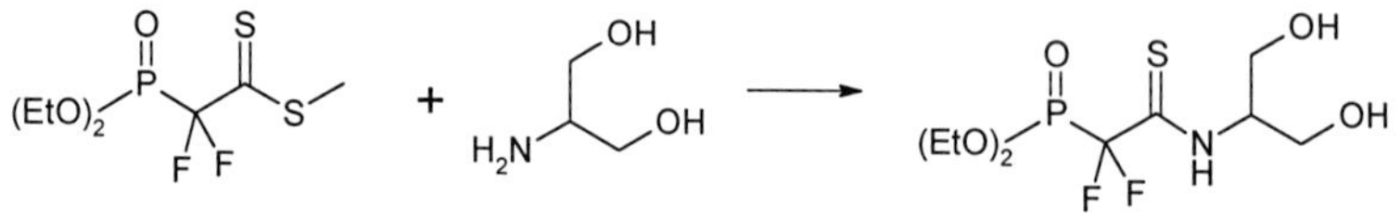

79 + → **80**

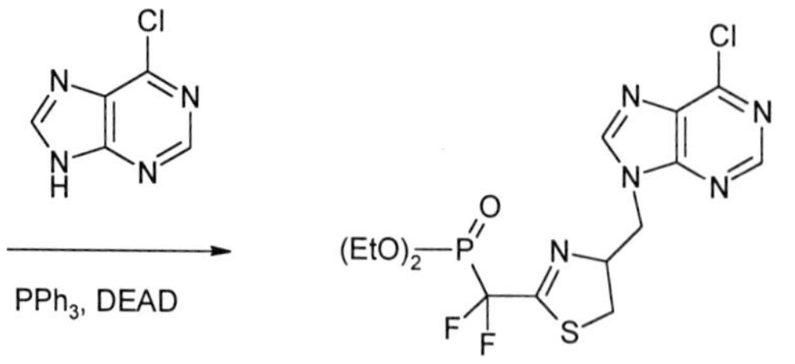

→ (PPh₃, DEAD) → **81**

The synthesis of thiazoline based nucleosides has been reported. The synthesis starts with methyldifluorodiethoxyphosphonodithioacetate (**79**), which is coupled with β-aminoalcohols to give thioamides such as **80**. These thioamides could be cyclodehydrated to thiazolines using thionyl chloride/pyridine. If the diol **80** is treated under Mitsunobu conditions with 6-chloropurine the nucleoside **81** is afforded in 66% yield <02JOC843>.

The synthesis of natural products containing a thiazole ring continues to attract a lot of interest. The epothilones are a promising class of anti-tumor agents and a number of papers have appeared describing the total synthesis of members of this family of compounds <02TL2895, 02JACS9825, 02T6413, 02ACIE1381>.

The total synthesis of the bisthiazole macrocycle cystothiazole A has been described <02TL643>. The synthesis of the thiazole unit of the anti-inflammatory metabolite halipeptin C has also been reported <02TL5707>.

5.5.3 THIADIAZOLES

5.5.3.1 1,2,3-Thiadiazoles

Heating solutions of 5-aryloxy-1,2,3-thiadiazoles **82** (X = O) in DMF at 100 °C in the presence of excess sodium hydride afforded 1,4-benzoxathiins **84** (X = O). The transformation was proposed to proceed *via* initial cleavage of the thiadiazole ring with subsequent N_2 elimination to give the intermediate **83**. Subsequent intramolecular rearrangement of intermediates gave the 1,4-benzoxathiins **84** (X = O) <02H483>.

The corresponding 5-arylthio-1,2,3-thiadiazoles **82** (X = S) underwent the same transformation in DMF in the presence of NaH at 120° C to give **84** (X = S).

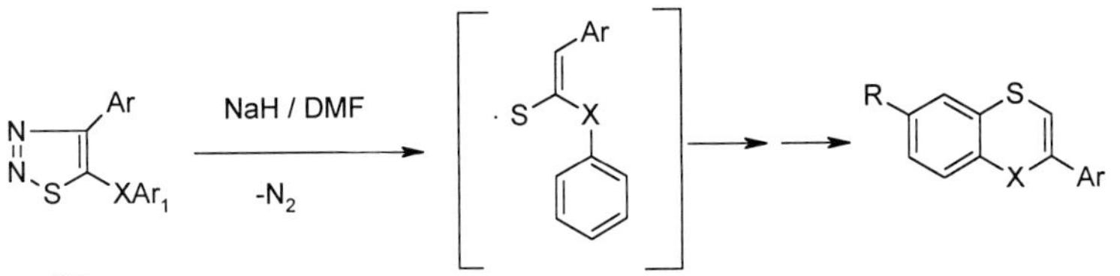

Ar = Ph, Thiophen-2-yl; Ar^1 = 4-MeOC_6H_4, 4-MeC_6H_4, 3-ClC_6H_4; X = O, S; Yields = 11-60%

1,2,3-Thiadiazole-4-carbohydrazides **85** undergo ring cleavage with liberation of N_2 when treated with *t*-BuOK in DMSO at 20° C. Recyclisation and subsequent alkylation then allows isolation of 5-thiopyrazolones, 6-thiomethylidene-1,3,4-oxadiazin-5-ones or 5-thio-7*H*-pyrazolo[5,1-*b*][1,3]thiazine-2,7-diones, depending on how many equivalents of base was used in the reaction <02TL1015>.

5.5.3.2 1,2,4-Thiadiazoles

A thorough search of the literature in 2002 revealed very few interesting references to the 1,2,4-thiadiazole ring system. The authors considered that none of the work described in these references warranted inclusion in this chapter.

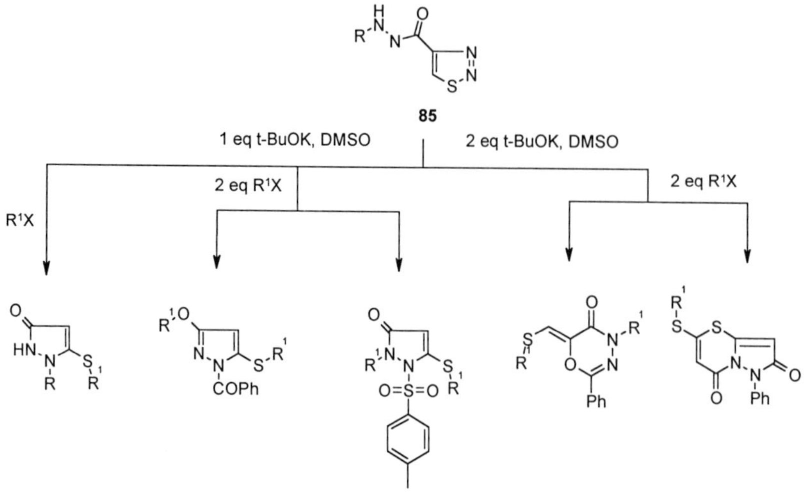

[Scheme showing compound 85 and its reactions]

85

1 eq t-BuOK, DMSO   2 eq t-BuOK, DMSO

2 eq R¹X   2 eq R¹X

R¹X

R = PhCO, 1,2,3-thiadiazole-4-carbonyl, Ts; R¹X = MeI, C₁₆H₃₃Br, PhCH₂Cl; Yields: 20-51%

5.5.3.3 1,2,5-Thiadiazoles

A thorough search of the literature in 2002 revealed very few interesting references to the 1,2,5-thiadiazole ring system. The authors considered that none of the work described in these references warranted inclusion in this chapter.

5.5.3.4 1,3,4-Thiadiazoles

Huisgen and co-workers reported, in three papers (02T507, 02T4185 and 02HCA1523> the conversion of various 1,3,4-thiadiazolines **86** into thiocarbonyl ylides **87** *via* extrusion of N_2. They then described cycloaddition reactions of these ylides **87** with various α,β-unsaturated esters and nitriles and postulated reaction mechanisms for the regioselectivity and stereochemistries observed in the transformations.

In a similar manner, Mloston *et al.* prepared the thiadiazoline **88** and treated it with electron-deficient acetylenes giving 2,5-dihydrothiophenes **89** in generally good yields. Reaction of **89** with TFA at room temperature, resulted in the formation of thiophene **90** *via* isomerisation and ring-opening of the cyclobutane ring <02HCA451>.

Cyclocondensation of 2-amino-5-(ethylsulfonyl)thiadiazole (**91**) with the sterically demanding base **92** afforded the 1,3,4-thiadiazolo[2,3-*b*]quinazoline **93** in 60% yield <02JCS(P1)555>.

R^1 = CN, MeOCO, PhCO, H_2NCO, ${}^tBuOCO$
R^2 = MeOCO, CN, PhCO, H_2NCO, H
Yields (25-75%)

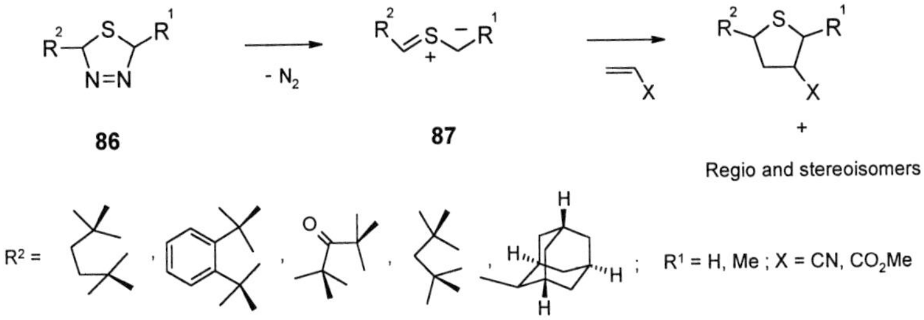

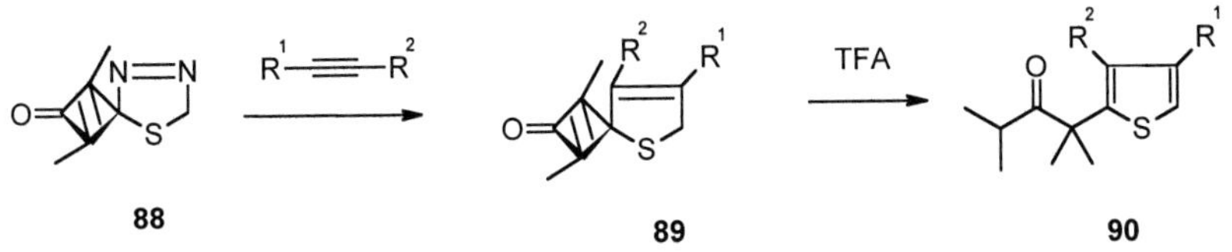

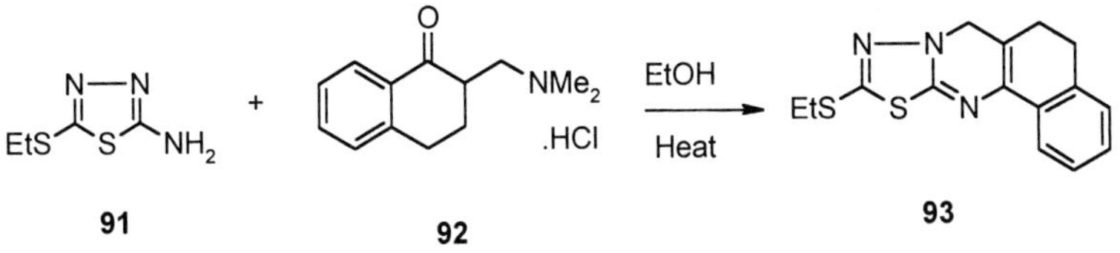

5.5.4 SELENAZOLES AND SELENADIAZOLES

The preparation of 5-alkylideneselenazoline-2-ones **96** has been described. Aminoalkynes **94** were treated with carbon monoxide and selenium to give an intermediate carbamoselenoate **95**. The intermediate carbamoselenoate **95**, which is highly nucleophilic at selenium, then undergoes an intermolecular cycloaddition reaction to give **96**. When R^1 is a primary or secondary alkyl group the reaction proceeds in high yield (95% R^1 = Bu); however, a tertiary alkyl or phenyl substituent severely effects or stops carbon monoxide absorption, presumably due to steric hindrance in the former case and lower nucleophilicity in the latter case. Substituents at R^2 that can stabilise an α anion at the alkynyl carbon such as phenyl facilitate the cycloaddition step; bulky groups such as trimethylsilyl at this centre resulted in a poor yield <02JOC6275>.

94 **95** **96**

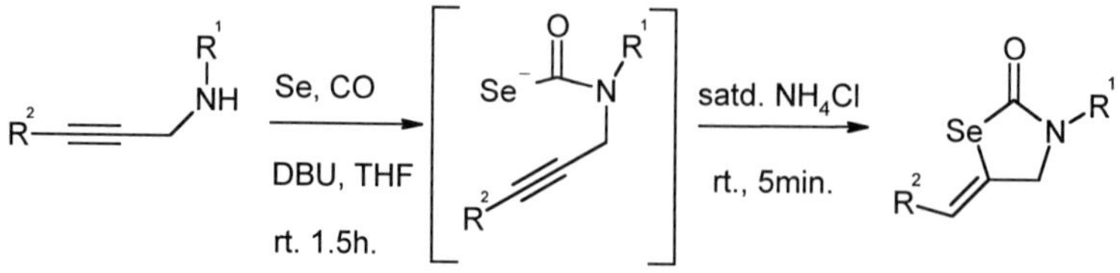

2-Imino-1,3-selenazolidin-4-one derivatives **99** have been synthesised in very high yields by the reaction of *N,N*-disubstituted selenoureas **97** with α-haloacyl halides **98** in the presence of pyridine. In contrast monosubstituted selenoureas such as the piperidine derivative **100** gave the selenazolone **101** in low yield <02S195>.

97 **98** **99**

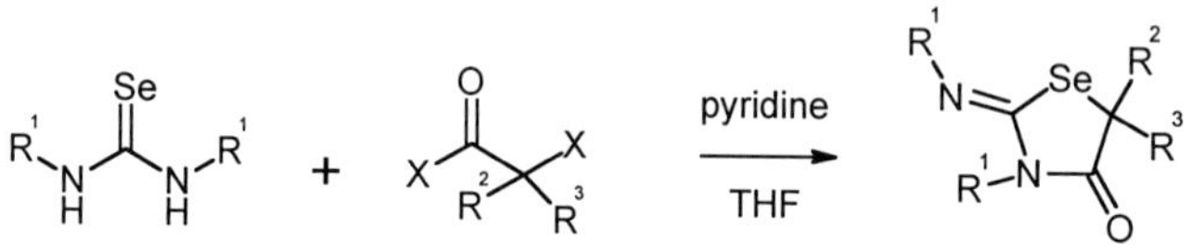

R^1 = alkyl or aryl; R^2 and R^3 = H or alkyl

100 **101**

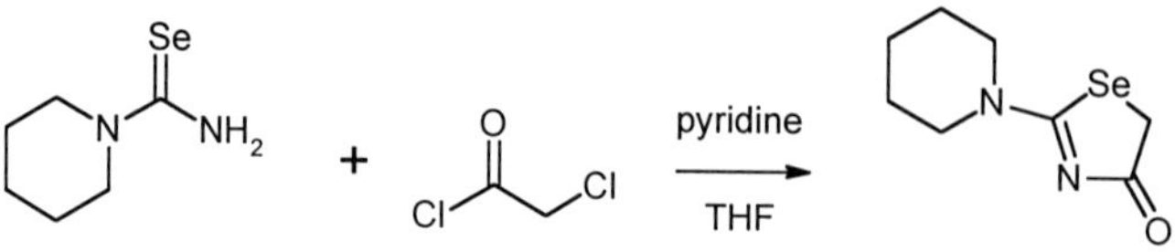

1,2,3-Selenadiazoles are useful intermediates for the preparation of alkenes because they can be easily decomposed with the loss of nitrogen and a selenium atom under free radical conditions. However, if 1,2,3-selenadiazoles such as **102** are treated with allyltributyl stannane/AIBN in the presence of an olefin or diene dihydroselenophenes such as **103** are formed provided the 1,2,3-selenadiazole has been derived from a cyclic ketone. Under similar conditions 1,2,3-selenadiazoles prepared from linear or aromatic ketones afford alkynes as the sole products <02JOC1520>.

102 **103**

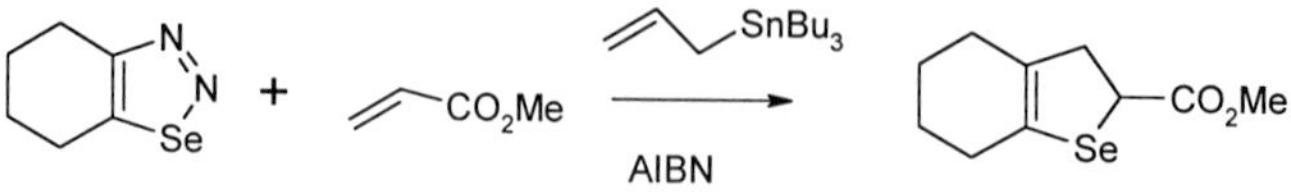

1,2-Diaza-1,3-butadienes **104** react with selenoureas **105** to give 2-selenazolin-4-ones **106**. Nucleophilic addition of the selenium atom at the terminal carbon of the heterodiene is followed by intramolecular ring closure. **104** also reacts with selenobenzamides **107** to give the 2-selenazolines **108** and **109**. It was found that the isomer **109** could be aromatised to **110** but not isomer **108**, suggesting that the aromatisation process involves an anti-elimination <02TL5707>.

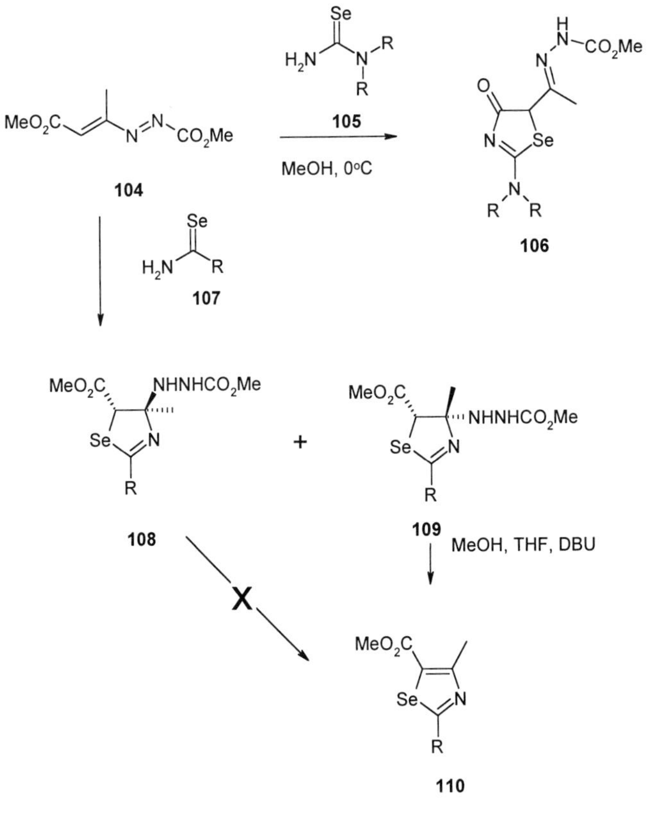

5.5.5 REFERENCES

02ACEI1381	J. Sun, C.S. Sinha, *Angew. Chem. Int. Ed.* **2002**, *41*, 1381.
02CC1454	H. Miyabe, A. Nishimura, M. Veda, T. Naito, *Chem. Commun.* **2002**, 1454.
02EJOC2953	M. Grimstrup, F. Zaragoza, *Eur. J. Org. Chem.* **2002**, 2953.
02H393	A. Gebert, A. Linden, G. Mloston, H. Heimgartner, *Heterocycles* **2002**, *56*, 393.
02H483	A.R. Katritzky, G.N. Nikonov, D.O. Tymoshenko, E.L. Moyano, P.J. Steel, *Heterocycles* **2002**, *56*, 483.
02HCA183	K. Taubert, J. Sieler, L. Hennig, M. Findeisen, B. Schulze, *Helv. Chim. Acta* **2002**, *85*, 183.
02HCA451	T. Gendek, G. Mloston, A. Linden, H. Heimgartner, *Helv. Chim. Acta* **2002**, *58*, 451.
02HCA1523	H. Giera, R. Huisgen, E. Langhals, K. Polborn, *Helv. Chim. Acta* **2002**, *85*, 1523.

02IJC(B)593	L.D.S. Yadav, S. Dubey, *Indian J. Chem. Sect. B,* **2002**, *41B,* 593.
02JACS9825	K. Biswas, H. Lin, J.T. Njardarson, M.D. Chapell, T-C, Chou, Y. Guan, W.P. Tong, L. He, S.B. Horwitz, S.J. Danishefsky, *J. Am. Chem. Soc.* **2002**, *124,* 9825.
02JCS(P1)302	Z. Liu, N. Shibata, Y. Takeuchi, *J. Chem. Soc., Perkin Trans. 1* **2002**, 302.
02JCS(P1)555	J. Quiroga, P. Hernandez, B. Insuasty, R. Abonia, J. Cobo, A. Sanchez, M. Nogueras, J.N. Low, *J. Chem. Soc., Perkin Trans. 1* **2002**, 555.
02JCS(P1)652	D. Brickute, F.A. Slok, C. Romming, A. Sackus, *J. Chem. Soc. Perkin Trans.1* **2002**, 652.
02JCS(P1)1213	N.C.P. Araujo, P.M.M. Barraca, J.F. Bickley, A.F. Brigas, M.L.S. Cristiano, R.A.W. Johnstone, R.M.S. Loureiro, P.C.A. Pena, *J. Chem. Soc., Perkin Trans. 1* **2002**, 1213.
02JCS(P1)1314	J.G. Lee, K.Il Choi, A. Nim Pae, H.Y. Koh, Y. Kang, Y.S. Cho, *J. Chem, Soc., Perkin Trans. 1* **2002**, 1314.
02JHC149	S. Hamilakis, D. Kontonassios, A. Tsolomitis, *J. Heterocycl. Chem.* **2002**, *39,* 149.
02JOC1520	Y. Nishiyama, Y. Hada, M. Anjiki, K. Miyake, S. Hanita, N. Sonoda, *J. Org. Chem.,* **2002**, 1520.
02JOC2335	V.K. Aggarwal, E. Alonso, M. Ferrara, S.E. Spey, *J. Org. Chem.,* **2002**, *67,* 2335.
02JOC2375	L. Bunch, P. Krogsgaard-Larsen, U. Madsen, *J. Org. Chem.* **2002**, *67,* 2375.
02JOC2919	J.L. Garcia Ruano, C. Alemparte, F.R. Clemente, L. Gonzalez Gutierrez, R. Gordillo, A.M. Martin Castro, J.H. Rodriguez Rames, *J. Org. Chem.* **2002**, *67,* 2919.
02JOC4045	T.M.V.D. Pinho e Melo, M.I.L. Soares, A.M. d'A. Rocha Gonsalves, J.A. Paixao, A.M. Beja, M.R. Silva, L. Alte da Veiga, J.C. Pessoa , *J. Org. Chem.* **2002**, *67,* 4045.
02JOC4595	B. Lui, R. Davis, B. Joshi, D.W. Reynolds, *J. Org. Chem.* **2002**, *67,* 4595.
02JOC5375	D.J. Lee, B.S. Kim, K. Kim, *J. Org. Chem.* **2002**, *67,* 5375.
02JOC5789	T. Bach, S. Heuser, *J. Org. Chem.* **2002**, *67,* 5789.
02JOC6275	S. Fujiwara, Y. Shikano, T. Shinike, N. Kambe, N. Sonoda, *J. Org. Chem.* **2002**, *67,* 6275.
02OL843	E. Pfund, T. Lequeux, S. Masson, M. Vazeux, *Org. Lett.* **2002**, *4,* 843.
02OL1363	K.J. Hodgetts, M.T. Kershaw, *Org. Lett.* **2002**, *4,* 1363.
02PHA384	L. Soubh, A. Besch, H.H. Otto, *Pharmazie* **2002**, 384.
02RCB187	N.V. Voskoboev, A.I. Gerasyuto, S.G. Zlotin, *Russ. Chem. Bull.* **2002**, *51,* 187.
02S195	M. Koketsu, F. Nada, H. Ishihara, *Synthesis* **2002**, 195.
02T507	R. Huisgen, G. Mloston, H. Giera, E. Langhals, *Tetrahedron* **2002**, *58,* 507.
02T2137	A. Noack, H. Hartmann, *Tetrahedron* **2002**, *58,* 2137.
02T3779	M. Shimizu, Y. Sugano, T. Konakahara, Y. Gama, I. Shibuya, *Tetrahedron* **2002**, *58,* 3779.
02T4185	G. Mloston, R. Huisgen, H. Giera, *Tetrahedron* **2002**, *58,* 4185.
02T6413	K.C. Nicolaou, A. Ritzen, K. Namoto, R.M. Buey, J. Fernando Diaz, J.M. Andreu, M. Wartmann, K-H, Altmann, A, O'Brate, P. Giannakakou, *Tetrahedron* **2002**, *58,* 6413.
02TA429	M. Tiecco, L. Testaferri, C. Santi, C. Tomassini, F. Marini, L. Bagnoli, A. Temperini, *Tetrahedron: Asymmetry* **2002**, *13,* 429.
02TL643	K. Kato, A. Nishimura, Y. Yamamoto, H. Akito, *Tetrahedron Lett.* **2002**, *43,* 643.
02TL1015	A. Hameurlaine, M.A. Abramov, W. Dehaen, *Tetrahedron Lett.* **2002**, *43,* 1015.
02TL1717	M. Kodomari, T. Aoyama, Y. Suzuki, *Tetrahedron Lett.* **2002**, *43,* 1717.
02TL2483	H. Chabane, C. Lamazzi, V. Thiery, G. Guillaumet, T. Besson, *Tetrahedron Lett.* **2002**, *43,* 2483.
02TL2895	M.S. Ermolenko, P. Potier, *Tetrahedron Lett.* **2002**, *43,* 2895.
02TL3193	S. El Kazzouli, S. Berteina-Raoin, A. Mouaddib, G. Guillaumet, *Tetrahedron Lett.* **2002**, *43,* 3193.
02TL4099	C. Landreau, D. Deniaud, A. Reliquet, J.C. Meslin, *Tetrahedron Lett.* **2002**, *43,* 4099.
02TL5707	C.D. Monica, A. Randazzo, G. Bifulco, P. Cimino, M. Aquino, I. Izzo, F. DeRiccardis, L. Gomez-Paloma, *Tetrahedron Lett.* **2002**, *43,* 5707.

Chapter 5.6

Five-Membered Ring Systems: With O & S (Se, Te) Atoms

R. Alan Aitken* and Stephen J. Costello
University of St. Andrews, UK
raa@st-and.ac.uk

5.6.1 1,3-DIOXOLES AND DIOXOLANES

The reaction of epoxides with acetone to form 1,3-dioxolanes **1** can be catalysed by tin(IV) tetraphenylporphyrin <01JCR(S)365>, bismuth(III) salts <01SC3411> or cobalt polyoxometalates such as $K_5CoW_{12}O_{40}$ <01MI205>. The reaction of aldehydes and ketones with ethanediol to form 1,3-dioxolanes is accelerated by the combination of a Dean and Stark apparatus with microwave irradiation <01SC3323> and reaction of catechol with aldehydes and ketones to form benzodioxoles **2** is catalysed by a solid super-acid catalyst based on ZrO_2/SO_4^{2-} <01JCR(S)289>. Improved preparations of dioxolanes **3** and **4** have been reported <02SC449> and a synthesis of the perfumery ingredient "apple ester β" **5** uses ketalisation catalysed by $H_8[Si(W_2O_7)_6]$ <01MI208>.

A variety of improved methods for reaction of epoxides with CO_2 to give 1,3-dioxolan-2-ones **6** have appeared including reaction with 1 atmosphere of CO_2 in an ionic liquid <02OL2561>, electrocatalytic reaction in an ionic liquid <02CC274>, catalysis by $ZnBr_2$/pyridine <02USP13477> and reaction in DMF at 120 °C <02JAP53573>. Treatment of phthalimidoalkyl-substituted epoxides with $BF_3 \cdot Et_2O$ in chlorobenzene at 130 °C results in dimerisation to give dioxolanes **7** <02T7065>.

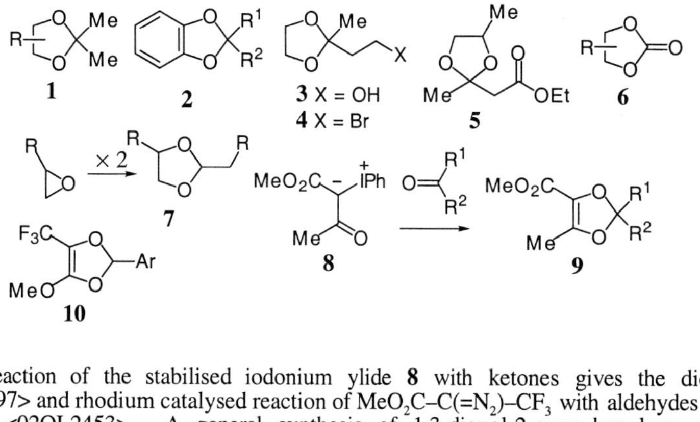

Reaction of the stabilised iodonium ylide **8** with ketones gives the dioxoles **9** <02TL5997> and rhodium catalysed reaction of $MeO_2C-C(=N_2)-CF_3$ with aldehydes similarly gives **10** <02OL2453>. A general synthesis of 1,3-dioxol-2-ones has been described

<02TL1161>. The synthesis of monoglyceride ketals **11** has been patented <02JAP69068> and the synthesis and reactivity of dioxolanes **12** derived from glycerol monoallyl ether has been examined <01JGU542>. The chiral dioxolane alcohol **13**, prepared in enantiomerically pure form, is a key intermediate in a synthesis of frontalin, <02TA155> and a series of alditol cyclic thionocarbonates (1,3-dioxolane-2-thiones) have been used as precursors for thiaheterocycles <02TL815>. The spiro bis(dioxolane) **14** is obtained as a stable crystalline product from degradation of PTFE with accelerated electrons followed by treatment of the resulting perfluoroalkene mixture with ethanediol <01JFC(111)129>. The dioxolane functions play a key role in synthesis of the highly pyramidalised alkenes **15** and **16** <02T10081>. Theoretical and experimental studies on the structure of simple bis(dioxolanes) including the X-ray structure of **17** have appeared <02NJC1686>.

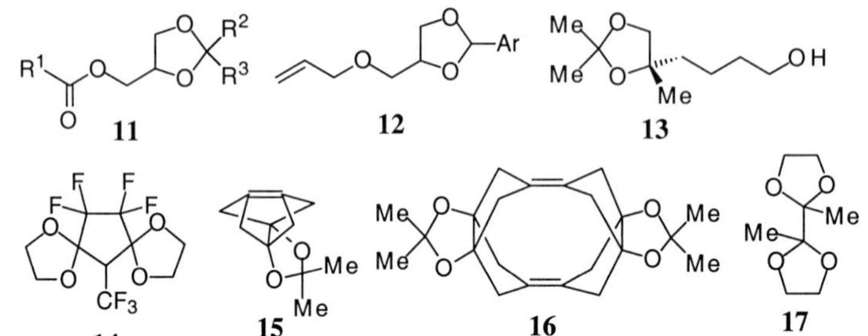

Treatment of 2-phenyl-1,3-dioxolane with IN_3 results in cleavage to give **18** <02SL1111> and while dioxolane **19** reacts with Me_3SiNEt_2 and MeI to give **20**, treatment with Et_3SiH and MeI gives the completely unexpected product **21** resulting from combination of two molecules of **19** with cleavage of the cyclohexane ring in one of them <02JOC5170>. Anodic monofluorination of 1,3-dioxolane and 1,3-dioxolan-2-one gives **22** and **23** respectively, <02TL1503>. The course of ozonolysis of 2-substituted 1,3-dioxolanes has been examined and the primary products established to be **24** <02JA11260>. Reaction of 2,2-dimethyl-4-phenyldioxolane with dichlorocarbene under phase-transfer conditions gives the insertion product **25** <02JAP47229>. Preparation of a range of spiro acetals such as **26** has been patented <02JAP69072>.

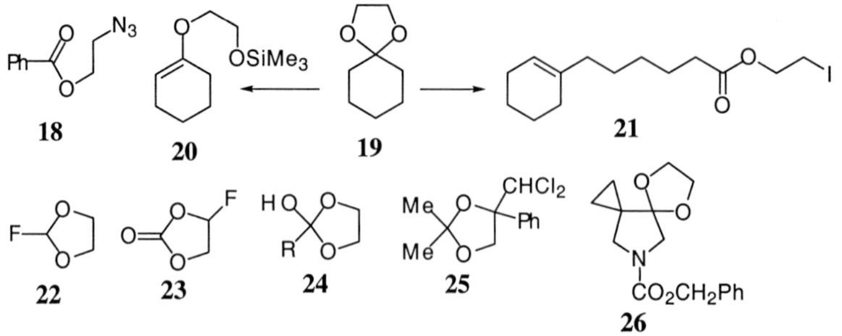

There have again been a large number of reports involving chiral dioxolane compounds. The chiral benzoquinone monoketal **27** has been used to obtain a variety of chiral building-blocks <01TA3077> and ultrasound-promoted conjugate addition of alkyl iodides in the presence of Zn/CuI to substrates such as **28** has been examined <02SL1435>. A new type of dioxolane-containing bicyclic amino acid **29** has been prepared <02T9865> and chiral dioxolanylaziridines such as **30** have been prepared as synthetic intermediates <02MIP85893>. The enantiomerically pure nitrone **31** can be prepared in a convenient one-pot procedure <02JOC1678>. Preparation of dioxolanone **32** from lactic acid and pivalaldehyde in the presence of scandium triflate and chiral binaphthol has been patented <02JAP105071>.

27 **28** **29** **30** **31** **32** **33** **34**

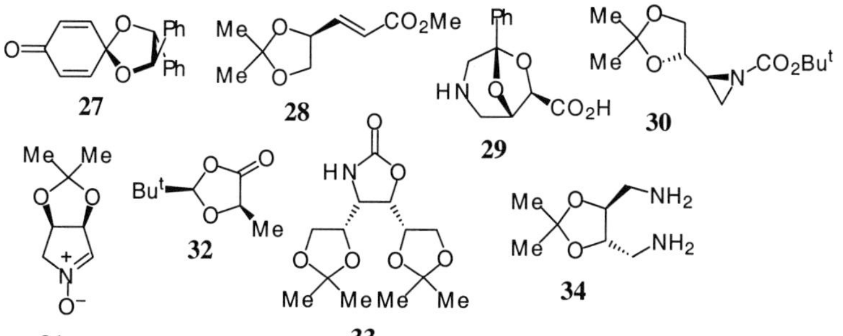

The bis(dioxolanyl)oxazolidinone **33** has been prepared from D-mannitol and evaluated as a chiral auxiliary <02MI749> and the diamine **34** has been examined as a ligand for rhodium catalysed asymmetric hydrogenation of diethyl itaconate <02JOU104>. Deracemisation of 2-benzylcyclohexanone by formation of an inclusion complex with the TADDOL compound **35** has been described and the mechanism clarified by X-ray structure determination of the complex <02T3401>. A production process for the bis(phosphine oxide) **36** has been patented <02USP6472539>.

35 **36** **37** **38** **39** **40** **41**

42 R = CH_2OH
43 R = CO_2H

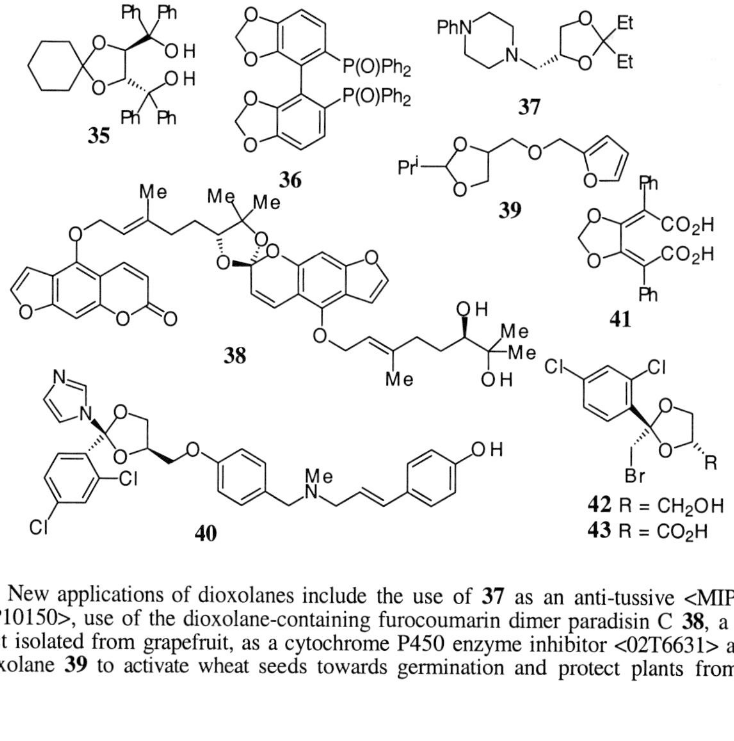

New applications of dioxolanes include the use of **37** as an anti-tussive <MIP10149, 02MIP10150>, use of the dioxolane-containing furocoumarin dimer paradisin C **38**, a natural product isolated from grapefruit, as a cytochrome P450 enzyme inhibitor <02T6631> and use of dioxolane **39** to activate wheat seeds towards germination and protect plants from water stress <00URP2152942>. Compounds such as **40** have fungicidal activity <02MIP90354> while a toxic natural product from a poisonous mushroom has been identified as **41** <02CC1384> and lipase-catalysed kinetic resolution has been used to obtain intermediates such as **42** and **43** important for ketoconazole synthesis <02TA2501>.

5.6.2 1,3-DITHIOLES AND DITHIOLANES

The synthesis and reactions of 1,3-dithioles and dithiolanes have been reviewed <01S1747>. Catalysts for the conversion of aldehydes and ketones into 1,3-dithiolanes with ethanedithiol include iodine <01JOC7527>, MoCl$_5$ <01PS(175)207> and indium triflate <02T7897>. Iodine and indium triflate are also effective in effecting transthioacetalisation of dioxolanes to give dithiolanes; indium chloride has also been used for this purpose <02SL727>. Conversion of aldehydes, but not ketones, into 1,3-dithiolanes is possible using *N*-bromosuccinimide <02SL474>. Benzyne generated by diazotisation of anthranilic acid, reacts with CS$_2$ in isoamyl alcohol giving benzodithiole **44** (77%) <02JCR(S)11>.

Treatment of 2-substituted 1,3-dithiolanes with *N*-bromosuccinimide and ethanediol results in conversion into the corresponding 1,3-dioxolanes <02T4513> and cleavage of 2-substituted 1,3-dithiolanes to give aldehydes can be achieved using thionyl chloride-treated silica in DMSO <02JOC2572>. Application of the same conditions to 2,2-disubstituted dithiolanes results in ring expansion to give dihydro-1,4-dithiins. An unusual method for sulfoxide reduction involves reaction with 1,3-dithiolane and a catalyst resulting in formation of formaldehyde from the sulfoxide oxygen and the 2-CH$_2$ of the dithiolane <02JOC2826>. The use of silicon chemistry in 2-functionalisation of 1,3-dithiolanes has been reported <02SL1447> and the synthesis of cyanoethyl dithiolethiones **45** and **46** has been optimised <02S2177>. The preparation and use in asymmetric synthesis of C$_2$-symmetric bis(sulfoxides) such as **47** has been reviewed <02EJO3507> and diastereoselective oxidation with Bu^{t}OOH and Cp$_2$TiCl$_2$ has been used to form the dithiolane monosulfoxide **48** <02S505>.

The presence of the dithiolane function in **49** plays a key role in allowing its photochemical [2+2] cyclisation to afford **50** <02CC736>. Theoretical and experimental studies on the structure of simple bis(dithiolanes) including the X-ray structure of **51** have appeared <02NJC1686>. Preparative biocatalytic hydrolysis and *S*-methylation has been reported for the thienodithiolethione **52** <02T2589>.

Work on tetrathiafulvalenes (TTFs) and related compounds has continued at a high level. TTF itself has been used for reduction of silver nitrate to produce silver dendritic nanostructures

<02CC1300> and donor/acceptor compounds such as **53** show second order non-linear optical properties <01JOC8872>. A variety of halogenated TTFs have been prepared including tetraiodo-TTF <01JMAC2181> and other simple substituted TTFs prepared include **54** <02JHC691>, **55** <02JOC3160> and the selenium/tellurium analogues **56** and **57** <01JMAC2431>. The doubly ^{13}C-labelled tetraselenafulvalene **58** has been prepared starting from $^{13}CH_2Cl_2$ <01MI1035>. The unexpected cycloaddition of a thiazolium ylide to tetramethoxycarbonyl-TTF has been observed giving product **59** <02TL3879>.

54

55 Het = 2-pyridyl, 2-pyrazinyl, 1,3-dioxolan-2-yl

56

58

59

57

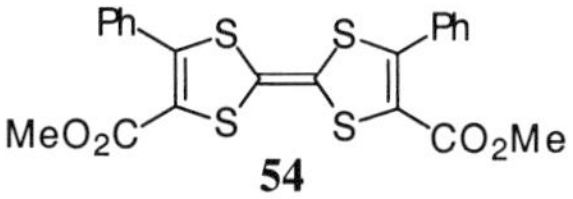

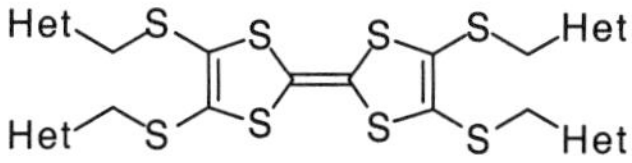

A novel metal-insulator phase transition has been observed in a salt of **60** <02JMAC2600>. Other ring-fused TTF donors whose properties have been examined include **61** <02JMAC1640>, **62** <02JMAC159>, **63** <02SM(130)129>, **64** <02SM(128)155>, **65** <02SM(128)273> and **66** <02SM(128)325>. Many studies have involved BEDT-TTF **67** and derivatives, including synthesis of chlorinated and fluorinated derivatives <01JCS(P1)3399>, formation of a variety of new crystalline salts <02SM(131)41, 02NJC490, 02MI1318> and synthesis of all six possible selenium analogues **68** <02JOC4218>.

60

61

62

63

64

65

66

67

68 X,Y,Z = S or Se

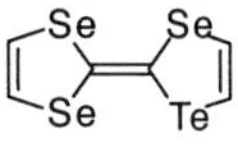

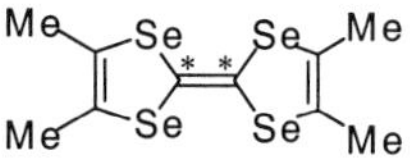

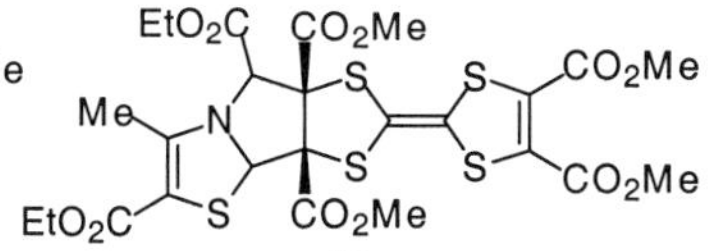

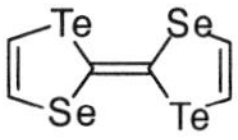

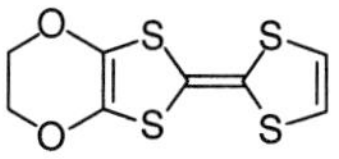

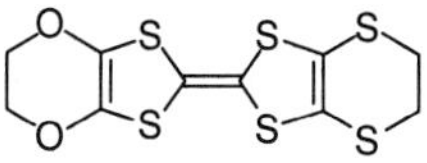

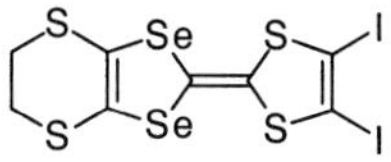

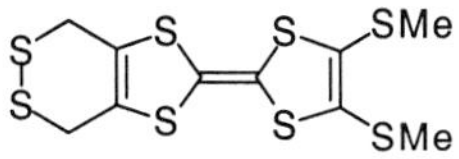

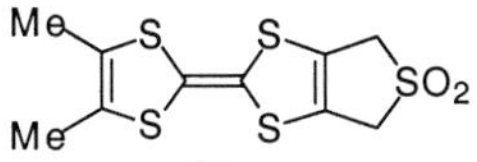

Other new donors of interest include **69** <02T1119>, the radicals **70** <02CL1048> and **71** <01TL7991> and the perpendicularly-fused TTF dimers **72** <02OL961>. Synthesis and properties of pyrazine-fused TTFs such as **73** <01ZN(B)297> and **74** <01ZN(B)963> have been reported and compounds **75** with a metal binding site have been prepared <02CL592>.

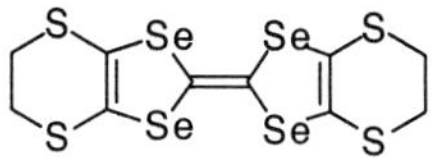

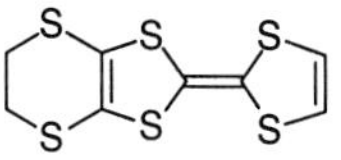

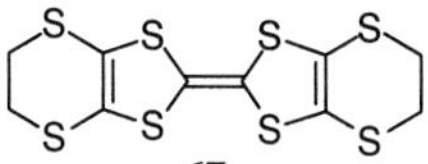

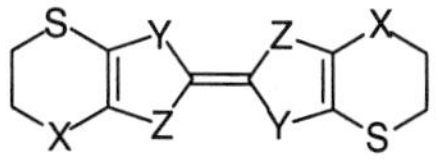

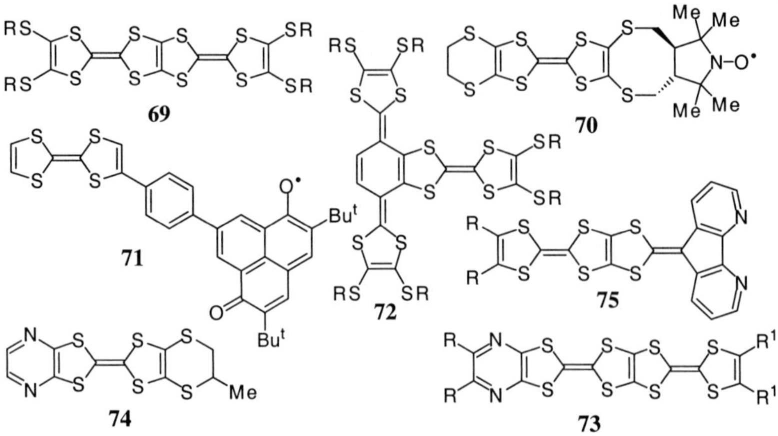

[Structures 69, 70, 71, 72, 73, 74, 75 shown]

A range of extended TTF derivatives have been described including examples with TTFs separated by an acetylenic linker <02CL590> and an aromatic or heteroaromatic ring <02CL1002> and tris- and tetrakis-TTFs have also been prepared <02SM(130)99>. Compounds reported with dithiole rings separated by a spacer include **76** <02CEJ3601>, **77** <01JCR(S)482>, **78** <02JMAC2696>, and **79** and **80** <02CEJ784>. The diselenole dialdehyde **81** has been prepared and used to make dendralenes such as **82** <01JOC7757> and the first dendralene-like TTF derivative containing a 1,3-ditellurole ring **83** has been reported <02OL2581>.

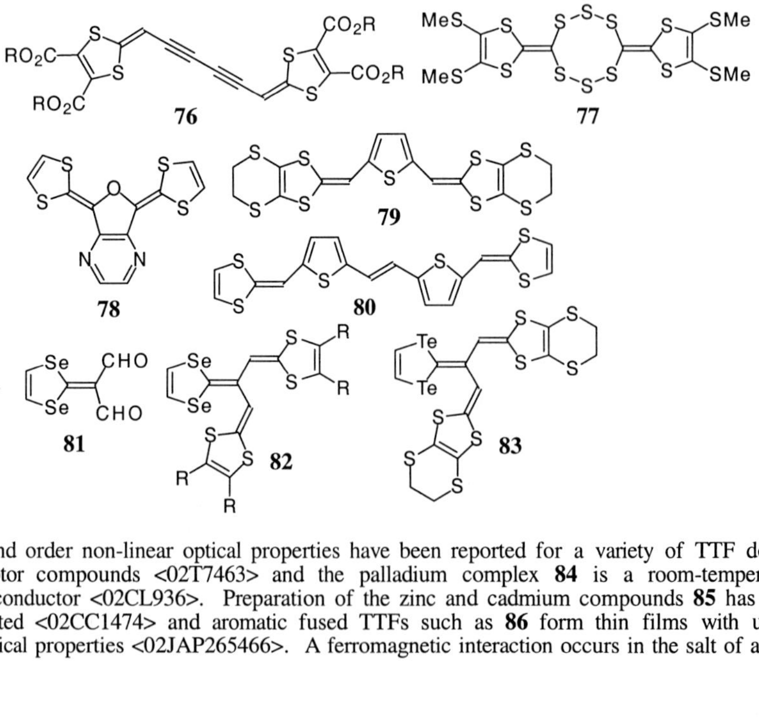

[Structures 76, 77, 78, 79, 80, 81, 82, 83 shown]

Second order non-linear optical properties have been reported for a variety of TTF donor-acceptor compounds <02T7463> and the palladium complex **84** is a room-temperature semiconductor <02CL936>. Preparation of the zinc and cadmium compounds **85** has been reported <02CC1474> and aromatic fused TTFs such as **86** form thin films with useful electrical properties <02JAP265466>. A ferromagnetic interaction occurs in the salt of a TTF

Five-Membered Ring Systems: With O & S (Se, Te) Atoms   255

cyclophane <02CL910>. TTF-containing dendrimers with up to 21 TTF units have been reported <02CC2950, 02JMAC27>. Self-assembled monolayers of redox-switchable TTF crown ether hybrids on a gold surface <02NJC1320> and a TTF molecular belt <02OL1327> have been reported. Several new complex structures containing both TTF and C_{60} units have been examined <02CC2968, 02JMAC2100, 02SM(126)263, 02SM(131)87>.

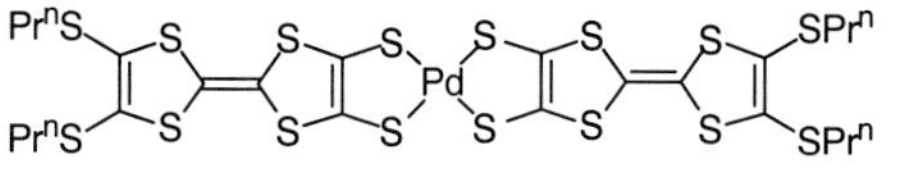

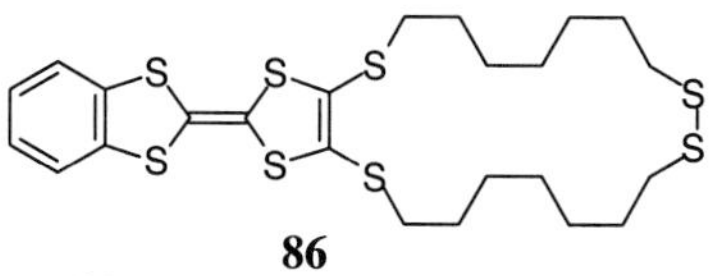

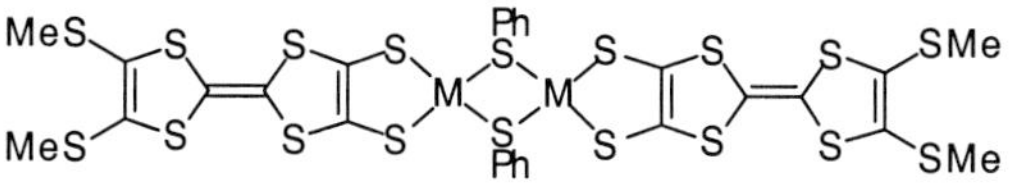

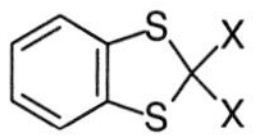

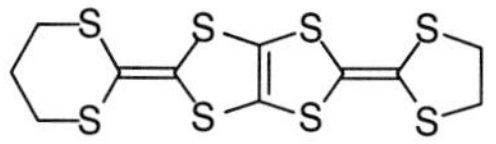

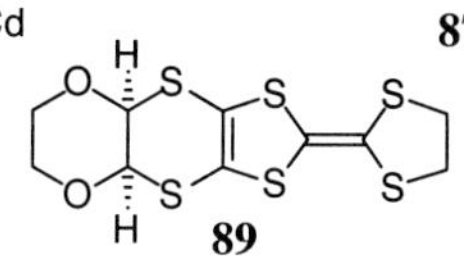

The silyl and stannyl benzodithioles **87** have been evaluated as donors <01BCJ1717> and new dihydro-TTF compounds include **88** which forms conducting salts which remain metallic down to 2K <02CC1118> and **89** whose salts are superconducting <02JA730>.

5.6.3 1,3-OXATHIOLES AND OXATHIOLANES

New catalysts for the formation of 1,3-oxathiolanes from aldehydes and ketones with mercaptoethanol include indium triflate <02SL1535>, *N*-bromosuccinimide <02TL6947>, tetrabutylammonium tribromide <02TL2843> and aminopropyl hydrochloride-functionalised silica which is effective for α,β-unsaturated ketones <02TI0455>. The reaction of α-ethoxyacrolein with mercaptoethanol takes a complex course resulting eventually in formation of bis(oxathiolane) **90** <01JOU1693> and *N*-bromosuccinimide catalyses reaction of 2-substituted 1,3-oxathiolanes with ethanediol to give the corresponding dioxolanes <02T4513>. A novel synthesis of the fluorinated oxathiolanones **91** has been reported <02SL996>.

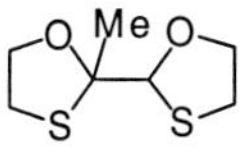

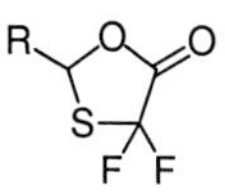

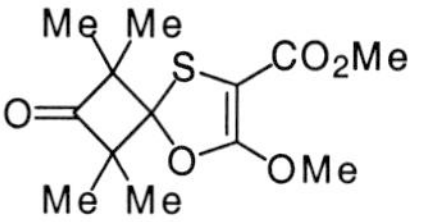

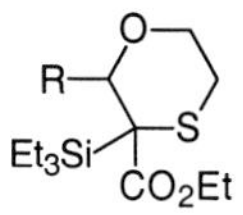

Other new oxathiolane syntheses include reaction of an epoxide with a stable thioketone <01HCA3319>. Rhodium catalysed reaction of dimethyl diazomalonate with a thioketone has been used to prepare oxathiole **92** <02PJC551> and the iodonium ylide **8** reacts with thioketones or carbon disulfide to form oxathioles <02TL5997>.

Cleavage of 2-substituted 1,3-oxathiolanes to give aldehydes and ketones can be achieved using V_2O_5 combined with H_2O_2 and NH_4Br in a two-phase system <02JCS(P1)1026> while treatment of 2-substituted oxathiolanes with $Et_3Si–C(=N_2)–CO_2Et$ results in ring expansion to give 1,4-oxathianes **93** <02CC346>. The use of 2-propenyl-, 2-(2-furyl)- and 2-phenyl-1,3-oxathiolanes as flavourings has been described <01MI57>.

5.6.4 1,2-DITHIOLES AND DITHIOLANES

The synthesis and reactions of 1,2-dithioles and dithiolanes have been reviewed <01S1747>. Treatment of the corresponding dithiols with bromine on hydrated silica has been used to prepare 1,2-dithiolanes **94** and **95** <02TL6271>. The behaviour of all ten possible naphthodichalcogenoles **96** as strong carbon bases in the gas phase has been reported <02NJC1747>. The synthesis and reactions of fluorinated 1,2-dithiole-3-thiones **97** and **98** have been described <02TL5809>. The regioselectivity of reaction of aminodithiolethiones **99** with electrophiles has been examined <02MI55>. The cycloaddition reactions of compound **100**, formed by reaction of Pr^iNEt_2 with S_2Cl_2, with acetylenic compounds has been examined <02T9785, 02JOC6439> and by altering the ratio of components in the reaction which forms **100** the alternative products **101** and **102** can be isolated <01MC165>.

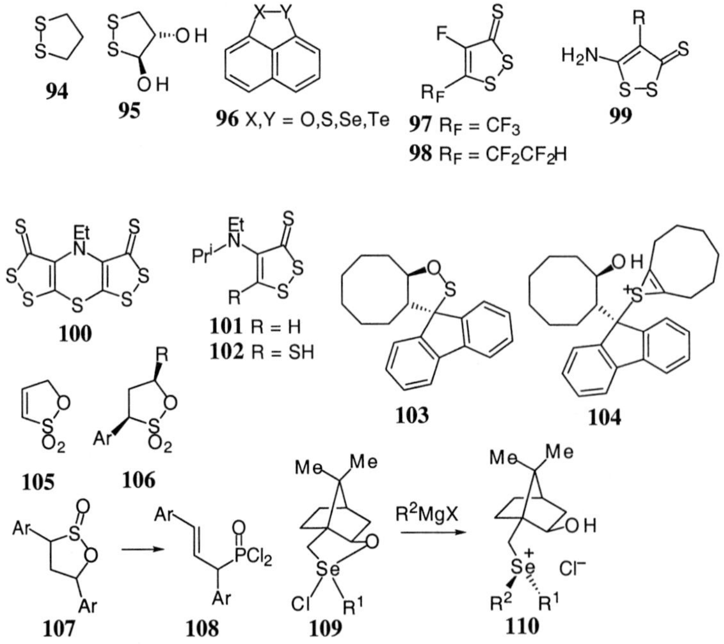

5.6.5 1,2-OXATHIOLES AND OXATHIOLANES

The unusual 1,2-oxathiolane **103** is an effective reagent for sulfur transfer onto alkenes to form thiiranes <02OL599> and it reacts with cyclooctyne in the presence of TFA to give the trifluoroacetate of the thiirenium ion **104** <02JA8316>. Ring-closing metathesis of allyl vinylsulfonate using a second-generation Grubbs catalyst gives the oxathiole S,S-dioxide **105** <02SL2019>. A series of chiral γ-sultones **106** have been prepared using a carbohydrate-based chiral auxiliary <02SL1727> and reaction of the sultine **107** with PCl_3 results in ring-opening to give **108** <01CHE649>. The 1,2-oxaselenolane **109** reacts with Grignard reagents as shown to give chiral selenonium salts **110** <01SC2441>.

5.6.6 THREE HETEROATOMS

A study of the effect of steric hindrance on the direction of cleavage of 1,2,3-trioxolanes in ozonolysis of highly hindered alkenes has appeared <02T891> and ozonolysis of cyclic dienes in the presence of carbonyl compounds results in formation of cross ozonides <02MI423>. The products from photosensitised oxidation of tetraarylbutadiene monoepoxides have been identified as 1,2,4-trioxolanes **111** rather than trioxepines as previously reported <01TL9203>. Treatment of either 6-ring or 5-ring monosulfoxides of the benzotrithiole **112** with sulfuric acid results in formation of a dication in solution <02CL540>. The first tellurium-containing trichalcogenoles **113** have been prepared and X-ray structures determined for X = S and Y = Se and for X = Y = Se <02CC1918>.

111

112

113 X,Y = S,Se

5.6.7 REFERENCES

00URP2152942	T.G. Dedikova, L.A. Badovskaya, N.I. Nen'ko, T.P. Kosulina, *Russ. Pat.* 2,152,942 (**2000**) [*Chem. Abstr.* **2002**, *136*, 294819].
01BCJ1717	H. Li, K. Nishiwaki, K. Itami, J. Yoshida, *Bull. Chem. Soc. Jpn.* **2001**, *74*, 1717.
01CHE649	E.V. Grigor'ev, L.G. Saginova, *Chem. Heterocycl. Compd. (Engl. Transl.)* **2001**, *37*, 649 [*Chem. Abstr.* **2002**, *136*, 216798].
01HCA3319	C. Fu, A. Linden, H. Heimgartner, *Helv. Chim. Acta* **2001**, *84*, 3319.
01JCR(S)289	T.-S. Jin, S.-L. Zhang, X.-F. Wong, T.-J. Guo, T.-S. Li, *J. Chem. Res. (S)* **2001**, 289 [*Chem. Abstr.* **2002**, *136*, 37540].
01JCR(S)365	S. Tangestaninejad, M.H. Habibi, V. Mirkhani, M. Moghadam, *J. Chem. Res. (S)* **2001**, 365 [*Chem. Abstr.* **2002**, *136*, 200123].
01JCR(S)482	J. Dai, M.-Y. Zhou, G.-Q. Bian, X. Wang, Q.-F. Xu, M. Munakata, M. Maekawa, *J. Chem. Res. (S)* **2001**, 482.
01JCS(P1)3399	O.J. Dautel, M. Fourmigué, *J. Chem. Soc., Perkin Trans. 1* **2001**, 3399.
01JFC(111)129	D. Prescher, J. Schulze, B. Costisella, K. Seppelt, *J. Fluorine Chem.* **2001**, *111*, 129.
01JGU542	Kh. A. Kerimov, *Russ. J. Gen. Chem. (Engl. Transl.)* **2001**, *71*, 542 [*Chem. Abstr.* **2002**, *136*, 102309].
01JMAC2181	A.S. Batsanov, M.R. Bryce, A. Chesney, J.A.K. Howard, D.E. John, A.J. Moore, C.L. Wood, H. Gershtenman, J.Y. Becker, V.Y. Khodorkovsky, A. Ellern, J. Bernstein, I.F. Perepichka, V. Rotello, M. Gray, A. O. Cuello, *J. Mater. Chem.* **2001**, *11*, 2181.
01JMAC2431	A. Morikami, K. Takimiya, Y. Aso, T. Otsubo, *J. Mater. Chem.* **2001**, *11*, 2431.
01JOC7527	H. Firouzabadi, N. Iranpoor, H. Hazarkhani, *J. Org. Chem.* **2001**, *66*, 7527.
01JOC7757	R.R. Amaresh, D. Liu, T. Konovalova, M.V. Lakshmikantham, M.P. Cava and L.D. Kispert, *J. Org. Chem.* **2001**, *66*, 7757.
01JOC8872	M. González, J.L. Segura, C. Seoane, N. Martín, J. Garín, J. Orduna, R. Alcalá, B. Villacampa, V. Hernández, J. T. López-Navarrete, *J. Org. Chem.* **2001**, *66*, 8872.
01JOU1693	N.A. Keiko, E.A. Funtikova, L.G. Stepanova, Yu. A. Chuvashev, A.I. Albanov, M.G. Voronkov, *Russ. J. Org. Chem.* **2001**, *37*, 1693 [*Chem. Abstr.* **2002**, *137*, 109224].
01MC165	L.S. Konstantinova, O.A. Rakitin, C.W. Rees, *Mendeleev Commun.* **2001**, 165 [*Chem. Abstr.* **2002**, *136*, 294754].

01MI57	Y. Yang, F. Zheng, B. Sun, F. Ding, Y. Liu, Y. Ren, *Chemical Journal on Internet* **2001**, *3*, 57 [http://www.chemistrymag.org/cji/2001/03b057nc.htm] [*Chem. Abstr.*, **2002**, *137*, 93705].
01MI205	M.H. Habibi, S. Tangestaninejad, V. Mirkhani, B. Yadollahi, *Catalysis Lett.* **2001**, *75*, 205 [*Chem. Abstr.* **2002**, *136*, 200124].
01MI208	J. Wang, *Huaxue Yanjiu Yu Yingyong* **2001**, *13*, 208 [*Chem. Abstr.* **2002**, *136*, 263114].
01MI1035	J.-B. Christensen, K. Bechgaard, G. Paquignon, *J. Labelled Compd. Radiopharm.* **2001**, *44*, 1035 [*Chem. Abstr.* **2002**, *136*, 232253].
01PS(175)207	H. Firouzabadi, B. Karimi, *Phosphorus, Sulfur, Silicon Relat. Elem.* **2001**, *175*, 207 [*Chem. Abstr.* **2002**, *136*, 232217].
01S1747	G.H. Elgemeie, S.H. Sayed, *Synthesis* **2001**, 1747.
01SC2441	J. Zhang, S. Saito, T. Koizumi, *Synth. Commun.* **2001**, *31*, 2441.
01SC3323	G.V. Salmoria, A. Neves, E.L. Dall'Oglio, C. Zucco, *Synth. Commun.* **2001**, *31*, 3323.
01SC3411	I. Mohammadpoor-Baltork, A.R. Khospour, H. Aliyan, *Synth. Commun.* **2001**, *31*, 3411.
01TA3077	F. Busqué, P. de March, M. Figueredo, J. Font, S. Rodríguez, *Tetrahedron Asymmetry* **2001**, *12*, 3077.
01TL7991	Y. Morita, J. Kawai, N. Haneda, S. Nishida, K. Fukui, S. Nakazawa, D. Shiomi, K. Sato, T. Takui, T. Kawakami, K. Yamaguchi, K. Nakasuji, *Tetrahedron Lett.* **2001**, *42*, 7991.
01TL9203	M. Kamata, K. Komatsu, R. Akaba, *Tetrahedron Lett.* **2001**, *42*, 9203.
01ZN(B)297	G.C. Papavassiliou, Y. Misaki, K. Tokahashi, J. Yamada, G.A. Mousdis, T. Sharahata, T. Ise, *Z. Naturforsch., Teil B* **2001**, *56*, 297.
01ZN(B)963	G.C. Papavassiliou, A. Terzis, C.P. Raptopoulou, *Z. Naturforsch., Teil B* **2001**, *56*, 963.
02CC274	H. Yang, Y. Gu, Y. Deng, F. Shi, *Chem. Commun.* **2002**, 274.
02CC346	M. Ioannou, M.J. Porter, F. Saez, *Chem. Commun.* **2002**, 346.
02CC736	B.K. Joseph, B. Verghese, C. Sudharsanakumar, S. Deepa, D. Viswam, P. Chandran, C.V. Asokan, *Chem. Commun.* **2002**, 736.
02CC1118	J. Yamada, M. Watanabe, T. Toita, H. Akutsu, S. Nakatsuji, H. Nishikawa, I. Ikemoto, K. Kikuchi, *Chem. Commun.* **2002**, 1118.
02CC1300	X. Wang, K. Naka, H. Itoh, S. Park, Y. Chujo, *Chem. Commun.* **2002**, 1300.
02CC1384	Y. Sano, K. Sayama, Y. Arimoto, T. Inakuma, K. Kobayashi, H. Koshino, H. Kawagishi, *Chem. Commun.* **2002**, 1384.
02CC1474	G.-Q. Bian, J. Dai, G.-Y. Zhu, W. Yang, Z.-M. Yan, M. Munakata, M. Maekawa, *Chem. Commun.* **2002**, 1474.
02CC1918	S. Ogawa, Y. Soshimura, N. Nagahora, Y. Kawai, Y. Mitaki, R. Sato, *Chem. Commun.* **2002**, 1918.
02CC2950	A. Beeby, M.R. Bryce, C.A. Christensen, G. Cooke, F.M.A. Duclairoir, V.M. Rotello, *Chem. Commun.* **2002**, 2950.
02CC2968	M.A. Herranz, N. Martín, J. Ramey, D.M. Guldi, *Chem. Commun.* **2002**, 2968.
02CEJ784	P. Frère, M. Allain, E.H. Elandaloussi, E. Levillain, F.-X. Sauvage, A. Riou, J. Roncali, *Chem. Eur. J.* **2002**, *8*, 784.
02CEJ3601	M.B. Nielsen, N.F. Utesch, N.N.P. Moonen, C. Boudon, J.-P. Gisselbrecht, S. Concilio, S. P. Piotto, P. Seiler, P. Günter, M. Gross, F. Diederich, *Chem. Eur. J.* **2002**, *8*, 3601.
02CL540	T. Kimura, S. Ito, T. Sasaki, S. Niizuma, S. Ogawa, R. Sato, Y. Kawai, *Chem. Lett.* **2002**, 540.
02CL590	M. Iyoda, M. Hasegawa, J. Takano, K. Hara, Y. Kuwatani, *Chem. Lett.* **2002**, 590.
02CL592	K. Sako, Y. Misaki, M. Fujiwara, T. Maitani, K. Tanaka, H. Tatemitsu, *Chem. Lett.* **2002**, 592.
02CL910	A. Izuoka, J. Tanabe, T. Sugawara, T. Kudo, T. Saito, Y. Kawada, *Chem. Lett.* **2002**, 910.
02CL936	W. Suzuki, E. Fujiwara, A. Kobayashi, A. Hasegawa, T. Miyamoto, H. Kobayashi, *Chem. Lett.* **2002**, 936.
02CL1002	K. Takahashi, H. Tanioka, H. Fueno, Y. Misaki, K. Tanaka, *Chem. Lett.* **2002**, 1002.
02CL1048	H. Fujiwara, E. Fujiwara, H. Kobayashi, *Chem. Lett.* **2002**, 1048.
02EJO3507	B. Delouvrié, L. Fensterbank, F. Nájera, M. Malacria, *Eur. J. Org. Chem.* **2002**, 3507.
02JA730	H. Nishikawa, T. Morimoto, T. Kodama, I. Ikemoto, K. Kikuchi, J. Yamada, H. Yoshino, K. Murata, *J. Am. Chem. Soc.* **2002**, *124*, 730.
02JA8316	W. Adam, S.G. Bosio, B. Fröhling, D. Leusser, D. Stalke, *J. Am. Chem. Soc.* **2002**, *124*, 8316.
02JA11260	B. Plesnicar, J. Cerkovnik, T. Tuttle, E. Kraka, D. Cremer, *J. Am. Chem. Soc.* **2002**, *124*, 11260.
02JAP47229	Y. Masaki, *Jpn. Pat.* 47,229 (**2002**) [*Chem. Abstr.* **2002**, *136*, 167365].
02JAP53573	Y. Ikushima, H. Kawanami, K. Torii, *Jpn. Pat.* 53,573 (**2002**) [*Chem. Abstr.* **2002**, *136*, 167366].

02JAP69068	T. Imanaka, T. Tanaka, H. Tahara, H. Nagumo, *Jpn. Pat.* 69,068 (**2002**) [*Chem. Abstr.* **2002**, *136*, 232285].
02JAP69072	M. Muto, Y. Tani, N. Ota, T. Akiba, *Jpn. Pat.* 69,072 (**2002**) [*Chem. Abstr.* **2002**, *136*, 232286].
02JAP105071	N. Hirayama, *Jpn. Pat.* 105,071 (**2002**) [*Chem. Abstr.* **2002**, *136*, 294820].
02JAP265466	S. Ogawa, K. Ogura, N. Yoshimoto, Y. Hiroi, *Jpn. Pat.* 265,466 (**2002**) [*Chem. Abstr.* **2002**, *137*, 247701].
02JCR(S)11	A.E.W. Sarhan, T. Izumi, *J. Chem. Res. (S)* **2002**, 11.
02JCS(P1)1026	E. Mondal, P.R. Sahu, G. Bose, A.T. Khan, *J. Chem. Soc., Perkin Trans. 1* **2002**, 1026.
02JHC691	A.A.O. Sarhan, M. Murakami, T. Izumi, *J. Heterocycl. Chem.* **2002**, *39*, 691.
02JMAC27	N. Godbert, M. R. Bryce, *J. Mater. Chem.* **2002**, *12*, 27.
02JMAC159	T. Imakubo, N. Tajima, M. Tamura, R. Kato, Y. Nishio, K. Kajita, *J. Mater. Chem.* **2002**, *12*, 159.
02JMAC1640	G. Saito, H. Sasaki, T. Aoki, Y. Yoshida, A. Otsuka, H. Yamochi, O. O. Drozdova, K. Yakuchi, H. Kitagawa, T. Mitani, *J. Mater. Chem.* **2002**, *12*, 1640.
02JMAC2100	G. Kodis, P.A. Liddell, L. de la Garza, A.L. Moore, T.A. Moore, D. Gust, *J. Mater. Chem.* **2002**, *12*, 2100.
02JMAC2600	A. Ota, H. Yamaochi, G. Saito, *J. Mater. Chem.* **2002**, *12*, 2600.
02JMAC2696	M. Uruichi, K. Yakuchi, T. Shirahata, K. Takahashi, T. Mori, T. Nakamura, *J. Mater. Chem.* **2002**, *12*, 2696.
02JOC1678	S. Cicchi, M. Corsi, A. Brandi, A. Goti, *J. Org. Chem.* **2002**, *67*, 1678.
02JOC2572	H. Firouzabadi, N. Iranpoor, H. Hazarkhani, B. Karimi, *J. Org. Chem.* **2002**, *67*, 2572.
02JOC2826	N. Iranpoor, H. Firouzabadi, H. R. Shaterian, *J. Org. Chem.* **2002**, *67*, 2826.
02JOC3160	S.-X. Liu, S. Dolder, M. Pilkington, S. Decurtins, *J. Org. Chem.* **2002**, *67*, 3160.
02JOC4218	K. Takimiya, T. Jigami, M. Kawashima, M. Kodani, Y. Aso, T. Otsubo, *J. Org. Chem.* **2002**, *67*, 4218.
02JOC5170	A. Iwata, H. Tang, A. Kunoi, J. Ohshita, Y. Yamamoto, C. Matui, *J. Org. Chem.* **2002**, *67*, 5170.
02JOC6439	S. Barriga, P. Fuertes, C.F. Marcos, O. A. Rakitin, T. Torroba, *J. Org. Chem.* **2002**, *67*, 6439.
02JOU104	B.A. Shainyan, M.V. Ustinov, V.K. Bel'skii, L.O. Nindakova, *Russ. J. Org. Chem.* **2002**, *38*, 104 [*Chem. Abstr.* **2002**, *137*, 185439].
02MI55	R. Cmelik, J. Marek, P. Pazdera, *Heterocycl. Commun.* **2002**, *8*, 55 [*Chem. Abstr.* **2002**, *137*, 247628]
02MI423	S.H. Park, T.S. Huh, *Bull. Korean Chem. Soc.* **2002**, *23*, 423 [*Chem. Abstr.* **2002**, *137*, 63210].
02MI749	S.-M. Kim, H. Jin, J.-G. Jun, *Bull. Korean Chem. Soc.* **2002**, *23*, 749.
02MI1318	Z. Lin, Q. Fang, W.-T. Yu, G. Xue, M.-H. Jiang, B. Zhang, J.-B. Zhang, D. B. Zhu, *Huaxue Xuebao* **2002**, *60*, 1318 [*Chem. Abstr.* **2002**, *137*, 337842].
02MIP10149	M. Allegreti, M.C. Cesta, R. Curti, L. Pellegrini, G. Melillo, *PCT Int. Appl.* WO 10,149 (**2002**) [*Chem. Abstr.* **2002**, *136*, 167390].
02MIP10150	M. Allegreti, M.C. Cesta, R. Curti, L. Nicolini, *PCT Int. Appl.* WO 10,150 (**2002**) [*Chem. Abstr.* **2002**, *136*, 167391].
02MIP85893	J.S. Kang, S.-K. Chang, K.-M. Seol, M.-K. Kim, *PCT Int. Appl.* WO 85,893 (**2002**) [*Chem. Abstr.* **2002**, *137*, 337874].
02MIP90354	D. Babin, J. Weston, *PCT Int. Appl.* WO 90,354 (**2002**) [*Chem. Abstr.* **2002**, *137*, 353025].
02NJC490	C. Jia, D. Zhang, C.-M. Liu, W. Xu, H. Hu, D. Zhu, *New J. Chem.* **2002**, *26*, 490.
02NJC1320	G. Trippé, M. Oçafrain, M. Besbes, V. Monroche, J. Lyskawa, F. Le Derf, M. Sallé, J. Becher, B. Colonna, L. Echegoyen, *New J. Chem.* **2002**, *26*, 1320.
02NJC1686	W. Chen, Y.-L. Lam, M.W. Wong, H.H. Huang, E. Liang, *New J. Chem.* **2002**, *26*, 1686.
02NJC1747	P. Sanz, O. Mó, *New J. Chem.* **2002**, *26*, 1747.
02OL599	W. Adam, B. Fröhling, *Org. Lett.* **2002**, *4*, 599.
02OL961	N. Gautier, N. Gallego-Planas, N. Mercier, E. Levillain, P. Hudhomme, *Org. Lett.* **2002**, *4*, 961.
02OL1327	K. Nielsen, J.O. Jeppesen, N. Thorup, J. Becher, *Org. Lett.* **2002**, *4*, 1327.
02OL2453	B. Jiang, X. Zhang, Z. Luo, *Org. Lett.* **2002**, *4*, 2453.
02OL2561	V. Calo, A. Nacci, A. Monopoli, A. Fanizzi, *Org. Lett.* **2002**, *4*, 2561.
02OL2581	D. Rajagopal, M.V. Lakshmikantham, M.P. Cava, *Org. Lett.* **2002**, *4*, 2581.
02PJC551	G. Mloston, J. Romanski, H. Heimgartner, *Pol. J. Chem.* **2002**, *76*, 551.
02S505	G. Della Sala, S. Labano, A. Lattanzi, C. Tedesco, A. Scettri, *Synthesis* **2002**, 505.
02S2177	C. Jia, D. Zhang, X. Guo, S. Wan, W. Xu, D. Zhu, *Synthesis* **2002**, 2177.
02SC449	R. J. Petroski, *Synth. Commun.* **2002**, *32*, 449.

02SL474	A. Kamal, G. Chouhan, *Synlett* **2002**, 474.
02SL727	B.C. Ranu, A. Das, S. Samanta, *Synlett* **2002**, 727.
02SL996	S. Gouault, J.-C. Pommelet, T. Lequeux, *Synlett* **2002**, 996.
02SL1111	M. Baruah, M. Bols, *Synlett* **2002**, 1111.
02SL1435	R.M. Suárez, J. P. Sestelo, L. A. Sarandeses, *Synlett* **2002**, 1435.
02SL1447	A. Capperucci, V. Cerè, A. Degl'Innocenti, T. Nocentini, S. Pollicino, *Synlett* **2002**, 1447.
02SL1535	K. Kazahaya, N. Hamada, S. Ito, T. Sato, *Synlett* **2002**, 1535.
02SL1727	D. Enders, W. Harnying, N. Vignola, *Synlett* **2002**, 1727.
02SL2019	S. Karsch, P. Schwab, P. Metz, *Synlett* **2002**, 2019.
02SM(126)263	I. Olejniczak, A. Graja, A. Bogucki, M. Golub, P. Hudhomme, A. Gorgues, D. Kreher, M. Cariou, *Synth. Met.* **2002**, *126*, 263.
02SM(128)155	M. Mas-Torrent, J. Llacay, K. Wurst, V. Laukhin, J. Vidal-Gancedo, J. Veciana, C. Rovira, *Synth. Met.* **2002**, *128*, 155 [*Chem. Abstr.* **2002**, *137*, 262986].
02SM(128)273	V. Gritsenko, E. Fujiwara, H. Fujiwara, H. Kobayashi, *Synth. Met.* **2002**, *128*, 273.
02SM(128)325	L.V. Zorina, M. Gener, S.S. Khasanov, R.P. Shibaeva, E. Canadell, L.A. Kushch, E.B. Yagubskii, *Synth. Met.* **2002**, *128*, 325.
02SM(130)99	C. Carcel, J.-M. Fabre, *Synth. Met.* **2002**, *130*, 99.
02SM(130)129	S. Le Moustarder, N. Mercier, P. Hudhomme, N. Gallego-Planas, A. Gorgues, A. Riou, *Synth. Met.* **2002**, *130*, 129.
02SM(131)41	S.S. Khasanov, L.V. Zorina, R.P. Shibaeva, S.I. Pesotskii, M.V. Kartsovnik, L.F. Veiros, E. Canadell, *Synth. Met.* **2002**, *131*, 41.
02SM(131)87	D.V. Konarev, I.S. Neretin, Yu. L. Slovokhotov, A.L. Litvinov, A. Otsuka, R.N. Lyubovskaya, G. Saito, *Synth. Met.* **2002**, *131*, 87.
02T891	S. Kawamura, H. Yamakoshi, M. Abe, A. Masuyama, M. Nojima, *Tetrahedron* **2002**, *58*, 891.
02T1119	S. Kimura, H. Kurai, T. Mori, *Tetrahedron* **2002**, *58*, 1119.
02T2589	W. Kroutil, A.A. Stämpfli, R. Dahinden, M. Jörg, U. Müller, J.P. Pachlatko, *Tetrahedron* **2002**, *58*, 2589.
02T3401	H. Kaku, S. Takaoka, T. Tsunoda, *Tetrahedron* **2002**, *58*, 3401.
02T4513	B. Karimi, H. Seradj, J. Maleki, *Tetrahedron* **2002**, *58*, 4513.
02T6631	T. Ohta, T. Maruyama, M. Nagahashi, Y. Miyamoto, S. Hosoi, F. Kiuchi, Y. Yamazoe, S. Tsukamoto, *Tetrahedron* **2002**, *58*, 6631.
02T7065	S. Kanoh, T. Nishimura, M. Naka, M. Motoi, *Tetrahedron* **2002**, *58*, 7065.
02T7463	M. Otero, M.A. Herranz, C. Seoane, N. Martín, J. Garín, J. Orduna, R. Alcalá, B. Villacampa, *Tetrahedron* **2002**, *58*, 7463.
02T7897	S. Muthusamy, S.A. Babu, C. Gunanathan, *Tetrahedron* **2002**, *58*, 7897.
02T9785	S. Barriga, C.F. Marcos, O. Riant, T. Torroba, *Tetrahedron* **2002**, *58*, 9785.
02T9865	A. Guarna, I. Bucelli, F. Machetti, G. Menchi, E.G. Occhiato, D. Scarpi, A. Trabocchi, *Tetrahedron* **2002**, *58*, 9865.
02T10081	P. Camps, X. Pujol, S. Vázquez, *Tetrahedron* **2002**, *58*, 10081.
02T10455	S. Kerverdo, L. Lizzani-Cuvelier, E. Duñach, *Tetrahedron* **2002**, *58*, 10455.
02TA155	S. Jew, D.-Y. Lim, J.-Y. Kim, S. Kim, E. Roh, H.-J. Yi, J.-M. Ku, B. Park, B. Jeong, H. Park, *Tetrahedron Asymmetry* **2002**, *13*, 155.
02TA2501	Y.H. Kim, C.S. Cheong, S.H. Lee, S.J. Jun, K.S. Kim, H.-S. Cho, *Tetrahedron Asymmetry* **2002**, *13*, 2501.
02TL815	S. Halila, M. Benazza, G. Demailly, *Tetrahedron Lett.* **2002**, *43*, 815.
02TL1161	C.-Q. Sun, P.T.W. Cheng, J. Stevenson, T. Dejneka, B. Brown, T.C. Wang, J.A. Robl, M.A. Poss, *Tetrahedron Lett.* **2002**, *43*, 1161.
02TL1503	M. Hasegawa, H. Ishii, T. Fuchigami, *Tetrahedron Lett.* **2002**, *43*, 1503.
02TL2843	E. Mondal, P.R. Sahu, G. Bose, A.T. Khan, *Tetrahedron Lett.* **2002**, *43*, 2843.
02TL3879	R. Toplak, P. Benard-Rocherullé, D. Lorcy, *Tetrahedron Lett.* **2002**, *43*, 3879.
02TL5809	V.M. Timoshenko, J.-P. Bouillon, Yu.G. Shermolovich, C. Portella, *Tetrahedron Lett.* **2002**, *43*, 5809.
02TL5997	C. Batsila, G. Kostakis, L.P. Hadjiarapoglou, *Tetrahedron Lett.* **2002**, *43*, 5997.
02TL6271	M.H. Ali, M. McDermott, *Tetrahedron Lett.* **2002**, *43*, 6271.
02TL6947	A. Kamal, G. Chouhan, K. Ahmed, *Tetrahedron Lett.* **2002**, *43*, 6947.
02USP13477	H.S. Kim, J.J. Kim, B.G. Lee, H.G. Kim, *U.S. Pat. Appl.* 13,477 (**2002**) [*Chem. Abstr.* **2002**, *136*, 134746].
02USP6472539	T. Yokozawa, T. Saito, N. Sayo, T. Ishizaki, *U.S. Pat.* 6,472,539 (**2002**) [*Chem. Abstr.* **2002**, *137*, 325510].

Chapter 5.7

Five-Membered Ring Systems: With O & N Atoms

Stefano Cicchi, Franca M. Cordero, Donatella Giomi
Università di Firenze, Italy
donatella.giomi@unifi.it

5.7.1 ISOXAZOLES

The interest in this heterocyclic system is still growing and from a synthetic viewpoint significant improvements have been obtained through the applications of new techniques to classical synthetic approaches. In particular, 1,3-dipolar cycloadditions (1,3-DC) of nitrile oxides continue to be one of the main strategies for the synthesis of isoxazoles and isoxazolines and the first soluble polymer-supported synthesis of structurally different derivatives has been described: a soluble poly(ethylene glycol) (PEG)-supported alkyne **1** was allowed to react with nitrile oxides, generated *in situ* from chlorooximes **2**, affording isoxazoles **4** in good yield and purity after cleavage from the resin <02S1663>.

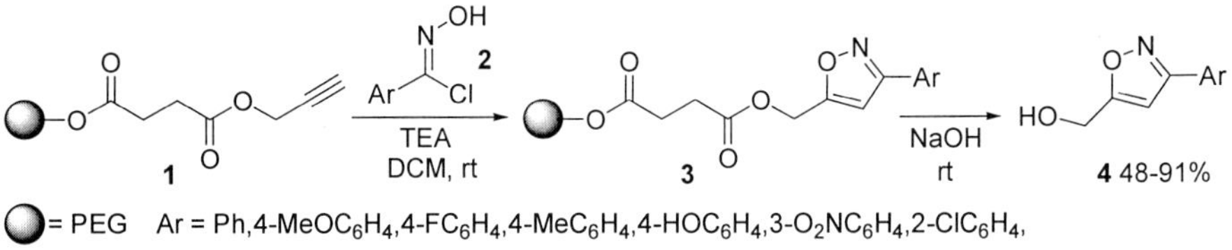

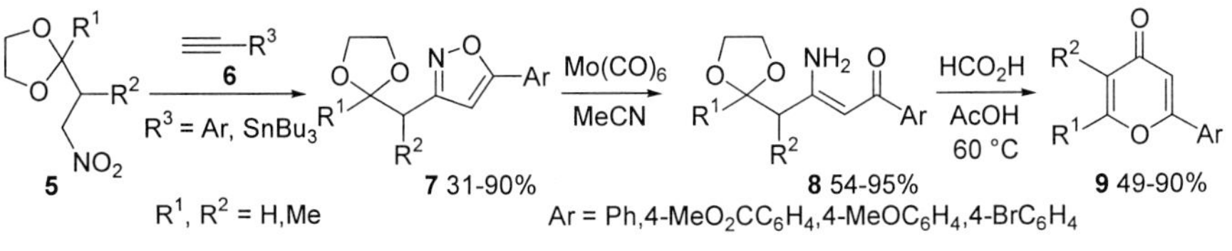

3-Acetyl- and 3-benzoylisoxazoles have been conveniently obtained by one-pot reactions of alkynes with ammonium cerium(IV) nitrate or ammonium cerium(III) nitrate tetrahydrate in acetone or acetophenone; these processes probably involve 1,3-DC of nitrile oxides formed by nitration of the carbonyl compound by cerium salts <02TL7035>.

1,3-DC of acetylenes **6** and nitrile oxides, generated *in situ* from nitro compounds **5**, afforded isoxazoles **7** directly (R^3 = Ar) or through Stille coupling with aryl iodides of the 5-

stannyl isoxazole precursors (R^3 = SnBu$_3$); reductive ring-opening of **7** with Mo(CO)$_6$ furnished enamino ketones **8**, converted with aqueous formic acid into substituted pyran-4-ones **9** <02TL3565>.

Alkynyliodonium salts easily react with 2,4,6-trimethylbenzonitrile oxide to give the corresponding (isoxazol-4-yl)iodonium salts <02JOM196>. 3-Aryl-5-alkylisoxazoles have been synthesised in high yields by regioselective 1,3-DC of arylnitrile oxides with free enolate ions regioselectively obtained by metallation of alkyl methyl ketones with LDA <02T2659>.

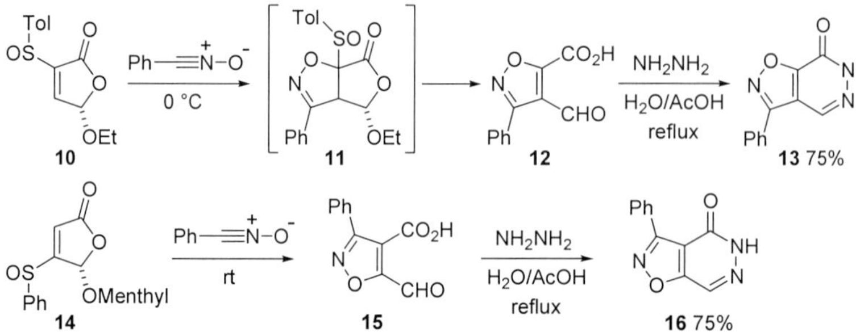

Completely regioselective additions of benzonitrile oxide to arylsulfinyl-5-alkoxyfuran-2(5H)-ones **10** and **14** gave rise to regioisomeric 4,5-difunctionalised isoxazoles **12** and **15**, after spontaneous transformations of the primary adducts through desulfinylation and opening of the lactone ring; the subsequent condensation with hydrazine yielded isoxazolopyridazinones **13** and **16**, which were more easily obtained in 75% overall yield in a one-pot two-step synthetic sequence <02SL73>.

Novel antiviral nucleoside analogues **19**, consisting of isoxazole rings as modified sugars and nucleobases (thymine, uracil, and 5-fluorouracil) joined by a methylene linker, have been synthesised in satisfactory yields: N-propargyl pyrimidines **17** were allowed to react with N-Boc amino aldoximes **18** in the presence of a commercial bleaching agent (containing 4% NaOCl) for the generation of nitrile oxides <02BMCL1395>.

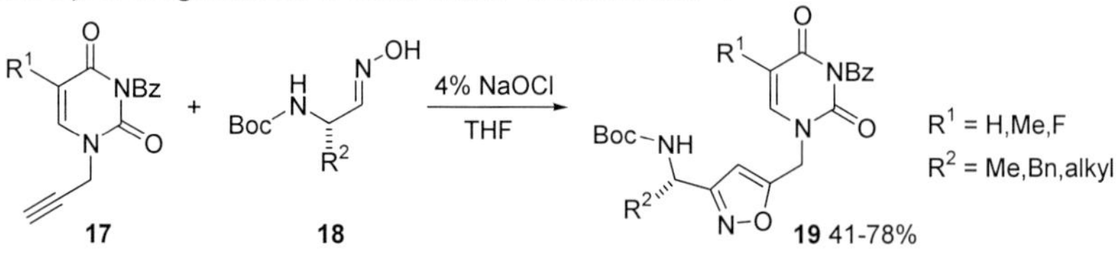

Nucleophilic attack of hydroxylamine at position 6 of 2H-pyran-2-ones **20** afforded stereoselectively, through pyran ring-opening and subsequent cyclisation, (β-isoxazol-4-yl)-α,β-didehydroamino acids **21** mainly as (Z)-isomers <02JCS(P1)675>. In a similar way, a short glycal-mediated synthesis of new enantiomerically pure 5-substituted isoxazoles **23a,b** has been accomplished by treatment of enones **22a,b**, derived from D-galactal and D-glucal, with hydroxylamine and subsequent dehydration of intermediate epimeric isoxazolines; the final products were selectively converted into derivatives with different O-protections in the glycerol side-chain <02TL4613>.

An analogous ring-ring transformation was observed in the reaction of 2-methylthio-4-

nitrothiophene (**24**) with pyrrolidine and silver nitrate leading by an initial ring-opening to the highly functionalised building-block **25**: a three-step sequence involving chemoselective replacement of the pyrrolidine group with an aryl residue, reduction of the nitrovinyl moiety, and final cyclisation of the resulting oximes accompanied by methanethiol elimination, afforded 3-arylmethyl-5-(methylthio)isoxazoles **26** in good yields <02T3379>.

20 R = Me,Ph **21** 76-95% **22 a** R^1=OBn R^2=H **22 b** R^1=H R^2=OBn **23a** 81% **23b** 70%

24 (*E*)-**25** 70% **26** 49-81%

Ar = Ph,2-MeC_6H_4,4-MeC_6H_4,1-naphthyl,1-thienyl

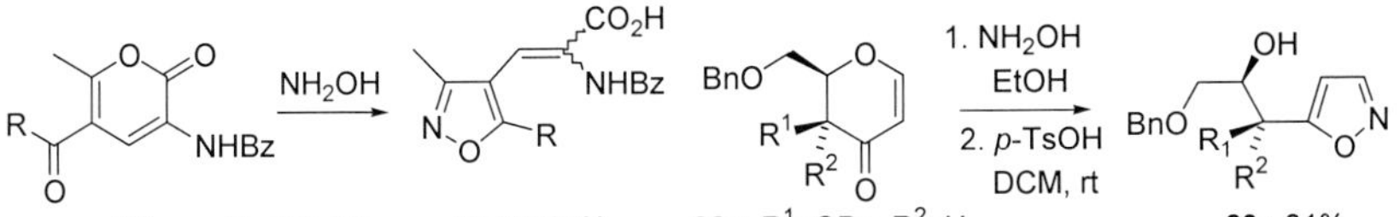

In a study of new nicotinic acetylcholine receptor ligands, analogues to the alkaloid epibatidine and presenting the isoxazole nucleus, compounds **27** and **28** have been synthesised by condensation of the 1,4-dilithium salt of acetone oxime with the appropriate ester derivative <02EJMC163, 02T4505>. The same procedure allowed the synthesis of the isoxazole moiety of bicyclic dioxetanes **29**, obtained by singlet oxygenation of the corresponding dihydrofuran precursors; these systems exhibit remarkably high chemiluminescence efficiency even in aqueous media <02TL8955>.

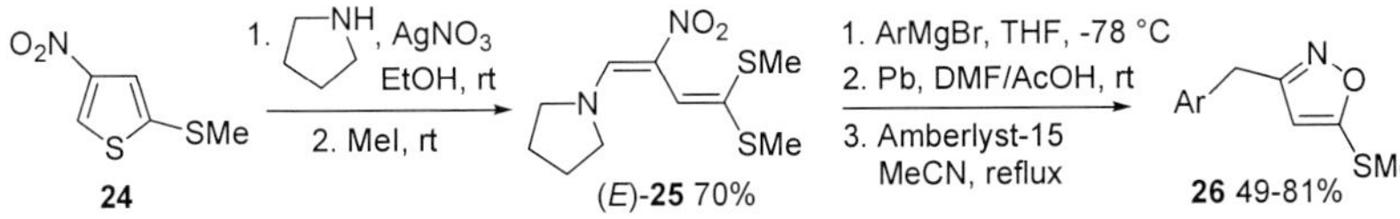

A novel class of 4-arylalkyl substituted 3-isoxazole $GABA_A$ antagonists **30** has been synthesised by condensation of the appropriate β-oxo esters with hydroxylamine <02JMC2454>.

Thermolysis of 2-halo-2*H*-azirines, as well as of β-azido-α,β-unsaturated ketones and esters, gave rise to new 4-haloisoxazoles in high yields <02S605>.

27 n = 0,1 Ar = Ph,pyridyl,pyrazinyl **28** R = H,Me **29** R = H,CF_3,*t*Bu **30** R = $(CH_2)_n$Ph,$CHPh_2$, $(CH_2)_n$naphthyl, CH_2anthracenyl

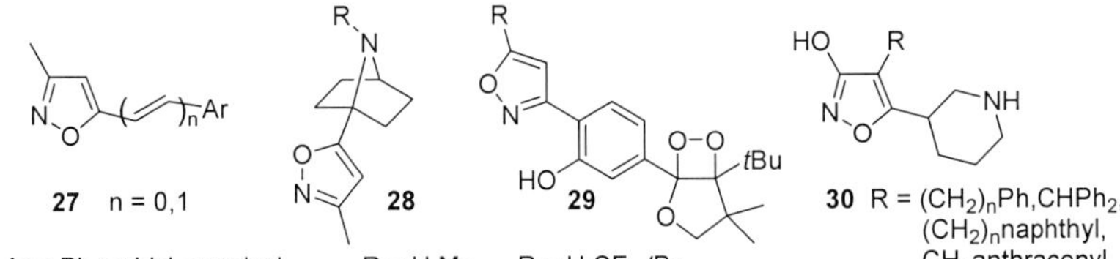

3-Aryl-5-isoxazolecarboxaldehydes **31** have been identified as activated aldehydes for the generation of isoxazole-based combinatorial libraries on solid phase through automation. Baylis-Hillman (BH) reaction of **31** with supported acrylate **32** led to BH adducts **33** which, through Michael addition of primary amines and cleavage from the resin, furnished various derivatives **34** as diastereomeric mixtures in excellent yields and purity. In another synthetic strategy, reductive amination of the resin bound amino acids **35** with **31** followed by alkylation of the NH group afforded highly functionalised isoxazoles **36** <02BMCL1905>. A synthetic and mechanistic study of the $TiCl_4$-promoted BH reactions of **31** with cycloalkenones was also reported <02JOC5783>.

Novel phenanthro[9,10-*d*]isoxazoles **37** have been prepared by intramolecular Stille-Kelly stannylation/coupling of *o,o'*-diiodo-4,5-diarylisoxazoles and by PIFA-mediated non-phenolic oxidative coupling of the corresponding non-halogenated substrates <02T3021>.

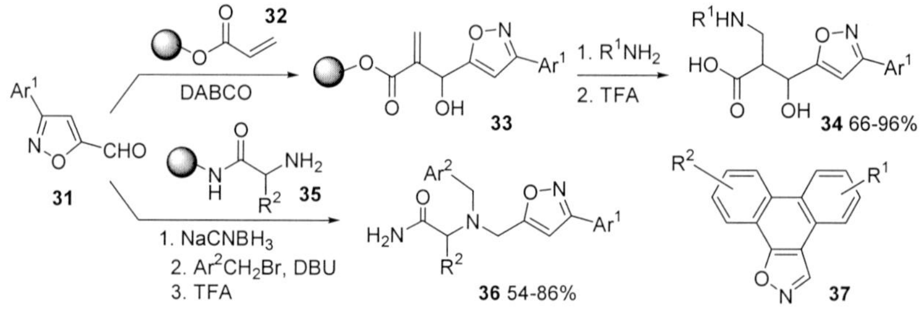

5.7.2 ISOXAZOLINES

Solid supported isoxazolines **40** were prepared starting from a sulfinate-functionalised resin **38**. Oxidation of the resin linked cyclobutanols **40**, with concomitant cleavage of the sulfone linker, produced isoxazolinocyclobutenones **41** in 34-38% overall yield (4 steps) <02OL741>. A five-step solid phase synthesis of isoxazolino-pyrrole-2-carboxylates that employing the same traceless sulfone linker strategy has also been reported <02JOC6564>.

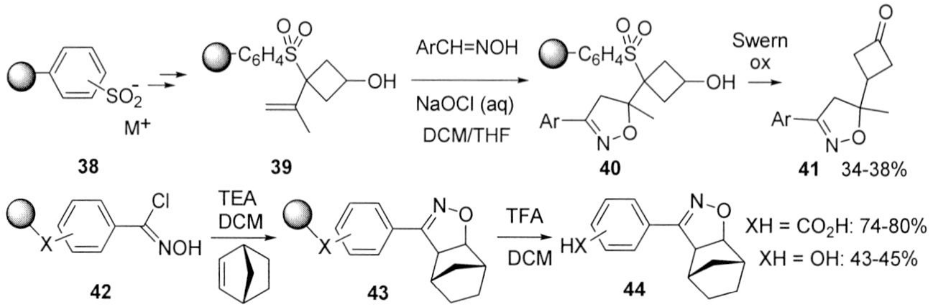

Wang resin supported nitrile oxides, generated from resins **42**, displayed increased stability and could be isolated or directly trapped with a variety of dipolarophiles. The cleaved adducts, such as **44**, were recovered by treatment of the resin **43** with TFA <02EJO1175>.

The chiral induction of carbohydrate enol ethers in 1,3-DC with some aromatic nitrile oxides was investigated. The highest diastereoseletivity (28:1 dr) was achieved with a 3-*O*-vinyl-β-D-fructopyranose derivative and 2,4,6-trimethylbenzonitrile oxide <02TA2535>.

Exo-methyleneprolinate **45** reacted as a dipolarophile with nitrile oxides to generate spiroisoxazolinoprolinates **46** in good yields (70-75%) and with *ca*. 1:4 *cis*:*trans* diastereoselectivity <02JOC5673>.

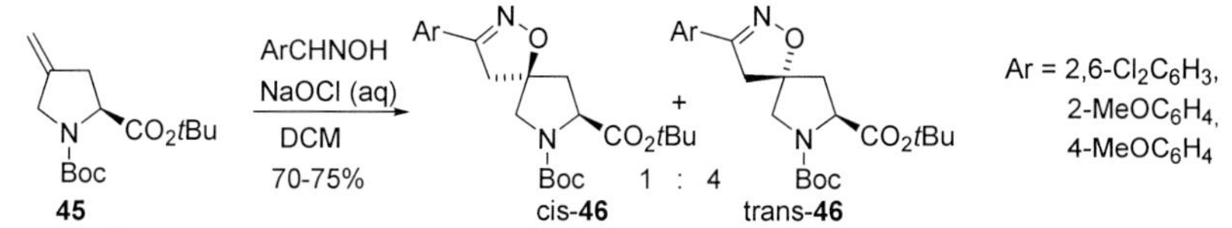

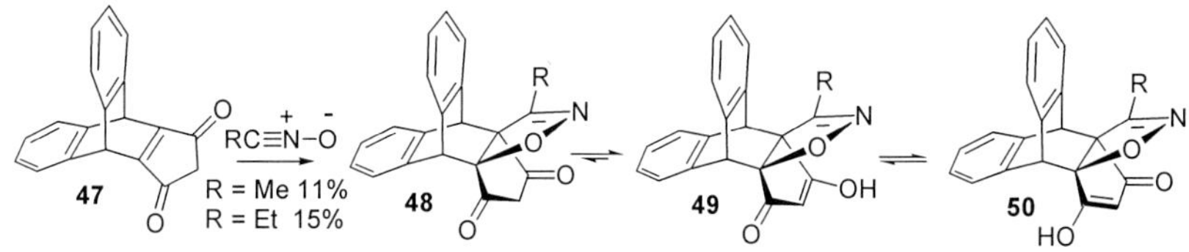

Triptycene cyclopentenedione **47** underwent 1,3-DC with nitrile oxides affording polycyclic isoxazolines **48** bearing the triptycene moiety. The adducts **48** exist in solution in their two enolic forms **49** and **50**, in equimolecular amounts <02JOC4612>.

Substituted isoxazoline *N*-oxides **52** were synthesised by condensation of α-nitro-esters **51** and aldehydes catalysed by zinc complexes of amino acids. In DMSO the selectivity of the reaction was high and products **52** were obtained as single *trans* diastereomers in 77-88% yields <02TL5287>.

The cyclic sulfate isoxazoline **53** was converted into the polyhydroxylated piperidine **54** through one-pot reduction and cyclisation reactions. The process was highly stereo- and regioselective and gave **54** as a single isomer in 82% yield <02SL1359>.

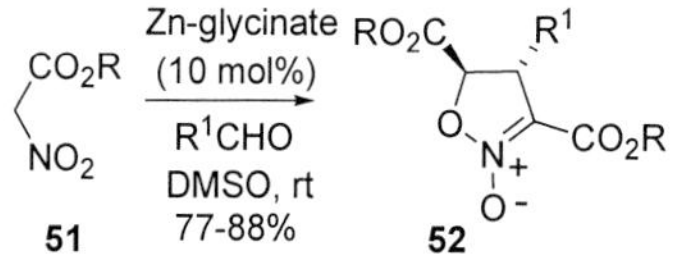

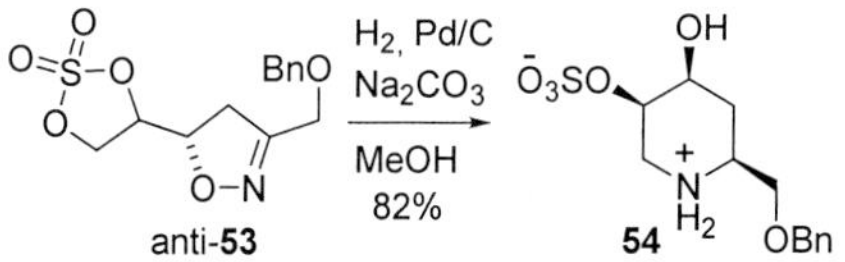

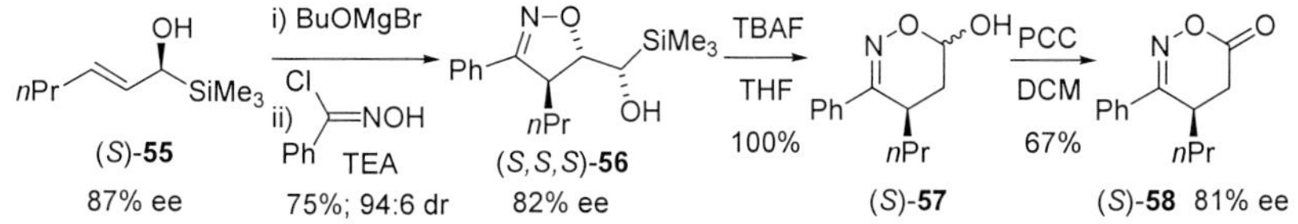

Magnesium ion–mediated nitrile oxide 1,3-DC reactions to allylic alcohols have been reviewed <02SL1371>. New examples have been recently reported, in particular, cycloadditions of aromatic and aliphatic nitrile oxides with optically active α-silylallyl alcohols in the presence of magnesium cations. The substituted isoxazolines, which were obtained with high diastereo- and enantioselectivity, were smoothly converted to [1,2]-oxazine derivatives by treatment with TBAF. For example, oxazin-3-one (*S*)-**58** was obtained in 81% ee starting from dipolarophile (*S*)-**55** <02T9613>.

The bicyclic isoxazolines **59** underwent molybdenum-mediated tandem reductive N–O bond cleavage-retroaldol reactions to provide the compounds **60** or **61** as mixtures of stereoisomers <02OL4101>.

The isoxazoline *syn*-**62** underwent iodoetherification by iodine monochloride to give tetrahydrofuran derivative **63** with good diastereosectivity (29:1 *trans:cis* ratio) <02SL1691>.

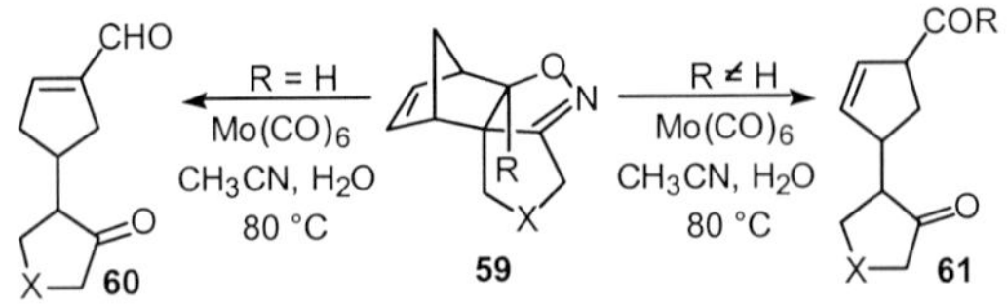

R	X	yield%
60a H	CH$_2$	62
60b H	CH$_2$CH$_2$	60
60c H	O	53
61a C$_6$H$_{13}$	CH$_2$	73
61b CH$_2$OMe	CH$_2$	67

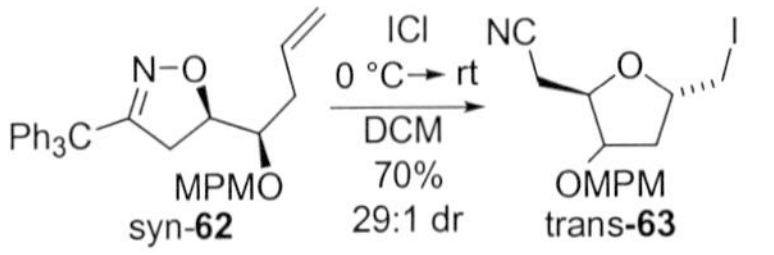

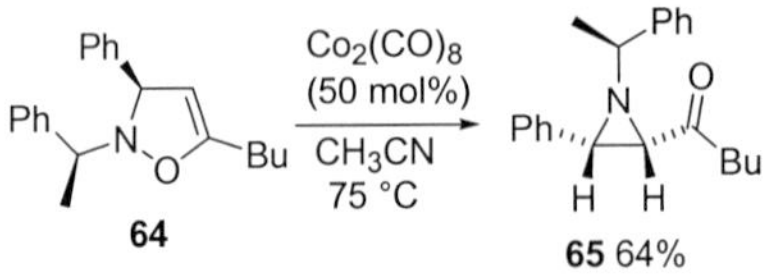

Dicobalt octacarbonyl in anhydrous acetonitrile promoted the rearrangement of 4-isoxazolines to isomeric 2-acylaziridines with yields ranging from 39 to 92%. The optically pure isoxazoline **64** underwent a totally diastereoselective rearrangement to give aziridino ketone **65** as a single isomer <02OL1907>.

A new approach to a variety of α-branched alkynes **70** was achieved by a Knoevenagel type condensation of 4-unsubstituted isoxazolin-5-ones **66** with aldehydes or ketones **67**, followed by conjugate addition of an organometallic species and nitrosative cleavage of the heterocyclic ring <02SL1257>.

The 13-membered macrolide fungal metabolite (+)-brefelin A (**73**) was synthesised through an intramolecular nitrile oxide cycloaddition <02JOC764, 02JOC772>.

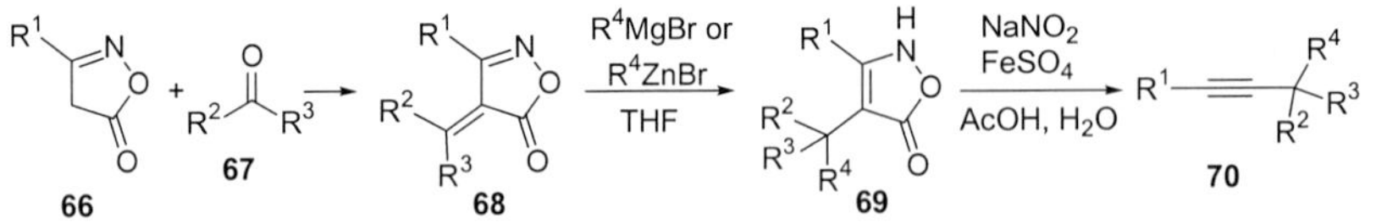

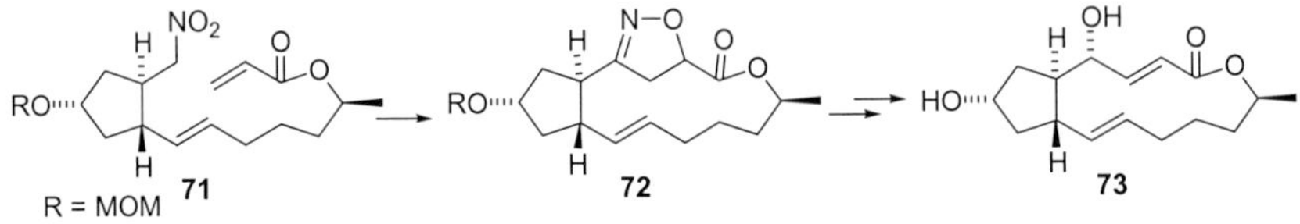

5.7.3 ISOXAZOLIDINES

Isoxazolidines have been used as key intermediates in the syntheses of a variety of natural compounds and unnatural analogues of biological interest. The isoxazolidine moiety has also been introduced in some target compounds. For example, the diastereoselective synthesis of a new class of nucleosides analogues **74** was achieved through 1,3-DC of C-ethoxycarbonyl N-methyl nitrone with 2-acetyloxyacrylate, followed by nucleosidation with silylated bases and reduction of ester moieties <02EJO1206, 02T581>. Other isoxazolidinyl nucleosides were prepared from sugar-derived chiral nitrones and N-vinylated bases <02SL1113>. Azetidinone-tethered nitrones were used to synthesise isoxazolidinyl-β-lactams, –

indolizidines and -quinolines <02JOC7004, 02SL85>. Glycoconjugated isoxazolidine-fused chlorins, such as **75**, and bacteriochlorins were prepared by 1,3-DC of porphyrin derivatives and glycosyl nitrones <02TL603>.

The bicyclic isoxazolidine **76** was prepared by nucleophilic addition of 2-trimethylsilyloxyfuran to *N*-gulosyl-nitrones and employed in a formal synthesis of Polyoxin C <02OL1111>.

74a **74b**

B = thymine, cytosine, 5-fluorouridine
B' =adenine

75

R = TBDPS
76

R, R' = Ph, H; CO_2Et, H ; CN, CN

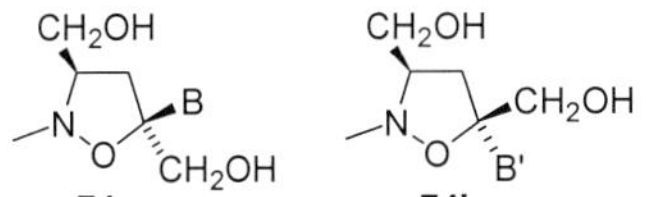

Styrene, ethyl acrylate and fumaronitrile have been used to mask (3*S*)-pentacyclic nitrone **77**. After a suitable elaboration of isoxazolidines **78** affording **79**, the nitrone moiety was restored by thermally induced cycloreversion (145–180 °C), and the tricyclic intermediate **80** was obtained directly through intramolecular 1,3-DC (31-84% yield). This approach was applied to the synthesis of stereodifferentiated polyhydroxyindolizidines such as **81** <02EJO1941>.

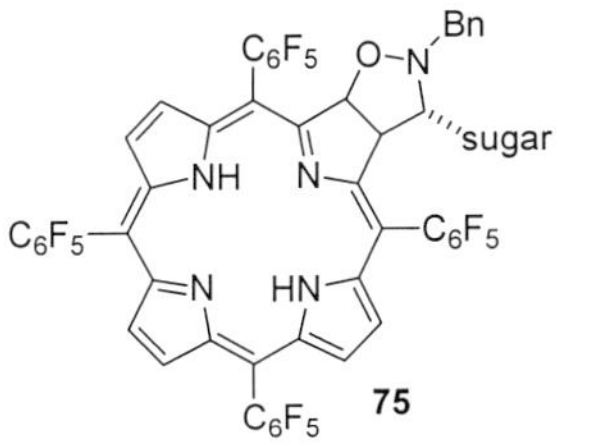

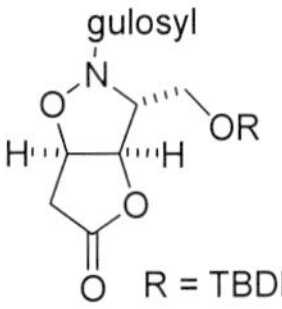

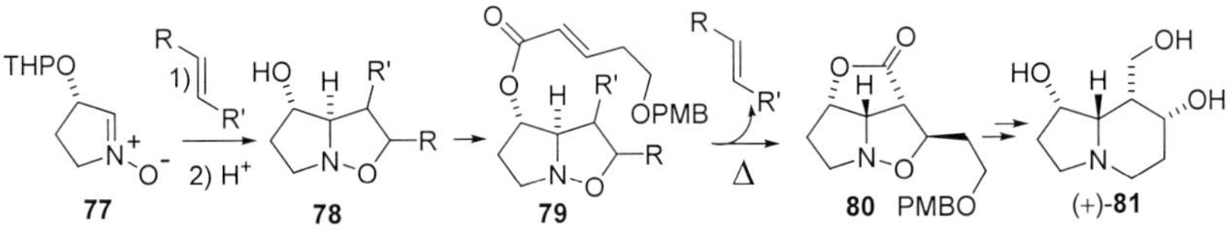

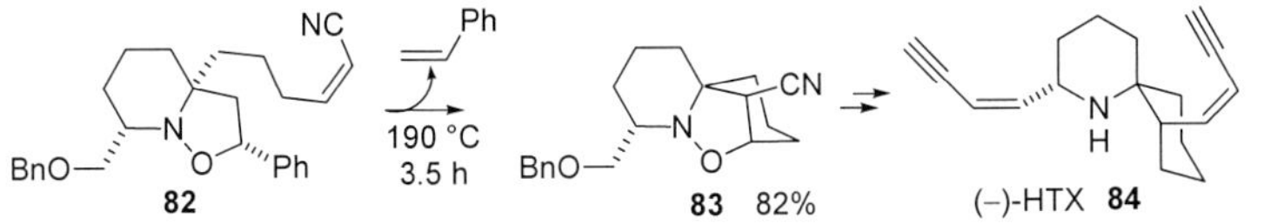

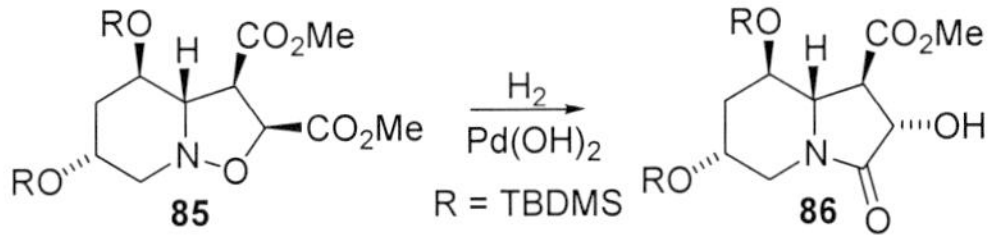

The nitrone protection strategy has also been used in the total synthesis of (–)-histrionicotoxin (**84**) and some related alkaloids. The bicyclic isoxazolidine **82** underwent the cycloreversion-cycloaddition reaction at 190 °C to give the key intermediate **83** in high yield (82%) <02JCS(P1)1494>.

The enantiopure bicyclic isoxazolidine **85** was converted into the protected polyhydroxyindolizidinone **86** in high yield (88%) <02TL9357>.

The syntheses of sedum and related alkaloids using isoxazolidines as key intermediates have been reviewed <02T5957>.

1,3-DC of nitrones is the most versatile and reliable synthetic route to isoxazolidines. Many efforts have been devoted to the preparation of enantiopure cycloadducts, either starting from a chiral non-racemic dipolarophile or nitrone, and through asymmetric versions of the 1,3-DC reaction in the presence of enantiopure catalysts. A highly diastereoselective intramolecular nitrone cycloaddition onto a chiral ketene equivalent was applied to the synthesis of the antifungal antibiotic cispentacin and to its enantiomer **89** <02OL1227>.

The double diastereoselection of cycloaddition of chiral nitrones with Oppolzer's sultam acrylamide was studied and the adducts were converted into 4-hydroxy-pyroglutamic acid derivatives <02TA167, 02TA173>.

The syntheses of isoxazolines and isoxazolidines through 1,3-DC reactions of carbohydrate derived dipoles or dipolarophiles have been reviewed <02JCS(P1)2419>, and some new examples have been reported recently <02TL2741, 02SL1344>.

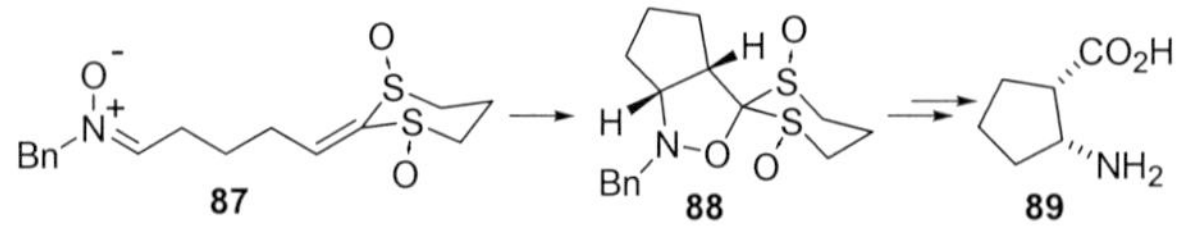

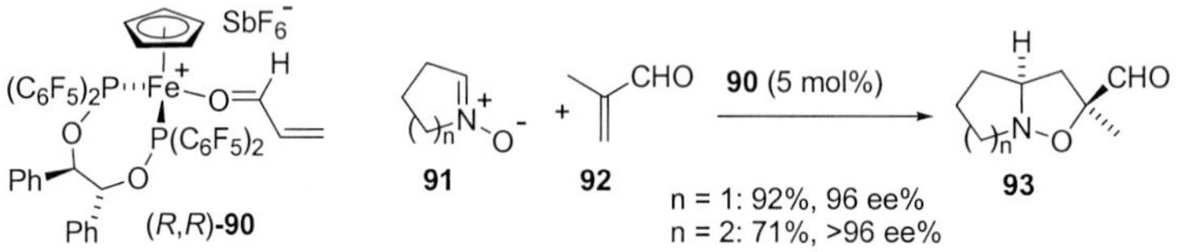

Chiral iron complex **90** catalysed asymmetric 1,3-DC reactions between nitrones and α,β-unsaturated aldehydes. The best results were achieved with cyclic nitrones **91** and methacrolein (**92**) which afforded the CHO-endo isoxazolidines **93** in good yield and with high ee <02JA4968>.

The Lewis acid ATPH **97** was effective both in rate enhancement and in control of the regiochemistry in the cycloaddition reactions between C,N-diphenyl nitrone (**94**) and α,β-unsaturated carbonyl compounds. For example, in the presence of a catalytic amount (10 mol%) of **97**, the regiochemistry of the 1,3-DC of **94** and **92** was reversed and the isoxazolidine-4-carbaldehyde **95** was obtained as the major cycloadduct in high yield <02TL657>.

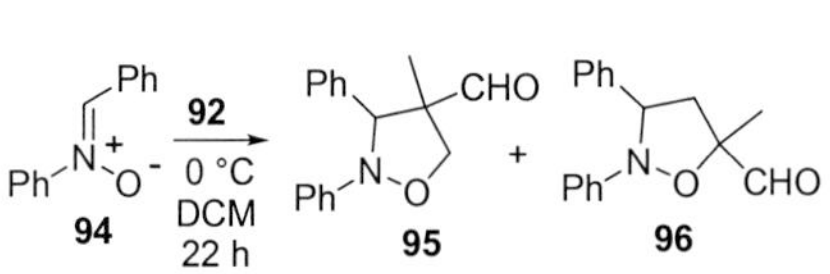

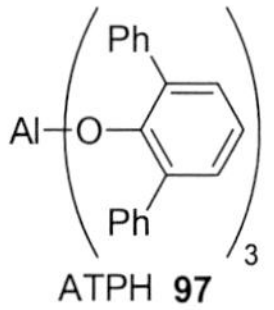

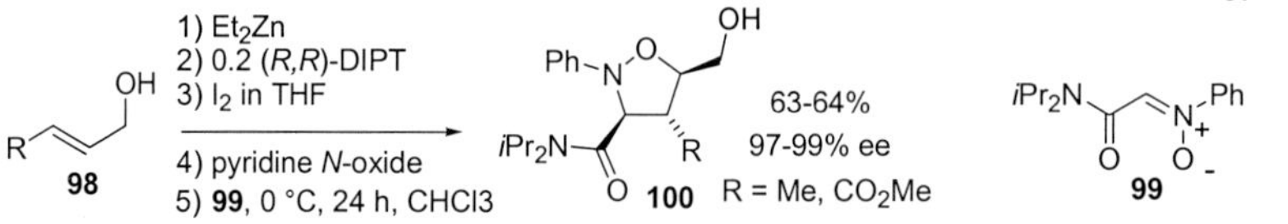

A catalytic asymmetric 1,3-DC of nitrone **99** with γ-substituted allylic alcohols was achieved by using diisopropyl tartrate as chiral ligand. Isoxazolidines **100** were formed with high ee <02CL302>.

Isoxazolidines **104** were produced in high yields with excellent endo selectivities and high enantioselectivities by 1,3-DC of α,β-unsaturated aldehyde **102** and nitrones **101** in the presence of catalytical amount of the optically active cobalt(III) complex **103** <02OL2457>.

Some optically active 3-alkoxycarbonyl-2-methylisoxazolidines were obtained by asymmetric decomposition in the presence of catalytic amount of Pd-BINAP complex. For instance, the kinetic resolution of racemic **105** by **106** afforded (+)-**105** in 48% yield and with 99% ee <02EJO3855>.

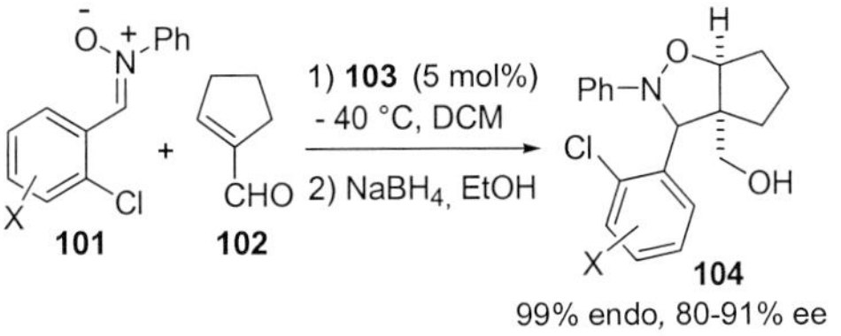

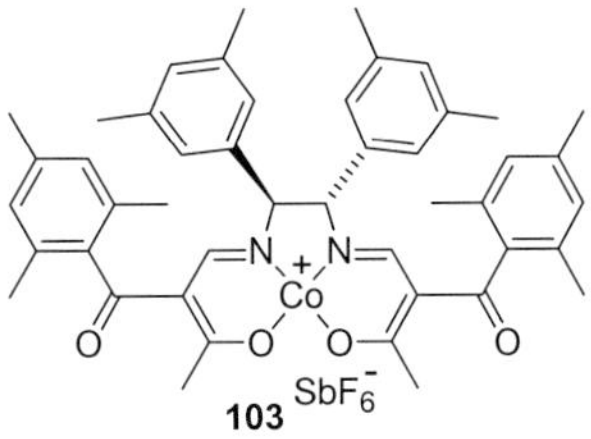

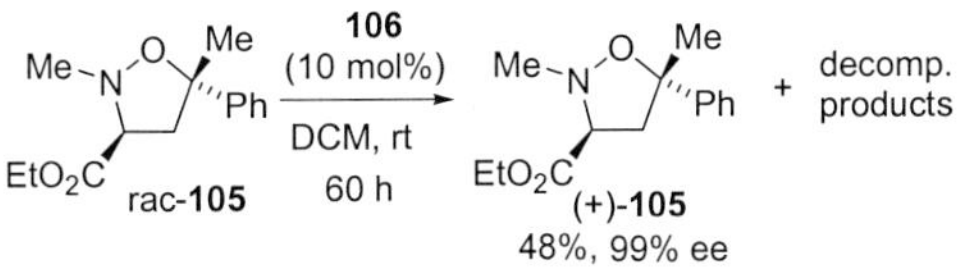

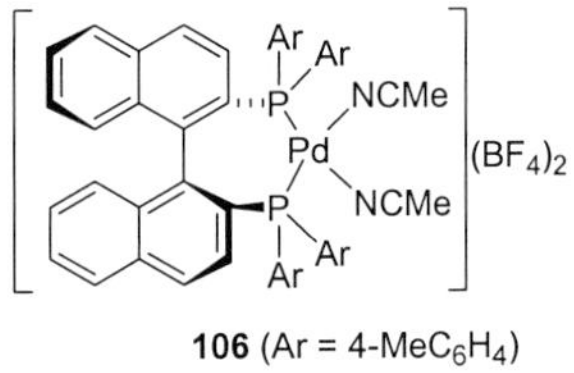

Six fused isoxazolidines of general structure **110** were synthesised through a new palladium-catalysed allene insertion-intramolecular 1,3-DC cascade reaction. The *cis*-ring junction stereochemistry of **110** was established by X-ray analysis <02CC1754>.

R^1	R^2	X	yield%
Ph	H	O	87
Me	Me	O	80
H	Ph	CbzN	92

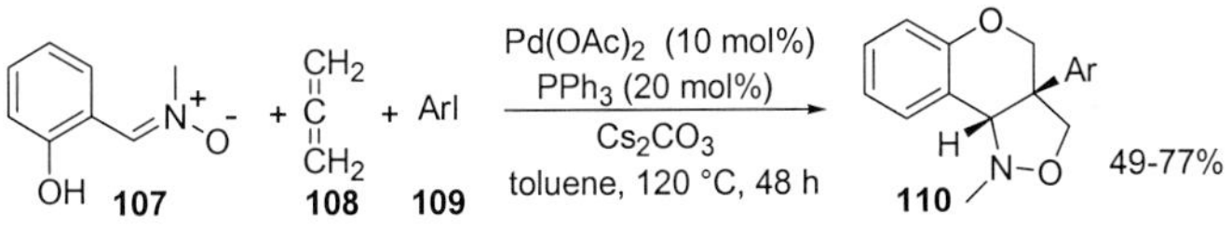

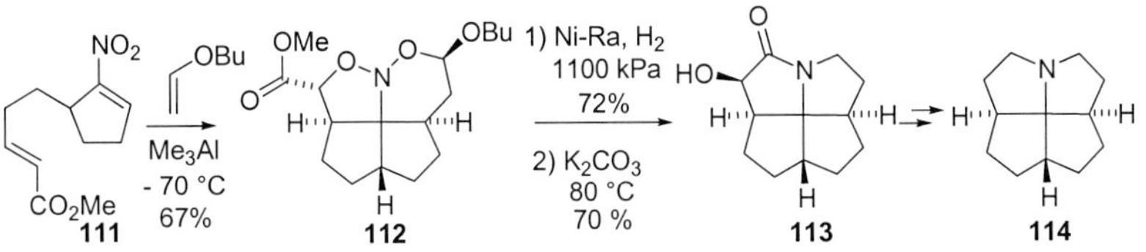

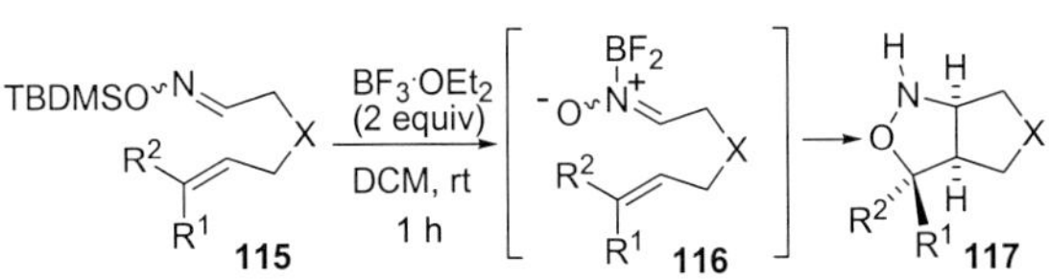

The tandem [4+2]/[3+2] cycloaddition of nitroalkenes was applied to the synthesis of the strained *cis,cis,cis,cis*-[5.5.5.5]-1-azafenestrane **114**, starting from butyl vinyl ether and **111**. The structure of **114** was confirmed by X-ray analysis of its adduct with borane (**114**.BH_3) <02AG(E)4122 >.

N-Unsubstituted isoxazolidines **117** were obtained by treating *O-tert*-butyldimethylsilyloximes **115** with two equiv of $BF_3 \cdot OEt_2$. Probably, the reaction goes through an intramolecular cycloaddition of the *in situ* generated *N*-borano-nitrone **116** <02CC1128>.

5.7.4 OXAZOLES

The Robinson-Gabriel synthesis has been one of the most useful procedures for the preparation of oxazoles through cyclodehydration of α-acylamino ketones. On this basis, a general approach to 2,4-disubstituted oxazoles **120** has been evolved employing the reagent combination triphenylphosphine/hexachloroethane for an easier cyclodehydration of α-acylamino aldehydes **119**, produced from Weinreb amides of Boc-protected α-amino acids **118** <02OL2665>. An analogous procedure applied to keto-amides **122**, obtained by ring-opening of phosphorylated 2*H*-azirines **121** with *N*-protected amino acids, afforded optically active phosphorylated oxazoles **123** containing amino acid residues. This strategy has been extended to simple carboxylic acids as well as to *N*-protected peptides to give the first family of optically active oxazoles with peptide side-chains <02TA2541, 02JOC7283>.

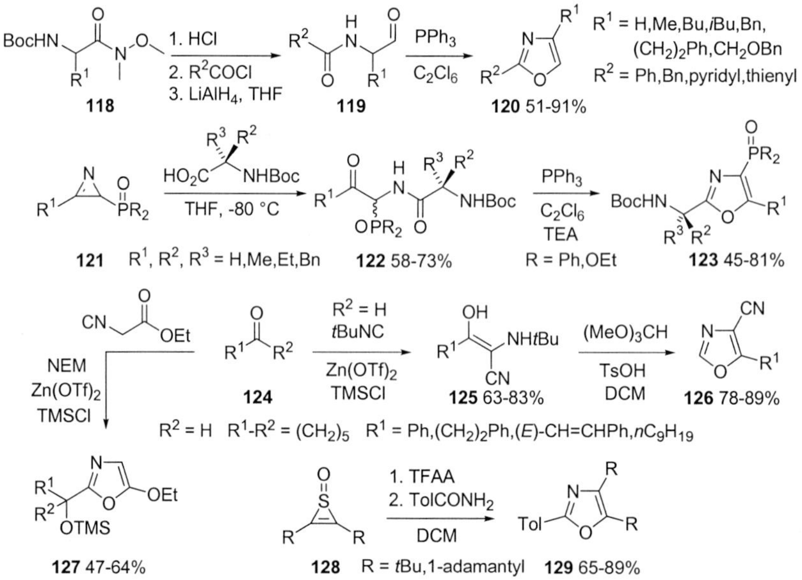

When the Brønsted acid in the classical Passerini reaction was replaced by zinc triflate and chlorotrimethylsilane, different functionalised oxazoles were produced in satisfactory yields; in particular, carbonyl compounds **124** (R^2=H) reacted with 2 equiv of *tert*-butyl isonitrile leading to 4-cyanooxazoles **126** through cyanoenamines **125**, while ethyl isocyanoacetate gave 2-(α-silyloxyalkyl)-5-alkoxy derivatives **127** <02S1969, 02OL1631>. Starting from α-isocyanoacetamides, a multi-component Ugi-type reaction affording 5-amino-oxazoles was efficiently accelerated by addition of ammonium chloride in toluene <02JA2560>.

Thiirene-1-oxides **128** furnished oxazoles **129** with two bulky alkyl substituents at vicinal positions by treatment with trifluoroacetic anhydride and *p*-toluamide <02CL314>.

Using the rhodium carbenoid N-H insertion strategy developed by Moody, an array of trisubstituted oxazoles has been prepared starting from polymer bound β-ketoesters. Alternative approaches, respectively employing insoluble JandaJel resins and soluble microgel polymers as supports, allowed the preparation of final compounds in similar yields and purity <02TL5407, 02JOC3045>.

When *N*-protected indole-3-carboxaldehydes were allowed to react with tosylmethyl isocyanide, 3-(oxazol-5-yl)-indoles were obtained in satisfactory yields and applied to the synthesis of novel indole-based IMPDH inhibitors <02BMCL3305>. In the same context, a modified approach to 2-(*N*-aryl)oxazoles employing an iminophosphorane/ isothiocyanate-mediated methodology and its application to the synthesis of the potent IMPDH inhibitor BMS-337197, was reported <02OL2091>.

o-Iodoxybenzoic acid (IBX), highly effective in carrying out oxidations adjacent to carbonyl functionalities and at benzylic carbon centres, was also efficient in the transformation of methyloxazoles **130** into formyl derivatives **131** in a process probably initiated by single electron transfer from the aryl moiety of the substrate to IBX <02JA2245>.

The addition of carbonylated electrophiles to the 2-lithio derivative of 4-oxazolinyloxazole **132** allowed the efficient preparation of 5-phenyloxazoles **134** bearing a variety of hydroxyalkyl groups at C-2 position and a carboxyl (or formyl) function at C-4. This protocol suppresses the troublesome electrocyclic ring-opening reaction and allows access to the target compounds by simple chemical transformation of the oxazoline moiety of **133** <02JOC3601>. A direct chemoselective C-2 silylation of oxazoles was performed by treatment of the lithiated parent compounds with silyl triflates <02TL935>.

The readily available chloro ester **135** proved to be a versatile scaffold for the synthesis of disubstituted- and trisubstituted-oxazoles: Suzuki, Stille, and Negishi Pd-catalysed coupling reactions were all successfully used to install substituents at the C-2 position affording compounds **136**. Following bromination, a second Pd-catalysed coupling (including a Sonogashira reaction) led to oxazoles **137** which, through the 4-bromo derivatives, were converted into **138** <02OL2905>. Analogously, 2-, 4-, and 5-trifloyl oxazoles were subjected to Sonogashira cross-coupling with variously substituted terminal alkynes <02OL2485>.

A novel strategy for the enantioselective synthesis of polyhydroxypiperidines **141**, which can be viewed as amino 2,6-dideoxyazasugars, was developed based on 5-bromo-4-bromomethyloxazole **139** and *N*-protected α-amino esters: alkylation and bromine-lithium exchange gave rise to bicyclic derivatives **140** which after reduction, alkylation, and oxazole ring-opening afforded the final products (all-*cis* relationship) in a highly diastereo- (de>90%) and enantioselective way with an overall maximum yield of 21%. Inclusion of an additional Mitsunobu step in the synthetic sequence allowed the preparation of derivatives with inverted configurations at positions 3 and 4 (*trans-cis-trans* relationship) <02JOC3184>.

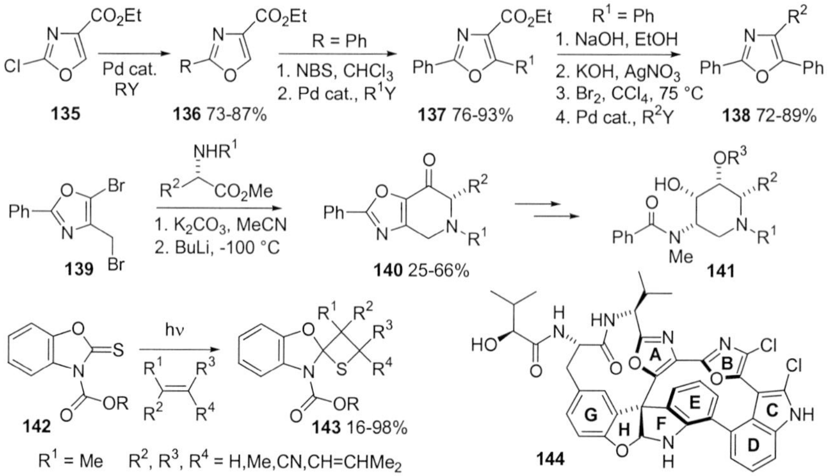

Photocycloaddition reactions of alkyl and aryl 2-thioxo-3*H*-benzoxazole-3-carboxylates **142** to alkenes afforded stable isolable spirocyclic aminothietanes **143** <02HCA2383>; similar reactions with both electron-poor and electron-rich alkenes were also performed on 2-methyloxazolo[5,4-*b*]pyridine <02EJO4211>.

Nicolaou and co-workers have reported a total synthesis of diazonamide A (**144**) which confirmed its highly unusual molecular architecture containing an additional ring (ring H) and a nitrogen atom leading to a structurally unique aminal moiety <02AG(E)3495>.

Oxazole and thiazole-based amino acids were shown to undergo novel cyclooligomerisations in the presence of pentafluorophenyl diphenyl phosphinate (FDPP) producing the natural hexapeptide dendroamide A <02H(58)521>.

The first intramolecular Diels-Alder (IMDA) reactions of simple dienes **145** featuring an *N*-substituted oxazolone as dienophile have been performed: **145** (n=1) afforded a mixture of endo and exo cycloadducts **147** and **148**, while **145** (n=0) reacted exclusively in overall endo fashion leading to **146** <02CC438>. Chiral pyrrolo[1,2-*c*]thiazoles **151** were obtained as single enantiomers, even if in moderate yields, via intramolecular 1,3-DC of bicyclic münchnones **150** derived from cyclodehydration of 2-substituted *N*-acylthiazolidine-4-carboxylic acids **149** <02JOC4045>. The intramolecular Pauson-Khand reaction of 2-oxazolone derivatives with a suitable pentynyl appendage were also reported <02OL4301>. A highly diastereoselective multicomponent synthesis of unsymmetrical imidazolines was performed probably through 1,3-DC of oxazol-5-ones with intermediate imines <02OL3533>.

(+)-(*R*)-[2.2]Paracyclophane[4,5-*d*]oxazol-2(3*H*)-one (**152**) exhibiting planar chirality has been used as chiral auxiliary in asymmetric Diels-Alder (DA) and Michael reactions via α,β-unsaturated carboxy imides and in asymmetric aldol additions via enolate imides. The endo/exo- and face-diastereoselectivity were good and the chiral auxiliary can easily be removed and quantitatively recovered <02JOC2665>.

146 65% **145** **147** 65% **148** 16%

149 R = Ph,Me,COAr **150** n = 0,1 **151** 17-42%

152 **153a** **153b**

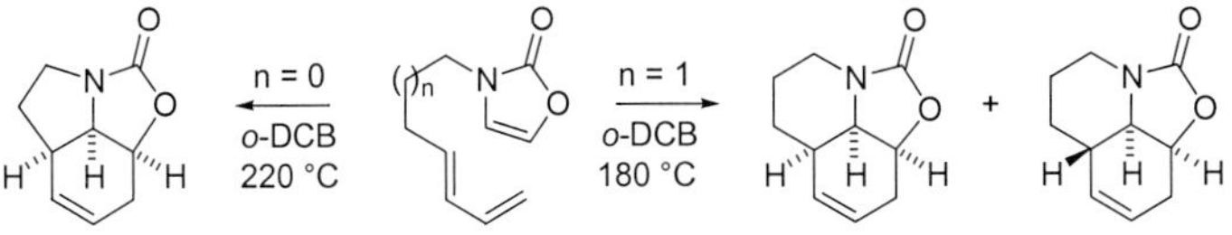

4-Substituted benzoxazol-2(3*H*)-ones behaved as achiral templates for enantioselective DA reactions as a result of a chiral relay effect of the substituent at C-4 position. Upon complexation with a chiral Lewis acid, *N*-acryloylbenzoxazol-2(3*H*)-ones should take up the two distereomeric conformations **153a** and **153b** because of the presence of a chiral axis in the substrate. The best results were obtained with benzyl derivatives. <02OL39>.

154 **155** **156** **157** 90%

158 **159** **160**

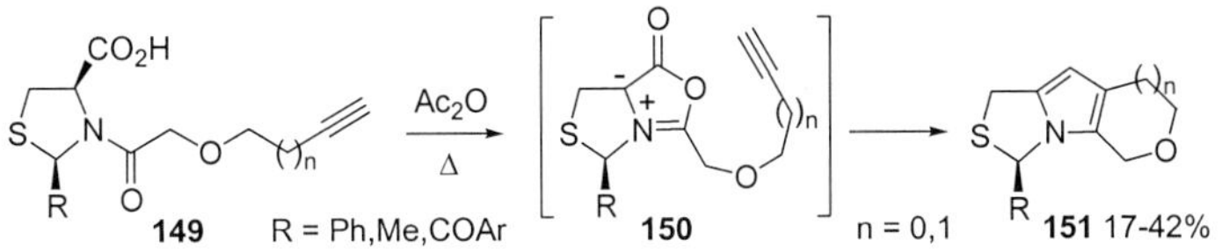

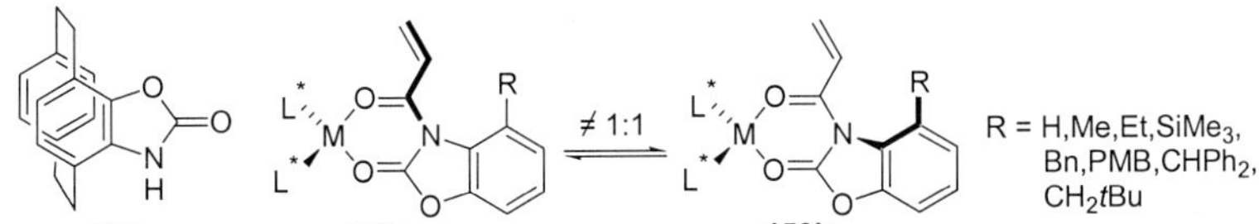

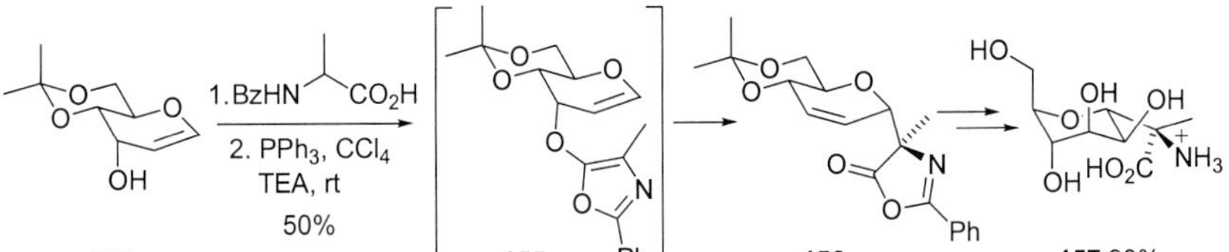

α-D-*C*-Mannosyl-(*R*)-alanine **157** was synthesised in only four steps starting from the acetonide derivative **154** and racemic *N*-benzoylalanine, through the oxazolone (4*R*)-**156**; the key C–C bond formation between the sugar and the amino acid moiety was effected through a Claisen rearrangement of the oxazole intermediate **155** <02T9381>.

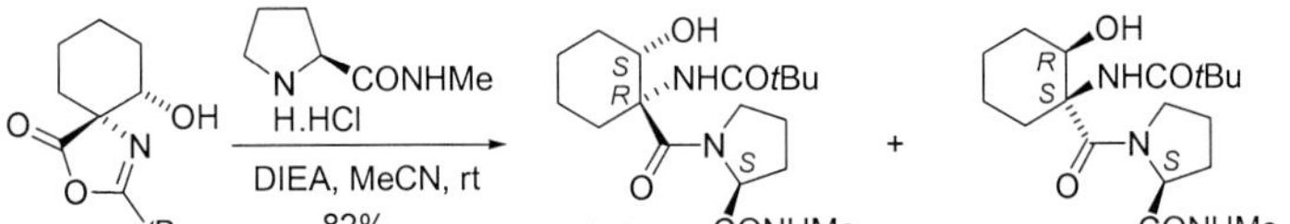

A resolution method involving the aminolysis of a racemic oxazolone intermediate **158** with L-Pro-NHMe hydrochloride allowed the incorporation of enantiomerically pure (1*R*,2*S*)- and (1*S*,2*R*)-1-amino-2-hydroxycyclohexane-1-carboxylic acids (c$_6$Ser) into Xaa-Pro dipeptide, affording diastereomers **159** and **160** separated by column chromatography; the strained cis-c$_6$-Ser was capable of inducing a type 1 β-turn or a non-folded structure in the dipeptide <02TL1429>.

5.7.5 OXAZOLINES

New syntheses of the oxazoline ring have appeared in the literature. The use of a zeolite (E-4) resulted in an efficient preparation of 2-substituted oxazoline derivatives **163** starting from carboxylic acids and β-aminoalcohols <02TL3985>. The same reaction can be performed starting from β-aminoalcohols and Fmoc-protected α-amino acids using CCl$_4$, PPh$_3$, and DIPEA in a one-pot fashion, affording very efficiently amine functionalised oxazolines **166** <02OL3399>.

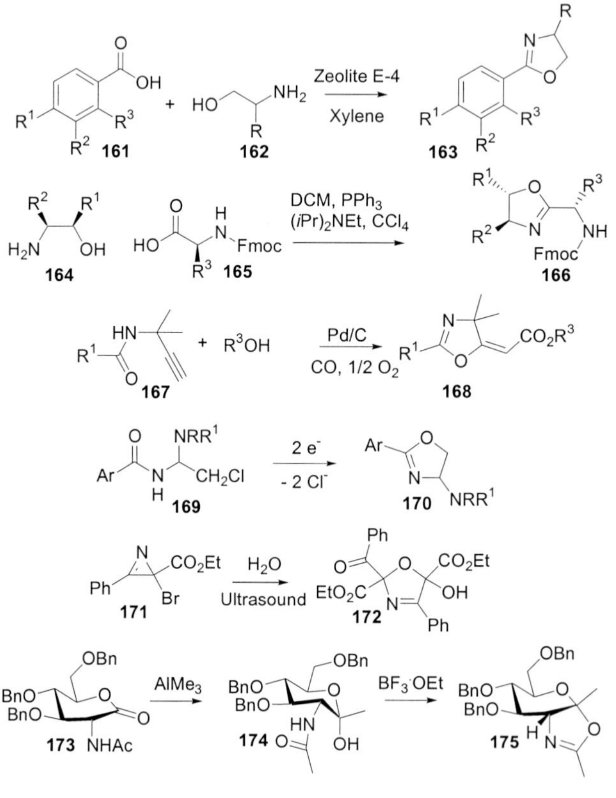

Oxazolines **168** and bisoxazolines can be synthesised starting from a prop-2-ynylamide in a reaction catalysed by Pd/C or PdI$_2$ in the presence of O$_2$ and CO <02JOC4450>. 4-Amino-2-aryl-2-oxazolines **170** can be synthesised electrochemically starting from *N*-(1-amino-2,2-

dichloroethyl)benzamides **169** <02T9853>. A study on the reactivity of 2-halo-2*H*-azirines showed that two molecules of compound **171** are hydrolysed using ultrasound to afford the 3-oxazoline **172** <02JOC66>. Glycosyl oxazolines are useful glycosyl donors. A new synthesis of **175** was described by addition of trimethylaluminum to glyconolactone **173** and subsequent treatment with a Lewis acid <02TL7101>.

A new and mild method for opening sugar oxazolines, mediated by Cu(II) salts, was described, affording 2-acetamido-2-deoxy-β-D-glucopyranosides **177** in good yield <02EJO1363>.

Sugar oxazolines can be oxidized with *m*-CPBA and subsequently with Dess-Martin periodinane into nitroglycal derivatives<02TL347>.

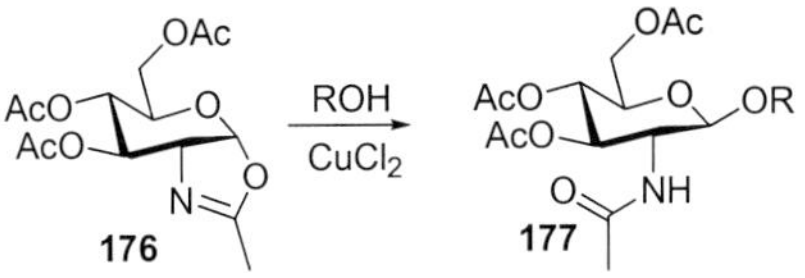

The presence of an oxazoline intermediate was proved in the mechanism of C-2-amidoglycosidation of glycals <02JA9789>.

Several studies on the lithiation of oxazoline derivatives were published. 4,4-Dimethyl-2-(*o*-tolyl)oxazoline **178** undergoes both lateral and *ortho* lithiation depending on the reaction conditions <02TL9069>. The use of TMEDA, together with a different lithiating agent, allowed the diastereoselective lithiation of chromium arene complexes substituted with an oxazoline moiety <02AG(E)3884>. Lithiation of 2-chloromethyl-2-oxazoline **181** afforded derivatives that behave like chlorocarbenes and tend to "dimerise" or "trimerise" to afford 1,2,3-tri(oxazolinyl)cyclopropane (**182**) <02JOC759>. The β-lithiation of oxazolinyloxirane **183** occured diastereoselectively and the intermediate reacted with carbonyl compounds to afford γ-butyrolactone derivatives **184** <02OL1551>.

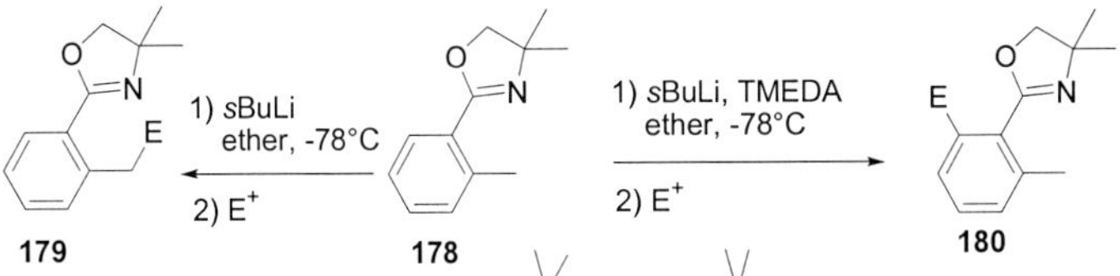

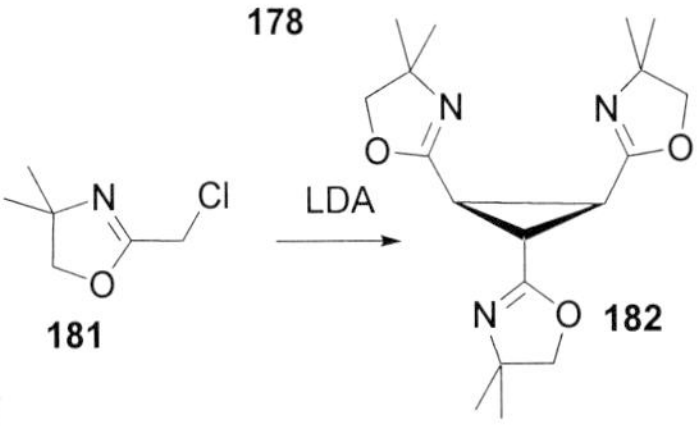

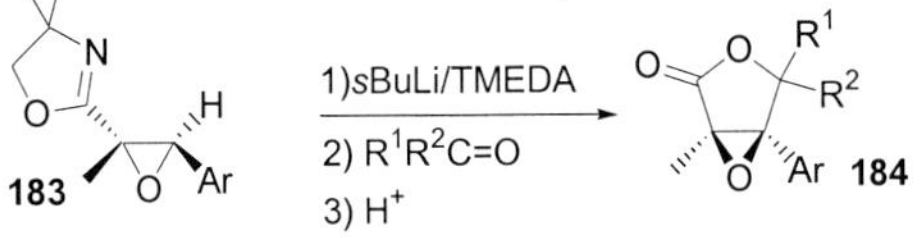

N-(2-oxoalkyl)oxazolinium salts reacted with ammonia to afford 1,2,4-trisubstituted imidazoles <02TL6997>. Imidazolium salts **187** were obtained from bioxazolines **185** and silver triflate/chloromethylpivalate (**186**). These salts were finally transformed in enantiopure palladium carbene complexes <02CC2704>.

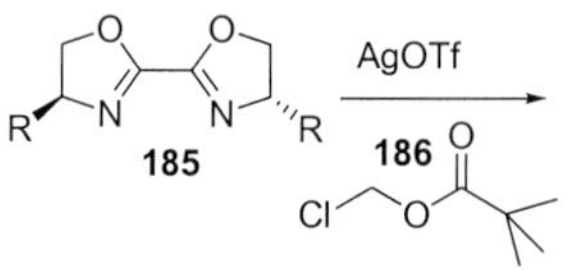

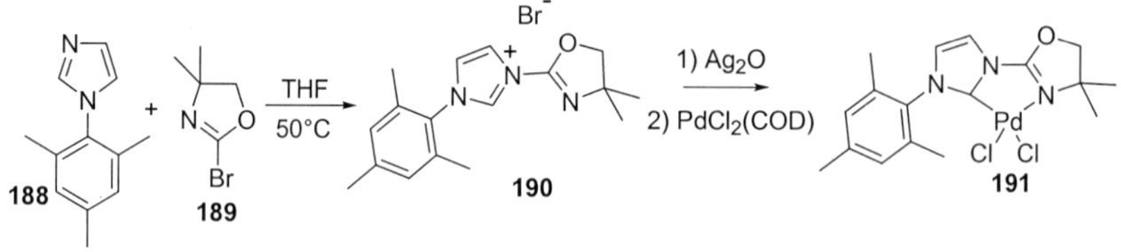

Through the direct coupling of 2-bromo-4,4-dimethyloxazoline **189** and 1-mesityl imidazole (**188**) the corresponding imidazolium salt **190** was obtained and used for the preparation of a mono-carbene-palladium complex **191** active as a catalyst in Heck and Suzuki C–C coupling reactions <02OM5204>.

A particular use of oxazoline ketones as chiral auxiliaries, was described in the synthesis of (−)-rhazinilam, an antitumor agent <02JA6900>.

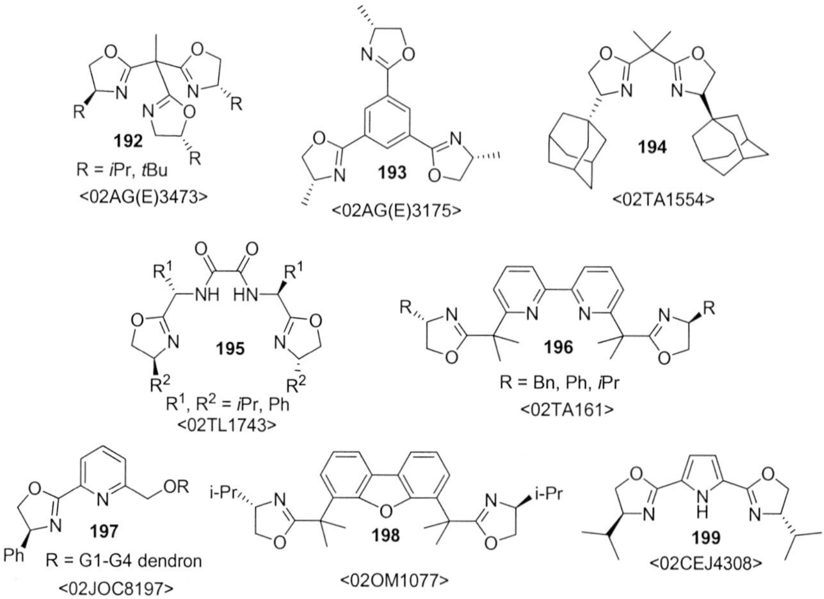

Several new C_3 <02AG(E)3473, 02AG(E)3175, 02JA591> and C_2 <02OM1077, 02TL1743, 02CEJ4308, 02TA161, 02TA1554> symmetrical ligands **192-199** were synthesised and the properties of their complexes studied. Ligand **199** afforded palladium complexes characterised by enantiopure double helices while ligand **193** afforded a propeller shaped supramolecular capsule around three silver ions.

The dendritic oxazoline ligand **197** <02JOC8197> was used for the formation of palladium complexes able to catalyse enantioselective allylic alkylation reactions.

Ligands, already demonstrated as useful in solution, were immobilised on a solid phase <02OL3927, 02TL5587>. New ligands were synthesised, such as phosphino oxazoline ligands <02TL2811>, hydroxymethyloxazolines <02TL1535, 02TA2497> and have found, between others, application in asymmetric hydrogenation of ketones <02TA469>, in the enantioselective conjugate addition of 1,3-dicarbonyl compounds to nitroalkenes <02JA13097>, in the asymmetric Michael addition of indoles to alkylidene malonates <02JA9030> and in the palladium-catalysed asymmetric DA reaction <02TA1841>.

5.7.6 OXAZOLIDINES

(−)-Statine was synthesised from the oxazolidine derivative **201** obtained through a stereoselective conjugate addition of an hydroxymethyl group in a basic methanolic solution. The conjugate addition proceeded faster than the dehydroxymethylation<02OL1213>.

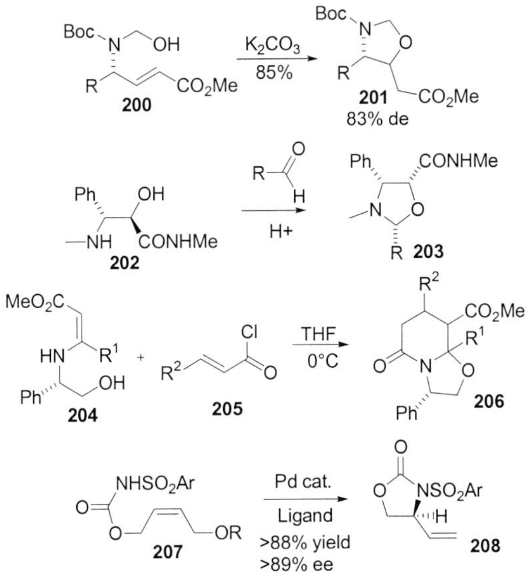

While the reaction of amide **202** with 1,1-dihaloalkanes or paraformaldehyde afforded the six membered hexahydro-4-pyrimidinone, reaction with each of 2,2-dimethoxypropane, benzaldehyde, and acetaldehyde afforded chemoselectively oxazolidines **203** <02T3281>.

Oxazololactams **206** can be obtained easily by reaction of β-enaminoesters with acryloyl derivatives <02TL2521>.

Vinyl-substituted 2-oxazolidinones **208** were synthesised by a catalytic <02JOC974> and also asymmetric intramolecular aminopalladation of readily available derivatives of (Z)-2-buten-1,4-diol **207** <02JA12>.

A Diels-Alder reaction of 4,5-diethylidene-oxazolidinone **209** afforded bicyclic compound **210** <02HCA464>.

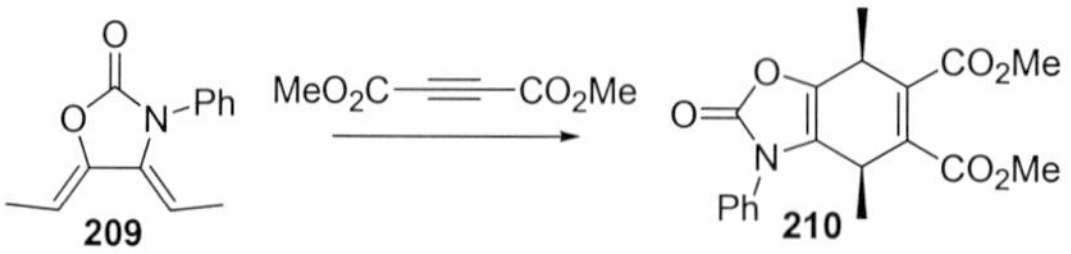

The reactions of propargylamines with carbon dioxide have been systematically examined in the presence of transition metals; $Pd(OAc)_2$ was the best catalyst for the formation of the corresponding oxazolidinone **212** <02JOC16>.

The reduction under free radical conditions of cyclic carbamate **213** effected a stereoselective ring contraction to provide oxazolidinone **214** <02JOC1972>.

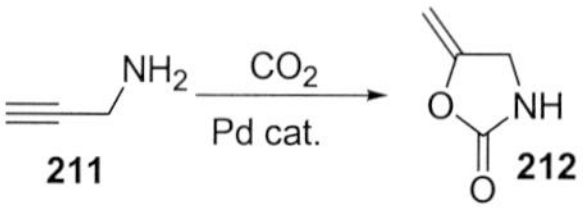

An improved electrochemical synthesis of chiral oxazolidin-2-ones from 1,2-aminoalcohols was achieved by reaction with CO_2 and electrogenerated acetonitrile anion <02TL5863>. The reaction of N-protected amino acids with paraformaldehyde was greatly accelerated by microwave irradiation <02TL9461>. Oxazolidin-2-ones were synthesised by sulfur assisted thiocarbonylation of 2-aminoethanols with carbon monoxide and subsequent oxidative cyclisation with molecular oxygen <02T7805>.

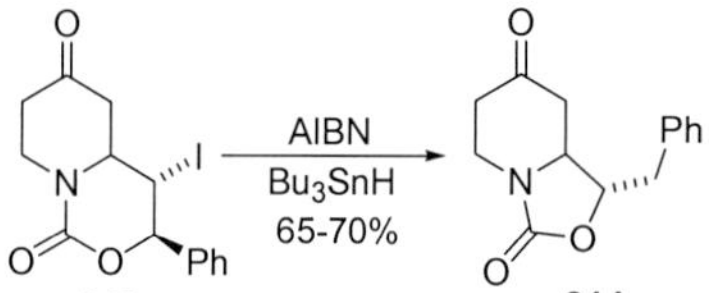

The reaction of (S)-phenylglycinol-derived oxazolidines **215** with allyltrimethylsilane and propynyltrimethylsilane lithium reagents, afforded β-aminoalcohols with a vinylsilane or alkynylsilane appendage **216** and **217**. The same reaction, performed in the presence of $Ti(OiPr)_4$ gave allylsilane and allenylsilane derivatives **218** and **219** <02JOC1496>.

The diastereoselective synthesis of a dipeptide isostere **221** was performed, through an organocopper mediated anti-S_N2' reaction on substituted oxazolidinone **220** <02OL1055>.

An enantiodivergent route to 2-arylpiperidines was described starting from bicyclic lactam **222** with different reducing agents <02CC256>.

Treatment of N-acyloxazolidinones with hydroxylamines using samarium triflate as a Lewis acid, afforded the corresponding hydroxamic acids in good yields <02OL3343>.

A 1:2 mixture of Me_3SiNEt_2 and MeI induces the ring cleavage of oxazolidines to afford the corresponding N-[2-(trimethylsiloxy)ethyl]imine <02JOC5170>.

A new spirocyclic ketone **225** was proposed for the asymmetric epoxidation of olefins <02JOC2435>.

Five-Membered Ring Systems with O & N Atoms   279

A polymer-supported oxazolidine aldehyde **226** was developed for asymmetric chemistry <02JOC6646> as well as a soluble polymer-bound Evan's chiral auxiliary <02TA333> and 3,5-disubstituted oxazolidin-2-one **227** anchored on a solid phase <02TL8327>.

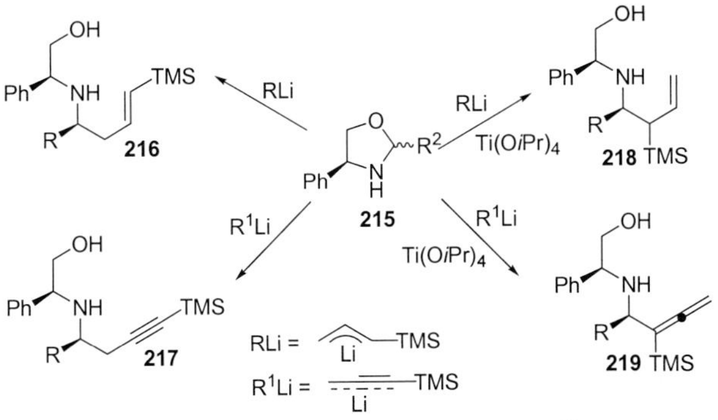

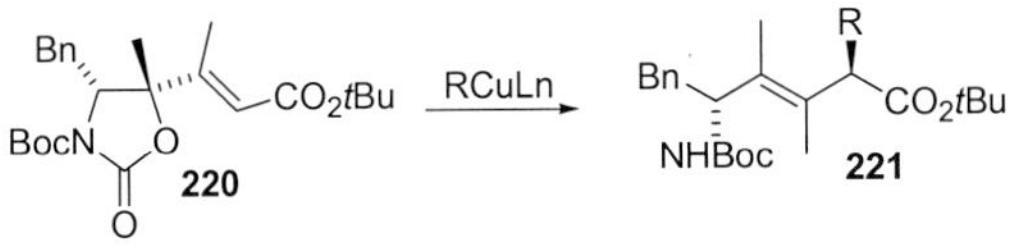

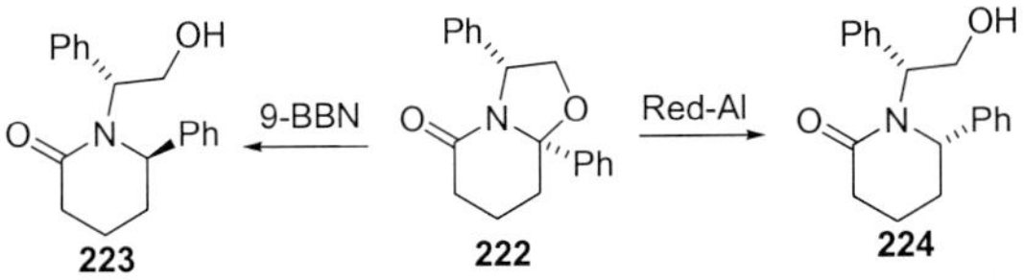

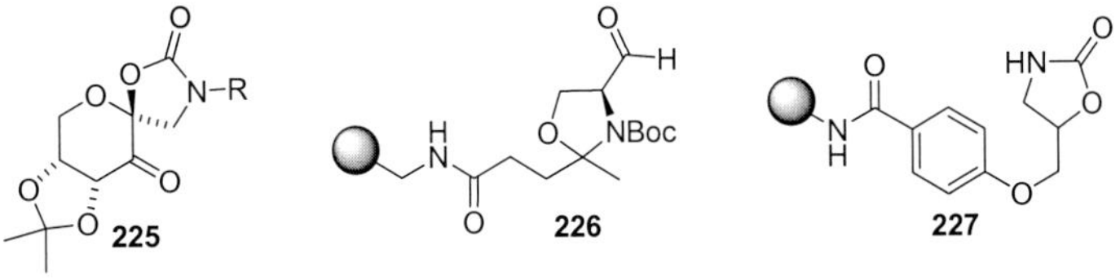

5.7.7 OXADIAZOLES

A solid phase synthesis of substituted 3-alkylamino-1,2,4-oxadiazoles has been developed. The synthesis started from *N*-acyl-1*H*-benzotriazole-1-carboximidamides **228**. Treatment with hydroxylamine afforded the oxadiazole skeleton **229** in high yield <02TL5043>.

280 S. Cicchi, F.M. Cordero, and D. Giomi

228 → **229**

Tandem intramolecular DA/1,3-DC reactions of 1,3,4-oxadiazoles were reported. The DA reaction step produced the intermediate **231** which was not isolated but evolved, through the loss of nitrogen and a subsequent 1,3-DC, to the final product **232** <02JA11292>

230 — DA → **231** — $-N_2$, 1,3-DC → **232**

The irradiation of some 5-alkyl-3-amino-1,2,4-oxadiazoles **233** at 254 nm in methanol, resulted in ring-photoisomerization to 2-alkyl-5-amino-1,3,4-oxadiazoles **234** and 3-alkyl-5-amino-1,2,4-oxadiazoles **235** <02JOC6253>.

233 — hν → **234** + **235**

5.7.8 REFERENCES

02AG(E)3174	H.-J. Kim, D. Moon, M.S. Lah, J.-I. Hong, *Angew. Chem. Int. Ed.* **2002**, *41*, 3174.
02AG(E)3473	S. Bellemin-Laponnaz, L.H. Gade, *Angew. Chem. Int. Ed.* **2002**, *41*, 3473
02AG(E)3495	K.C. Nicolaou, M. Bella, D.Y.-K. Chen, X. Huang, T. Ling, S.A. Snyder, *Angew. Chem. Int. Ed.* **2002**, *41*, 3495.
02AG(E)3884	L. E. Overman, C.E. Owen, G.G. Zipp, *Angew. Chem. Int. Ed.* **2002**, *41*, 3884.
02AG(E)4122	S.E. Denmark, L.A. Kramps, J.I. Montgomery, *Angew. Chem. Int. Ed.* **2002**, *41*, 4122.
02BMCL1395	Y.-S. Lee, B.H. Kim, *Bioorg. Med. Chem. Lett.* **2002**, *12*, 1395.
02BMCL1905	S. Batra, T. Srinivasan, S.K. Rastogi, B. Kundu, A. Patra, A.P. Bhaduri, *Bioorg. Med. Chem. Lett.* **2002**, *12*, 1905.
02BMCL3305	T.G.M. Dhar, Z. Shen, C.A. Fleener, K.A. Rouleau, J.C. Barrish, D.L. Hollenbaugh, E.J. Iwanowicz, *Bioorg. Med. Chem. Lett.* **2002**, *12*, 3305.
02CC256	M. Amat, M. Cantò, N. Llor, J. Bosch, *Chem. Commun.* **2002**, 526.
02CC438	S.P. Fearnley, E. Market, *Chem. Commun.* **2002**, 438.
02CC1128	T. Tamura, T. Mitsuya, H. Ishibashi, *Chem. Commun.* **2002**, 1128.
02CC1754	T. Aftab, R. Grigg, M. Ladlow, V. Sridharan, M. Thornton-Pett, *Chem. Commun.* **2002**, 1754.
02CC2704	F. Glorius, G. Altenhoff, R. Goddard, C. Lehmann, *Chem. Commun.* **2002**, 2704.
02CEJ4308	C. Mazet, L.H. Gade, *Chem. Eur. J.* **2002**, *8*, 4308
02CL302	X. Ding, Y. Ukaji, S. Fujinami, K. Inomata, *Chem. Lett.* **2002**, 302.
02CL314	Y. Ono, Y. Sugihara, A. Ishii, J. Nakayama, *Chem. Lett.* **2002**, 314.
02EJMC163	N.M. Silva, J.L.M. Tributino, A.L.P. Miranda, E.J. Barreiro, C.A.M. Fraga, *Eur. J. Med. Chem.* **2002**, *37*, 163.
02EJO1175	G. Faita, M. Mella, A. Mortoni, A. Paio, P. Quadrelli, P. Seneci, *Eur. J. Org. Chem.* **2002**, 1175.
02EJO1206	U. Chiacchio, A. Corsaro, V. Pistarà, A. Rescifina, D. Iannazzo, A. Piperno, G. Romeo, R.

	Romeo, G. Grassi, *Eur. J. Org. Chem.* **2002**, 1206.
02EJO1363	V. Wittmann, D. Lennartz, *Eur. J. Org. Chem.* **2002**, 1363.
02EJO1941	F.M. Cordero, F. Pisaneschi, M. Gensini, A. Goti, A. Brandi, *Eur. J. Org. Chem.* **2002**, 1941.
02EJO3855	T. Ohta, F. H. Kamizono, A. Kawamoto, K. Hori, I. Furukawa, *Eur. J. Org. Chem.* **2002**, 3855.
02EJO4211	D. Donati, S. Fusi, F. Ponticelli, *Eur. J. Org. Chem.* **2002**, 4211.
02H(58)521	A. Bertram, G. Pattenden, *Heterocycles*, **2002**, *58*, 521.
02HCA464	R. Martinez, H.A. Jimenez-Vazquez, A. Reyes, J. Tamariz, *Helv. Chim. Acta* **2002**, *85*, 464.
02HCA2383	T. Nishio, K. Shiwa, M. Sakamoto, *Helv. Chim. Acta* **2002**, *85*, 2383.
02JA12	L.E. Overman, T.P. Remarchuk, *J. Am. Chem. Soc.* **2002**, *124*, 12.
02JA591	S.-G. Kim, K.-H. Kim, J. Jung, S. K. Shin, K.H. Ahn, *J. Am. Chem. Soc.* **2002**, *124*, 591.
02JA2245	K.C. Nicolaou, T. Montagnon, P.S. Baran, Y.-L. Zhong, *J. Am. Chem. Soc.* **2002**, *124*, 2245.
02JA2560	P. Janvier, X. Sun, H. Bienaymé, J. Zhu, *J. Am. Chem. Soc.* **2002**, *124*, 2560.
02JA4968	F. Viton, G. Bernardinelli, E.P. Kundig,, *J. Am. Chem. Soc.* **2002**, *124*, 4968.
02JA6900	J.A. Johnson, N. Li, D. Sames, *J. Am. Chem. Soc.* **2002**, *124*, 6900.
02JA9030	J. Zhou, Y. Tang, *J. Am. Chem. Soc.* **2002**, *124*, 9030.
02JA9789	J. Liu, D.Y. Gin, *J. Am. Chem. Soc.* **2002**, *124*, 9789.
02JA11292	G.D. Wilkie, G.I. Elliot, B.S.J. Blagg, S.E. Wolkenberg, D.R. Soenen, M.M. Miller, S. Pollack, D.L. Boger, *J. Am. Chem. Soc.* **2002**, *124*, 11292.
02JA13097	D.M. Barnes, J. Ji, M.G. Fickes, M.A. Fitzgerald, S.A. King, H.E. Morton, F.A. Plagge, M. Preskill, S.H. Wagaw, S.J. Wittenberger, J. Zhang, *J. Am. Chem. Soc.* **2002**, *124*, 13097.
02JCS(P1)675	L. Vraničar, A. Meden, S. Polanc, M. Kočevar, *J. Chem. Soc., Perkin Trans. I* **2002**, 675.
02JCS(P1)1494	E.C. Davison, M.E. Fox, A.B. Holmes, S. D. Roughley, C.J. Smith, G.M. Williams, J.E. Davies, P.R. Raithby, J.P. Adams, I.T. Forbes, N.J. Press, M. Thompson, *J. Chem. Soc., Perkin Trans. I* **2002**, 1494.
02JCS(P1)2419	H.M.I. Osborn, N. Gemmell, L.M. Harwood, *J. Chem. Soc., Perkin Trans. I* **2002**, 2419.
02JMC2454	B. Frølund, A.T. Jørgensen, L. Tagmose, T.B. Stensbøl, H.T. Vestergaard, C. Engblom, U. Kristiansen, C. Sanchez, P. Krogsgaard-Larsen, T. Liljefors, *J. Med. Chem.* **2002**, *45*, 2454.
02JOC16	M. Shi, Y.-M. Shen, *J. Org. Chem.* **2002**, *67*, 16
02JOC66	T.M.V.D. Pinho e Melo, C.S.J. Lopes, A.M. d'A. Rocha Gonsalves, A.M. Beja, J.A. Paixao, M.R. Silva, L.A. da Vega, *J. Org. Chem.* **2002**, *67*, 66.
02JOC759	V. Capriati, S. Florio, R. Luisi, M.T. Rocchetti, *J. Org. Chem.* **2002**, *67*, 759.
02JOC764	D. Kim, J. Lee, P.J. Shim, J.I. Lim, H. Jo, S. Kim, *J. Org. Chem.* **2002**, *67*, 764.
02JOC772	D. Kim, J. Lee, P.J. Shim, J.I. Lim, T. Doi, S. Kim, *J. Org. Chem.* **2002**, *67*, 772.
02JOC974	A. Lei, G. Liu, X. Lu, *J. Org. Chem.* **2002**, *67*, 974.
02JOC1496	C. Agami, S. Comesse, C. Kadouri-Puchot, *J. Org. Chem.* **2002**, *67*, 1496.
02JOC1972	A.L. Williams, T.A. Grillo, D.L. Comins, *J. Org. Chem.* **2002**, *67*, 1972.
02JOC2435	H. Tian, X. She, H. Yu, L. Shu, Y. Shi, *J. Org. Chem.* **2002**, *67*, 2435
02JOC2665	A. Cipiciani, F. Fringuelli, O. Piermatti, F. Pizzo, R. Ruzziconi, *J. Org. Chem.* **2002**, *67*, 2665.
02JOC3045	C. Spanka, B. Clapham, K.D. Janda, *J. Org. Chem.* **2002**, *67*, 3045.
02JOC3184	S. Swaleh, J. Liebscher, *J. Org. Chem.* **2002**, *67*, 3184.
02JOC3601	A. Couture, P. Grandclaudon, C. Hoarau, J. Cornet, J.-P. Hénichart, R. Houssin, *J. Org. Chem.* **2002**, *67*, 3601.
02JOC4045	T.M.V.D. Pinho e Melo, M.I.L. Soares, A.M.d'A. Rocha Gonsalves, J.A. Paixão, A.M. Beja, M. Ramos Silva, L. Alte da Veiga, J. Costa Pessoa, *J. Org. Chem.* **2002**, *67*, 4045.
02JOC4450	A. Bacchi, M. Costa, B. Gabriele, G. Pelizzi, G. Salerno, *J. Org. Chem.* **2002**, *67*, 4450
02JOC4612	S. Spyroudis, N. Xanthopoulou, *J. Org. Chem.* **2002**, *67*, 4612.
02JOC5170	A. Iwata, H. Tang, A. Kunai, *J. Org. Chem.* **2002**, *67*, 5170
02JOC5673	W.-C. Cheng, Y. Liu, M. Wong, M.M. Olmstead, K.S. Lam, M.J. Kurth, *J. Org. Chem.* **2002**, *67*, 5673.
02JOC5783	A. Patra, S. Batra, B.S. Joshi, R. Roy, B. Kundu, A.P. Bhaduri, *J. Org. Chem.* **2002**, *67*, 5783.
02JOC6253	S. Buscemi, A. Pace, I. Pibiri, N. Vivona, *J. Org. Chem.* **2002**, *67*, 6253.
02JOC6564	S.H. Hwang, M.J. Kurth, *J. Org. Chem.* **2002**, *67*, 6564.
02JOC6646	A.J. Wills, Y. Krishnan-Ghosh, S. Balasubramanian, *J. Org. Chem.* **2002**, *67*, 6646.
02JOC7004	B. Alcaide, P. Almendros, J.M. Alonso, M.F. Aly, C. Pardo, E. Sàez, M.R. Torres, *J. Org. Chem.* **2002**, *67*, 7004.
02JOC7283	F. Palacios, D. Aparicio, A.M. Ochoa de Retana, J.M. de los Santos, J.I. Gil, J.M. Alonso, *J. Org. Chem.* **2002**, *67*, 7283.

02JOC8197	M. Malkoch, K. Hallman, S. Lutsenko, A. Hult, E. Malstrom, C. Moberg, *J. Org. Chem.* **2002**, *67*, 8197.
02JOM196	T. Kitamura, Y. Mansei, Y. Fujiwara, *J. Organomet. Chem.* **2002**, *646*, 196.
02OL39	L. Quaranta, O. Corminboeuf, P. Renaud, *Org. Lett.* **2002**, *4*, 39.
02OL741	W.-C. Cheng, M. Wong, M.M. Olmstead, M.J. Kurth, *Org. Lett.* **2002**, *4*, 741.
02OL1055	S. Oishi, A. Niida, T. Kamano, Y. Odagaki, H. Tamamuro, A. Otaka, N. Hamanaka, N. Fujii, *Org. Lett.* **2002**, *4*, 1055.
02OL1111	N. Mita, O. Tamura, H. Ishibashi, M. Sakamoto, *Org. Lett.* **2002**, *4*, 1111.
02OL1213	D. Yoo, J.S. Oh, Y.G. Kim, *Org. Lett.* **2002**, *4*, 1213.
02OL1227	V.K. Aggarwal, S.J. Roseblade, J.K. Barrel, R. Alexander, *Org. Lett.* **2002**, *4*, 1227.
02OL1551	V. Capriati, L. DeGennaro, R. Favia, S. Florio, R. Luisi, *Org. Lett.* **2002**, *4*, 1551
02OL1631	Q. Xia, B. Ganem, *Org. Lett.* **2002**, *4*, 1631.
02OL1907	T. Ishikawa, T. Kudoh, J. Yoshida, A. Yasuhara, S. Manabe, S. Saito, *Org. Lett.* **2002**, *4*, 1907.
02OL2091	T.G.M. Dhar, J. Guo, Z. Shen, W.J. Pitts, H.H. Gu, B.-C. Chen, R. Zhao, M.S. Bednarz, E.J. Iwanowicz, *Org. Lett.* **2002**, *4*, 2091.
02OL2457	T. Mita, N. Ohtsuki, T. Ikeno, T. Yamada, *Org. Lett.* **2002**, *4*, 2457.
02OL2485	N.F. Langille, L.A. Dakin, J.S. Panek, *Org. Lett.* **2002**, *4*, 2485.
02OL2665	T. Morwick, M. Hrapchak, M. De Turi, S. Campbell, *Org. Lett.* **2002**, *4*, 2665.
02OL2905	K.J. Hodgetts, M.T. Kershaw, *Org. Lett.* **2002**, *4*, 2905.
02OL3343	M.P. Sibi, H. Gasegawa, S.R. Ghorpade, *Org. Lett.* **2002**, *4*, 3343.
02OL3399	S. Rajram, M.S. Sigman, *Org. Lett.* **2002**, *4*, 3399.
02OL3533	S. Peddibhotia, S. Jayakumar, J.J. Tepe, *Org. Lett.* **2002**, *4*, 3533.
02OL3927	A. Comejo, J.M. Fraile, J.I. Garcia, E. Garcia-Verdugo, M.J. Gil, G. Legarreta, S.V. Luis, V. Martinez-Merino, J.A. Mayoral, *Org. Lett.* **2002**, *4*, 3927.
02OL4101	G.K. Tranmer, W. Tam, *Org. Lett.* **2002**, *4*, 4101.
02OL4301	I. Nomura, C. Mukai, *Org. Lett.* **2002**, *4*, 4301.
02OM1077	M. Gomez, S. Jansat, G. Muller, M.A. Maestro, J. Mahia, *Organometallics* **2002**, *21*, 1077.
02OM5204	V. Cesar, S. Bellemin-Laponnaz, L.H. Gade, *Organometallics* **2002**, *21*, 5204.
02S605	T.M.V.D. Pinho e Melo, C.S.J. Lopes, A.M.d'A. Rocha Gonsalves, R.C. Storr, *Synthesis* **2002**, 605.
02S1663	Y.-J. Shang, Y.-G. Wang, *Synthesis* **2002**, 1663.
02S1969	Q. Xia, B. Ganem, *Synthesis* **2002**, 1969.
02SL73	J.L. García Ruano, F. Bercial, A. Fraile, M.R. Martín, *Synlett* **2002**, 73.
02SL85	B. Alcaide, C. Pardo, E. Sàez, *Synlett* **2002**, 85
02SL1113	R. Fischer, A. Drucková, L. Fišera, A. Rybár, C. Hametner, M.K. Cyrański, *Synlett* **2002**, 1113.
02SL1257	D. Renard, H. Rezaei, S.Z. Zard, *Synlett* **2002**, 1257.
02SL1344	O. Tamura, A. Toyao, H. Ishibashi, *Synlett* **2002**, 1344.
02SL1359	M. Lemaire, N. Veny, T. Gefflaut, E. Gallienne, R. Chenevert, J. Bolte, *Synlett* **2002**, 1359.
02SL1371	S. Kanemasa, *Synlett* **2002**, 1371.
02SL1691	H.C. Kim, S.W. Woo, M.J. Seo, D.J. Jeon, Z. No, H.R. Kim, *Synlett* **2002**, 1691.
02T581	D. Iannazzo, A. Piperno, V. Pistarà, A. Rescifina, R. Romeo, *Tetrahedron* **2002**, *58*, 581.
02T2659	L. Di Nunno, A. Scilimati, P. Vitale, *Tetrahedron* **2002**, *58*, 2659.
02T3021	R. Olivera, M. SanMartin, I. Tellitu, E. Domínguez, *Tetrahedron* **2002**, *58*, 3021.
02T3281	C. Hajji, M.L. Testa, E. Zaballos-Garcia, R.J. Zaragoza, J. Server-Carriò, J. Sepulveda-Arques, *Tetrahedron* **2002**, *58*, 3281.
02T3379	L. Bianchi, C. Dell'Erba, A. Gabellini, M. Novi, G. Petrillo, C. Tavani, *Tetrahedron* **2002**, *58*, 3379.
02T4505	A. Avenoza, J.H. Busto, C. Cativiela, A. Dordal, J. Frigola, J.M. Peregrina, *Tetrahedron* **2002**, *58*, 4505.
02T5957	R.W. Bates, K. Sa-Ei, *Tetrahedron* **2002**, *58*, 5957.
02T7805	T. Mizuno, J. Takahashi, A. Ogawa, *Tetrahedron* **2002**, *58*, 7805.
02T9381	L. Colombo, M. di Giacomo, P. Ciceri, *Tetrahedron* **2002**, *58*, 9381.
02T9613	A. Kamimura, Y. Kaneko, A. Ohta, K. Matsuura, Y. Fujimoto, A. Kakehi, S. Kanemasa, *Tetrahedron* **2002**, *58*, 9613.

02T9853	A. Guirado, R. Andreu, J. Galvez, P.G. Jones, *Tetrahedron* **2002**, *58*, 9853.
02TA161	Y.-Z. Zhu, Z.-P. Li, J.-A. Ma, F.-Y. Tang, L. Kang, Q.-L. Zhou, A.S.C. Chan, *Tetrahedron: Asymmetry* **2002**, *13*, 161.
02TA167	P. Merino, J. Revuelta, T. Tejero, U. Chiacchio, A. Rescifina, A. Piperno, G. Romeo, *Tetrahedron: Asymmetry* **2002**, *13*, 167.
02TA173	P. Merino, J.A. Mates, J. Revuelta, T. Tejero, U. Chiacchio, G. Romeo, D. Iannazzo, R. Romeo, *Tetrahedron: Asymmetry* **2002**, *13*, 173.
02TA333	G. Desimoni, G. Faita, A. Galbiati, D. Pasini, P. Quadrelli, F. Rancati, *Tetrahedron: Asymmetry*, **2002**, *13*, 333.
02TA469	Y.-B. Zhou, F.-Y. Tang, H.-D. Xu, X.-Y. Wu, J.-A. Ma, Q.-L. Zhou, *Tetrahedron: Asymmetry* **2002**, *13*, 469.
02TA1554	J. Clariana, J. Comelles, M. Moreno-Manas, A. Vallribera, *Tetrahedron: Asymmetry* **2002**, *13*, 1551.
02TA1841	K. Hiroi, K. Watanabe, *Tetrahedron: Asymmetry* **2002**, *13*, 1841.
02TA2497	M. Chrzanowska, *Tetrahedron: Asymmetry* **2002**, *13*, 2497.
02TA2535	M. Desroses, F. Chéry, A. Tatibouët, O. De Lucchi, P. Rollin, *Tetrahedron: Asymmetry* **2002**, *13*, 2535.
02TA2541	F. Palacios, A.M. Ochoa de Retana, J.I. Gil, J.M. Alonso, *Tetrahedron: Asymmetry* **2002**, *13*, 2541.
02TL347	M.A. Clark, Q. Wang, B. Ganem, *Tetrahedron Lett.* **2002**, *43*, 347.
02TL603	A.M.G. Silva, A.C. Tome, M.G.P.M.S. Neves, A.M.S. Silva, J.A.S. Cavaleiro, D. Perrone, A. Dondoni, *Tetrahedron Lett.* **2002**, *43*, 603.
02TL657	S. Kanemasa, N. Ueno, M. Shirahase, *Tetrahedron Lett.* **2002**, *43*, 657.
02TL935	R.A. Miller, R.M. Smith, S. Karady, R.A. Reamer, *Tetrahedron Lett.* **2002**, *43*, 935.
02TL1429	A. Avenoza, J.H. Busto, C. Cativiela, J.M. Peregrina, F. Rodriguez, *Tetrahedron Lett.* **2002**, *43*, 1429.
02TL1535	X. Zhang, W. Ling, L. Gang, A. Mi, X. Cui, Y. Jiang, M.C.K. Choi, A.S.C. Chan, *Tetrahedron Lett.* **2002**, *43*, 1535.
02TL1743	M. Pastor, H. Adolfsson, *Tetrahedron Lett.* **2002**, *43*, 1743.
02TL2521	C. Agami, L. Dechoux, S. Hebbe, *Tetrahedron Lett.* **2002**, *43*, 2521.
02TL2741	S. Cicchi, M. Corsi, M. Marradi, A. Goti, *Tetrahedron Lett.* **2002**, *43*, 2741.
02TL2811	G. Xu, S.R. Gilbertson, *Tetrahedron Lett.* **2002**, *43*, 2811.
02TL3565	C.-S. Li, E. Lacasse, *Tetrahedron Lett.* **2002**, *43*, 3565.
02TL3985	A. Cwik, Z. Hell, A. Hegedus, Z. Finta, Z. Horvath, *Tetrahedron Lett.* **2002**, *43*, 3985.
02TL4613	H.-G. Weinig, P. Passacantilli, M. Colapietro, G. Piancatelli, *Tetrahedron Lett.* **2002**, *43*, 4613.
02TL5043	G.M. Makara, P. Schell, K. Hanson, D. Moccia, *Tetrahedron Lett.* **2002**, *43*, 5043.
02TL5287	A. Chatterjee, S. C. Jha, N. N. Joshi, *Tetrahedron Lett.* **2002**, *43*, 5287.
02TL5407	B. Clapham, S.-H. Lee, G. Koch, J. Zimmermann, K.D. Janda, *Tetrahedron Lett.* **2002**, *43*, 5407.
02TL5587	N. S. Shaikh, V. H. Deshpande, A. V. Bedekar, *Tetrahedron Lett.* **2002**, *43*, 5587.
02TL5863	M. Feroci, A. Gennaro, A. Ines, M. Orsini, L. Palombi, *Tetrahedron Lett.* **2002**, *43*, 5863.
02TL6997	M.P. John, S.A. Hermitage, J.R. Titchmarsh, *Tetrahedron Lett.* **2002**, *43*, 6997
02TL7035	K. Itoh, S. Takahashi, T. Ueki, T. Sugiyama, T.T. Takahashi, C.A. Horiuchi, *Tetrahedron Lett.* **2002**, *43*, 7035.
02TL7101	S. Knapp, C. Yang, T. Haimowitz, *Tetrahedron Lett.* **2002**, *43*, 7101.
02TL8327	S.K. Rastogi, G.K. Srivastava, S.K. Singh, R.K. Grover, R. Roy, B. Kundu, *Tetrahedron Lett.* **2002**, *43*, 8327.
02TL8955	M. Matsumoto, T. Sakuma, N. Watanabe, *Tetrahedron Lett.* **2002**, *43*, 8955.
02TL9069	N. Tahara, T. Fukuda, M. Iwao, *Tetrahedron Lett.* **2002**, *43*, 9069.
02TL9357	A. Brandi, S. Cicchi, V. Paschetta, D.G. Pardo, J. Cossy, *Tetrahedron Lett.* **2002**, *43*, 9357.
02TL9461	S.J. Tantry, Kantharaju, V.V.S. Babu, *Tetrahedron Lett.* **2002**, *43*, 9461.

Chapter 6.1

Six-Membered Ring Systems: Pyridines and Benzo Derivatives

D. Scott Coffey, Stanley P. Kolis and Scott A. May
Chemical Process Research & Development, Eli Lilly & Company, Indianapolis, IN, USA
coffey_scott@lilly.com, spk@lilly.com and may_scott_a@lilly.com

6.1.1 INTRODUCTION

Pyridines and their benzo-derivatives have played an important role in the synthesis of biologically active synthetic and natural substances. As a result, the construction of this molecular architecture has attracted the attention of a diverse array of synthetic methodologies. Notably, transition metal catalysis, radical reactions and cycloaddition chemistry-based methods have been developed for the construction of this important ring system. Detailed herein is a summary of the methods developed for the synthesis of pyridines, quinolines, isoquinolines and piperidines that were disclosed in the literature in 2002. Rather than survey all existing methods for the construction of these compound classes, this review will serve as a supplement and update to the review published last year in this series.

6.1.2 PYRIDINES

6.1.2.1 Preparation of Pyridines

The synthesis of pyridines and pyridine derivatives has been an active area of research over the past year. One particularly active area has been [2+2+2] cycloadditon reactions involving a nitrile and two acetylenes. Sato and co-workers report a titanium-mediated process wherein two different unsymmetrical acetylenes (**1** and **2**) combine to form titanacyclopentadiene **3**. Intermediate **3** undergoes [4+2] cycloaddition with a sulfonylnitrile regioselectively to afford bicyclic titanium complex **4**, which degrades under acidic conditions to afford the substituted pyridine (**5**→**6**) <02JA3518>. Later in the year, Takahashi reported a similar strategy for pyridine synthesis, but with zirconium as the metal of choice <02JA5059>. Additionally, Heller and co-workers have disclosed a photocatalyzed cobalt [2+2+2] cycloaddition reaction for the synthesis of 2-substituted pyridines <02JOC4414>.

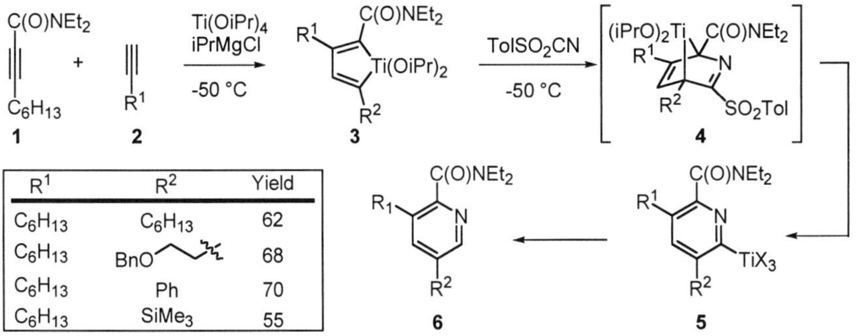

Scheme 1

R¹	R²	Yield
C$_6$H$_{13}$	C$_6$H$_{13}$	62
C$_6$H$_{13}$	BnO~~	68
C$_6$H$_{13}$	Ph	70
C$_6$H$_{13}$	SiMe$_3$	55

Several strategies for pyridine synthesis involving [4+2] cycloaddition reactions have been reported. Zhu and co-workers have disclosed full details of their ammonium chloride-promoted four-component synthesis of fused pyridines <02JA2560>. This work relies upon an intramolecular [4+2] cycloaddition and fragmentation to afford the desired product. In similar fashion, fused thienopyridines have been made through intramolecular [4+2] strategies involving 1,2,4-triazenes <02JCR(S)60, 02CPB463>.

Intermolecular [4+2] cycloaddition strategies have also been used successfully. Moody and co-workers have reported the synthesis of a core piece of the thiopeptide antibiotics through a [4+2] cycloaddition <02CC1760>. For example, 2-azadiene **7** and 2-thiazolyl dienophile **8** were submitted to microwave heating (180 °C) for 15 minutes. The substituted pyridine product **9** was isolated in modest yield. Palacios has also reported an intermolecular [4+2] approach involving 2-azadienes <02JOC2131>.

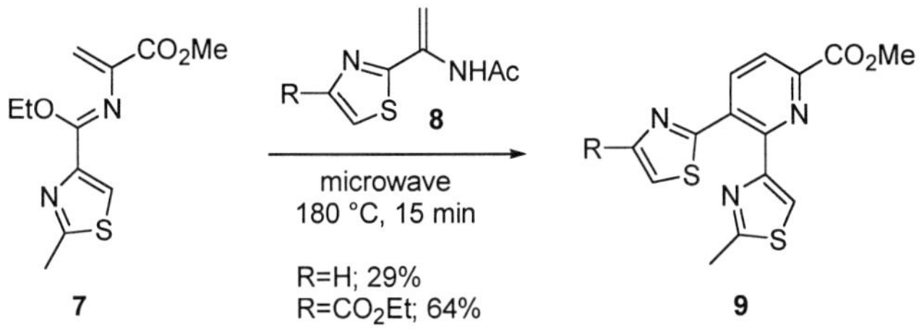

Scheme 2

Perhaps the most widely reported strategy in 2002 for pyridine synthesis was through cyclocondensation reactions. One aspect of this work is the cyclization of amino-heptadienones and aminoheptadienoate esters. Bagley has published variations on the Bohlmann-Rahtz reaction wherein the final cyclodehydration reaction can be promoted by acid catalysis <02JCS(P1)1663, 02CC1682> or microwave heating <02TL8331>. An impressive example of

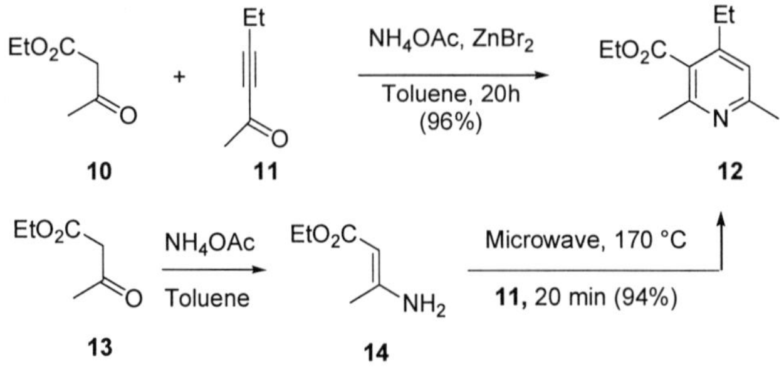

Scheme 3

this methodology is shown in Scheme 3. Treatment of ethyl acetoacetate (**10**) and two equivalents of hex-3-yn-2-one (**11**) in the presence of ammonium acetate and an acid catalyst in toluene affords the tetrasubstituted pyridine **12** in excellent yield. The same pyridine (**12**) can also be synthesized in high yield by heating ethyl β-aminocrotonate (**14**) and **11** under microwave conditions. Additionally, Koike reports the formation of 1,4-dihydropyridines from aminoheptadienoate esters <02CPB484> and Tomioka has published an approach toward fused dihydrofurano[2,3-*b*]pyridines <02JHC743>. Reactions involving the condensation of ammonia and 1,5-diones have also been reported <02H421, 02JHC1035>.

Müller reports a four component, one-pot synthesis of pyridines <02TL6907>. For example, aryl halide **15** and propargylic alcohol **16** were combined in the presence of copper and palladium to afford enone **17**. The addition of cyclic enamine **18** led to Michael addition and the subsequent cyclocondensation was achieved by adding ammonium chloride and acetic acid (**19→20**). Other multicomponent approaches to substituted pyridines have been reports by Litvinov <02RCBIE362>, Elkholy <02SC3493> and Veronese <02T9709>.

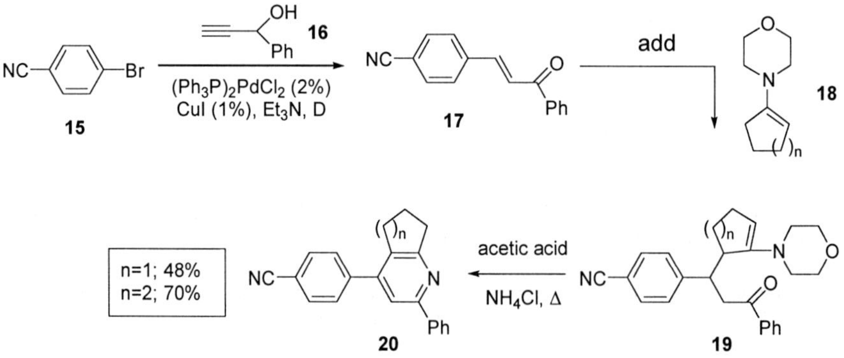

Scheme 4

Several reports have detailed the electrocyclization of aza-heptadienes. In two separate accounts, Brandsma described the formation of disubstituted pyridines by addition of lithiated allenic ethers to thioisocyanates and subsequent thermal cylization <02TL9679, 02RJOC917>. Formation of 2-methylthio substituted pyridines was also achieved under unusual Vilsmeier-Haack conditions <02TL2273>. The authors speculate that dehydration of **22** leads to butadiene **23**. Butadiene **23**, in turn, reacts with the Vilsmeier reagent to afford iminium salt **24** which cyclizes in modest yields to the pyridine **25** (Scheme 5). Other approaches to pyridines include an interesting rearrangement of amidines <02T1213> and fragmentation of Dewar pyridines <02S497>. Finally, new methods for oxidation of 1,4-dihydropyridines have also been reported <02T5069, 02SC793>.

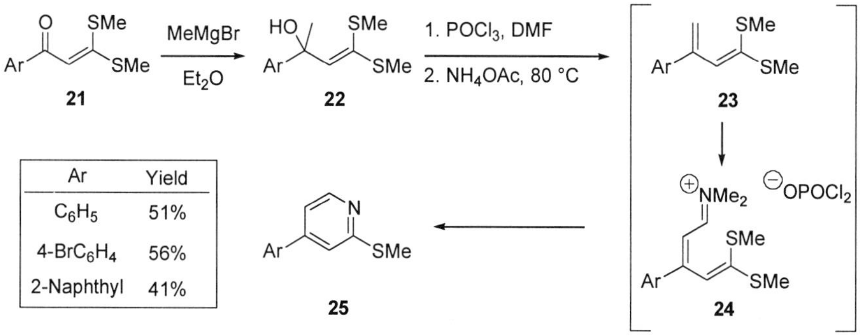

Ar	Yield
C_6H_5	51%
4-BrC_6H_4	56%
2-Naphthyl	41%

Scheme 5

6.1.2.2 Reactions of Pyridines

The use of substituted pyridines in organic synthesis has broad application. The activation of the pyridine ring toward nucleophilic attack is well known in the literature. The products of such reactions are often dihydropyridines which can serve as intermediates in more complex synthetic strategies. Rudler and co-workers have reported on the nucleophilic addition of bis(trimethylsilyl)ketene acetals to pyridine (**26**). The 1,4-addition product **27** was then cyclized with iodine to afford bicycle **28** in 90% overall yield <02CC940>. Yamada has elegantly shown that facial selectivity can be achieved and chiral 1,4-dihydropyridines accessed in high yield and de (**29**→**30**) <02JA8184>.

The pyridine nitrogen has also been used as an effective handle for additional functionalization of the 2-position. Fort has published several papers this past year detailing the use of the unimetal superbase BuLi-LiDMAE. This work includes a labeling study on the direct ortho metallation of 2-heterosubstituted pyridines <02JOC234>, the functionalization of 3,5-lutidine <02SL628> and the formation of 2-piperidylpiperazines <02OL1759>. The latter is of particular interest since it represents a solid phase approach. Polymer bound 2-pyridylpiperazine (2% cross-linked chloromethyl polystyrene) **31** was treated with BuLi-LiDMAE then cooled to –5 °C before treating with an electrophile of choice. The crude polymeric product was then treated with methyl chloroformate to afford the final products **32** in good overall yields.

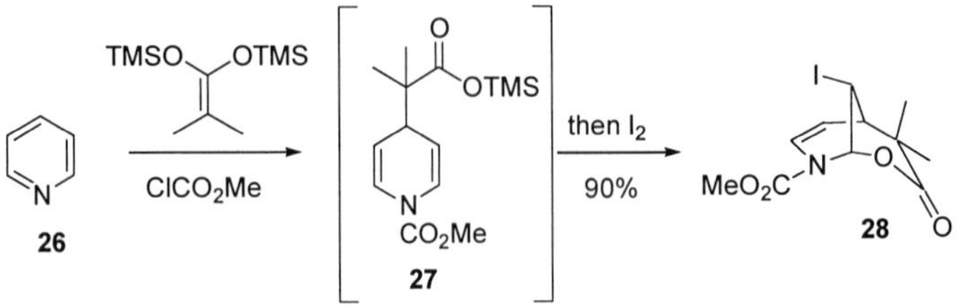

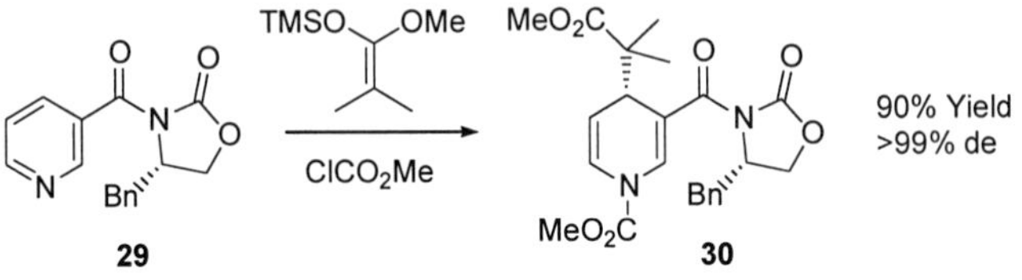

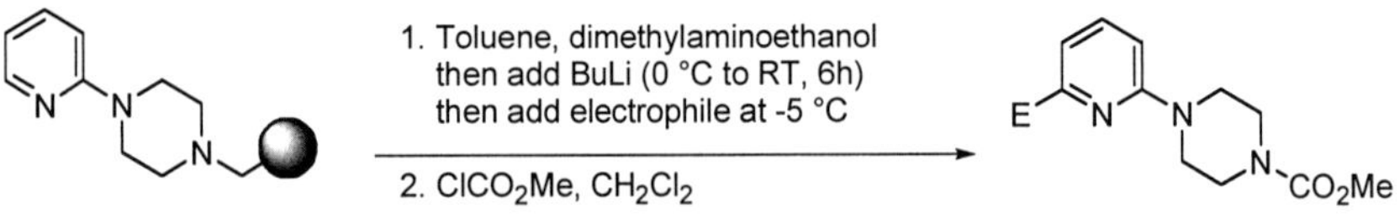

Other noteworthy examples of directing effects include Kelly's selective monacylation of a diamine <02OL2653> and Schlosser's silyl-mediated halogen exchange of 2-halopyridines <02EJO4181>.

The use of pyridines in cross coupling reactions was widespread in 2002. Conceptually this is not new, however the number of examples clearly illustrates the importance of the pyridine group in organic synthesis. For example, pyridines participated as cross-coupling partners for the Pd-catalyzed Negishi reaction <02SL808>, the Suzuki reaction <02JCS(P1)581>, Kumata-type cross-coupling <02T4429>, and Hartwig-Buchwald amination reactions <02TL7945>. Several reports detailed the preparation of pyridinyl boronic acids, which were later used in Suzuki reactions <02T4369, 02TL4258, 02JOC5394>. Other noteworthy examples of pyridines in metal-mediated reactions include a dissolving metal/radical cyclization reaction <02SL331> and a solid phase oxoalkylation reaction involving lead tetraacetate <02RCB1812>.

The pyridine functionality can also be part of a more structurally complex molecule. Oftentimes, pyridines serve as the starting materials for the preparation of fused or annulated target molecules <02TL1093, 02SL155, 02H21>. Other transformations such as the Diels-Alder reaction (**33→35**) <02T3655> and 1,3-dipolar cyloaddition reactions <02JCR(S)560> offer ready access to more complex products.

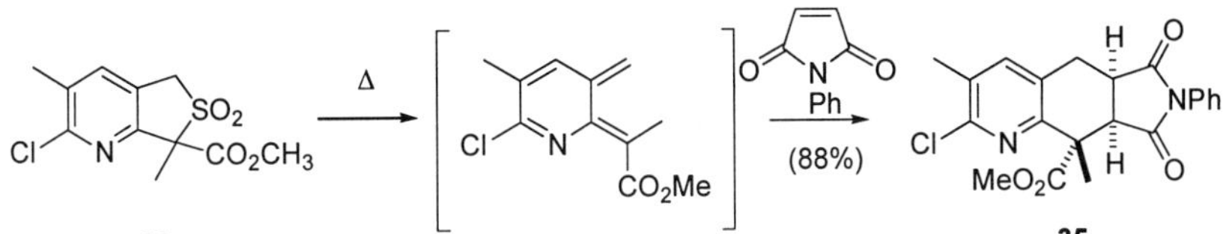

Finally, 2-halopyridines have been reacted with nucleophiles under various conditions <02JHC347>. For example, facile reaction between 2-fluoro, iodo, bromo or chloro pyridine and a variety of nucleophiles (**36→37**) can be promoted on brief exposure to microwave heating <02T4931>.

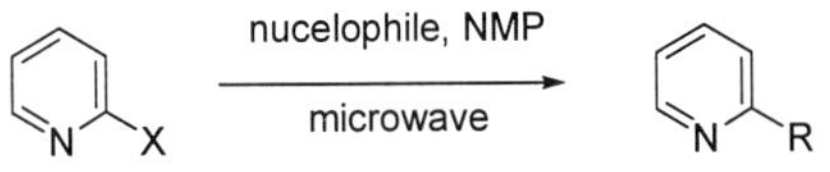

X	Nucleophile	Temp (°C)	R	Yield (%)
I	MeSNa	110	SMe	99
Br	BnOH	100	OBn	91
Cl	PhCH$_2$CN	110	CH(CN)Ph	68
F	PhONa	110	OPh	56

6.1.2.3 Pyridine *N*-Oxides and Pyridinium Salts

The synthesis and utilization of pyridinium salts and pyridine *N*-oxides has also been a topic of interest in 2002. Pyridine *N*-oxides have been synthesized from the corresponding pyridines using a number of different oxidants. Kulkarni reported a new method using 30% hydrogen peroxide in the presence of molecular sieve catalysts <02JMC(A)109>. Sain and co-workers detailed their use of catalytic ruthenium chloride and molecular oxygen to promote *N*-oxidation <02CC1040>. This exceptionally mild method involved simply stirring the reagents at room temperature in dichloroethane while bubbling oxygen through the solution. While metals have aided in oxidation of pyridines to *N*-oxides, they also have been used as effective catalysts for deoxygenation. For example, Yoo reports the efficient deoxygenation of *N*-oxide with bis(cyclopentadienyl)titanium(IV) dichloride/indium <02BKCS797>.

Pyridinium salts have been shown to be effective starting materials for synthesis. In Bennasar's total synthesis of (+)-16-epivinoxine and (-)-vinoxine, the chiral enolate of **38** was added to pyridinium **39** with modest facial control to provide dihydropyridine intermediate **40**. Cyclization afforded **41** in a single pot, albeit in modest yield (Scheme 6) <02TA95>.

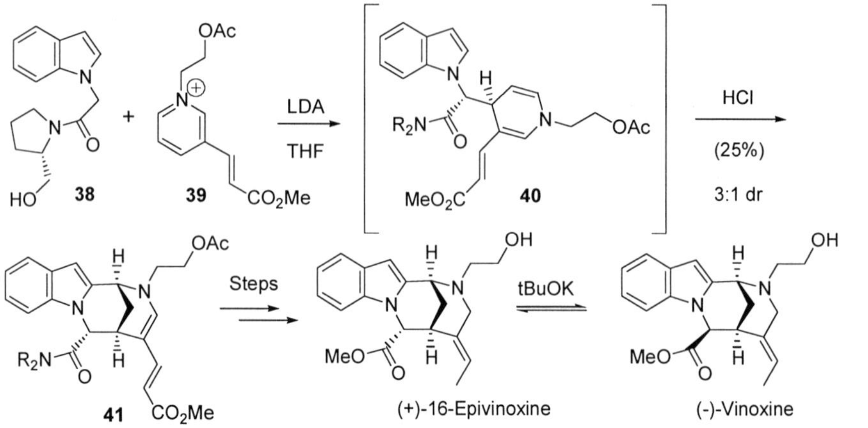

Scheme 6

Pyridinium salts were reduced to dihydropyridines *via* radical pathways induced by Zn-Cu and sonication <02CC850>. In a novel approach to 4-arylpyridines, pyridinium salt **42** was converted into dihydropyridine **43** on treatment with sulfite ion. After ring opening (**43**→**44**) a second equivalent of the pyridinium salt (**42**) was added with base to form aminoheptatriene **45**. Cyclization and re-aromatization (**46**→**47**) afforded the product in modest yield (Scheme 7)<02EJO4123>. The chemistry of dihydropyridines was reviewed in 2002 by Lavilla <02JCS(P1)1141>.

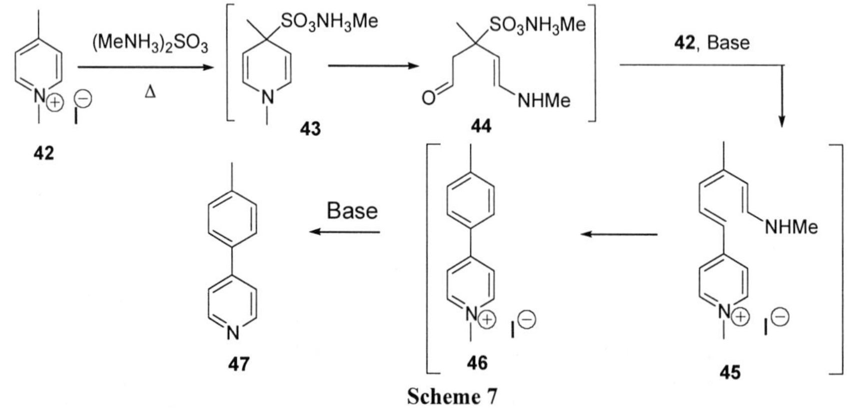

Scheme 7

Other reports include the formation of annulated pyridinium salts <02RJOC424, 02*EJO*375>, activation of carbocations by pyridinium salts <02TL6841>, and Stille reactions on pyridinium cations <02SL1904>. Finally two reports detail pyridine-based brominating reagents. Muathen details the use of pyridinium dichlorobromate <02S169> and Plumet reports on ipsobromodeformylation with pyridinium tribromide <02TL9303>.

6.1.3 QUINOLINES

6.1.3.1 Preparation of Quinolines

Organometallic reagents and methods continue to provide convenient access to quinolines and quinoline derivatives. Functionalized Grignard reagents prepared from *o*-iodoanilines by iodide-magnesium exchange were shown to be useful precursors to functionalized quinoline derivatives <02OL1819>. A zinc(II)-mediated alkynylation-cyclization of *o*-trifluoroanilines **48** provided an efficient route to 4-trifluoromethyl-substituted quinoline derivatives **49** <02JOC9449>. Preparation of 1,2,3,4-tetrahydroquinolines *via* an intramolecular amination of aryl halides mediated by copper(I) iodide and cesium acetate was reported as well <02SL231>. Samarium diiodide promoted reductive cyclization of 2-nitro-1,3-diphenyl-2-propen-1-one <02SC3617> and indium-mediated reductive cyclization of an analogous nitro propenal <02H467> were shown to provide quinoline derivatives. Nitroarenes undergo reductive cyclization with 3-amino-1-propanols in dioxane and water in the presence of a ruthenium catalyst and tin(II) chloride dihydrate to afford the corresponding quinolines <02JHC291>. Additionally, dicobaltoctacarbonyl was demonstrated to be an effective catalyst for the conversion of diallylanilines and arylimines in the presence of diallylaniline to the corresponding quinolines <02JMC(A)565>. Quinolines and isoquinolines were also prepared by a coupling reaction of zirconacyclopentadienes with the corresponding dihalopyridines <02JA576>.

Cycloaddition reactions continue to be a powerful tool for constructing quinoline derivatives. Ketene dithioacetals such as **50** were utilized as dienophiles in Sc(OTf)$_3$ catalyzed, aza-Diels-Alder reactions with *N*-arylimines to afford tetrahydroquinolines such as **51** with selectivities ranging from 1:1 to 100:0 anti:syn (Scheme 8). Further manipulation of **51** afforded 2,3-tetrahydroquinolines **52**, 4-quinolone **53** and 2,3-dihydroquinolone **54** <02OL4411>. The Lewis acid catalyzed [4 + 2] cycloaddition of *N*-acetyl-2-azetine with imines prepared from aromatic amines was reported to afford cycloadducts that would react with aromatic amines to afford 2,3,4-trisubstituted 1,2,3,4-tetrahydroquinolines with good selectivity <02CC444>. Another method to prepare 2,3,4-trisubstituted 1,2,3,4-tetrahydroquinolines using an InCl$_3$ catalyzed domino reaction of aromatic amines and cyclic enol ethers or 2-hydroxy cyclic ethers was also reported <02JOC3969>. Additionally, an approach to pyrrolo[3,3-*c*]quinolones *via* an intramolecular azomethine ylide alkene cycloaddition reaction was reported <02TL1171>.

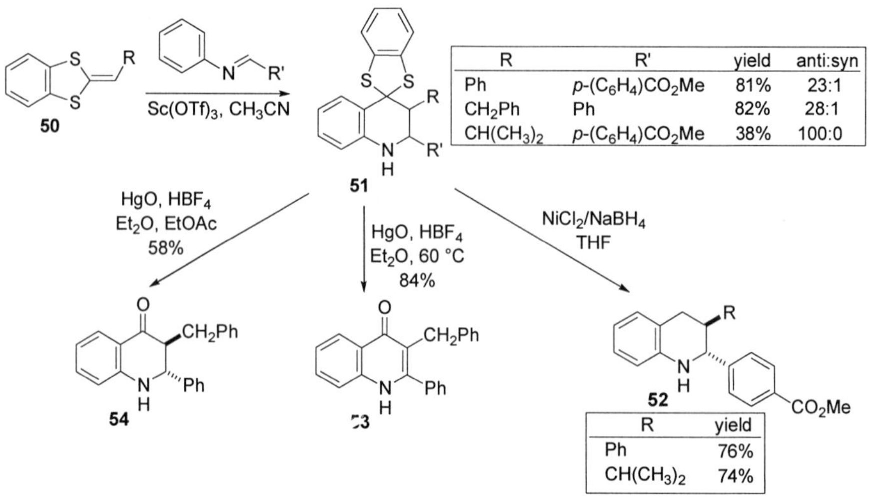

R	R'	yield	anti:syn
Ph	p-$(C_6H_4)CO_2Me$	81%	23:1
CH_2Ph	Ph	82%	28:1
$CH(CH_3)_2$	p-$(C_6H_4)CO_2Me$	38%	100:0

R	yield
Ph	76%
$CH(CH_3)_2$	74%

Scheme 8

Baylis-Hillman adducts such as **55** and **56** derived from 2-nitrobenzaldehydes were shown to function as useful precursors to functionalized (*1H*)-quinol-2-ones and quinolines. Treatment of **55** and **56** with iron and acetic acid at 110 °C afforded **57** and **58**, respectively <02T3693>. A variety of other cyclization reactions utilized in the preparation of the quinoline scaffold were also reported. An iridium-catalyzed oxidative cyclization of 3-(2-aminophenyl)propanols afforded 1,2,3,4-tetrahydroquinolines <02OL2691>. The intramolecular cyclization of aryl radicals to prepare pyrrolo[3,2-*c*]quinolines was studied <02T1453>. Additionally, photocyclization reactions of *trans-o*-aminocinnamoyl derivatives were reported to provide 2-quinolones and quinolines <02JHC61>. Enolizable quinone and mono- and diimide intermediates were shown to provide quinolines *via* a thermal 6π-electrocyclization <02OL4265>. Quinoline derivatives were also prepared from nitrogen-tethered 2-methoxyphenols. The corresponding 2-methoxyphenols were subjected to a iodine(III)-mediated acetoxylation which was followed by an intramolecular Michael addition to afford the quinoline

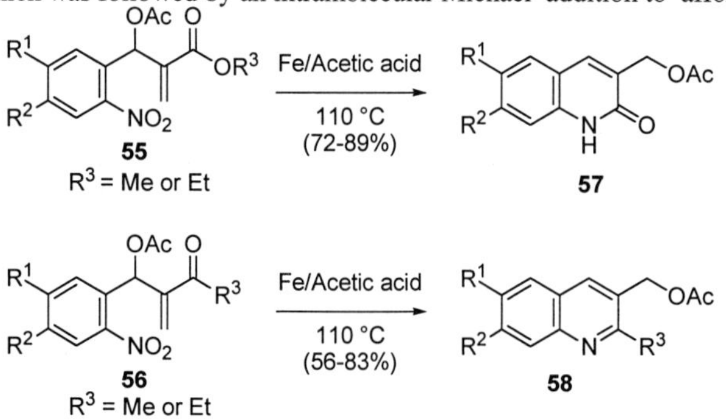

derivatives <02JOC3425>. Furthermore, preparation of 2,2,4-substituted 1,2-dihydroquinolines *via* a modified Skraup reaction that utilized lanthanide catalysis and microwave technology was also reported <02TL3907>.

6.1.3.2 Reactions of Quinolines

The intermolecular [2 + 2] photocycloaddition reactions of 2-quinolones such as **59** mediated by a chiral lactam host **60** or *ent*-**60** were reported with enantioselectivities of 93%. The intermolecular version of this reaction was also shown to be highly enantioselective. One example shown in Scheme 9 afforded cycloadduct **63** in 92% *ee* with a diastereoselectivity of >95:5 <02JA7982>.

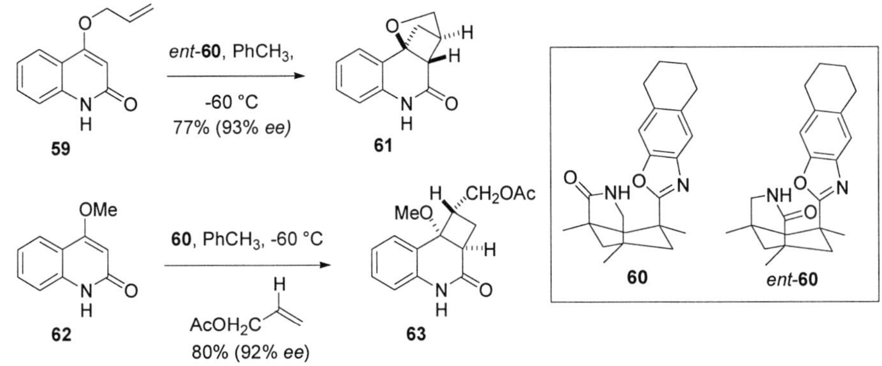

Scheme 9

The aerobic oxidation of methylquinolines using a N-hydroxyphthalimide/Co(OAc)$_2$/ Mn(OAc)$_2$ catalyst in the presence of nitrogen dioxide was reported <02CC180>. The catalytic hydrogenation of acetamidoquinolines and acetamidoisoquinolines with catalytic PtO$_2$ was reported to afford the corresponding amino-5,6,7,8-tetrahydroquinolines and tetrahydroisoquinolines <02JOC7890>. Additionally, a study on the directed lithiation of unprotected quinolinecarboxylic acids with lithium 2,2,6,6-tetramethylpiperidide (LTMP) was reported. Quinoline-2-carboxylic acid undergoes metallation at the 3 position at –25 °C while the metallation of 4-methoxyquinoline-2-carboxylic acid occurs at 0 °C. Quinoline-3- and -4-carboxylic acids are metallated at the 4 and 3 positions, respectively, at –50 °C <02TL767>.

There were several reports of cross coupling reactions and substitution reactions of halo-substituted quinolines. For example, the treatment of 5-iodoquinoline (**64**) and **65** with Pd(OAc)$_2$, tri-2-furylphosphine (TFP), norbornene and cesium carbonate to afford benzoxepine **66** was reported <02JOC3972>. Additionally, the copper-catalyzed amidation of 3-bromoquinoline using copper(I) iodide, a diamine ligand and K$_3$PO$_4$, K$_2$CO$_3$ or Cs$_2$CO$_3$ was reported <02JA7421>. Other selected examples include the iron-catalyzed arylation of 2- and 3- halo-substituted quinolines by Grignard reagents <02TL3547>, an iridium-catalyzed C-H coupling reaction at the 3-position of quinoline with bis(pinacolato)diboron <02TL5649>, the palladium catalyzed carbonylation of quinolyl bromides and triflates <02JCR(S)218> and nucleophilic substitution of 2-quinolylhalides using microwave technology <02T1125>.

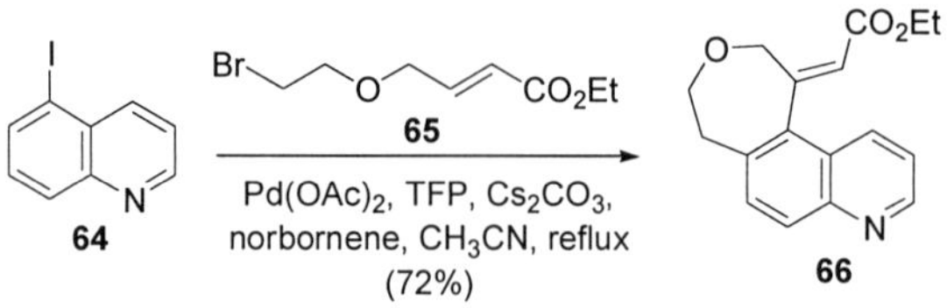

6.1.4 ISOQUINOLINES

6.1.4.1 Preparation of Isoquinolines

The isoquinoline scaffold is a common structural motif found in many natural products and other biologically interesting compounds. Larock and co-workers disclosed several approaches to functionalized isoquinolines starting from N-tert-butyl-o-(1-alkynyl)benzaldimines (**67**) (Scheme 10). For example, cyclization of **67** in the presence of an appropriate electrophile and base afforded 3,4-disubstituted isoquinolines **68** <02JOC3437>. The palladium-catalyzed carbonylative cyclization of N-tert-butyl-o-(1-alkynyl)benzaldimines **67** and aryl halides was shown to provide 3-substituted-4-aroylisoquinolines **69** <02OL193, 02JOC7042>. Additionally, the Pd(II)-catalyzed cyclization N-tert-butyl-o-(1-alkynyl)benzaldimines **67** and subsequent Heck reaction was reported to give 4-(1-alkenyl)-3-arylisoquinolines **70** <02TL3557>. A more in-depth investigation of the previously published palladium-catalyzed cyclization of N-(2-iodobenzylidene)-tert-butylamine and 1-phenyl-2-(trimethylsilyl)acetylene was also reported <02JOC86>.

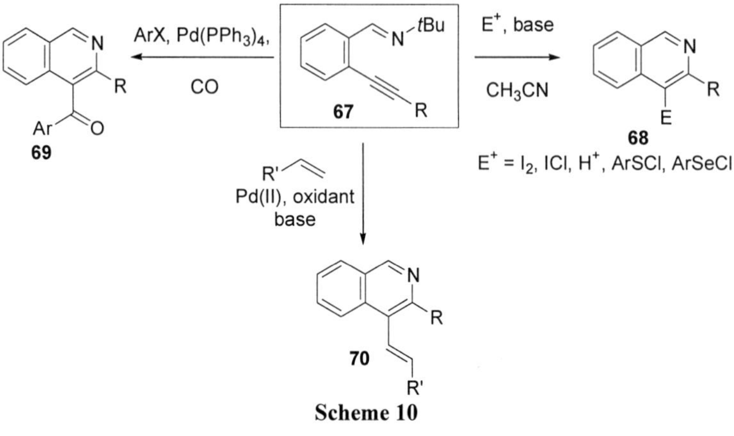

Scheme 10

The synthesis of 1-substituted tetrahydroisoquinolines using an intramolecular Pd-catalyzed α-enolate arylation was described. Treatment of α-amino esters such as **71** and **73** with LiOt-Bu, Pd$_2$(dba)$_3$ and ligand **75** or **76** afforded the corresponding isoquinolines **72** and **74**. Investigations to develop an asymmetric version of this reaction were reported to be ongoing <02JOC465>.

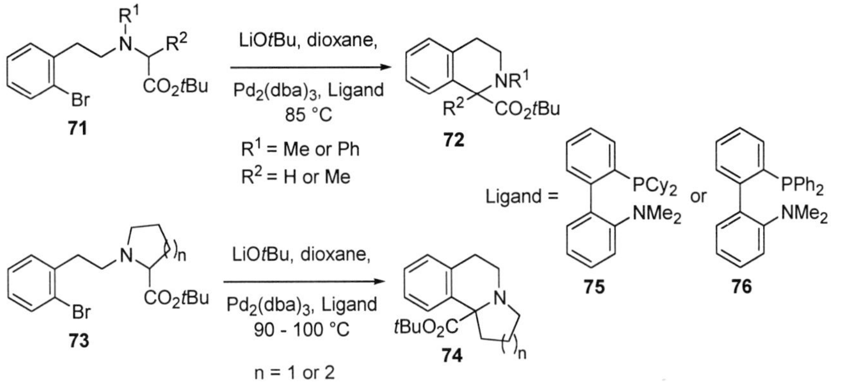

Scheme 11

The asymmetric syntheses of tetrahydroisoquinoline derivatives were also reported. Optically pure 3,4-disubstituted tetrahydroisoquinolines such as **78** were prepared by Friedel-Crafts cyclization of amino alcohols **77** <02TL1885>. Enantioselective syntheses of dihydropyrrolo[2,1-*a*]isoquinolines *via* a highly diastereoselective, chiral auxiliary assisted *N*-acyliminium cyclization was disclosed <02SL593>. The enantioselective synthesis (-)-tejedine, a seco-bisbenzyltetrahydroisoquinoline was also reported. One key step in this synthesis involved a chiral auxiliary-assisted diastereoselective Bischler-Napieralski cyclization <02OL2675>. Additionally, an asymmetric Bischler-Napieralski was reported for the preparation of 1,3,4-trisubstituted 1,2,3,4-tetrahydroisoquinolines <02JCS(P1)116>.

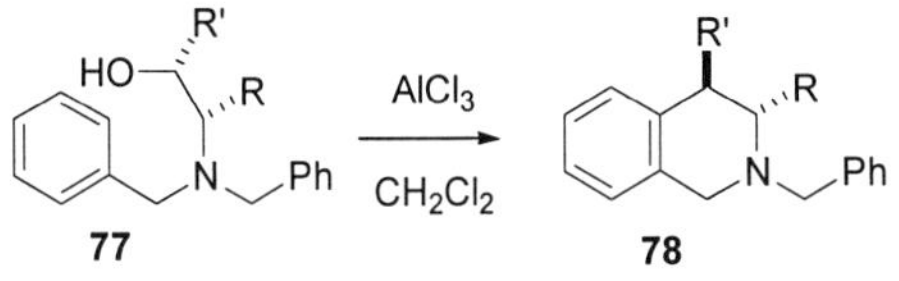

A palladium-catalyzed three-component reaction with 2-iodobenzoyl chloride or methyl 2-iodobenzoate, allene and primary aliphatic or aromatic amines to prepare *N*-substituted 4-methylene-3,4-dihydro-1(*2H*)-isoquinolin-1-ones was disclosed <02TL2601>. A synthesis of 1-substituted 1,2,3,4-tetrahydroisoquinolines *via* a Cp$_2$TiMe$_2$–catalyzed, intramolecular hydroamination/cyclization of aminoalkynes was also reported <02TL3715>. Additionally, a palladium-catalyzed one-atom ring expansion of methoxyl allenyl compounds **79** to prepare compounds **80** that can serve as precursors to isoquinolones was reported <02OL455, 02SL480>.

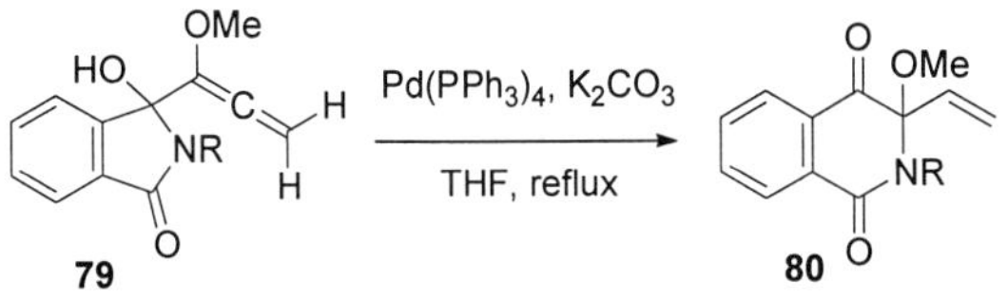

6.1.3.2 Reactions of Isoquinolines

Examples of the asymmetric functionalization of isoquinoline derivatives were reported in 2002. One example illustrated the enantioselective addition of ketene silyl acetal (**82**) to cyclic nitrone (**81**) catalyzed by a chiral titanium complex to prepare 1,2,3,4-tetrahydroisoquinoline (**83**) in 84% yield with 83% ee <02JA2888>.

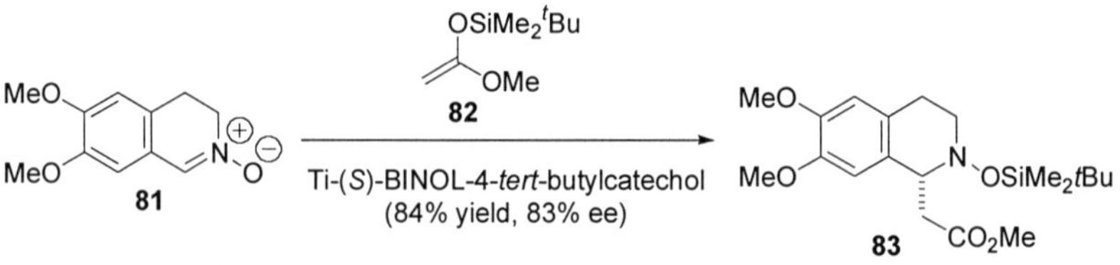

A diastereoselective three-component process to prepare 2H-pyrimido[2,1-a]-isoquinolines **84** from isoquinoline was reported <01OL3575>. Pyrrolo[2,1-a]isoquinolines were also prepared by an annelation reaction with 1,1-diiodo-2,2-dinitroethylene and isoquinolinium salts containing active methylene groups <02SL1547>.

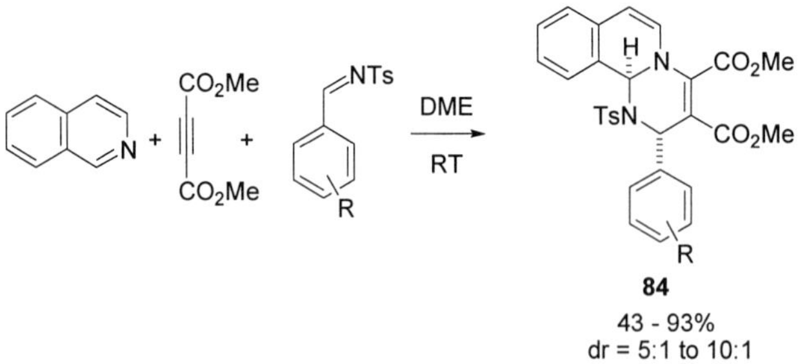

The preparation and reaction of 1-alkylthioisoquinolinium salts with active methylene compounds was studied <02CPB225>. The selective N- and O-alkylation of 1-isoquinolones using Mitsunobu conditions was also studied <02JCS(P1)335>. Additionally, the oxidation of isoquinoline derivatives to the corresponding isoquinolones using iodosobenzene and catalytic tetrabutylammonium iodide was reported <02HCA1069>.

6.1.5 PIPERIDINES

2002 was an active year for the publication of syntheses and reactivity of piperidines. The year was highlighted by review articles that either contained piperidine syntheses as their main topic or as a substantial part of the review material covered. Specifically, comprehensive reviews appeared on the subject of piperidine syntheses *via* aziridinium intermediates <02CT579>, the asymmetric synthesis of piperidines from chiral building blocks <02DOF143>, and on the synthesis of the *sedum* and related alkaloids <02T5957>, most of which contain the piperidine structure. A review by Puentes and Kouznetsov whose main focus was homoallylamine derivatives devoted a large part of its subject matter to syntheses of piperidines from these structures <02JHC595>.

6.1.5.1 Preparations of Piperidines

Intermolecular cycloaddition reactions constitute an important and convergent route to piperidines and related compounds. The research group of Franklin Davis at Temple University published a novel route to optically active piperidines that proceeded through an imino Diels-Alder reaction with enantiomerically enriched compound **84** (Scheme 16) <02OL655>. On reaction with *trans*-1,3-pentadiene, intermediate **85** was produced as a single diastereomer in 89% yield. Hydrogenation of this strained intermediate yielded the 2,6-disubstituted piperidine **86** in 75% yield.

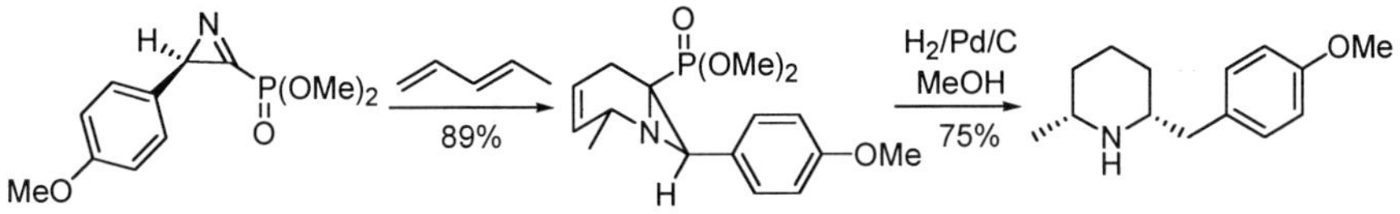

Scheme 16

A variety of other groups reported successful use of the imino Diels-Alder reaction to synthesize piperidines <02TL1071>. Asymmetric approaches to piperidines using imines modified with chiral esters <02TL1067> and *N*-substituted carbohydrate derivatives <02MFC571> were used to synthesize a variety of optically enriched piperidines. Barluenga and co-workers reported on an efficient solid-phase synthesis of polysubstituted piperidines <02OL3667>, and Lau and co-workers from Novo Nordisk published large-scale experimental procedures for the syntheses of 2-substituted 4-oxo piperidine derivatives in optically active form <02T7339>. The research group of Cheng used a formal [3+3] cycloaddition between an α-substituted sulfonyl acetamide and glutarimide to produce 3-aryl-*N*-substituted piperidines and demonstrated formal total syntheses of a variety of compounds from Johnson & Johnson <02JCCS1079, 02T3623>. Finally, dipolar cycloadditions were also used effectively by a number of research groups to achieve asymmetric <02JCS(P1)1494>, diastereoselective <02T3159> and racemic <02S771> syntheses of piperidine-containing natural and unnatural products.

The lactamization/reduction approach to synthesizing piperidines also continues to be of widespread utility <02JOC7573, 02JA11689, 02T1519>. Amat and co-workers reported a dynamic kinetic resolution (DKR) and desymmetrization process for the synthesis of a variety of enantiopure piperidines (Scheme 17)<02OL2787, 02JOC5343>. Compound **91** is synthesized as the major enantiomer from achiral, *meso* starting material **87**. The authors invoke two competing transition states for the cyclization reaction to produce the products in the isomer ratio shown. Lactamization takes place faster through the chair-like transition state **89** where the large substituent is in a psuedo-equatorial position. The bicycle **91** is transformed into piperidine **93** (and others) through well-established procedures. The utility of this chemistry is demonstrated by elaborating the lactams obtained from this reaction sequence to pharmaceutically important piperidine intermediates, as well as a formal total synthesis of preclamol. Amat and co-workers used a similar strategy (without dynamic kinetic resolution) to synthesize the piperidine alkaloid (−)-anabasine. The piperidine ring was formed via reduction of an intermediate lactam <02CC526>.

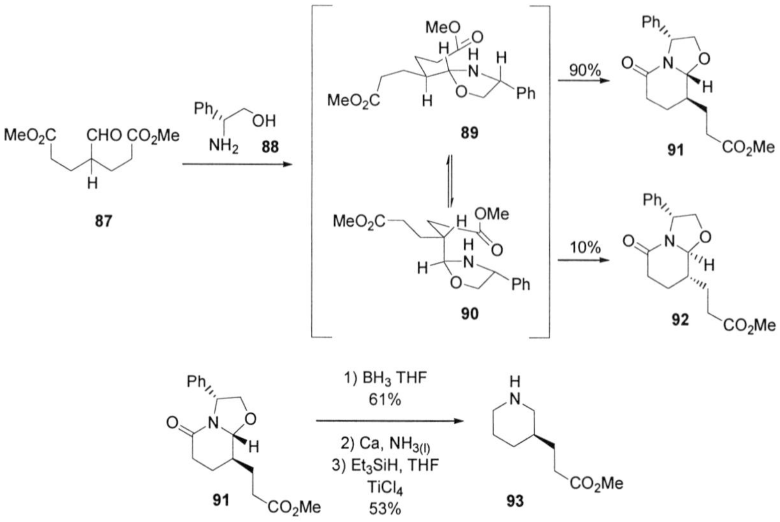

Scheme 17

Ring closure via reductive amination also continues to be used as an excellent method for synthesizing optically active piperidines. The source of asymmetry varies from case to case, with carbohydrates <02ARK91>, amino acids <02JOC865>, oxazolidinones <02TL7711> and SAMP-hydrazones <02TA587> serving as the source of chirality. In their synthesis of (+)-carpamic acid, Singh and co-workers used a silyl group as both a masked hydroxy substituent and as a handle to provide selectivity and efficiency in other portions of the synthesis <02TL7711>.

Transition metal catalysis continues to play an extremely important role in the synthesis of piperidines in both optically active and racemic form. Hu and co-workers at Procter and Gamble Pharmaceuticals synthesized a variety of amino piperidines in optically active form by using a protected D-serine molecule as the source of chirality, followed by ring-closing metathesis to generate the piperidine ring (Scheme 18) <02OL4499>. Cossy and co-workers used an enantioselective allyltitanation reaction followed by ring-closing metathesis to generate the piperidine alkaloids (+)-sedamine and (−)-prosophylline <02JOC1982>. Transition metal catalysis was also employed for the synthesis of racemic piperidines. Catalyst systems based on palladium <02CC720, 02ACE343, 02S87>, titanium <02TL3715>, cobalt (Pauson-Khand reaction) <02H1409>, and lanthanides (hydroamination/cyclization) <02JA7886> found utility for syntheses of piperidines via a variety of pathways.

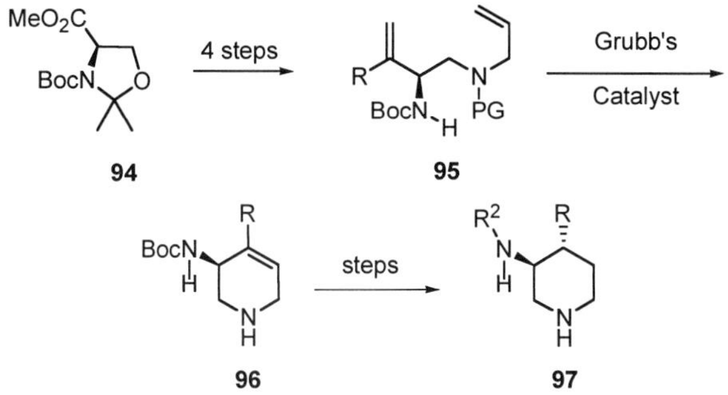

Scheme 18

Cyclizations via anionic <02TL225, 02JOC3184>, cationic <02OL1471> and radical pathways <02OL2573> also saw use in the synthesis of piperidines. Other work included cyclization under neutral conditions <02SL1359>, the intramolecular Michael reaction which led to the synthesis of 2,6-disubstituted piperidines <02T7983>, and the preparation of piperidine enamino esters via intramolecular cyclization on an activated alkyne <02TL3471>. Snaith and co-workers examined and developed a novel approach to 3,4-disubstituted piperidines that proceeded via carbonyl-ene or Prins-type cyclizations <02OL3727>. They were able to achieve diastereoselectivities of up to 98:2, and could select for the cis- vs. trans-substituted products depending upon the nature of the acid catalyst that was employed. For example, Lewis acids resulted in the formation of trans-disubstituted piperidines whereas Bronsted-acid catalysis resulted in the formation of cis-disubstituted piperidines (Scheme 19).

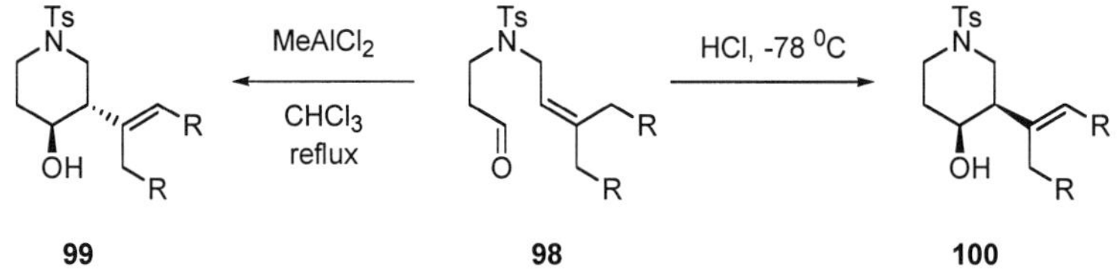

Scheme 19

6.1.5.2 Reactions of Piperidines

Succesful methods for elaborating existing piperidine rings are continually being developed <02S87, 02RJOC875, 02T1343>. In addition to the large amount of chemistry that surrounds reduction of the carbonyl group in 6-membered lactams, the addition of organometallic reagents at the 2-position of lactams <02T5301> and piperidinyl imines <02OL103> provide 2-alkylated piperidines in low yield. Matsumura and co-workers developed a catalytic, asymmetric addition of malonates to 2-alkoxy substituted piperidines **101** (Scheme 20, equation A). The reaction itself is at an early state of development, but certainly will provide a powerful method for assembling piperidine structures once further optimization has taken place <02TL3229>. Diez and co-workers undertook a systematic survey of additions of nucleophiles to optically pure aziridinyl lactam **103** (Scheme 20, equation B); a variety of heteroatomic nucleophiles added regioselectively to the 2-position of the compound, and the products were further elaborated to β-pseudopeptides <02TA995>. Takeuchi and Harayama also published a re-revision of the stereo structure of the piperidine lactone, a 2,3-disubstituted piperidine used in the synthesis of Febrifugine <02CPB1011>.

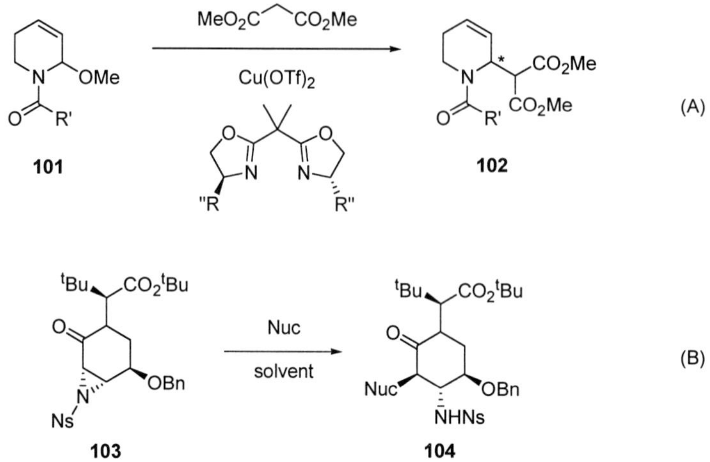

Scheme 20

Nuc = NaSMe, NaN$_3$, R$_3$N (R=H, alkyl, aryl), Me$_2$CuLi
Solvents = MeCN, DMF, THF
Yields = 63 - 82%

Novel methods for functionalizing piperidines at the 3- and 4-positions were also introduced. Mete and co-worker synthesized 3-diazo-piperidin-2-one and characterized its reactivity in transition-metal catalyzed reactions, particularly H-X insertion reactions and cyclopropanation reactions <02T3137>. Christoffers and co-workers developed an asymmetric Michael addition reaction with a chirally modified 4-piperidone-enamine. They were able to create a quaternary carbon center in >95% *de* and elaborate the compound on through classical means to the functionalized piperidine **107** (Scheme 21) <02EJO1505>.

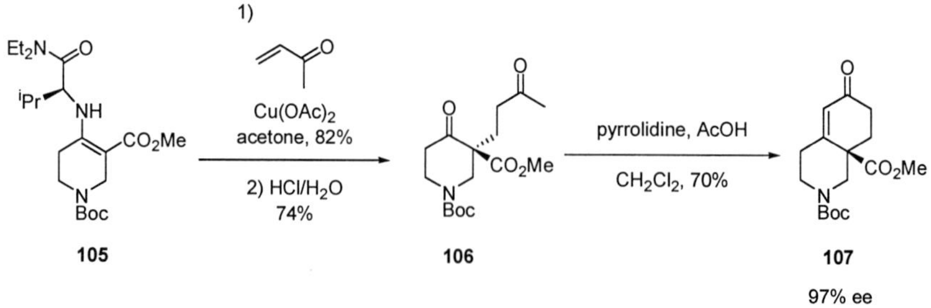

Scheme 21

Functionalization of piperidines at the 4-position is most commonly carried out by the addition of an organometallic reagent or other nucleophile to 4-piperidone <02OL423>, *via* Michael addition to substituted unsaturated lactams <02OL2787, 02TL1995, 02TL1991> or addition of nucleophiles to 3,4-aziridinyl piperidines <02TL5315>. Hannesian and co-workers were able to synthesize two diversely functionalized libraries of piperidines via Michael addition of nitromethanes and conjugate additions of organocuprate reagents (Scheme 22). The syn- to anti- ratios in the products varied depending upon the nature of the nucleophile used. The diastereomers could be separated later to provide optically pure compounds. Hu and co-workers demonstrated the scope of addition of a variety of heteroatom nucleophiles (alkoxides, thiols, carboxylates and halogens) and organometallic reagents (Grignards, alkylzincs) to piperidinyl aziridines (Scheme 23). The regioselectivity was extremely selective for addition at the 4-position, and in all cases the amino substituent was found *anti* to the newly added substituent <02TL4289>.

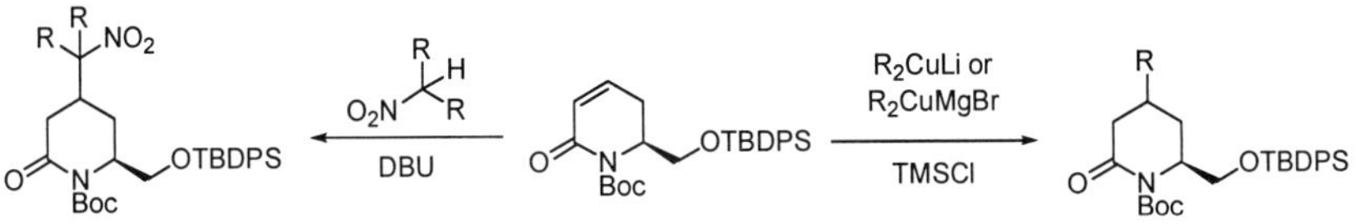

R=H, Me
109

108

R=aryl, Me
110

Scheme 22

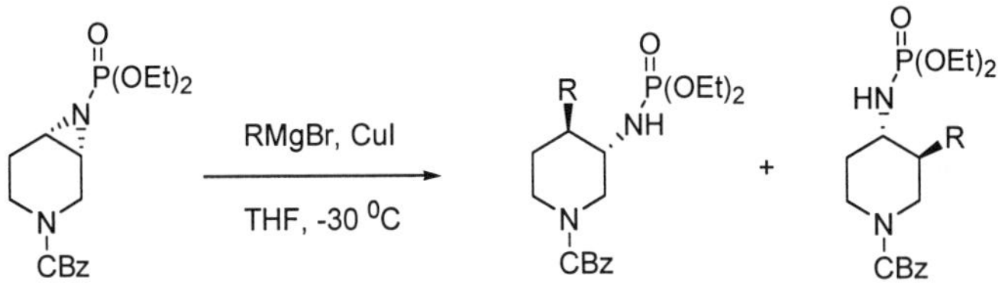

111 **112** **113**

R	Yield(%)	Ratio 112/113
Me	86	13:1
Et	78	8:1
Cyclopropyl	78	14:1
Cyclohexyl	82	12:1

Scheme 23

REFERENCES

02A91	D.D. Dhavale, V.N. Desai, N.N. Saha, J.N. Tilekar, *Arkivoc* **2002**, 91.
02AG(E)343	S.-K. Kang, Y.-H. Ha, B.-S. Ko, Y. Lim, J. Jung *Angew. Chem. Int. Ed.* **2002**, *41*, 343.
02BKCS797	B.W. Yoo, J.W. Choi, D.Y. Kim, S.K. Hwang, K.I. Choi, J.H. Kim, *Bull. Korean Chem. Soc.* **2002**, *23*, 797.
02CC180	S. Sakaguchi, A. Shibamoto, Y. Ishii, *Chem. Commun.* **2002**, 180.
02CC444	P.J. Stevenson, M. Nieuwenhuyzen, D. Osborne, *Chem. Commun.* **2002**, 444.
02CC526	M. Amat, M. Canto, N. Llor, J. Bosch, *Chem. Commun.* **2002**, 526.
02CC720	J. Helaja, R. Goettlich *Chem. Commun.* **2002**, 720.
02CC850	R. Lavilla, M.C. Bernabeu, E. Brillas, I. Carranco, J.L. Diaz, N. Llorente, M. Rayo, A. Spada, *Chem. Commun.* **2002**, 850.
02CC940	H. Rudler, B. Denise, A. Parlier, J.-C. Daran, *Chem. Commun.* **2002**, 940.
02CC1040	S.L. Jain, B. Sain, *Chem. Commun.* **2002**, 1040.
02CC1682	M.C. Bagley, J.W. Dale, J. Bower, *Chem. Commun.* **2002**, 1682.
02CC1760	C.J. Moody, R.A. Hughes, S.P. Thompson, L. Alcaraz, *Chem. Commun.* **2002**, 1760.
02CPB225	R. Fujita, N. Watanabe, H. Tomisawa, *Chem.Pharm. Bull.* **2002**, *50*, 225.
02CPB463	D. Branowska, S. Ostrowski, A. Rykowski, *Chem. Pharm. Bull.* **2002**, *50*, 463.
02CPB484	T. Koike, N. Takeuchi, *Chem.Pharm. Bull.* **2002**, *50*, 484.
02CPB1011	Y. Takeuchi, K. Azuma, H. Abe, K. Sasaki, T. Harayama *Chem. Pharm. Bull.* **2002**, *50*, 1011.
02CT579	J. Cossy, D. Gomez Pardo *Chemtracts* **2002**, *15*, 579.
02DOF143	N. Toyooka, H. Nemoto *Drugs of the Future* **2002**, *27*, 143.
02EJO375	J.-C. Berthet, M. Nierlich, M. Ephritikhine, *Eur. J. Org. Chem.* **2002**, 375.
02EJO1505	J. Christoffers, H. Scharl *Eur. J. Org. Chem.* **2002**, 1505.
02EJO4123	S.P. Gromov, N.A. Kurchavov, *Eur. J. Org. Chem.* **2002**, 4123.
02EJO4181	M. Schlosser, F. Cottet, *Eur. J. Org. Chem.* **2002**, 4181.
02H21	E. Moreau, J.-M. Chezal, C. Dechambre, D. Canitrot, Y. Blache, C. Lartigue, O. Chavignon, J.-C. Teulade, *Heterocycles* **2002**, *57*, 2.
02H421	H. Gorohmaru, T. Thiemann, T. Sawada, K. Takahashi, K. Nishi-i, N. Ochi, Y. Kosugi, S. Mataka, *Heterocycles* **2002**, *56*, 421.
02H467	B.K. Banik, I. Banik, L. Hackfeld, F.F. Becker, *Heterocycles* **2002**, *56*, 467.
02H1409	M. Ishizaki, M. Masamoto, O. Hoshino *Heterocycles* **2002**, *57*, 1409.
02HCA1069	W.-J. Huang, O.V. Singh, C.-H. Chen, S.-Y. Chiou, S.-S. Lee, *Helv. Chim. Acta* **2002**, *85*, 1069.
02JA576	T. Takahashi, Y. Li, P. Stepnicka, M. Kitamura,Y. Liu, K. Nakajima, M. Kotora, *J. Am. Chem. Soc.* **2002**, *124*, 576.
02JA2560	P. Janvier, X. Sun, H. Bienayme, J. Zhu, *J. Am. Chem. Soc.* **2002**, *124*, 2560.
02JA2888	S. Murahashi, Y. Imada, T. Kawakami, K. Harada, Yonemushi, N. Tomita, *J. Am. Chem. Soc.* **2002**, *124*, 2888.
02JA3518	D. Suzuki, R. Tanaka, H. Urabe, F. Sato, *J. Am. Chem. Soc.* **2002**, *124*, 3518.
02JA5059	T. Takahashi, F.-Y. Tsai, Y. Li, H. Wang, Y. Kondo, M. Yamanaka, K. Nakajima, M. Kotora, *J. Am. Chem. Soc.* **2002**, *124*, 5059-5067.
02JA7421	A. Klapars, X. Huang, S.L. Buchwald, *J. Am. Chem. Soc.* **2002**, *124*, 7421.
02JA7886	S. Hong, T.J. Marks *J. Am. Chem. Soc.* **2002**, *124*, 7886.
02JA7982	T. Bach, H. Bergmann, B. Grosch, K. Harms, *J. Am. Chem. Soc.* **2002**, *124*, 7982.
02JA8184	S. Yamada, C. Morita, *J. Am. Chem. Soc.* **2002**, *124*, 8184.
02JA11689	T.A. Johnson, D.O. Jang, B.W. Slafer, M.D. Curtis, P. Beak *J. Am. Chem. Soc.* **2002**, *124*, 11689.
02JCCS1079	M.-Y. Chang, Y.-C.L. John, S.-T. Chen, N.-C. Chang, *J. Chinese Chem. Soc.* **2002**, *49*, 1079.
02JCR(S)60	Y.A. Ibrahim, B. Al-Saleh, *J. Chem. Res.* **2002**, 60-61.
02JCR(S)218	C.W. Holzapfel, A.C. Ferreira, W. Marais, *J. Chem. Res.* **2002**, 218.
02JCR(S)560	G. Zhang, Y. Hu, . *J. Chem. Res.* **2002**, 560.
02JCS(P1)116	M. Nicoletti, D. O'Hagan, A.M.Z. Slawin, *J. Chem. Soc., Perkin Trans. 1* **2002**, 116.
02JCS(P1)335	S. Ferrer, D.P. Naughton, I. Parveen, M.D. Threadgill, *J. Chem. Soc., Perkin Trans. 1* **2002**, 335.
02JCS(P1)581	G.M. Chapman, S.P. Stanforth, B. Tarbit, M.D. Watson, *J. Chem. Soc., Perkin Trans. 1* **2002**, 581.

02JCS(P1)1141	R. Lavilla, *J. Chem. Soc., Perkin Trans. 1* **2002**, 1141.
02JCS(P1)1494	E.C. Davison, M.E. Fox, A.B. Holmes, S.D. Roughley, C.J. Smith, G.M. Williams, J.E. Davies, P.R. Raithby, J.P. Adams, I.T. Forbes, N. J. Press, M.J. Thompson *J. Chem. Soc., Perkin Trans. 1* **2002**, 1494.
02JCS(P1)1663	M.C. Bagley, C. Brace, J.W. Dale, M. Ohnesorge, N.G. Phillips, X. Xiong, J. Bower, *J. Chem. Soc., Perkin Trans. 1* **2002**, 1663.
02JHC61	T. Horaguchi, N. Hosokawa, K. Tanemura, T. Suzuki, *J. Heterocycl. Chem.* **2002**, *39*, 61.
02JHC291	C.S. Cho, T.K. Kim, T.-J. Kim, S.C. Shim, N.S. Yoon, *J. Heterocycl. Chem.* **2002**, *39*, 291.
02JHC347	T.J. Delia, J.B. Kanaar, E. Knefelkamp, *J. Heterocycl. Chem.* **2002**, *39*, 347.
02JHC595	C.O. Puentes, V. Kouznetsov, *J. Heterocycl. Chem.* **2002**, *39*, 595.
02JHC743	H. Maruoka, M. Yamazaki, Y. Tomioka, *J. Heterocycl. Chem.* **2002**, *39*, 743.
02JHC1035	B. Al-Saleh, M.M. Abdelkhalik, A.M. Eltoukhy, M.H. Elnagdi, *J. Heterocycl. Chem.* **2002**, *39*, 1035.
02JMOC(A)565	J. Jacob, C.M. Cavalier, W.D. Jones, S.A. Godleski, R.R. Valente, *J. Mol. Catal. A* **2002**, *186*, 565.
02JMOC(A)109	M.R. Prasad, G. Kamalakar, G. Madhavi, S.J. Kulkarni, K.V. Raghavan, *J. Mol. Catal. A* **2002**, *186*, 109.
02JOC86	K.R. Roesch, R.C. Larock, *J. Org. Chem.* **2002**, *67*, 86.
02JOC234	P. Gros, S. Choppin, J. Mathieu, Y. Fort, *J. Org. Chem.* **2002**, *67*, 234.
02JOC465	O. Gaertzen, S.L. Buchwald, *J. Org. Chem.* **2002**, *67*, 465.
02JOC865	C.-B. Xue, X. He, J. Roderick, R.L. Corbett, C.P. Decicco, *J. Org. Chem.* **2002**, *67*, 865-870.
02JOC1982	J. Cossy, C. Willis, V. Bellost, S.J.M. BouzBouz, *J. Org. Chem.* **2002**, *67*, 1982.
02JOC2131	F. Palacios, E. Herran, G. Rubiales, J.M. Ezpeleta, *J. Org. Chem.* **2002**, *67*, 2131.
02JOC3184	S. Swaleh, J. Liebscher *J. Org. Chem.* **2002**, *67*, 3184.
02JOC3425	L. Pouysegu, A.-V. Avellan, S. Quideau, *J. Org. Chem.* **2002**, *67*, 3425.
02JOC3437	Q. Huang, J.A. Hunter, R.C. Larock, *J. Org. Chem.* **2002**, *67*, 3437.
02JOC3972	M. Lautens, J.-F. Paquin, S. Piguel, *J. Org. Chem.* **2002**, *67*, 3972.
02JOC3969	J. Zhang, C.-J. Li, *J. Org. Chem.* **2002**, *67*, 3969.
02JOC4414	B. Heller, B. Sundermann, H. Buschmann, H.-J. Drexler, J. You, U. Holzgrabe, E. Heller, G. Oehme, *J. Org. Chem.* **2002**, *67*, 4414.
02JOC5343	M. Amat, M. Canto, N. Llor, C. Escolano, E. Molins, E. Espinosa, J. Bosch, *J. Org. Chem.* **2002**, *67*, 5343.
02JOC5394	L. Wenjie, D.P. Nelson, M.S. Jensen, R.S. Hoerrner, H. Cai, R.D. Larsen, P.J. Reider, *J. Org. Chem.* **2002**, 67, 5394.
02JOC7042	G. Dai, R.C. Larock, *J. Org. Chem.* **2002**, *67*, 7042.
02JOC7573	C. Agami, L. Dechoux, C. Menard, S. Hebbe, *J. Org. Chem.* **2002**, *67*, 7573.
02JOC7890	K.A. Skupinska, E.J. McEachern, R.T. Skerlj, G.J. Bridger, *J. Org. Chem.* **2002**, *67*, 7890.
02JOC9449	B. Jiang, Y.-G. Si, *J. Org. Chem.* **2002**, *67*, 9449.
02MC571	M. Weymann, M. Schulz-Kukula, S. Knauer, H. Kunz *Monatsh. Chem.* **2002**, *133*, 571.
02OL103	B.G. Davis, M.A. Maughan, T.M. Chapman, R. Villard, S. Courtney *Org. Lett.* **2002**, *4*, 103.
02OL185	N. Langlois *Org. Lett.* **2002**, *4*, 185.
02OL193	G. Dai, R.C. Larock, *Org. Lett.* **2002**, *4*, 193.
02OL423	A.M. Shestopalov, Y.M. Emeliyanova, A.A. Shestopalov, L.A. Rodinovskaya, Z.I. Niazimbetova, D.H Evans. *Org. Lett.* **2002**, *4*, 423.
02OL455	Y. Nagao, A. Ueki, K. Asano, S. Tanaka, S. Sano, M. Shiro, *Org. Lett.* **2002**, *4*, 455.
02OL655	F.A. Davis, Y. Wu, H. Yan, K. R. Prasad, W. McCoull *Org. Lett.* **2002**, *4*, 655.
02OL1471	B. Schlummer, J.F. Hartwig *Org. Lett.* **2002**, *49*, 1471.
02OL1759	P. Gros, F. Louerat, Y. Fort, *Org. Lett.* **2002**, *4*, 1759-61.
02OL1819	D.M. Lindsay, W. Dohle, A.E. Jensen, F. Kopp, P. Knochel, *Org. Lett.* **2002**, *4*, 1819.
02OL2573	D. Crich, S. Neelamkavil *Org. Lett.* **2002**, *4*, 2573.
02OL2653	T.R. Kelly, M. Cavero, *Org. Lett.* **2002**, *4*, 2653.
02OL2675	Y.-C. Wang, P.E. Georghiou, *Org. Lett.* **2002**, *4*, 2675.
02OL2691	K. Fujita, K. Yamamoto, R. Yamaguchi, *Org. Lett.* **2002**, *4*, 2691.
02OL2787	M. Amat, M. Perez, N. Llor, J. Bosch *Org. Lett.* **2002**, *4*, 2787.
02OL3575	V. Nair, A.R. Sreekanth, N. Abhilash, M.M. Bhadbhade, R.C. Gonnade, *Org. Lett.* **2002**, *4*, 3575.

02OL3667	J. Barluenga, C. Mateos, F. Aznar, C. Valdes *Org. Lett.* **2002**, *4*, 3667.
02OL3727	J.T. Williams, P.S. Bahia, J.S. Snaith *Org. Lett.* **2002**, *4*, 3727.
02OL4265	K.A. Parker, T.L. Mindt, *Org. Lett.* **2002**, *4*, 4265.
02OL4411	D. Cheng, J. Zhou, E. Saiah, G. Beaton, *Org. Lett.* **2002**, *4*, 4411.
02OL4499	X.E. Hu, N.K. Kim, B. Ledoussal, *Org. Lett.* **2002**, *4*, 4499.
02RCB(IE)362	V.V. Dotsenko, S.G. Krivokolysko, V.P. Litvinov, A.N. Chernega, *Russ. Chem. Bull.* **2002**, *51*, 362.
02RCB(IE)1812	G.I. Nikishin, L.L. Sokova, N.I. Kapustina, *Russ. Chem. Bull.* **2002**, *51*, 1812.
02RJOC424	A.N. Karaseva, V.F. Mironov, V.V. Karlin, A.I. Konovalov, O.V. Tsepaeva, E.R. Yunusov, *Russ. J. Org. Chem.* **2002**, *38*, 424.
02RJOC875	I.L. Lysenko, A.V. Bekish, O.G. Kulinkovich *Russ. J. Org. Chem.* **2002**, *38*, 875.
02RJOC917	N.A. Nedolya, L. Brandsma, N.I. Shlyakhtina, *Russian J. Org. Chem.* **2002**, *38*, 917.
02S87	L.S. Santos, R.A. Pilli *Synthesis* **2002**, 87.
02S169	H.A. Muathen, *Synthesis* **2002**, 169.
02S771	A. Varlamov, V. Kouznetsov, F. Zubkov, A. Chernyshev, O. Shurupova, L.Y.V. Mendez, A.P. Rodriguez, J.R. Castro, A.J. Rosas-Romero *Synthesis* **2002**, 771.
02SC793	D.-P. Cheng, Z.-C. Chen, *Synth. Commun.* **2002**, *32*, 793.
02SC3493	F.A. Abu-Shanab, Y.M. Elkholy, M.H. Elnagdi, *Synth. Commun.* **2002**, *32*, 3493.
02SC3617	X. Wang, Y. Zhang, *Synth. Commun.* **2002**, *32*, 3617.
02SL155	M.-J. Perez-Perez, E.-M. Priego, M.-L. Jimeno, M.-J. Camarasa, *Synlett* **2002**, 155.
02SL231	K. Yamada, T. Kubo, H. Tokuyama, T. Fukuyama, *Synlett* **2002**, 231.
02SL331	T.J. Donohoe, L. Mace, M. Helliwell, O. Ichihara, *Synlett* **2002**, 331.
02SL480	Y. Nagao, S. Tanaka, A. Ueki, I.-Y. Jeong, S. Sano, M. Shiro, *Synlett* **2002**, 480.
02SL593	I. Gonzalez-Temprano, N. Sotomayor, E. Lete,. *Synlett* **2002**, 593.
02SL628	P. Gros, C. Viney, Y. Fort, *Synlett* **2002**, 628.
02SL808	G. Karig, N. Thasana, T. Gallagher, *Synlett* **2002**, 808.
02SL1359	M. Lemaire, N. Veny, T Gefflaut, E. Gallienne, R. Chenevert, J. Bolte *Synlett* **2002**, 1359.
02SL1547	E. Boultadakis, B. Chung, M.R.J. Elsegood, G.W. Weaver, *Synlett* **2002**, 1547.
02SL1904	D. Garcia-Cuadrado, A.M. Cuadro, J. Alvarez-Builla, J.J. Vaquero, *Synlett* **2002**, 1904.
02S497	J. Nikolai, J. Schlegel, M. Regitz, G. Maas, *Synthesis* **2002**, 497.
02T1125	Y.-J. Cherng, *Tetrahedron* **2002**, *58*, 1125.
02T1213	E.M. Beccalli, A. Contini, P. Trimarco, *Tetrahedron* **2002**, *58*, 1213.
02T1343	A. Stehl, G. Seitz, K. Schulz *Tetrahedron* **2002**, *58*, 1343.
02T1453	C. Escolano, K. Jones, *Tetrahedron* **2002**, *58*, 1453.
02T1519	M.T. Barros, M.A. Januario-Charmier, C.D. Maycock, T. Michaud *Tetrahedron* **2002**, *58*, 1519.
02T3137	I.S. Hutchinson, S.A. Matlin, A. Mete *Tetrahedron* **2002**, *58*, 3137.
02T3159	Y. Liu, A. Maden, W.V. Murray *Tetrahedron* **2002**, *58*, 3159.
02T3623	M.-Y. Chang, S.-T. Chen, N.-C. Chang *Tetrahedron* **2002**, *58*, 3623.
02T3655	S.L. Cappelle, I.A. Vogels, T.C. Govaerts, S.M. Toppet, F. Compernolle, G.J. Hoornaert, *Tetrahedron* **2002**, *58*, 3655.
02T3693	D. Basavaiah, R.M. Reddy, N. Kumaragurubaran, D.S. Sharada, *Tetrahedron* **2002**, *58*, 3693.
02T4369	A. Bouillon, J.-C. Lancelot, V. Collot, P.R. Bovy, S. Rault, *Tetrahedron* **2002**, *58*, 4369.
02T4429	V. Bonnet, F. Mongin, F. Trecourt, G. Quéguiner, P. Knochel, *Tetrahedron* **2002**, *58*, 4429.
02T4931	Y.-J. Cherng, *Tetrahedron* **2002**, *58*, 4931.
02T5069	M. Anniyappan, D. Muralidharan, P.T. Perumal, *Tetrahedron* **2002**, *58*, 5069.
02T5301	Z. Li, Y. Zhang *Tetrahedron* **2002**, *58*, 5301.
02T5957	R.W. Bates, K. Sa-Ei *Tetrahedron* **2002**, *58*, 5957.
02T7339	J.F. Lau, T. Kruse Hansen, J. Paul Kilburn, K. Frydenvang, D. Holsworth, Y. Ge, R. T Uyeda, L.M. Judge, H. Sune Andersen *Tetrahedron* **2002**, *58*, 7339.
02T7983	S. Sengupta, S. Mondal *Tetrahedron* **2002**, *58*, 7983.
02T9709	A.C. Veronese, C.F. Morelli, M. Basato, *Tetrahedron* **2002**, *58*, 9709.
02TA95	M.L. Bennasar, E. Zulaica, Y. Alonso, B. Vidal, J.T. Vazquez, J. Bosch, *Tetrahedron: Asymmetry* **2002**, *13*, 95.
02TA587	D. Enders, B. Nolte, J. Runsink *Tetrahedron: Asymmetry* **2002**, *13*, 587.

02TA995	J. Piro, P. Forns, J. Blanchet, M. Bonin, L. Micouin, A. Diez, *Tetrahedron: Asymmetry* **2002**, *13*, 995.
02TL225	F.-X. Felpin, J. Lebreton *Tetrahedron Lett.* **2002**, *43*, 225.
02TL767	A.-S. Rebstock, F. Mongin, F. Trecourt, G. Quéguiner, *Tetrahedron Lett.* **2002**, *43*, 767.
02TL1067	P.D. Bailey, P.D. Smith, F. Pederson, W. Clegg, G.M. Rosair, S.J. Teat *Tetrahedron Lett.* **2002**, *43*, 1067.
02TL1071	P.D. Bailey, P.D. Smith, K.M. Morgan, G.M. Rosair, *Tetrahedron Lett.* **2002**, *43*, 1071.
02TL1093	B.A. Trofimov, L.V. Andriyankova, S.A. Zhivet'ev, A.G. Mal'kina, V.K. Voronov, *Tetrahedron Lett.* **2002**, *43*, 1093.
02TL1171	Y. He, H. Mahmud, B.R. Wayland, H.V.R. Dias, C.J. Lovely, *Tetrahedron Lett.* **2002**, *43*, 1171.
02TL1885	S. Chandrasekhar, N.R. Reddy, M.V. Reddy, B. Jagannadh, A. Nagaraju, A. Ravi Sankar, A.C. Kunwar, *Tetrahedron Lett.* **2002**, *43*, 1885.
02TL1991	S. Hanessian, M. Seid, I. Nilsson *Tetrahedron Lett.* **2002**, *43*, 1991.
02TL1995	S. Hanessian, W.A.L. van Otterlo, I., Nilsson, U. Bauer *Tetrahedron Lett.* **2002**, *43*, 1995.
02TL2273	A.D. Thomas, C.V. Asokan, *Tetrahedron Lett.* **2002**, *43*, 2273.
02TL2601	R. Grigg, T. Khamnaen, S. Rajviroongit, V. Sridharan, *Tetrahedron Lett.* **2002**, *43*, 2601.
02TL3229	O. Onomura, Y. Kanda, Y. Nakamura, T. Maki, Y. Matsumura, *Tetrahedron Lett.* **2002**, *43*, 3229.
02TL3471	O. David, M.-C. Fargeau-Bellassoued, G. Lhommet *Tetrahedron Lett.* **2002**, *43*, 3471.
02TL3547	J. Quintin, X. Franck, R. Hocquemiller, B. Figadere, *Tetrahedron Lett.* **2002**, *43*, 3547.
02TL3557	Q. Huang, R.C. Larock, *Tetrahedron Lett.* **2002**, *43*, 3557.
02TL3715	I. Bytschkov, S. Doye, *Tetrahedron Lett.* **2002**, *43*, 3715.
02TL3907	M.-E. Theoclitou, L.A. Robinson, *Tetrahedron Lett.* **2002**, *43*, 3907.
02TL4285	D. Cai, R.D. Larsen, P.J. Reider, *Tetrahedron Lett.* **2002**, *43*, 4285.
02TL4289	X.E. Hu, N.K. Kim, B. Ledoussal, A.-O. Colson *Tetrahedron Lett.* **2002**, *43*, 4289.
02TL5315	X.E. Hu *Tetrahedron Lett.* **2002**, *43*, 5315.
02TL5649	J. Takagi, K. Sato, J.F. Hartwig, T. Ishiyama, N. Miyaura, *Tetrahedron Lett.* **2002**, *43*, 5649.
02TL6841	Y. Zhang, D.A. Klumpp, *Tetrahedron Lett.* **2002**, *43*, 6841.
02TL6907	N.A.M. Yehia, K. Polborn, T.J.J. Muller, *Tetrahedron Lett.* **2002**, *43*, 6907.
02TL7711	R. Singh, S.K. Ghosh *Tetrahedron Lett.* **2002**, *43*, 7711.
02TL7945	Y. Miyazaki, T. Kanbara, T. Yamamoto, *Tetrahedron Lett.* **2002**, *43*, 7945.
02TL8331	M.C. Bagley, R. Lunn, X. Xiong, *Tetrahedron Lett.* **2002**, *43*, 8331.
02TL9303	R. Cordoba, J. Plumet, *Tetrahedron Lett.* **2002**, *43*, 9303.
02TL9679	N.A. Nedolya, N.I. Schlyakhtina, L.V. Klyba, I.A. Ushakov, S.V. Fedorov, L. Brandsma,. *Tetrahedron Lett.* **2002**, *43*, 9679.

Chapter 6.2

Six-Membered Ring Systems: Diazines and Benzo Derivatives

Michael P. Groziak
San José State University, San José, CA, USA
mgroziak@science.sjsu.edu

6.2.1 INTRODUCTION

Once again, the diazines pyridazine, pyrimidine, pyrazine, and their benzo-fused derivatives cinnoline, phthalazine, quinazoline, quinoxaline, and phenazine were the central focus in a great many chemical and biological investigations. Progress on the syntheses of these heterocycles and/or their use as key intermediates in broader synthetic routes was abundant, and a significant number of studies relied on solid-phase, microwave irradiation, or novel metal-assisted strategies. In addition, some reports were mainly of an X-ray, computational, spectroscopic, natural product, or biological nature. To a large extent, reports of a similar nature have been grouped together.

6.2.2 REVIEWS AND GENERAL STUDIES

One review focused on the oxidative addition, Suzuki and Stille coupling, and Buchwald-Hartwig amination reactions of chloropyridazines <02JHC535>, another on the solid-phase synthetic strategies for generating benzannelated nitrogen heterocycles <02BMC2415>, and still another on Cu- and Co-complexing macrocyclic pyridazines like **1** <02EJI2535>. Reviews of a biological nature included the preparation of pyridazines and their use as analgesics and aldose reductase inhibitors <02JHC545> and the syntheses and antibacterial and antifungal activities of pyridazino[3,4-*b*]quinoxalin-4-ones <02JHC551>. Quinoline, quinazoline, and acridone alkaloids from plant, microbial, and animal sources were also reviewed <02NPR742>.

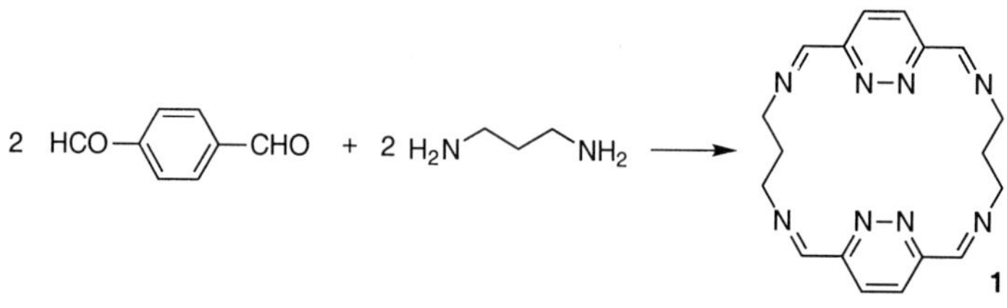

Some investigations pertained to more than one type of diazine and so were more general in nature. For instance, Pd-catalyzed cross-coupling reactions of 2-chloropyrimidine (**2**), 2-chloropyrazine (**3**), and 5-bromopyrimidine (**4**) gave phenyldiazines **5**, **6**, and **7**, respectively, under extremely mild conditions (< 0 °C) <02T4429>. A Ni-catalyzed cross-coupling between aryl Grignard reagents and various fluorodiazines **8** gave arylated diazines **9** in excellent fashion <02JOC8991>, and another Ni-catalyzed reaction afforded pyridyldiazines <02SL1008>.

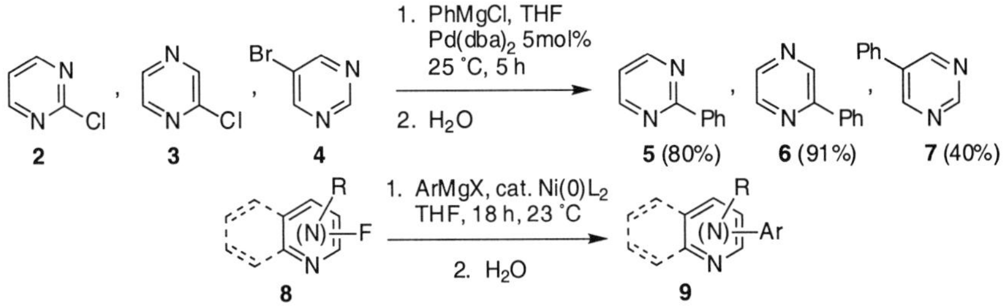

A Suzuki coupling of 5-chloro-2-methyl-6-phenyl-2H-pyridazin-3-one (**10**) ultimately led to diazino-fused indole **11** and cinnoline **12** and allowed access to a novel pyrimidoisoquinoline ring system in a one-pot fashion <02T10137>. Mn(II)-azido networks of the type [Mn(N$_3$)$_2$(L)]$_n$ like **13** with new 3-D topologies were obtained using both pyridazine and pyrimidine ligands <02CC64>.

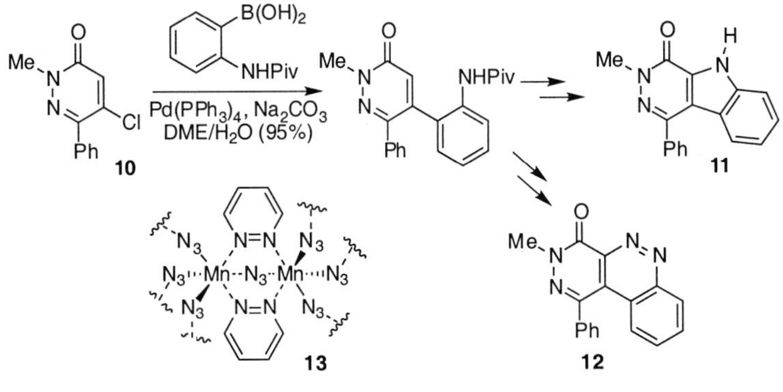

Diazinyl-substituted carbodiimides were used for the synthesis of novel guanidines, isothioureas, and isoureas <02JHC695>, and K$_2$CO$_3$ was used as a support for a "green" synthesis of 2-amino-1,3,4-thia/oxadiazines and 2-hydrazino-1,3,4-thiadiazines <02JHC1045>. A reversed-dipole, or 'umpolung', nitration of various diazines was found to proceed moderately well <02ARK19>. The structure, stability, and reactivity of new diazinium ylides were investigated <02ARK73>, as was the electrochemical behavior of benzo[c]cinnoline and some of its bromo derivatives <02TJC617>. Two 1,10-phenanthroline units symmetrically appended to a central diazine ring gave new bis-tridentate chelators **14**, **15**, and **16** for ruthenium <02OL1253>. Thiazolodiazines obtained from N-diazinylcarboxamides with Lawesson's reagent were used in metalation reactions <02JHC1077>, and novel diaza-metallacycles like **17** were obtained by inserting tungsten(II) aryloxides into the aromatic rings of the diazines pyridazine, benzo[*c*]cinnoline and phthalazine <02CC2482>. Finally, enantiopure diazine analogues of novel nicotinic acetylcholine receptor (nAChR) ligands were found to have very high binding affinities <02JMC1064>.

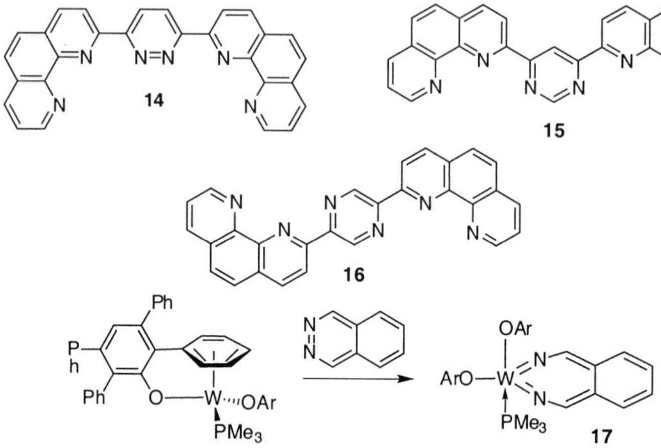

Computational studies continue to help define the physicochemical properties of the diazines. The heats of formation for twelve monocyclic azines were derived in a Gaussian-3 study <02JSC257>, and the effect of hydrogen bonding solvents on the electronic and magnetic properties of diazines in dilute solution were examined by *ab initio* methods <02JA1506>. The structural non-rigidity of heteroaromatic rings has been investigated by quantum chemical non-empirical methods at the MP2 level <02JMS159>, and the role of steric hindrance in solute-solvent interactions exhibited by 2-substituted 4,5-dimethoxypyridazin-3(2*H*)-ones **18** has been studied by both semiempirical and *ab initio* methods <02CCC1790>. An AM1(RHF) level theoretical study of the 1,3-dipolar cycloaddition reactions of C_{60} with 3-phenylphthalazinium-1-olate was reported <02TC165>. A zwitterionic dehydrocinnolinium species was predicted by D-functional theory to be an intermediate in the thermal cyclization of (2-ethynylphenyl)triazenes to cinnolines **19** <02JA1572, 02JA13463, 02JOC6395>. A quantum-mechanical computational analysis of 5-substituted-6-phenyl-3(2*H*)-pyridazinones **20** revealed that the 5-substituent exerts a great effect on the reactivity of these heterocycles <02T2389>.

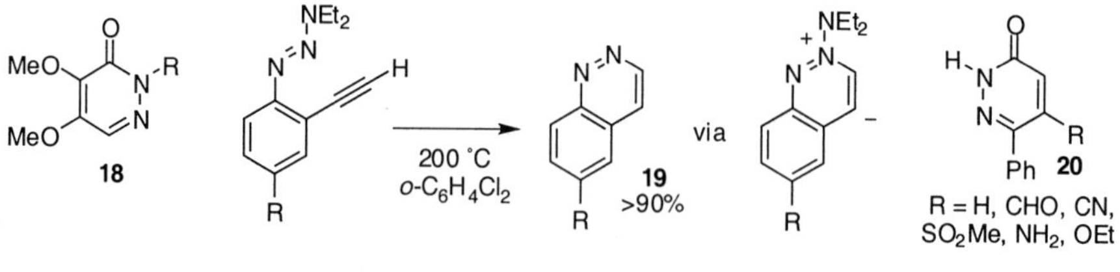

6.2.3 PYRIDAZINES AND BENZO DERIVATIVES

Carbohydrate-based tetrahydropyridazines **22** were prepared by hetero-Diels-Alder reaction of chiral 1,2-diaza-1,3-butadienes **21** with acrylonitrile, and the stereochemistry of the major cycloadduct was established by X-ray crystallography <02JOC2241>. The X-ray crystal structure of 2-phenyl-3*H*-pyridazino[6,1-*b*]quinazoline-3,10(5*H*)-dione **23a** revealed that it exists as one of four possible tautomers **23a-d** <02EJO133>. Crystal structures of the related *O*-methylated (**24**) and *N*-methylated (**25**) compounds were also obtained. The nucleophilic *ipso* and *cine* substitution of bromine by morpholine in pyridazinones was investigated, and the X-ray crystal structures of derivatives **26-29** was invaluable <02PJC45>. Bactericidal metal ion complexes of the antibacterial agents cinoxacin (**30**) and ciprofloxacin were studied by spectroscopic and X-ray methods <02JIB65>. The structure of $\{[Cd(Cx)_3][Cd(Cx)_3](H_2O)\} \cdot 12H_2O$ (Cx = cinoxacin) was determined.

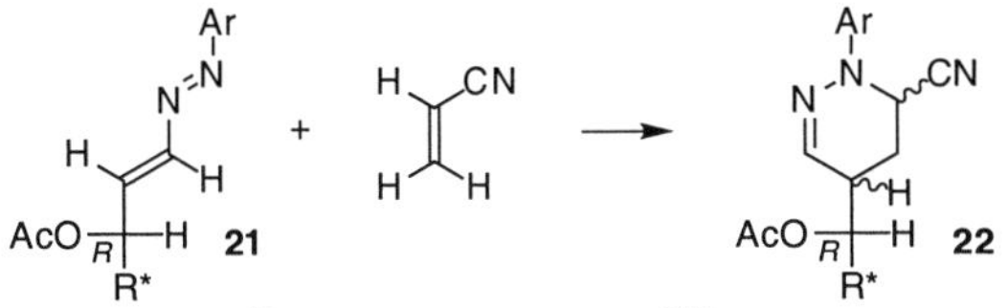

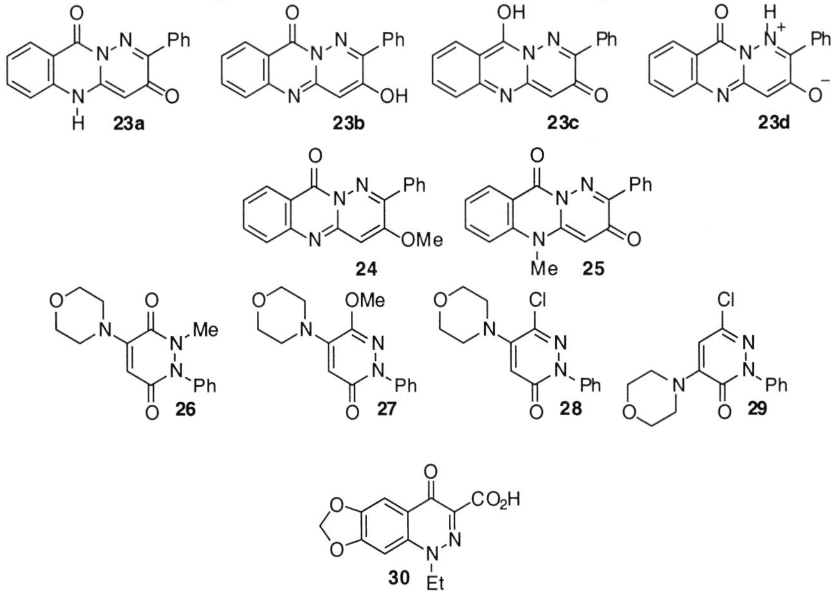

X-ray crystal structures of 3,6-di(thiophen-2-yl)pyridazine (**31**) <02AX(C)640>, 4-methyl-2-(8-quinolyl)phthalazin-1-one (**32**) <02AX(E)1051>, 1,4-di-(2'-pyridylamino)phthalazine (**33**) <02ZK61>, and bis(phthalazino)azine <02ZK203> and a di(phenanthroline) dineodymium(III)

complex <02ZK292> were reported, as were those of an imidazo[4,5-*d*]pyridazino[1,2-*a*]-pyridazin-4-one-based angiotensin II receptor antagonist <02CPB1022>, and *N*-substituted *cis*-tetra- and *cis*-hexahydrophthalazinone-based selective PDE4-inhibiting *in vitro* antiinflammatory agents <02JMC2526>. Coordination compounds of Cp_2Yb with phenazine, 2,2'-bipyrimidine, and other bridging ligands gave compounds that underwent antiferromagnetic exchange coupling across the ligand. The crystal structure of $(Cp_2Yb)_2(\mu$-bipyrimidine) was determined <02OM4622>.

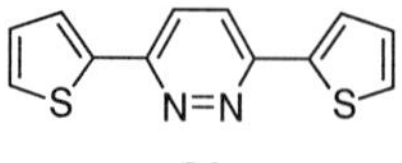

NMR and other spectroscopic techniques were used to gain a better understanding of pyridazine-based heterocycles. For example, NMR was used to study the stereoisomerism and ring-chain tautomerism in pyrazolo[1,2-*a*]pyridazine-5,8-diones and pyrazolo[1,2-*b*]phthalazine-5,10-diones **34** <02EJO2046>, and also in 1-hydroxy-2,3-dihydro-1*H*-pyrazolo[1,2-*a*]pyridazine-5,8-diones and 1-hydroxy- and 1-amino-2,3-dihydro-1*H*-pyrazolo[1,2-*b*]phthalazine-5,10-diones **35** <02EJO3447>. The selectivity of methylation of [1,2,3]triazolo[4,5-*d*]pyridazines was studied by ^{1}H-^{15}N NMR methods <02JMS73>, and the ring currents and π-electron effects in heterocycles including pyrimidine, pyrazine, and pyridazine, and their methylated derivatives were studied by ^{1}H NMR <02JCS(P2)1081>. By 1D and 2D NMR, the unsymmetrical diamine 1,2-dihydro-2-(4-aminophenyl)-4-[4-(4-aminophenoxy)-4-phenyl]-(2*H*)phthalazin-1-one exists as a phthalazinone rather than a phthalazine ether <02MRC738>.

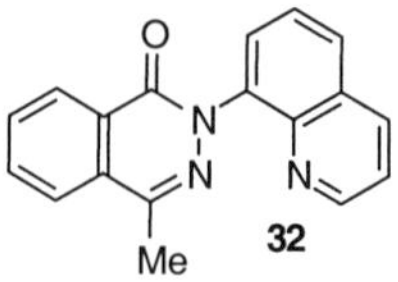

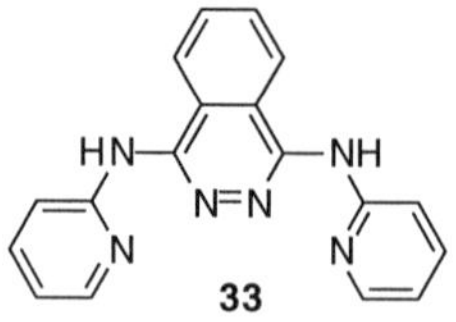

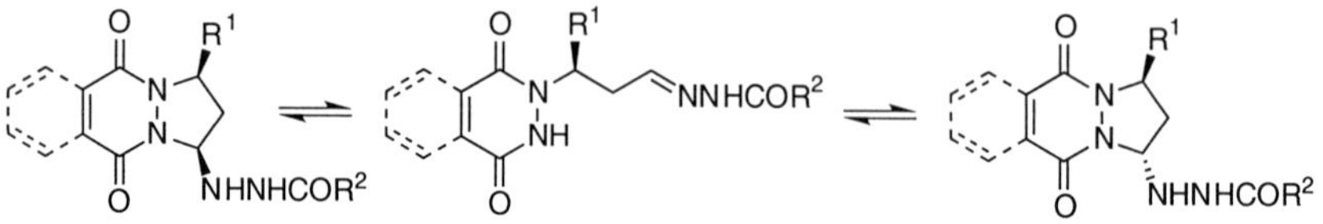

34, R^1 = H, Me; R^2 = H, Me, Ph

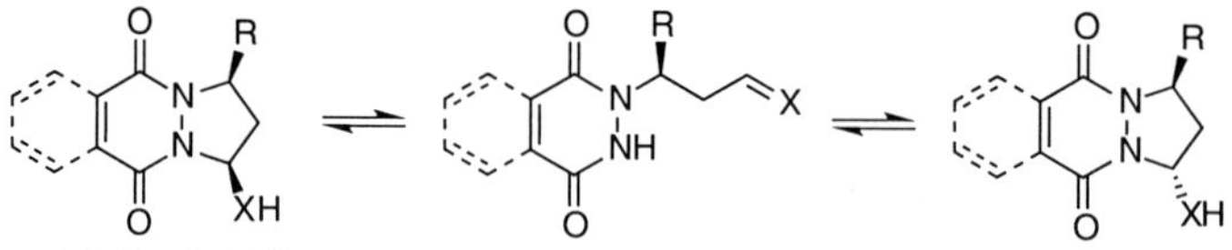

35, X = O, NR^2

6.2.3.1 Syntheses

Condensation routes are a common way to synthesize pyridazines. Pyridazino-, furo-, and polysubstituted pyridazines can all be accessed via a common condensation route <02SC989>.

Ring-fused pyridazin-3-ones are prepared in one-pot fashion by cyclocondensation of hydrazines and ketocarboxylic acids <02H(57)723>, and bicyclic [*b*]-heteroannulated pyridazines like 7,8-dihydro-3,8,8-trimethyltriazolo[4,3-*b*]pyridazin-6(5*H*)-one are obtained via acid-catalyzed cleavage of the condensation adduct formed between tetrahydropyridazine-3,6-dione 3-hydrazones and β-keto esters <02PJC1577>. New pyridazino cinnoline derivatives are obtained by condensing 2,5-dibromo-3,6-(dihydrazino)-1,4-benzoquinones and active methylene reagents <02JHC853>, and pyridazines can be condensed with PhNCS and then hetero-cyclized with EtO_2CCH_2Cl to give pyridino[2,3-*d*]pyridazines, pyrazolo[3,4-*d*]pyridazines, and isoxazolo[4,5-*d*]pyridazines <02HAC258>.

(2-Quinolyl)pyrazolo[3,4-*d*]pyridazines were prepared by cyclocondensation reactions of an anilino-pyrazole acid with acetaldehyde <02JHC869>. Cyclopentadienyl-derived γ-diketones and arylhydrazines condensed to 4-(1,4-diaryl-2*H*-cyclopent[*d*]pyridazin-2-yl)-benzenesulfonamides <02H(57)2383>. Sulfonated poly(phthalazinone ether sulfone)s **37** were prepared by polycondensation of 4-(hydroxyphenyl)phthalazinone **36** with various ratios of disodium 5,5'-sulfonylbis(2-fluorobenzenesulfonate) and bis(4-fluorophenyl)sulfone <02P5335>.

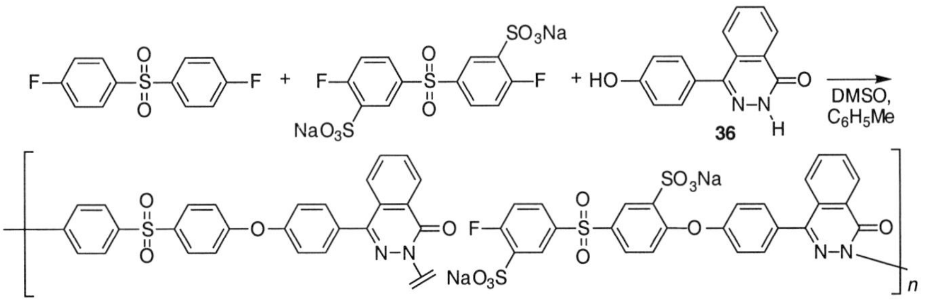

New linear and angular pyridazine furocoumarins have been obtained <02S43>, some by a regioselective inverse electron demand Diels-Alder reaction <02S475>. Isoxazolo[4,5-*d*]-pyridazin-4(5*H*)- and 7(6*H*)-ones were prepared in two-step, one-pot fashion from functionalized vinyl sulfoxides <02SL73>. 3,6-Diarylpyridazines **38**, as well as 3,5-diarylpyrroles were formed from 2,2'-sulfonylbis(1,3-diarylprop-2-en-1-ones) and hydrazine <02T2227>. 6,7-Disubstituted pyrrolo[1,2-*b*]pyridazines **39** were prepared via regioselective $AlCl_3$-mediated C–C bond formation <02T9933>.

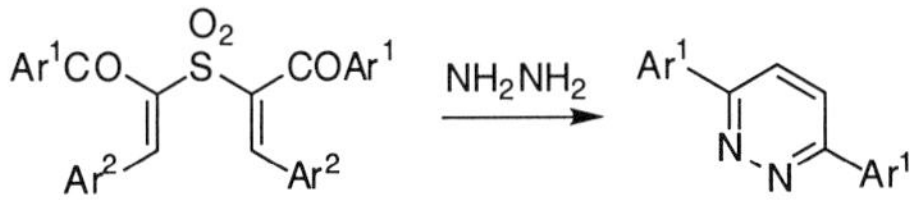

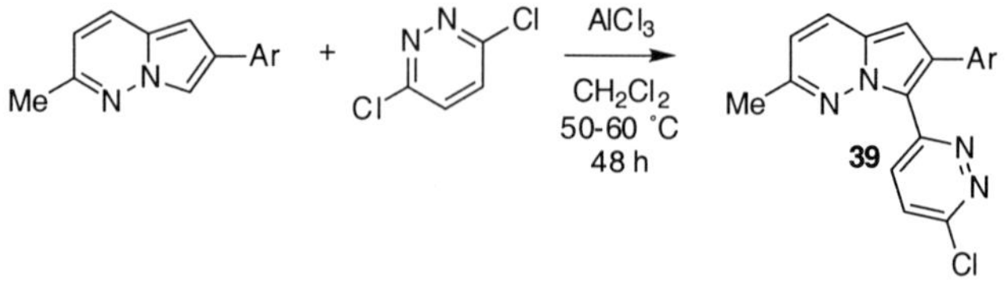

Fused carbazoles related to pyrido[4,3-*b*]carbazole alkaloids were prepared by a Diels-Alder route, and a 3-aza bioisostere of the antitumor alkaloid olivacine was synthesized <02CPB1479>. Indoloid [3.3]cyclophane **40a** gave the pentacyclic indoloid **41a** upon heating <02OL127>. This led to a concise formal total synthesis of (±)-strychnine in 12 facile steps from tryptamine when a similar transannular inverse-electron-demand Diels-Alder reaction of indoloid [3.3]cyclophane **40b** gave **41b** <02AG(E)3261>.

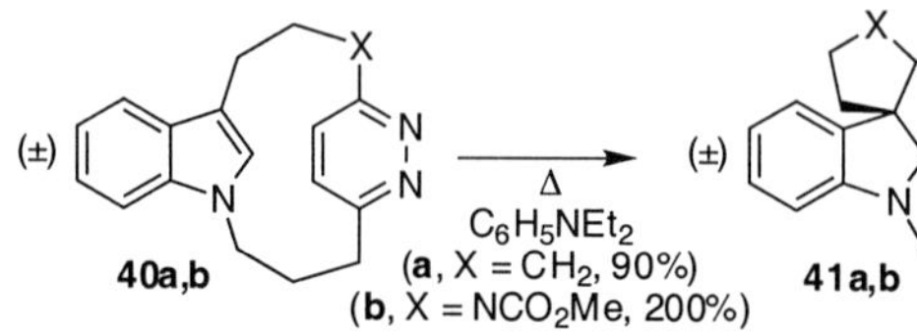

α,β-Didehydroglutamates **42** were diastereoselectively transformed via intermediates **43** into 6-oxoperhydropyridazine-3-carboxylic acids **44**, a new class of cyclic amino acid derivatives <02JOC2789>. In a related study, a diastereoselective synthesis of α-methylpyroglutamates from α,β-didehydro α-amino acids relied on pyridazinones **45** as key intermediates <02EJO4190>.

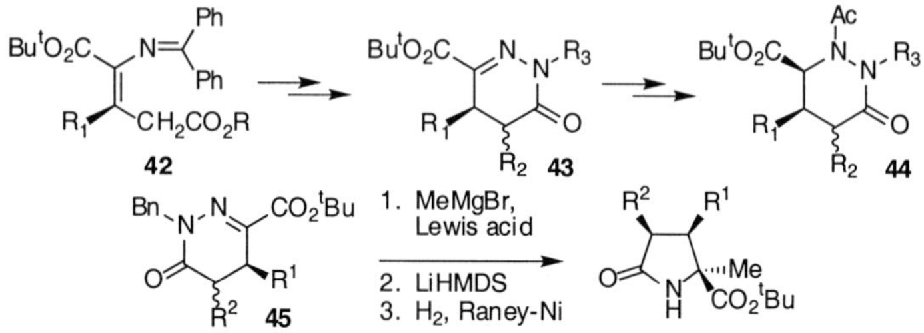

New thieno[2,3-*c*]pyridazines <02BKC1715>, nitrogen bridgehead heterocycles containing a pyridazinone/triazole moiety <02IJHC225>, and 4-phenyl-1-phthalazinylhydrazones were reported <02RJO1309>. The synthesis and properties of 6-amino-7-hetaryl-5-R-5*H*-pyrrolo-[2,3-*b*]pyrazine-2,3-dicarbonitriles **46** were described <02CHE336>, and functionalized 8,9-diazatricyclo[4.4.0][0.1.5]decanes, members of a new tricyclic system containing the pyridazine ring, were accessed <02H(57)523>. [1,4]Benzodioxino[2,3-*c* and 2,3-*d*]pyridazinones were obtained by reaction of chloropyridazin-3-ones with catechol under basic conditions <02JHC685>, and methylthiomaleimides were used to prepare pyridazines and polycyclic pyridazinediones as chemiluminophors <02JHC571>.

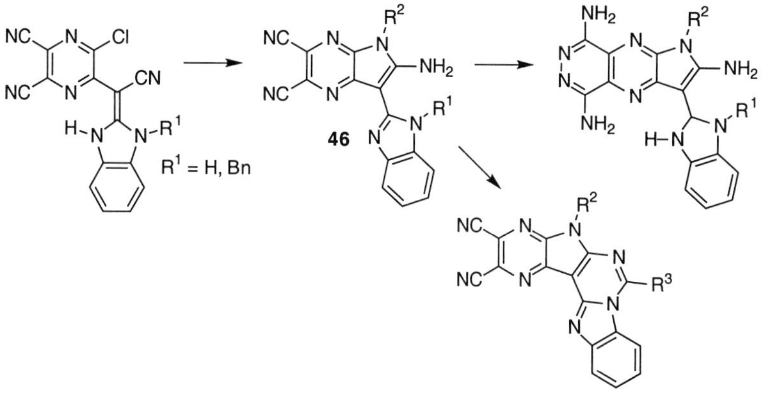

Pyridazinones and pyridazin-6-imines were obtained via benzenediazonium chloride coupling reactions <02SC481> and pyridazino[4,5-c]isoquinolinones **48** were prepared from pyridazin-3(2H)-ones **47** via Suzuki cross-coupling <02T5645>. The treatment of benzotriazolylacetone with C and N electrophilic reagents generated benzotriazolyl-cinnolines <02HAC141>. Hydroxy-substituted hexahydrophthalazinones **49-52** were obtained via treatment of cyclohexane- and norbornanelactones or ketallactones with hydrazine hydrate <02M241>.

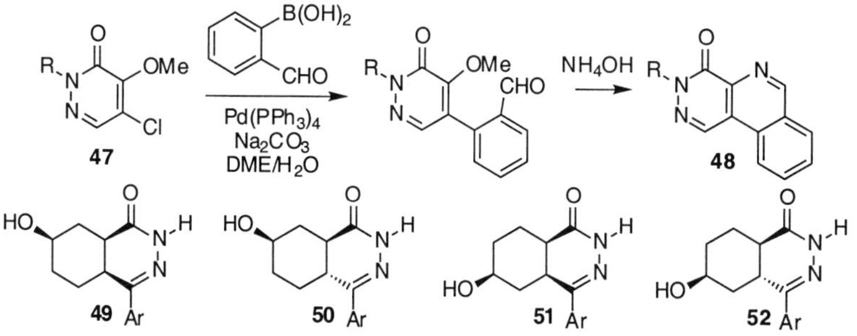

4-Diarylmethyl-1-(2H)phthalazinones were obtained from 2,2-diaryl-1,3-indanediones <02SL823>, and pyrrolo[2,1-a]phthalazines were obtained by a rearrangement following an intramolecular 1,1-cycloaddition in triene-conjugated diazo-compounds <02ARK67>. New 1,2-dihydrophthalazines were synthesized from 6-chloro- and 6,7-dichloro-4-(4-nitrophenyl)-phthalazine <02JHC989>. 6,7-Dichloro-5,8-phthalazinedione (**54**) was obtained in excellent yield via chloroxidation of 5,8-diaminophthalazine (**53**), and **54** was further transformed into **55** with ethanolic pyridine (83%), **56** with aniline and CeCl$_3$ (97%), and **57a** and **57b** with 2-aminopyridine in MeOH (53%) and EtOH (48%), respectively <02BKC1425>. The crystal structure of **57b** was determined.

57a, R = Me; b, R = Et

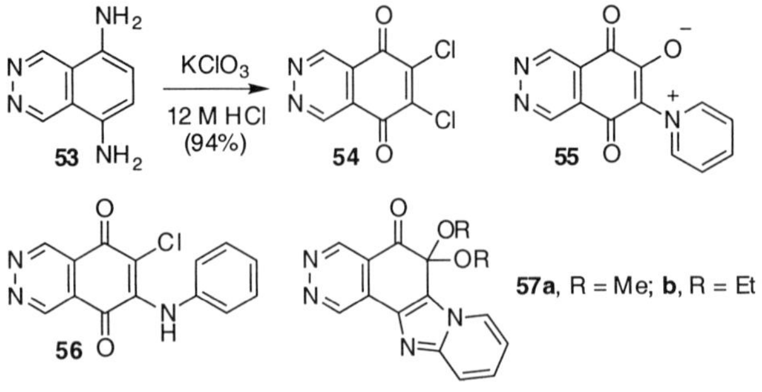

6.2.3.2 Reactions

The palladium-catalyzed amination and alkylation of 3-iodo-6-arylpyridazines was shown to be simple and efficient <02SL1123>, and the dicyclopalladation of new mesogenic pyridazines followed by reaction with β-diketones afforded some novel metallomesogens (chiral liquid crystals) **58** <02JOM246>. Pyridazinethiocarboxamides like **59** (prepared from their carboxamides) were metalated and condensed with electrophiles to give various polysubstituted pyridazines **60** <02T2743>. It was surprising to find this substitution occurring regioselectively *meta* to the thiocarboxamide group, but a mechanistic explanation was offered.

58, $R^1 = C_nH_{2n+1}$ (n = 4-8,10); R^2 = Me, Bu

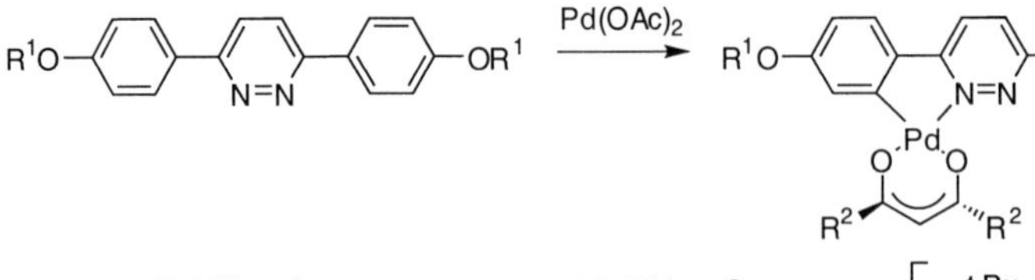

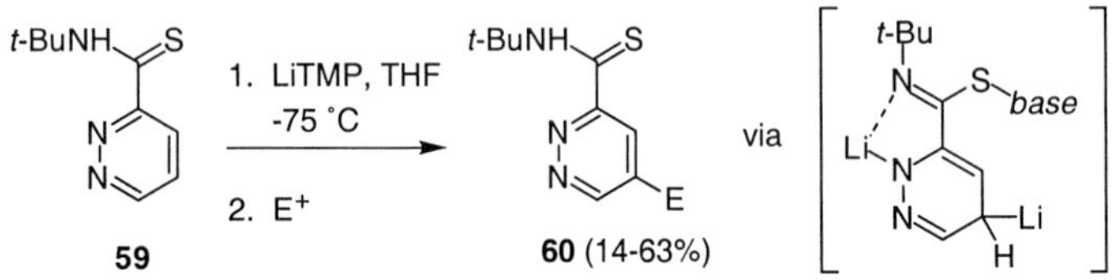

Pyridazines readily participate in cycloaddition reactions. For example, 4,5-dicyanopyridazine (**61**) undergoes Diels-Alder cycloaddition to the C-2/C-3 double bond of pyrroles to give, after loss of N_2 and oxidation, indoles like **62** <02T8067>. They also react with indoles to give substitution products like **63**. The influences of water and solvent-free conditions on the 1,3-dipolar cycloaddition of pyridazinium dicyanomethanide and phthalazinium-2-dicyanomethanide (**64**) with dipolarophiles surprisingly revealed a strong rate enhancement by water <02JCS(P2)1807>.

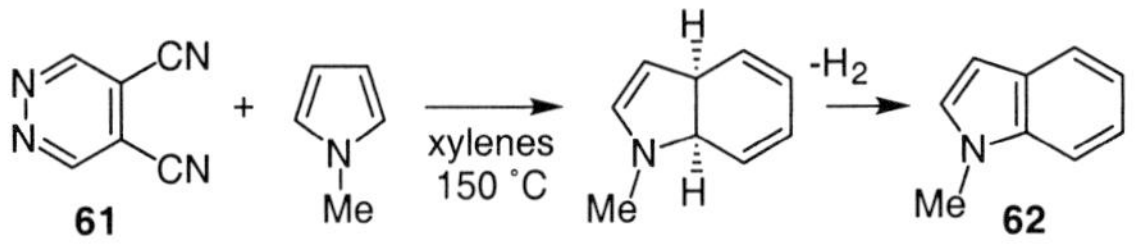

The 4-carboxamides of some pyridazino[4,5-*b*]indoles and pyridazino[4,5-*b*]benzo[*b*]furans were prepared via regioselective amidation followed by Krapcho dealkoxycarbonylation <02SL2095>. The *N*(2)-oxide and 3-amino derivatives of 6,8-dimethylpyrimido[4,5-*c*]pyridazine-5,7(6*H*,8*H*)-dione were shown to react with alkylamines in the presence of an oxidant to generate condensed imidazolines or imidazoles <02MC157>. The reaction of 2-benzenesulfonyl-4,5-dichloropyridazin-3-ones with aliphatic amines under neutral conditions was shown to generate the corresponding 5-(alkylamino) compounds <02JHC203>. Pyridazine-3-hydrazidic acids were used in a one-pot synthesis of thiazole derivatives <02PS2661>.

Pd-catalyzed coupling reactions at the 5-position of 6-phenyl-3(2*H*)-pyridazinones using a retro-ene transformation have been reported <02SL2062>, and the Pd-catalyzed arylation of 4-bromo-6-chloro-3-phenylpyridazine has been shown to be efficient and regioselective <02SL223>. The *N*-methylation of substituted 3(2*H*)-pyridazinones with *N,N*-dimethylformamide dimethylacetal has been explored <02SC1675>, as have the preparation of 3-nitro-, -nitroso- and –chloro-derivatives of 2-substituted imidazo[1,2-*b*]pyridazines via electrophilic substitution <02JHC737>, and the reactivity of 5-alkynyl-4-chloro- and 4-alkynyl-5-chloro-pyridazin-3(2*H*)-ones toward O- and S-nucleophilic reagents was investigated <02H(57)2115>.

Treatment of quinoxaline *N*-oxides with 4,4,4-trifluoroacetoacetate gave 1-methyl-3-trifluoromethylpyridazino[3,4-*b*]quinoxalin-4(1*H*)-ones <02H(56)291>, and treatment with diethyl ethoxymethylenemalonate gave 1-methyl-pyridazino[3,4-*b*]quinoxalin-4(1*H*)-ones <02H(58)359>. 3-Aminopyridazines were used in the synthesis of 4-oxo-4*H*-pyrimido[1,2-*b*]-pyridazine and 1-(substituted pyridazin-3-yl)-1*H*-1,2,3-triazoles <02ARK143>, and 2-substituted 4-aryl-5-hydroxy- and 5-aryl-4-hydroxypyridazin-3(2*H*)-ones were used to prepare 4,5-diaryl-pyridazin-3(2*H*)-ones (**65**) and isochromeno[3,4-*d*]pyridazinediones, respectively <02T9713>. Oxidation of 1,4-phthalazinedione (**66**) in the presence of furfural or 5-methylfurfural led to the formation of [5,6]benza-3a,7a-diazaindanes **67** via skeletal rearrangement <02OL773>. Phenazine, pyrimidine, and pyrazine gave adducts **68**, **69**, and **70**, respectively, upon reaction with 2-(chloroseleno)benzoyl chloride <02PJC953>.

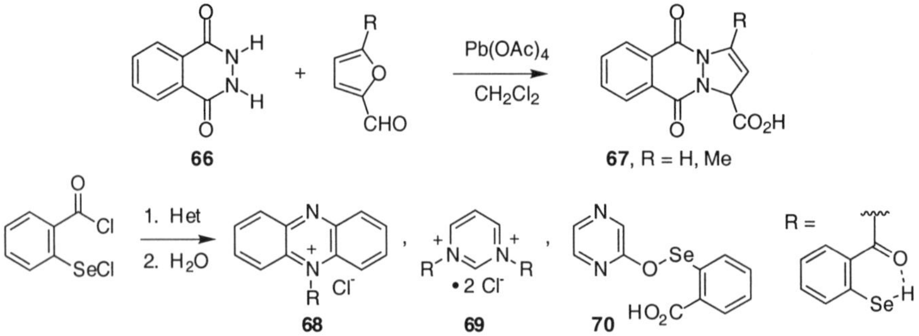

6.2.3.3 Applications

The variety of applications for pyridazines is widening. 2-Acyl-4,5-dichloropyridazin-3-ones are mild and chemoselective acylating agents for amines under neutral conditions <02S733>. Heteroaromatic chromophores experience a dramatic enhancement of fluorescence when pyridazine- or quinoxaline-based spacers **71-75** are inserted into oligo(triacetylene)s prepared from monomer **76** <02HCA2195>. When incorporated into diiron(II) complexes, the sterically hindered bridging phthalazine ligands **77a,b** help in the construction of small molecular models of the active sites in nonheme carboxylate-bridged diiron enzymes <02ICA212>.

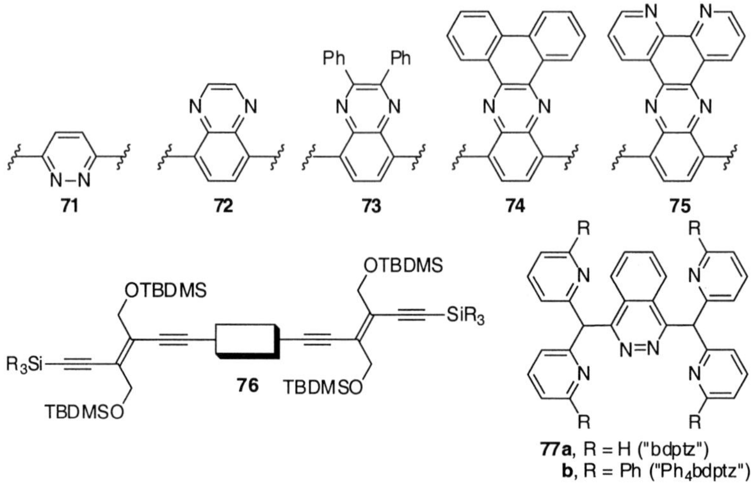

Biologically active pyridazines and benzo-fused derivatives were still avidly sought. For example, pyridazines were prepared as inhibitors of a protein tyrosine phosphatase <02BMC3197>, trisubstituted pyridazines as inhibitors of p38 MAPK <02BMCL689>, and phthalazines as antimicrobials <02SUL183>. 5-Substituted- <02BMC2873> and 5-alkylidene- <02BMCL1575> 6-phenyl-3(2H)-pyridazinones were studied as platelet aggregation inhibitors, and 4,5-disubstituted-6-phenyl-3(2H)-pyridazinones with antiplatelet activity were accessed <02CPB1574>. Pyridazine-fused carbazoles were prepared as antitumor agents <02JHC511>, 4,5-dihydropyridazin-3-(2H)ones were prepared as antibacterials and antifungals <02RRC649>, and 6-(1,2,4-thiadiazol-5-yl)-3-aminopyridazines as antiangiogenics <02BMCL589>. Imidazo-[4,5-d]pyridazine nucleosides were designed as modulators of dsDNA unwinding by a nucleoside triphosphatase/helicase <02AAC1231>, and the interaction of polycyclic derivatives of phenanthro and acenaphtho azaquinolizinium salts with DNA was shown to involve mainly the carbocyclic system rather than to the heterocyclic moiety <02PAC37>.

Other biologically active pyridazines included 3-acetamido-6-arylpyridazine Y5 receptor antagonists <02CPB636>, 1,2,3-triazolo[4,5-d]1,2,4-triazolo[3,4-b]pyridazine adenosine A1 and A2A receptor binding agents <02JHC889>, 4-amino-5-vinyl-3(2H)-pyridazinone antinociceptive agents <02JHC523>, 3-O-substituted benzyl pyridazinone cyclooxygenase inhibitors <02EJM339>, and 2-substituted-3-nitroimidazo[1,2-b]pyridazines <02JHC173>. Still others included α-substituted 3-bisarylthio-N-hydroxypropionamide MMP inhibitors <02BMC531>, imidazo[2,1-a]phthalazine p38 MAP kinase inhibitors <02AP7>, triazolo- and imidazophthalazine non-competitive AMPA antagonists <02MCR39>, and mono- and di-substituted 5,6-diphenyl-3-alkylaminopyridazine acyl-CoA:cholesterol acyltransferase inhibitors <02H(57)39>. 6-Chloropyridazin-3-yl-based compounds <02JMC4011> and pyridazine-based analogues **80a-c** of anabasine (**78**) obtained from tetrazines **79** <02T1343> were studied as nAChRs binding agents. 3-Amino-6-phenylpyridazine was discovered to possess anti-neuroinflammatory activity <02JMC563>. N-Substituted cis-tetra- and cis-hexahydrophthalazinones were developed as selective PDE4-inhibiting in vitro antiinflammatory agents <02JMC2520>, and 1-substituted-4-hydroxyphthalazines were developed as sympathetic blockers against electroshock-induced seizures <02EJM793>.

6.2.4 PYRIMIDINES AND BENZO DERIVATIVES

X-ray crystallography and NMR have been very helpful in defining the structures of pyrimidines. In a new synthesis of 5-(benzylsulfonyl)-6-(polyfluoroalkyl)uracils **81** by treating polyfluoro-1,1-dihydroalkyl sulfones with NaOCN/Et₃N, the X-ray structure of the hexafluoropropyl compound ($n = 3$) was key <02EJO1619>. In the thermal rearrangement of N-alkoxycarbonyl imidazole acryl azides, the size of the alkoxy group determines if imidazo[1,5-c]-pyrimidinones or imidazo[4,5-c]pyridinones are obtained. The crystal structure determination of 6-ethylimidazo[1,5-c]pyrimidin-5-one assisted in this investigation <02TL5879>. Finally, useful NMR structures of pyrazolo[3,4-d]pyrimidinone derivatives were established <02JMS41>.

318   M.P. Groziak

$H(CF_2)_nCF_2CH_2SO_2Bn + 2\ NaOCN \xrightarrow{Et_3N}$ [structure **81**: pyrimidine with H(CF$_2$)$_n$, SO$_2$Bn, H-N, =O •Et$_3$N substituents]

6.2.4.1 Syntheses

Solid-phase syntheses of pyrimidines continue to appear at a rapid pace. Solid-phase syntheses of olomoucine from 4,6-dichloro-2-(methylthio)-5-nitropyrimidine <02TL8071> and of 2,6-disubstituted 4(3*H*)-quinazolinones from 2,4-dichloro-6-hydroxyquinazoline <02TL2971> have been developed. 2-(Arylamino)quinazolinones **83** were accessed by a traceless parallel solid-phase route from 2-nitrobenzamides **82** <02JOC5831>. A solid-phase synthesis of quinazolin-4(3*H*)-ones with 3-point diversity relied on the use of immobilized arylguanidines <02TL5579>.

[Scheme: compound **82** (R^1-substituted 2-nitrobenzamide with N-H-P) → compound **83** (quinazolinone with R^1, R^2, R^3 substituents)]

Microwave irradiation is becoming increasingly important to organic synthesis, and its application to pyrimidine synthesis is particularly attractive. The FeCl$_3$-catalyzed synthesis of 5-alkoxycarbonyl-4-aryl-3,4-dihydropyrimidin-2(1*H*)-ones <02SC147> and the conversion of enaminoketones formed *in situ* into pyrimidines and other heterocycles <02S1669> are both greatly facilitated by microwave irradiation. Sometimes, the use of solventless conditions in combination with microwave irradiation gives beneficial results. For example, both the Biginelli and Hantzsch syntheses of azines like the pyrimdinethiones **84** can be rendered environmentally benign in this manner <02JCS(P1)1845>. In other cases, however, the combination gives unexpected results. The reaction of 6-aminopyrimidines and 3-formylchromone was shown to be 6-hydroxy-6,9-dihydrobenzopyranopyrido[2,3-*d*]pyrimidin-8-ones in all of the cases examined <02TL9061>.

[Scheme: ArCHO + Me-C(O)-CH$_2$-CO$_2$Et + H$_2$N-C(S)-NH$_2$ $\xrightarrow{MW}$ compound **84** (dihydropyrimidinethione with EtO$_2$C, Ar, Me, =S substituents)]

Metal-assisted syntheses are also becoming more widely utilized in pyrimidine chemistry. A new Ni(II)-mediated aromatic C-F activation was applied to the synthesis of pyrimidines like 5-chloro-2,6-difluoro-4-iodopyrimidine (**86**) from its 2,4,6-trifluoro counterpart **85** <02JCS(D)297>. The nucleophilic addition of Grignard reagents at the 2-position of 2-cyanopyrimidinones has led to polysubstituted pyrimidinones <02TL8901>. Highly functionalized amine-containing Grignard reagents derived from diiodo benzenes **87** were elaborated to polyfunctional indoles, quinolines, and quinazolinones like **88** <02OL1819>, and a convergent synthesis of trifunctionalized pyrimidines **89** was achieved by an alkylative-annulation reaction of α,α-dibromo oxime ethers with Grignard reagents <02JA9032>.

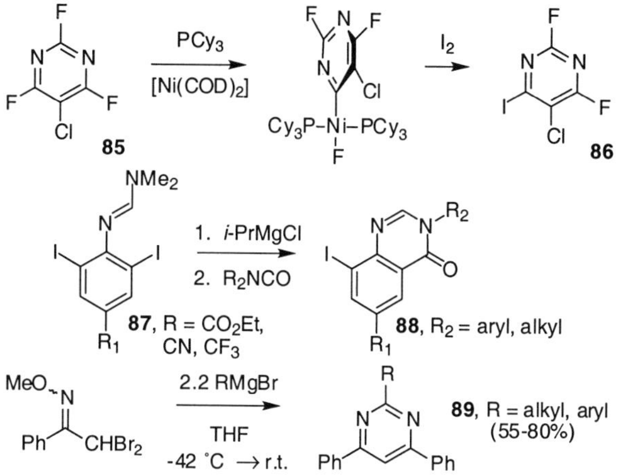

A mild reduction of azides to amines using FeCl₃/NaI was employed in a synthesis of fused [2,1-*b*]quinazolinones <02TL6861>. The use of SmI₂ enabled the development of a one-pot synthesis of 2-aryl-2,3-dihydro-4(1*H*)-quinazolinones <02JHC1271>. Palladium-catalyzed cyclocarbonylation of bis(o-trifluoroacetamidophenyl)acetylene with aryl or vinyl halides and triflates led to 12-acylindolo[1,2-*c*]quinazolines **90** <02OL1355>, and palladium-catalyzed couplings afforded 5-arylpyrrolo[1,2-*c*]pyrimidin-1(2*H*)-ones **92** from stannylated 2*H*-pyrrolo[1,2-*c*]pyrimidin-1-one **91** <02JCS(P1)471>. Facile variation of the C-5 substituent in 1,2,3-triazolo[1,5-*a*]quinazolines **93** was accomplished via Pd-catalyzed cross-coupling <02T9973>. Finally, microwave irradiation facilitated the synthesis of aminopyrimidines via nucleophilic aromatic substitution and Suzuki cross-coupling <02TL5739>.

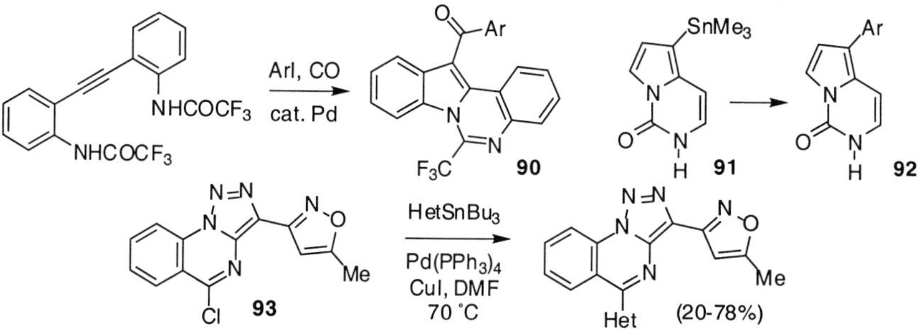

Pyrimidine rings are components in natural nucleoside aglycons, and therefore there are a great many investigations that formally involve pyrimidine chemistry. Although most are beyond the scope of this review, a handful deserve mention here. A new class of adenosine and guanosine analogues **94** featuring fusion-separated imidazole and pyrimidine fragments was reported <02JOC3365>. New 5',6-oxomethylene transglycosidically tethered pyrimidine nucleoside monophosphates **95** have been accessed <02JOC2152>. A new route to mono-*O*-protected *anti-*

conformationally constrained pyrimidine acyclic nucleosides was developed <02CPB1028>, and a series of 9-substituted 9-deazapurines were designed as purine nucleoside phosphorylase (PNPase) inhibitors <02CPB364>. Thiocyanate-substituted nucleoside mimics were prepared by an aminothiocyanation involving the addition of ClSCN to a mixture of a silylated heterocycle and an electron rich alkene <02SC343>. Finally, the 5-hydroxymethylation of pyrimidines and their nucleosides was shown to be greatly facilitated under microwave irradiation <02SL2043>.

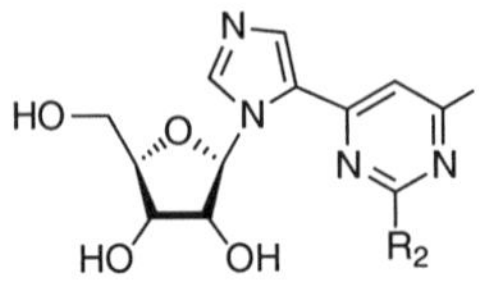

Condensation syntheses of pyrimidines and their benzo-fused analogues continue to be particularly useful. 5H-Thiazolo[3,2-a]pyrimidin-5-ones are obtained via condensation of β-alkoxyvinyl trichloromethyl ketones with 2-aminothiazole <02TL9315>, and β-amino-β-(pyrid-4-yl)acrylonitrile (97) condenses with various reagents to afford fused 1,2,4-triazines 96 and pyrimidines 98 containing a pyridyl group <02T9423>. 7-Arylpyrido[2,3-d]pyrimidines 100 can be prepared from pyrimidinone 99 via regioselective cyclocondensation <02T4873>. The condensation of 6-hydrazinouracils with isocyanates now constitutes a facile one-pot synthesis of pyrazolo[3,4-d]pyrimidines <02TL895>. Zwitterionic pyrimidinylium and 1,3-diazepinylium derivatives condense with isocyanates and thiophosgene to give unusually substituted [1,3]oxazolo[3,4-a]pyrimidines <02SL1831>.

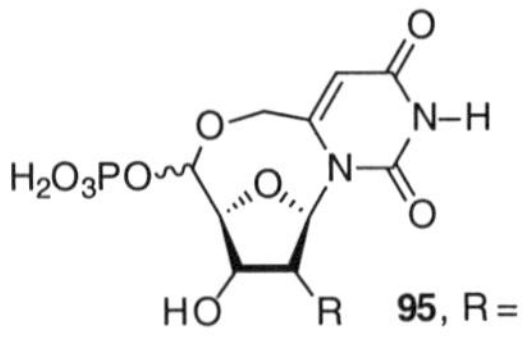

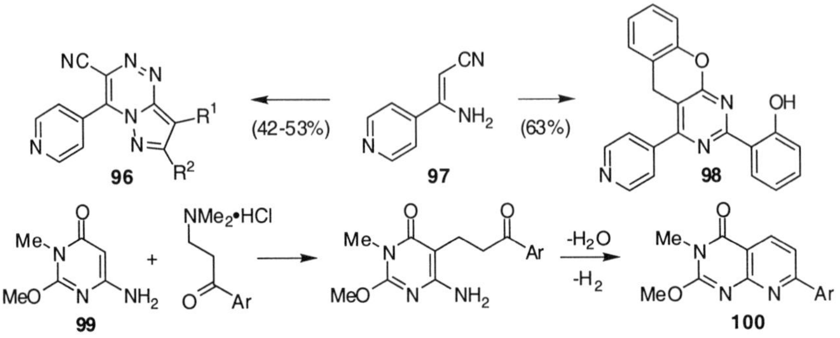

Cyclocondensation routes also provide access to pyrimidines. 2,3-Disubstituted pyrido[2,3-h]-quinazolin-4(3H)-ones are obtained via cyclocondensation of 5-aminoquinoline-6-carboxylic acid with acid chlorides <02SC235>. 5,6,8-Trialkyl-7-methoxy-2-aminoquinazolines are obtained from 1,3-dimethoxybenzenes via cyclocondensation of intermediate dihydrobenzenes with guanidine carbonate <02TL3295>. Diastereoselective intramolecular hetero Diels-Alder cyclization of a pyrazole carboxaldehyde condensed onto 1,3-dimethylbarbituric acid (101) gave polycyclic heterocycle 102 <02T531>. An efficient one-step synthesis of cyclobutene-annelated pyrimidinones 103 from methyl 2-chloro-2-cyclopropylideneacetate and amidines has been developed, and these were used to prepare 2-alkyl-5,6,7,8-tetrahydro-3*H*-quinazolin-4-ones **104** <02OL839>. 5-(Cyclohex-2-enyl)-1,3-dimethyl-6-hydroxyuracil was shown to undergo regioselective heterocyclization via halonium ions **105** to afford fused tricyclic heterocycles <02M1187>. Benzimidazo[1,2-*c*]quinazoline-6(5*H*)-thiones can be obtained by cyclization of 3-(2-aminophenyl)quinazoline-2-thioxo-4-ones <02HEC233>.

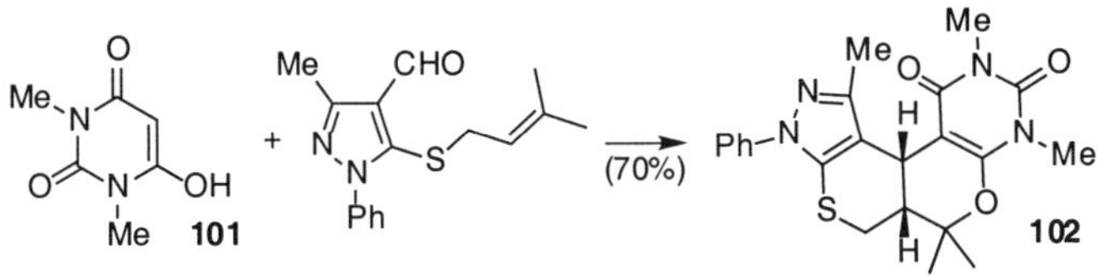

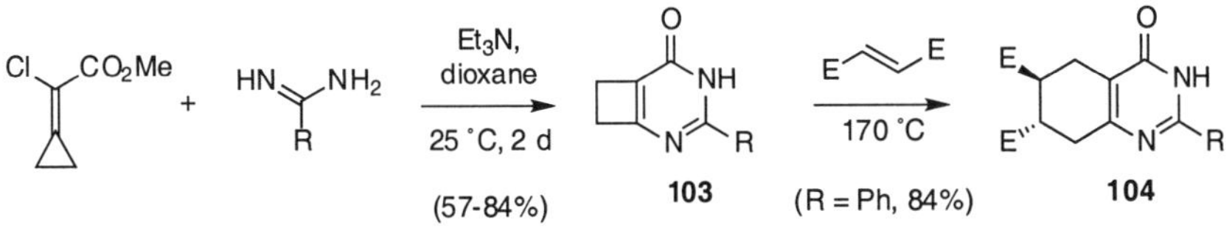

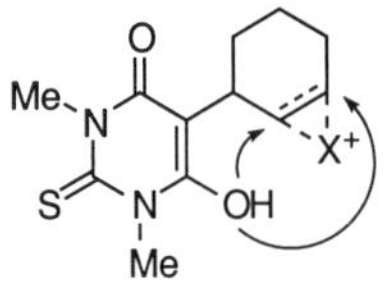

A facile route to 6-hetaryl-pyrrolo[1,2-*a*]thieno[3,2-*e*]pyrimidines **106** was developed <02CCC365>, and pyrrolo[3,2-*d*]pyrimidines were prepared in regioselective fashion from 6-azidouracils and ylide phosphoranes <02HAC357>. 2-Substituted pyrimidine-5-carboxylates were prepared from amidinium salts <02S720>, and 2-trihalogenomethyl-3,4-dihydrofuro[2,3-*d*]-pyrimidin-4-ones were prepared from trihalogenoacetamidines <02SC3749>. A one-pot synthesis of pyrido[2,1-*f*]purine-2,4-diones from 6-aminouracils <02SL155>, and a two-step synthesis of 4-phenyl-2-chloropyrimidines from acetophenone cyanoimines under Vilsmeier-Haack conditions <02SC3011> were both reported.

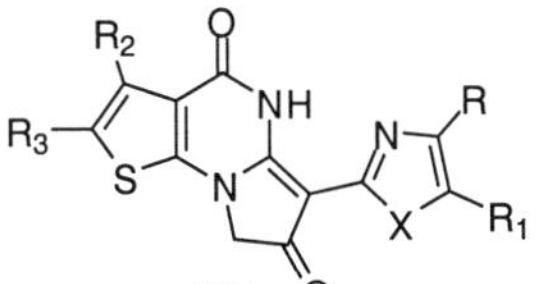

The inverse electron demand Diels-Alder reaction of 1,3,5-triazines **107** with 2-amino-4-cyano-pyrroles **108** constitutes a one-pot synthesis of highly substituted pyrrolo[2,3-*d*]-pyrimidines **109** <02JOC8703>. A high-yielding and unequivocal synthesis of 4-methyl-2-oxo-(2*H*)-pyrido-[1,2-*a*]pyrimidines is now available <02SC741>. A synthesis of heterocycles **110** containing a partially hydrogenated spiro[isoquinoline-4,4'-(2*H*)-pyran] fragment was accomplished <02CHE300>.

Purine mimetical 2,4,6,8-tetrasubstituted pyrimido[5,4-*d*]-pyrimidines were prepared via controlled stepwise nucleophilic substitutions on 2,4,6,8-tetrachloropyrimido[5,4-*d*]pyrimidine <02JCS(P1)108>. An inexpensive and safe route to 3-(1-carboxyalkyl)pyrido[2,3-*d*]pyrimidinediones **111** from 2,3-pyridinedicarboxylic acid has been developed <02CHE306>. 2-Aminomethyl-5-chloropyrimidine was synthesized from *S*-methylthiouronium sulfate and mucochloric acid <02SC153>. Electrochemical oxidation of 3,4-dihydroxybenzoic acid in the presence of 1,3-dimethylbarbituric acids and 1,3-diethyl-2-thiobarbituric acid (**112a,b**) was shown to generate benzofuro[2,3-*d*]pyrimidines **113a,b** <02JOC5036>. Substituted 4-oxo-4*H*-pyrido[1,2-*a*]pyrimidines have been prepared as "exotic" amino acids <02CHE836>, and a one-pot synthesis of imidazo[1,2-*c*]pyrimidines **114a,b** from thioureido-acetamides and benzaldehydes and ethyl cyanoacetate or malononitrile has been developed <02T7241>.

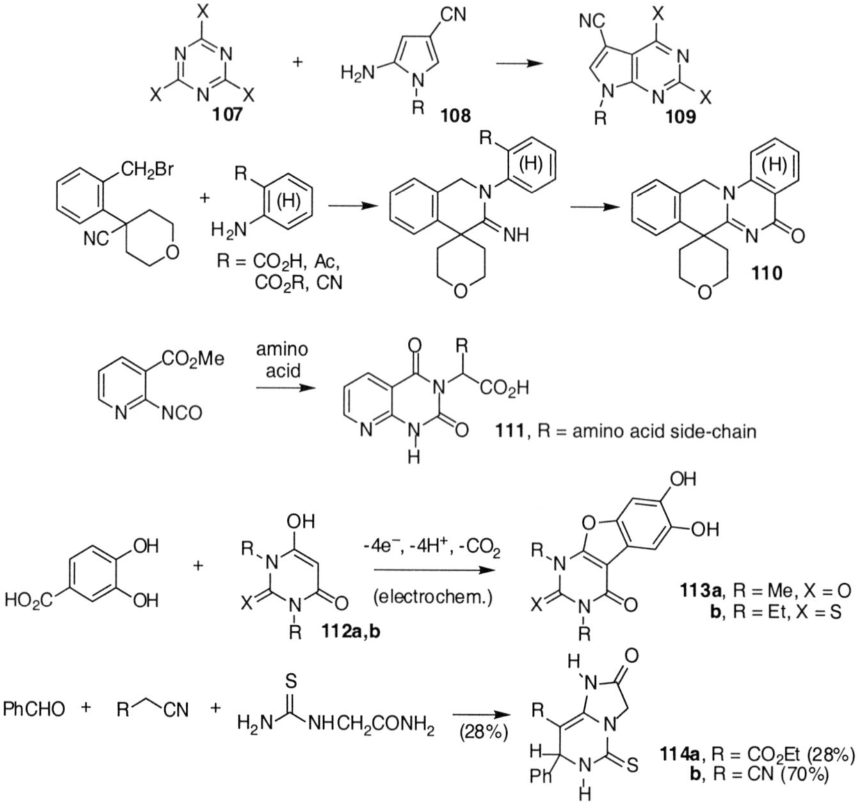

The chemical fixation of CO_2 in the presence of DBU or DBN led to 1*H*-quinazoline-2,4-diones **115** <02T3155>, 5-acylated thiouracils **116a,b** were obtained from pyridin-3- and 4-yl thioureas <02AJC287>, and dihydropyrimidines were obtained by an improved Biginelli reaction protocol

<02SC1847>. The substituted imidazo[1,2-*a*]pyrimidine products expected from the multi-component reactions between aldehydes, isonitriles, and 2-aminoazines have been critically evaluated <02TL4267>. Pyrimidines like **117** bearing two polyhaloalkyl groups were prepared from aromatic methyl ketimines and trichloroacetonitrile, trifluoroacetonitrile, or 2,2,3,3-tetrafluoropropionitrile <02T1375>, and various 2-substituted 4,6-di-2-pyridylpyrimidines **118a-f** were prepared <02CL628>. Compounds **118a** and **b** were found to fluoresce in solution. New tetracyclic azolopyrido[4',3':4,5]thieno[2,3-*d*]pyrimidines were accessed <02HAC280>. Tröger's-base analogues **120** bearing fused pyrimidine rings were synthesized from 6-aminouracils **119** <02JCS(P1)1588>.

115 (97%)

116a, R = pyridin-3-yl (69%)
b, R = pyridin-4-yl (64%)

117

118a, R = CH_3; **b**, R = NH_2; **c**, R = Cl; **d**, R = CH_2Br; **e**, R = $CHBr_2$; **f**, R = CBr_3

119 **120**

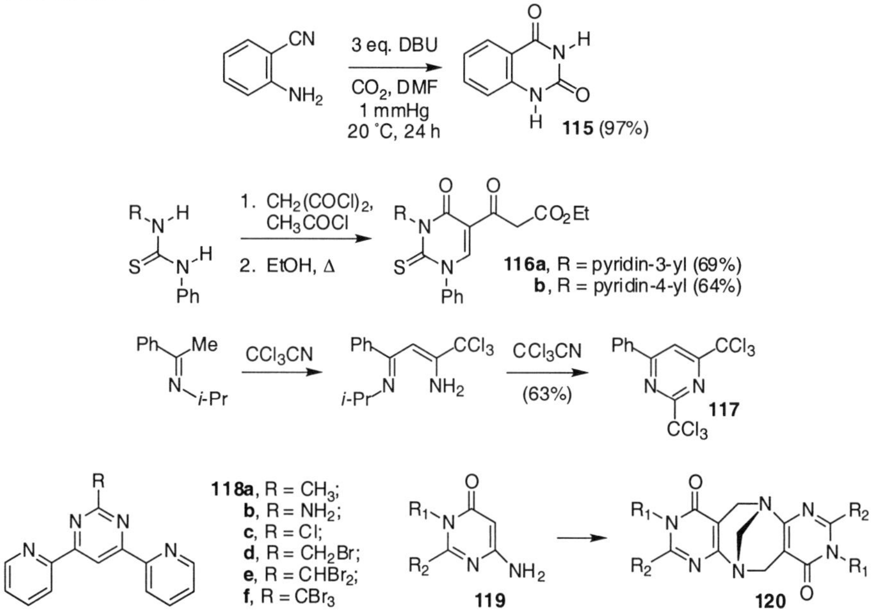

An unanticipated synthesis of 4-trifluoromethyl-2-aminopyrimidines and 5-trifluoroethyl-2,4-diaminopyrimidines analogous to trimethoprim and pyrimethamine, respectively, was reported <02TL9233>. Convenient syntheses of 2-arylidene-5*H*-thiazolo[2,3-*b*]quinazoline-3,5(2*H*)-diones and their benzoquinazoline derivatives are now available <02JHC1153>. The synthesis, basicity, and hydrolysis of 4-amino-2-(*N,N*-diethylamino)quinazolines were investigated <02JHC1289>. Benzimidazo[1,2-*a*]quinazoline-5(7*H*)-one was used for the construction of novel polyheterocyclic frameworks <02JHC1007>. A new route to tetracyclic systems containing the quinazolin-4(3*H*)-one ring system (benzazepino[2,3-*b*]quinazolinones, isoquino[1,2-*b*]quinazolinones, thienopyrimidinones, and isoquino[1,2-*c*]quinazoline-6-carbonitriles) was developed <02H(57)1471>, as was a new quinazoline ring synthesis based on the cycloaddition of *N*-arylketenimines to *N,N*-disubstituted cyanamides <02CPB426>. 2-Substituted 4(3*H*)-quinazolinones were obtained by directed lithiation of 3-*t*-butoxycarbonyl-4(3*H*)-quinazolinone using LDA <02H(57)323>, and heavily substituted 3-hetaryl-1,2,4,5-tetrahydropyrrolo[1,2-*a*]-quinazoline-2,5-diones **121** have been prepared <02CHE324>.

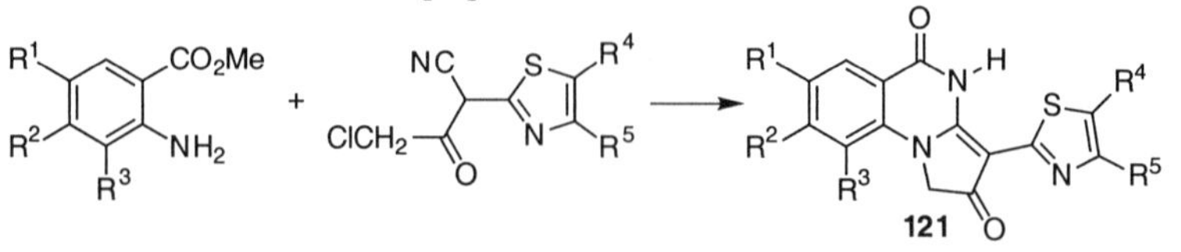

6.2.4.2 Reactions

The reaction of 6-amino-2-thiouracil (**122**) with hydrazonoyl halides was shown to give 7-amino-1,3-disubstituted-1,2,4-triazolo[4,3-*a*]pyrimidines **123** in a regioselective manner <02M1297>. The base-catalyzed coupling of active [(4-oxo-6-phenyl-3*H*-pyrimidin-2-yl)thio]-methine compounds with diazotized anilines gave rise to [1,2,4]triazolo[4,3-*a*]pyrimidines <02HAC136>. 6-Alkoxy-8,9-dialkylpurines **125** were obtained via reaction of 5-amino-4-chloro-6-alkylaminopyrimidines **124** with *N*,*N*-dimethylalkaneamides and alkoxide ions <02T7607>. An electrophilic *N*-amination of quinazoline-2,4-diones with substituted (nitrophenyl)hydroxylamines to give **126a,b** was optimized <02OPRD230>.

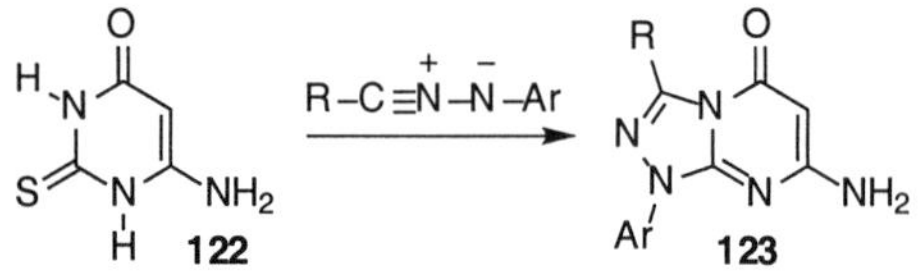

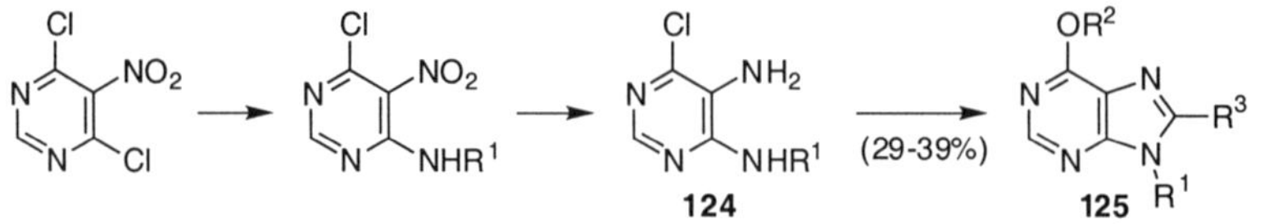

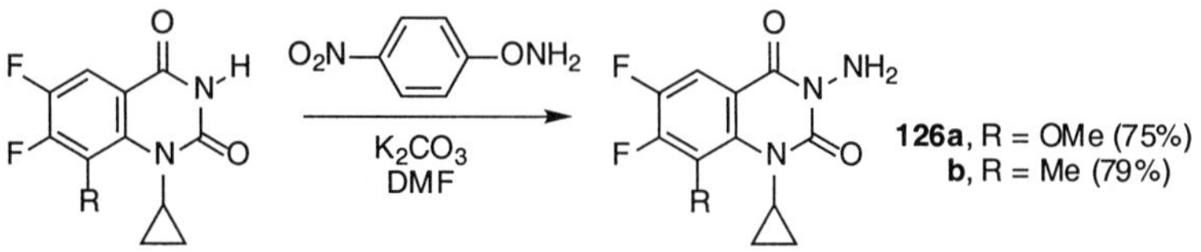

The regioselectivity of bromination of quinazoline, quinoxaline, and other heterocycles in strong acid was investigated <02S83>. Barbituric acid (**127**) was nitrated first to **128**, then again to 5,5-*gem*-dinitropyrimidine-4,6-dione **129**, whose structure and reactivity were studied, leading to 1,3-diacylureas **130** <02JOC7833>. Regioselective cross-coupling of halopyrimidines with 3-bromopyridine and 4-bromoanisole succeeded in generating diazine-containing biaryl derivatives **131** <02JCS(P1)1847>.

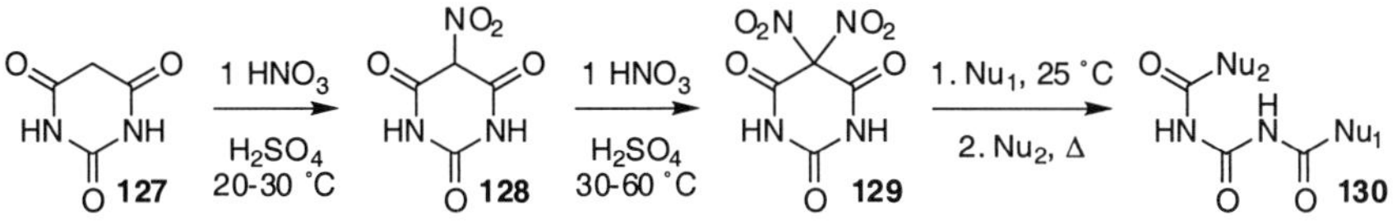

6.2.4.3 Applications

In the pursuit of bioactive pyrimidine heterocycles, 3-arylpyrazolo[4,3-*d*]pyrimidines have been prepared as corticotropin-releasing factor antagonists <02BMCL2133>, and 7-substituted 5-amino-2-(2-furyl)pyrazolo[4,3-*e*]-1,2,4-triazolo[1,5-*c*]pyrimidines have been designed to be A2A adenosine receptor antagonists <02JMC115>. Chloropyrimidines were synthesized as antimicrobial agents <02BMC869>, and annelated 1,2,3-triazolo[1,5-*a*]pyrimidines **132** were synthesized as DNA-binding compounds <02T9723>. A new natural product from *L. vulgaris* was shown to be the pyrroloquinazoline alkaloid **133** <02CPB1393>, and a novel spiro pyrrolo[4,5-*b*]quinazoline alkaloid designated as trisulcusine was obtained from *Anisotes trisulcus* <02IJC2385>.

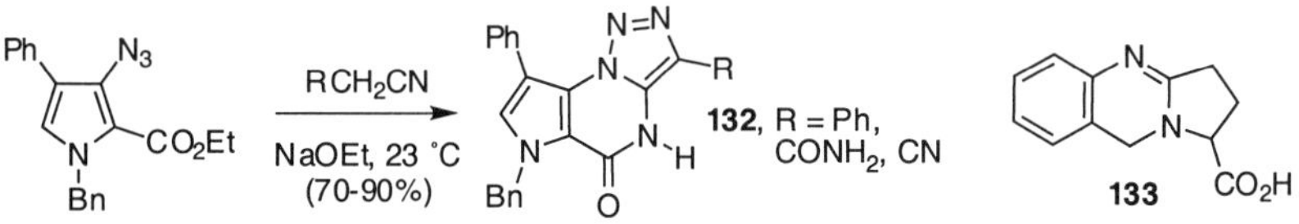

6.2.5 PYRAZINES AND BENZO DERIVATIVES

The X-ray crystal structures of pyrazine *N,N'*-dioxide (**134**) <02AX(E)1253>, the β-polymorph of phenazine (**135**) <02AX(C)181>, cobalt(III) complexes of pyrazine-2,6- and pyridine-2,6-dicarboxylic acids <02JIC458>, and bis-urea-substituted phenazines <02ZN(B)937> were reported. Fluorescent pyrido[1,2-*a*]quinoxalines **136** prepared as pH indicators were examined by X-ray crystallography <02JCS(P2)181>, as were macrocyclic quinoxaline-bridged porphyrinoids obtained from the condensation of dipyrrolylquinoxalines **137** and 1,8-diaminoanthracene

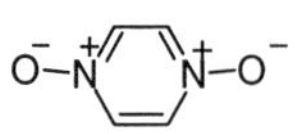

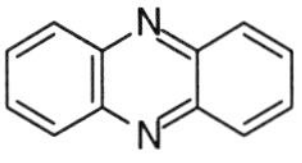

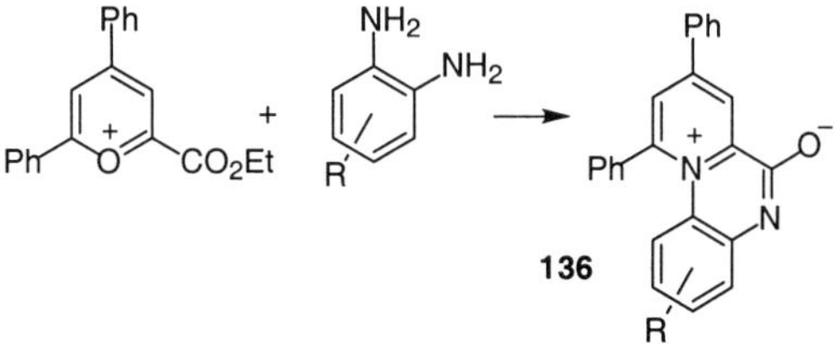

<02JA13474>. A trypanocidal phenazine derived from β-lapachone was found to be active against *Trypanosoma cruzi*, the microorganism that causes Chagas' disease, and its structure was determined in part by X-ray crystallography <02JMC2112>.

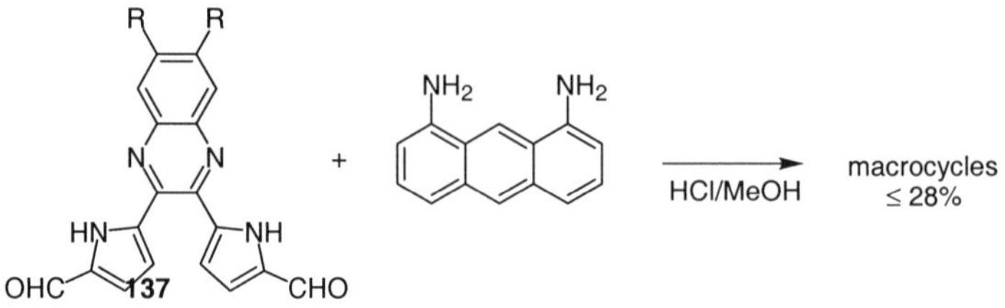

An NMR structural study of 3-benzoylpyrrolo[2,3-*b*]quinoxalin-2(4*H*)-ones in solution confirmed the presence of only one enaminone isomer <02H(57)1413>. The regioselectivity of lithiation and functionalization of dihydrodipyridopyrazines was investigated <02SL1356>, and the Ag(I)-catalyzed reaction of 2-chloroquinoxaline with phenoxide gave quinoxalines <02SC813>.

6.2.5.1 Syntheses

Microwave irradiation has been applied to the synthesis of pyrazines and their fused analogues. It has been shown to facilitate the nucleophilic substitution reactions of pyrimidyl and pyrazyl halides like 5-bromopyrimidine (**138**), 2-bromo- and 2–chloropyrimidine (**139a,b**), and 2-chloropyrazine (**140**) <02T887>, and the intramolecular hetero-Diels-Alder reactions in functionalized 2(1*H*)-pyrazinones **141** to give bicyclo adducts **142** were found to undergo a significant rate enhancement using controlled microwave irradiation <02JOC7904>. The solvent-less, microwave-assisted condensation of pyrazine o-quinodimethanes and non-activated dienophiles was found to constitute a viable route to quinoxalines <02SL2037>.

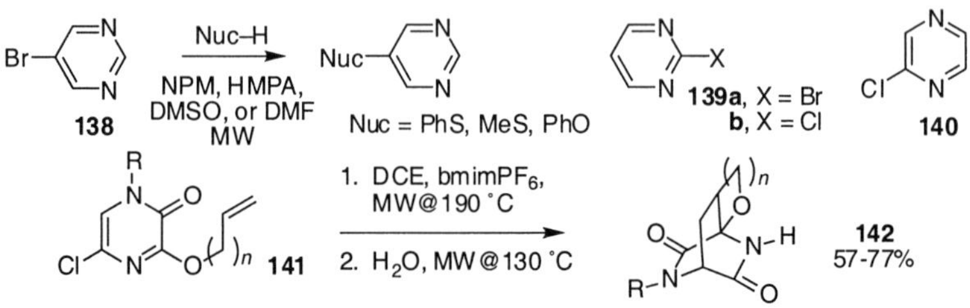

Condensation and cyclization routes continue to be popular for the construction of pyrazines. 2,3-Dihetaryl-6-methylquinoxalines were obtained by condensation of 4-methyl-*o*-phenyl-enediamine with 1,2-diketones <02JCSP42>. 2,3-Dicyano-5,6-diethyl-1,4-pyrazine (**144**) was condensed with 3,6-diphenylphthalonitrile (**143**) to afford donor-acceptor phthalocyanines **145** displaying adjacently-fused pyrazine rings <02CL866>. Dipyrido[3,2-*a*:2',3'-*c*]quinolino[2,3-*h*]-phenazines, angular nitrogen polyheterocycles, were prepared via regioselective condensation of 5,6-phendione and 3,4-diaminoacridines <02TL7883>. A simple synthesis of [1,2,3]triazolo[1,5-*a*]-pyrazines was developed using a cyclization route <02H(56)353>. 1,2-Diazepino[3,4-*b*]-

quinoxalines **146** were prepared via 1,3-dipolar cycloaddition and then subjected to oxidative ring contraction to give pyridazino[3,4-*b*]quinoxalines **147** <02BKC511>. 6,8-Disubstituted-5,7-difluoro-3,4-dihydro-1*H*-quinoxalin-2-ones were prepared via reductive cyclization of 2,4,6-substituted-3,5-difluoronitrobenzenes <02TL6435>, and intramolecular cyclization of *N*-H and *N*-alkyl quaternary salts of quinoxaline-2-carboxaldehyde hydrazones was found to give pyrazolo-[3,4-*b*]quinoxalines in good yields <02MC68>. 2,3-Dihydro-1*H*-pyrimido[1,2-*a*]-quinoxaline 6-oxides were obtained via heterocyclization <02S2687>.

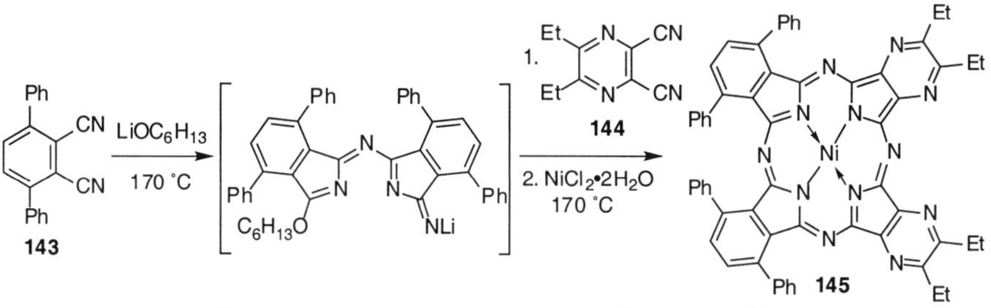

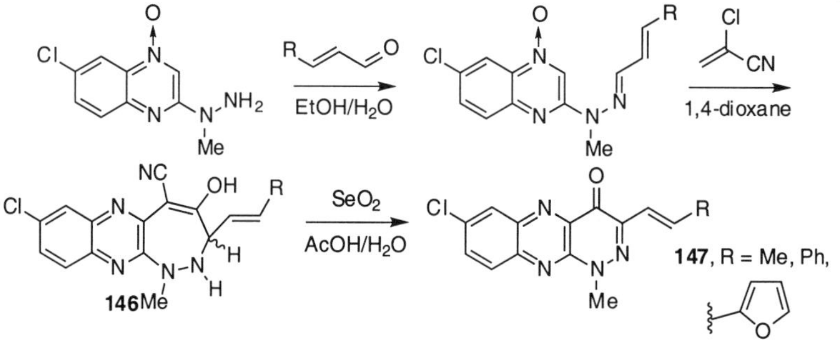

Metal-assisted chemistry was used to prepare a variety of pyrazines and fused analogues. Regioselective metalation of 2-fluoropyrazine (**148**) allows access to various 5- and 6-iodofluoropyrazines **149a-c**, used for Suzuki and Negishi couplings <02T283>. A new synthesis of quinoxalines via Bi-catalyzed oxidative coupling of epoxides and ene-1,2-diamines was reported <02TL3971>, and 2-(oxoalkyl)pyrazines **150** were obtained by oxoalkylation of pyrazine with 1-methylcycloalkanols and Pb(OAc)$_2$ upon mechanical activation <02CHE494>. 5,7-Disubstituted quinoxalino[2,3-*b*]phenazines were prepared via hydrogenation and then air oxidation of dinitrobenzenediamines <02ARK175>.

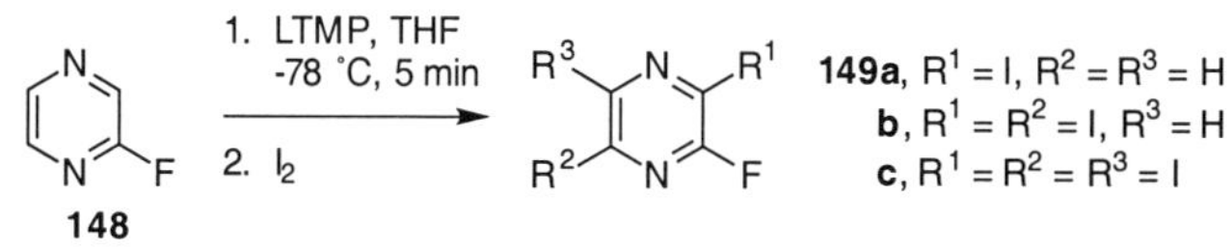

150

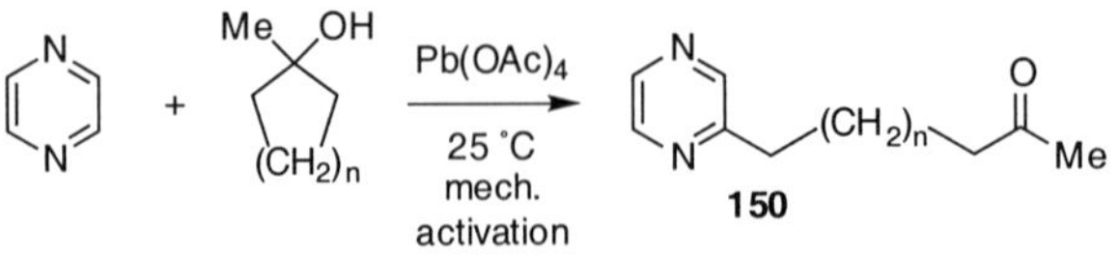

Substituted pyrazino[2,1-*b*]quinazolinediones were prepared and their methylenations explored <02T6163>. Pyrazine-phosphonates and -phosphine oxides **152** were obtained thermally from 2*H*-azirines **151** and oximes **153** <02OL2405>. A general route to 2,3-disubstituted thieno[3,4-*b*]-pyrazines **156** from 3,4-diaminothiazole (**154**) and 1,2-dicarbonyl compounds **155** was developed <02JOC9073>. 1,2,4-Triazolo[1,5-*a*]quinoxalines **157** were obtained by treating benzyl bromide or benzyl cinnamate with *N*-(benzotriazol-1-ylmethyl)arylimidoyl chlorides and t-BuOK <02JOC3118>, and a solid-phase route to asymmetrically disubstituted furano[3,4-*b*]pyrazines was developed <02TL4741>.

151, R = OEt, Ph; R_1 = H, Me, Ph

152, R = OEt, Ph; R_1 = H, Me, Ph; R_2 = H, PO(OEt)$_2$

153, X = Ts, H; R = OEt, Ph

154 **155**

156, R = H, Me, Ph, C_6H_{13}, C_8H_{17}, $C_{10}H_{21}$, $C_{12}H_{25}$

157, R = aryl, heteroaryl

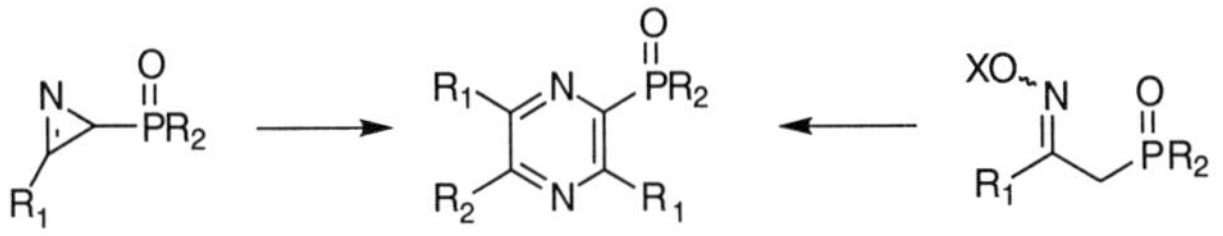

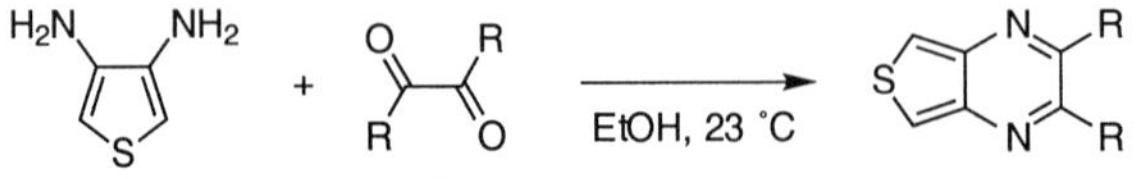

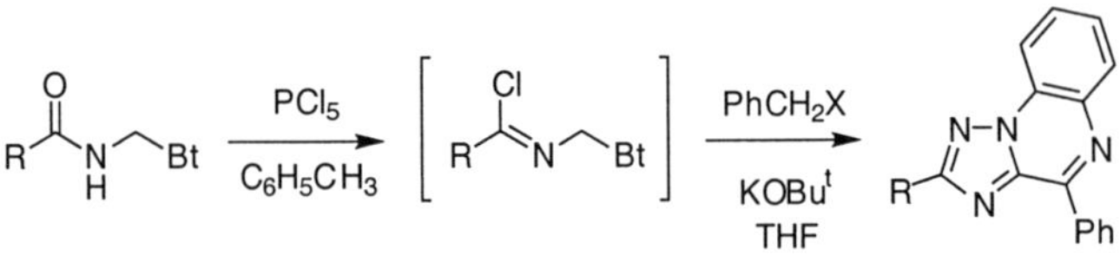

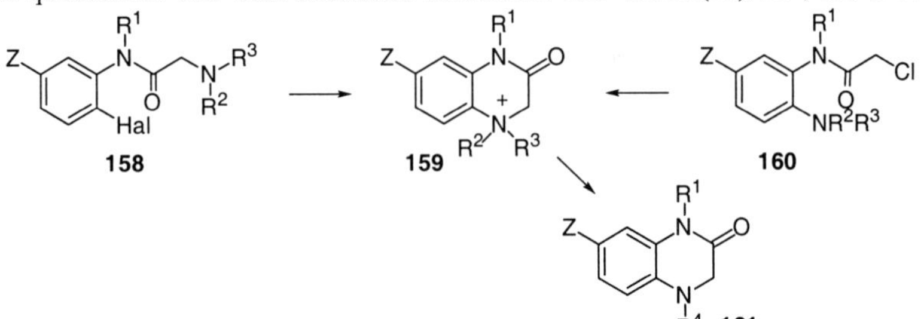

Imidazo[2',1':2,3]thiazolo[4,5-*b*]quinoxalines were obtained in one step from imidazole precursors <02IJH257>. Intramolecular quaternization reactions of either **158** or **159** ultimately afforded quinoxalines **161** from substituted acetanilides **160** <02JCS(P1)790>, and a two-step

158 **159** **160** **161**

synthesis of quinoxalinones utilizing a Ugi/de-Boc/cyclization strategy was developed <02TL1637>. The synthesis and reactivity of pyrimido- (**162**) and oxadiazolyl- (**163**) thienoquinoxalines was reported <02BKC567>.

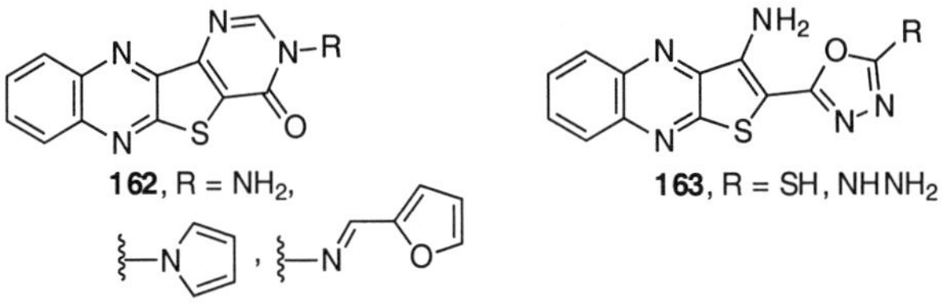

162, R = NH$_2$,

163, R = SH, NHNH$_2$

A new set of Mannich bases of 10H-indolo[3,2-b]quinoxaline was prepared <02IJH157>. A versatile synthetic route to N-protected L-amino acid (3-benzylquinoxalin-2-yl)hydrazides was developed <02LPS49>. The properly substituted bis(indole) pyrazine **165** from the marine alkaloid dragmacidin D was prepared from 2,5-dibromo-3-methoxypyrazine (**164**) <02JOC9392>.

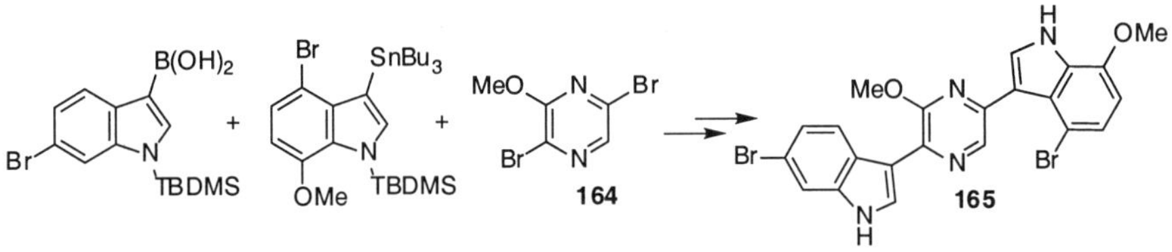

6.2.5.2 Reactions

The regio- and diastereoselective C-4 alkylation of 1-alkyl-2,4-dihydro-1H-pyrazino[2,1-b]-quinazoline-3,6-diones glycine templates was used in a synthesis of Fiscalin B <02TA3387>. A palladium-catalyzed N-arylation of 2-aminopyridines **166** and other amino heterocycles **167** was developed using Xantphos as a ligand. The nature of the base and solvent was found to be crucial in this preparation of diheteroaryl amines **168** <02OL3481>. The reductive decyanation of pyrazinecarbonitriles using H$_2$ and Pt/C under acidic conditions was reported <02TL6747>, and the oxidation of 4-amino-(4H,8H)-bisfurano[3,4-b:3',4'-e]pyrazines with electropositive halogens unexpectedly gave a stable nitrogen-centered biradical <02MC66>.

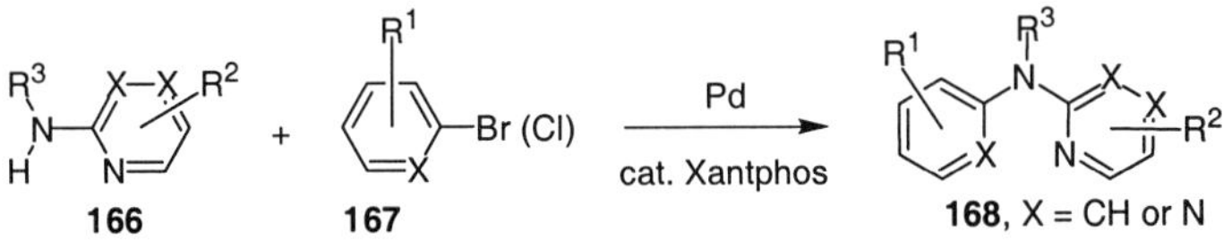

168, X = CH or N

The stereoselective intramolecular Diels-Alder reaction of 3-alkenyl(oxy)-2(1H)-pyrazinones leading to tricyclic ring systems was investigated <02TL447>. A one-pot 7-alkoxylation of 6-arylpyrazino[2,3-c][1,2,6]thiadiazine 2,2-dioxides **169** was accomplished by using N-halosuccinimides <02EJO2109>, and the nitration of 2-(5-methyl-2-furyl)quinoxaline was shown to proceed at C-4' <02CHE783>. Condensation of 2,3-dichloroquinoxalines with alkyl cyanoacetates was unexpectedly found to generate 2-N,N-dimethylamino-3-(2-alkylcyanoacetato)quinoxalines <02IJH99>. Amino- and hydrazino-substituted quinoxalines were prepared, and their ability to complex iron(III) and zinc(II) was investigated <02ZN(B)946>.

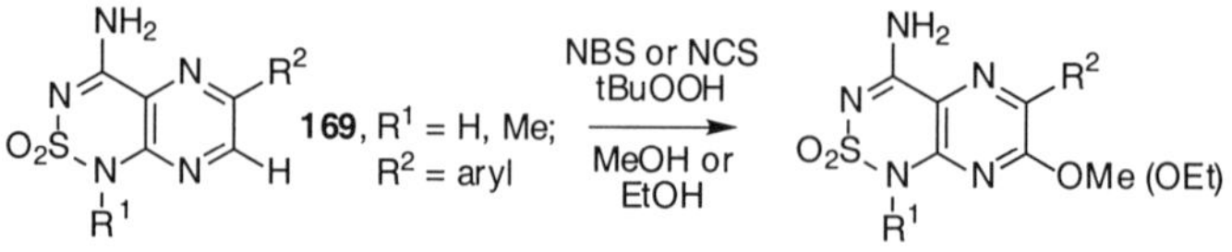

6.2.5.3 Applications

Quinoxalinium dichromate (QxDC, **170**) was shown to be a new and efficient reagent for the oxidation of primary and secondary alcohols and oximes, as well as of anthracene <02MC1417>. 2-Styryl-6,7-dichlorothiazolo[4,5-*b*]quinoxaline (**171**) was used to prepare fluorescent dyes, and their coloristic, fluorophoric, and polyester dyeing properties were studied <02JHC303>. Two families (represented by **172** and **173**) of ditopic *N*-heterocyclic ligands containing a central pyrazine ring and annelated terpene fragments were accessed for use in self-assembly reactions <02JCS(P1)1881>.

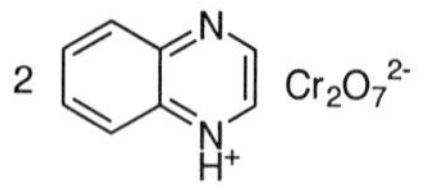

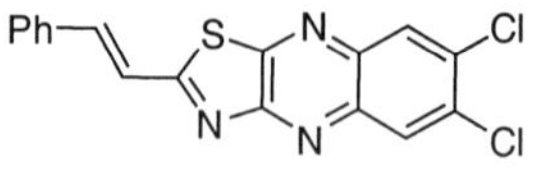

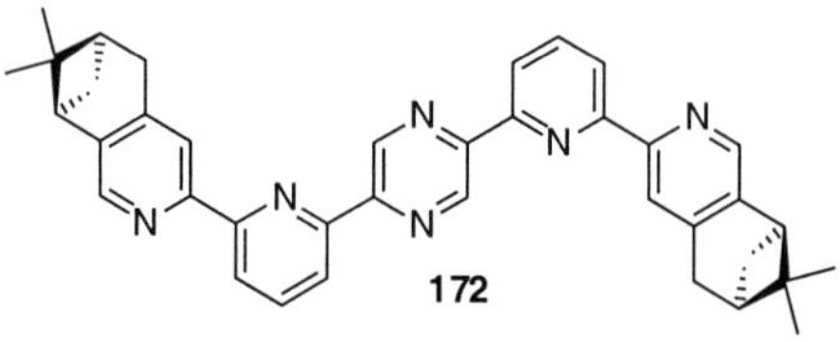

As for biologically active pyrazines, thieno[2,3-*b*]pyrazines have been prepared as bioisosteres for quinoxaline-based reverse transcriptase inhibitors <02H(57)97>, and pyrido[2,3-*b*]pyrazines and 6,7-ethylenedioxyquinoxalines like **174** have been examined as antineoplastic agents <02T5241>. 2,3-Bifunctionalized quinoxalines were synthesized as anticancer, antituberculosis, and antifungal agents <02MOL641>, and substituted quinoxaline 1,4-dioxides were prepared as antimycobacterials <02EJM355>. Quinoxaline-carbohydrate hybrids like **175** were examined as photo-induced GG-selective DNA cleaving agents <02CC212>, 5-oxy-pyrido[2,3-*b*]quinoxaline-9-carboxylic acids were examined as potential cytotoxic DNA intercalators <02JHC1173>, and pyrazines and quinoxalines were examined as antimycobacterials <02JMC5604>.

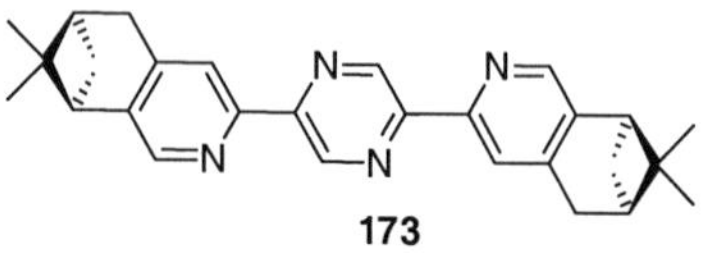

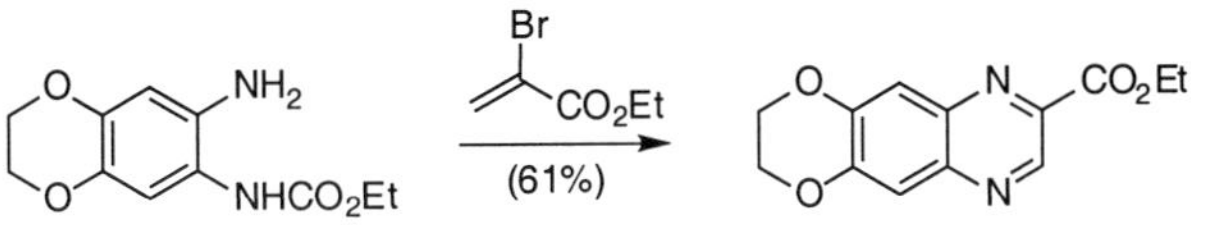

2-Amino-3-benzyl-5-(4-hydroxyphenyl)pyrazine (**176**), a precursor of Watasenia preluciferin (coelenterazine, a natural product found in marine bioluminescent animals), was synthesized in 3 steps from 4-hydroxyphenylglyoxal aldoxime in a 56% overall yield <02CPB301>.

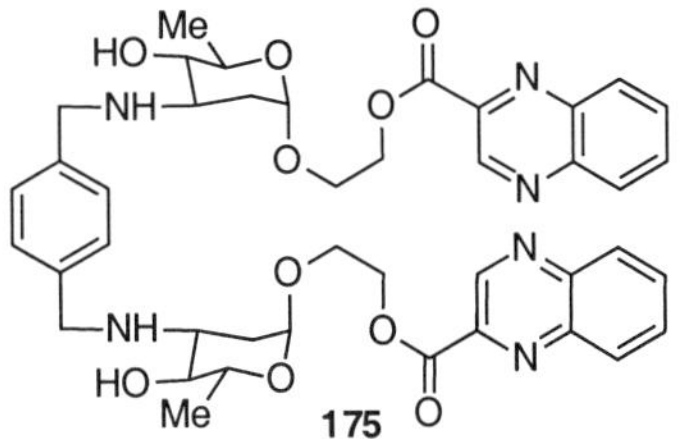

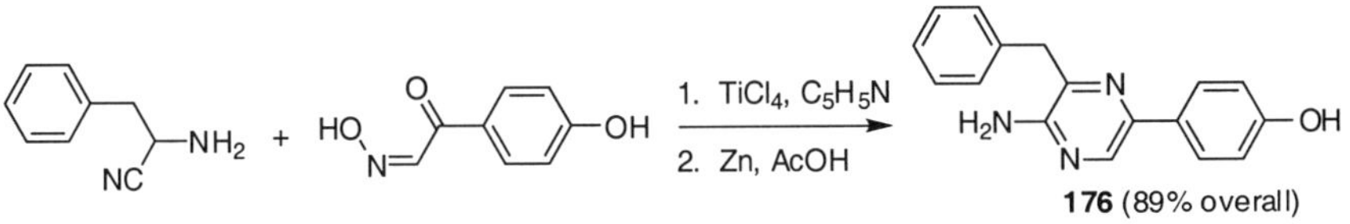

6.2.6 REFERENCES

02AAC1231	P. Borowski, M. Lang, A. Haag, H. Schmitz, J. Choe, H.-M. Chen, R.S. Hosmane *Antimicrob. Agents Chemother.* **2002**, *46*, 1231.
02AG(E)3261	G.J. Bodwell, J. Li *Angew. Chem. Int. Ed.* **2002**, *41*, 3261.
02AJC287	L.W. Deady, D. Ganame, A.B. Hughes, N.H. Quazi, S.D. Zanatta *Aust. J. Chem.* **2002**, *55*, 287.
02AP7	S. Mavel, I. Thery, A. Gueiffier *Arch. Pharm. (Weinheim, Ger.)* **2002**, *335*, 7.
02ARK19	R.W. Millar, R.P. Claridge, J.P.B. Sandall, C. Thompson *Arkivoc* **2002**, 19.
02ARK67	J.T. Sharp, A.G. Cessford, A.R. Stewart *Arkivoc* **2002**, 67.
02ARK73	I.I. Mangalagiu, M.C. Caprosu, G.C. Mangalagiu, G.N. Zbancioc, M.G. Petrovanu *Arkivoc* **2002**, 73.
02ARK143	T. Kocar, S. Recnik, J. Svete, B. Stanovnik *Arkivoc* **2002**, 143.
02ARK175	P.A. Koutentis *Arkivoc* **2002**, 175.
02AX(C)181	W. Jankowski, M. Gdaniec *Acta Crystallogr., Sect. C* **2002**, *C58*, 181.
02AX(C)640	B. Ackers, A.J. Blake, S.J. Hill, P. Hubberstey *Acta Crystallogr., Sect. C* **2002**, *C58*, 640.
02AX(E)1051	D.E. Lynch, I. McClenaghan *Acta Crystallogr., Sect. E* **2002**, *E58*, 1051.
02AX(E)1253	C. Näther, P. Kowallik, I. Jess *Acta Crystallogr., Sect. E* **2002**, *E58*, 1253.
02BKC511	H.S. Kim, S.U. Lee, H.C. Lee, Y. Kurasawa *Bull. Korean Chem. Soc.* **2002**, *23*, 511.
02BKC567	O.S. Moustafa, M.Z.A. Badr, T.I. El-Emary *Bull. Korean Chem. Soc.* **2002**, *23*, 567.
02BKC1425	J.S. Kim, K.J. Shin, D.C. Kim, Y.K. Kang, D.J. Kim, K.H. Yoo, S.W. Park *Bull. Korean Chem. Soc.* **2002**, *23*, 1425.
02BKC1715	E.A. Bakhite, O.S. Mohamed, S.M. Radwan *Bull. Korean Chem. Soc.* **2002**, *23*, 1715.

02BMC531	A.-M. Chollet, T. Le Diguarher, N. Kucharczyk, A. Loynel, M. Bertrand, G. Tucker, N. Guilbaud, M. Burbridge, P. Pastoureau, A. Fradin, et al. *Biorg. Med. Chem.* **2002**, *10*, 531.
02BMC869	N. Agarwal, P. Srivastava, S.K. Raghuwanshi, D.N. Upadhyay, S. Sinha, P.K. Shukla, V. Ji Ram *Biorg. Med. Chem.* **2002**, *10*, 869.
02BMC2415	S. Brase, C. Gil, K. Knepper *Biorg. Med. Chem.* **2002**, *10*, 2415.
02BMC2873	E. Sotelo, N. Fraiz, M. Yanez, V. Terrades, R. Laguna, E. Cano, E. Ravina *Biorg. Med. Chem.* **2002**, *10*, 2873.
02BMC3197	C. Liljebris, J. Martinsson, L. Tedenborg, M. Williams, E. Barker, J.E.S. Duffy, A. Nygren, S. James *Biorg. Med. Chem.* **2002**, *10*, 3197.
02BMCL589	J.-P. Bongartz, R. Stokbroekx, M. Van der Aa, M. Luyckx, M. Willems, M. Ceusters, L. Meerpoel, G. Smets, T. Jansen, W. Wouters, et al. *Biorg. Med. Chem. Lett.* **2002**, *12*, 589.
02BMCL689	C.J. McIntyre, G.S. Ponticello, N.J. Liverton, S.J. O'Keefe, E.A. O'Neill, M. Pang, C.D. Schwartz, D.A. Claremon *Biorg. Med. Chem. Lett.* **2002**, *12*, 689.
02BMCL1575	E. Sotelo, N. Fraiz, M. Yanez, R. Laguna, E. Cano, J. Brea, E. Ravina *Biorg. Med. Chem. Lett.* **2002**, *12*, 1575.
02BMCL2133	J. Yuan, M. Gulianello, S. De Lombaert, R. Brodbeck, A. Kieltyka, K.J. Hodgetts *Biorg. Med. Chem. Lett.* **2002**, *12*, 2133.
02CC64	A. Escuer, R. Vicente, F.A. Mautner, M.A.S. Goher, M.A.M. Abu-Youssef, *Chem. Commun.* **2002**, 64.
02CC212	K. Toshima, R. Takano, T. Ozawa, S. Matsumura *Chem. Commun.* **2002**, 212.
02CC2482	M.R. Lentz, P.E. Fanwick, I.P. Rothwell *Chem. Commun.* **2002**, 2482.
02CCC365	Y.M. Volovenko, E.V. Resnyanska, A.V. Tverdokhlebov *Collect. Czech. Chem. Commun.* **2002**, *67*, 365.
02CCC1790	M. Samalikova, A. Perjessy, Q. Liu, D. Loos, V. Konecny, H. Dehne, S. Sokolowski, Z. Sustekova *Collect. Czech. Chem. Commun.* **2002**, *67*, 1790.
02CHE300	V.M. Kisel, E.O. Kostyrko, M.O. Platonov, V.A. Kovtunenko *Chem. Heterocycl. Compd. (Engl. Transl.)* **2002**, *38*, 300.
02CHE306	M. Saoud, F.B. Benabdelouahab, F. El Guemmout, A. Romerosa *Chem. Heterocycl. Compd. (Engl. Transl.)* **2002**, *38*, 306.
02CHE324	Y.M. Volovenko, E.V. Resnyanskaya, A.V. Tverdokhlebov *Chem. Heterocycl. Compd. (Engl. Transl.)* **2002**, *38*, 324.
02CHE336	Y.M. Volovenko, G.G. Dubinina *Chem. Heterocycl. Compd. (Engl. Transl.)* **2002**, *38*, 336-343.
02CHE494	G.I. Nikishin, L.L. Sokova, V.D. Makhaev, L.A. Petrova, N.I. Kapustina *Chem. Heterocycl. Compd. (Engl. Transl.)* **2002**, *38*, 494.
02CHE783	N.O. Saldabol, J. Popelis, V. Slavinska *Chem. Heterocycl. Compd. (Engl. Transl.)* **2002**, *38*, 783.
02CHE836	I. Ravina, D. Zicane, M. Petrova, E. Gudriniece, U. Kalejs *Chem. Heterocycl. Compd. (Engl. Transl.)* **2002**, *38*, 836.
02CL628	S. Takagi, T. Sahashi, K. Sako, K. Mizuno, M. Kurihara, H. Nishihara *Chem. Lett.* **2002**, 628.
02CL866	T. Fukuda, N. Kobayashi *Chem. Lett.* **2002**, 866.
02CPB301	H. Kakoi *Chem. Pharm. Bull.* **2002**, *50*, 301.
02CPB364	H. Shih, H.B. Cottam, D.A. Carson *Chem. Pharm. Bull.* **2002**, *50*, 364.
02CPB426	M. Shimizu, A. Oishi, Y. Taguchi, Y. Gama, I. Shibuya *Chem. Pharm. Bull.* **2002**, *50*, 426.
02CPB636	S. Guery, Y. Rival, C.-G. Wermuth, P. Renard, J.-A. Boutin *Chem. Pharm. Bull.* **2002**, *50*, 636.
02CPB1022	H. Ishii, K. Yamaguchi, H. Seki, S. Sakamoto, Y. Tozuka, T. Oguchi, K. Yamamoto *Chem. Pharm. Bull.* **2002**, *50*, 1022.
02CPB1028	C.-T. Liu, T.-C. Tu, L.-Y. Hsu *Chem. Pharm. Bull.* **2002**, *50*, 1028.
02CPB1393	H. Hua, M. Cheng, X. Li, Y. Pei *Chem. Pharm. Bull.* **2002**, *50*, 1393.
02CPB1479	N. Haider, E. Sotelo *Chem. Pharm. Bull.* **2002**, *50*, 1479.
02CPB1574	E. Sotelo, N. Fraiz, M. Yanez, R. Laguna, E. Cano, E. Ravina *Chem. Pharm. Bull.* **2002**, *50*, 1574.
02EJI2535	S. Brooker *Eur. J. Inorg. Chem.* **2002**, 2535.
02EJM339	V.K. Chintakunta, V. Akella, M.S. Vedula, P.K. Mamnoor, P. Mishra, S.R. Casturi, A. Vangoori, R. Rajagopalan *Eur. J. Med. Chem.* **2002**, *37*, 339.

02EJM355	A. Carta, G. Paglietti, M.E. Rahbar Nikookar, P. Sanna, L. Sechi, S. Zanetti *Eur. J. Med. Chem.* **2002**, *37*, 355.
02EJM793	R. Sivakumar, S. Kishore Gnanasam, S. Ramachandran, J. Thomas Leonard *Eur. J. Med. Chem.* **2002**, *37*, 793.
02EJO133	D. Csanyi, G. Hajos, G. Timari, Z. Riedl, A. Kotschy, T. Kappe, L. Parkanyi, O. Egyed, M. Kajtar-Peredy, S. Holly *Eur. J. Org. Chem.* **2002**, 133.
02EJO1619	V.M. Timoshenko, Y.V. Nikolin, A.N. Chernega, Y.G. Shermolovich *Eur. J. Org. Chem.* **2002**, 1619.
02EJO2046	J. Sinkkonen, V. Ovcharenko, K.N. Zelenin, I.P. Bezhan, B.A. Chakchir, F. Al-Assar, K. Pihlaja *Eur. J. Org. Chem.* **2002**, 2046.
02EJO2109	N. Campillo, J.A. Paez, P. Goya *Eur. J. Org. Chem.* **2002**, 2109.
02EJO3447	J. Sinkkonen, V. Ovcharenko, K.N. Zelenin, I.P. Bezhan, B.A. Chakchir, F. Al-Assar, K. Pihlaja *Eur. J. Org. Chem.* **2002**, 3447.
02EJO4190	C. Alvarez-Ibarra, A.G. Csaky, C.G. De la Oliva *Eur. J. Org. Chem.* **2002**, 4190.
02H(56)291	Y. Kurasawa, I. Matsuzaki, W. Satoh, Y. Okamoto, H.S. Kim *Heterocycles* **2002**, *56*, 291.
02H(56)353	T. Jug, M. Polak, T. Trcek, B. Vercek *Heterocycles* **2002**, *56*, 353.
02H(57)39	L. Toma, M.P. Giovannoni, V. Dal Piaz, B.-M. Kwon, Y.-K. Kim, A. Gelain, D. Barlocco *Heterocycles* **2002**, *57*, 39.
02H(57)97	T. Erker, K. Trinkl *Heterocycles* **2002**, *57*, 97.
02H(57)323	O. Sugimoto, Y. Yamauchi, K.-I. Tanji *Heterocycles* **2002**, *57*, 323.
02H(57)523	X. Cachet, B. Deguin, F. Tillequin, M. Koch, A. Chiaroni *Heterocycles* **2002**, *57*, 523.
02H(57)723	K. Suzuki, A. Senoh, K. Ueno *Heterocycles* **2002**, *57*, 72.
02H(57)1413	K. Ostrowska, M. Zylewski, K. Walocha *Heterocycles* **2002**, *57*, 1413.
02H(57)1471	P.K. Mohanta, K. Kim *Heterocycles* **2002**, *57*, 1471.
02H(57)2115	O. R'Kyek, B.U.W. Maes, G.L.F. Lemiere, R.A. Dommisse *Heterocycles* **2002**, *57*, 2115.
02H(57)2383	M. Blankenbuehler, S. Parkin *Heterocycles* **2002**, *57*, 2383.
02H(58)359	Y. Kurasawa, J. Takizawa, Y. Maesaki, A. Kawase, Y. Okamoto, H.S. Kim *Heterocycles* **2002**, *58*, 359.
02HAC136	A.S. Shawali, M.A. Abdallah, M.A.N. Mosselhi, T.A. Farghaly *Heteroatom Chem.* **2002**, *13*, 136.
02HAC141	B. Al-Saleh, M.M. Abdel-Khalik, E. Darwich, O. Abdel-Motaleb Salah, M.H. Elnagdi *Heteroatom Chem.* **2002**, *13*, 141.
02HAC258	R.M. Mohareb, D.H. Fleita *Heteroatom Chem.* **2002**, *13*, 258.
02HAC280	E.K. Ahmed *Heteroatom Chem.* **2002**, *13*, 280.
02HAC357	W.M. Abdou, A.F.M. Fahmy, A.A. Kamel *Heteroatom Chem.* **2002**, *13*, 357.
02HCA2195	M.J. Edelmann, J.-M. Raimundo, N.F. Utesch, F. Diederich *Helv. Chim. Acta* **2002**, *85*, 2195.
02HEC233	A. Ivachtchenko, S. Kovalenko, O. Drushlyak *Heterocycl. Comm.* **2002**, *8*, 233.
02ICA212	J. Kuzelka, B. Spingler, S.J. Lippard *Inorg. Chim. Acta* **2002**, *337*, 212.
02IJC2385	A.J. Al-Rehaily, K.A. El-Sayed, M.S. Al-Said, B. Ahmed *Indian J. Chem., Sect. B* **2002**, *41B*, 2385.
02IJH99	P.K. Dubey, S. Vijaya, B. Babu, V.S.H. Krishnan, K.S. Chowdary *Indian J. Heterocycl. Chem.* **2002**, *12*, 99.
02IJH157	S.K. Sridhar, C. Roosewelt, J.T. Leonard, N. Anbalagan *Indian J. Heterocycl. Chem.* **2002**, *12*, 157.
02IJH257	J. Mohan, D. Khatter, S. Mahalaxmi *Indian J. Heterocycl. Chem.* **2002**, *11*, 257.
02IJHC225	Z.A. Filmwala, S.M. Bhalekar, J.P. D'Souza, P.S. Fernandes *Indian J. Heterocycl. Chem.* **2002**, *11*, 225.
02JA1506	B. Mennucci *J. Am. Chem. Soc.* **2002**, *124*, 1506.
02JA1572	D.B. Kimball, R. Herges, M.M. Haley *J. Am. Chem. Soc.* **2002**, *124*, 1572.
02JA9032	H. Kakiya, K. Yagi, H. Shinokubo, K. Oshima *J. Am. Chem. Soc.* **2002**, *124*, 9032.
02JA13463	D.B. Kimball, T.J.R. Weakley, R. Herges, M.M. Haley *J. Am. Chem. Soc.* **2002**, *124*, 13463.
02JA13474	J.L. Sessler, H. Maeda, T. Mizuno, V.M. Lynch, H. Furuta *J. Am. Chem. Soc.* **2002**, *124*, 13474.
02JCS(D)297	M.I. Sladek, T. Braun, B. Neumann, H.-G. Stammler *J. Chem. Soc., Dalton Trans.* **2002**, 297.

02JCS(P1)108 J.S. Northen, F.T. Boyle, W. Clegg, N.J. Curtin, A.J. Edwards, R.J. Griffin, B.T. Golding *J. Chem. Soc., Perkin Trans. 1* **2002**, 108.
02JCS(P1)471 M. Alvarez, D. Fernandez, J.A. Joule *J. Chem. Soc., Perkin Trans. 1* **2002**, 471.
02JCS(P1)790 S. de Castro, R. Chicharro, V.J. Aran *J. Chem. Soc., Perkin Trans. 1* **2002**, 790.
02JCS(P1)1588 R. Abonia, A. Albornoz, H. Larrahondo, J. Quiroga, B. Insuasty, H. Insuasty, A. Hormaza, A. Sanchez, M. Nogueras *J. Chem. Soc., Perkin Trans. 1* **2002**, 1588.
02JCS(P1)1845 M. Kidwai, S. Saxena, R. Mohan, R. Venkataramanan *J. Chem. Soc., Perkin Trans. 1* **2002**, 1845.
02JCS(P1)1847 N.M. Simkovsky, M. Ermann, S.M. Roberts, D.M. Parry, A.D. Baxter *J. Chem. Soc., Perkin Trans. 1* **2002**, 1847.
02JCS(P1)1881 T. Bark, H. Stoeckli-Evans, A. von Zelewsky *J. Chem. Soc., Perkin Trans. 1* **2002**, 1881.
02JCS(P2)181 K.J. Duffy, R.C. Haltiwanger, A.J. Freyer, F. Li, J.I. Luengo, H.-Y. Cheng *J. Chem. Soc., Perkin Trans. 2* **2002**, 181.
02JCS(P2)1081 R.J. Abraham, M. Reid *J. Chem. Soc., Perkin Trans. 2* **2002**, 1081.
02JCS(P2)1807 R.N. Butler, A.G. Coyne, W.J. Cunningham, L.A. Burke *J. Chem. Soc., Perkin Trans. 2* **2002**, 1807.
02JCSP42   N. Ansar *J. Chem. Soc. Pakistan* **2002**, *24*, 42.
02JHC173   T. Terme, C. Galtier, J. Maldonado, M.P. Crozet, A. Gueiffier, P. Vanelle *J. Heterocycl. Chem.* **2002**, *39*, 173.
02JHC203   D.-H. Kweon, H.-K. Kim, J.-J. Kim, H.A. Chung, W.S. Lee, S.-K. Kim, Y.-J. Yoon *J. Heterocycl. Chem.* **2002**, *39*, 203.
02JHC303   N.D. Sonawane, D.W. Rangnekar *J. Heterocycl. Chem.* **2002**, *39*, 303.
02JHC511   N. Haider *J. Heterocycl. Chem.* **2002**, *39*, 511.
02JHC523   V. Dal Piaz, S. Pieretti, C. Vergelli, M.C. Castellana, M.P. Giovannoni *J. Heterocycl. Chem.* **2002**, *39*, 523.
02JHC535   B.U.W. Maes, J. Kosmrlj, G.L.F. Lemiere *J. Heterocycl. Chem.* **2002**, *39*, 535.
02JHC545   G. Cignarella, D. Barlocco *J. Heterocycl. Chem.* **2002**, *39*, 545.
02JHC551   Y. Kurasawa, H.S. Kim *J. Heterocycl. Chem.* **2002**, *39*, 551.
02JHC571   Y. Tominaga, Y. Shigemitsu, K. Sasaki *J. Heterocycl. Chem.* **2002**, *39*, 571.
02JHC685   H.-A. Chung, J.-J. Kim, S.-D. Cho, S.-G. Lee, Y.-J. Yoon, S.-K. Kim *J. Heterocycl. Chem.* **2002**, *39*, 685.
02JHC695   G. Heinisch, B. Matuszczak, D. Rakowitz, K. Mereiter *J. Heterocycl. Chem.* **2002**, *39*, 695.
02JHC737   M. Hervet, C. Galtier, C. Enguehard, A. Gueiffier, J.-C. Debouzy *J. Heterocycl. Chem.* **2002**, *39*, 737.
02JHC853   A.M. Soliman *J. Heterocycl. Chem.* **2002**, *39*, 853.
02JHC869   A. Sener, R. Kasimogullari, M.K. Sener, I. Bildirici, Y. Akcamur *J. Heterocycl. Chem.* **2002**, *39*, 869.
02JHC889   G. Biagi, F. Ciambrone, I. Giorgi, O. Livi, V. Scartoni, P.L. Barili *J. Heterocycl. Chem.* **2002**, *39*, 889.
02JHC989   G. Lukacs, G. Simig *J. Heterocycl. Chem.* **2002**, *39*, 989.
02JHC1007   A. Da Settimo, G. Primofiore, F. Da Settimo, G. Pardi, F. Simorini, A.M. Marini *J. Heterocycl. Chem.* **2002**, *39*, 1007.
02JHC1045   M. Kidwai, R. Venkataramanan, B. Dave *J. Heterocycl. Chem.* **2002**, *39*, 1045.
02JHC1077   C. Fruit, A. Turck, N. Ple, G. Queguiner *J. Heterocycl. Chem.* **2002**, *39*, 1077.
02JHC1153   A.I. Khodair *J. Heterocycl. Chem.* **2002**, *39*, 1153.
02JHC1173   A. Varvaresou, K. Iakovou *J. Heterocycl. Chem.* **2002**, *39*, 117.
02JHC1271   G. Cai, X. Xu, Z. Li, W.P. Weber, P. Lu *J. Heterocycl. Chem.* **2002**, *39*, 1271.
02JHC1289   W. Zielinski, A. Kudelko *J. Heterocycl. Chem.* **2002**, *39*, 1289.
02JIB65   M.P. Lopez-Gresa, R. Ortiz, L. Perello, J. Latorre, M. Liu-Gonzalez, S. Garcia-Granda, M. Perez-Priede, E. Canton *J. Inorg. Biochem.* **2002**, *92*, 65.
02JIC458   G.S. Sanyal, R. Ganguly, P.K. Nath, R.J. Butcher *J. Indian Chem. Soc.* **2002**, *79*, 458.
02JMC115   P.G. Baraldi, B. Cacciari, R. Romagnoli, G. Spalluto, A. Monopoli, E. Ongini, K. Varani, P.A. Borea *J. Med. Chem.* **2002**, *45*, 115.

02JMC1064	H. Gohlke, D. Guendisch, S. Schwarz, G. Seitz, M.C. Tilotta, T. Wegge *J. Med. Chem.* **2002**, *45*, 1064.
02JMC563	S. Mirzoeva, A. Sawkar, M. Zasadzki, L. Guo, A.V. Velentza, V. Dunlap, J.-J. Bourguignon, H. Ramstrom, J. Haiech, L.J. Van Eldik, et al. *J. Med. Chem.* **2002**, *45*, 563.
02JMC2112	C. Neves-Pinto, V.R.S. Malta, M.d.C.F.R. Pinto, R.H.A. Santos, S.L. de Castro, A.V. Pinto *J. Med. Chem.* **2002**, *45*, 2112.
02JMC2520	M. Van der Mey, H. Boss, A. Hatzelmann, I.J. Van der Laan, G.J. Sterk, H. Timmerman *J. Med. Chem.* **2002**, *45*, 2520.
02JMC2526	M. Van der Mey, H. Boss, D. Couwenberg, A. Hatzelmann, G.J. Sterk, K. Goubitz, H. Schenk, H. Timmerman *J. Med. Chem.* **2002**, *45*, 2526.
02JMC4011	L. Toma, P. Quadrelli, W.H. Bunnelle, D.J. Anderson, M.D. Meyer, G. Cignarella, A. Gelain, D. Barlocco *J. Med. Chem.* **2002**, *45*, 4011.
02JMC5604	L.E. Seitz, W.J. Suling, R.C. Reynolds *J. Med. Chem.* **2002**, *45*, 5604.
02JMS41	F. Miklos, I. Kanizsai, P. Sohar, G. Stajer *J. Mol. Struct.* **2002**, *610*, 41.
02JMS73	A. Csampai, P. Kover, G. Hajos, Z. Riedl *J. Mol. Struct.* **2002**, *616*, 73.
02JMS159	O.V. Shishkin, K.Y. Pichugin, L. Gorb, J. Leszczynski *J. Mol. Struct.* **2002**, *616*, 159.
02JOC2152	M.P. Groziak, D.W. Thomas *J. Org. Chem.* **2002**, *67*, 2152.
02JOC2241	M. Avalos, R. Babiano, P. Cintas, F.R. Clemente, R. Gordillo, J.L. Jimenez, J.C. Palacios *J. Org. Chem.* **2002**, *67*, 2241.
02JOC2789	C. Alvarez-Ibarra, A.G. Csakye, C. Gomez de la Oliva *J. Org. Chem.* **2002**, *67*, 2789.
02JOC3118	A.R. Katritzky, T.-B. Huang, O.V. Denisko, P.J. Steel *J. Org. Chem.* **2002**, *67*, 3118.
02JOC3365	K.L. Seley, L. Zhang, A. Hagos, S. Quirk *J. Org. Chem.* **2002**, *67*, 3365.
02JOC5036	D. Nematollahi, H. Goodarzi *J. Org. Chem.* **2002**, *67*, 5036.
02JOC5831	Y. Yu, J.M. Ostresh, R.A. Houghten *J. Org. Chem.* **2002**, *67*, 5831.
02JOC6395	D.B. Kimball, T.J.R. Weakley, M.M. Haley *J. Org. Chem.* **2002**, *67*, 6395.
02JOC7833	A. Langlet, N.V. Latypov, U. Wellmar, U. Bemm, P. Goede, J. Bergman, I. Romero *J. Org. Chem.* **2002**, *67*, 7833.
02JOC7904	E. Van der Eycken, P. Appukkuttan, W. De Brggraeve, W. Dehaen, D. Dallinger, C.O. Kappe *J. Org. Chem.* **2002**, *67*, 7904.
02JOC8703	Q. Dang, J.E. Gomez-Galeno *J. Org. Chem.* **2002**, *67*, 8703.
02JOC8991	F. Mongin, L. Mojovic, B. Guillamet, F. Trecourt, G. Quéguiner *J. Org. Chem.* **2002**, *67*, 8991.
02JOC9073	D.D. Kenning, K.A. Mitchell, T.R. Calhoun, M.R. Funfar, D.J. Sattler, S.C. Rasmussen *J. Org. Chem.* **2002**, *67*, 9073.
02JOC9392	C.-G. Yang, G. Liu, B. Jiang *J. Org. Chem.* **2002**, *67*, 9392.
02JOM246	J.W. Slater, D.P. Lydon, J.P. Rourke *J. Organomet. Chem.* **2002**, *645*, 246.
02JSC257	M.-F. Cheng, H.-O. Ho, C.-S. Lam, W.-K. Li *J. Serbochem. Soc.* **2002**, *67*, 257.
02LPS49	A. El-Faham, A.M. El Massry, A. Amer, Y.M. Gohar *Lett. Peptide Sci.* **2002**, *9*, 49.
02M241	J.A. Szabo, P. Sohar, A. Csampai, G. Stajer *Monatsh. Chem.* **2002**, *133*, 241.
02M1187	K.C. Majumdar, S.K. Samanta *Monatsh. Chem.* **2002**, *133*, 1187.
02M1297	M.A.N. Mosselhi *Monatsh. Chem.* **2002**, *133*, 1297.
02M1417	N. Degirmenbasi, B. Oezguen *Monatsh. Chem.* **2002**, *133*, 1417.
02MC66	A.B. Sheremetev, I.L. Yudin *Mendeleev Commun.* **2002**, 66.
02MC68	M.G. Ponizovsky, A.M. Boguslavsky, M.I. Kodess, V.N. Charushin, O.N. Chupakhin *Mendeleev Commun.* **2002**, 68.
02MC157	A.V. Gulevskaya, A.F. Pozharskii, D.V. Besedin, O.V. Serduke, Z.A. Starikova *Mendeleev Commun.* **2002**, 157.
02MCR39	S. Solyom, T. Hamori, A.P. Borosy, I. Tarnawa, P. Berzsenyi, I. Pallagi *Med. Chem. Res.* **2002**, *11*, 39.
02MOL641	M.J. Waring, T. Ben-Hadda, A.T. Kotchevar, A. Ramdani, R. Touzani, S. Elkadiri, A. Hakkou, M. Bouakka, T. Ellis *Molecules* **2002**, *7*, 641.
02MRC738	T. Wang, C. Lin, T. Zhang, H. Yuan, S. Mao *Magn. Reson. Chem.* **2002**, *40*, 738.
02NPR742	J.P. Michael *Nat. Prod. Rep.* **2002**, *19*, 742.
02OL127	G.J. Bodwell, J. Li *Org. Lett.* **2002**, *4*, 127.
02OL773	A.S. Amarasekara, S. Chandrasekara *Org. Lett.* **2002**, *4*, 773.

02OL839	M.W. Noetzel, K. Rauch, T. Labahn, A. de Meijere *Org. Lett.* **2002**, *4*, 839.
02OL1253	D. Brown, S. Muranjan, Y. Jang, R. Thummel *Org. Lett.* **2002**, *4*, 1253.
02OL1355	G. Battistuzzi, S. Cacchi, G. Fabrizi, F. Marinelli, L.M. Parisi *Org. Lett.* **2002**, *4*, 1355.
02OL1819	D.M. Lindsay, W. Dohle, A.E. Jensen, F. Kopp, P. Knochel *Org. Lett.* **2002**, *4*, 1819.
02OL2405	F. Palacios, A.M. Ochoa de Retana, J.I. Gil, R. Lopez de Munain *Org. Lett.* **2002**, *4*, 2405.
02OL3481	J. Yin, M.M. Zhao, M.A. Huffman, J.M. McNamara *Org. Lett.* **2002**, *4*, 3481.
02OM4622	D.J. Berg, J.M. Boncella, R.A. Andersen *Organometallics* **2002**, *21*, 4622.
02OPRD230	D.C. Boyles, T.T. Curran, R.V.I.V. Parlett, M. Davis, F. Mauro *Org. Process Res. Dev.* **2002**, *6*, 230.
02P5335	G. Xiao, G. Sun, D. Yan, P. Zhu, P. Tao *Polymer* **2002**, *43*, 5335.
02PAC37	M.A. Martin, A.S. Bouin, S. Munoz-Botella, B. del Castillo *Polycyclic Aromatic Compounds* **2002**, *22*, 37.
02PJC45	A.A. Katrusiak, A. Katrusiak, S. Baloniak, K. Zielinska *Pol. J. Chem.* **2002**, *76*, 45.
02PJC953	M. Osajda, J. Mlochowski *Pol. J. Chem.* **2002**, *76*, 953.
02PJC1577	K. Wejroch, J. Karolak-Wojciechowska, J. Lange, J.G. Sosnicki *Pol. J. Chem.* **2002**, *76*, 1577.
02PS2661	W.W. Wardakhan, Y.M. Elkholy *Phosphorus, Sulfur Silicon Relat. Elem.* **2002**, *177*, 2661.
02RJO1309	F.V. Bagrov, T.V. Vasil'eva *Russ. J. Org. Chem. (Transl. of Zh. Org. Khim.)* **2002**, *38*, 1309.
02RRC649	R.R. Kassab, G.H. Sayed, A.M. Radwan, N.A. El-Azzez *Rev. Roum. Chim.* **2002**, *46*, 649.
02S43	J.C. Gonzalez-Gomez, L. Santana, E. Uriarte *Synthesis* **2002**, 43.
02S83	W.D. Brown, A.-H. Gouliaev *Synthesis* **2002**, 83.
02S475	J.C. Gonzalez, J. Lobo-Antunes, P. Perez-Lourido, L. Santana, E. Uriarte *Synthesis* **2002**, 475.
02S720	P. Zhichkin, D.J. Fairfax, S.A. Eisenbeis *Synthesis* **2002**, 720.
02S733	Y.-J. Kang, H.-A. Chung, J.-J. Kim, Y.-J. Yoon *Synthesis* **2002**, 733.
02S1669	V. Molteni, M.M. Hamilton, L. Mao, C.M. Crane, A.P. Termin, D.M. Wilson *Synthesis* **2002**, 1669.
02S2687	M.B. Garcia, L.R. Orelli, M.L. Magri, I.A. Perillo *Synthesis* **2002**, 2687.
02SC147	S.-J. Tu, J.-F. Zhou, P.-J. Cai, H. Wang, J.-C. Feng *Synth. Commun.* **2002**, *32*, 147.
02SC153	Y.-M. Liang, S.-J. Luo, Z.-X. Zhang, Y.-X. Ma *Synth. Commun.* **2002**, *32*, 153.
02SC235	T.A. Kumari, M.S. Reddy, P.J.P. Rao *Synth. Commun.* **2002**, *32*, 235.
02SC343	J. Chmielewski, M. Haun, K. Topmiller, J. Ward, K.M. Church *Synth. Commun.* **2002**, *32*, 343.
02SC481	S.M. Sayed, M.A. Khalil, M.A. Ahmed, M.A. Raslan *Synth. Commun.* **2002**, *32*, 481.
02SC741	O.P. Suri, K.A. Suri, B.D. Gupta, N.K. Satti *Synth. Commun.* **2002**, *32*, 741.
02SC813	A. Rizzo, G. Campos, A. Alvarez, A. Cuenca *Synth. Commun.* **2002**, *32*, 813.
02SC989	A.A. Shalaby, M.M. El-Shahawi, N.A. Shams, S. Batterjee *Synth. Commun.* **2002**, *32*, 989.
02SC1675	E. Sotelo, E. Ravina *Synth. Commun.* **2002**, *32*, 1675.
02SC1847	T. Jin, S. Zhang, T. Li *Synth. Commun.* **2002**, *32*, 1847.
02SC3011	S.J. Cuccia, L.B. Fleming, D.J. France *Synth. Commun.* **2002**, *32*, 3011.
02SC3749	M.V. Vovk, A.V. Bol'but, V.I. Dorokhov, V.V. Pyrozhenko *Synth. Commun.* **2002**, *32*, 3749.
02SL73	J.L. Garcia Ruano, F. Bercial, A. Fraile, M. Rosario Martin *Synlett* **2002**, 73.
02SL155	M.-J. Perez-Perez, E.-M. Priego, M.-L. Jimeno, M.-J. Camarasa *Synlett* **2002**, 155.
02SL223	E. Sotelo, E. Ravina *Synlett* **2002**, 223.
02SL823	S.K. Kundu, A. Pramanik, A. Patra *Synlett* **2002**, 823.
02SL1008	V. Bonnet, F. Mongin, F. Trecourt, G. Breton, F. Marsais, P. Knochel, G. Queguiner *Synlett* **2002**, 1008.
02SL1123	I. Parrot, G. Ritter, C.G. Wermuth, M. Hibert *Synlett* **2002**, 1123.
02SL1356	S. Blanchard, I. Rodriguez, P. Caubere, G. Guillaumet *Synlett* **2002**, 1356.
02SL1831	B. Zaleska, M. Karelus *Synlett* **2002**, 1831.
02SL2037	A. Diaz-Ortiz, A. De la Hoz, A. Moreno, P. Prieto, R. Leon, M.A. Herrero *Synlett* **2002**, 2037.
02SL2043	A.A.H. Abdel-Rahman, E.S.H. El Ashry *Synlett* **2002**, 2043-.
02SL2062	A. Coelho, E. Ravina, E. Sotelo *Synlett* **2002**, 2062.
02SL2095	J.C. Gonzalez-Gomez, E. Uriarte *Synlett* **2002**, 2095.
02SUL183	Y.A. Issac, E.G. El-Karim, S.G. Donia, M.S. Behalow *Sulfur Lett.* **2002**, *25*, 183.
02T283	F. Toudic, N. Ple, A. Turck, G. Queguiner *Tetrahedron* **2002**, *58*, 283.

02T531	E. Ceulemans, M. Voets, S. Emmers, K. Uytterhoeven, L. Van Meervelt, W. Dehaen *Tetrahedron* **2002**, *58*, 531.
02T887	Y.-J. Cherng *Tetrahedron* **2002**, *58*, 887.
02T1343	A. Stehl, G. Seitz, K. Schulz *Tetrahedron* **2002**, *58*, 1343.
02T1375	V.Y. Sosnovskikh, B.I. Usachev, G.-V. Roschenthaler *Tetrahedron* **2002**, *58*, 1375.
02T2227	M. Gnanadeepam, S. Selvaraj, S. Perumal, S. Renuga *Tetrahedron* **2002**, *58*, 2227.
02T2389	E. Sotelo, N.B. Centeno, J. Rodrigo, E. Ravina *Tetrahedron* **2002**, *58*, 2389.
02T2743	C. Fruit, A. Turck, N. Ple, L. Mojovic, G. Queguiner *Tetrahedron* **2002**, *58*, 2743.
02T3155	T. Mizuno, Y. Ishino *Tetrahedron* **2002**, *58*, 3155.
02T4429	V. Bonnet, F. Mongin, F. Trecourt, G. Queguiner, P. Knochel *Tetrahedron* **2002**, *58*, 4429.
02T4873	J. Quiroga, H. Insuasty, B. Insuasty, R. Abonia, J. Cobo, A. Sanchez, M. Nogueras *Tetrahedron* **2002**, *58*, 4873.
02T5241	M. Mateu, A.S. Capilla, Y. Harrak, M.D. Pujol *Tetrahedron* **2002**, *58*, 5241.
02T5645	Z. Riedl, B.U.W. Maes, K. Monsieurs, G.L.F. Lemiere, P. Matyus, G. Hajos *Tetrahedron* **2002**, *58*, 5645.
02T6163	M.L. Heredia, E. de la Cuesta, C. Avendano *Tetrahedron* **2002**, *58*, 6163.
02T7241	J. Schmeyers, G. Kaupp *Tetrahedron* **2002**, *58*, 7241.
02T7607	P.G. Baraldi, A.U. Broceta, M.J.P.d.l. Infantas, J.J.D. Mochun, A. Espinosa, R. Romagnoli *Tetrahedron* **2002**, *58*, 7607.
02T8067	D. Giomi, M. Cecchi *Tetrahedron* **2002**, *58*, 8067.
02T9423	S.A.S. Ghozlan, A.Z.A. Hassanien *Tetrahedron* **2002**, *58*, 9423.
02T9713	B.U.W. Maes, K. Monsieurs, K.T.J. Loones, G.L.F. Lemiere, R. Dommisse, P. Matyus, Z. Riedl, G. Hajos *Tetrahedron* **2002**, *58*, 9713.
02T9723	A. Lauria, P. Diana, P. Barraja, A. Montalbano, G. Cirrincione, G. Dattolo, A.M. Almerico *Tetrahedron* **2002**, *58*, 9723.
02T9933	M. Pal, V.R. Batchu, S. Khanna, K.R. Yeleswarapu *Tetrahedron* **2002**, *58*, 9933.
02T9973	P. Jones, M. Chambers *Tetrahedron* **2002**, *58*, 997.
02T10137	P. Tapolcsanyi, G. Krajsovszky, R. Ando, P. Lipcsey, G. Horvath, P. Matyus, Z. Riedl, G. Hajos, B.U.W. Maes, G.L.F. Lemiere *Tetrahedron* **2002**, *58*, 10137.
02TA3387	F. Hernandez, F.L. Buenadicha, C. Avendano, M. Sollhuber *Tetrahedron: Asymmetry* **2002**, *12*, 3387.
02TC165	L. Turker *Theochem* **2002**, *588*, 165.
02TJC617	N. Secken, M.L. Aksu, A.O. Solak, E. Kilic *Turkish J. Chem.* **2002**, *26*, 617.
02TL447	W.M. De Borggraeve, F.J.R. Rombouts, B.M.P. Verbist, E.V. Van der Eycken, G.J. Hoornaert *Tetrahedron Lett.* **2002**, *43*, 447.
02TL895	P.J. Bhuyan, H.N. Borah, J.S. Sandhu *Tetrahedron Lett.* **2002**, *43*, 895.
02TL1637	T. Nixey, P. Tempest, C. Hulme *Tetrahedron Lett.* **2002**, *43*, 1637.
02TL2971	C. Weber, A. Bielik, G.I. Szendrei, I. Greiner *Tetrahedron Lett.* **2002**, *43*, 2971.
02TL3295	Y. Bathini, I. Sidhu, R. Singh, R.G. Micetich, P.L. Toogood *Tetrahedron Lett.* **2002**, *43*, 3295.
02TL3971	S. Antoniotti, E. Dunach *Tetrahedron Lett.* **2002**, *43*, 3971-3973.
02TL4267	G.S. Mandair, M. Light, A. Russell, M. Hursthouse, M. Bradley *Tetrahedron Lett.* **2002**, *43*, 4267.
02TL4741	E. Fernandez, S. Garcia-Ochoa, A. Huss, A. Mallo, J.M. Bueno, F. Micheli, A. Paio, E. Piga, P. Zarantonello *Tetrahedron Lett.* **2002**, *43*, 4741.
02TL5579	A.P. Kesarwani, K. Srivastava, S.K. Rastogi, B. Kundu *Tetrahedron Lett.* **2002**, *43*, 5579.
02TL5739	G. Luo, L. Chen, G.S. Poindexter *Tetrahedron Lett.* **2002**, *43*, 5739.
02TL5879	Y. Jiao, E. Valente, S.T. Garner, X. Wang, H. Yu *Tetrahedron Lett.* **2002**, *43*, 5879.
02TL6435	R.J. Holland, I.R. Hardcastle, M. Jarman *Tetrahedron Lett.* **2002**, *43*, 6435.
02TL6747	J. Albaneze-Walker, M. Zhao, M.D. Baker, P.G. Dormer, J. McNamara *Tetrahedron Lett.* **2002**, *43*, 6747.
02TL6861	A. Kamal, K.V. Ramana, H.B. Ankati, A.V. Ramana *Tetrahedron Lett.* **2002**, *43*, 6861.
02TL7883	R. Dinica, F. Charmantray, M. Demeunynck, P. Dumy *Tetrahedron Lett.* **2002**, *43*, 7883.
02TL8071	L.G.J. Hammarstrom, D.B. Smith, F.X. Talamas, S.S. Labadie, N.E. Krauss *Tetrahedron Lett.* **2002**, *43*, 8071.

02TL8901	D. Zhang, K. Sham, G.-Q. Cao, R. Hungate, C. Dominguez *Tetrahedron Lett.* **2002**, *43*, 8901.
02TL9061	J. Quiroga, A. Rengifo, B. Insuasty, R. Abonia, M. Nogueras, A. Sanchez *Tetrahedron Lett.* **2002**, *43*, 9061.
02TL9233	H. Berber, M. Soufyane, M. Santillana-Hayat, C. Mirand *Tetrahedron Lett.* **2002**, *43*, 9233.
02TL9315	H.G. Bonacorso, R.V. Lourega, A.D. Wastowski, A.F.C. Flores, N. Zanatta, M.A.P. Martins *Tetrahedron Lett.* **2002**, *43*, 9315.
02ZK61	E. Balogh-Hergovich, G. Speier, M. Reglier, M. Giorgi *Z. Kristallogr.* **2002**, *217*, 61.
02ZK203	S. Ianelli, M. Carcelli *Z. Kristallogr.* **2002**, *217*, 203.
02ZK292	X. Li, Q.H. Jin, Y.Q. Zou, K.B. Yu *Z. Kristallogr.* **2002**, *217*, 292.
02ZN(B)937	T. Fricke, A. Dickmans, U. Jana, M. Zabel, P.G. Jones, I. Dix, B. Konig, R. Herges *Z. Naturforsch., Teil B* **2002**, *57*, 937.
02ZN(B)946	O. Hampel, C. Rode, D. Walther, R. Beckert, H. Gorls *Z. Naturforsch., Teil B* **2002**, *57*, 946.

Chapter 6.3

Triazines, Tetrazines and Fused Ring Polyaza Systems*

Carmen Ochoa and Pilar Goya
Instituto de Química Médica (CSIC), Madrid, Spain
carmela@iqm.csic.es, iqmg310@iqm.csic.es

6.3.1. TRIAZINES

The synthesis and structural characterization of a copper(II) complex of 3-acetyl-5-benzyl-1-phenyl-4,5-dihydro-1,2,4-triazin-6-one oxime have been described <02ZN(B)547>. Three new enantiomerically pure bifunctional chiral auxiliaries for enantioselective HPLC, belonging to a family of biselector 1,3,5-triazine systems, have been synthesized and their enantiodiscriminating capability studied <02TA1805>. A novel microtubule destabilizing entity from orthogonal synthesis of a triazine library and zebrafish embryo screening has been reported <02JA11608>. The formation and photophysical properties of a stable concave-convex supramolecular complex of C_{60} and a substituted 1,3,5-triazine derivative have been studied <02CC2538>. The design of coordination polymer gels containing melamine derivatives, as stable catalytic systems, has been reported <02CEJ5028>. Heterogeneous enantioselective epoxidation of olefins catalyzed by unsymmetrical Mn(III) complexes supported on amorphous or MCM-41 silica through a new 1,3,5-triazine-based linker has been reported <02CC716>. Heterocumulene metathesis by iridium guanidinate and ureylene complexes-catalysis involving reversible insertion to form metalla-perhydro-1,3,5-triazines has been reported <02JA9010>. An unprecedented ladder coordination polymer-based on a pentanuclear copper(II) 2,4,6-tris(dipyridin-2-ylamino)-1,3,5-triazine building-block has been reported <02CC1488>. The preparation and X-ray structure determination of organomercury derivatives of 2,4,6-trimercapto-1,3,5-triazine have been described <02JOM101>. Two new cyclotriveratrylenes bearing complementary H-bond donor-acceptor substituents such as melamine or cyanuric acid have been described <02JCR(S)359>. Two novel disc-shaped 1,3,5-triazine derivatives with strong intramolecular hydrogen bonds have been prepared and characterized <02CC1350>. A study of the kinetic stabilities of hydrogen-bonded double, tetra and hexarosette assemblies bearing melamine units <02JOC4808>, and diastereoselective noncovalent synthesis of hydrogen-bonded double rosette assemblies with dimelamine or cyanurate components have been reported <02CEJ2288>. A superstructural example from an alkylated 2,6-diamino-1,3,5-

---

*In Memoriam of Prof. Manfred Stud.

triazine and the formation of one-dimensional hydrogen-bonded arrays or cyclic rosettes have been examined <02MI15>. New polydentate and polynucleating *N*-donor ligands from amines and 2,4,6-trichloro-1,3,5-triazine have been reported <02TL6783>. 2,4,6-Tris(4-nitrophenoxy)-1,3,5-triazine forms a honeycomb network of molecular and nitro-trimer synthons with guest species included in the hexagonal voids <02CC952>. New classes of star-shaped discotic liquid-crystal containing a 1,3,5-triazine unit as a core have been described <02BCJ615, 02TL3863>. The first example of a diamino-1,3,5-triazine receptor that exhibits controllable allostery using redox chemistry has been reported <02CC178>. Triazinyl-amino acids have been reported as new building-blocks for pseudopeptides <02SL557>. For the first time, rhenium complexes based on a cyclotriphosphazane scaffold with exocyclic pyrazolyl substituents have been prepared <02CC1386>. A new Pd(0) complex of a pendant cyclophosphazene-containing cross-linked polymer has been found to be an effective heterogeneous catalyst for the Heck arylation reaction <02OL2113>. Structure-alkali metal cation complexation relationships for macrocyclic PNP-lariat ether ligands have been studied <02JCS(P2)442>.

6.3.1.1. Synthesis

3,5-*Bis*(5-carboxy-6-azauracil-1-yl)aniline and 1,3,5-*tris*(5-carboxy-6-azauracil-1-yl)benzene have been prepared from 3-amino-5-nitroacetanilide <02JHC357>. Reaction between nitrilimines, generated *in situ* from the corresponding hydrazonoyl chlorides **1**, and ethyl isocyanoacetate (**2**) with an excess of triethylamine in boiling benzene yielded, through two competing pathways, 2,3-dihydro-1,2,4-triazines **3** and 1,3-oxazoles **5**. *In situ*, cycloaddition of unreacted nitrilimine with the triazines **3** gave rise to the bicyclic triazolotriazines **4** <02MI165>.

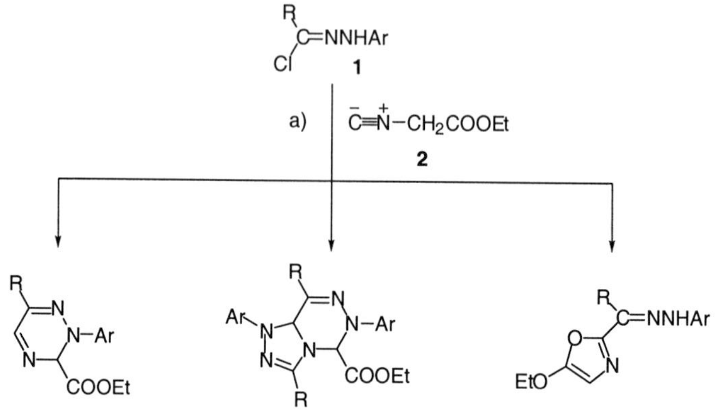

a: R = Ar = Ph; **b**: R = 4-MeOC$_6$H$_4$, Ar = Ph; **c**: R = Ph, Ar = 4-MeOC$_6$H$_4$; **d**: R = Ph, Ar = 4-O$_2$NC$_6$H$_4$; **e**: R = Ac, Ar = Ph; **f**: R = Ac, Ar = 4-MeC$_6$H$_4$; **g**: R = EtOOC, Ar = 4-BrC$_6$H$_4$.
a) Et$_3$N (3-4 eqv.), benzene, reflux 1 h.

Reaction of benzoylacetylene with 1,5-diphenyl-2,4-dithiobiuret in methanol yielded a hexahydro-1,3,5-triazine-2,4-dithione derivative, whilst this reaction in benzene afforded a dihydro-1,3,5-thiadiazine derivative <02KGS81>. The synthesis of hexadentate hexahydro-1,3,5-triazine-based ligands and their copper(I) complexes have been described <02M1157>. An efficient approach to 1,3,5-tris-aryl-hexahydro-1,3,5-triazines **7** by reaction of anilines **6** with 1,3,6,8-tetrazatricyclo[4.4.1.1(3.8)]dodecane (TATD) has been described <02SC1407>.

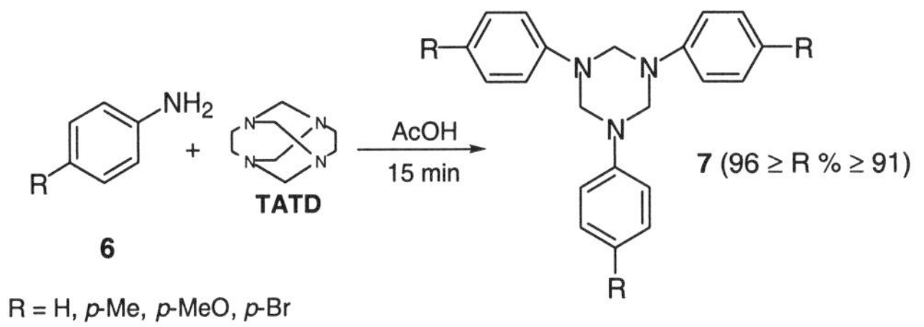

An efficient synthesis of 1-aryl-4,6-diamino-1,2-dihydro-1,3,5-triazines from an acid-catalyzed reaction between the corresponding arylbiguanidine and carbonyl compounds has been reported <02SC2089>. A new family of functionalized terpyridine-like ligands containing a central 1,3,5-triazine, (**8a**) and (**8b**) has been synthesized. These compounds exhibit room-temperature luminiscence as their Ru(II) complexes (**9a**), (**9b**) and (**9c**) <02CC1356>.

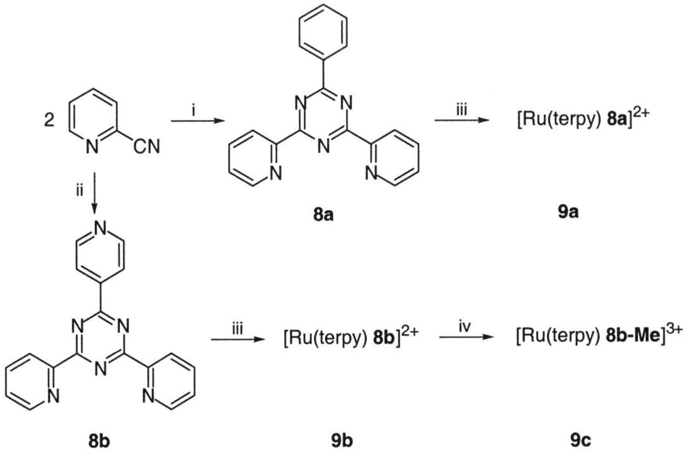

i) lithium *N,N*-dimethylbenzamidinate, diethyl ether, r.t., 30 min
ii) 4-cyanopyridine, NaH, 180 °C, 30 min
iii) Ru(tpy)Cl$_3$, ethylene glycol, reflux, 10 min
iv) MeI, r.t., 24 h (**8b-Me** is ligand **8b** methylated in the 4-pyridyl position)

The reaction of ethylideneisopropylamine (**10**) with halonitriles has been studied. Trichloroacetonitrile did not react with **10**, whilst difluoroacetonitrile reacted, either at room temperature or in a sealed tube at 80 °C, to give a complex mixture of at least three products from which the crystalline previously unknown triazine (**11**) was isolated <02T1375>.

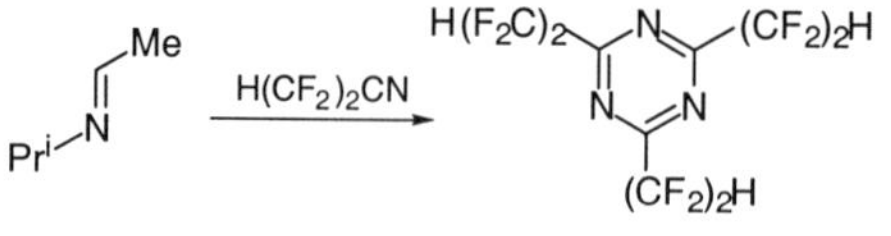

A novel sorbitol dehydrogenase inhibitor bearing a 1,3,5-triazine moiety has been described <02JMC4398>. On refluxing isatin-3-thiosemicarbazone and 1-*p*-chlorophenyl-3-phenylprop-2-yn-1-one in ethanol a mixture of 2,4,6-tris(2-oxo-2*H*,3*H*-benzo[*b*]pyrrolidine-3-iminyl)amino-1,3,5-triazine and Z,Z-3,3'-thiodi(1-*p*-chlorophenyl-3-phenylprop-2-en-1-one) was obtained <02PS173>. Complex (**12**) reacted with two equivalents of alkyl isocyanates yielding the complexes (**13-15**) with six-membered cyclic ligands which were characterized spectroscopically and, for **15**, by X-ray diffraction analysis. Reaction of **13-15** with triflic acid (HOTf) cleanly gave demetalated 1,3,5-triazine-2,4-diones (**16-18**) <02AG(E)3858>.

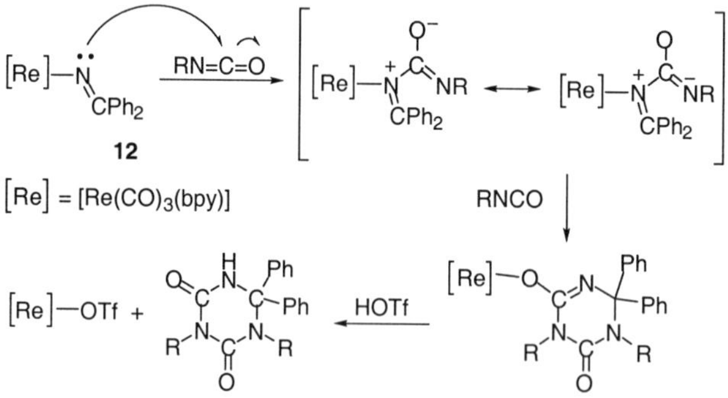

The synthesis and evaluation of *N*-(triazin-1,3,5-yl)phenylalanine derivatives, as VLA-4 integrin antagonists, have been reported <02BMCL1591>. Rapid synthesis of 1,3,5-triazine inhibitors of inosine monophosphate dehydrogenase has been carried out <02BMCL2137>. The solid phase synthesis of 1,3-disubstituted and 1,3,5-trisubstituted 1,3,5-triazine-2,4,6-triones from MBHA and Wang resin has been described <02JCO484>. The synthesis and antitumor activities of novel 4-substituted 2-amino-4-(3,5,5-trimethyl-2-pyrazolino)-1,3,5-triazines have been reported <02EJM709>.

6.3.1.2. Reactions

Deoxygenative *vs* vicarious nucleophilic substitution of hydrogen in reactions of 1,2,4-triazine-4-oxides with α-halocarbanions has been described <02EJO1412>. Tandem vicarious nucleophilic substitution of hydrogen/intramolecular Diels-Alder reaction of 1,2,4-triazine (**19**) into functionalized cycloalkenopyridines **21** has been reported <02CPB463>.

19 **20** **21**

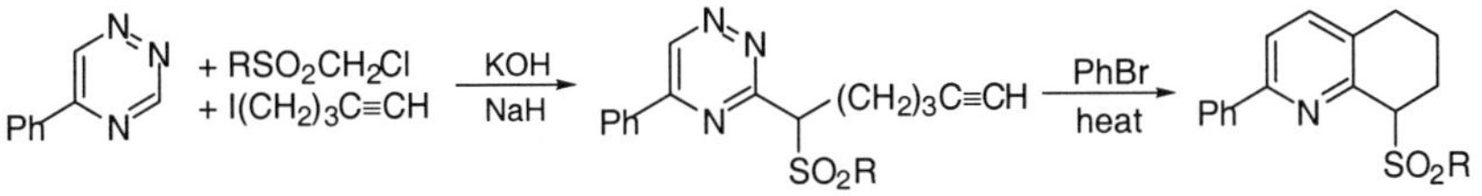

The reaction of 3-propargylmercapto-1,2,4-triazin-5-one and an aromatic iodide *via* palladium catalysis yielded substituted thiazolo[2,3-*c*][1,2,4]triazinones <02PS2491>. Synthetic approaches towards benzopyrano[4,3-*b*]thieno[2,3-*e*]pyridines *via* intramolecular Diels-Alder reactions of 1,2,4-triazines have been described <02JCR(S)60>. The palladium-catalyzed *N*-(hetero)arylation of a number of heteroarylamines including 3-amino-1,2,4-triazine (**22**) has been carried out using Xantphos as the ligand. Choice of the base and solvent is critical for the success of these reactions to give compounds such as (**23**) <02OL3481>.

22 Xantphos **23** (86%)

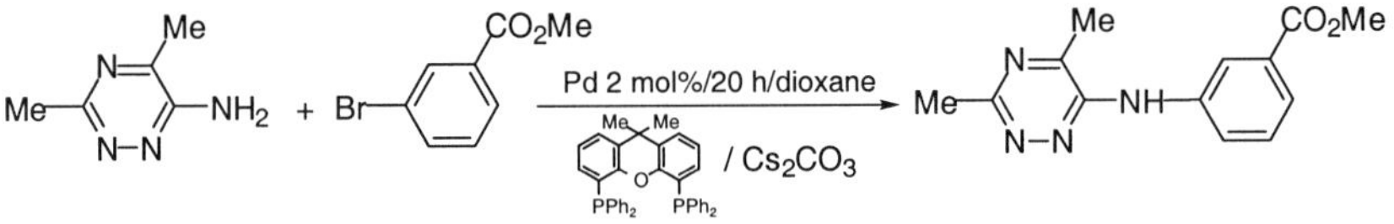

Some transformations of 5-cyano-1,2,4-triazines in reactions with nucleophiles have been reported <02MI744>. The palladium-catalyzed coupling reaction of 3-thiomethyl-1,2,4-triazine with different organoboron compounds in the presence of copper(I) 3-methylsalicylate afforded the corresponding 3-substituted-1,2,4-triazines in good yield <02SL447>. Benzyl-3-(arylmethylidenhydrazino)-1,2,4-triazin-5(4*H*)-ones **24** underwent regioselective cyclization upon treatment with either bromine in acetic acid/sodium acetate or with ferric chloride in refluxing ethanol to give the respective triazolo[4,3-*b*][1,2,4]triazin-7(8*H*)-ones **25** in overall good yields <02T8559>.

	R
a	Ph
b	4-MeC_6H_4
c	4-ClC_6H_4
d	4-BrC_6H_4
e	2-furyl
f	2-thienyl

24(a-f) **25(a-f)**

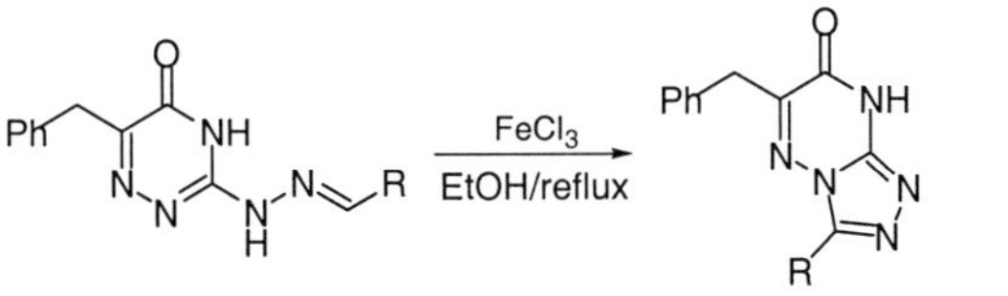

Syntheses of 5-[2-(halogenobenzylthio)-ethyl]thio-6-azauracil derivatives, as potential antiviral agents, have been achieved from 5-mercapto-6-azauracil <02RRC1007>. Synthesis of some novel *N,N*'-bis-(1,2,4-triazin-4-yl)dicarboxylic acid amides and some fused rings starting from 4-amino-6-methyl-5-oxo-3-thioxo-2,3,4,5-tetrahydro-1,2,4-triazine has been described <02PS587>. New complexes of some bivalent metals with new Schiff bases derived from the condensation of 4-amino-3-mercapto-6-methyl-5-oxo-1,2,4-triazine with aromatic aldehydes have been prepared <02SRI171>. A high yield synthesis of polysubstituted 2-azafluorenones from 1,2,4-triazines using metalation and intramolecular Diels-Alder reaction has been described <02S2532>. Reactions of π-electron rich 1,2,4-triazines with organolithium nucleophiles have been studied <02JCS(P1)2549>. Preparation of macrocycle (**27**) from 5-ethoxy-5,6-diphenyl-4,5-dihydro-2*H*-1,2,4-triazine (**26a**) and chromiun chloride using a high dilution technique <02IJC(A)1629> or starting from (**26b**) and mercury salts <02T1525> has been reported.

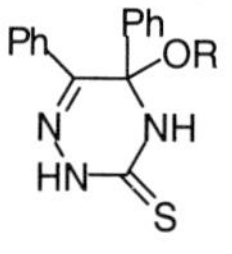

26a, R= Et
26b, R=Me

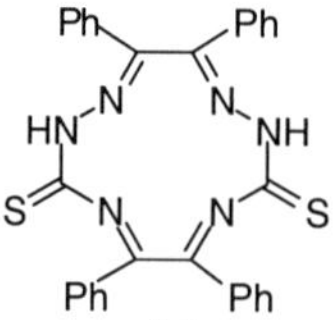

27

Inverse electron demand Diels-Alder reaction between enaminones and 1,3,5-triazine yielded functionalized quinazolinones as intermediates in the synthesis of CNS agents <02TL3551>. Hydration with mercuric acetate and then reduction with 9-BBN-H of 2-(1-alkenyl)-4,6-dimethyl-1,3,5-triazines have been described <02JOC3202>. 2-Chloro-4,6-disubstituted-1,3,5-triazines have been converted into their quaternary methylammonium salts, and their capability to be transformed in other derivatives studied <02KGS197, 02KGS326>. 2,4-Dichloro-6-alkylamino-1,3,5-triazine derivatives have been used as starting materials to obtain a new series of estrogen-receptor modulators <02JMC5492>. Selective amination of cyanuric chloride in the presence of 18-crown-6 has been reported <02EJO1551>. Synthetic duplex oligomers from cyanuric chloride <02T721> and 2-amino-4,5-dichloro-1,3,5-triazine <02JA5074> have been prepared. A novel artificial hydrogen-bonding receptor (**30**) which possesses a three-point triaminotriazine donor-acceptor-donor site to bind imides (e.g. uracil), has been reported. Receptor (**30**) was prepared from cyanuric chloride and anilino derivative (**28**) as outlined below <02OL881>.

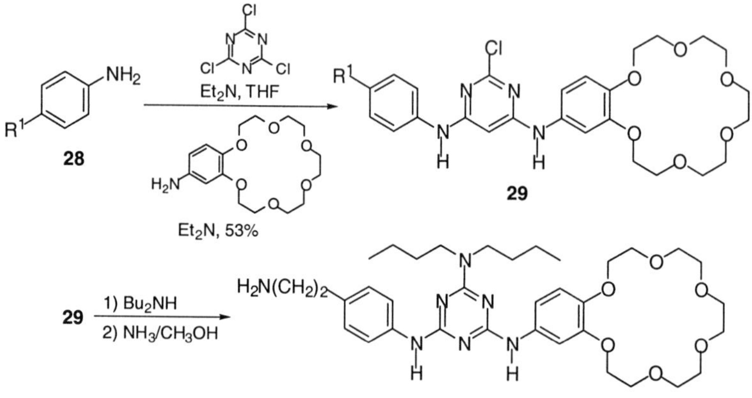

A simple procedure for the isolation of 2-hydroxy-4,6-dimethoxy-1,3,5-triazine (HO-DMT) a co-product arising from dehydrating condensation using 4-(4,6-dimethoxy-1,3,5-triazin-2-yl)-4-methylmorpholinium chloride (DMT-MM) has been established <02TL3323>. Benzyl 4,6-dimethoxy-1,3,5-triazinyl carbonate has been found to be a useful reagent for the introduction of the benzyloxycarbonyl group into amines <02CL66>. Some 2-benzylmercapto-1,3,5-triazine-6-thione derivatives on treatment with ammonia/hydrazine hydrate afforded the corresponding 2-amino/hydrazino derivatives which, on treatment with aryl isothiocyanates/aryl aldehydes, gave the related thiocarbamides/thiosemicarbazides or Schiff bases which were tested as antibacterials <02JIC176>. Starting from 2,4-diamino-6-phenyl-1,3,5-triazines, novel N-alkyl and N-acyl-(7-substituted)-1,3,5-triazin-4-yl-amines were prepared as adenosine receptor antagonists <02JMC5030>. Silylation of amino-1,3,5-triazines afforded new ligands with potential multi-coordination modes <02MI461>. A novel synthesis of hexa-aryloxy-cyclotriphosphazenes from hexachlorotriphosphazenes has been reported <02SC203>. Reversible skeletal substitution reactions involving group-13 heterophosphazenes have been described <02CC1102>. Thermolysis of N-trialkyl borazines at 500 °C produced homogeneous, amorphous boron carbonitride phases, whose compositions were dependent upon the borazine substituents and whose structures are similar to that of icosahedral boron carbide B_4C <02CC718>.

6.3.2. TETRAZINES

6.3.2.1. Synthesis

Reaction of 2-cyanophenanthroline (**31**) with hydrazine in refluxing ethanol provided dihydrotetrazine (**32**) in good yield <02OL1253>.

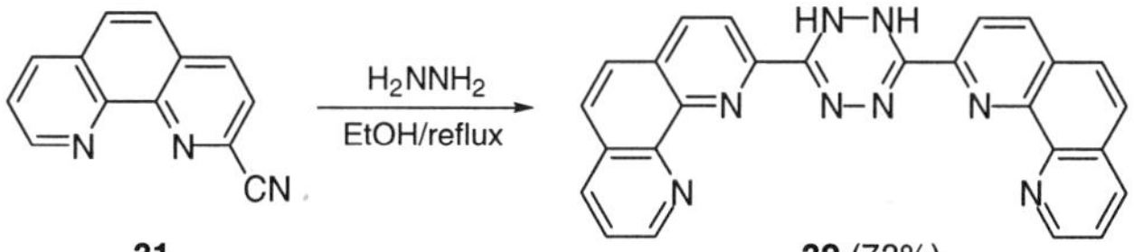

The synthesis and characterization of a new 6-phosphoverdazyl radical incorporated into a spirocyclic framework containing a cyclotriphosphazane ring has been reported <02CJC1501>. Verdazyl-3-carboxylic acid (**35**) was prepared by condensation of carbonic acid bis(1-methylhydrazide) (**33**) with glyoxylic acid, to give tetrazine (**34**), and subsequent oxidation with benzoquinone. The coordination chemistry of this water-soluble verdazyl derivative was explored <02CC1688>.

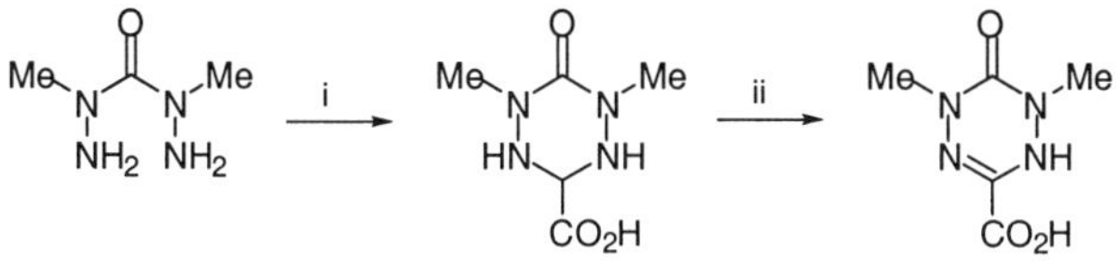

i) glyoxylic acid, H_2O, 0.5 h, r.t.;
ii) p-benzoquinone, EtOAc/MeOH, 1.5 h, 45 °C

Dipolar cycloadditions between nitrilimines **37**, obtained *in situ* from **36**, and alkenyl dipolarophiles **38** afford a mixture of pyrazoles **39** and **41** and tetrazines **40**. Tetrazine by-products **40** were obtained in good yields using 1-hexene as dipolarophile, Cs_2CO_3 as base and with aryl as p-tolyl (50%) or as p-Br-phenyl (67%) <02NJC1340>.

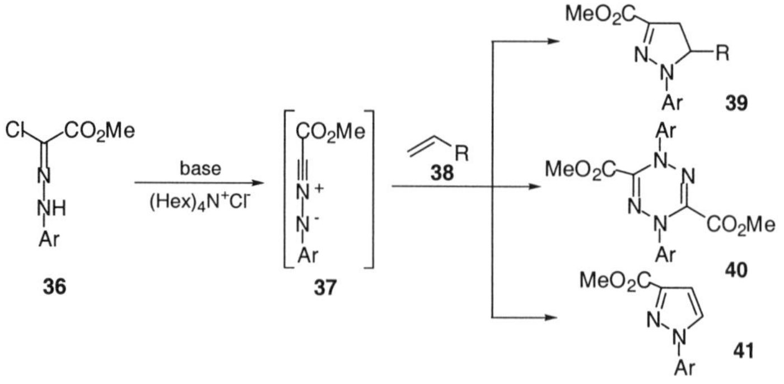

6.3.2.2. Reactions

The syntheses of spirocycloalkane-condensed-hydrotetrazines **43** and **44** have been described starting from 1,2,4,5-tetraazaspiro-cycloalkane-3-thiones **42** <02IJC(B)400>.

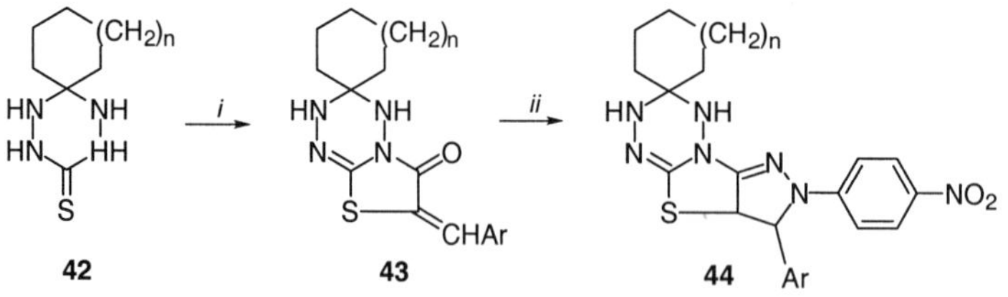

a, n = 3; **b**, n = 2; **c**, n = 1
i) ClCH$_2$COOEt, pyridine, ArCHO, piperidine; ii) 4-O$_2$NC$_6$H$_4$NHNH$_2$

Dihydrotetrazine **32** was oxidized with nitric acid in acetic acid to yield tetrazine **45** which suffered cycloaddition with acetylene in refluxing DMF followed by the extrusion of nitrogen to give pyridazine **46** which was used as bis-tridentate bridging ligand <02OL1253>.

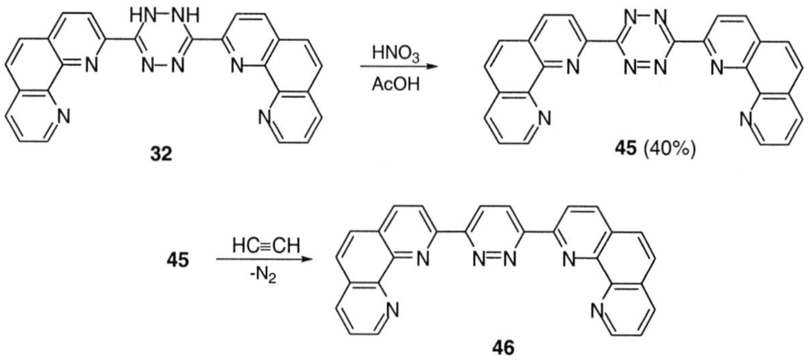

A simple and unexpected acid-catalyzed cleavage of tetrahydrotetrazines leading to 1,2-bis(hydrazones) has been reported <02JOC2378>. Inverse demand Diels-Alder reaction of suitable dihydrofuran-3-one derivatives with 3,6-bis(trifluoromethyl)1,2,4,5-tetrazine has been used to obtain linear and angular pyridazine furocoumarins <02S43>. Regioselective inverse electron demand Diels-Alder reaction of **47** with 3,6-bis-substituted 1,2,4-tetrazines **48** afforded angular pyridazinopyrrolocoumarins **49** in good yield <02S475>.

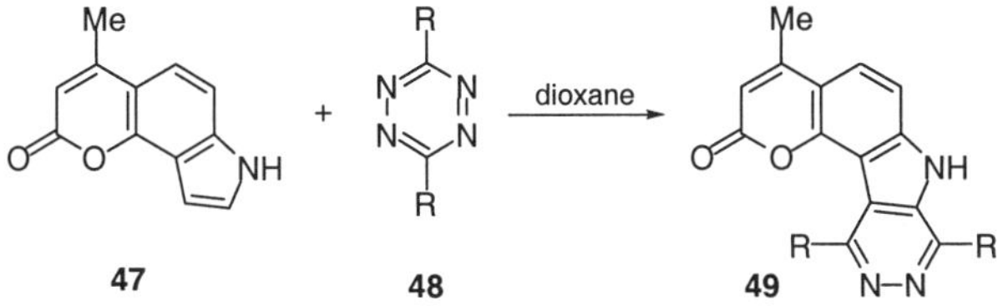

R = CO₂Me, CF₃

A cycloaddition-cycloelimination sequence transformed 1,2,4,5-tetrazines **50** on reaction with 3,3'-bicyclopropenyl (**51**) into 3,4-diazanorcaradienes, which, on gentle heating in inert solvents, were transformed into semi-bullvalenes **52** <02EJO791>.

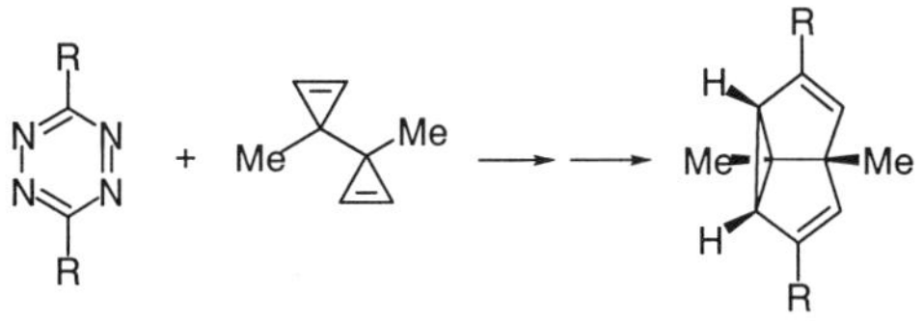

R = 2-thiazolyl, 2-pyridyl, 4-pyridyl, 2-pyrazinyl, (thiazolyl)

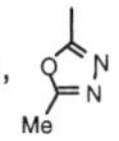

6.3.3. FUSED [6]+[5] POLYAZA SYSTEMS

6.3.3.1. Synthesis

The reaction of 5-amino-1,2,4-triazoles with isothiourea derivatives to give isomeric 5,7-diamino-1,2,4-triazolo[1,5-*a*][1,3,5]triazines has been reinvestigated <02JHC327>. 2-Aryl substituted imidazo[5,1-*f*][1,2,4]triazin-4(3*H*)-ones have been described as a new class of potent cGPM-PDE5 inhibitors. From them, vardenafil (BAY 38-9456) has been selected for clinical studies for erectile dysfunction <02BMCL865>. New derivatives of 1,3,4-thiadiazolo[2,3-*c*][1,2,4]triazinones with antibacterial and antitumor activity have been described <02FES693>. One-pot synthesis of some new biologically active 5-aryl-2-furyl-thiadiazolotriazine-3-ones has been reported <02IJC(B)2690>. 1,3,5-Trisubstituted pyrazoles have been used to prepare <02MI602>.

A new synthesis of the antitumor imidazotetrazine, temozolomide, has been reported involving an efficient condensation between nitrosoimidazoles and phenyl *N*-methylcarbazate. This yields phenyloxycarbonyl substituted triazenylimidazoles which through isomerization of the triazene N=N bond provides temozolomide in high yield

<02JCS(P1)1877>. Two rapid radiosynthetic routes for the preparation of this compound labeled with ^{11}C have also been reported <02JMC5448>.

An efficient one-pot synthesis of 6-alkoxy-8,9-dialkylpurines has been accomplished by cyclization of the corresponding 5-amino-4-chloro-6-(alkylamino)pyrimidines promoted by alkoxides and N,N-dimethylamides where the latter act as solvent and reagent <02T7607>. A series of O-6-substituted guanines were prepared to probe the ATP ribose binding domain of kinases CDK1 and CDK2 <02JMC3381>. 9[(Hydroxymethyl)phenyl]adenines have been synthesized and studied for their adenosine deaminase (ADA) substrate activity <02BMCL1489>. Several derivatives of xanthines substituted at the 1-, 3- ,7- and 8-positions <02JMC2131> and 1,8-disubstituted <02JMC1500> have been synthesized as adenosine A(2B) receptor antagonists. A three-step method for the preparation of tetra-substituted xanthines from an easily prepared imidazole precursor has been reported <02CPB1379>. In order to provide a novel route for a scaleable synthesis of the chiral xanthine CVT-124, a method for the late stage pyrimidine ring closure of the corresponding nitrogen-protected endo 2-norbornenyl imidazole has been developed <02JOC188>. Several publications have dealt with xanthine analogs as inhibitors of PDE5 including 8-aryl derivatives <02BMCL2587>, derivatives modified at the C-8 position <02BMCL3149> and a solid phase approach towards xanthines using Mitsunobu chemistry <02BMCL1973>.

A facile one-pot synthesis of pyrazolo[3,4-d]pyrimidines **55** from 6-hydrazino-uracils **53** with isocyanates **54** has been reported <02TL895>.

	R^1	R^2
55a	Me	Ph
55b	H	Ph
55c	Me	Bn
55d	H	Bn

Pyrazolo[4,3-d]pyrimidine and pyrazolo[3,4-d]pyrimidine sulfonamides have been synthesized as selective calcitonin inducers <02JMC2342>. A series of para-substituted 3-arylpyrazolo[3,4-d]pyrimidines have been synthesized and evaluated as inhibitors of lck kinase <02BMCL1687). The synthesis of 3-aryl-pyrazolo[4,3-d]pyrimidines as potential corticotropin-releasing factor (CRF-1) antagonists has been described <02BMCL2133>. Cyclization of 4-chloro-5-heteroimine-1,2,3-thiadiazoles furnished pyrazolo[3,4-d]pyrimidines, triazolo[4,5-d]pyrimidines and purines <02BMC449>.

A novel approach to 1,2,3-triazolo[1,5-a]pyrazines involving the cyclization of (4,5-dicyano-1,2,3-triazol-1-yl)propenoates has been reported <02H353>. The synthesis of 1,2,3-triazolo[1,5-a]pyrazinium-3-olate has been described <02KGS1302>. The synthesis of new fluorine containing triazolo and tetrazolopyrimidines has been reported <02S901>.

Several papers have dealt with 1,2,4-triazolo[3,4-b][1,3,4]thiadiazines including synthesis of new derivatives and tautomeric studies <02SC3455; 02ZN(B)552; 02PS487> and a one-pot synthesis starting from 4-amino-5-substituted-1,2,4-triazole-3-thiones involving a solid acid-induced cyclocondensation <02PS2403>.

A new solid-phase synthesis of families of asymmetrically disubstituted furazano[3,4-b]-pyrazines, which overcomes selectivity problems found in solution, has been achieved by stepwise displacement of the two chlorine atoms of 5,6-dichlorofurano[3,4-b]pyrazines with nucleophiles <02TL4741>. Pyrazolo[3,4-b]quinoxalines as a new class of cyclin-dependent kinase inhibitors have been described <02BMC2177>.

A stereoselective synthesis of 3-(1,2,4-triazolo[4,3-x]azin-3-yl)-bicyclo[2.2.1]heptanones **58** starting from (1*R*)-(+)camphor **56** has been described. The reaction comprises oxidative cyclization of the intermediate hydrazones **57** with methanolic bromine <02TA821>.

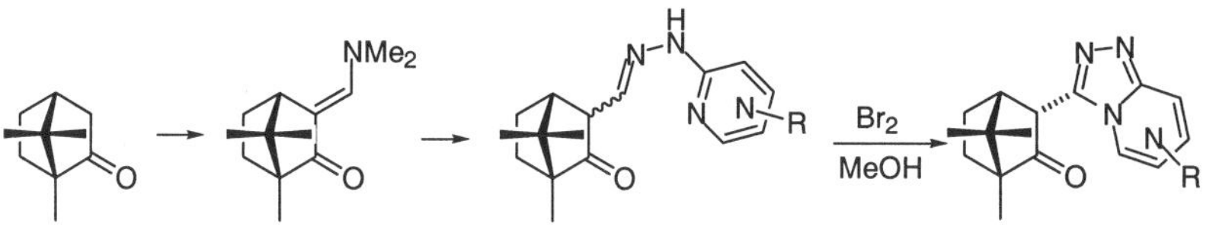

6.3.3.2. Reactions

N-Adamantyl derivatives of 1,2,4-triazolo[5,1-*c*][1,2,4]triazin-4-ones have been described <02MI272>. The synthesis and antifungal activity of 4-substituted pyrazolo[3,4-*c*][1,2,4]triazines have been reported <02IJC(B)664>.

In order to obtain dual-action agents, the imidazotetrazines, mitozolomide and temozolomide combined with a β-lactam antibiotic were synthesized <02EJM323>. Carboxylic acids derived from the amido groups of these imidazotetrazines have been conjugated to simple aminoacids and peptides by carbodiimide coupling <02JMC5458>.

Dimethylpyridin-4-ylamine (DMAP) catalyzed reactions of 2-amino-*N*,*N*,*N*-trimethyl-9*H*-purin-6-ylammonium chloride (**59**) with various alkoxides in DMSO, gave the corresponding substitution products in moderate to good yields. The procedure can be used to obtain *O*-6-substituted guanine derivatives such as (**60**) <02S538>.

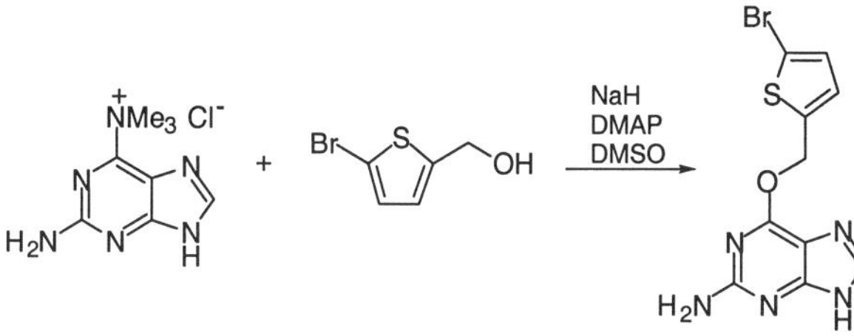

Stille coupling between suitable 6-chloropurines and aryl(tributyl)tin afforded 6-arylpurines with a variety of substituents in the 9-position which were screened for antimycobacterial activity <02JMC1383>. The Stille cross-coupling reactions of 1,4- and 1,3-bis(trialkylstannyl)benzenes with 9-benzyl-6-chloropurine gave 1,4- and 1,3-bis(purin-6-yl)benzenes and 4-(purin-9-yl)benzenes <02T7431>.

Reaction of several purines with 1-benzyl-4-(chloromethyl)piperidine has been studied. It proceeds through the initial formation of 1-benzyl-1-azoniabicyclo[2.2.1]heptane system which undergoes nucleophilic attack at two different carbons yielding *N*-benzylpiperidine and *N*-benzylpyrrolidine substituted purines <02S911>.

Self complementary mesomeric betaines of guanine obtained from guanine *N*-oxide have been described and their base pairing properties studied <02CL222>. The stereoselective reaction of adenine with methylglyoxal, in water, under mild conditions, afforded a new family of heterocycles, and can be used for understanding the biological effects of methylglyoxal <02CC1114>. Xanthines with an 8-halogen were reacted with phenyl- and styrylboronic acids, in a Suzuki cross coupling reaction to give 8-substituted 3,7-dihydro-purine-2,6-dione (xanthine) derivatives <02H871>.

Traceless solid-phase synthesis of 2,6,9-trisubstituted purines from resin-bound 6-thiopurines <02T7911>, and microwave assisted solid-phase synthesis of 2,6,9-trisubstituted purines <02TL6169> have been described. A resin-capture and release strategy toward combinatorial libraries of 2,6,9-trisubstituted purines has been reported <02JCO183>. Alkylated purines chlorinated at the 6,8- or 2,6,8-positions can be captured onto a solid support and further elaborated by aromatic substitution or *via* palladium catalyzed cross-coupling reactions <02JA1594>.

7-Hydrazido derivatives of 1,2,3-triazolo[4,5-*d*]pyrimidines when heated in Dowtherm underwent intramolecular cyclization to give tricyclic 1,2,3-triazolo[4,5-*e*]1,2,4]triazolo[4,3-*c*]pyrimidine derivatives <02JHC885>. A highly resigoselective anodic mono- and difluorination of 1,2,4-triazolo[3,4-*b*][1,3,4]thiadiazine derivatives **61** to give **62** and **63** has been described <02TL273>.

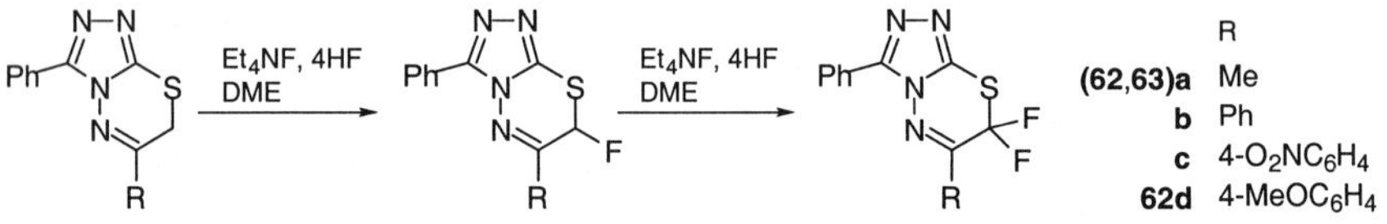

Et4NF, 4HF
DME
Et4NF, 4HF
DME
R
(62,63)a Me
b Ph
c 4-O2NC6H4
62d 4-MeOC6H4

61 **62(a-d)** **63(a-d)**

6.3.4. FUSED [6]+[6] POLYAZA SYSTEMS

Three novel pteridine alkaloids bearing two pteridine units and a tryptophan core have been isolated from the sponge *Clathria* sp <02T4481>. Structural requirements for inhibition of neural nitric oxide synthase (NOS-I) and 3D-QSAR analysis of 4-oxo-pteridine-based and 4-amino-pteridine-based inhibitors have been reported <02JMC2923>.

6.3.4.1. Synthesis

Synthesis of 3-phenyloctahydropyrimido[1,2-*a*][1,3,5]triazines was performed via iminodimethylation of dialkylated 2-aminopyrimidinedione synthons by substituted primary arylamines <02JHC663>. Synthesis of various 9,9-dialkylated octahydropyrimido[3,4-*a*][1,3,5]-triazines, as antifungals, by an iminodimethylation reaction between a 5,5-dialkyl-

+ SCNCO2Et
64a, R1 = H
64b, R1 = Me
65
R2R3NH
HgCl2
DMF
R2,R3=H, Me,
Et, Pr, i-Pr,
cyclohexyl
66
66
HCl/dioxane
67

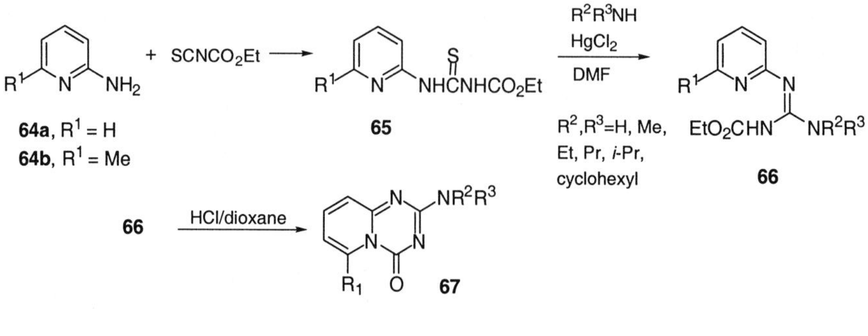

6-aminopyrimidine-2,4-dione, a substituted aniline, and two mol equivalents of formaldehyde has been achieved <02FES109>. Synthesis of new 2-aminopyrido[1,2-a][1,3,5]triazin-4-ones **67** starting from 2-aminopyridine (**64a**) or 6-aminopicoline (**64b**) through intermediates **65** and **66** has been reported <02JHC1061>.

The synthesis and NMR study of benzo[1,2,3,4]tetrazine 1,3-dioxides have been described <02EJO2342>. A convenient one-pot synthesis and antimicrobial activity of pyrimido[1,2-b][1,2,4,5]tetrazines **68** bearing different substituents have been reported <02JHC45>.

An unusual rearrangement following cyclization of 2-anilino-2-ethoxy-3-oxothiobutanoic acid with aromatic 1,2-diamines yielded fused heterocyclic systems such as pteridine, quinoxaline or pyrido[2,3-b]pyrazine <02JOC4526>. The total synthesis of methotrexate-γ-tris-fatty acid conjugates has been reported <02AJC635>. Design, synthesis and evaluation of 6-carboxyalkyl and 6-phosphonoxyalkyl derivatives of 7-oxo-8-ribitylaminolumazines as inhibitors of riboflavin synthase and lumazine synthase have been described <02JOC5807>. Syntheses of labeled vitamers of folic acid, to be used as internal standards in stable-isotope dilution assays, have been reported <02MI4760>. Solid-phase and solution-phase protocols for the synthesis of pyrimido[4,5-d]pyridazines, furo[3,4-d]pyrimidines and pyrrolo[3,4-d]pyrimidines have been studied <02JCO501>. The Michael addition-cyclodehydration of 6-aminouracils **69** and alkynone (**70**) afforded 5-deazapterin derivatives **71** with total control of regiochemistry <02SL1332>.

	R^1	R^2
71a	H	H
71b	Me	H
71c	Me	Me

Synthesis of pyrido[2,3-b]pyrazines as intermediates in the preparation of antineoplastic agents has been described <02TL5241>. The reaction of β-methylsulfanylacroleins with 6-aminouracils or 2,6-diaminouracil provided a new synthetic route to 6-substituted pyrido[2,3-d]pyrimidines <02H2081>. Pyrido[2,3-d]pyrimidines have been synthesized *via* a selective cyclocondensation reaction between 6-aminopyrimidines and some propiophenone hydrochloride derivatives <02TL4873>. Different syntheses of novel pyrido[2,3-d]pyrimidines have been studied <02IJC(B)804, 02IJC(B)430>. Microwave-assisted synthesis of 5-deaza-5,8-dihydropterins has been performed <02SL1718>.

6.3.4.2. Reactions

The introduction of a *tert*-butyl carbonate moiety at the 3-position of 3,4-dihydro-3-hydroxy-4-oxo-1,2,3-benzotriazine yielded an efficient active carbonate <02TL2529>.

Carbethoxypyrazol-5-yl-, indazol-3-yl- and indazol-5-yl-2-iodobenzamides have been synthesized from the corresponding 1,2,3-benzotriazinones <02FES183>. The reduction of benzo-1,2,3,4-tetrazine 1,3 dioxides **72** afforded benzotriazoles **74** *via* intermediates **73** <02OL3227>.

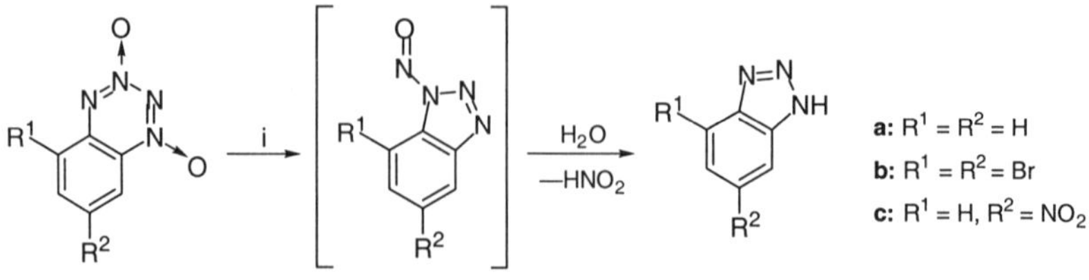

i) Na$_2$S$_2$O$_4$ (4 equiv), EtOAc/H$_2$O, 24 °C, 3 min, **72a→74a** (90%); **72b→74b** (84%); **72c→74c** (66%); SnCl$_2$•2H$_2$O (4 equiv), EtOAc/EtOH, **72a→74a** (97%); **72b→74b** (95%); **72c→74c** (98%)

Regioselective preparation of pterin 6-triflate and its application to 6-substituted pterin synthesis was achieved by reaction of 4-butoxypteridine 8-oxide with trifluoromethanesulfonic anhydride. The triflate group could be replaced by various nucleophiles <02H1841>. Reactions of folic acid with reducing sugars and sugar degradation products have been studied <02MI1647>. Reductive alkylation of *N*-2-*i*-Bu-6-formylpterin yielded γ-cysteamine modified folic acid <02TL4439>. Controlled stepwise conversion of 2,4,6,8-tetrachloropyrimido[4,5-*d*]pyrimidine into 2,4,6,8-tetrasubstituted pyrimido[5,4-*d*]pyrimidines has been carried out <02JCS(P1)108>. Pyrimido[5',4':5,6]-pyrido[2,8-*d*]pyrimidine derivatives **77** was synthesized, as new antiviral agents, from 7-amino-5-aryl-6-cyanopyrido[2,3-*d*]pyrimidines **75** through intermediates **76** <02AP289>.

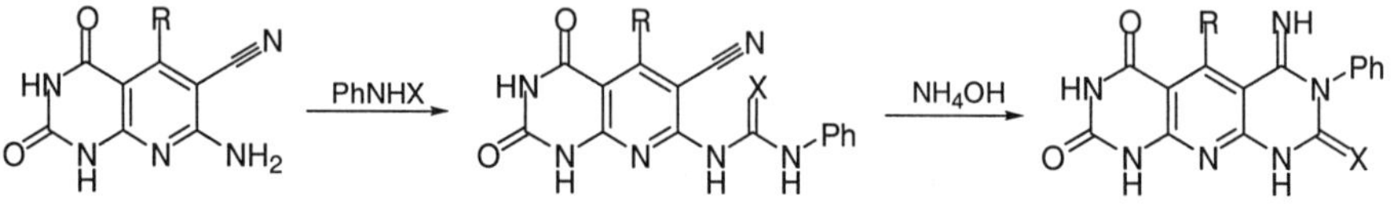

5,7-Disubstituted 2-thioxo-pyrido[2,3-*d*]pyrimidinones condensed with aldehydes in the presence of chloroacetic acid to yield thiazolo[2,3-*a*]pyrido[2,3-*d*]pyrimidinones <02PS45>. One-pot alkoxylation at the 7-position of 6-arylpyrazino[2,3-*c*][1,2,6]thiadiazine 2,2-dioxides with NBS or NCS in the appropriate alcohol has been studied <02EJO2109>.

6.3.5. MISCELLANEOUS FUSED RING POLYAZA SYSTEMS

6.3.5.1. Synthesis

New syntheses of pyrido[2,3-*b*]thieno[3,2-*d*]-pyrimidine and -1,2,3-triazine have been achieved <02SC3493>. Intramolecular conversion of 1-aryl-3,3-dialkyltriazenes into tetrahydrobenzo-1,2,3-triazine condensed derivatives has been described <02AG484>. A new heteroaromatic system, benzo[4,5]imidazo[1,2-c]pyrido[3',2'/4,5]thieno[2,3-e][1,2,3]triazine, has been described <02KGS713>. o-Diaminopyrazolo[4,3-c]pyridine derivatives have been used to build up a third heterocyclic ring, such as, dihydro-1,2,4-triazine, triazole and tetrazole through its interaction with some electrophilic reagents <02H1121>. General methods for the syntheses of mesoionic derivatives of [1,2,3]triazolo[1,5-a]quinoline, [1,2,3]triazolo[1,5-a]quinazoline, [1,2,3]triazolo[1,5-a]quinoxaline, and [1,2,3]triazolo[1,5-c]benzo[1,2,4]triazine have been studied <02T3185>. Condensation reactions of aceanthrene quinone with 2-aminoguanidine, semicarbazide, and thiosemicarbazide yielded aceanthryleno[1,2-e][1,2,4]triazines and with 4,5,6-triamino-pyrimidine derivatives afforded aceanthryleno[1,2-g]pteridines <02M79>. Synthesis of some new tricyclic, 1,4-benzothiazinones fused to a 1,2,4-triazine ring, has been reported <02JIC472>. Synthesis and antimicrobial activity of thiazolo[3',2':2,3][1,2,4]triazino[5,6-b]indoles have been reported <02IJC(B)628, 02IJC(B)2364>. 5-Chlorosulfonyl-3,3-dichloro-2-indolinone are promising scaffolds for the solid- and liquid-phase syntheses of new combinatorial libraries of various heterocycles including indol-1,2,4-triazine derivatives <02JCO419>. An efficient method for the solid-phase synthesis of trisubstituted 1,3,5-triazino[1,2-a]benzimidazole-2,4-diones from resin-bound amino acids has been described <02JCO345>. 6-Nitroquinoline (**78**) underwent direct cyclocondensation with aromatic keto hydrazones **79** to give 3,3-disubstituted 2,3-dihydro[1,2,4]triazino[6,5-f]quinoline 4-oxides **80** in moderate to good yields <02JCS(P1)696>.

	R^1	R^2
81a	Ph	H
81b	C_6H_4Me-4	Me
81c	Me	H
81d	Me	Me
81e	Me	Br

Unprecedented dihydrodipyridopyrazines were easily obtained by hetarynic dimerization of 2-alkylamino-3-halopyridines <02T3513>. The reaction of hexamethylphosphotriamide with amidines derived from N-benzimidazoyl imidates **81** afforded benzimidazolo-1,3,5,2-triazaphosphorine 2-oxides **83** in good yield. When the condensation was carried out at room temperature N-phosphonic amidines **82** could be isolated <02PS1033>.

	R^1	R^2		R^1	R^2
81 - 83 (a)	Me	Ph	81 - 83 (d)	Et	Ph
81 - 83 (b)	Me	Bn	81 - 83 (e)	Et	Bn
81 - 83 (c)	Me	$(CH_2)_{11}Me$	81 - 83 (f)	Et	$(CH_2)_{11}Me$

A facile one-pot synthesis of 1,4,7,8,9,10-hexahydro-6-*H*-[1]benzothieno[2',3':4,5]pyrimido-[1,2-b][1,2,4,5]tetrazine-6-ones has been described <02ZN(B)699>. An efficient method for the synthesis of 4,7-dihydro-1*H*-imidazo[4,5-

e][1,2,4]triazepin-8-one derivatives was accomplished by the use of microwave methodology <02SL519>. Tricyclic imidazo[2,1-*c*]purinones and ring-enlarged analogues have been prepared as adenosine receptor (AR) antagonists <02JMC3440>. Imidazo[1,2-*c*]pyrimido[5,4-*e*]pyrimidinones, tetraazaacenaphthene-3,6-diones and tetra-azaphenalene 1,7-dione have been prepared from cyclic ketene aminals through tandem addition-cyclization reactions <02BMC1275>. Domino reaction of 3-azidopyrroles and methylene active nitriles afforded annelated 1,2,3-triazolo[1,5-*a*]pyrimidines, which are new tricyclic systems of biological interest <02T9723>. A new and efficient one-pot synthesis of [1,2,3]triazolo[1,5-*a*][1,4]benzodiazepine-6-(4*H*)-ones starting from readily available anthranilic acids has been described <02BMCL1881>. Novel derivatives of pyrido[2,3-*d*][1,2,4]triazolo[4,3-*a*]pyrimidine-5,6-dione have been synthesized <02M1297>. Derivatives of 6,9-dihydro- and 6,7,8,9-tetrahydro-2-thia-3,5,6,7,9-pentaazabenz[*cd*]azulenes representing a new heterocyclic system have been prepared <02TL695>.

6.3.5.2. Reactions

The synthesis and binding study to benzodiazepine receptors of new 8-chloropyrazolo[5,1-*c*][1,2,4]benzotriazine 5-oxide 3-esters, obtained from the corresponding acids, have been reported <02JMC5710>. Regiospecific formation of the linear 1,2,4-triazolo[4'.3':2,3][1,2,4]triazino[5,6-*b*]indoles from the corresponding 3-hydrazino-1,2,4-triazino[5,6-*b*]indoles and aldehydes <02JCR(S)314> or acids <02PHA438> have been studied. Hexamethylenetetramine has been used to synthesize novel β-substituted indol-3-yl ethylamido melatoninergic analogues <02JPP147>. Nonpeptidic tryptamine derivatives have been prepared, among them 3-(2-aminoacetyl)-1-phenylsulfonylindole hydrochloride, which was synthesized via α-bromination of 3-acetyl-1-phenylsulfonylindole followed by amination with hexamethylenetetramine, strongly stimulated cell growth <02H519>. Regioselective and chemoselective 1,3-dipolar cycloaddition of nitrilimines to 1,3,4-benzotriazepin-5-one derivatives yielded a novel heterocyclic ring system, [1,2,4]triazolo[1,3,4]benzotriazepines <02NJC1545>.

6.3.4. REFERENCES

02AG(E)484	K. Nishiwaki, T. Ogawa, K. Matsou, *Angew. Chem. Int. Ed* **2002**, *41*, 484.
02AG(E)3858	E. Hevia, J. Pérez, V. Riera, D. Miguel, *Angew. Chem. Int. Ed.* **2002**, *41*, 3858.
02AJC635	C.L. Francis, Q. Yang, N.K. Hart, F. Widmer, M.K. Manthey, H.M. Hewilliams, *Aust. J. Chem.* **2002**, *55*, 635.
02AP289	M.N. Nasr, M.M. Gineinah, *Arch. Pharm.* **2002**, *335*, 289.
02BCJ615	C.H. Lee, T. Yamamoto, *Bull. Chem. Soc. Jpn.* **2002**, *75*, 615.
02BMC449	P.G. Baraldi, M.G. Pavani, M.C. Nuñez, P. Brigidi, B. Vitali, R. Gambari, R. Romagnoli, *Bioorg. Med. Chem.* **2002** *10*, 449.
02BMC1275	V.J. Ram, A. Goel, S. Sarkhel, P.R. Maulik, *Bioorg. Med. Chem.* **2002**, *10*, 1275.
02BMC2177	M.A. Ortega, M.E. Montoya, B. Larranz, A. Jaso, I. Aldana, S. Leclerc, L. Meijer, A. Monge, *Bioorg. Med. Chem.* **2002**, *10*, 2177.
02BMCL865	H. Haning, U. Niewöhner, T. Schenke, M. Es-Sayed, G. Schmidt, T. Lampe, E. Bischoff, *Bioorg. Med. Chem. Lett.* **2002**, *12*, 865.
02BMCL1489	M. Brakta, D. Murthy, L. Ellis, S. Phadtare, *Bioorg. Med. Chem. Lett.* **2002**, *12*, 1489.
02BMCL1591	J.R. Portter, S.C. Archibald, J.A. Brown, K. Childs, D. Critchley, J.C. Head, B. Hutchinson, T.A.H. Parton, M.K. Robinson, A. Shock, G.J. Warrellow, A. Zomaya, *Bioorg. Med. Chem. Lett.* **2002**, *12*, 1591.
02BMCL1687	A.F. Buchat, D.J. Calderwood, M.M. Friedman, G.C. Hirst, B. Li, P. Rafferty, K. Ritter, B.S. Skinner, *Bioorg. Med. Chem. Lett.* **2002**, *12*, 1687.

02BMCL1881	A.W. Thomas, *Bioorg. Med. Chem. Lett.* **2002**, *12*, 1881.
02BMCL1973	D. Beer, G. Bhalay, A. Dusntan, A. Glen, S. Haberthuer, H. Moser *Bioorg. Med. Chem. Lett.* **2002**, *12*, 1973.
02BMCL2133	J. Yuan, M. Gulianello, S. De Lombaert, R. Brodbeck, A. Kieltyka, K.J. Hodgetts, *Bioorg. Med. Chem. Lett.* **2002**, *12*, 2133.
02BMCL2137	W.J. Pitts, J. Guo, T.G.M. Dhar, Z. Shen, H.H. Gu, S.H. Watterson, M.S. Bednarz, B.-C. Chen, J.C. Barrish, D. Bassolino, D. Cheney, C.A. Fleener, K.A. Rouleau, D.L. Hollenbaugh, E.J. Iwanowicz, *Bioorg. Med. Chem. Lett.* **2002**, *12*, 2137.
02BMCL2587	R. Arnold, D. Beer, G. Bhalay, U. Baettig, S.P. Collingwood, S. Craig, N. Devereux, A. Dunstan, A. Glen, S. Gomez, S. Haverthuer, T. Howe, S. Jelfs, H. Moser, R. Naef, P. Nicklin, D. Sandham, R. Stringer, K. Turner, S. Watson, M. Zurini, *Bioorg. Med. Chem. Lett.* **2002**, *12*, 2587.
02BMCL3149	U. Wang, S. Chackalamanil, Z. Hu, C.D. Boyle, C.M. Lankin, Y. Sia, R. Xu, T. Asberom, D. Pissarnitski, A.W. Stamfor, W.J. Greenlee, J. Skell, S. Kurowski, S. Vemulapalli, J. Palamanda, M. Chintala, P. Wu, J. Myers, P. Wang, *Bioorg. Med. Chem. Lett.* **2002**, *12*, 3149.
02CC178	M.H. Al-Sayah, N.R. Branda, *Chem. Commun.* **2002**, 178.
02CC716	F. Bigi, L. Moroni, R. Maggi, G. Sartori, *Chem. Commun.* **2002**, 716.
02CC718	R. Brydson, H. Daniels, M. A. Fox, R. Greatrex, C. Workman, *Chem. Commun.* **2002**, 718.
02CC952	R.K.R. Jetti, P.K. Thallapally, A. Nangia, C.K. Lam, T.C.V. Mak, *Chem. Commun.* **2002**, 952.
02CC1102	A.R. McWilliams, E. Rivard, A.J. Lough, I. Manners, *Chem. Commun.* **2002**, 1102.
02CC1114	C. Routaboul, L. Dumas, I. Gautierluneau, J. Vergne, M.C. Maurel, J.L. Decout, *Chem. Commun.* **2002**, 1114.
02CC1350	W.M. Shu, S. Valiyaveettil, *Chem. Commun.* **2002**, 1350.
02CC1356	M.I.J. Polson, N.J. Taylor, G.S. Hanan, *Chem. Commun.* **2002**, 1356.
02CC1386	M. Harmjanz, B.L. Scott, C.J. Burns, *Chem. Commun.* **2002**, 1386.
02CC1488	P. Gamez, P. Dehoog, O. Roubeau, M. Lutz, W.L. Driessen, A.L. Spek, J. Reedijk, *Chem. Commun.* **2002**, 1488.
02CC1688	T.M. Barclay, R.G. Hicks, M.T. Lemaire, L.K. Thompson, Z.Q. Xu *Chem. Commun.* **2002**, 1688.
02CC2538	D.I. Schuster, J. Rosenthal, S. Macmahon, P.D. Jarowski, C.A. Alabi, D.M. Guldi, *Chem. Commun.* **2002**, 2538.
02CJC1501	T.M. Barclay, R.G. Hicks, A.S. Ichimura, G.W. Patenaude, *Can. J. Chem.* **2002**, *80*, 1501.
02CL66	K. Hioki, M. Fujiwara, S. Tani, M. Kunishima, *Chem. Lett.* **2002**, 66.
02CL222	A. Schmidt, N. Kobakhidze, *Chem. Lett.* **2002**, 222.
02CEJ2288	L.J. Prins, R. Hulst, P. Timmerman, D.N. Reinhoudt, *Chem.-Eur. J.* **2002**, *8*, 2288.
02CEJ5028	B.G. Xing, M.F. Choi, B. Xu, *Chem.-Eur. J.* **2002**, *8*, 5028.
02CPB463	D. Branowska, S. Otrowski, A. Rykowski, *Chem. Pharm. Bull.* **2002**, *50*, 463.
02CPB1379	A.R. Hergueta, M.J. Figueira, C. López, O. Caamaño, F. Fernández, J.E. Rodríguez-Borges, *Chem. Pharm. Bull.* **2002**, *50*, 1379.
02EJM323	Y.F. Wang, P. Lambert, L.X. Zhao, D. Wang, *Eur. J. Med. Chem.* **2002**, *37*, 323.
02EJM709	Z. Brzozowski, F. Saczewski, *Eur. J. Med. Chem.* **2002**, *37*, 709.
02EJO791	J. Sauer, P. Bauerlein, E. Ebenbeck, J. Schuster, I. Sellner, H. Sichert, H. Stimmelmayr, *Eur. J. Org. Chem.* **2002**, 791.
02EJO1412	D.N. Kozhevnikov, V.L. Rusinov, O.N. Chupakhin, M. Makosza, A. Rykowski, E. Wolinska, *Eur. J. Org. Chem.* **2002**, 1412.
02EJO1551	S. Samaritani, P. Peluso, C. Malanga, R. Menicagli, *Eur. J. Org. Chem.* **2002**, 1551.
02EJO2109	N. Campillo, J.A. Páez, P. Goya, *Eur. J. Org. Chem.* **2002**, 2109
02EJO2342	A.M. Churakov, O.Y. Smirnov, S.L. Ioffe, Y.A. Strelenko, V.A. Tartakovsky, *Eur. J. Org. Chem.* **2002**, 2342.
02FES109	A. Ghaib, S. Menager, P. Verite, O. Lafont, *Farmaco* **2002**, *57*, 109.
02FES183	D. Raffa, G. Daidone, F. Plescia, D. Schillaci, B. Maggio, L. Tortal, *Farmaco* **2002**, *57*, 183.
02FES693	B.S. Holla, B.S. Rao, R. Gonsalves, B.K. Sarojini, K. Shridhara, *Farmaco* **2002**, *57*, 693.
02H353	T. Jug. M. Polak, T. Trcek, B. Vercek, *Heterocycles* **2002**, *56*, 353.
02H519	H. Suzuki, T. Furukawa, C. Yamada, M. Kurimi, T. Yokoma, Y. Murakami, *Heterocycles* **2002**, *56*, 519.
02H871	K. Vollmann, C.E. Muller, *Heterocycles* **2002**, *57*, 871.
02H1121	M.M. Kandeel, R.A. Ahmed, S.K. Youssef, *Heterocycles* **2002**, *57*, 1121.
02H1841	M. Kujime, K. Kudoh, S. Murata, *Heterocycles* **2002**, *57*, 1841.
02H2081	T. Kuwada, K. Harada, J. Nobuhiro, T. Choshi, S. Hibino, *Heterocycles* **2002**, *57*, 2081.

02IJC(A)1629	S. Chandra, A. Sangeetika, *Indian J. Chem. (A)* **2002**, *41*, 1629.
02IJC(B)400	J. Mohan, Anapuma, A. Kumar, D. Khatter, *Indian J. Chem. (B)* **2002**, *41*, 400.
02IJC(B)430	G. Singh, G. Singh, A.K. Yadav, A.K. Mishra, *Indian J. Chem. (B)* **2002**, *41*, 430.
02IJC(B)628	J. Mohan, Anapuma, *Indian J. Chem. (B)* **2002**, *41*, 628.
02IJC(B)664	A.K. Tewari, L. Mishra, H.N. Verma, A. Mishra, *Indian J. Chem. (B)* **2002**, *41,* 664.
02IJC(B)804	P. Saikia, A.J. Takur, D. Prajapati, J. S. Sandhu, *Indian J. Chem. (B)* **2002**, *41*, 804.
02IJC(B)2364	J. Mohan, A. Kumar, *Indian J. Chem. (B)* **2002**, *41*, 2364.
02IJC(B)2690	B.S. Holla, M.K. Shivananda, B. Veerendra, *Indian J. Chem. (B)* **2002**, *41*, 2690.
02JA1594	S. Ding, N.S. Gray, X. Wu, Q. Ding, P.G. Schultz, *J. Am. Chem. Soc.* **2002**, *124,* 1594.
02JA5074	E.A. Archer, M.J. Krische, *J. Am. Chem. Soc.* **2002**, *124*, 5074.
02JA9010	A.W. Holland, R.G. Bergman, *J. Am. Chem. Soc.* **2002**, *124*, 9010.
02JA11608	H.S. Moon, E.M. Jacobson, S.M. Khersonsky, M.R. Luzung, D.P. Walsh, W.N. Xiong, J.V. Lee, P.B. Parikh, J.C. Lam, T.V. Kang, G.R. Rosania, A.F. Schier, Y.T. Chang, *J. Am. Chem. Soc.* **2002**, *124*, 11608.
02JCO183	S. Ding, N.S. Gray, Q. Ding, X. Wu, P.G. Schultz, *J. Comb. Chem.* **2002**, *4*, 183.
02JCO419	A.V. Ivachtchenko, A.P. Ilyin, V.V. Kobak, D.A. Zolotarev, L.V. Boksha, A.S. Trifilenko, D.M. Ugoleva, *J. Comb. Chem.* **2002**, *4*, 419.
02JCO484	Y.P. Yu, J.M. Ostresh, R.A. Houghten, *J. Comb. Chem.* **2002**, *4*, 484.
02JCO345	G. Klein, A.N. Acharya, J.M. Ostresh, R.A. Houghten, *J. Comb. Chem.* **2002**, *4*, 345.
02JCO501	R. Pére, T. Beryozkina, O.I. Zbruyev, W. Haas, C.O. Kappe, *J. Comb. Chem.* **2002**, *4*, 501.
02JCR(S)60	Y.A. Ibrahim, B. Alsaleh, *J. Chem. Res. (S)* **2002**, 60.
02JCR(S)314	E.S.H. El Ashry, H.A. Hamid, A. Mousaad, E.S. Ramadan, *J. Chem. Res. (S)* **2002**, 314.
02JCR(S)359	M. Darzac, S. Montesinos, A. Collet, J.P. Dutasta, *J. Chem. Res. (S)* **2002**, 359.
02JCS(P1)108	J.S. Northen, F.T. Boyle, W. Clegg, N.J. Curtin, A.J. Edwards R.J. Griffin, B.T. Golding, *J. Chem. Soc. Perkin Trans. 1* **2002**, 108.
02JCS(P1)696	K. Uehata, T. Kawakami, H. Suzuki, *J. Chem. Soc. Perkin Trans. 1* **2002**, 696.
02JCS(P1)1877	M.J. Wanner, G.J. Koomen, *J. Chem. Soc. Perkin Trans. 1* **2002**, *1*, 1877.
02JCS(P1)2549	J. Szczepkowskasztolcman, A. Katrusiak, H. Wojtowiczrajchel, K.Golankiewicz, *J. Chem. Soc. Perkin Trans. 1* **2002**, 2549.
02JCS(P2)442	R.A. Bartsch, E.K. Lee, S.K. Chun, N. ElKarim, K. Brandt I. Porwolikczomperlik, M. Siwy, D. Lach, J. Silberring, *J. Chem. Soc. Perkin Trans. 2* **2002**, 442.
02JHC45	A.S. Shawali, M.A. Abdallah, M.M. Zayed, *J. Heterocycl. Chem.* **2002**, *39*, 45.
02JHC327	B.L. Pongo, I. Kovesdi, J. Reiter, *J. Heterocycl. Chem.* **2002**, *39*, 327.
02JHC357	P. Bilek, J. Slouka, *J. Heterocycl. Chem.* **2002**, *39*, 357.
02JHC663	L. Lucry, F. Enoma, F. Estour, H. Oulyadi, S. Menager, O. Lafont, *J. Heterocycl. Chem.* **2002**, *39*, 663.
02JHC885	G. Biagi, I. Giorgi, O. Livi, F. Pacchini, V. Scartoni, *J. Heterocycl. Chem.* **2002**, *39,* 885.
02JHC1061	M. Kopp, J.C. Lancelot, S. Dagdag, H. Miel, S. Rault, *J. Heterocycl. Chem.* **2002**, *39*, 1061.
02JIC176	S. Mehrotra, V. Pandey, R. Lakhan, P.K. Srivastava, *J. Indian Chem. Soc.* **2002**, *79*, 176.
02JIC472	M.B. Deshmukh, S.S. Patil, A.R. Mulik, *J. Indian Chem. Soc.* **2002**, *79*, 472.
02JMC1383	L.L. Gundersen, J. Nissenmeyer, B. Spilsberg, *J. Med. Chem.* **2002**, *45,* 1383.
02JMC1500	A.M. Hayallah, J. Sandoval-Ramírez, U. Reith, U. Schobert, B. Preiss, B. Schumacher, J.W. Daly, C. Muller, *J. Med. Chem.* **2002**, *45*, 1500.
02JMC2131	S.A. Kim, M.A. Marshall, N. Melman, H.S. Kim, C.E. Muller, J. Linden, K.A. Jacobson, *J. Med. Chem.* **2002**, *45*, 2131.
02JMC2342	A.M. Gilbert, S. Caltabiano, F.E. Koehn, Z.J. Chen, G.D. Francisco, J.W. Ellingboe, Y, Kharode, A. Mangine, R. Francis, M. Trailsmith, D. Gralnick, *J. Med. Chem.* **2002**, *45,* 2342.
02JMC2923	H. Matter, P. Kotsonis, O. Klingler, H. Strobel, L.G. Frohlich, A. Frey, W. Pfleiderer, H. H.H.W. Schmidt, *J. Med. Chem.* **2002**, *45*, 2923.
02JMC3381	A.E. Gibson, C.E. Arris, J. Bentley, F.T. Boyle, N.J. Curtin, T.G. Davies, J.A. Endicott, B.T. Golding, S. Grant, R.J. Griffin, P. Jewsbury, L.N. Johnson, V. Mesguiche, D.R. Newell, M.E.M. Noble, J.A. Tucker, H.J. Whitfield, *J. Med. Chem,* **2002**, *45*, 3381.
02JMC3440	C.E. Muller, M. Thorand, R. Qusirhi, M. Diekmann, K.A. Jacobson, W.L. Padgett, J.W. Daly, *J. Med. Chem.* **2002**, *45,* 3440.
02JMC4398	B.L. Milary, P.J. Oates, W.J. Zembrowski, D.A. Deebe, E.L. Conn, J.B. Coutcher, M.T. Ogorman, M.C. Linhares, G.J. Withbroe, *J. Med. Chem.* **2002**, *45*, 4398.
02JMC5030	E. Novellino, E. Abignente, B. Cosimelli, G. Greco, M. Iadanza, S. Laneri, A. Lavecchia, M. G. Rimoli, *J. Med. Chem.* **2002**, *45*, 5030.
02JMC5448	G.D. Brown, S.K. Luthra, C.S. Brock, M.F.G. Stevens, P.M. Price, F. Brady, *J. Med. Chem.* **2002**, *45*, 5448.

02JMC5458	J. Arrowsmith, S.A. Jennings, A.S. Clark, M.F.G. Stevens, *J. Med. Chem.* **2002**, *45*, 5458.
02JMC5492	B.R. Henke, T.G. Consler, N. Go, R.L. Hale, D.R. Hohman, S.A. Jones, A.T. Lu, L.B. Moore, J.T. Moore, L.A. Orbandmiller, R.G. Robinett, J. Shearin, P.K. Spearing, E.L. Stewart, P.S. Turnbull, S.L. Weaver, S.P. Williams, G.B. Wisely, M.H. Lambert, *J. Med. Chem.* **2002**, *45*, 5492.
02JMC5710	A. Constanzo, G. Guerrini, G. Ciciani, F. Bruni, C. Costagli, S. Selleri, F. Besnard, B. Costa, C. Martini, P. Malmbergaiello, *J. Med. Chem.* **2002**, *45*, 5710.
02JOC188	R.J. Herr, P.F. Vogt, H. Meckler, M.P. Trova, S.R. Schow, R.C. Petter, *J. Org. Chem.* **2002**, *67*, 188.
02JOC2378	M. Avalos, R. Babiano, P. Cintas, F.R. Clemente, R. Gordillo, M.B. Hursthouse, J.L. Jiménez, M.E. Light, J.C. Palacios, *J. Org. Chem.* **2002**, *67*, 2378.
02JOC3202	H.L. Nyquist, E.A. Beeloo, L.S. Hurlbut, R. Watsonclark, D.E. Harwell, *J. Org. Chem.* **2002**, *67*, 3202.
02JOC4526	B. Zaleska, T. Bazanek, R. Socha, M. Karelus, J. Grochowski, P. Serda, *J. Org. Chem.* **2002**, *67*, 4526.
02JOC4808	L.J. Prins, E.E. Neuteboom, V. Paraschiv, M. Gregocalama, P. Timmerman, D.N. Reinhoudt, *J. Org. Chem.* **2002**, *67*, 4808.
02JOC5807	M. Cushman, D.L. Yang, S. Gerhardt, F. Huber, M. Fischer, K. Kis, A. Bacher, *J. Org. Chem.* **2002**, *67*, 5807.
02JOM101	C.T. Quian, W.L. Nie, Y.F. Chen, J. Sun, *J. Organomet. Chem.* **2002**, *645*, 101.
02JPP147	K. Iakovou, A. Varvaresou, A.P. Kourounakis, K. Stead, D. Sugden, A. Tsotinis, *J. Pharm. Pharmacol.* **2002**, *54*, 147.
02KGS81	T.E. Glotova, A.S. Nakhmanovich, A.I. Alvanov, N.I. Protsuk, T.V. Nizovtseva, V.A. Lopyrev, *Khim. Geterotsikl. Soedin.* **2002**, 81.
02KGS197	A.A. Chesnyiuk, S.N. Mikhailichenko, K.S. Zavodnov, V.N. Zaplishny, *Khim. Geterotsikl. Soedin.* **2002**, 197.
02KGS326	S.N. Mikhailichenko, A.A. Chesnyuk, V.E. Zavodnik, S.I. Firgang, L.D. Konyushkin, V.N. Zaplishny, *Khim. Geterotsikl. Soedin.* **2002**, 326.
02KGS713	V.K. Vasilin, E.A. Kaigorodova, M.M. Lipunov, G.D. Krapivin, *Khim. Geterotsikl. Soedin.* **2002**, 713.
02KGS1302	J.L. Nein, Y.Y. Morzherin, Y.A. Rozin, V.A. Bakulev, *Khim Geterotsikl. Soedin.* **2002**, *9*, 1302.
02M79	A.M. Amer, M. Elmobayed, A.M. Ateya, T.S. Muhdi, *Monatsh. Chem.* **2002**, *133*, 79.
02M1297	M.A.N. Mosselhi, *Monatsh. Chem.* **2002** *133*, 1297.
02M1157	G. Kickelbick, D. Rutzinger, T. Gallauner, *Monatsh. Chem.* **2002**, *133*, 1157.
02M1165	D. Moderhack, A. Daoud, P.G. Jones, *Monatsh. Chem.* **2002**, *133*, 1165.
02MI15	M.L. Highfill, A. Chandrasekaran, D.E. Lynch, D.G. Hamilton, *Cryst. Growth Des.* **2002**, *2*, 15.
02MI272	E.N. Ulomskii, S.L. Deev, A.V. Tkachev, I.K. Moiseev, V.L. Rusinov, *Russ. J. Org. Chem.* **2002**, *38*, 272.
02MI461	U. Abram, C.C. Gatto, E. Bonfada, E.S. Lang, *Inorg. Chem. Commun.* **2002**, *5*, 461
02MI602	Z. Turgut, N. Ocal, *Russ J. Org. Chem.* **2002**, *38*, 602.
02MI744	D.N. Kozhevnikov, V.N. Kozhevnikov, I.S. Kovalev, V.L. Rusinov, O.N. Chupakhin, G.G. Aleksandrov, *Russ. J. Org. Chem.* **2002**, *38*, 744.
02MI1647	M. Schneider, M. Klotzsche, C. Werzinger, J. Hegele, R. Waibel, M. Pischetsrieder, *J. Agr. Food Chem.* **2002**, *50*, 1647.
02MI4760	A. Freisleben, P. Schieberle, M. Rychlik, *J. Agr. Food Chem.* **2002**, *50*, 4760.
02NJC1340	G. Molteni, A. Ponti, M. Orlandi, *New J. Chem.* **2002**, *26*, 1340.
02NJC1545	R. Jalal, M. Elmessaoudi, *New J. Chem.* **2002**, *26*, 1545.
02OL881	M.H. Alsayah, N.R. Branda, *Org. Lett.* **2002**, *4*, 881.
02OL1253	D. Brown, S. Muranjan, Y. Jang, R. Tummel, *Org. Lett.* **2002**, *4*, 1253.
02OL2113	V. Chandrasekhar, A. Athimoolam, *Org. Lett.* **2002**, *4*, 2113.
02OL3227	M.O. Ratnikov, D.L. Lipilin, A.M. Churakov, Y.A. Strelenko, V.A. Tartakovsky, *Org. Lett.* **2002**, *4*, 3227.
02OL3481	J.J. Yin, M.M. Zhao, M.A. Huffman, J.M. McNamara, *Org. Lett.* **2002**, *4*, 3481.
02PHA438	B. Dundar, O. Bozdagdundar, B. Kaneke, T. Coban, M. Iscan, E. Bayukbingole, *Pharmazie* **2002**, *57*, 438.
02PS45	A.B.A. Elgazzar, A.M. Gaafar, A.S. Aly, *Phosphorus Sulfur* **2002**, *177*, 45.
02PS173	A.S.A. Youssef, *Phosphorus Sulfur* **2002**, *177*, 173.

02PS487	M.A.N. Mosselhi, M.A. Abdallah, Y.F. Mohamed, S. Shawali, *Phosphorus Sulfur* **2002**, *177*, 487.
02PS587	M.A. Elbadawi, A.A. Elbarbary, Y.M. Loksha, M. Eldaly, *Phosphorus Sulfur* **2002**, *177*, 587.
02PS1033	R. Abderrahim, B. Baccar, M.L. Benkhoud, *Phosphorus Sulfur* **2002**, *177*, 1033.
02PS2403	M.M. Heravi, M. Bakherad, M. Rahimzadeh, M. Bakavoli, *Phosphorus Sulfur* **2002**, *177*, 2403.
02PS2491	M.M. Heravi, A. Kivanloo, M. Rahimizadeh, M. Bakavoli, M. Ghassemzadeh, *Phosphorus Sulfur* **2002**, *177*, 2491.
02PS2497	F.M.A. Ellatif, E.A.E. Rady, M.A. Khalil, *Phosphorus Sulfur* **2002**, *177*, 2497.
02RRC1007	F. Czobor, C. Cristescu, *Rev. Roum. Chim.* **2002**, *46*, 1007.
02S43	J.C. González-Gómez, L. Santana, E. Uriarte, *Synthesis* **2002**, 43.
02S475	J.C. González, J. Lobo Antunes, J. Pérez Lourido, L. Santana, E. Uriarte, *Synthesis* **2002**, 475.
02S538	R. Schirrmacher, B. Wangler, E. Schirrmacher, T. August, F. Rosch, *Synthesis* **2002**, 538.
02S901	A.L. Krasovsky, A.M. Moiseev, V.G. Nenajdenko, E.S. Balenkova, *Synthesis* **2002**, 901.
02S911	M.I. Rodríguez-Franco, M.I. Fernández-Bachiller, *Synthesis* **2002**, 911.
02S2532	T. Hundsdorf, E.V. Blyumin, H. Neunhoeffer, *Synthesis* **2002**, 2532.
02SC203	C.F. Ye, Z.F. Zhang, W.M. Liu, *Syn. Commun.* **2002**, *32*, 203.
02SC1407	A. Rivera, O.L. Torres, J.D. Leiton, M.S. Morales-Ríos, P. Josephnathan, *Synth. Commun.* **2002**, *32*, 1407.
02SC2089	N. Saesaengseerung, T. Vilaivant, Y. Thebtaranonth, *Synth. Commun.* **2002**, *32*, 2089.
02SC3455	Y Xiong, L.X. Zhang, A.J. Zhang, D.J. Xu, *Synth. Commun.* **2002**, *32*, 3455.
02SC3493	F.A. Abushanab, Y.M. Elkholy, M.H. Elnagdi, *Synth. Commun.* **2002**, *32*, 3493.
02SL447	F.A. Alphonse, F. Suzenet, A. Keromnes, B. Lebret, G. Guillaumet, *Synlett* **2002**, 447.
02SL519	P. Raboisson, B. Norberg, J.R. Casimir, J.J. Bourguignon, *Synlett* **2002**, 519.
02SL557	J.A. Zerkowski, L.M. Hensley, D. Abramowitz, *Synlett* **2002**, 557.
02SL1332	D.D. Hughes, M.C. Bagley, *Synlett* **2002**, 1332.
02SL1718	M.C. Bagley, N. Singh, *Synlett* **2002**, 1718.
02SRI171	G. Singh, P.A. Singh, K. Singh, S.N. Dubey, R.N. Handa, J.H. Choi, *Syn. Reactiv. Inorg. Metal-Org.* **2002**, *32*, 171.
02T721	E.A. Archer, D.F. Cauble, V. Lynch, M.J. Krische, *Tetrahedron* **2002**, *58*, 721.
02T1375	V.Y. Sosnovskikh, B.I. Usachev, G.V. Roschenthaler, *Tetrahedron* **2002**, *58*, 1375.
02T1525	M.A. Blanco, E. López Torres, M.A. Mendiola, E. Brunet, M.T. Sevilla, *Tetrahedron* **2002**, *58*, 1525.
02T3185	P.A. Abbott, R.V. Bonnert, M.V. Caffrey, P.A. Cage, A.J. Cooke, D.K. Donald, M. Furber, S. Hill, J. Withnall, *Tetrahedron*, **2002**, *58*, 3185.
02T3513	S. Blanchard, I. Rodríguez, C. Kuehmcaubere, P. Renard, B. Pfeiffer, G. Guillaumet, P. Caubere, *Tetrahedron*, **2002**, *58*, 3513.
02T4481	I.A. Zuleta, M.L. Vitelli, R. Baggio, M.T. Garland, A.M. Seldes, J.A. Palermo, *Tetrahedron*, **2002**, *58*, 4481.
02T7431	M. Havelkova, D. Dvorak, M. Hocek, *Tetrahedron* **2002**, *58*, 7431.
02T7607	P.G. Baraldi, A.U. Broceta, M.J.P. Desasinfantas, J.J.D. Mochun, A. Espinosa, R. Romagnoli, *Tetrahedron* **2002**, *58*, 7607.
02T7911	V. Brun, M. Legraverend, D.S. Grierson, *Tetrahedron* **2002**, *58*, 7911.
02T8559	A.S. Shawali, S.M. Gomha, *Tetrahedron* **2002**, *58*, 8559.
02T9723	A. Lauria, P. Diana, P. Barraja, A. Montalbo, G. Cirrincione, G. Dattolo, A.M. Almerico, *Tetrahedron* **2002**, *58*, 9723.
02TA821	U. Groselj, S. Recnik, J. Svete, A. Meden, B. Stanovnik, *Tetrahedron: Asymmetry* **2002**, *13*, 821.
02TA1805	A. Iuliano, E. Attolino, P. Salvadori, *Tetrahedron: Asymmetry* **2002**, *13*, 1805.
02TL273	M.R. Shaaban, T. Fuchigami, *Tetrahedron Lett.* **2002**, *43*, 273.
02TL695	S. Tumkevicius, L.A. Agrofoglio, A. Kaminskas, G. Urbelis, T.A. Zevaco, O. Walter, *Tetrahedron Lett.* **2002**, *13*, 695.
02TL895	P.J. Bhuyan, H.N. Borah, J.S. Sandhu, *Tetrahedron Lett.* **2002**, *43*, 895.
02TL2529	Y. Basel, A. Hassner, *Tetrahedron Lett.* **2002**, *13*, 2529.
02TL3323	M. Kunishima, K. Hioki, A. Wada, H. Kobayashi, S. Tani, *Tetrahedron Lett.* **2002**, *43*, 3323.
02TL3551	E.R. Bilbao, M. Alvarado, C.F. Maseguer, E. Raviña, *Tetrahedron Lett.* **2002**, *43*, 3551.
02TL3863	C.J. Lee, S.J. Lee, J.Y. Chang, *Tetrahedron Lett.* **2002**, *43*, 3863.
02TL4439	C.J. Lee, S.J. Lee, J.Y. Chang, *Tetrahedron Lett.* **2002**, *43*, 4439.
02TL4741	E. Fernández, S. García-Ochoa, A. Huss, A. Mallo, J.M. Bueno, F. Micheli, A. Paio, E. Piga, P. Zarantonello, *Tetrahedron Lett.* **2002**, *43*, 4741.

02TL4873	J. Quiroga, H. Insuasty, B. Insuasty, R. Abonia. J. Cobo, A.Sánchez, M. Nogueras, *Tetrahedron Lett.* **2002**, *43*, 4873.
02TL5241	M. Mateu, A.S. Capilla, Y. Harrak, M.D. Pujol, *Tetrahedron Lett.* **2002**, *43*, 5241.
02TL6169	R.E. Austin, J.F. Okonya, D.R.S. Bond, F. Alobeidi, *Tetrahedron Lett,* **2002**, *43*, 6169.
02TL6783	P. Dehoog, P. Gámez, W.L. Driessen, J. Reedijk, *Tetrahedron Lett.* **2002**, *43*, 6783.
02ZN(B)547	M.M. Elabadelah, A.S. Abushamleh, C.M. Mossmer, W. Voelter, *Z. Naturforsch. (B)* **2002**, *57*, 547.
02ZN(B)552	A.S. Shawali, M.A. Abdallah, M.A.N. Mosselhi, Y.J. Mohamed, *Z. Naturforsch (B)* **2002**, *57*, 552.
02ZN(B)699	M.A. Abdallah, *Z. Naturforsch. (B)* **2002**, *57*, 699.

Chapter 6.4

Six-Membered Ring Systems: With O and/or S Atoms

John D. Hepworth
James Robinson Ltd., Huddersfield, UK
j.d.hepworth@tinyworld.co.uk

B. Mark Heron
Department of Colour Chemistry
University of Leeds, Leeds, UK
b.m.heron@leeds.ac.uk

6.4.1 INTRODUCTION

Reviews of saturated O-heterocycles <02JCS(P1)2301>, the value of pyran-2-ones in the synthesis of arenes and heteroarenes <01COC571> and of the polyacylated anthocyanin flower pigments <02H(56)633> have been published. A review of o-quinone methides includes an account of their value in pyran chemistry <02T5367> and reviews of the annular tautomerism of six-membered heterocycles <02AHC(81)253>, biologically active carbazole alkaloids <02CR4303> and cyclisation reactions of 1,3-bis-silyl enol ethers <02S441> contain material pertinent to this chapter.

Strategies adopted for the total synthesis of marine macrolides that bind with the actin cytoskeleton and their biological properties have been reviewed <02AG(E)4633>. Total syntheses of (-)-laulimalide feature an asymmetric acyl halide-aldehyde cyclocondensation <02JA13654>, a diastereoselective allylstannane addition and a Mitsunobu macrolactonisation <02JA5958>, an intermolecular asymmetric Sakurai reaction to couple aldehyde and allyl silane moieties followed by a Yamaguchi macrolactonisation <02JA4956>, and a Lewis acid-promoted coupling of an allyl silane with a chiral acetal, again followed by a Yamaguchi cyclisation <02TL3381>. An enantioselective synthesis of callipeltoside A has been reported <02JA5654>.

A Symposia-in-Print is dedicated to the methodology and total synthesis of marine natural products containing polycyclic ethers <02T1779>. Additional reports on individual polycyclic ethers include those on the ciguatoxins <02JOC3301, 02OL2771, 02OL4551, 02SL1835> and gambierol <02JA14983>. A combination of Lewis acid-mediated intramolecular allylation of a-acetoxy ethers and a ring closing metathesis (RCM) of the products offers a convergent synthesis of polycyclic ethers <02JA3562>, as does a Pd-catalysed coupling of ketene acetal triflates with a Zn homoenolate and subsequent hydroboration and cyclisation of the resulting enol ethers <02JOC3494>. Trans-fused polyclic ethers can be obtained via initial reaction between a THP-aldehyde and a pyran-based Li alkynide. A hetero Michael addition features in the remaining steps <02OL2739>.

This chemistry features in a synthesis of the left hand fragment of yessotoxin and adriatoxin <02OL3943>. Assembly of the F-N ring component of gymnomycin A, involves a b-alkyl Suzuki-Miyaura coupling reaction <02OL1747>.

Work on spiroketals includes that on the spongistatins <02BMC2039, 02JA5661, 02OL3719, 02OL3723, 02TL3285>, the pectenotoxins <02AG(E)4569, 02AG(E)4573>, bistriamide C <02CEJ1670>, didemnaketals <02JCS(P1)565, 02T1697> and pinnatoxin A <02T10375>. Key steps in an enantioselective synthesis of (-)-preussomerin G and other naphthalene-based spiroketals are a chiral epoxidation of an enone and an oxidative spirocyclisation <02JOC2735>. A total synthesis of the marine pyran-4-one polypropionate, siphonarin B, has been reported <02OL391>.

The chemistry and biology of the leptomycin family of natural products, which contain a 5,6-dihydro-2H-pyran-2-one unit, has been reviewed <02S981>. Two total syntheses of (-)-callystatin A have been reported. One creates the stereogenic centres using SAMP/RAMP methodology <02CEJ4272> and the other involves an efficient sp^2-sp^2 Suzuki coupling <02JOC2751>. Total syntheses of (-)- <02TL8791> and (±)- arisugacin A <02OL367>, (-)-malyngolide <02T8929> and solanapyrones <02JOC5969> have been published.

6.4.2 HETEROCYCLES CONTAINING ONE OXYGEN ATOM

6.4.2.1 Pyrans

4-Methylenepyran Fischer-carbene complexes undergo self-dimerisation in the presence of a Pd(0) catalyst, giving electron-rich bispyrans. Oxidation leads to a methylenepyran fulvene (Scheme 1) <02TL3967>.

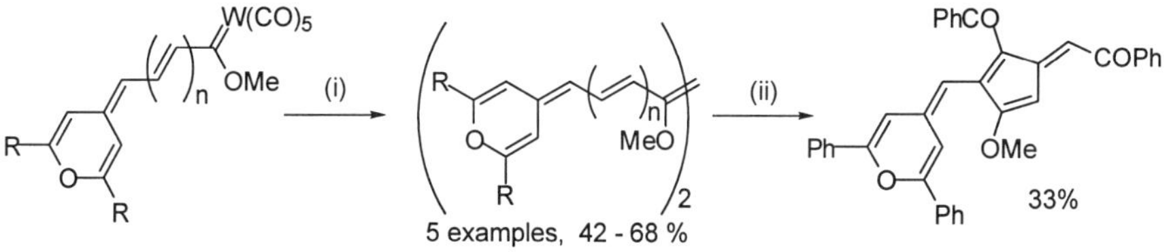

Reagents: (i) either 10 mol% Pd(PPh$_3$)$_4$ or Pd(PPh$_3$)$_2$Cl$_2$, Et$_3$N, THF, 70 °C; (ii) O$_2$, SiO$_2$, CH$_2$Cl$_2$

Scheme 1

Treatment of 3-bromo-4H-pyran with base in the presence of 18-crown-6 generates $3\Delta^2$-pyran **1** which is trapped, even when furan or styrene are present in the reaction mixture, by tert-BuO$^-$, yielding 4-tert-butoxy-4H-pyran. The preference for nucleophilic attack over cycloaddition implies a significant contribution of the zwitterion **2** to the ground state of **1**. Calculations indicate **2** to be the most stable planar structure and to be the ground state in THF solution <02JA287>.

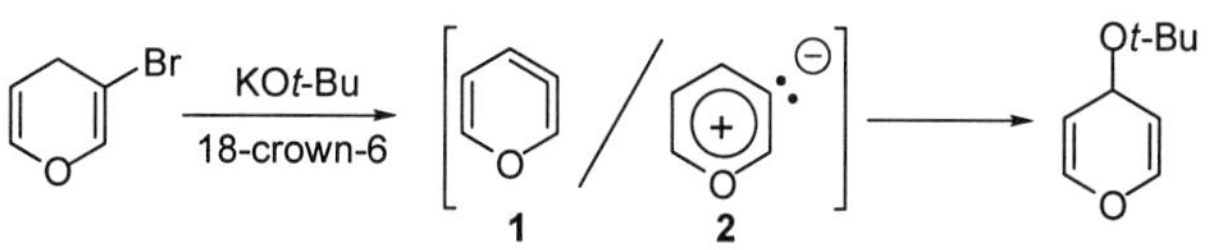

New additions to the hetero Diels-Alder (hDA) approach to dihydropyrans include the reaction between enol ethers and 3-aryl-2-cyanoprop-2-enoates that under high pressure yields substituted 3,4-dihydro-2*H*-pyrans with high *endo*- and complete regio- selectivity. The products can be transformed into a wide variety of other compounds <02EJO3126>.

Enol ethers react with γ-substituted β-unsaturated α-oxo esters in the presence of a Lewis acid, the choice of which can, along with the structure of the reagents, influence the steric course of the reaction. Of particular interest is the total *exo* selectivity shown in the reaction of methyl (*E*)-*tert*-butoxymethylenepyruvate with 1-methoxycyclohexene <02EJO514>.

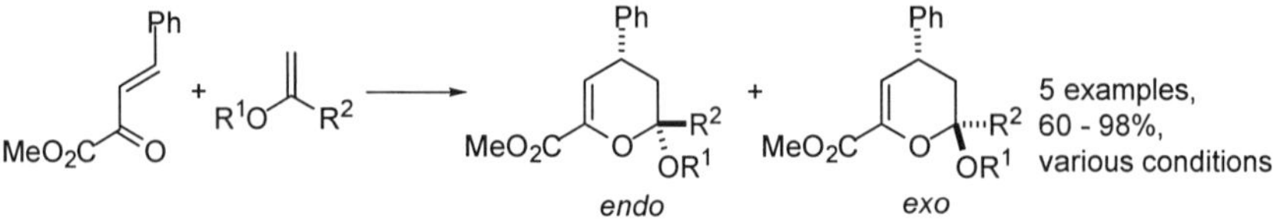
Ph
MeO2C
O
R1O
R2
endo
exo
OR1
5 examples,
60 - 98%,
various conditions

Trans-fused pyranopyrans are formed in good yield and with high ee when butenoate esters react with pent-4-en-1-ols in the presence of chiral Cu oxazoline complexes. An initial transesterification is followed by an intramolecular hDA <02TL9397>.

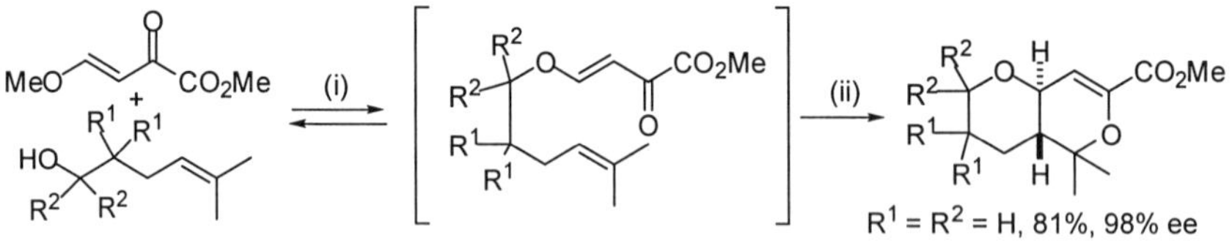
MeO
CO2Me
HO
R1
R2
(i)
(ii)
H
R1 = R2 = H, 81%, 98% ee

Reagents: (i) chiral Lewis acid cat., 40°C, -MeOH; (ii) Ti-TADDOLate cat., -40 °C, 24 h, CH_2Cl_2, mol. sieve.

Phosphonoacrolein, $(EtO)_2P(O)C(CHO){=}CH_2$, takes part in hDA reactions with alkenes and cyclic conjugated dienes, to give phosphono-substituted 3,4-dihydro-2*H*-pyrans and their annulated derivatives. The reaction with alkynes gives the 1:1 adduct, a 4*H*-pyran, initially but this undergoes a second cycloaddition leading to a tetrahydropyrano[3,2-*b*]pyran **3** <02JOC7303>.

The ether-tethered allenyne **4** is converted into a cross-conjugated triene, the 5,6-dihydro-2*H*-pyran **5**, in a Rh-catalysed allenic Alder ene reaction <02JA15186> and allylic halides undergo a Pd-catalysed coupling reaction with 3,4-allen-1-ols to give 5,6-dihydro-2*H*-pyrans (Scheme 2) <02JOC6104>.

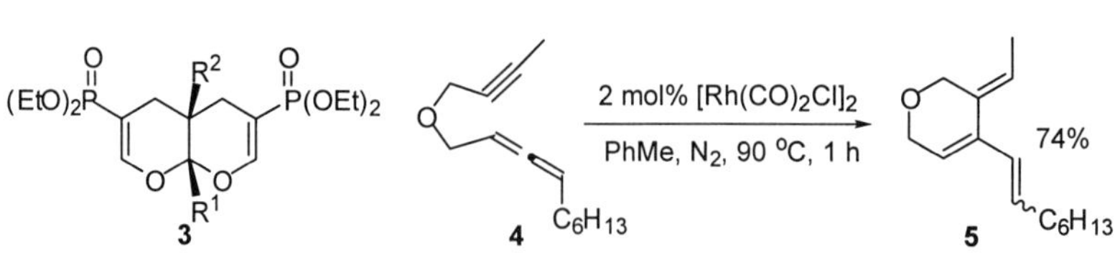
(EtO)2P
P(OEt)2
R2
R1
3
4
5
C6H13
2 mol% [Rh(CO)2Cl]2
PhMe, N2, 90 °C, 1 h
74%

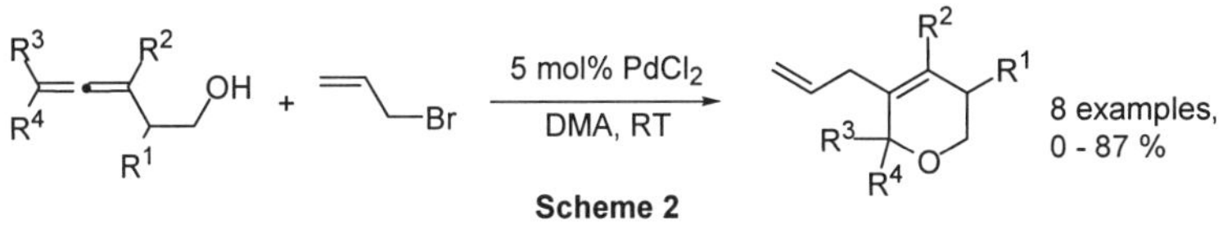

Scheme 2

The mechanism of both the uncatalysed and W-catalysed cycloisomerisation of pent-4-yn-1-ol has been studied in detail. The *endo* reaction is complex but is favoured mainly as a result of stabilisation of a vinylidene intermediate <02JA4149>. The W-mediated cyclisation of alkynols has been used to synthesise glycals of the antibiotics vancosamine and saccharosamine <02OL749> and *trans*-fused THP derivatives of 5*H*-dibenzo[*a,d*]cycloheptene <02TL8697>.

Ru-catalysed enyne metathesis offers a short approach to chiral derivatives of 3-vinyl-5,6-dihydro-2*H*-pyrans. Some epimerisation can occur at the pyranyl C atom at elevated temperatures (Scheme 3) <02T5627>. The bispropargyloxynorbornene derivative **6** undergoes a cascade of metathesis reactions in the presence of alkenes and Grubbs' catalyst incorporating an enyne-RCM that leads to fused cyclic dienes. A dienophile can be added to the reaction mixture, resulting in Diels-Alder reactions and the formation of functionalised polycyclic products <02TL1561>.

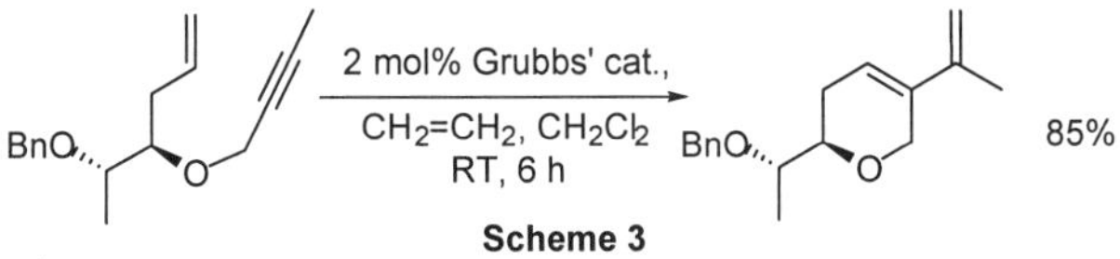

Scheme 3

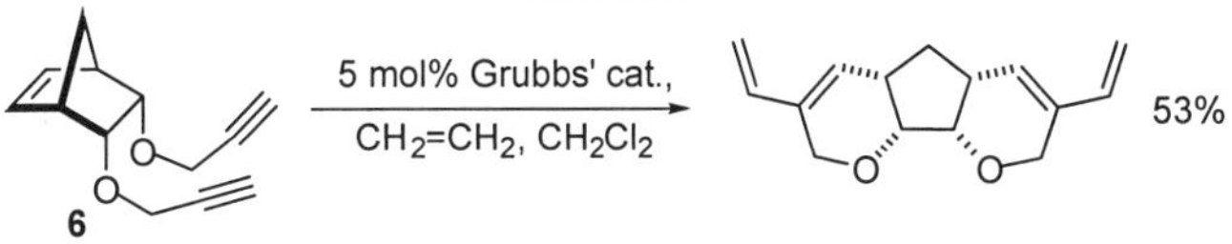

Sequential one-pot RCM-cyclisation of dienes and alkene isomerisation have been achieved through conversion of the Ru-alkylidene RCM catalyst into a Ru-hydride catalyst, providing an efficient route to 3,4-dihydro-2*H*-pyrans from acyclic dienes <02JA13390>.

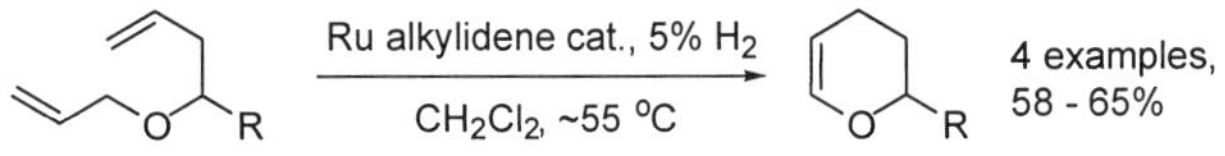

The acid-catalysed rearrangement of 2,4,5-triols leads to the thermodynamic product, a tetrahydrofuran. However, cyclisation using trimethyl orthoacetate and pyridinium toluene-*p*-sulfonate gives a mixture of the THF and a tetrahydropyran, in which the former predominates, by attack of the primary hydroxy group at different ends of the episulfonium ion intermediate. This route nevertheless provides a useful route to THPs through equilibration of this reaction mixture <02JCS(P1)2646, 02JCS(P1)2652>.

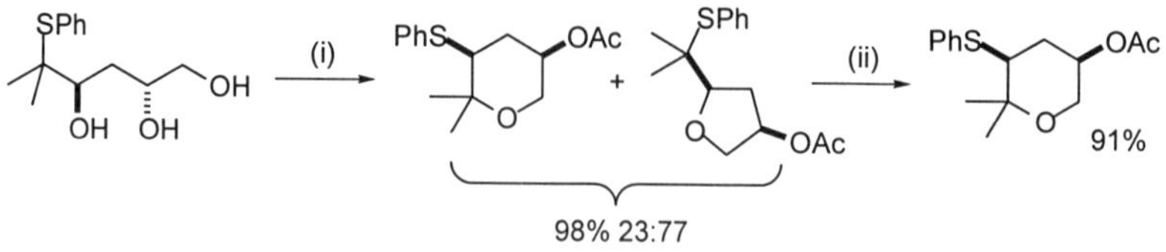

Reagents: (i) (MeO)$_3$CMe, C$_5$H$_6$N$^+$ TsO$^-$, CH$_2$Cl$_2$, RT, 24 h.; (ii) TsOH, CH$_2$Cl$_2$, 40 °C, 24 h.

4-Methylene-2,6-disubstituted tetrahydropyrans are formed from the In-catalysed reaction of aldehydes with homoallylic alcohols which proceeds by an intramolecular oxonium-ene cyclisation and exhibits excellent diastereoselectivity <02TL7193>.

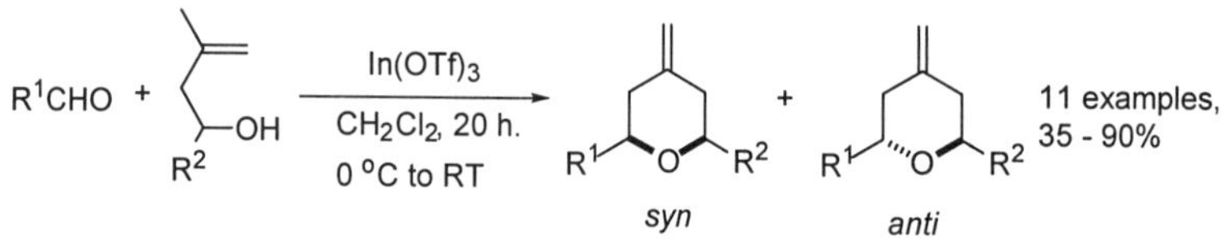

Spirocyclic 4-substituted tetrahydropyrans are readily obtained through the Prins reaction involving cyclic ketones, homoallylic alcohols and MeSO$_3$H <02H(58)659>. The cationic species generated when alkyne-Co complexes derived from δ-valerolactone are treated with SnCl$_4$ undergo a double cyclisation to yield the oxaspiro[5.5]undecane <02T2755>.

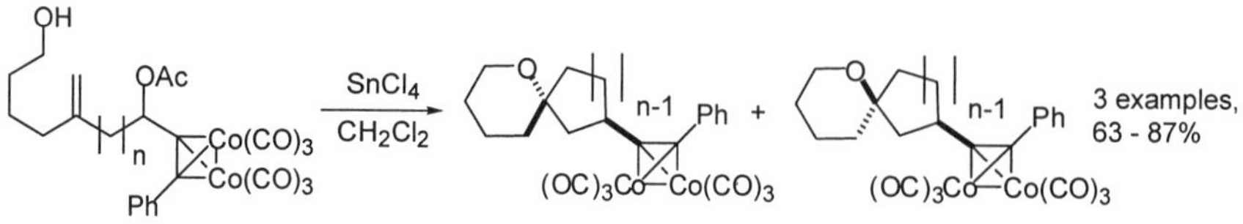

6.4.2.2 [1]Benzopyrans and Dihydro[1]benzopyrans (Chromenes and Chromans)

Examples of the formation of 2H-[1]benzopyrans by the thermal rearrangement of aryl propargyl ethers include a synthesis of the 1,7-dihydropyrano[2,3-g]indole ring system which is present in the paraherquamide alkaloids <02TL2149> and a commercial synthesis of (+)-calanolide A, a potent inhibitor of HIV-1 reverse transcriptase, which involves a novel disulfonate monodeprotection process <02TL2899>. In contrast, prop-2-yn-1-ols undergo a cycloaddition reaction with electron-rich phenols in the presence of thiolate-bridged diruthenium complexes to give the isomeric 4H-[1]benzopyrans. The reaction is thought to proceed through nucleophilic attack at the two electrophilic centres in the allenylidene ligand <02JA7900>.

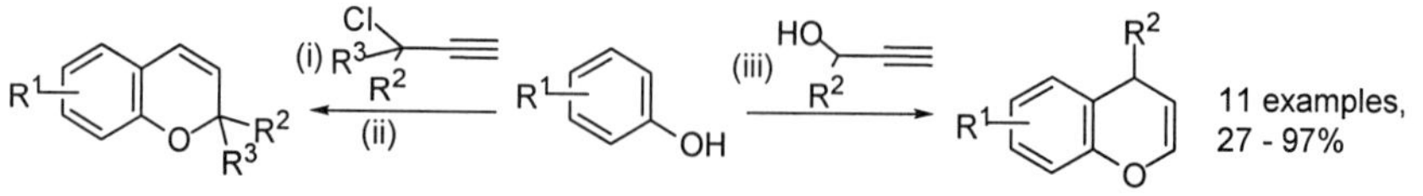

Reagents: CuCl$_2$.2H$_2$O, DBU, MeCN, 0 °C, 5h; (ii) o-C$_6$H$_4$Cl$_2$, heat;
(iii) 5 mol% [Cp*RuCl(μ$_2$-SMe)$_2$RuCp*Cl], NH$_4$BF$_4$, ClCH$_2$CH$_2$Cl, 60 °C

There has been a number of developments in the use of salicylaldehydes as precursors of both chromenes and chromans. Alkenes activated by acyl, formyl, nitrile and phenylsulfonyl groups react with 2-hydroxybenzaldehydes and 2-hydroxy-1-naphthaldehyde under Bayliss-Hillman conditions to yield 3-substituted chromenes *via* the *in situ* dehydration of the initially formed chroman-4-ol <02JCS(P1)1318>. In like manner, β-nitrostyrenes yield 2- and 2,2-substituted derivatives of 3-nitrochromenes <02H(57)1033>. A simple route to 2-phenyl-2*H*-chromenes starting from salicylaldehyde and utilising a Pd(0)-catalysed cyclisation of an allylic acetate has been described <02SC3667>.

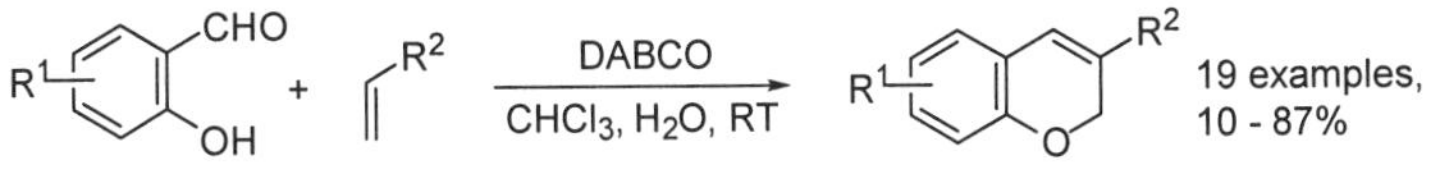

$$R^1\text{-salicylaldehyde} + R^2\text{-alkene} \xrightarrow{\text{DABCO, CHCl}_3, \text{H}_2\text{O, RT}} R^1\text{-chromene-}R^2 \quad \text{19 examples, 10 - 87\%}$$

A sequence of allylation, epoxidation and an acid-mediated 6-*exo* cyclisation converts salicylaldehydes into 2-hydroxymethyl-2-methyl-2*H*-[1]benzopyrans. A bicyclic chroman arising from attack of the hydroxymethyl group on the intermediate benzylic cation has been isolated <02SL322>. A twelve-step enantioselective synthesis of a 2-hydroxymethyl-2-methylchroman with an overall yield of 48% uses related methodology and introduces the chirality through an asymmetric Sharpless epoxidation <02JCS(P1)496>.

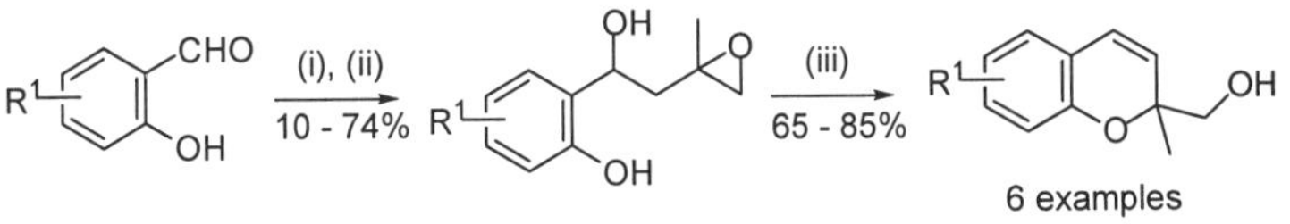

$$R^1\text{-CHO} \xrightarrow[10-74\%]{\text{(i), (ii)}} R^1\text{-intermediate} \xrightarrow[65-85\%]{\text{(iii)}} R^1\text{-product}$$

6 examples

Reagents: (i) CH$_2$=C(Me)CH$_2$MgCl, THF, -20 °C; (ii) *m*-CPBA, CH$_2$Cl$_2$, 0 °C; (iii) 4 mol%, 4-TsOH, PhH, heat

Triflates of transition metals catalyse the reaction between alkenes, salicylaldehydes and trimethyl orthoformate (TMOF) that is particularly useful for the synthesis of fused chromans. Generation of an *o*-quinone methide and its capture by the alkene is the key feature. The examples in Scheme 4 are illustrative. *Trans*-fused pyrano[3,2-*c*]benzopyrans

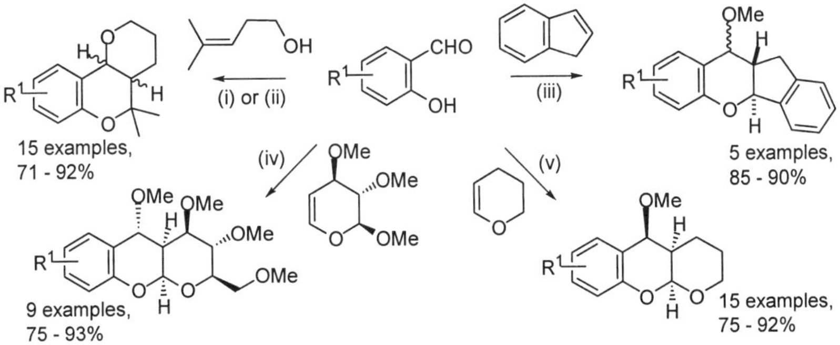

15 examples, 71 - 92%

9 examples, 75 - 93%

5 examples, 85 - 90%

15 examples, 75 - 92%

Reagents: (i) I$_2$, TMOF, CH$_2$Cl$_2$, RT <02JCS(P1)1401>; (ii) Sc(OTf)$_3$, TMOF, CH$_2$Cl$_2$, RT <02S217>; (iii) Sc(OTf)$_3$, TMOF, CH$_2$Cl$_2$, RT <02TL2999>; (iv) Sc(OTf)$_3$, TMOF, CH$_2$Cl$_2$, RT <02TL4527>; (iv) Yb(OTf)$_3$, TMOF, CH$_2$Cl$_2$, RT <02JCS(P1)165>

Scheme 4

are readily formed by the reaction of salicylaldehydes with unsaturated alcohols and TMOF, which proceeds through an I_2-catalysed hDA reaction <02JCS(P1)1401>.

The reaction between salicylaldehydes, aromatic amines and enol ethers, which yields 4-aminobenzopyran derivatives, is catalysed by $PPh_3.HClO_4$. The method is easily adapted to the synthesis of pyrano- and furano- cis-fused analogues <02T10301>.

The one-pot reaction of O-BOC protected salicylaldehydes and salicyl alcohols with electron-rich alkenes and a Grignard reagent involves a diastereoselective cycloaddition to an o-quinone methide and offers access to a wide range of 4-substituted chromans <02JOC6911>.

Two specific o-quinone methides merit mention. Spectral data for the o-quinone methide derived from α-tocopherol by oxidation with Ag_2O has been obtained from its complex with N-methylmorpholine N-oxide. The amine oxide extends the lifetime of the methide from about 2 s to several minutes <02OL4285>. Tropolone o-quinone methide, generated from a 2,4-dihydroxy-5-phenylthiomethyltropone, reacts with the sesquiterpene humulene in a hDA reaction to give the fungus-derived deoxyepolone B <02OL3009>.

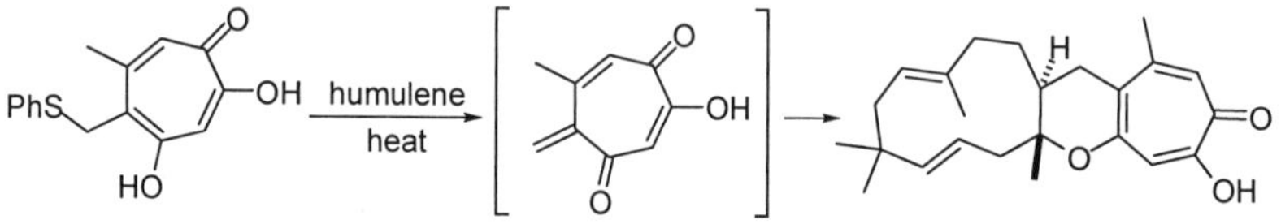

The cyclobutanone ring in the ether **7**, derived by ring expansion of an oxaspiropentane, intramolecularly alkylates the activated aromatic system on treatment with PTSA, leading to 2H-cyclobuta[c]chroman-4-ols **8** (X = OH). Subsequent fission of the four-membered ring yields 3,4-disubstituted 2H-chromenes <02SL796>. A variation of this approach allows the synthesis of the 4β-phenylsulfanyl derivative of **8** (X = SPh), oxidation of which affords the cyclopropa[c]chroman entity <02OL2565>.

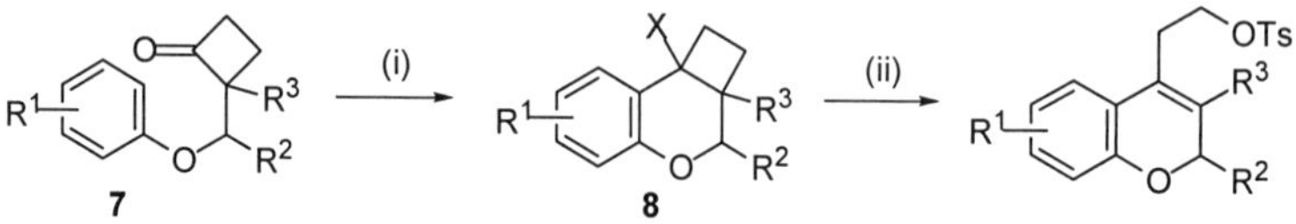

6 examples, 30 - 70%
Reagents: (i) 4-TsOH, PhH, heat, 6 h; (ii) 4-TsOH (equimolar), PhH, heat, 30 min

Stable palladacycles **9**, derived from o-substituted iodobenzenes, react with activated alkynes to give good yields of 2H-[1]benzopyrans. Generally, unsymmetrical alkynes insert in a regioselective manner <02OL3679>.

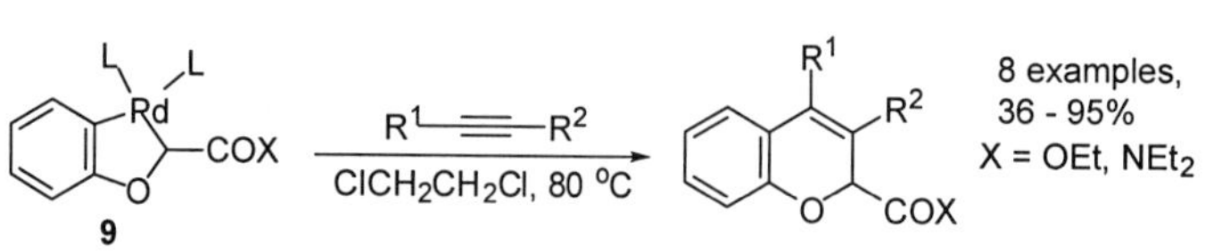

8 examples, 36 - 95%
X = OEt, NEt_2

The cyclohexadienone **10** undergoes an intramolecular asymmetric Heck reaction in the presence of a chiral monodentate phosphoramidate ligand to give the benzo[c]chromene derivative **11** with excellent enantioselectivity and conversion <02JA184>.

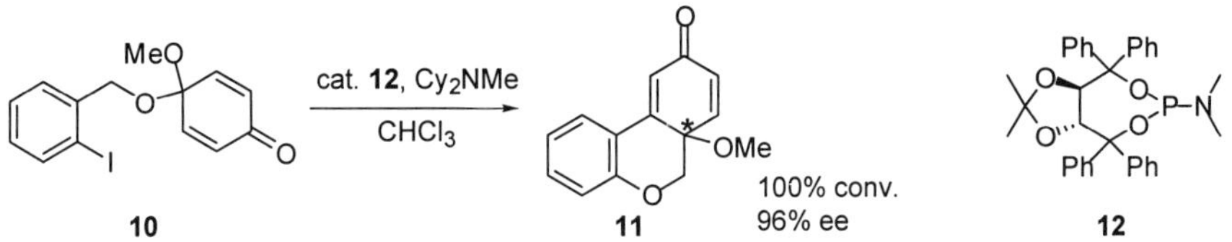

3-Nitrocoumarins undergo regio- and endo- selective [4+2]-cycloadditions with electron-rich dienophiles to yield nitronates **13**, hydrolysis of which yields a mixture of chroman-2-ols. The one-pot process can be carried out under aqueous conditions and is readily adapted to the synthesis of *cis*-fused furo- and pyrano- [2,3-*b*]chromans <02JOC7238>.

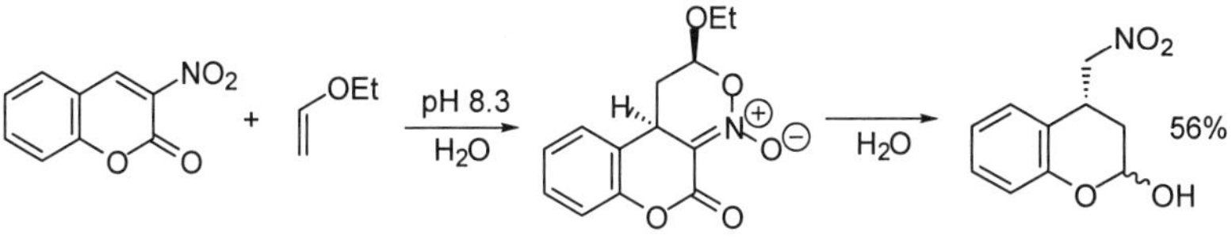

Optically pure Δ^9-THC has been obtained *via* the 1,4-addition of [Ar$_2$Cu(CN)Li$_2$] to α-iodocyclohexenones (Scheme 5) <02JOC8771>. Syntheses of (-)-Δ^9-tetrahydrocannabivarins deuterated at specific sites in the side-chain have been reported <02JCS(P1)2544>.

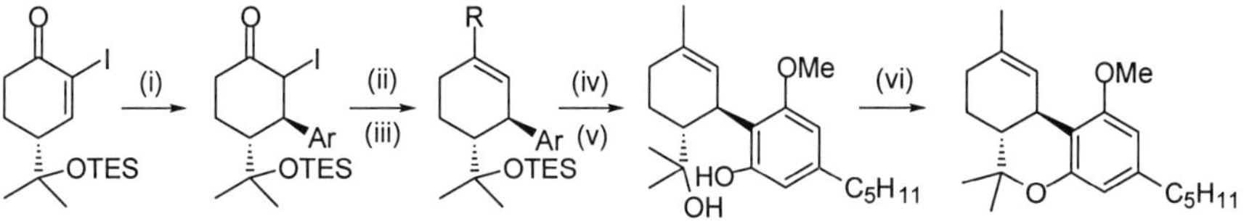

Reagents: (i) Ar$_2$Cu(CN)Li$_2$ (62%); (ii) EtMgBr; (iii) ClP(O)(OEt)$_2$ (66%); (iv) MeMgCl, Ni(acac)$_2$ (93%); (v) NaSEt (73%); (vi) ZnBr$_2$, MgSO$_4$ (84%)

Scheme 5

Protected 7-hydroxychromenes are phenylsulfonylated at C-4 and undergo conjugate addition of PhLi to give the *cis*-diastereomer. Desulfinylation by metal reduction provides access to isoflavans and elimination affords the isoflavene system <02TL6893>.

The bis-chroman **14** has been dimerised and subsequently cyclodimerised to give a 3,6-linked calix[4]naphthalene derivative <02JOC904>. Calixerene-like macrocycles are

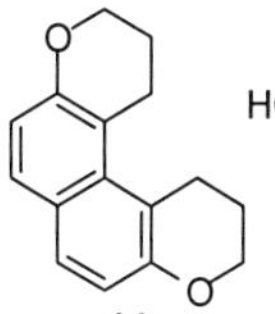

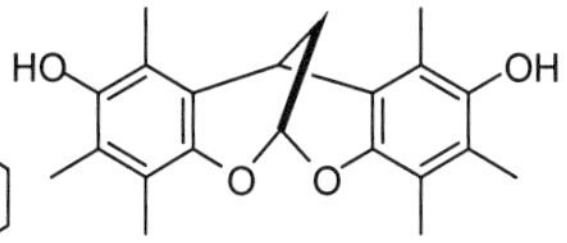

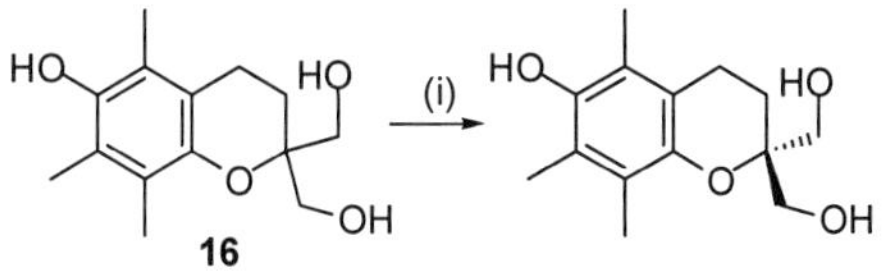

Reagents: (i) Candida antarctica lipase, vinyl acetate, Et$_2$O, 2 h

formed when **15** undergoes repeated oxidation – hDA – reduction sequences involving an *o*-quinone methide <02AG(E)1171>.

A synthesis of (*S*)-α-tocotrienol, in which the phytyl group of tocopherol is replaced by a farnesyl side chain, (six steps, 19%; ee = 98%) is based on the enzymatic desymmetrisation of the achiral 2,2-di(hydroxymethyl)chroman **16** <02TL7971>.

6.4.2.3 [2]Benzopyrans and Dihydro[2]benzopyrans (Isochromenes and Isochromans)

2-Iodobenzyl 2-butynyl ethers undergo a Pd-mediated cyclisation with terminal alkynes to give 1*H*-[2]benzopyrans. The absence of a Cu(I) co-catalyst allows an intramolecular 6-*exo-dig* cyclisation to compete successfully with an intermolecular Sonogashira aryl-alkynyl coupling <02T9007>.

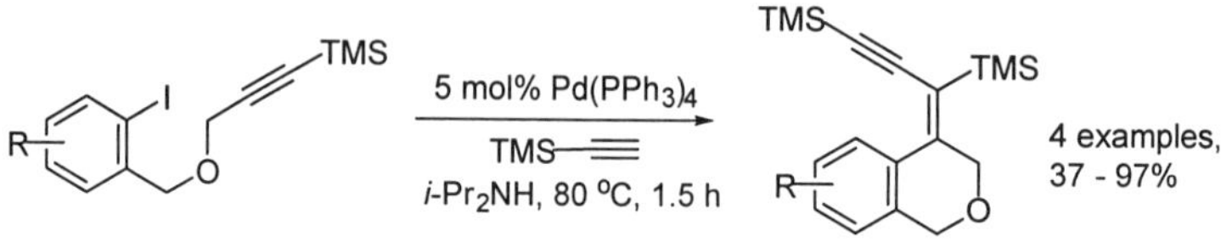

2-Alkynylbenzaldehydes yield isochromenes in a Pd-catalysed reaction with alcohols. The Pd plays a dual role, behaving as a Lewis acid to effect the formation of a hemiacetal and coordinating to the alkyne moiety to promote cyclic ether formation <02JA764>.

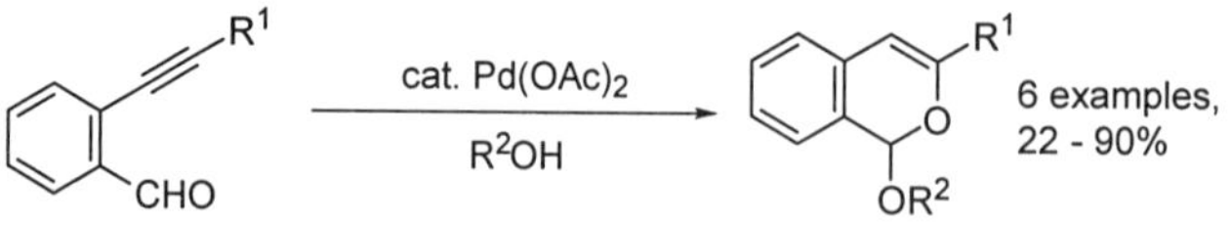

The trisethers derived from 1,3,5-tribromo-2,4,6-tris(bromomethyl)benzene by reaction with 2-alkenols in the presence of NaH undergo a triannulation reaction by way of a triple intramolecular Heck reaction to yield 3,7,11-trioxa-1,2,3,4,5,6,7,8,9,10,11,12-dodecahydro-triphenylenes <02JOC8280>.

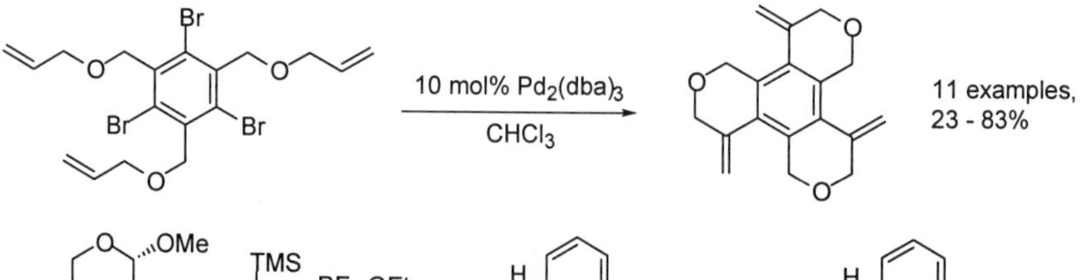

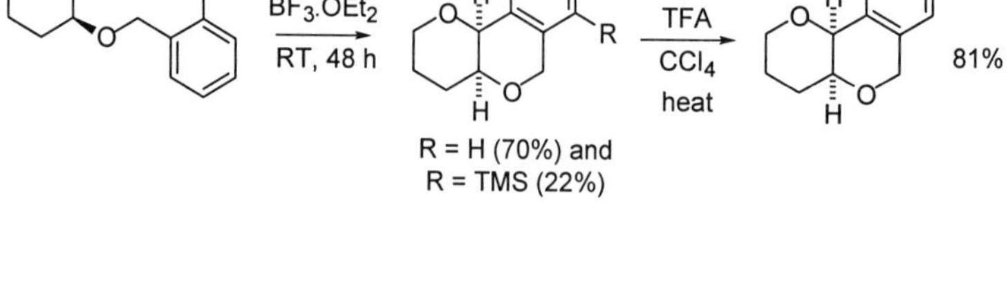

A trimethylsilyl group facilitates the Friedel-Crafts intramolecular cyclisation of 2-O-benzyl ethers to isochromans. The bulk of the TMS function may force the molecule into an alignment suitable for cyclisation <02OL3797>.

A total synthesis of (1R,3S)-thysanone, an isochromanquinone which shows potent activity against human rhinovirus 3C-protease, has been achieved from ethyl (S)-lactate <02JCS(P1)938>.

6.4.2.4 Pyrylium Salts

Cleavage of the pyran ring occurs during the hydrolysis of pyrano[4,3,2-de][1]benzopyrylium pigments, contributors to the colour of red wine, to their aglycone pyranoanthocyanidins and this results in a rearrangement to the furo[2,3-c][1]benzopyrylium ring system <02TL715>.

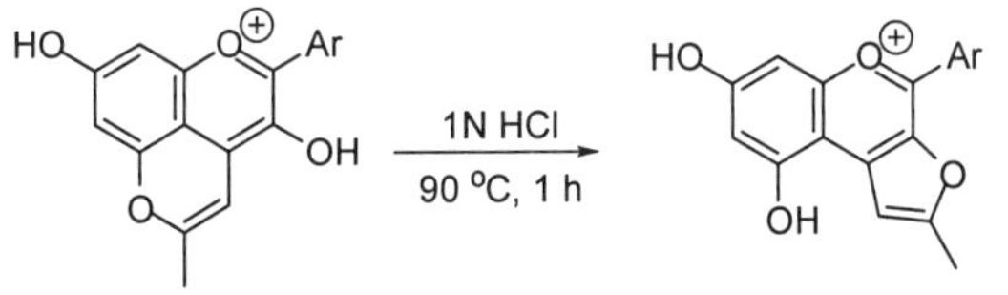

The conversion of pyrylium salts into arenes features in a synthesis of functionalised phenyl-ethynyl macrocycles <02OL4269>.

6.4.2.5 Pyranones

6-Alkyl-2H-pyran-2-ones are formed when (Z)-5-alkyl-2-en-4-ynoic acids, available by a Pd-catalysed coupling of alkynylzinc compounds with (Z)-3-iodopropenoic acid, are treated with $ZnBr_2$. When cyclisation is effected by Ag_2CO_3, 5-alkylidenefuran-2(5H)-ones are produced selectively <02TL5673>.

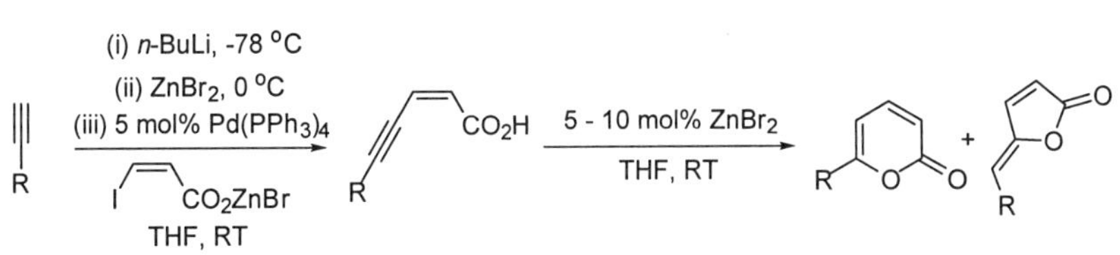

Methyl 2-en-4-ynoates undergo a regioselective iodocyclisation to 5-iodopyran-2-ones under mild conditions (Scheme 6) <02T5023>, augmenting the related cyclisation of 5-substituted (E)-2-bromo-2-en-4-ynoic acids <02EJO1063>. Application of this methodology to 2-(arylethynyl)benzoates yields isocoumarins. A further advantage of this route is that alkynyltin compounds can be coupled at the 5-position in a one-pot process. In a further example of the synthesis of pyran-2-ones by the Stille reaction, vinylstannanes are stereoselectively annulated by acyl chlorides <02JOC3941>.

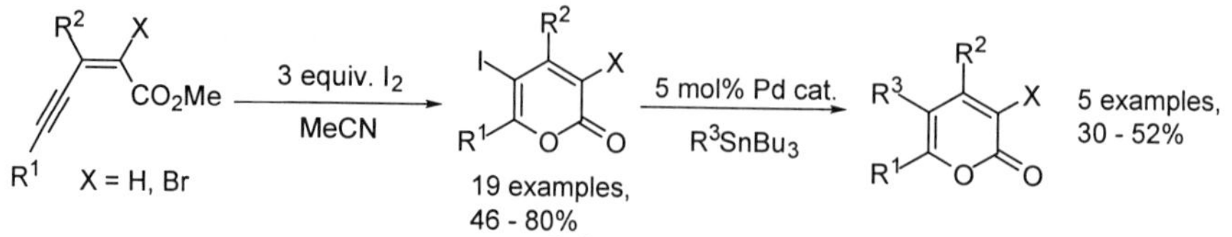

Scheme 6

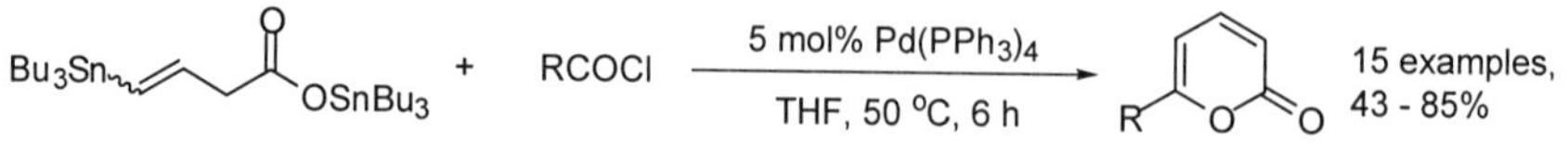

Substituted pyran-4-ones can be prepared from isoxazoles through reductive cleavage with $Mo(CO)_6$ and acid-catalysed cyclisation of the generated enaminoketone <02TL3565>.

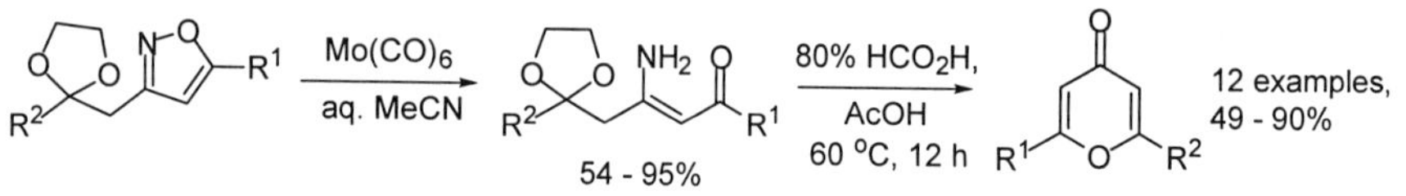

Fused 2*H*-pyran-2-ones are formed in excellent yields and under mild conditions through a Ni-catalysed [2+2+2] cycloaddition of diynes and CO_2 <02JA15188>. Oxidative demetalation of (η^3-allyl)Mo complexes of pyran with pyridinium dichromate (PDC) introduces a carbonyl function at the allylic terminus offering access to dihydropyranones of high enantiopurity <02JOC5773>.

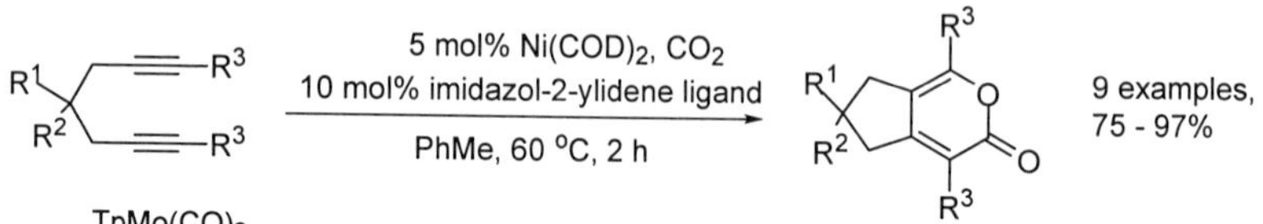

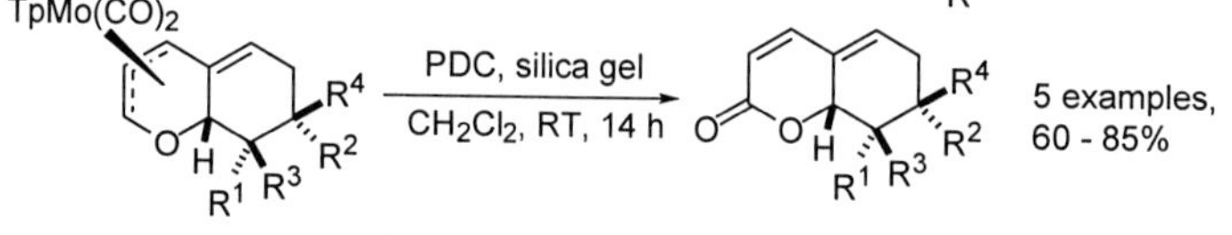

α-Substituted cyclopropylideneacetic acids and esters are cleaved at the proximal C–C bond when heated with $CuBr_2$ and subsequent cyclisation affords 4-bromo-5,6-dihydro-2*H*-pyran-2-ones <02OL4419>. The carbonylation of cyclopropenone can be accomplished via a Ru-catalysed C–C bond cleavage and gives pyranopyrandiones. In the presence of internal alkynes, a cross-carbonylation leads to unsymmetrically substituted derivatives <02JA6824>.

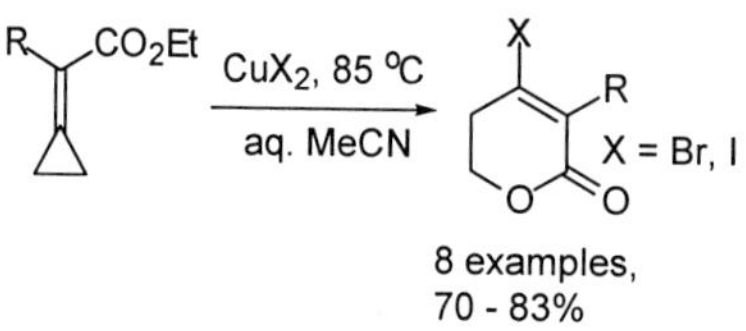

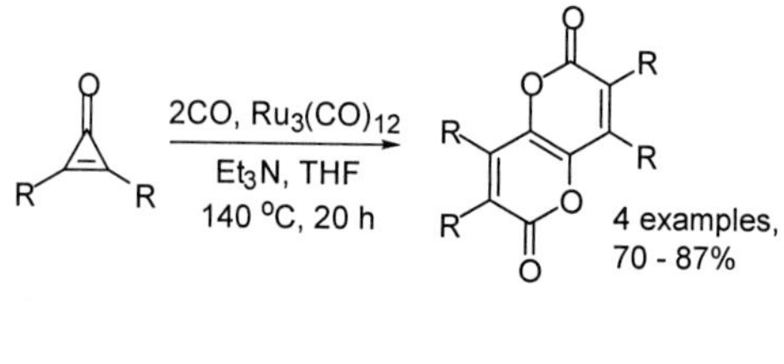

The 4-vinyloxetan-2-ones generated in a Pd-catalysed [2+2] cycloaddition between ketene and α,β-unsaturated carbonyl compounds undergo a spontaneous allylic rearrangment. A zwitterionic intermediate is proposed that cyclises to a 3,6-dihydro-2*H*-pyran-2-one, but which may alternatively decarboxylate to a diene <02T5215>.

Several reports relate to stereocontrol in the hDA synthesis of 2,3-dihydro-4*H*-pyran-4-ones. The reaction of aldehydes with Danishefsky's diene occurs with high *trans*- and enantio- selectivities when catalysed by binol complexes <02JOC2175, 02OL122>. A Ti-binol catalyst is efficient under solvent-free conditions <02JA10> and hydrogen-bonding solvents promote the hDA reaction with unactivated ketones such as cyclohexanone, when spiro-linked pyranones are formed in good yield <02JA9662>. The use of chiral aldehydes with a chiral Cr-salen catalyst enables any of the four possible stereoisomers to be prepared <02OL1795>.

6.4.2.6 Coumarins

A Suzuki coupling reaction carried out under aqueous conditions features in a four-step synthesis of 3,4-fused coumarins from tetralones and chroman-4-one <02TL1213>. 4-Tosylcoumarin is a precursor of 4-arylcoumarins using Suzuki methodology <02TL4395>.

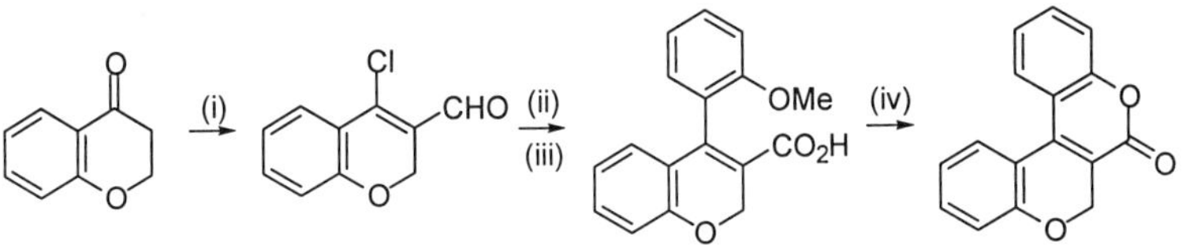

Reagents: (i) POCl$_3$, DMF (96%); (ii) 2-methoxyphenylboronic acid, Pd(OAc)$_2$, Bu$_4$NBr, K$_2$CO$_3$, H$_2$O, 45 °C (98%); (iii) NaClO$_2$, 30% H$_2$O$_2$, MeCN, RT (69%); (iv) SOCl$_2$, Et$_3$N, Et$_2$O, heat then AlCl$_3$, CH$_2$Cl$_2$, RT (96%)

An improvement in the synthesis of coumarins from phenols and β-keto esters results from the use of Zn to mediate a transesterification <02TL8583> and InCl$_3$ is an efficient catalyst for the Pechmann reaction <02TL9195>.

Naturally occurring 6-prenylcoumarins have been synthesised from 2-prenyloxy-benzaldehydes through tandem Wittig reaction and Claisen rearrangement. Several of the

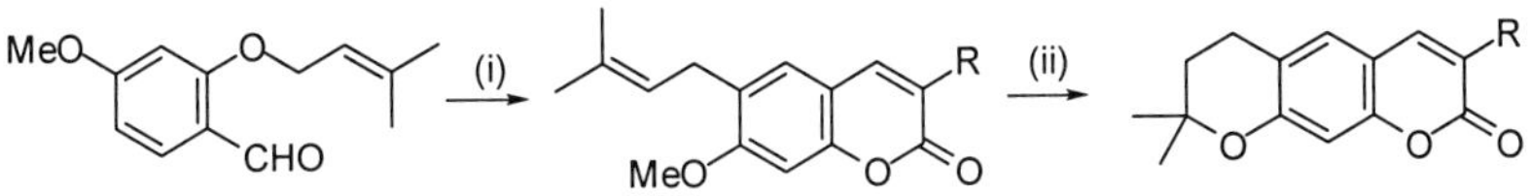

Reagents: (i) Ph$_3$P=C(R)CO$_2$Et, PhNMe$_2$, N$_2$, heat; (ii) py.HCl, N$_2$, heat

products have been converted into pyranocoumarins <02JCS(P1)371>. A combination of ozonolysis and chiral stationary phase-GC-MS is recommended for the assignment of absolute configuration to coumarins derived from prenyl epoxides <02CC3070>

A one-pot synthesis of dibenzo[b,d]pyranones from the aryl propenoate **17** proceeds via sequential Sonogashira coupling and an intramolecular benzannulation. <02JOC5138>.

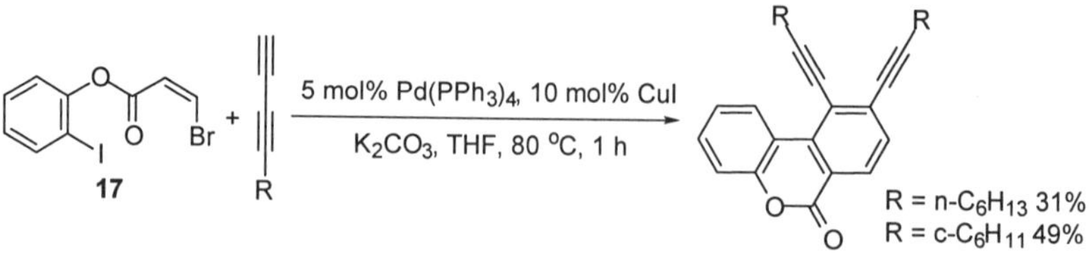

A simple conversion of hydroxycoumarins into the 2-oxoethoxy derivatives has been described <02T4851>. 4-Hydroxycoumarins have been arylated at C-3 through reaction of their phenyliodonium zwitterions with boronic acids <02OL3333>. Reaction of 4-hydroxycoumarin with O-prenylated aromatic aldehydes results in a Knoevenagel-hDA sequence which leads to a mixture of chromeno-fused pyrano[3,2-c]coumarins and pyrano[2,3-b]chromones through involvement of the keto and lactone carbonyl group respectively in the cycloaddition. Under microwave irradiation, the proportion of the major product, the coumarin, is significantly increased (Scheme 7) <02T997>. The zwitterion formed from DMAD and cyclohexyl isocyanide is trapped by 4-hydroxycoumarin to produce the pyrano[3,2-c]benzopyran ring system <02H(58)147>. In the presence of a Ru catalyst, the stable 3-diazobenzopyran-2,4(3H)-dione inserts into the O-H bond of alcohols and phenols to give 3-alkoxy- and 3-aryloxy- 4-hydroxycoumarins <02TL3637>.

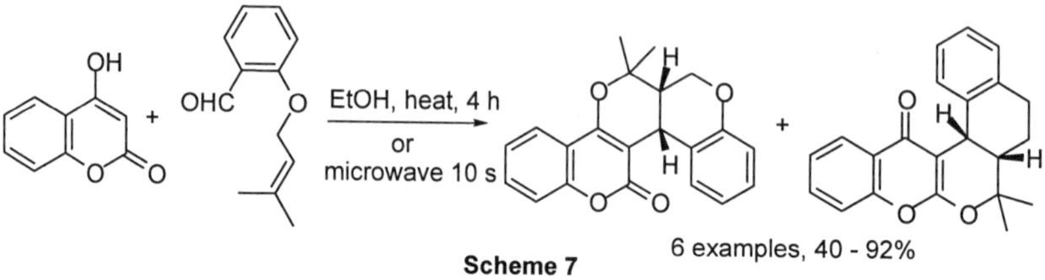

Scheme 7

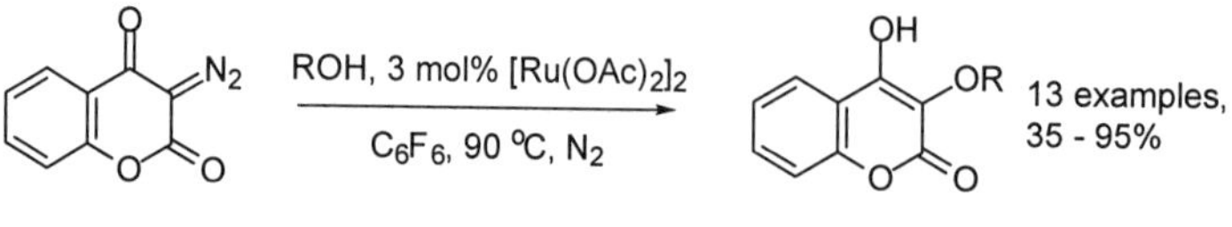

Iodocyclisation of 2-(1-alkynyl)benzoates under mild conditions leads to 3-substituted 4-iodoisocoumarins. Application of the method to (Z)-2-alken-4-ynoates gives 6-substituted

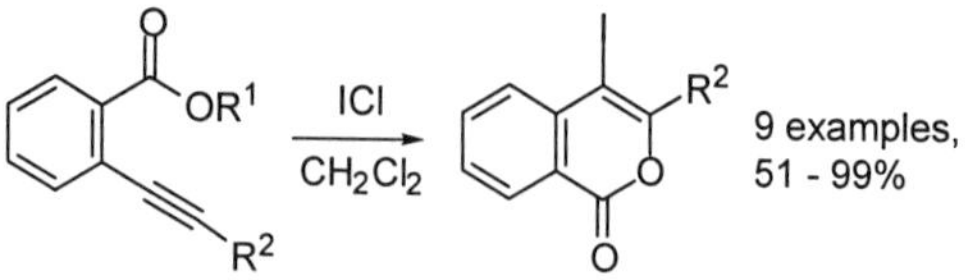

5-iodopyran-2-ones. In both series, elaboration of the products is possible through the iodo functionality <02TL7401>.

6.4.2.7 Chromones

Benz[*b*]indeno[2,1-*e*]pyrandiones are readily available by the acylation of *o*-hydroxyacetophenone with monomethyl phthalate and subsequent Baker-Venkataraman rearrangement <02TL4515>.

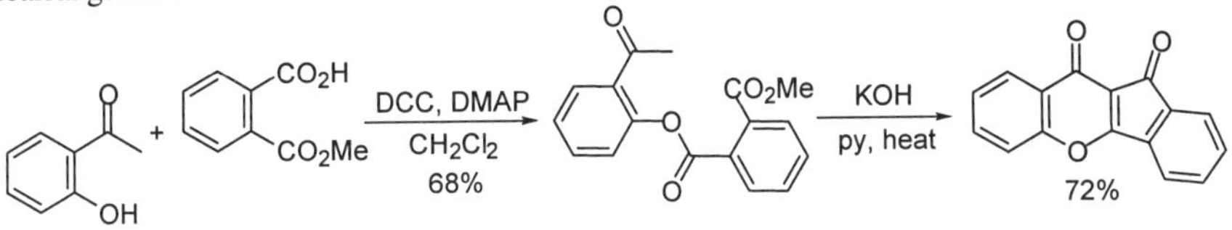

Polymer-supported diacetoxyiodobenzene can be used advantageously in the synthesis of isoflavones from 2-hydroxychalcones <02S2490>. Treatment of deoxybenzoins with CS_2 and an alkyl halide under phase transfer conditions leads to 2-(alkylthio)isoflavones <02TL6113>.

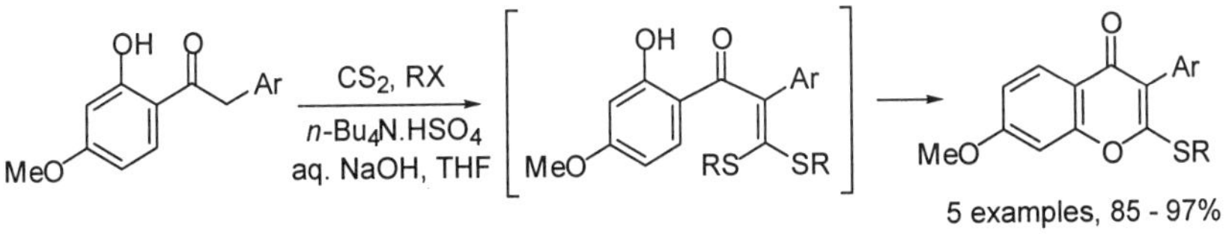

Furo[3,4-*b*][1]benzopyran-9-ones are formed directly by deprotection of the acetal **18**. The corresponding thiochromones behave similarly <02TL4507>. Incorporation of a trifluoromethyl group into chroman-4-ones can be effected through reaction of 2-trifluoromethyl-4*H*-chromen-4-imines with malonic acid; the acid serves as a methylating agent <02S2341>.

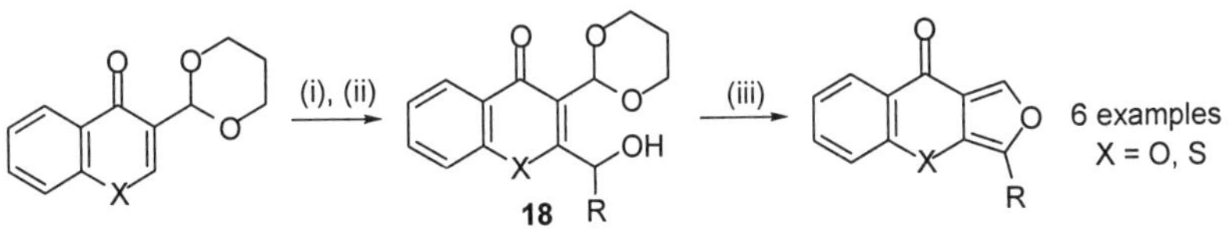

Reagents: (i) LTMP, THF, -78 °C; (ii) RCHO; (iii) 4-TsOH, PhMe, 50 °C

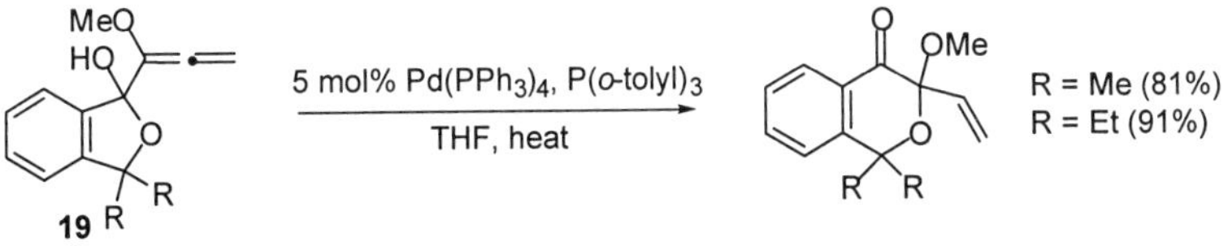

Ring expansion of the phthalans **19** to isochromanones is effected by Pd(0) complexes <02OL455>.

6.4.2.8 Xanthones and Xanthenes

An inter- followed by an intra- molecular double Heck reaction features in the synthesis of the xanthene **20** from 2-bromophenyl ether and ethyl acrylate <02TL8559>.

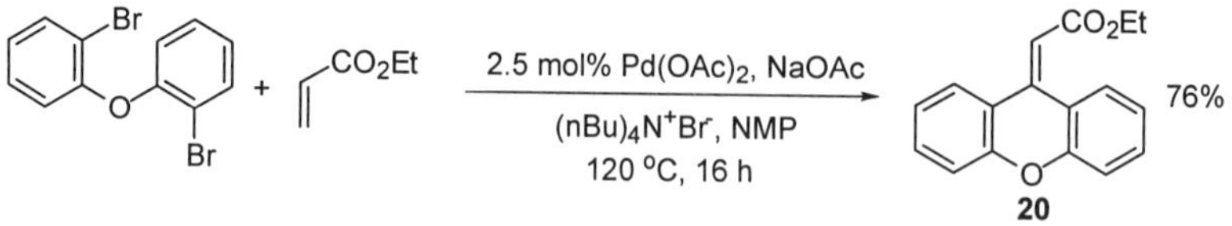

Benzo[a] and benzo[c] xanthene and dibenzo[a,c]xanthene have been synthesised by the Pd-catalysed cyclisation of aromatic 2-(arylmethyl)phenol triflate esters through S-O bond cleavage <02T5927>.

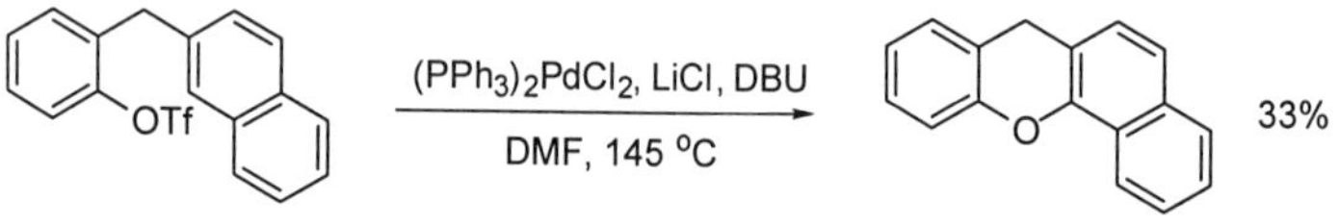

Benzo[b]xanthene-6,11,12-triones are readily available through the photo-induced 1,4-acylation of naphthoquinone with 2-hydroxybenzaldehydes followed by oxidation <02H(57)1915>. Chromone-3-carboxaldehyde reacts with o-benzoquinodimethane in a DA reaction with concomitant deformylation to give a diastereoisomeric mixture of tetrahydrobenzo[b]xanthones; oxidation can be accomplished with DMSO/I_2 <02T105>.

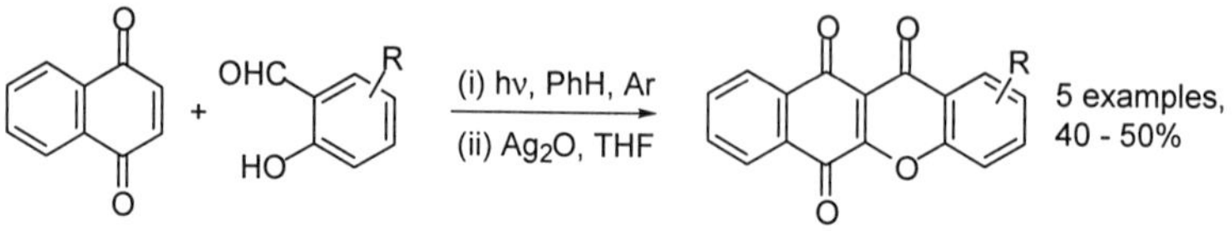

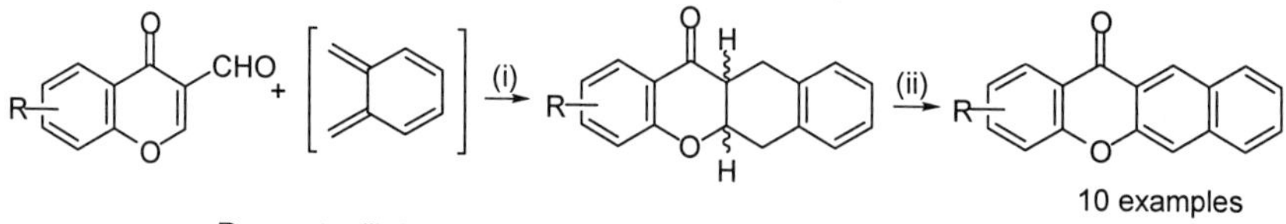

Reagents: (i) 1,2,4-trichlorobenzene, 250 °C, N_2; (ii) I_2, DMSO, heat

6.4.3 HETEROCYCLES CONTAINING ONE SULFUR ATOM

6.4.3.1 Thiopyrans and analogues

1,6-Diynes react with CS_2 and isothiocyanates in a Ru-catalysed [2+2+2] cycloaddition to yield cyclopenta[c]thiopyran 2-thiones and 2-imines respectively <02JA28>. Cycloaddition of CS$_2$ with 1,4-dilithio-1,3-dienes gives thiopyran-2-thiones, but cleavage of a C=S bond and formation of thiophene derivatives competes <02TL3533>.

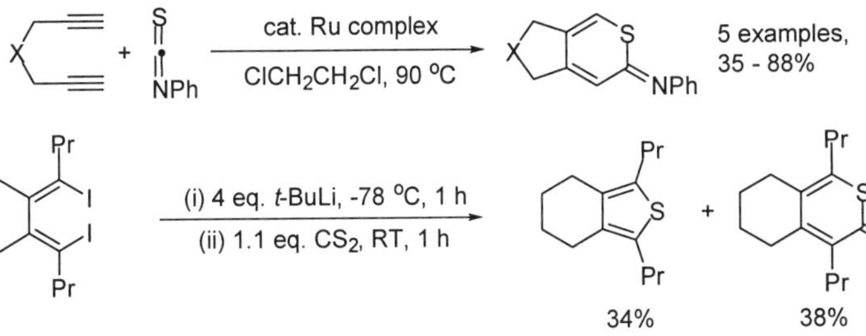

Bisenamination followed by heterocyclisation with SCl$_2$ has been used to synthesise 1-methyl-2-thiaadamantane **21**, found in petroleum, <02CC1750> and the 9-thiatricyclo[4.3.1.1]undecane ring system **22**, the *anti, anti* 2,7-diol derivative of which is the first heterocycle to form ellipsoidal clathrates <02JOC3221>.

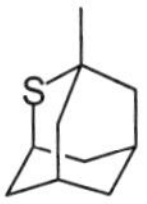

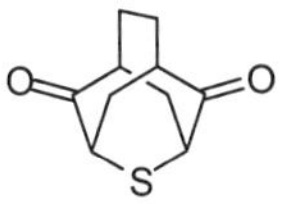

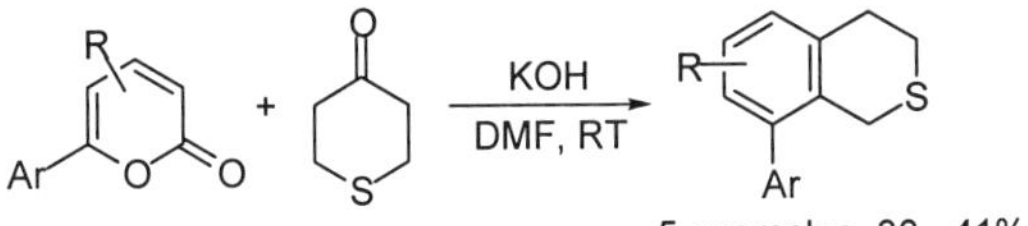

Pyran-2-ones react with the anions derived from tetrahydrothiopyran-4-one and thiochroman-4-one probably at C-6, initiating a ring transformation that culminates in isothiochromans (Scheme 8) <02JCS(P1)1426>.

Conditions have been elucidated to control the stereochemistry of the aldol reaction between tetrahydro-4*H*-thiopyran-4-one and keto-protected tetrahydro-4-oxothiopyran-3-aldehydes; a 4-ketal group is a valuable control element <02JOC1618>.

3-Hydroxyalkyl derivatives of seleno- and thio- chromones can be formed by the reaction of aldehydes with appropriate propynones in which an intramolecular Michael reaction is followed by an aldol condensation (Scheme 9) <02TL7039>.

Various substituted naphtho[2,1-*b*]thiopyran-1'-ylidene-9*H*-thioxanthene derivatives **23** have been synthesised using diazo-thioketone coupling as the key step. The compounds function as light-driven molecular motors and studies of the photochemical and thermal isomerisation processes indicate a four-step unidirectional 360° rotation about the double bond of the upper half of the molecule with respect to the lower half <02JA5037>.

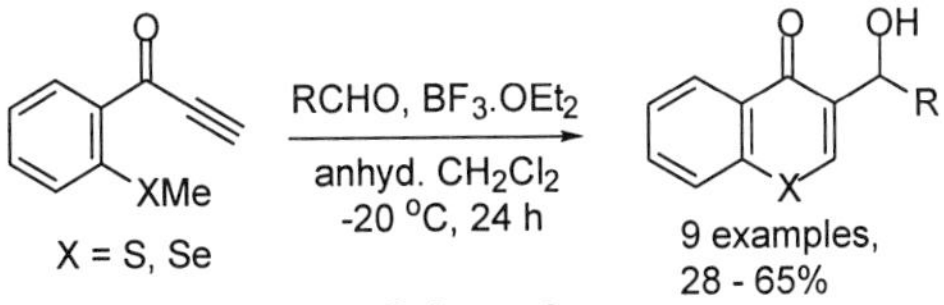

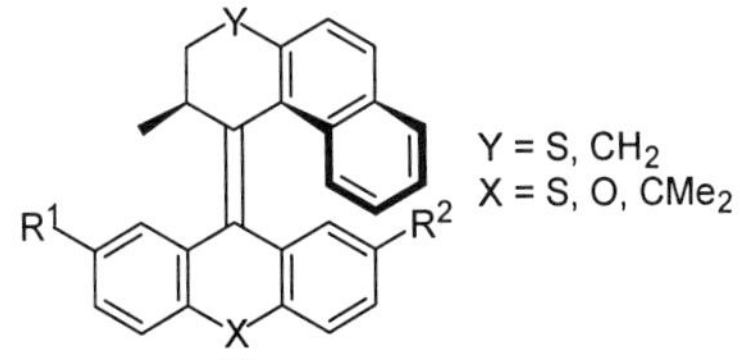

6.4.4 HETEROCYCLES CONTAINING TWO OR MORE OXYGEN ATOMS

6.4.4.1 Dioxins and dioxanes

A synthesis of the 2,3-dioxabicyclo[3.3.1]nonane system **24**, which is present in the antimalarial yingzhaosu, involves an initial thiol-limonene co-oxygenation reaction followed by reduction of a hydroperoxide by PPh_3 and a sulfenyl → sulfonyl oxidation <02T2449>.

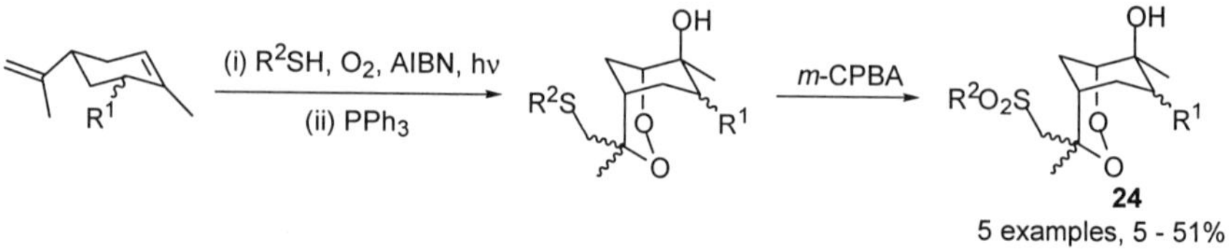

The cyclic peroxide system was constructed by photo-addition of O_2 to the diene **25** in a total synthesis of plakortic acid derivatives isolated from a marine sponge <02OL485> and the key step in a synthesis of related marine products is an intramolecular Michael addition of a peroxy hemiacetal <02BMCL69>.

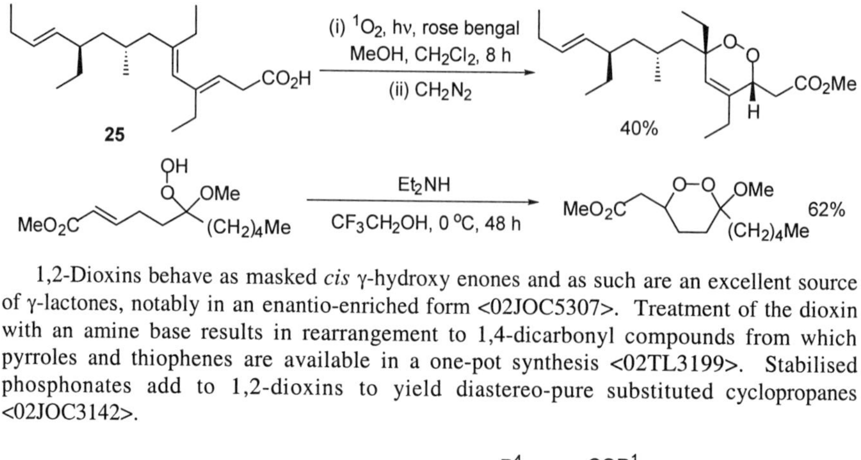

1,2-Dioxins behave as masked *cis* γ-hydroxy enones and as such are an excellent source of γ-lactones, notably in an enantio-enriched form <02JOC5307>. Treatment of the dioxin with an amine base results in rearrangement to 1,4-dicarbonyl compounds from which pyrroles and thiophenes are available in a one-pot synthesis <02TL3199>. Stabilised phosphonates add to 1,2-dioxins to yield diastereo-pure substituted cyclopropanes <02JOC3142>.

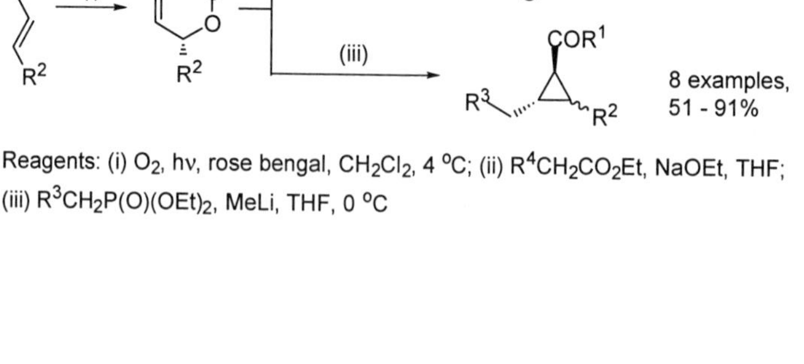

Reagents: (i) O_2, hv, rose bengal, CH_2Cl_2, 4 °C; (ii) $R^4CH_2CO_2Et$, NaOEt, THF; (iii) $R^3CH_2P(O)(OEt)_2$, MeLi, THF, 0 °C

The low-temperature, Mn(III)-oxidation of unsaturated derivatives of Meldrum's acid provides a route to cyclohexenes <02T25> and the readily available 5-substituted Meldrum's acids are a convenient source of α-substituted acrylate esters <02JOC7365>. Flash vacuum thermolysis of 5-aminomethylene derivatives of Meldrum's acid **26** generates iminopropadienones, RN=C=C=C=O <02JOC8558>.

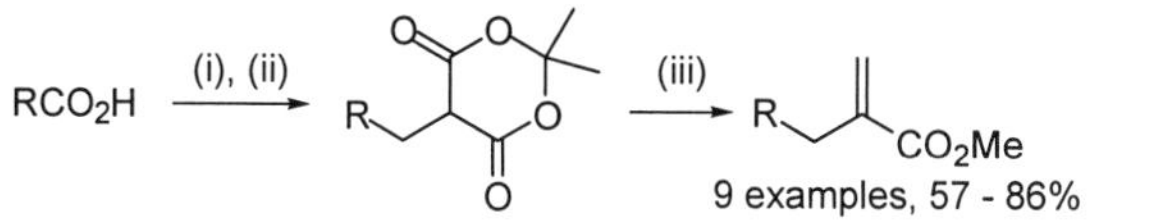

Reagents: (i) Meldrum's acid, DCC, DMAP, CH$_2$Cl$_2$, -5 °C; (ii) NaBH$_4$, AcOH, CH$_2$Cl$_2$, -5 °C; (iii) Me$_2$N$^+$=CH$_2$ I$^-$, MeOH, 65 °C

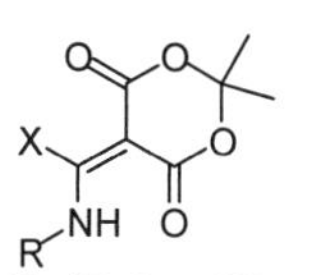

X = MeS or NMe$_2$
26

6.4.4.2 Trioxanes

Attention continues to be focused on 1,2,4-trioxanes because of their antimalarial properties, with much of the work being devoted to interconversions within the artemisinins.

10-Trifluoromethyldihydroartemisinin **27** is a source of 10α-(trifluoromethyl)-deoxoartemisinin by reduction of the 10-bromo derivative and of 10-trifluoromethyl-anhydrodihydroartemisinin **28** <02OL757>. This glycal undergoes contraction of the D-ring through rearrangement of the derived epoxide in hexafluoropropan-2-ol (HFIP), affording the trifluoromethyl ketone **29** as a single stereoisomer <02JOC1253>. Allylic bromination of both this glycal and its non-fluorinated analogue is facile and the allylic bromine atom is readily displaced by C-, N- and O-nucleophiles giving novel 16-substituted artemisinin derivatives <02TL7837>.

The 10-sulfones derived from the reaction of dihydroartemisinin with thiophenol and subsequent oxidation react with organozinc reagents to give 10β-substituted

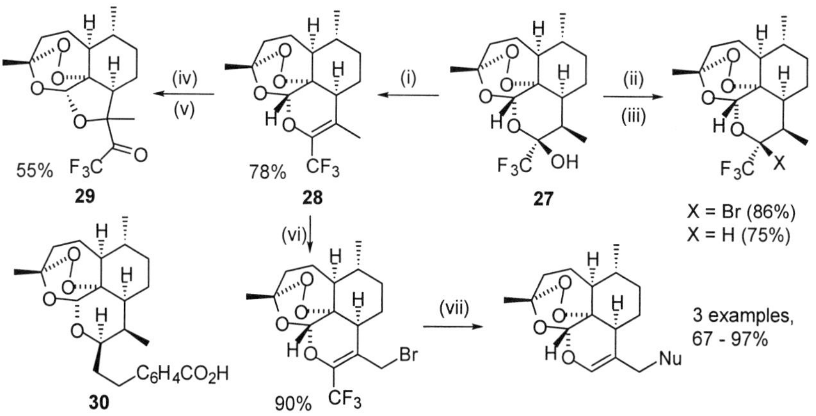

Reagents: (i) SOCl$_2$, py, 0 °C; (ii) SOBr$_2$, py, -30 °C; (iii) HSnBu$_3$, PhMe, heat; (iv) *m*-CPBA, CH$_2$Cl$_2$; (v) HFIP, RT, 1 h; (vi) NBS, CCl$_4$, heat; (vii) NuH, THF

deoxoartemisinins <02TL2891>. A variety of C-10 esters and ethers derived from dihydroartemisinin have been reported and the structures of 10α- and 10β- artesunate established <02EJO113>.

Conversion of artemisinin acid into (+)-deoxoartemisininelinic acid **30** which is water-soluble, hydrolytically stable and shows high antimalarial activity, involves epoxidation by an *S*-ylide and a photooxidative cyclisation <02JMC4940>.

Discussions of the mode of action of artemisinins have been published <02ACR167, 02ACR255> and the mechanism of the spontaneous autooxidation of dihydroartemisinic acid has been shown to involve four discrete steps of which three are assisted by the 12-carboxyl group <02T897, 02T909>. Studies of the reaction of simple 1,2,4-trioxanes related to artemisinin with Mn(II) tetraphenylporphyrin suggest that alkylation is a key factor in the antimalarial activity of the endoperoxide drugs <02JOC609>. The peroxide function of artemisinin is activated by iron(II) heme and generates a radical which then alkylates the heme <02CC414>.

Cyclohexanone is the starting point for the synthesis of thermally stable 3-carboxyphenyltrioxanes that are water-soluble and exhibit good antimalarial properties. Photooxidative cyclisation generates the trioxane ring system with almost exclusive formation of the 12α-stereoisomers <02JMC3824>. Photo-oxygenation of geraniol gives rise to hydroperoxides from which 1,2,4-trioxanes can be synthesised through treatment with ketones <02BMCL1913>. The β-hydroxyperoxyalcohol **31**, is peroxyacetalised on treatment with ketones in the presence of BF_3 etherate, yielding *trans*-5,6-disubstituted trioxanes. Aldehydes yield the trimer admixed with some 1,2,4-trioxane <02OL4193>.

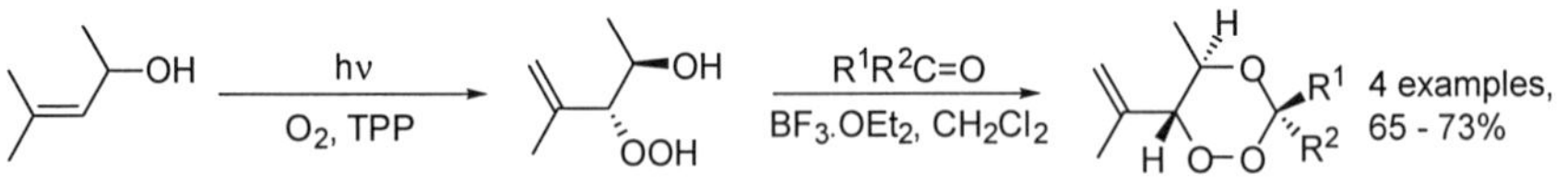

6.4.4.3 Tetraoxanes

The Baeyer-Villiger oxidation of cyclohexanone to ε-caprolactone using H_2O_2 in $(CF_3)_2CHOH$ proceeds *via* 7,8,15,16-tetraoxadispiro[5.2.5.2]hexadecane **32**. The tetraoxane can be isolated and rearranges to the lactone on treatment with catalytic amounts of 4-TsOH <02AG(E)4481>.

1,2,4,5-Tetraoxanes that show good antimalarial activity are formed when the steroid-derived *geminal*-dihydroperoxide **33** is treated with ketones <02JMC3331>.

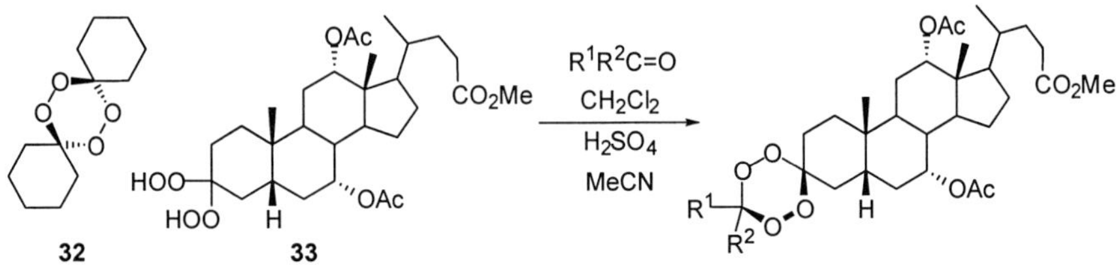

6.4.5 HETEROCYCLES CONTAINING TWO OR MORE SULFUR ATOMS

6.4.5.1 Dithianes and Trithianes

Fischer carbene complexes of Cr and W insert into the 3,4-bond of the 1,2-dithiole-3-thiones **34** to give 1,2-dithiin derivatives <02TL8037>.

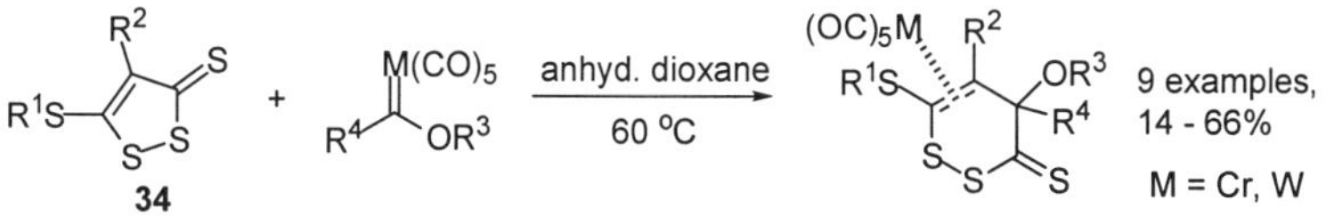

Ring expansion and concomitant chlorination of 1,3-dithiolanes with an enolisable methyl group occurs in DMSO on reaction with silica pre-treated with $SOCl_2$, giving dihydro-1,4-dithiin derivatives. Spirocyclic dithioacetals yield the annulated dithiin <02JOC2572>.

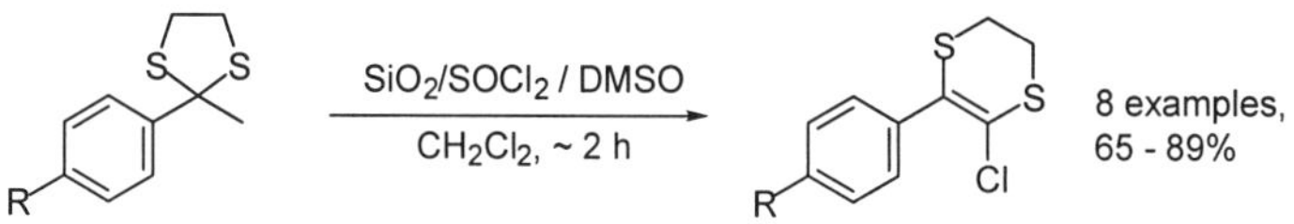

The central ring of both dibenzo[1,4]- dithiins and -oxathiins is cleaved on treatment with Li and a catalytic amount of 4,4″-di-*tert*-butylbiphenyl (DTBB) to afford thiols after reaction of the dilithio intermediate with electrophiles. In certain instances, the initial product can be cyclised to the dibenzo- dithiepine and -oxathiepine <02CL726>. The dilithio salt from thianthrene reacts sequentially with two different carbonyl compounds to give a 1,2-di(hydroxyalkyl)benzene. When CO_2 is used as the second electrophile, a phthalan results <02TL7205>.

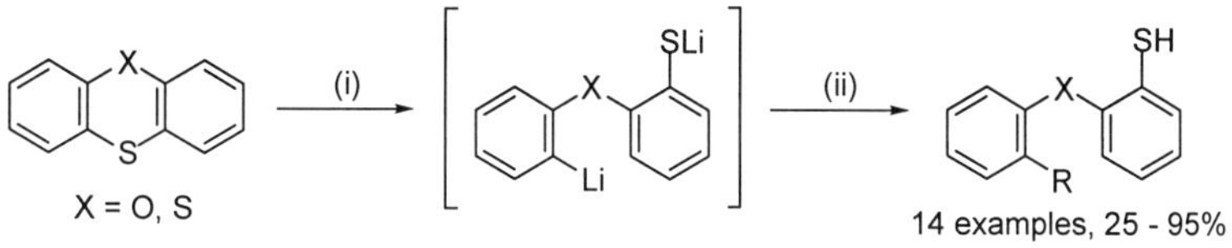

Reagents: (i) Li, 7.5 mol% DTBB, THF, -90 °C, (ii) electrophile, -78 °C then aq. HCl, RT

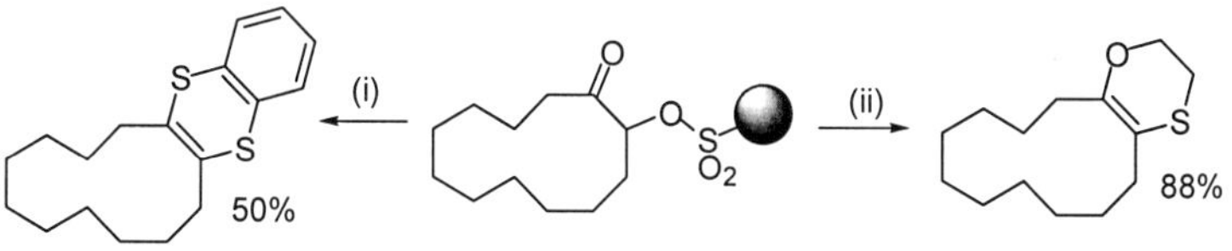

Reagents: (i) 1,2-benzenedithiol, PPTS, PhH, heat; (ii) $HOCH_2CH_2SH$, PPTS, PhH, heat

The reaction of bisnucleophiles with α-sulfonyloxy ketones bound to a resin results in the release of heterocycles; 1,2-benzenedithiol gives 1,4-benzodithiins, 2-mercaptoethanol yields 1,4-oxathiins and 1,2-diols afford 1,4-dioxins <02JA5718>.

6.4.6 HETEROCYCLES CONTAINING BOTH OXYGEN AND SULFUR IN THE SAME RING

6.4.6.1 Oxathianes

4-Mercapto-1,3-diols are deprotonated at S by Et_3N and *in situ* tosylation at S results in cyclisation through S–O bond formation, yielding substituted 1,2-oxathianes as a single diastereomer <02JCS(P1)2282>.

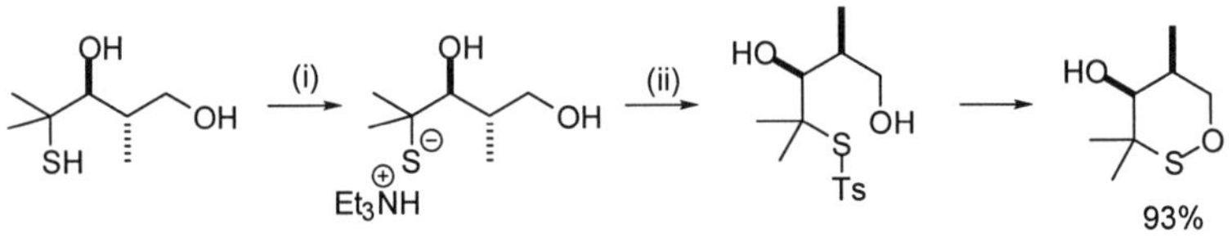

Reagents: (i) Et_3N, CH_2Cl_2; (ii) TsCl

Unsaturated sultones are obtained in high yield through RCM of unsaturated sulfonates using Grubbs' Ru catalysts (Scheme 10) <02SL2019>.

The diastereomerically pure camphor-based 1,3-oxathiane **36**, formed by the reaction of α,β-unsaturated aldehydes with the hydroxythiol **37**, undergoes a stereospecific ketene Claisen rearrangement with dichloroketene to give the stereo-pure 10-membered thiolactone <02CC2534>.

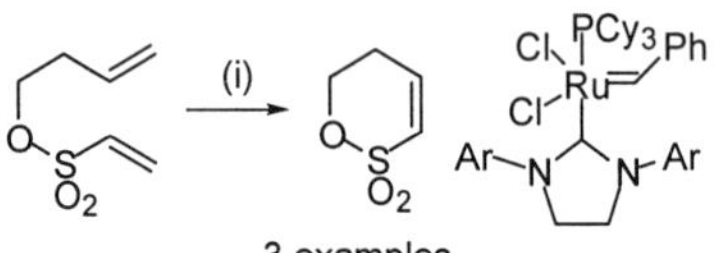

3 examples, 69 - 99%

35

Reagents: (i) catalyst **35**, CH_2Cl_2, 40 °C

Scheme 10

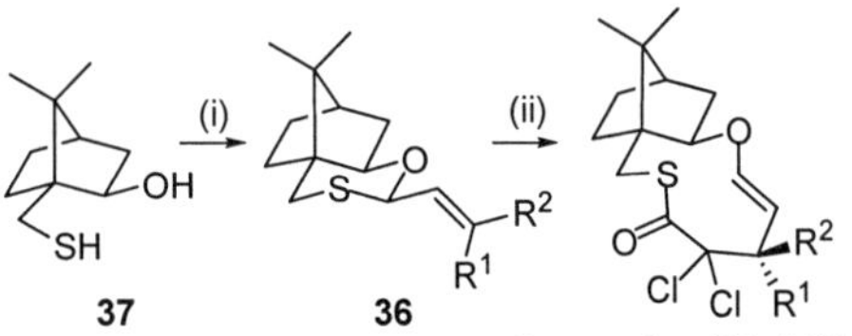

37 **36**

6 examples, 67 - 96%

Reagents: (i) R^1R^2C=CHCHO, PPTS, CH_2Cl_2, RT; (ii) CCl_2HCOCl, Et_3N, hexane, RT

o-Thioquinones undergo [4+2] cycloadditions with alkynes and arylalkenes <02H(56)471> and with fulvenes <02T3235> to give 1,4-benzoxathiins.

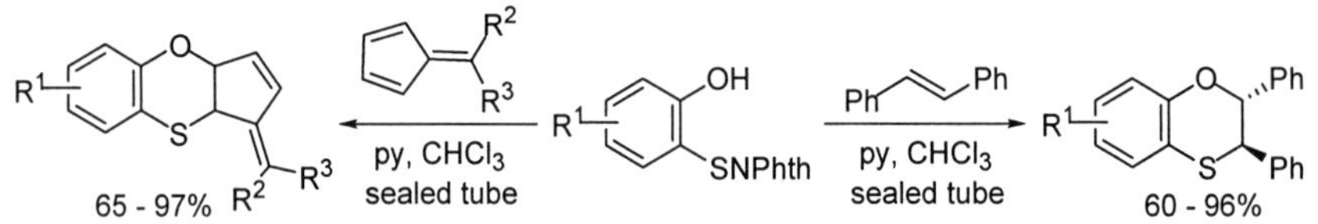

The Cu-catalysed ring expansion of 2-substituted 1,3-oxathiolanes with ethyl (triethylsilyl)diazoacetate affords 2,3-disubstituted 1,4-oxathianes with variable diastereoselectivity; the intermediacy of a Cu carbene and of a S-ylide is proposed <02CC346>.

6.4.6 REFERENCES

01COC571	V.J. Ram, P. Srivastava, *Curr. Org. Chem.* **2001**, *5*, 571.
02ACR167	A. Robert, O. Dechy-Cabaret, J. Cazelles, B. Meunier, *Acc. Chem. Res.* **2002**, *35*, 167.
02ACR255	Y. Wu, *Acc. Chem. Res.* **2002**, *35*, 255.
02AG(E)1171	T. Rosenau, A. Potthast, A. Hofinger, P. Kosma, *Angew. Chem., Int. Ed.* **2002**, *41*, 1171.
02AG(E)4481	A. Berkessel, M.R.M. Andreae, H. Schmickler, J. Lex, *Angew. Chem., Int. Ed.* **2002**, *41*, 4481.
02AG(E)4569	D.A. Evans, H.A. Rajapakse, D. Stenkamp, *Angew. Chem., Int. Ed.* **2002**, *41*, 4569.
02AG(E)4573	D.A. Evans, H.A. Rajapakse, A. Chiu, D. Stenkamp, *Angew. Chem., Int. Ed.* **2002**, *41*, 4573.
02AG(E)4633	K.-S. Yeung, I. Paterson, *Angew. Chem., Int. Ed.* **2002**, *41*, 4633.
02AHC(81)253	B. Stanovnik, M. Tisler, A.R. Katritzky, O. V. Denisko, *Adv. Heterocycl. Chem.* **2002**, *81*, 253.
02BMCL69	N. Murakami, M. Kawanishi, S. Itagaki, T. Horii, M. Kobayashi, *Bioorg. Med. Chem. Lett.* **2002**, *12*, 69.
02BMCL1913	C. Singh, N. Gupta, S.K. Puri, *Bioorg. Med. Chem. Lett.* **2002**, *12*, 1913.
02BMCL2039	A.B. Smith, III, R.M. Corbett, G.R. Pettit, J.-C. Chapuis, J. M. Schmidt, E. Hamel, M. K. Jung, *Bioorg. Med. Chem. Lett,* **2002**, *12*, 2039.
02CC346	M. Ioannou, M.J. Porter, F. Saez, *Chem. Commun.* **2002**, 346.
02CC414	A. Robert, Y. Coppel, B. Meunier, *Chem. Commun.* **2002**, 414.
02CC1750	S. Hanin, P. Adam, I. Kowalewski, A.-Y. Huc, B. Carpentier, P. Albrecht, *Chem. Commun.* **2002**, 1750.
02CC2534	V.K. Aggarwal, A. Lattanzi, D. Fuentes, *Chem. Commun.* **2002**, 2534.
02CC3070	D.R. Boyd, N.D. Sharma, P.L. Loke, J.F. Malone, W.C. McRoberts, J.T.G. Hamilton, *Chem. Commun.* **2002**, 3070.
02CEJ1670	P. Wipf, Y. Uto, S. Yoshimura, *Chem. Eur. J.* **2002**, *8*, 1670.
02CEJ4272	D. Enders, J.L. Vicario, A. Job, M. Wolberg, M. Müller, *Chem. Eur. J.* **2002**, *8*, 4272.
02CL726	M. Yus, F. Foubelo, J. V. Ferrández, *Chem. Lett.* **2002**, 726.
02CR4303	H.-J. Knölker, K.R. Reddy, *Chem. Rev.* **2002**, *102*, 4303.
02EJO113	R.K. Haynes, H.-W. Chan, M.-K. Cheung, W.-L. Lam, M.-K. Soo, H.-W. Tsang, A. Voerste, I.D. Williams, *Eur. J. Org. Chem.* **2002**, 113.
02EJO514	A. Martel, S. Leconte, G. Dujardin, E. Brown, V. Maisonneuve, R. Retoux, *Eur. J. Org. Chem.* **2002**, 514.
02EJO1063	M. Biagetti, F. Bellina, A.Carpita, S. Viel, L. Mannina, R. Rossi, *Eur. J. Org. Chem.* **2002**, 1063.
02EJO3126	R.W.M. Aben, R. de Gelder, H.W. Scheeren, *Eur. J. Org. Chem.* **2002**, 3126.
02H(56)471	V. Nair, B. Mathew, *Heterocycles* **2002**, *56*, 471.
02H(56)633	T. Honda, N. Saito, *Heterocycles* **2002**, *56*, 633.
02H(57)1033	M.-C. Yan, Y.-J. Jang, W.-U. Kuo, Z. Tu, K.-H. Shen, T.-S. Cuo, C.-H. Ueng, C.-F. Yao, *Heterocycles* **2002**, *57*, 1033.
02H(57)1915	K. Kobayashi, A. Matsunaga, M. Mano, O. Morikawa, H. Konishi *Heterocycles* **2002**, *57*, 1915.
02H(58)147	V. Nair, A.U. Vinod, R. Ramesh, R.S. Menon, L. Varma, S. Mathew, A. Chiaroni, *Heterocycles* **2002**, *58*, 147.
02H(58)659	A.K. Ghosh, D. Shin, G. Schiltz, *Heterocycles* **2002**, *58*, 659.
02JA10	J. Long, J. Hu, X. Shen, B. Ji, K. Ding, *J. Am. Chem. Soc.* **2002**, *124*, 10.
02JA28	Y. Yamamoto, H. Takagishi, K. Itoh, *J. Am. Chem. Soc.* **2002**, *124*, 28.

02JA184	R. Imbos, A.J. Minnaard, B.L. Feringa, *J. Am. Chem. Soc.* **2002**, *124*, 184.
02JA287	B. Engels, J. C. Schoneboom, A. F. Münster, S. Groetsch, M. Christi, *J. Am. Chem. Soc.* **2002**, *124*, 287.
02JA764	N. Asao, T. Nogami, K. Takahashi, Y. Yamamoto, *J. Am. Chem. Soc.* **2002**, *124*, 764.
02JA3562	I. Kadota, A. Ohno, K. Matsuda, Y. Yamamoto, *J. Am. Chem. Soc.* **2002**, *124*, 3562.
02JA4149	Y. Sheng, D.G. Musaev, K.S. Reddy, F.E. McDonald, K. Morokuma, *J. Am. Chem. Soc.* **2002**, *124*, 4149.
02JA4956	P.A. Wender, S.G. Hegde, R.D. Hubbard, L. Zhang, *J. Am. Chem. Soc.* **2002**, *124*, 4956.
02JA5037	N. Koumura, E.M. Geertsema, M.B. van Gelder, A. Meetsma, B.L. Feringa, *J. Am. Chem. Soc.* **2002**, *124*, 5037.
02JA5654	D.A. Evans, E. Hu, J.D. Burch, G. Jaeschke, *J. Am. Chem. Soc.* **2002**, *124*, 5654.
02JA5661	M.T. Crimmins, J.D. Katz, D.G. Washburn, S.P. Allwein, L.F. McAtee, *J. Am. Chem. Soc.* **2002**, *124*, 5661.
02JA5718	K.C. Nicolaou, T. Montagnon, T. Ulven, P.S. Baran, Y.-L. Zhong, F. Sarabia, *J. Am. Chem. Soc.* **2002**, *124*, 5718.
02JA5958	M.T. Crimmins, M.G. Stanton, S.P. Allwein, *J. Am. Chem. Soc.* **2002**, *124*, 5958.
02JA6824	T. Kondo, Y. Kanako, Y. Taguchi, A. Nakamura, T. Okada, M. Shiotsuki, Y. Ura, K. Wada, T. Mitsudo, *J. Am. Chem. Soc.* **2002**, *124*, 6824.
02JA7900	Y. Nishibayashi, Y. Inada, M. Hidai, S. Uemura, *J. Am. Chem. Soc.* **2002**, *124*, 7900.
02JA9662	Y. Huang, V.H. Rawal, *J. Am. Chem. Soc.* **2002**, *124*, 9662.
02JA13390	A.E. Sutton, B.A. Seigal, D.F. Finnegan, M.L. Snapper, *J. Am. Chem. Soc.* **2002**, *124*, 13390.
02JA13654	S.G. Nelson, W.S. Cheung, A.J. Kassick, M.A. Hilfiker, *J. Am. Chem. Soc.* **2002**, *124*, 13654.
02JA14983	H. Fuwa, N. Kainuma, K. Tachibana, M. Sasaki, *J. Am. Chem. Soc.* **2002**, *124*, 14983.
02JA15186	K.M. Brummond, H. Chen, P. Sill, L. You, *J. Am. Chem. Soc.* **2002**, *124*, 15186.
02JA15188	J. Louie, J.E. Gibby, M.V. Farnworth, T.N. Tekavec, *J. Am. Chem. Soc.* **2002**, *124*, 15188.
02JCS(P1)165	J.S. Yadav, B.V.S. Reddy, M. Aruna, C. Venugopal, T. Ramalingam, S.K. Kumar, A.C. Kunwar, *J. Chem. Soc., Perkin Trans. 1* **2002**, 165.
02JCS(P1)371	R.S. Mali, P.P. Joshi, P.K. Sandhu, A. Manekar-Tilve, *J. Chem., Soc. Perkin Trans. 1* **2002**, 371.
02JCS(P1)496	J.-Y. Goujon, A. Duval, B. Kirschleger, *J. Chem. Soc., Perkin Trans. 1* **2002**, 496.
02JCS(P1)565	Y. Jia, X. Li, P. Wang, B. Wu, X. Zhao, Y. Tu, *J. Chem. Soc., Perkin Trans. 1* **2002**, 565.
02JCS(P1)938	C.D. Donner, M. Gill, *J. Chem. Soc., Perkin Trans. 1* **2002**, 938.
02JCS(P1)1318	P.T. Kaye and X.W. Nocanda, *J. Chem. Soc., Perkin Trans. 1* **2002**, 1318.
02JCS(P1)1401	J.S. Yadav, B.V.S. Reddy, C.V. Rao, K.V. Rao, *J. Chem. Soc., Perkin Trans. 1* **2002**, 1401.
02JCS(P1)1426	V.J. Ram, N. Agarwal, A.S. Saxena, S. Farhanullah, A. Sharon, P.R. Maulik, *J. Chem. Soc., Perkin Trans. 1* **2002**, 1426.
02JCS(P1)2282	J. Eames, N. Kuhnert, S. Warren, *J. Chem. Soc., Perkin Trans. 1* **2002**, 2282.
02JCS(P1)2301	M.C. Elliott, *J. Chem. Soc., Perkin Trans. 1* **2002**, 2301.
02JCS(P1)2544	S.P. Nikas, G.A. Thakur, A. Makriyannis, *J. Chem. Soc., Perkin Trans. 1* **2002**, 2544.
02JCS(P1)2646	L. Caggiano, D.J. Fox, D. House, Z.A. Jones, F. Kerr, S. Warren, *J. Chem. Soc., Perkin Trans. 1* **2002**, 2646.
02JCS(P1)2652	D. House, F. Kerr, S. Warren, *J. Chem. Soc., Perkin Trans. 1* **2002**, 2652.
02JMC3331	B.A. Solaja, N. Terzic, G. Pocsfalvi, L. Gerena, B. Tinant, D. Opsenica, W.K. Milhous, *J. Med. Chem.* **2002**, *45*, 3331.
02JMC3824	G.H. Posner, H.B. Jeon, P. Ploypradith, I-H. Paik, K. Borstnik, S. Xie, T.A. Shapiro, *J. Med. Chem.* **2002**, *45*, 3824.
02JMC4940	M. Jung, K. Lee, H. Kendrick, B.L. Robinson, S.L. Croft, *J. Med. Chem.* **2002**, *45*, 4940.
02JOC609	J. Cazelles, A. Robert, B. Meunier, *J. Org. Chem.* **2002**, *67*, 609.
02JOC904	B.J. Shorthill, R.G. Granucci, D.R. Powell, T.E. Glass, *J. Org. Chem.* **2002**, *67*, 904.
02JOC1253	F. Grellepois, F. Chorki, B. Crousse, M. Ourévitch, D. Bonnet-Delpon, J.-P. Bégué, *J. Org. Chem.* **2002**, *67*, 1253.
02JOC1618	D.E. Ward, M. Sales, C.C. Man, J. Shen, P.K. Sasmal, C. Guo, *J. Org. Chem.* **2002**, *67*, 1618.
02JOC2175	B. Wang, X. Feng, Y. Huang, H. Liu, X. Cui, Y. Jiang, *J. Org. Chem.* **2002**, *67*, 2175.
02JOC2572	H. Firouzabadi, N. Iranpoor, H. Hazarkhani, B. Karimi, *J. Org. Chem.* **2002**, *67*, 2572.
02JOC2735	A.G.M. Barrett, F. Blaney, A.D. Campbell, D. Hamprecht, T. Meyer, A. J. P. White, D. Witty, D.J. Williams, *J. Org. Chem.* **2002**, *67*, 2735.
02JOC2751	J.A. Marshall, M.P. Bourbeau, *J. Org. Chem.* **2002**, *67*, 2751.
02JOC3142	M.C. Kimber, D.K. Taylor, *J. Org. Chem.* **2002**, *67*, 3142.
02JOC3221	S. Kim, R. Bishop, D.C. Craig, I.G. Dance, M.L. Scudder, *J. Org. Chem.* **2002**, *67*, 3221.
02JOC3301	M. Sasaki, T. Noguchi, K. Tachibana, *J. Org. Chem.* **2002**, *67*, 3301.

02JOC3494	I. Kadota, H. Takamura, K. Sato, Y. Yamamoto, *J. Org. Chem.* **2002**, *67*, 3494.
02JOC3941	J. Thibonnet, M. Abarbri, J.-L. Parrain, A. Duchêne, *J. Org. Chem.* **2002**, *67*, 3941.
02JOC5138	T. Kawasaki, Y. Yamamoto, *J. Org. Chem.* **2002**, *67*, 5138.
02JOC5307	B.W. Greatrex, M.C. Kimber, D.K. Taylor, G. Fallon, E.R.T. Tiekink, *J. Org. Chem.* **2002**, *67*, 5307.
02JOC5773	A. Alcudia, R.G. Arrayàs, L.S. Liebeskind, *J. Org. Chem.* **2002**, *67*, 5773.
02JOC5969	H. Hagiwara, K. Kobayashi, S. Miya, T. Hoshi, T. Suzuki, M. Ando, T. Okamoto, M. Kobayashi, I. Yamamoto, S. Ohtsubo, M. Kato, H. Uda, *J. Org. Chem.* **2002**, *67*, 5969.
02JOC6104	S. Ma, W. Gao, *J. Org. Chem.* **2002**, *67*, 6104.
02JOC6911	R.M. Jones, C. Selenski, T.R.R. Pettus, *J. Org. Chem.* **2002**, *67*, 6911.
02JOC7238	D. Amantini, F. Fringuelli, F. Pizzo, *J. Org. Chem.* **2002**, *67*, 7238.
02JOC7303	S. Arimori, R. Kouno, T. Okauchi, T. Minami, *J. Org. Chem.* **2002**, *67*, 7303.
02JOC7365	B. Hin, P. Majer, T. Tsukamoto, *J. Org. Chem.* **2002**, *67*, 7365.
02JOC8280	S. Ma, B. Ni, *J. Org. Chem.* **2002**, *67*, 8280.
02JOC8558	M. Shtaiwi, C. Wentrup, *J. Org. Chem.* **2002**, *67*, 8558.
02JOC8771	A.D. William, Y. Kobayashi, *J. Org. Chem.* **2002**, *67*, 8771.
02OL367	T. Sunazuka, M. Handa, K. Nagai, T. Shirahata, Y. Harigaya, K. Otoguro, I. Kuwajima, S. Omura, *Org. Lett.* **2002**, *4*, 367.
02OL391	I. Paterson, D.Y.-K. Chen, A.S. Franklin, *Org. Lett.* **2002**, *4*, 391.
02OL455	Y. Nagao, A. Ueki, K. Asano, S. Tanaka, S. Sano, M. Shiro, *Org. Lett.* **2002**, *4*, 455.
02OL485	G. Yao, K. Steliou, *Org. Lett.* **2002**, *4*, 485.
02OL749	W.W. Cutchins, FE. McDonald, *Org. Lett.* **2002**, *4*, 749.
02OL757	F. Chorki, F. Grellepois, B. Crousse, V.D. Huang, N.V. Hung, D. Bonnet-Delpon, J.-P. Bégué, *Org. Lett.* **2002**, *4*, 757.
02OL1221	Y. Yamashita, S. Saito, H. Ishitani, S. Kobayashi, *Org. Lett.* **2002**, *4*, 1221.
02OL1747	M. Sasaki, C. Tsukano, K. Tachibana, *Org. Lett.* **2002**, *4*, 1747.
02OL1795	G.D. Joly, E.N. Jacobsen, *Org. Lett.* **2002**, *4*, 1795.
02OL2565	A.M. Bernard, E. Cadoni, A. Frongia, P.P. Piras, F. Secci, *Org. Lett.* **2002**, *4*, 2565.
02OL2739	K. Suzuki, T. Nakata, *Org. Lett.* **2002**, *4*, 2739.
02OL2771	H. Takakura, M. Sasaki, S. Honda, K. Tachibana, *Org. Lett.* **2002**, *4*, 2771.
02OL3009	R.M. Adlington, J.E. Baldwin, A.V.W. Mayweg, G.J. Pritchard, *Org. Lett.* **2002**, *4*, 3009.
02OL3333	G. Zhu, J. Wu, R. Fathi, Z. Yang, *Org. Lett.* **2002**, *4*, 3333.
02OL3679	J.L.Portscheller, H.C. Malinakova, *Org. Lett.* **2002**, *4*, 3679.
02OL3719	E.B. Holson, W.R. Roush, *Org. Lett.* **2002**, *4*, 3719.
02OL3723	E.B. Holson, W.R. Roush, *Org. Lett.* **2002**, *4*, 3723.
02OL3797	S.P. Fearnley, M.W. Tidwell, *Org. Lett.* **2002**, *4*, 3797.
02OL3943	K. Susuki, T. Nakata, *Org. Lett.* **2002**, *4*, 3943.
02OL4193	A.G. Griesbeck, T.T. El-Idreesy, M. Fiege, R. Brun, *Org. Lett.* **2002**, *4*, 4193.
02OL4269	S. Höger, S. Rosselli, A.-D. Ramminger, V. Enkelmann, *Org. Lett.* **2002**, *4*, 4269.
02OL4285	T. Rosenau, A. Potthast, T. Elder, P. Kosma, *Org. Lett.* **2002**, *4*, 4285.
02OL4419	X. Huang, H. Zhou, *Org. Lett.* **2002**, *4*, 4419.
02OL4551	M. Inoue, H. Uehara, M. Maruyama, M. Hirama, *Org. Lett.* **2002**, *4*, 4551.
02S217	J.S. Yadav, B.V.S. Reddy, M. Aruna, M. Thomas, *Synthesis* **2002**, 217.
02S441	P. Langer, *Synthesis* **2002**, 441.
02S981	M. Kalesse, M. Christmann, *Synthesis* **2002**, 981.
02S2341	V.Y. Sosnovskikh, B.I. Usachev, *Synthesis* **2002**, 2341.
02S2490	Y. Kawamura, M. Maruyama, T. Tokuoka, M. Tsukayama, *Synthesis* **2002**, 2490.
02SC3667	J.-R. Labrosse, P. Lhoste, D. Sinou, *Synth. Commun.* **2002**, *32*, 3667.
02SL322	J.Y. Goujon, F. Zammattio, S. Pagnoncelli, Y. Boursereau, B. Kirschleger, *Synlett* **2002**, 322.
02SL796	A.M. Bernard, C. Floris, A. Frongia, P. P. Piras, *Synlett* **2002**, 796.
02SL1835	K. Fujiwara, Y. Koyama, K. Kawai, H. Tanaka, A. Murai, *Synlett* **2002**, 1835.
02SL2019	S. Karsch, P. Schwab, P. Metz, *Synlett* **2002**, 2019.
02T25	B.B. Snider, R.B. Smith, *Tetrahedron* **2002**, *58*, 25.
02T105	A. Sandulache, A.M.S. Silva, J.A.S. Cavaleiro, *Tetrahedron* **2002**, *58*, 105.
02T897	L.-K. Sy, G.D. Brown, *Tetrahedron* **2002**, *58*, 897.
02T909	L.-K. Sy, G.D. Brown, *Tetrahedron* **2002**, *58*, 909.
02T997	M. Shanmugasundaram, S. Manikandan, R. Raghunathan, *Tetrahedron* **2002**, *58*, 997.
02T1697	Y.X Jia, X. Li, B. Wu, X.Z. Zhao, Y.Q. Tu, *Tetrahedron* **2002**, *58*, 1697.
02T1779	M. Hirama, J.D. Rainier, *Tetrahedron* **2002**, *58*, 1779.

02T2449	E.E. Korshin, R. Hoos, A.M. Szpilman, L. Konstantinovski, G.H. Posner, M.D. Bachi, *Tetrahedron* **2002**, *58*, 2449.
02T2755	C. Mukai, H. Yamashita, M. Sassa, M. Hanaoka, *Tetrahedron* **2002**, *58*, 2755.
02T3235	V. Nair, B. Mathew, R.S. Menon, S. Mathew, M. Vairamani, S. Prabhakar, *Tetrahedron* **2002**, *58*, 3235.
02T4851	S. Chimichi, M. Boccalini, B. Cosimelli, *Tetrahedron* **2002**, *58*, 4851.
02T5023	M. Biagetti, F. Bellina, A. Carpita, P. Stabile, R. Rossi, *Tetrahedro* **2002**, *58*, 5023.
02T5215	T. Hattori, Y. Suzuki, Y. Ito, D. Hotta, S. Miyano, *Tetrahedron* **2002**, *58*, 5215.
02T5367	R.W. Van De Water, T.R.R. Pettus, *Tetrahedron* **2002**, *58*, 5367.
02T5627	H. Guo, R.J. Madhushaw, F.-M. Shen, R.-S. Liu, *Tetrahedron* **2002**, *58*, 5627.
02T5927	J.-Q. Wang, R.G. Harvey, *Tetrahedron* **2002**, *58*, 5927.
02T8929	H. Mizutani, M. Watanabe, T. Honda, *Tetrahedron* **2002**, *58*, 8929.
02T9007	F. Teply, I.G. Stará, I. Stary, A. Kollárovic, D. Saman, P. Fiedler *Tetrahedron*, **2002**, *58*, 9007.
02T10301	M. Anniyappan, D. Muralidharan, P.T. Perumal, *Tetrahedron* **2002**, *58*, 10301.
02T10375	S. Nakamura, J. Inagaki, T. Sugimoto, Y. Ura, S. Hashimoto, *Tetrahedron* **2002**, *58*, 10375.
02TL715	Y. Lu, L.Y. Foo, *Tetrahedron Lett.* **2002**, *43*, 715.
02TL1213	S. Hesse, G. Kirsch, *Tetrahedron Lett.* **2002**, *43*, 1213.
02TL1561	D. Banti, M. North, *Tetrahedron Lett.* **2002**, *43*, 1561.
02TL2149	R.J. Cox, R.M. Williams, *Tetrahedron Lett.* **2002**, *43*, 2149.
02TL2891	S. Lee, S. Oh, *Tetrahedron Lett.* **2002**, *43*, 2891.
02TL2899	M.E. Fox, I.C. Lennon, G. Meek, *Tetrahedron Lett.* **2002**, *43*, 2899.
02TL2999	J.S. Yadav, B.V.S. Reddy, C. Parisse, P. Carvalho, T.P. Rao, *Tetrahedron Lett.* **2002**, *43*, 2999.
02TL3199	C.E. Hewton, M.C. Kimber, D.K. Taylor, *Tetrahedron Lett.* **2002**, *43*, 3199.
02TL3285	I. Paterson, M.J. Coster, *Tetrahedron Lett.* **2002**, *43*, 3285.
02TL3381	J. Mulzer, M. Hanbauer, *Tetrahedron Lett.* **2002**, *43*, 3381.
02TL3533	J. Chen, Q. Song, Z. Xi, *Tetrahedron Lett.* **2002**, *43*, 3533.
02TL3565	C.-S. Li, E. Lacasse, *Tetrahedron Lett.* **2002**, *43*, 3565.
02TL3637	S.Cenini, G. Cravotto, G.B. Giovenzana, G. Palmisano, A. Penoni, S. Tollari, *Tetrahedron Lett.* **2002**, *43*, 3637.
02TL3967	F.R.-L. Guen, P. Le Poul, B. Caro, N. Faux, N. Le Poul, S. J. Green, *Tetrahedron Lett.* **2002**, *43*, 3967.
02TL4395	J. Wu, L. Wang, R. Fathi, Z. Yang, *Tetrahedron Lett.* **2002**, *43*, 4395.
02TL4507	G.E. Daia, C.D. Gabbutt, J.D. Hepworth, B.M. Heron, D.E. Hibbs, M.B. Hursthouse, *Tetrahedron Lett.* **2002**, *43*, 4507.
02TL4515	N. Thasana, S. Ruchirawat, *Tetrahedron Lett.* **2002**, *43*, 4515.
02TL4527	J.S. Yadav, B.V.S. Reddy, L. Chandraiah, B. Jagannadh, S.K. Kumar, A.C. Kunwar, *Tetrahedron Lett.* **2002**, *43*, 4527.
02TL5673	L. Anastasia, C. Xu, E. Negishi, *Tetrahedron Lett.* **2002**, *43*, 5673.
02TL6113	Y.-W. Kim, R.W. Breuggemeier, *Tetrahedron Lett.* **2002**, *43*, 6113.
02TL6893	A.B.C. Simas, L.F.O. Furtado, P.R.R. Costa, *Tetrahedron Lett.* **2002**, *43*, 6893.
02TL7039	T. Kataoka, H. Kinoshita, S. Kinoshita, T. Iwamura, *Tetrahedron Lett.* **2002**, *43*, 7039.
02TL7193	T.-P. Loh, J.-Y. Yang, L.-C. Feng, Y. Zhou, *Tetrahedron Lett.* **2002**, *43*, 7193.
02TL7205	M. Yus, F. Foubelo, J.V. Ferrández, *Tetrahedron Lett.* **2002**, *43*, 7205.
02TL7401	T. Yao, R.C. Larock, *Tetrahedron Lett.* **2002**, *43*, 7401.
02TL7837	F. Grellepois, F. Chorki, M. Ourévitch, B. Crousse, D. Bonnet-Delpon, J.-P. Bégué, *Tetrahedron Lett.* **2002**, *32*, 7837.
02TL7971	R. Chenevert, G. Courchesne, *Tetrahedron Lett.* **2002**, *43*, 7971.
02TL8037	A.M. Granados, J. Kreiker, R.H. de Rossi, *Tetrahedron Lett.* **2002**, *43*, 8037.
02TL8559	M. Prashad, Y. Liu, X.Y. Mak, D. Har, O. Repic, T.J. Blacklock, *Tetrahedron Lett.* **2002**, *43*, 8559.
02TL8583	S.P. Chavan, K. Shivasankar, R. Sivappa, R. Kale, *Tetrahedron Lett.* **2002**, *43*, 8583.
02TL8697	H. Mao, M. Koukni, T. Kozlecki, F. Compernolle, G.J. Hoornaert, *Tetrahedron Lett.* **2002**, *43*, 8697.
02TL8791	K.P. Cole, R.P. Hsung, *Tetrahedron Lett.* **2002**, *43*, 8791.
02TL9195	D.S. Bose, A.P. Rudradas, M.H. Babu, *Tetrahedron Lett.* **2002**, *43*, 9195.
02TL9397	E. Wada, H. Koga, G. Kumaran, *Tetrahedron Lett.* **2002**, *43*, 9397.

Chapter 7

Seven-Membered Rings

John B. Bremner
Institute for Biomolecular Science, Department of Chemistry
University of Wollongong, Wollongong, NSW 2522, Australia
john_bremner@uow.edu.au

7.1 INTRODUCTION

This review covers the years 2001-2002. Seven-membered heterocycles continued to be a focus of considerable activity over this period, with fused systems a particular highlight. Ring closing metathesis methodology has been adapted to further syntheses of seven-membered rings.

A major review on the synthesis and chemistry of 1,4-, 4,1-, and 1,5-benzoxazepines, covering the literature to the end of 2000, was published [01JHC1011]. The use of domino Wittig-pericyclic reactions in the synthesis of a range of bioactive heterocycles including 7-membered rings has also been reviewed [02MI1181].

Divisions in this Chapter are again based on the number of ring heteroatoms, with an emphasis on nitrogen, oxygen and sulfur. Sections on systems of pharmacological interest or significance, and on future directions, are also included.

7.2 SEVEN-MEMBERED SYSTEMS CONTAINING ONE HETEROATOM

While activity in this particular area has not been especially high, a number of interesting preparations and reactions have been described. As previously, the following material is divided into non-fused and fused examples.

7.2.1. Azepines and derivatives

A concise approach to substituted dihydroazepines **3** has been described involving the reaction of ylides, generated *in situ* from styryldiazoacetates (e.g. **1**), with imines **2**. In the case of **1** and the imine **2** (Ar = Ph, R = Me) the yield of the corresponding azepine **3** was 73% [01OL3741].

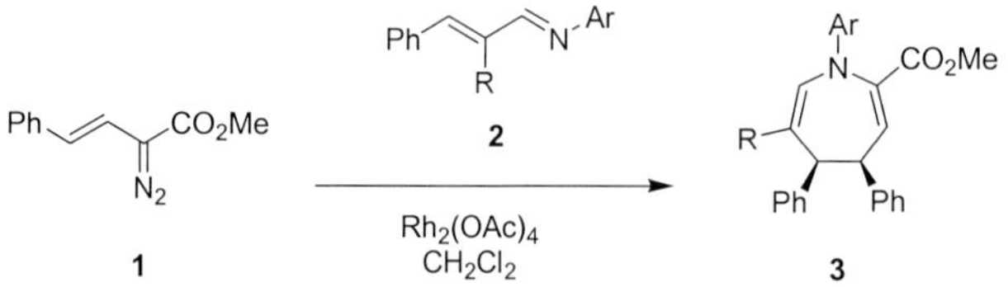

Heterocyclic phosphorus ylides (e.g. **5**, R = Me) have also been prepared, although in low yield, by flash vacuum pyrolysis of the corresponding open-chain ylide precursors **4** [01TL141].

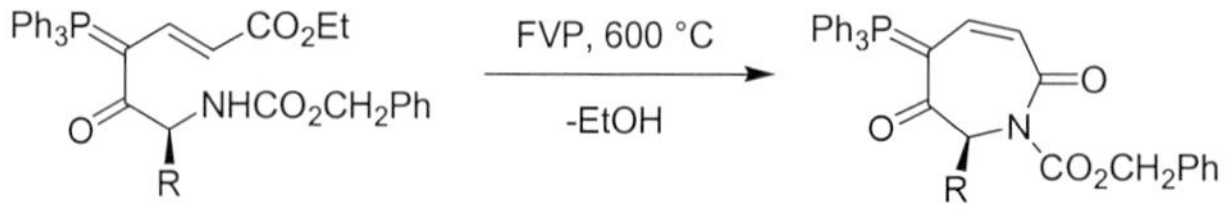

Access to more simply substituted azepine derivatives **7** and **8** has also been achieved by ruthenium-catalysed intramolecular hydroamination of the aminoalkyne **6** [01JOM149].

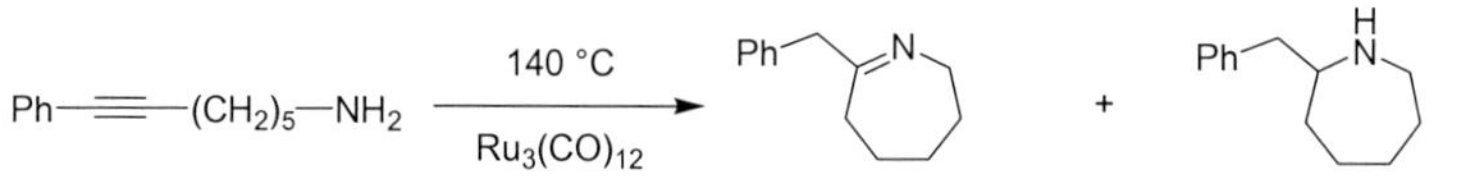

Ring closing metathesis using Grubbs' ruthenium catalyst has also been used in a novel preparation of the Fischer-type chromium carbene complex **10** from the precursor **9** in >98% yield [01SL757]. Similarly, the *N*-substituted tetrahydroazepine **12** could be accessed in near quantitative yield from **11** [01JOC3564].

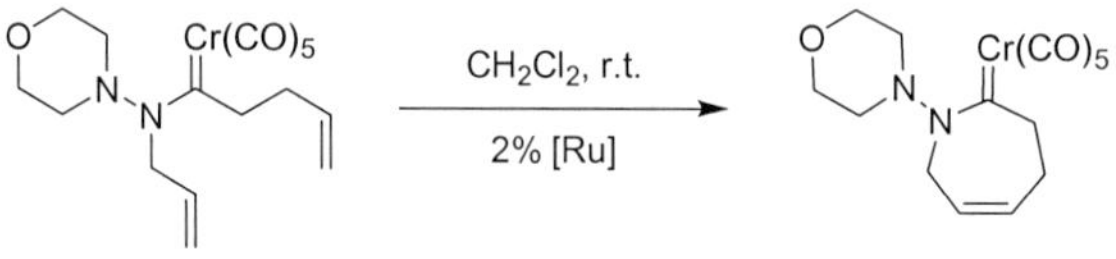

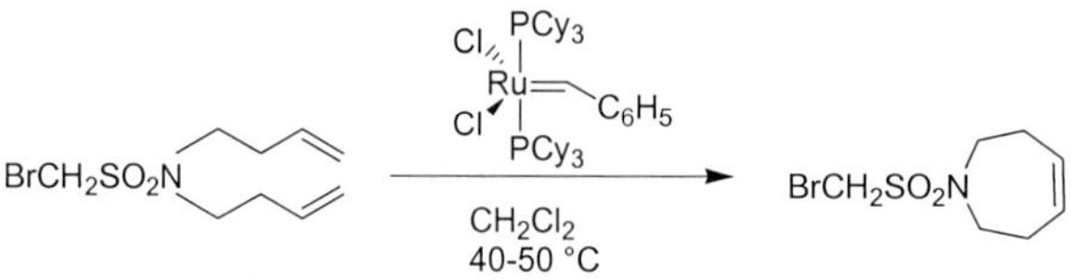

In an alternative approach to the azepine system, photolysis of **13** gave **15** in 14% yield, together with the fused derivative **14** [01H567].

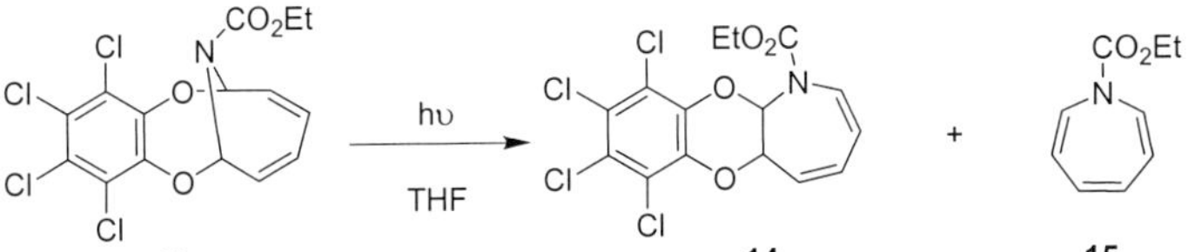

A nitrene-mediated ring expansion of the nitrobenzene **16** afforded the substituted 3*H*-azepine **17** in 74% yield; nucleophilic substitution by the appropriate alkoxide then gave the substituted analogues **18a-d** in yields ranging from 24 to 78% [02H223].

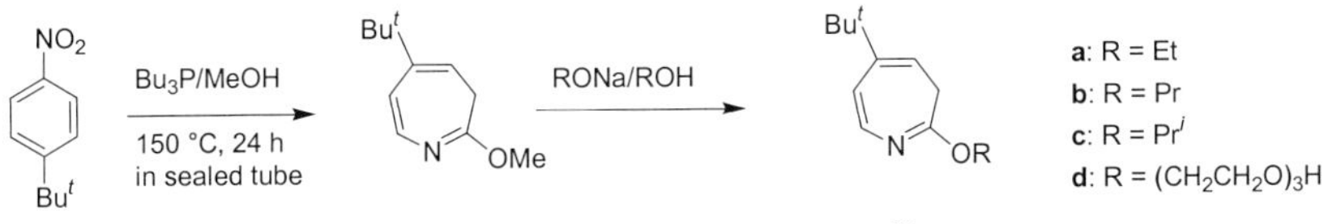

a: R = Et
b: R = Pr
c: R = Pr^i
d: R = (CH$_2$CH$_2$O)$_3$H

Other substituted azepinone derivatives (e.g. **20**, R = Me and **21**, R= Me) were made by Kim and co-workers from the 3-amino-azepanone **19** as part of a project on structure-property relationships with dermal penetration enhancers [01MI183].

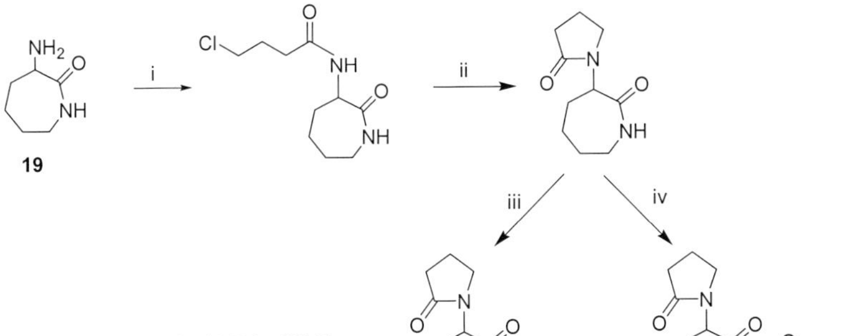

Reagents: (i), ClCO(CH$_2$)$_3$Cl, N(C$_2$H$_5$)$_3$, CH$_2$Cl$_2$;
(ii), KOC$_4$H$_9$(t), THF;
(iii), [(CH$_3$)$_2$Si]$_2$NK, toluene, R-Br;
(iv), [(CH$_3$)$_2$Si]$_2$NK, toluene, BrCH$_2$CO$_2$-R.

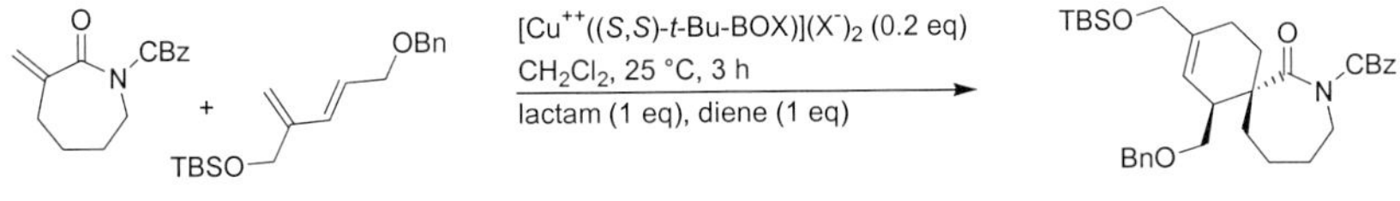

Murai *et al.* have also described the asymmetric synthesis of the spiroazepinone skeleton present in certain marine natural toxin by a Diels-Alder reaction. For example, **24** was prepared in 82% yield (96% ee; 99:1 *exo/endo* ratio) from **22** and the diene **23** with X = AsF$_6^-$ in the chiral copper complex [02SL403].

An elegant alternative approach to related spiro skeletons involved ring closing metathesis of **25a,b** to give **26a,b** in good yields using Grubbs' catalyst [02SL1827].

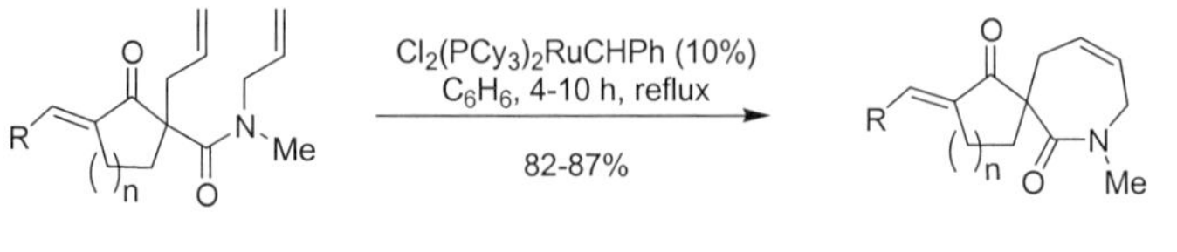

Cl$_2$(PCy$_3$)$_2$RuCHPh (10%)
C$_6$H$_6$, 4-10 h, reflux
82-87%

25a: n = 1; R = Phenyl
25b: n = 2; R = 2-Furyl

26a: n = 1; R = Phenyl
26b: n = 2; R = 2-Furyl

Amongst natural products with 7-membered nitrogen-containing rings, the azepinone bengamide Z (**27**) has been synthesised in chiral form by Boeckman *et al.* [02OL2109], the related ester, bengamide Q (**28**) together with other bengamides, has been isolated from a *Jaspis* species [01JOC1733], while the complex spirolide, 13-desmethyl-C (**29**), plus spirolides A and C, has been isolated from contaminated scallops and phytoplankton from Canada [01JNP308].

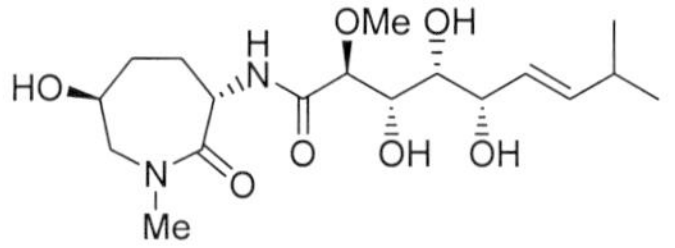

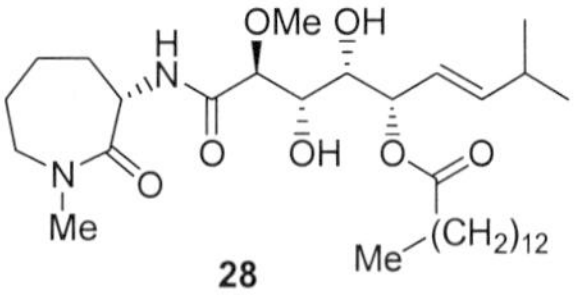

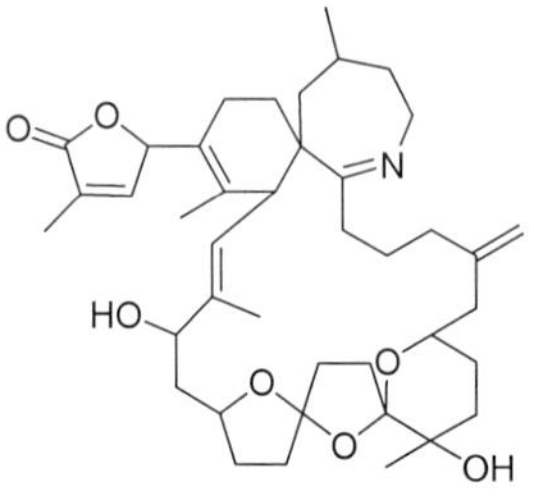

Control of regio- and stereochemistry in the synthesis of substituted azepanes has been achieved via piperidine ring expansion methodology and aziridinium ion intermediates. Thus, reaction of **30** with azide ion afforded **32** exclusively with backside attack by the azide ion at the methine carbon in the intermediate **31** being preferred [02JCS(P1)2080].

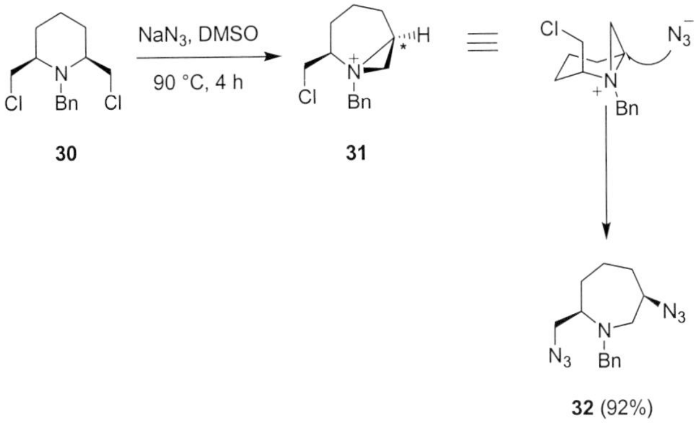

An alternative approach to functionalised azepine derivatives has been reported by Occhiato *et al.* [02JOC7144] involving a Pd-catalysed cross coupling reaction of the vinyl triflate **33** with the α-alkoxyboronate **34** (R = H) to give **35** in 45% yield; acid catalysed hydrolysis then afforded the azepine derivative **37** and the fused azepine **36** in low to fair yields respectively.

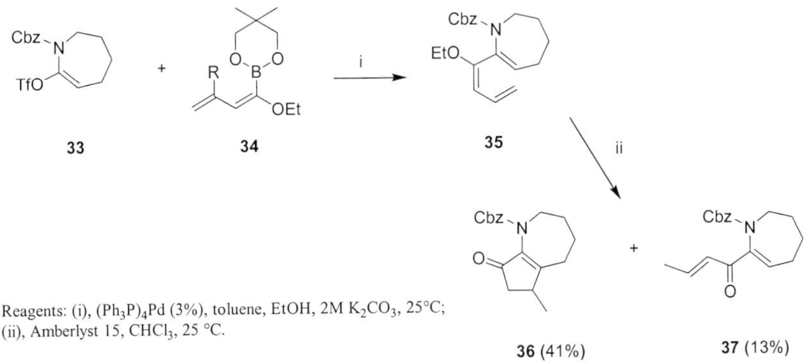

Reagents: (i), (Ph₃P)₄Pd (3%), toluene, EtOH, 2M K₂CO₃, 25°C; (ii), Amberlyst 15, CHCl₃, 25 °C.

7.2.2 Fused azepines and derivatives

The synthetic power of ring-closing metathesis technology is further illustrated by the synthesis of the cyclopentane-fused azepine **39** obtained in 78% yield from the triene precursor **38** using Grubbs' catalyst; the *N*-propargyl analogue of **38** could also be used, giving access to **40** in 63% yield [02SL1987].

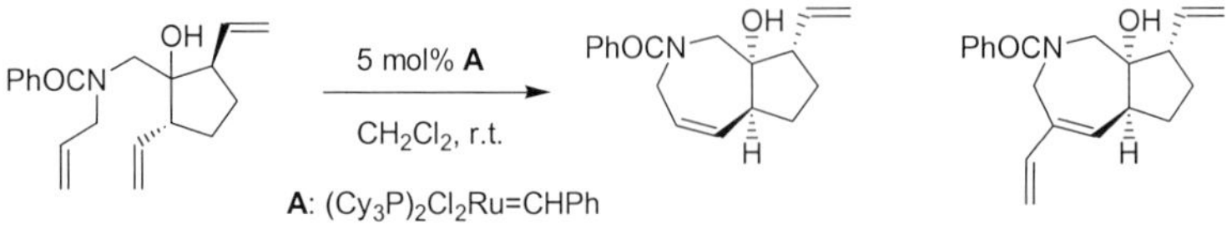

Ring construction approaches have also been used effectively in the diastereoselective synthesis of the fused azepine **43** from the pyrrolidinone **41** and (Z)-1,4-dichloro-2-butene **42** [01TA2205], and in the synthesis of the alkaloid 275A (**44**) from the Colombian poison frog *Dendrobates lehmanni* [01JNP421].

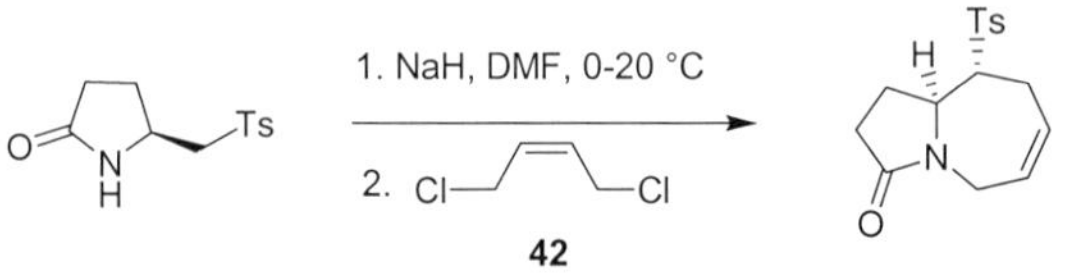

Eberbach and co-workers have reported [01EJO3313] a fascinating approach to benzazepinones **46** from the nitrone precursors **45**. Treatment of **45** with base under unusually mild basic conditions gave **46** in generally high yields e.g. **46**, $R^1 = R^3 = H$, $R^2 =$ Ph; 84%). The overall transformation is the result of a complex series of steps proposed to include allene formation, 1,7-dipolar cyclisation and a series of bond cleavage and formation steps (via a cyclopropanone intermediate).

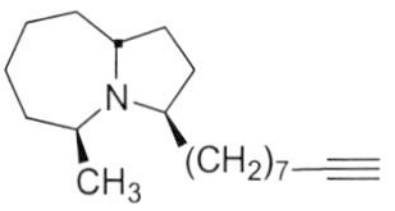

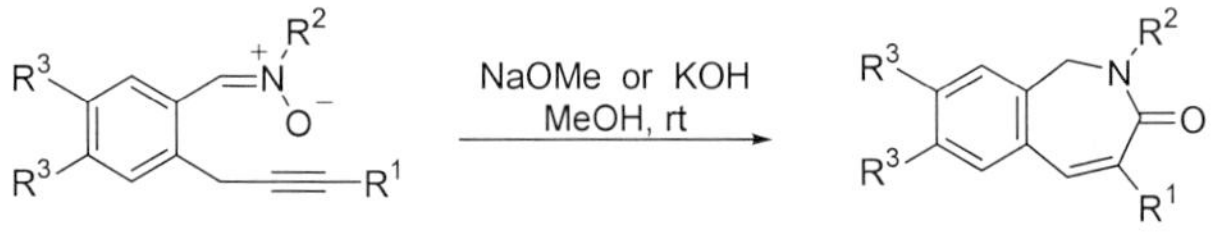

A more classical series of steps have been utilised by Sano *et al.* in the synthesis of the 2-benzazepines **53** from the aromatic aldehydes **47** and the amine **48** to give the imines **49**, followed by reduction to the amines **50**, *N*-formylation to **51**, and oxidation to the sulfoxides **52**. A modified Pummerer reaction on **52** (with trifluoroacetic acid and BF$_3$.Et$_2$O) was then used to complete the 7-membered ring in yields for this last step ranging from 45% to 78% (e.g. **53**, $R^1 = R^2 = H$, $R^3 = $ OMe, $R^4 = H$; 78%) [01H967].

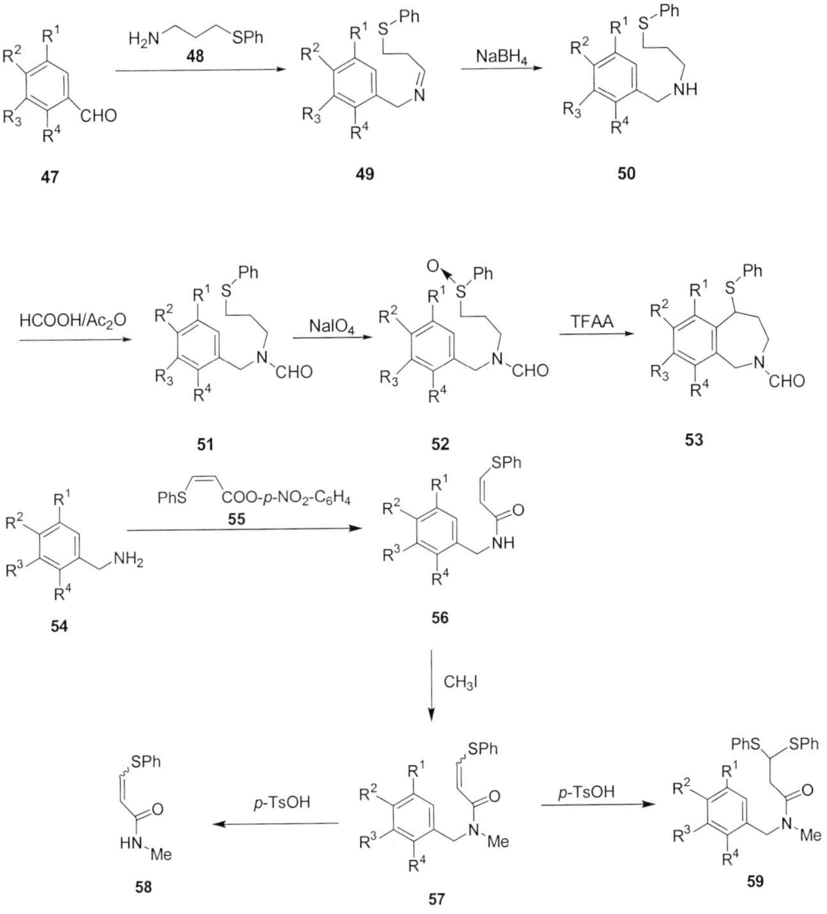

Seven-Membered Rings 391

	R¹	R²	R³	R⁴
a	H	H	H	H
b	H	H	H	OMe
c	H	OMe	H	H
d	H	H	OMe	H
e	OMe	H	H	H
f	OMe H	OMe	H
g	H	H	OMe	OMe
h	H	OMe	OMe	H
i	OMe	H	H	OMe
j	H	OMe	H	OMe

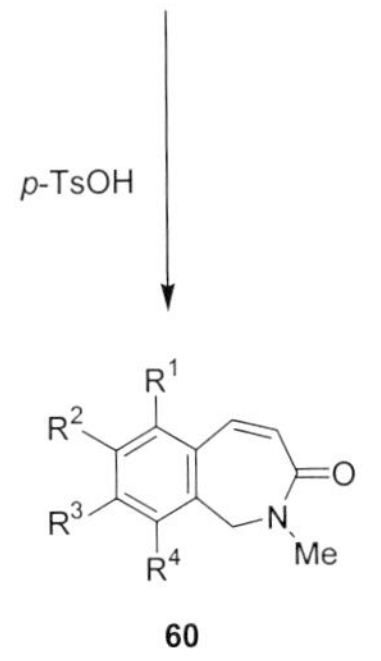

Acid-catalysed addition in toluene at reflux (followed by thiol elimination) has been used by Horiguchi and co-workers to access the benz[e]azepin-3-ones **60d-j** in moderate yields from the corresponding phenylsulfanylacrylamides **57**, which were prepared in turn from reaction of the benzylamines **54** with the ester **55** to afford the amides **56**, and then N-methylation with iodomethane in the presence of potassium hydroxide and tetraethylammonium bromide as a phase transfer catalyst. Other non-cyclized products ((E)- and (Z)-**58** and **59**, as well as (E)-**57**) were also observed depending on the structure of the N-aryl methyl group in **57** and on the solvent [02H1063].

An interesting solvent effect was observed in the palladium-catalysed cyclisation (10 mol% PPh$_3$, 2.5 eq. n-Bu$_4$NOAc, 85 °C) of the butenamides **61**. In anhydrous DMF, 6-membered ring formation was preferred, while in DMF/H$_2$O the 7-membered ring heterocycles **62**, **62a** and **62b** were formed, together with the 8-membered ring analogues **63** and **63a**. It is suggested that water decreases the medium basicity, thus reducing isomerisation of the double bond in the side-chain into conjugation with the amide carbonyl group; this latter isomer is the precursor for the 6-membered ring products [02SL1860].

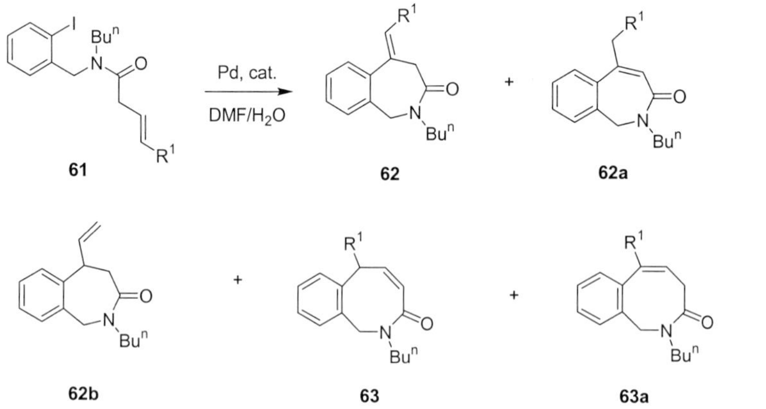

R^1 = H: **62** + **62a** + **63** + **63a**
R^1 = Me: **62** + **62a** + **62b** + **63** + **63a**
R^1 = Ph: **62** + **62a**

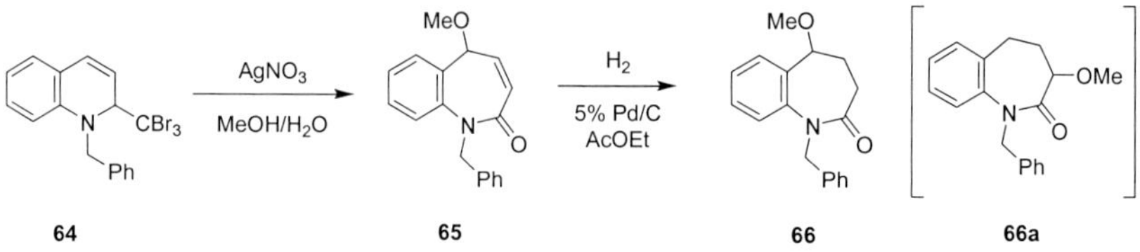

The benzo[b]azepin-2-one system continues to be a target for synthesis because of the pharmacological activities of compounds with this skeletal unit. In this context, Guingant *et al.* have reported an intriguing ring enlargement (via an aziridinium ion) route to **65** on treatment of **64** with silver nitrate. The structure of the product was then confirmed by reduction of **65** to **66**, which was also made by an unequivocal independent route; the isomer **66a** was also made separately and shown to be different [02SL1350].

Lipase-catalysed transesterification of the racemic alcohol **67** has been used effectively to produce (S)-(+)-**67** (LipaseQL, 0-5 °C, 4h; 47% yield; >99% ee), which was then converted to a chiral precursor required for the synthesis of the non-peptide vasopressin V_2 receptor agonist, OPC-51803 [02H635]. The chiral synthesis of a 1-benzazepine-based antagonist (OPC-41061) at this receptor has also been described [02H123].

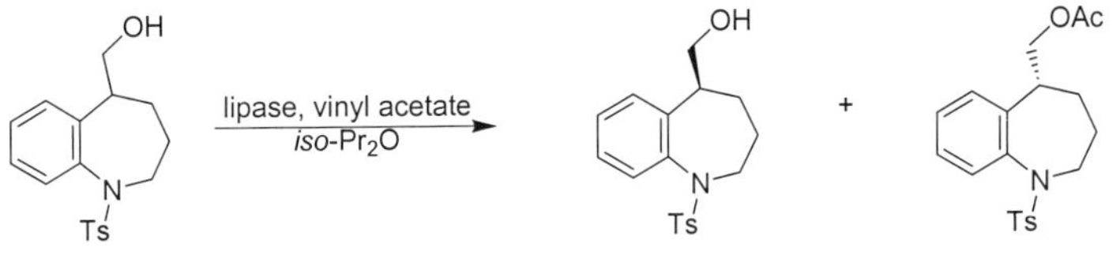

(±)-**67**   (+)-**67**   (-)-**68**

A variety of other ring fused azepines and azepinones have been described. Kocevar et al. have reported the first syntheses of the pyrano azepinediones **70** and **71** from **69** via the Schmidt reaction; at –15 °C to 0 °C (in $CHCl_3$ or CH_2Cl_2) the products **70** are favoured, while at 32 °C to 35 °C the percentage of the isomers **71** increased; overall yields were high [02H379].

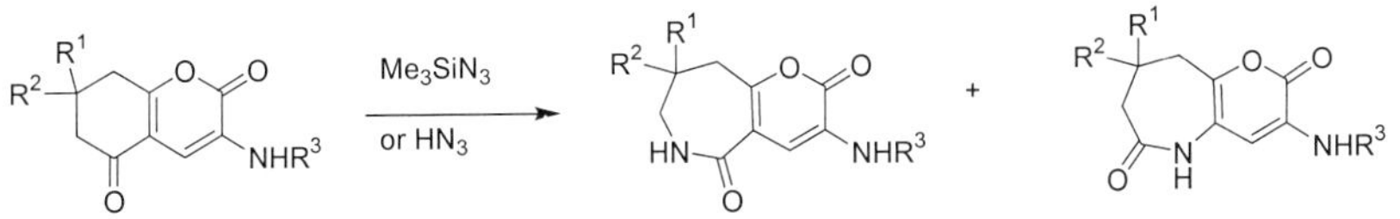

69   **70**   **71**

a: $R^1 = R^2 = R^3 = H$
b: $R^1 = R^3 = H, R^2 = Me$
c: $R^1 = R^2 = Me, R^3 = H$
d: $R^1 = R^2 = H, R^3 = COPh$
e: $R^1 = H, R^2 = Me, R^3 = COPh$
f: $R^1 = R^2 = Me, R^3 = COPh$

Lewis acid-catalysed rearrangement of the oxime tosylate **72** at 40 °C gave the oxazoloazepinone **73** in 70% yield [01JHC89].

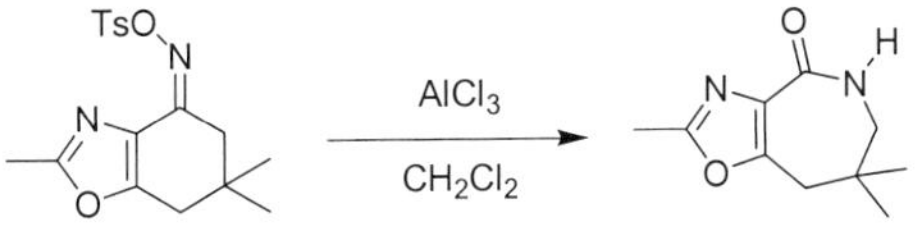

72   **73**

A new synthesis of the dibenz[b,e]azepin-11-one skeleton has been reported based on a lead tetraacetate-mediated oxidative cyclisation of the N-acylhydrazones **75**, obtained in turn from the benzophenone derivative **74**. The dibenzazepinones **76** were obtained in good yields (77% – 88%) [01H1057].

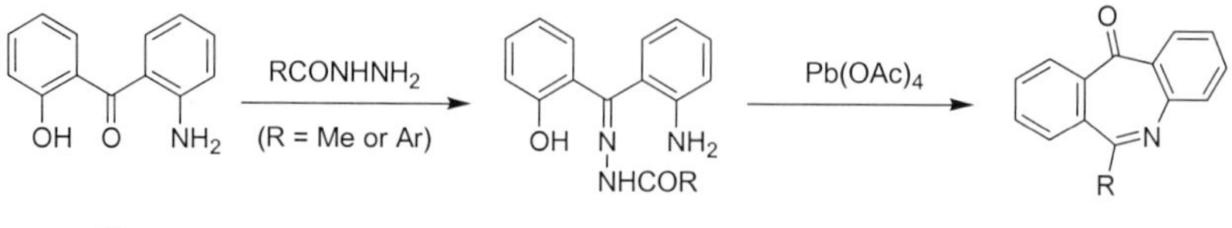

The dibenzazepinium salts **81** were prepared in moderate overall yields from the diol **77**, via **78**, the intermediate dibenzoxepine **79**, and the bromo aldehyde **80**. New catalysts for catalytic asymmetric epoxidation, for example **82** and **83**, were made in this way [02SL580].

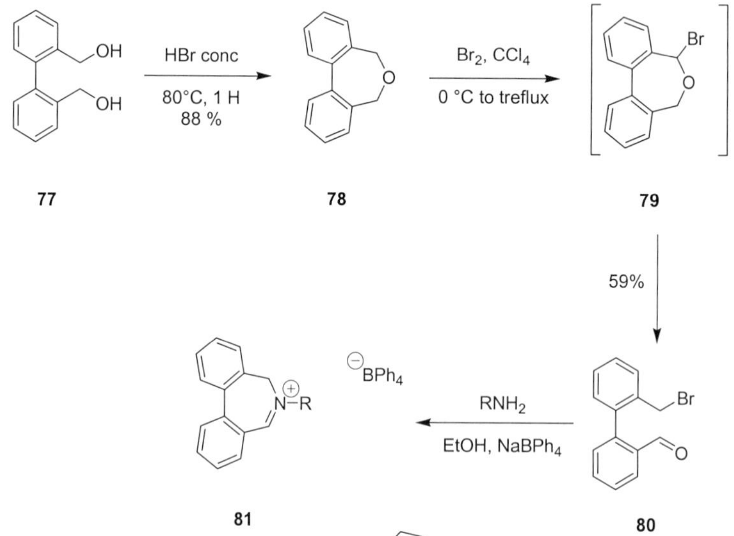

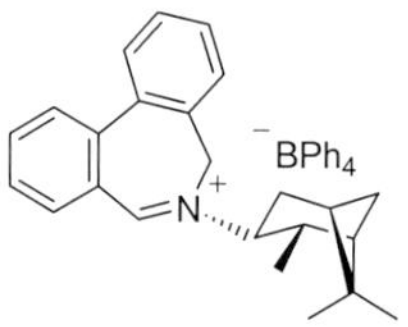

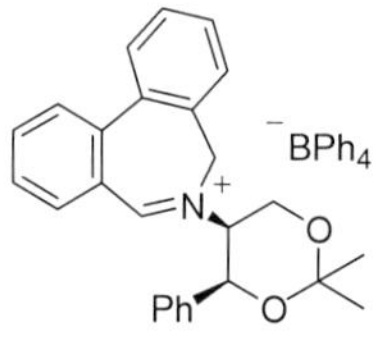

Indole-fused, or indole-benzo-fused azepinone derivatives have attracted synthetic attention and examples include the preparation of **85** in 84% yield from **84** by intramolecular Heck coupling [01SL848], as well as the preparation of paullone **87** (a CDK inhibitor) by cyclisation of **86** under basic conditions; borylation/Suzuki coupling technology was used to access **86** [02JOC1199]. Acid-catalysed cyclisation with polyphosphoric acid was used to prepare the racemic reduced azepino[4,5-b]indoles **92a,b** from the precursors **91**, which were obtained in turn from CDI-mediated coupling of **88** and **89**, followed by reduction of the amide with lithium aluminium hydride [01H1455].

84 **85**

86 **87** (51%)

88 **89** **90**

a: R = H; **b**: R = CH_3

92 **91**

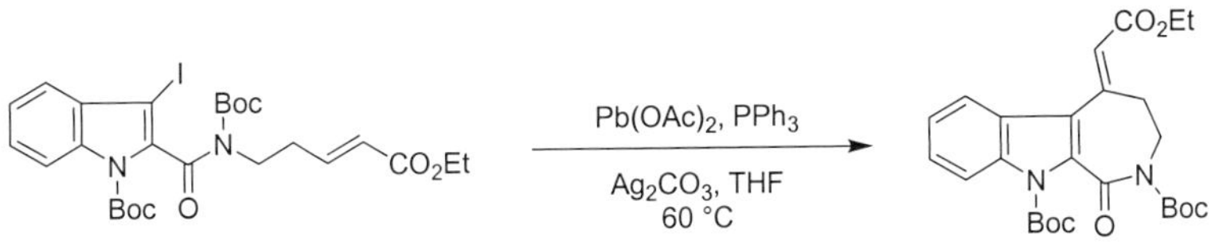

Other fused azepines, with aromatic (e.g. **94a,b**) or heteroaromatic fused rings (e.g. **94c,d**) have been prepared via the *N*-acyliminium ion intermediate **93** in high yields; lactam reduction then yielded the respective amines **95a,b** and **95c,d** [01H1519]. An intramolecular Friedel-Crafts reaction has been used to synthesise the fused azepinone **99** (*S,S* configuration) from **98**, the latter being prepared from the diastereopure α-hydroxylactam **96**, via the acid **97** [02H449].

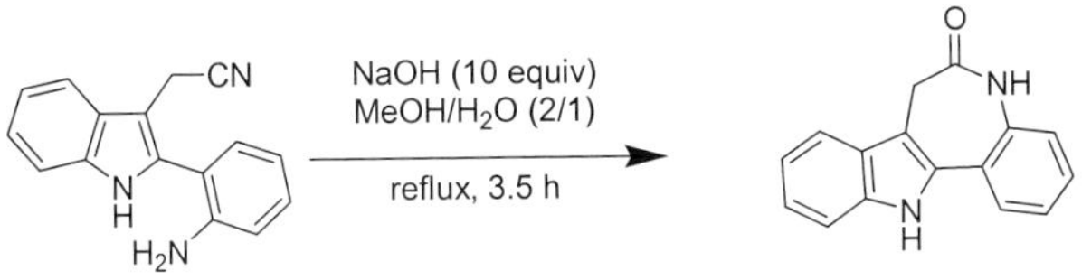

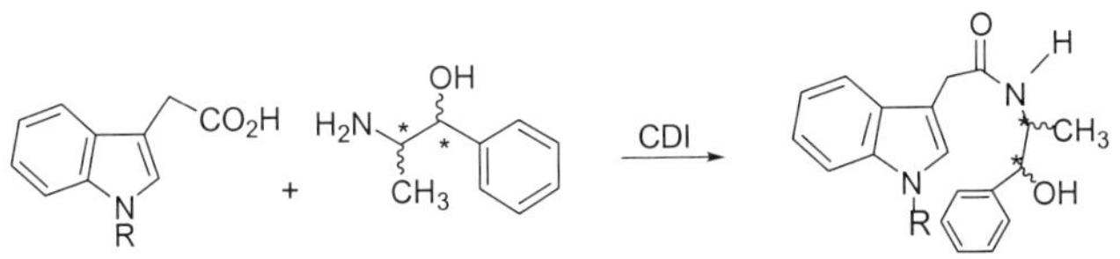

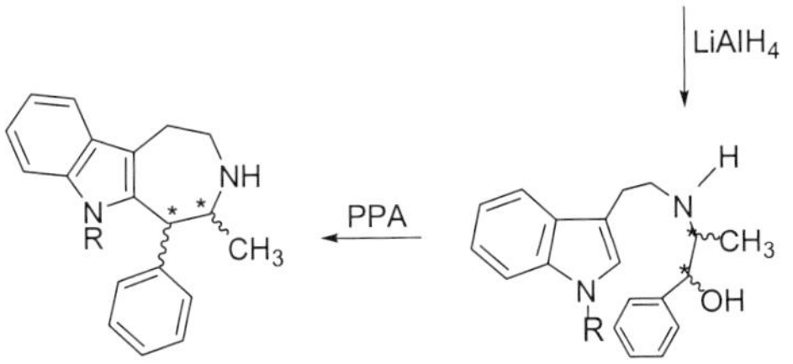

93

94a, n=1 (91%)
94b, n=2 (85%)

95a, n=1 (81%)
95b, n=2 (42%)

94c, n=1 (80%)
94d, n=2 (99%)

95c, n=1 (61%)
95d, n=2 (71%)

Reagents: (i), $BF_3.Et_2O$ and $BH_3.SMe_2$, THF, reflux.

96 **97** **98** **99**

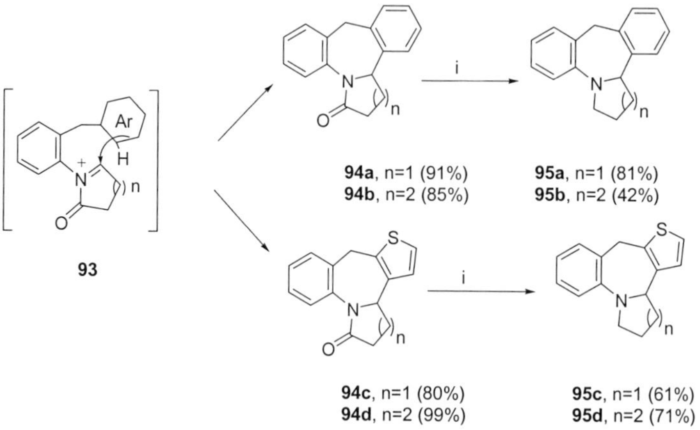

7.2.3 Oxepines and derivatives

An interesting approach to the substituted oxepine **102** has been described involving thermolysis of the bicyclic system **101**, obtained from the rhenium complex **100** (MeIm = *N*-methylimidazole; Tp = hydridotris(pyrazolyl)borate) [02JA7395].

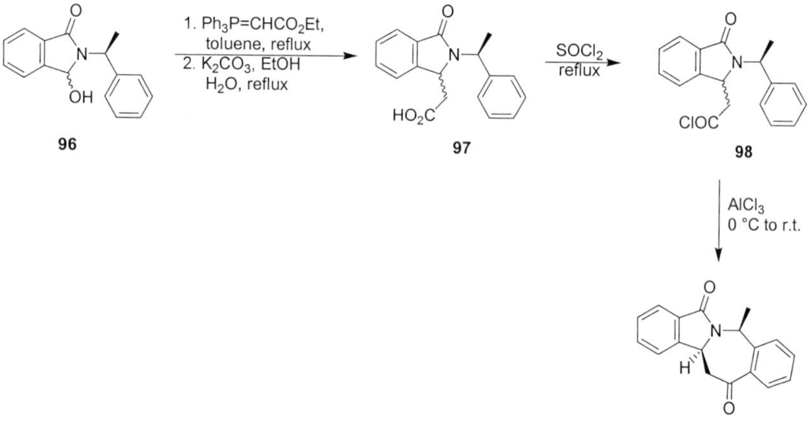

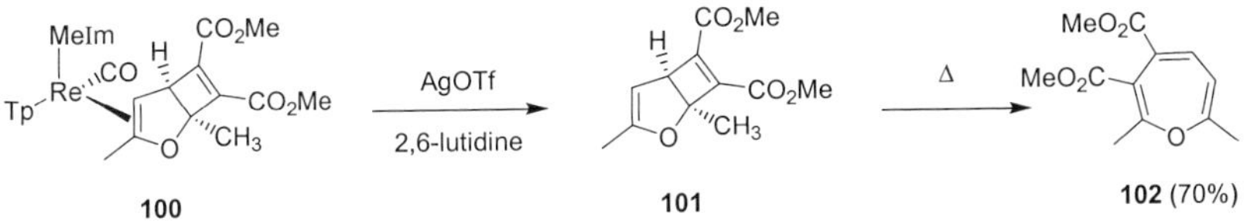

Trapping of the oxepine **104** ($R^1 = R^2 =$ Me; in equilibrium with the corresponding benzene-oxide **103**) with 4-phenyl-1,2,4-triazoline-3,5-dione **105** has been reported to give the *bis*-adduct **106**, whose structure was confirmed by X-ray analysis. Diels-Alder adducts **107** of the benzene-oxides were also observed [02CC1956].

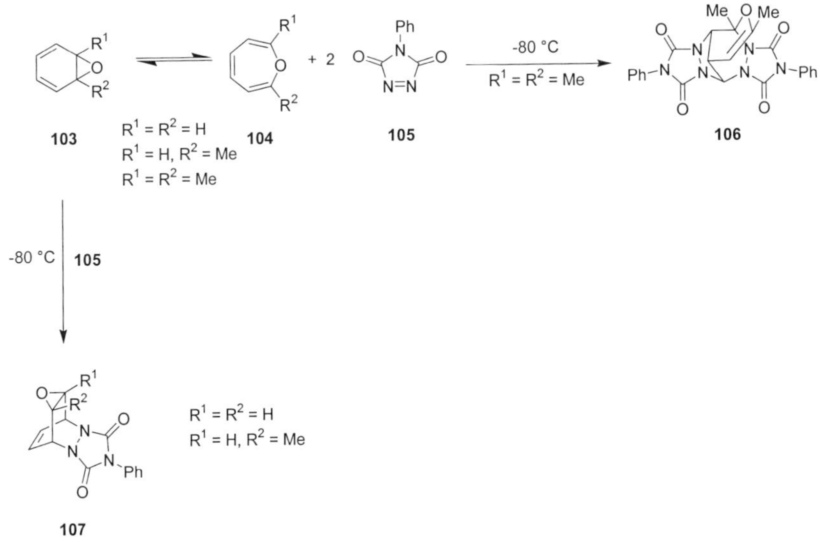

A three-component ring expansion reaction of cyclopropapyranones **108** with silyl enolate nucleophiles (Nu) and glyoxylate esters **109a,b** afforded the trisubstituted 4-oxepanones **110** in good yields (e.g. **110**, $R^1 =$ Ph, Nu = C(Me)$_2$COOMe, $R^2 =$ Et); further dehydration to compounds **111** was also undertaken [01H855]. Further exploration of this chemistry has also been reported by Sugita *et al.* [01TL1095].

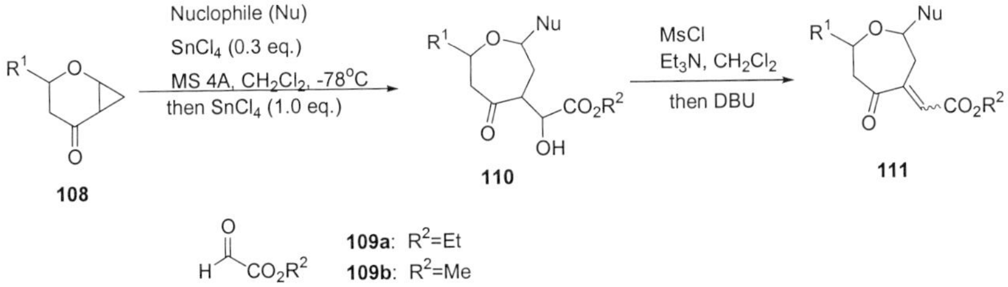

A study of the rearrangement-ring expansion and rearrangement-ring opening reactions of stereoisomeric tetrahydropyrans with zinc acetate has been undertaken. For example, the oxepane **113** was obtained from **112** in moderate yield, together with a small amount of the ring opened product **114** [02H113]. The first total synthesis of the marine natural product (+)-rogioloxepane A, containing a reduced oxepine ring, has also been reported [01TL1543].

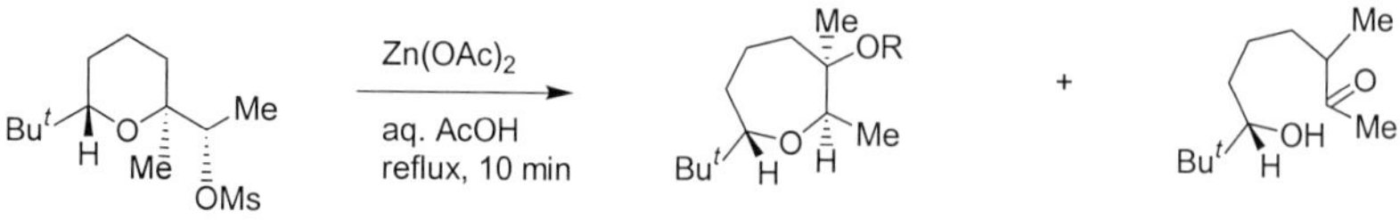

1',2-syn-2,6-anti
112

113 (58% (R=H, 49%; R=Ac, 9%))

114 (9%)

New conditions for the Baeyer-Villiger oxidation continue to be explored including selenium-catalysed oxidation with aqueous hydrogen peroxide (e.g. **115** to the oxepanone **117** in 95% yield) [01JOC2429] and tin-zeolite as a chemoselective heterogeneous catalyst [01NAT423].

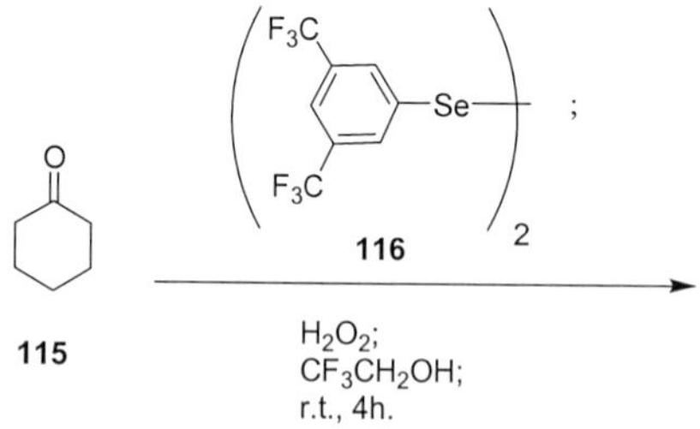

Seven-membered lactones can also be prepared in good yields by ruthenium-catalysed cyclocarbonylation of alkenyl alcohols (e.g. **119** from **118**) [01TL5459] or by ring closing metathesis using Grubbs' imidazolidine catalyst (e.g. **121** from **120**, n=2) [02H85].

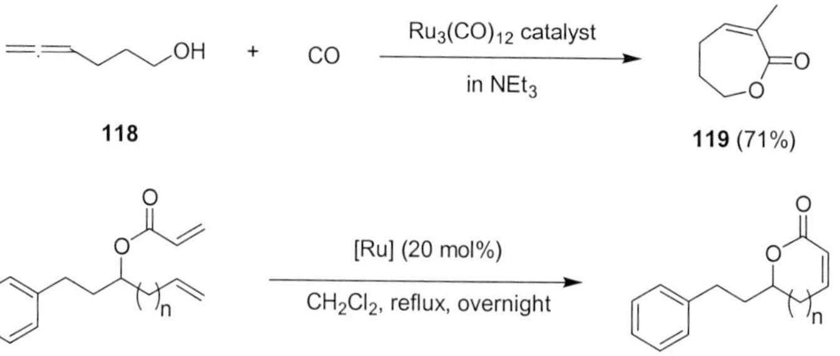

118

119 (71%)

120

121

Ring closing metathesis was also used to prepare the key oxepine derivative **123** starting from the allofuranose **122**; compound **123** then served as a precursor for the ultimate synthesis of the zoaptanol analogue **124** [01TL5749].

122

123

124

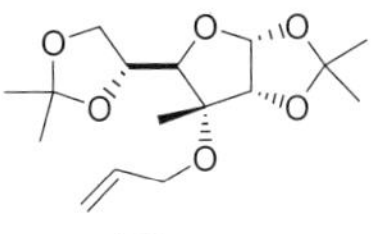

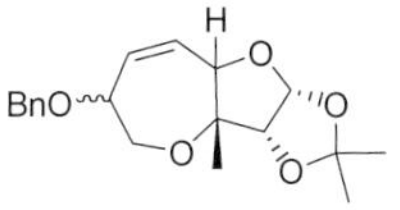

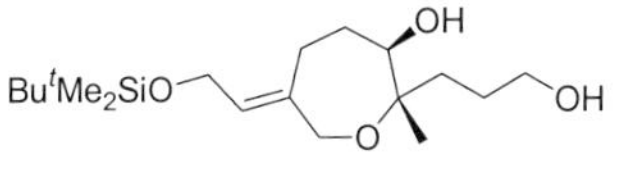

Increasing interest is being shown in the use of solid acid catalysts in potentially more environmentally acceptable syntheses. In this context the use of metal (IV) phosphates has been investigated for the cyclodehydration of diols to give cyclic ethers, including the conversion of the diol **125** to oxepane **126** (32% yield) [01GC143].

125 **126**

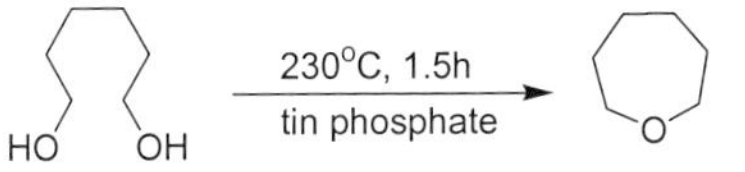

7.2.4 Fused oxepines and derivatives

The fused ring oxepine derivative **128** has been prepared in high yield (91%) by ring closing metathesis using Grubbs' catalyst (**A**) on the precursor **127** (7:3 ratio of diastereomers) [02SL1987]. The same catalyst was also used to prepare oxepine-annulated coumarins [02TL7781]. A convergent synthesis of the EFGH ring fragment of ciguatoxin CTX3C (containing a *bis*-fused oxepane ring) has been described [01TL6219], while a novel fused oxepane derivative, hypertricone **129**, has been isolated from the leaves of *Hypericum geminiflorum* [01HCA1976].

127 **A**: $(Cy_3P)_2Cl_2Ru{=}CHPh$ **128**

129

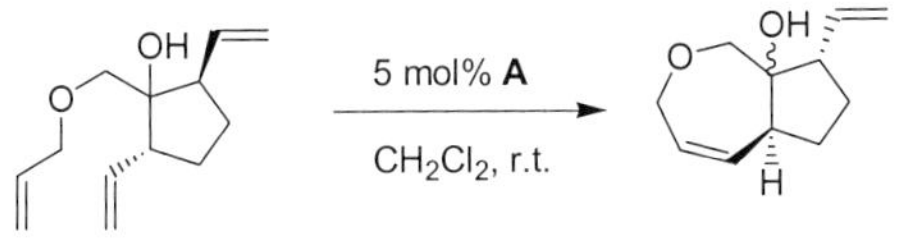

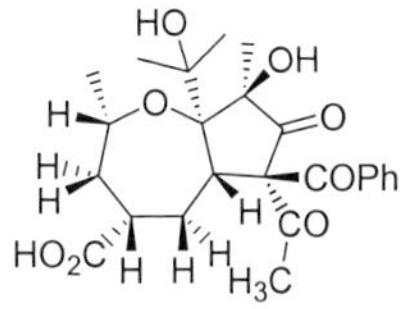

A neat hydroformylation – allylboration – hydroformylation reaction sequence reported by Hoffmann et al. afforded the pyrano-fused oxepanes **135** and **136** (48% total yield; ca 1:1 ratio) from the precursor **134**; the compound **134** required a number of steps for synthesis from the alcohol **130**, the ethynyl ether **131**, and the boronate **133** (by addition of **132**) in moderate overall yield. Dess-Martin oxidation of **135** and **136** then afforded the final lactones **137** and **138** [01NJC102].

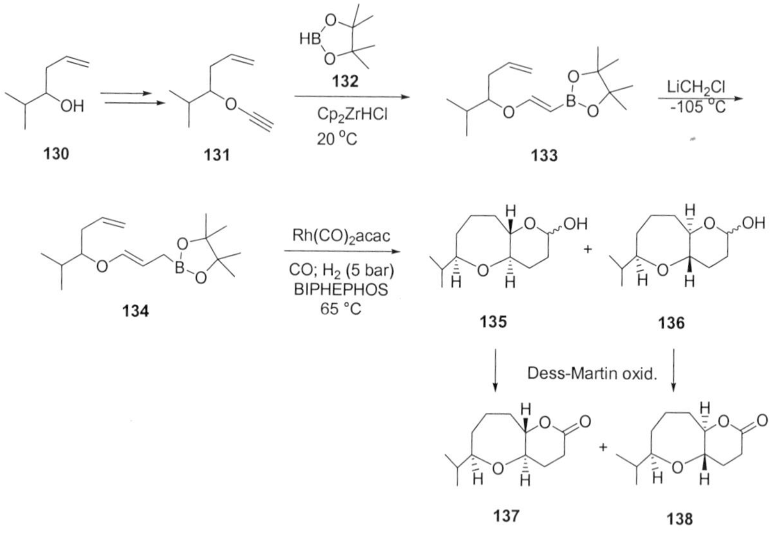

The structure and stereochemistry of an epoxydibenz[b,f]oxepinol derivative **141** has been described by Größnitzer and Wendelin [01MRC471]; the precursor **140** of compound **141** was accessed via the enone **139**.

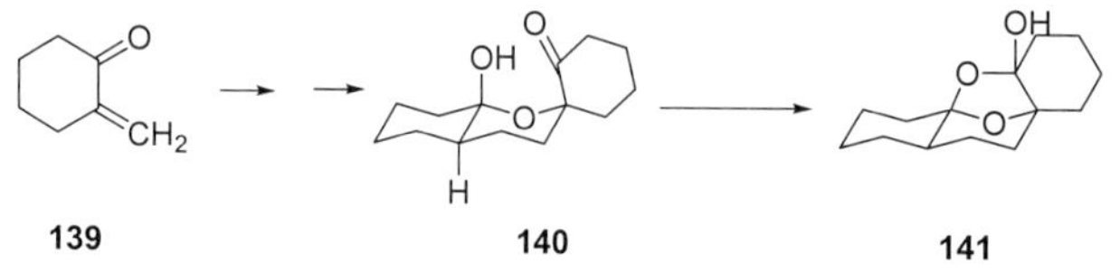

An enyne metathesis reaction using Grubbs' catalyst has been used by Pleixats et al. in an approach to benzoxepine derivatives capable of further ring construction via the Diels-Alder reaction. Formation of the propargyl ethers **143** from **142** set up the requirements for ring-closing metathesis to afford the benzoxepines **144**; reaction of **144** (Y = H) with diethyl azodicarboxylate, for example, then gave the tricyclic derivative **145** in 82% overall yield from **143** (Y = H). Other Diels-Alder reactions of **143** with maleimide, maleic anhydride, and p-benzoquinone were also described [01SL1784]. A similar ring-closing metathesis strategy has also been utilised to give the dihydrobenzo[b]oxepines **147**, **148**, and **149** in high yields from **146** [02H1997].

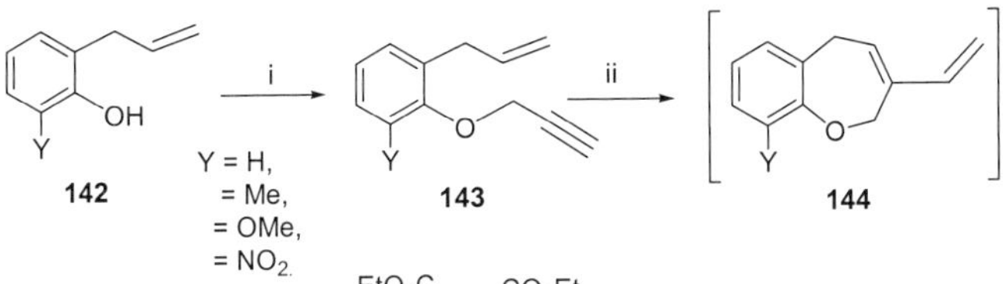

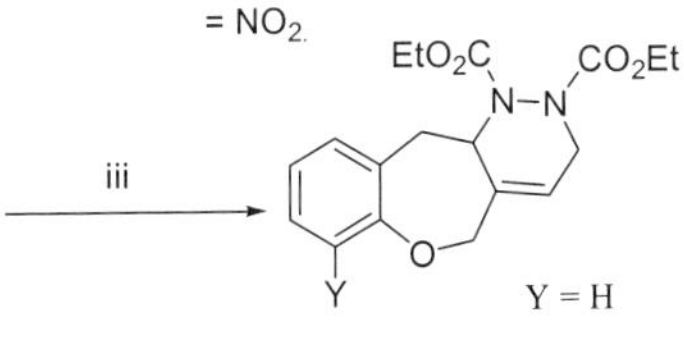

Reagents: (i), Propargyl bromide, K_2CO_3, acetone, reflux; (ii), $RuCl_2(PCy_3)_2$=CHPh (2-6% molar), CH_2Cl_2, r.t., 5h; (iii), EtO_2CN=NCO_2Et, CH_2Cl_2, reflux, 4 days.

145

146

147: R^1 = Me;
148: R^1 = Et;
149: R^1 = Bn

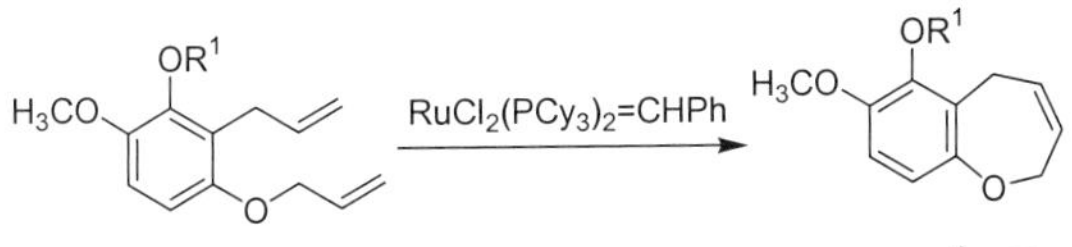

Interest continues in benzo[*b*]oxepine-ring containing natural products; a synthesis of racemic heliannuol D, **150** has been reported [02CC634], together with a synthesis of pterulone [01T7181], while the aglaforbesin derivative **151**, a new natural product, has been reported from the twigs of the plant *Aglaia oligophylla*. This plant contains insecticidal metabolites, but unfortunately **151** was not active as an insecticide against larvae of the insect pest *Spodoptera littoralis* [01JNP415].

150

151

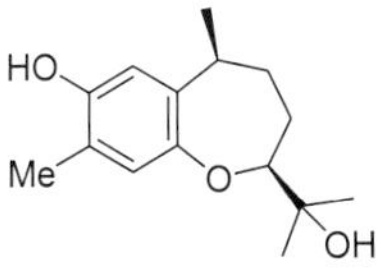

The synthesis of the highly hindered dibenzo[*c*,*e*]oxepine derivative **153** has been achieved by acid-catalysed dehydration of the diol **152** [02JCS(P1)2673].

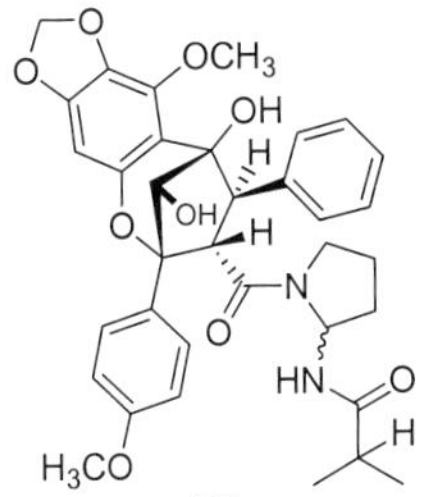

152

153 (70%)

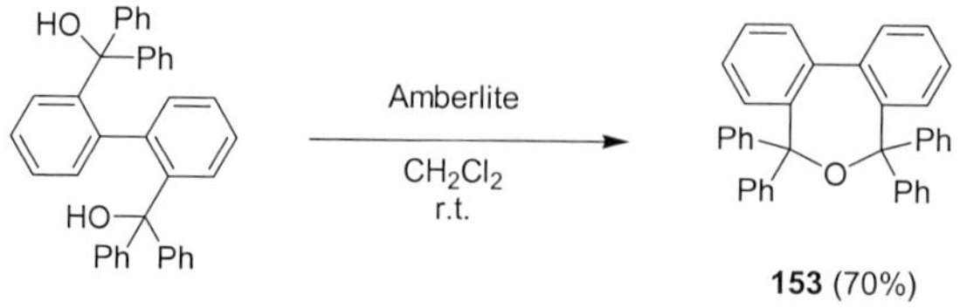

7.2.5 Thiepines and derivatives

Ab initio calculations at the B3LYP//6-31G(d) level indicate that Möbius-like conformations of hexafluorothiepine **154** do exist but that they are higher in energy than isomers with a plane of symmetry [02JCS(P2)388].

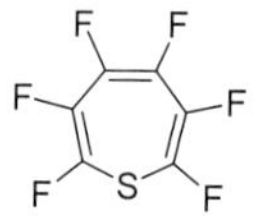

154

A synthesis of the tetrahydrothiepine **158** has also been reported [01JHC579] and is based on cyclisation of the lactol-derived precursor **155** via **156** or **157**; hydrogenation of **158** then gave **159** in high yield. The analogous thiopyran analogues ($X = CH_2$) were also synthesised in this work [01JHC579].

155 156 157 158 159

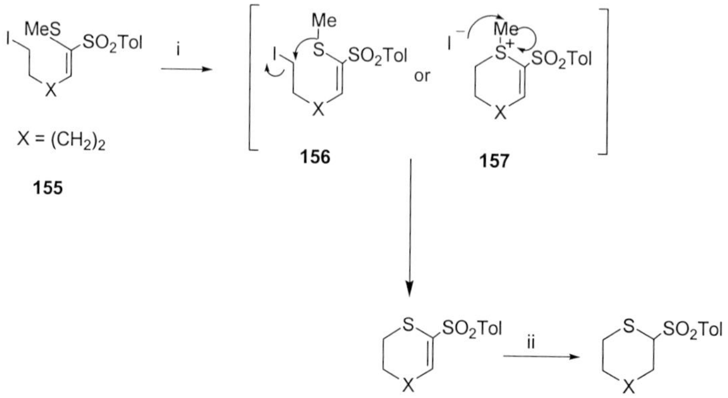

Reagents: (i), CH_3CN, NaI; (ii), 10% Pd/C, H_2, ethyl acetate, 20°C, 2 h, 80-87%.

7.2.6 Fused thiepines and derivatives

There has been more activity with fused thiepines of various types. Ring-closing metathesis (RCM) of the sulfoxide **160** gave the fused thiepene oxide derivative **161** as a 1:2 mixture of the *cis* and *trans* isomers; not unexpectedly, the thioether analogue of **160** did not cyclise under the RCM conditions due to chelation to the ruthenium centre in the catalyst [02SL1987].

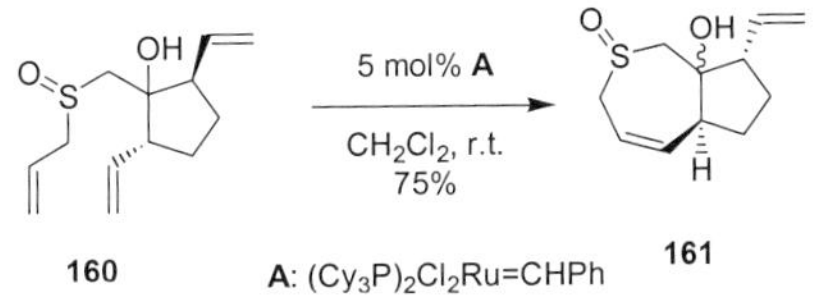

Coustard has reported the preparation of NMR-detectable cations of type **164**, including the 7-membered ring species (**164**, n = 2, R = Me or PhCH$_2$CH$_2$), by reaction of the nitroethenes **162** with trifluoromethanesulfonic acid; the presumed intermediates **163** were not observed. The cations **164** could be quenched with methanol to afford cyclic orthodithioester derivatives [01EJO1525].

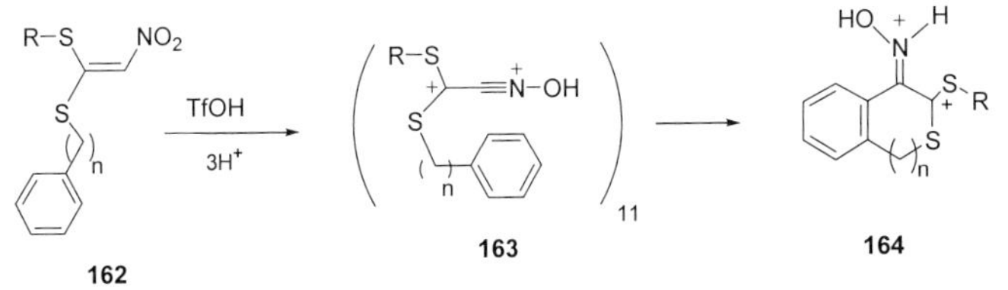

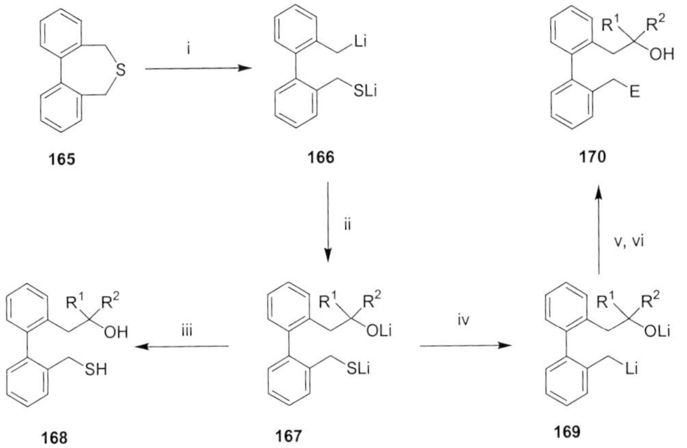

Reagents: (i), Li, DTBB (5 mol%), THF, -78 °C, 30 min; (ii), R^{1}R^{2}CO=tBuCHO, Ph(CH$_2$)$_2$CHO, PhCHO, Me$_2$CO, [Me(CH$_2$)$_4$]$_2$CO, (CH$_2$)$_5$CO, (CH$_2$)$_7$CO, (-)-menthone, -78 °C, 5 min; (iii), 3N HCl, -78 to 20 °C; (iv), 20 °C, 30 min; (v), E+ = Me$_2$CO, Et$_2$CO, (CH$_2$)$_5$CO, ClCO$_2$Et, -78 °C, 5 min; (vi), H$_2$O, -78 to 20 °C.

The two isosteric diastereomers of the calcium channel blocker UK-74,756 have been prepared based on the resolution of the isosteric enantiomers of the 6,11-dihydrodibenzo[*b*,*e*]thiepin-11-ol precursor [01TA975].

A study of the 4,4'-di-*tert*-butylbiphenyl(DTBB)-catalysed lithiation of dihydrodibenzothiepine **165** has been reported; lithiation results in **166**, which can then be converted into the alcohol derivatives **168** or **170** via the lithiated derivatives **167** or **169** respectively [01TL2469]. The preparation of the chiral pyrrolobenzo[*d*]thiepine **174** via intramolecular attack on the electrophile **173** (derived in turn from the hydroxylactam **171** via **172**) has been reported; the yield was 64% for **174**, n = 2 [01TL573].

171 **172** **173** n = 2 **174** n = 2

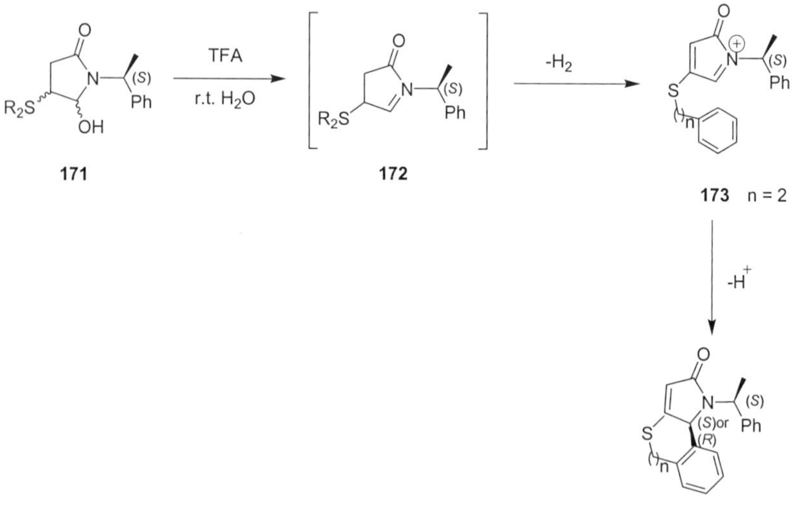

7.2.7 Miscellaneous systems with one heteroatom

Two further approaches to silicon-containing 7-membered ring systems have been described. The first involves RCM methodology and the Grubbs' catalyst to convert the dienes **175** to the reduced silepins **176a** and **176b** in moderate to low yields [01JOM160]. The second involves *endo* intramolecular radical cyclisation of the bromide **177** to **179**; yields of **179** increased with longer reaction times and slow addition of the Bu_3SnH, which decreased formation of the competing reduction product **178** [01MI363].

175

176a R = $(CH_2)_3Br$ (60%)
176b R = $(CH_2)_4CH{=}CH_2$ (35%)

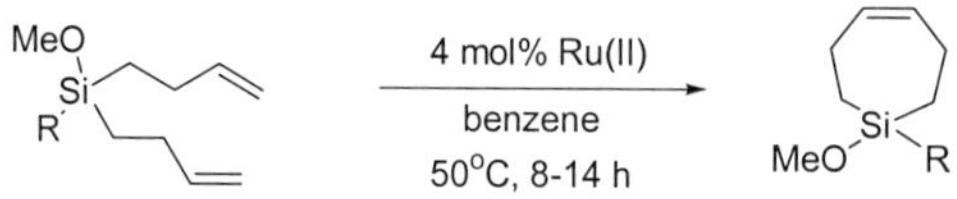

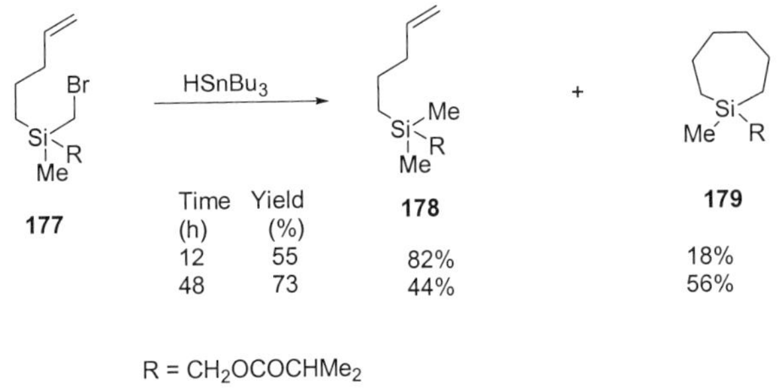

	Time (h)	Yield (%)	**178**	**179**
177	12	55	82%	18%
	48	73	44%	56%

R = CH$_2$OCOCHMe$_2$

Komatsu et al. have reported the first cyclic π-conjugated silatropylium ion by hydride removal from **184** on reaction with triphenylmethyl tetrakis(pentafluorophenyl)borate in dichloromethane at –50 °C for 0.5 hours. The cation was characterized by ^{1}H, ^{13}C and ^{29}Si NMR spectroscopy, with the observed chemical shifts suggestive of a degree of aromaticity not greatly less than that of the tropylium ion. The precursor **184** was made from **180** via the lithiated derivative **181**, the dichloride **182**, and the mesityl derivative **183** [01T3645]. An extensive report on the synthesis and chemistry of enantiomerically pure dihydrodibenzo[b,f]phosphine 5-oxides (e.g. **185**) has been published by Wyatt et al. [01JCS(P1)279].

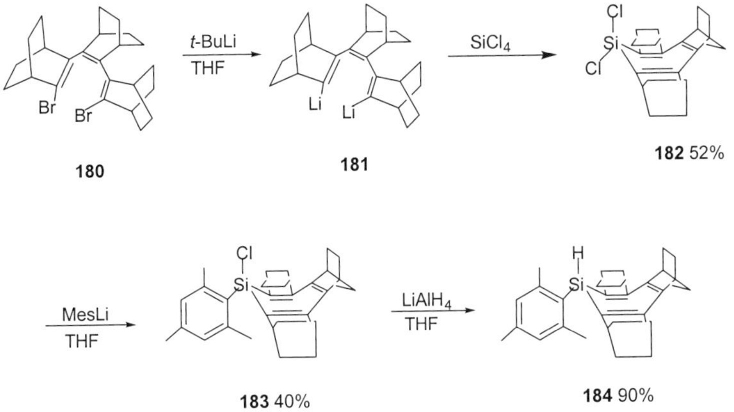

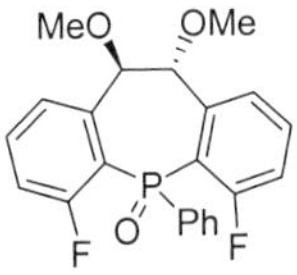

7.3 SEVEN-MEMBERED SYSTEMS CONTAINING TWO HETEROATOMS

7.3.1 Diazepines and fused diazepines and derivatives

A rich variety of chemistry is manifest with the diazepines and fused derivatives, and some selected examples are presented. The basicity and hydrolysis of 1,2-diaryl-1H-4,5,6,7-tetrahydro-1,3-diazepines has been reported by Fernandez et al. [01JHC895]. More work has also been carried out on 1,4-diazepine derivatives. Some, like compounds of type **189** (e.g. R^1 = PhCH$_2$, R^2 = H) are of interest as matrix metalloproteinase inhibitors; these diazepine derivatives were prepared by a reaction sequence starting with the orthogonally protected D-aspartic acid derivative **186**, followed by sulfonamide formation to give **187** followed by conversion to the hydroxyamides **188** and then a Mitsunobu reaction to give the heterocyclic ring and finally conversion of the methyl ester substituent to the hydroxamic acid in **189** [01BMCL1009].

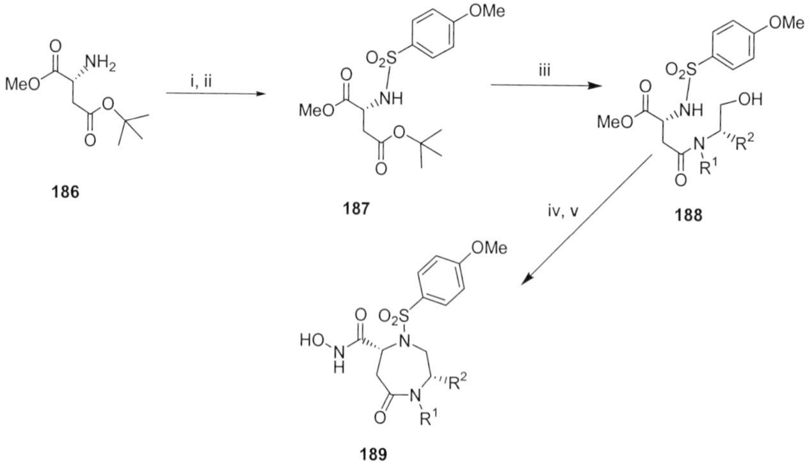

Reagents: (i), MeOC$_6$H$_4$SO$_2$Cl, Et$_3$N, CH$_2$Cl$_2$; (ii), TFA, CH$_2$Cl$_2$; (iii), R^{1}NHCHR^{2}CH$_2$OH, EDAC, HOBt, DMF; (iv), DEAD, Ph$_3$P, CH$_2$Cl$_2$; (v), NH$_2$OH, KOH, MeOH.

A new approach to 6-nitro-1H-[1,4]-diazepines **192** based on reaction of the formylated nitroenamine **190** with the 1,2-diamines **191** has been reported; yields were generally good in this ring construction sequence, and is applicable to sterically hindered diamines (e.g. **191**, R^1 = Et, R^2 = R^3 = H) [02H425].

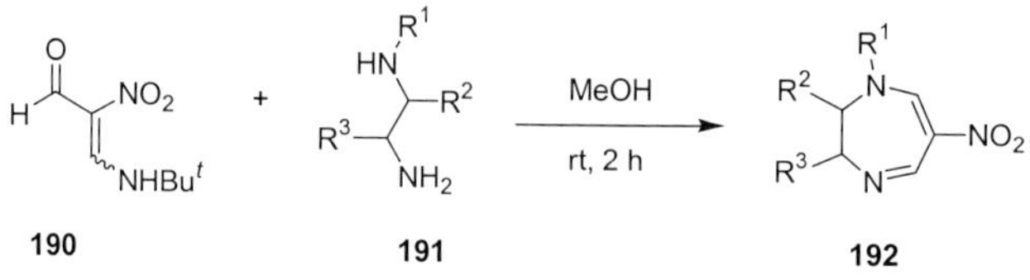

Stages involved in the reaction of 2,3-butanedione with propylenediamine have been monitored by NMR spectroscopy as well as by the isolation of products. The reduced 1,4-diazepine **194** was isolated, together with the macrocyclic dimeric product **196** and the tetrahydropyrimidine **195**. NMR spectroscopic evidence was obtained for the first formed imine **193** and the tautomeric diazepine **197** [02H11].

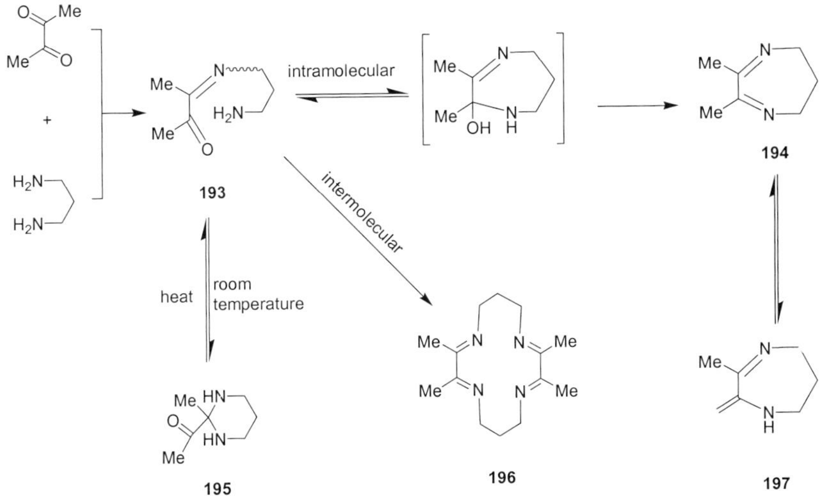

Natural products with a diazepine moiety present include the novel flavonoid alkaloid aquiledine, **198** [01JNP85] and the alkaloid TAN1251A, **199**, a muscarinic M_1 receptor antagonist [01OL1053].

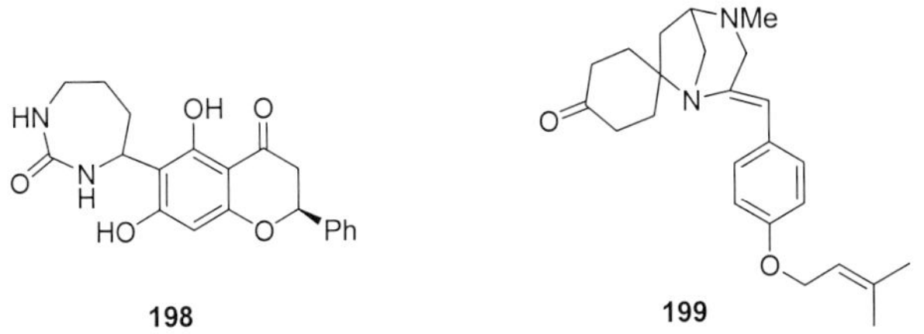

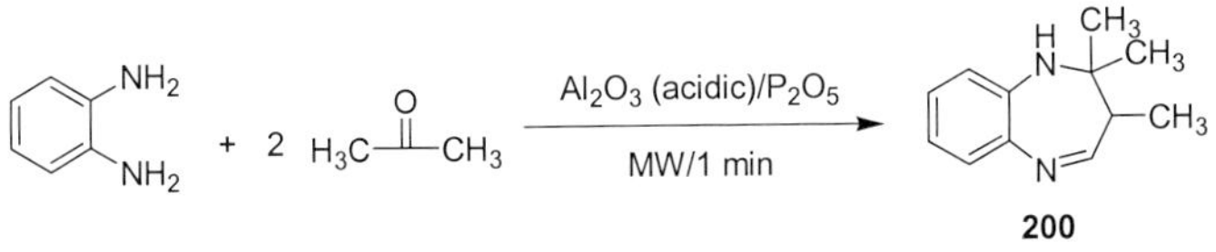

Microwave-induced or facilitated reactions are becoming increasingly significant in synthesis. One example is the efficient preparation of the 1,5-benzodiazepine **200** in 85%

yield from *o*-phenylenediamine and acetone in the presence of alumina/phosphorus pentoxide with microwave irradiation under solvent-free conditions [01H1443].

As part of a programme aimed at the synthesis of trifluoromethyl-substituted heterocycles, Kawase *et al.* have shown that the mesoionic compounds **201** react regiospecifically with *o*-phenylenediamine in dichloromethane at room temperature to give high yields of the 1,5-benzodiazepines **202** (e.g. R^1 = Me, R^2 = 4-MeOC$_6$H$_4$; 91%); the analogous reaction with *o*-aminothiophenol also gave the 1,5-benzothiazepines **203** in high yields [01H1919].

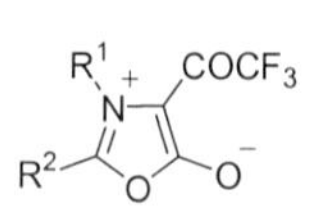

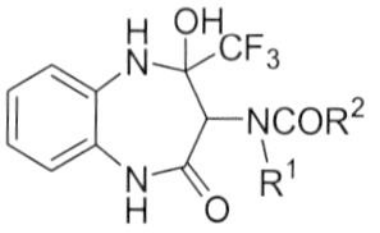

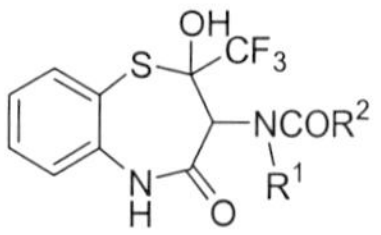

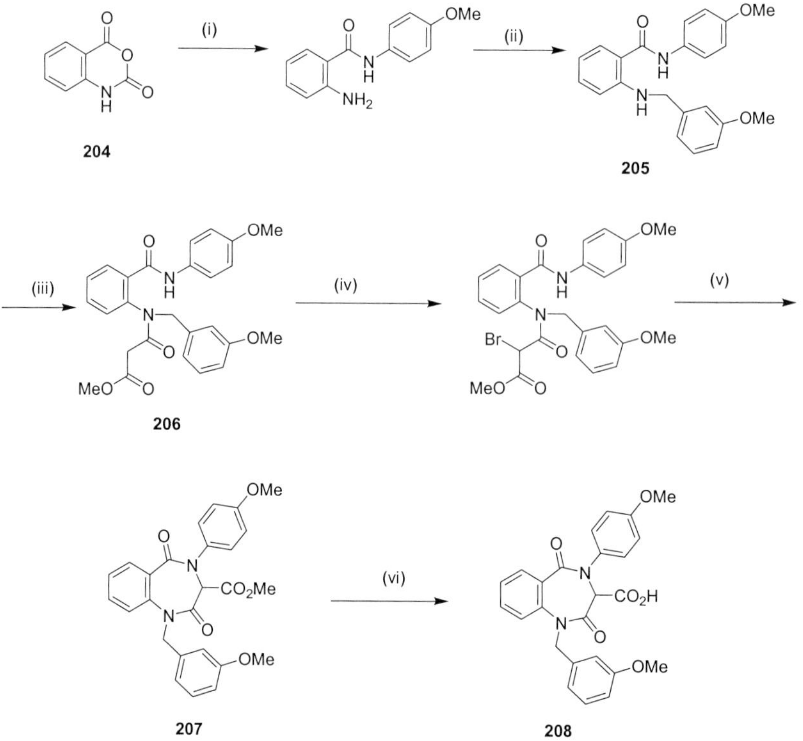

Reagents: (i), p-Anisidine / DMF; (ii), 3-Methoxybenzyl chloride / THF; (iii), Methyl malonylchloride / THF, 0 °C; (iv), PyHBr$_3$ / THF, H$^+$; (v), t-BuONa / DMF; (vi), LiOH, H$_2$O, 0 °C

The 1,4-benzodiazepine-2,5-dione system is of major pharmacological significance, with derivatives exhibiting a wide range of different activities. Efficient syntheses of specific derivatives thus continue to be of interest. Ho and co-workers have reported a concise route to a 3-carboxylic acid derivative **208**, which begins with isatoic anhydride **204** and then via the amino amide **205** to the key intermediate **206**, resulting from reaction of methyl malonylchloride with **205**. Bromination of **206** followed by base-induced intramolecular nucleophilic displacement then gave **207** in high yield (82%); hydrolysis then afforded **208** in 90% yield [02H1501].

The enantiopure tetrahydro-1,4-benzodiazepinone derivatives **210** have been prepared in moderate to good yields by a palladium-catalysed cyclisation of the chiral amine **209**; a range of bases and ligands were used with $Pd_2(dba)_3CHCl_3$ (10 mol%) [01SL803]. Copper-catalysed amination (CuI, 20 mol%) of an aryl bromide with L-aspartic acid (to give **211**) was used as a key early step in a concise synthesis of the chiral precursor **215** of lotrafiban, a platelet aggregation inhibitor. Later steps included methyl ester formation to give **212**, followed by conversion to the imine **213**, then reduction to the amine **214** and lactam ring formation to afford **215** in good overall yield [01SL1423].

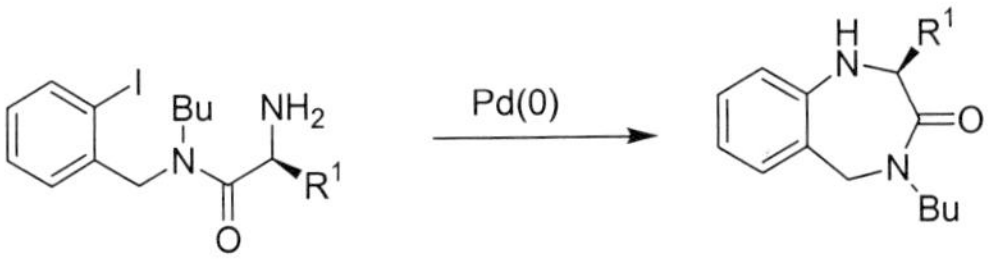

(*S*)-**209** → (*S*)-**210**

a R^1 = Me; **b** R^1 = Bn; **c** R^1 = Ph; **d** R^1 = CH_2CO_2Bn

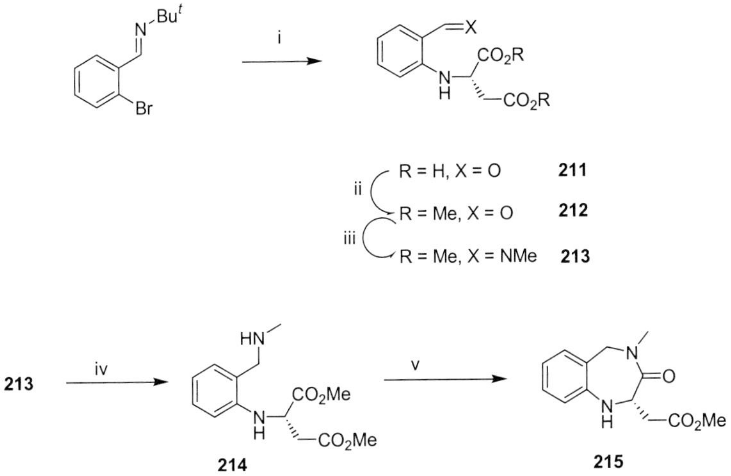

R = H, X = O   **211**
R = Me, X = O   **212**
R = Me, X = NMe **213**

213 → **214** → **215**

Reagents: (i), L-aspartic acid, tetrabutylammonium hydroxide, 1.8 eq., catalyst, MeCN, reflux; (ii), MeOH, AcCl, reflux; (iii), MeNH$_2$ in EtOH; (iv), Pd-C, MeOH, H$_2$; (v), stir in MeOH, r.t, 3 days.

An imidazo-fused 1,3-diazepine **218** has been accessed via nucleophilic substitution of 1-benzyl-5-nitroimidazole to give the dichloromethyl derivative **216** and then ultimately the aminoketone precursor **217**; cyclisation of **217** with triethyl orthoformate then gave in 81% yield the fused heterocyclic derivative **218**, which was isolated as a mono hydrochloride salt and mono DMSO solvate [02TL1595]. A novel ferrocenyl-substituted imidazo[1,2-a][1,3]diazepin-3-one derivative **220** has been prepared in 62% yield from the imidazolone precursor **219** via RCM methodology; removal of the N-Boc protecting groups then afforded **221** in 92% yield [02EJO3801].

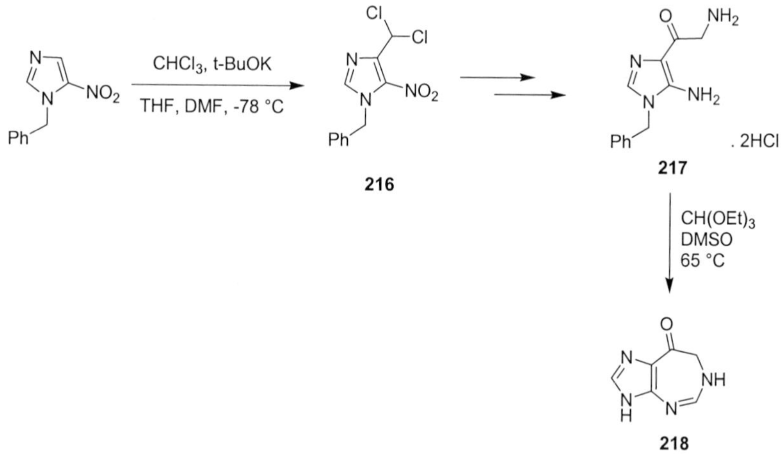

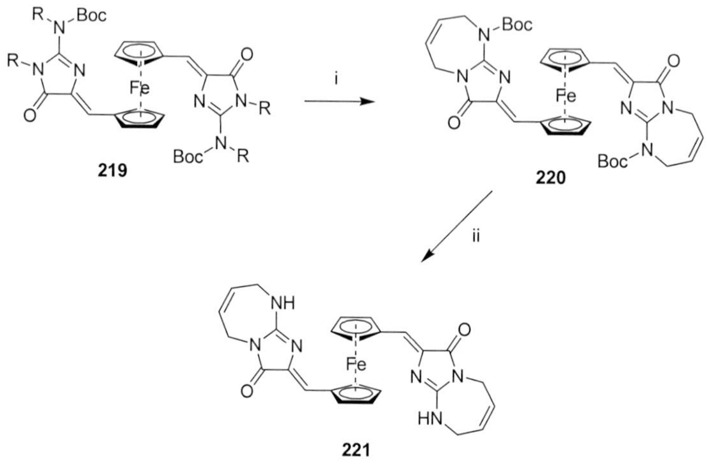

Reagents: (i), 10 mol% [RuCl$_2$(CHPh)(PCy$_3$)$_2$], reflux in CH$_2$Cl$_2$;
(ii), 30% F$_3$CCO$_2$H, reflux in CH$_2$Cl$_2$.

An attempted EDCI-mediated coupling of **222** and **223** afforded instead the tetrazolo[1,5-d][1,4]diazepin-6-one **224** in 93% yield; the structure of **224** was confirmed by X-ray crystallography and is of interest with respect to the design of *cis*-constrained peptidomimetics [01CC2080].

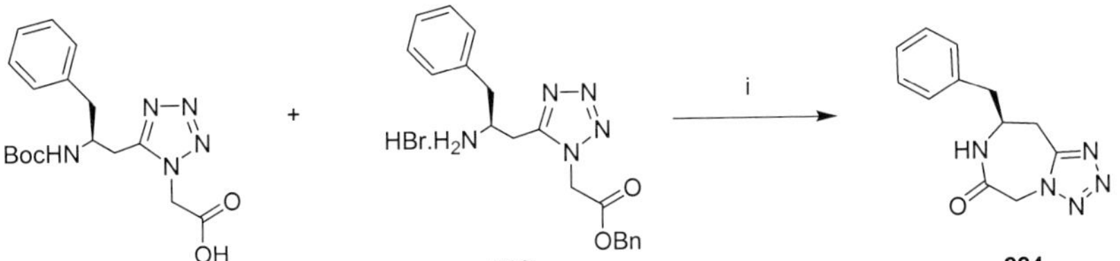

Reagents: (i), EDCI [1-(3-dimethyl-aminopropyl)-3-ethylcarbodiimide hydrochloride], HOBt (1-hydroxybenzotriazole hydrate), DIPEA (*N,N*-diisopropylethylamine), rt, 18 h.

A new route to dichlorofluoromethyl substituted heterocycles, including the 1,4-naphtho[1,8a,8-*ef*]diazepin-2(1*H*)-one **225**, has been described based on the reaction of bifunctional nucleophiles with 3-chloropentafluoropropane-1,2-oxide [01JFC1].

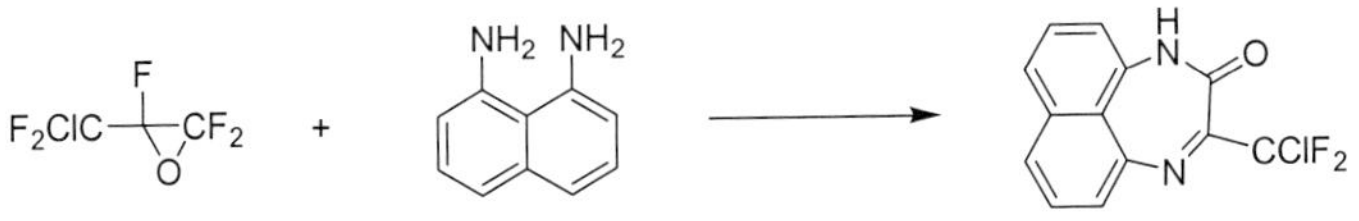

Other relevant papers in this general area include the formation of β-lactam derivatives of 2,3-dihydro-benzo[1,4]diazepines [01JHC1031], new scaffolds for potential drug discovery incorporating benzo[*e*][1,4]diazepin-5-one moieties [01OL1089], the synthesis of pyrrolo[3,4-*b*]hexahydro-1*H*-1,5-benzodiazepine derivatives [01SL1953], the isolation of circumdatin G (**226**), a new alkaloid from the fungus *Aspergillus ochraceus* [01JNP125], and the chiral syntheses of the related alkaloids circumdatin C and circumdatin F [01JOC2784], and the chiral synthesis of the pyrrolo-benzodiazepinone natural product DC-81 [01JOC2881].

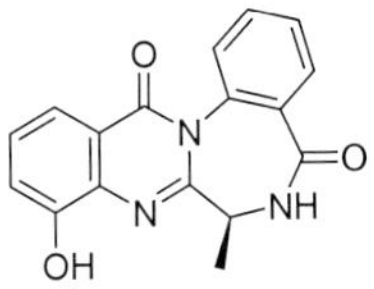

7.3.2 Dioxepines and fused dioxepines and derivatives

A serendipitous synthesis of the 1,4-dioxepin-5-one derivative **228** has been reported based on acid chloride formation from the acid **227**, followed by a ring opening and ring closing sequence on exposure of the acid chloride to 48% HBr solution [01TL2305].

227 **228** (75%)

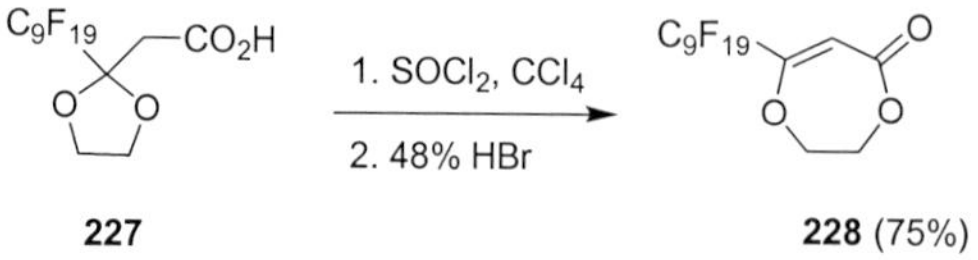

The chiral substituted 1,3-dioxepine derivative **231** has been prepared in 96% ee (50% conversion) by an asymmetric Heck reaction with the alkene **229** and the aryl triflate **230** [01OL161].

229 **230** **231** (96% ee)

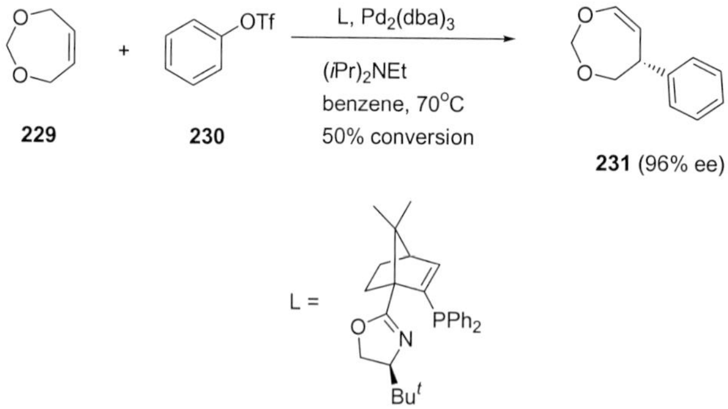

Enantioselective isomerisation of the 4,7-dihydro-1,3-dioxepine **232** to (*R*)-(-)-**233** has been achieved using (*R*,*R*)-(+)-CHIRAPHOS-(and (R,R)-(-)-Me-DuPHOS-) modified nickel complexes [01AG(E)177].

232 **233**

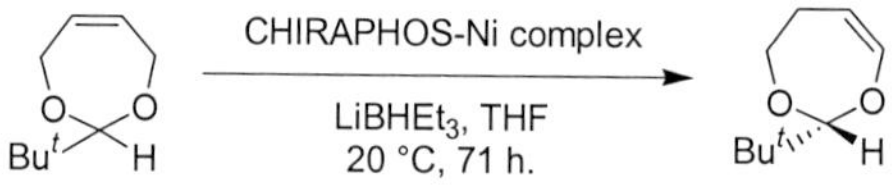

Nucleoside analogues **234** with a 1,4-dioxepane sugar-type moiety have been prepared and evaluated for antiviral activity, but no activity was found [01T3951]. One of five new depsidones from plants in the genus *Garcinia* has been shown to be the dibenzodioxepinone derivative **235**; the *Garcinia* species were investigated as part of a programme to locate new cancer chemopreventative agents but no biological data was given for **235** [01JNP147].

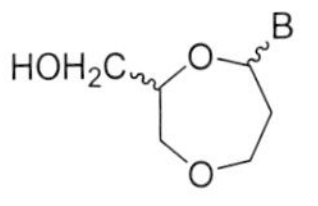

7.3.3. Miscellaneous derivatives with two heteroatoms

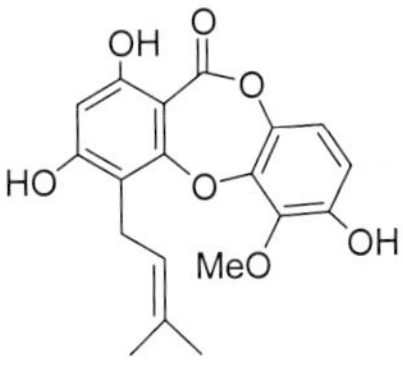

An efficient synthesis of 1,4-oxazepines has been developed based on iodine-sodium bicarbonate-promoted intramolecular cyclisation of the enamides **237**, which were derived in turn from the β-formyl enamides **236** by sodium borohydride reduction. Reductive removal of the iodo-substituent in **238** was readily achieved using sodium sulfite. The R^1 and R^2 groups were part of a ring system, for example a steroid system [01SC3281].

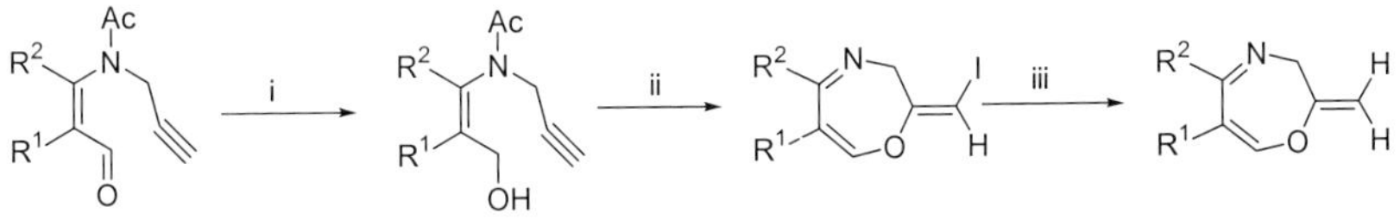

Reagents: (i), $NaBH_4$/MeOH/0-10 °C; (ii), I_2-$NaHCO_3$/Et_2O-H_2O/r.t.; (iii), Na_2SO_3-H_2O.

An alternative ring construction approach to 1,4-oxazepin-7-ones (e.g. **241**) utilises the Baylis-Hillman product **240**, subsequent reaction with a β-aminoalcohol, and ester hydrolysis followed by DCC-mediated intramolecular coupling to afford **241**. This sequence can be generalised to give a range of analogues [02S2232].

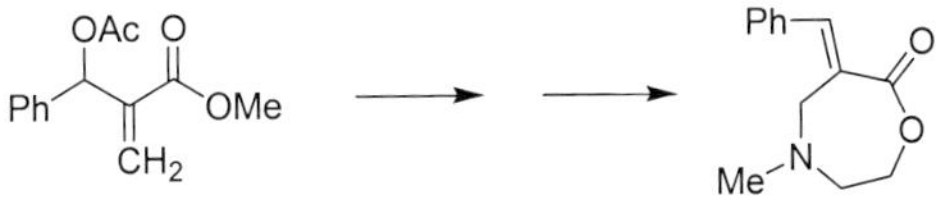

The fused 1,3-oxazepine derivative **244** is readily made from reaction of the amino alcohol **243** with the keto acid **242**, prepared in turn by a Grignard reaction with 3-methylglutaric anhydride. A 6:1 *trans:cis* diastereomeric ratio was obtained with **244**, which was then used in a new approach to a spiro heterocyclic system [01SL1506].

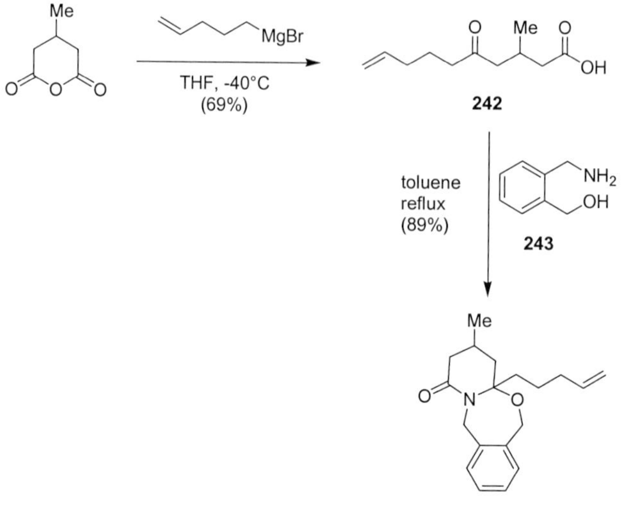

While TNT is well known as an explosive, it is not so commonly considered as a starting point for synthesis of new heterocyclic ring systems. Semenov *et al.* have used the acid chloride **245**, accessible from TNT, to prepare the dibenzo[*b,f*]-1,4-oxazepinone derivative **247** (together with other *bis*-fused 7-membered heterocyclic rings) in high overall yield [01MC109]. *Bis*-fused 1,5-benzoxazepines (e.g. **248**) are of interest as inducers of apoptosis in chronic myelogenous leukaemia cells and hence as potential anti-leukaemia therapeutics [01MI31].

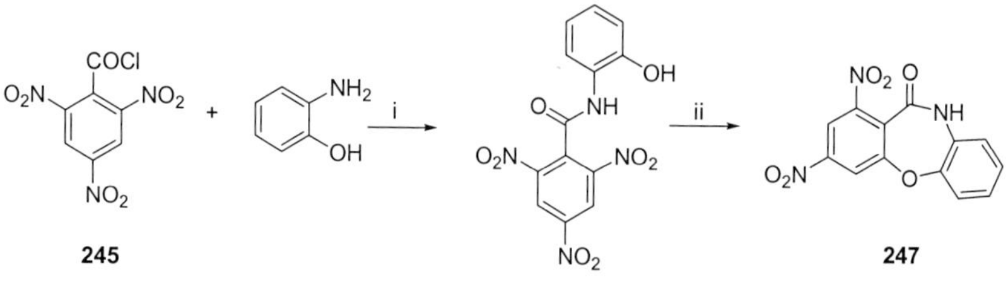

Reagents: (i), C_6H_6, reflux, 8 h, 81%; (ii), aqueous NH_3, EtOH-MeCN, 50°C, 60 h, 91%.

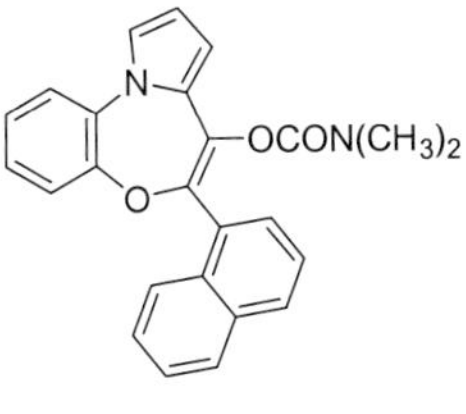

248

Reactivity studies on new types of tetracyclic 1,5-benzoxazepines (e.g. **249**, n = 1, X = O, R = H) have been undertaken by Sekhar *et al*. Reductive cleavage of the oxazepine ring (rather than imine bond reduction) occurred on catalytic hydrogenation in acetic acid, to give **250**, the structure of which was confirmed by alkaline hydrolysis to the phenol **251** [01JHC383]. Stereocontrolled reduction of the oxazepino-indolo[2,3-*a*]quinolizine derivative **253** to **254** (95% d.e.) was a key factor in the total synthesis of the indole alkaloid, (+)-tacamonine; the oxazepine **253** was made in high yield from the N-Boc protected enamine **252** [01TL7237]. The novel scaffold **256** upon which to base potential new drug leads was made in reasonable overall yield in a 3-step cyclisation sequence from **255** [01OL1089].

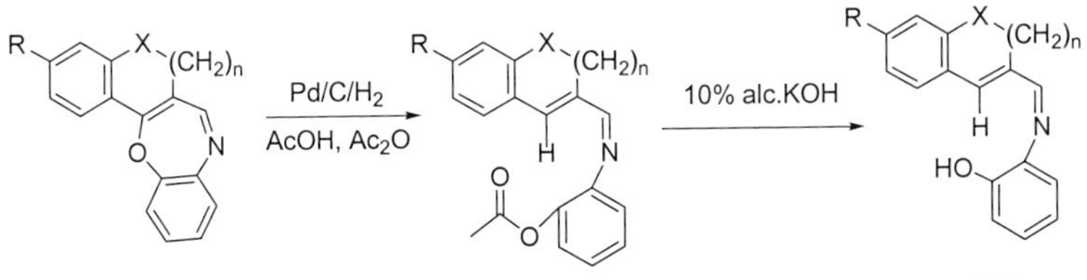

249 → **250** → **251**

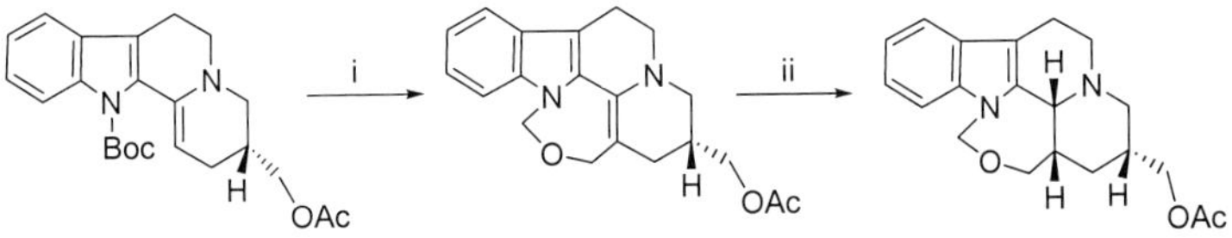

252 → **253** → **254**

Reagents: (i), (CH$_2$O)$_3$, HCO$_2$H, THF, reflux, 90%; (ii), H$_2$/PtO$_2$, dioxane, 75%.

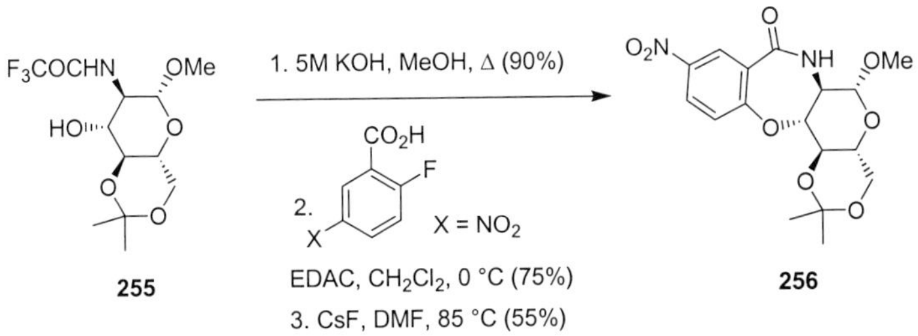

255 → **256**

1. 5M KOH, MeOH, Δ (90%)
2. X = NO$_2$, EDAC, CH$_2$Cl$_2$, 0 °C (75%)
3. CsF, DMF, 85 °C (55%)

Intramolecular free radical addition has been used to prepare the 1,2-thiazepine derivatives **258** and **259** from **257**; a further intramolecular radical cyclisation was then used to convert **258** to a new azabicyclic system with a bridgehead nitrogen [01JOC3564].

257 **258** (59%) **259** (14%)

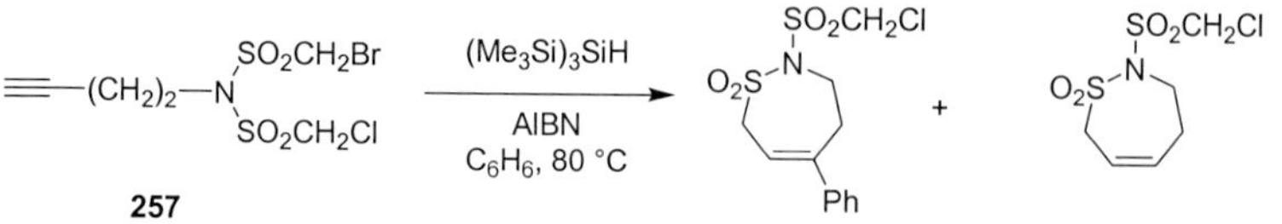

An unusual rhodium-catalysed cyclohydrocarbonylative ring expansion of acetylenic thiazoles (e.g. **260**) has been discovered providing access to 2-(*Z*)-6-(*E*)-4*H*-[1,4]-thiazepin-5-ones (e.g. **261**) in good yields. The process is a general one and a range of functional groups, including ether, ester, and chloro groups, are tolerated. Further study of this reaction will certainly be of interest [01JA1017].

260 **261** (86%)

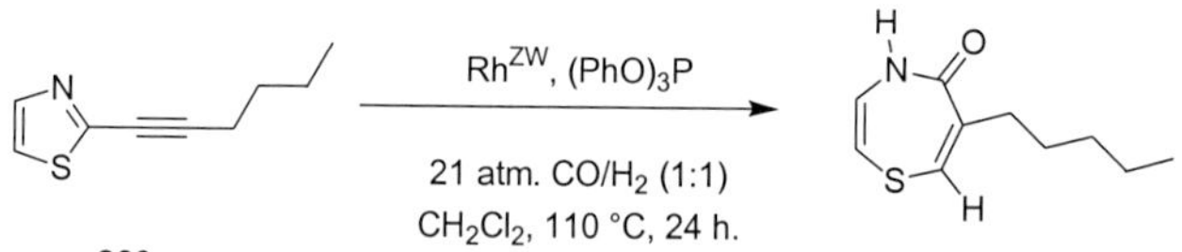

There have been a number of reports on the preparation and chemistry of fused thiazepine derivatives. Toda *et al.* have noted the stereocontrolled synthesis of tetrahydro-1,5-benzothiazepines, *trans*- and *cis*-**263**, on reaction of *o*-aminothiophenol with the epoxide **262** in the presence of magnesium perchlorate; the overall yield of **263** was 94% and the *trans:cis* ratio was 24:76 and the factors controlling the stereochemical outcome were discussed [01H1451].

(2*R**,1'*S**)-**262** *trans*-**263** *cis*-**263**

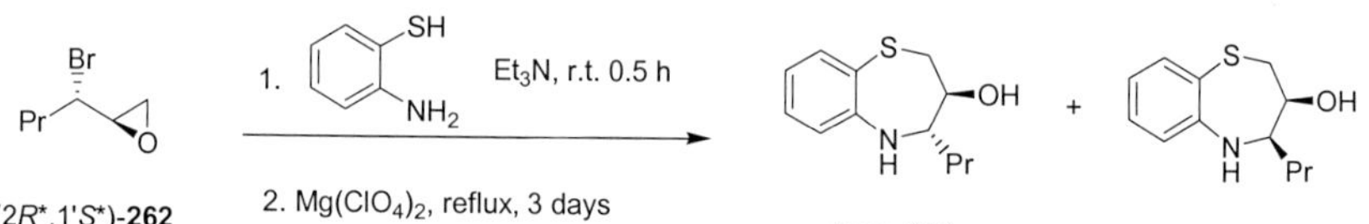

A combinatorial approach to 1,5-benzothiazepine derivatives **264** (as potential antibacterial agents) has also been reported. A ring construction approach is used to prepare these derivatives, with *o*-aminothiophenol being added in step iii [01JCO224].

With a view to combining two bioactive moieties in the one molecule, Wang and co-workers have reported the synthesis of the β-lactam fused 1,5-benzothiazepines **267** (e.g. R^1 = Ph, R^2 = 4-MeOC$_6$H$_4$, R^3 = H, R^4 = Cl) from **265** and the acid chlorides **266** in moderate yields [01JHC561].

264

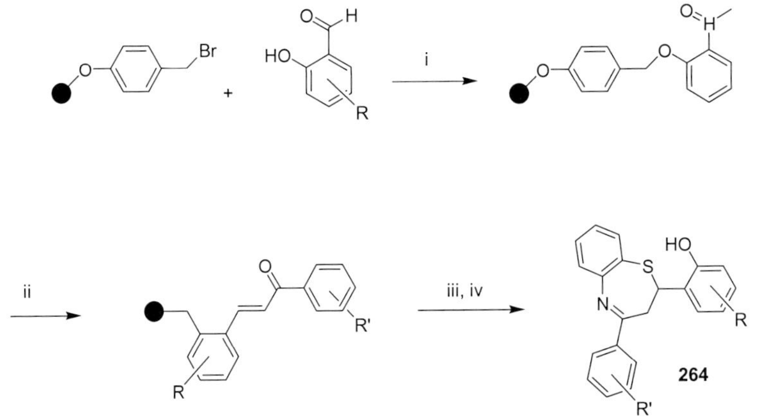

Reagents: (i), Cs_2CO_3, NaI, DMF, 50 °C; (ii), R'-COMe, NaOMe, MeOH, THF; (iii), THF or DMF, cat. AcOH, 60 °C, 5 h and then r.t. overnight; (iv), TFA, r.t, 3 h.

265 266 267

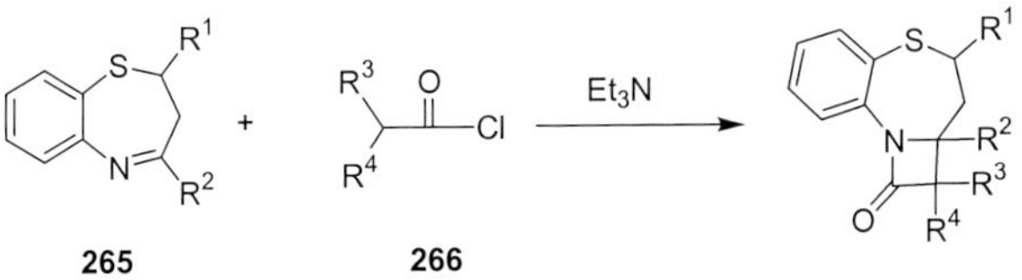

Stannous chloride-mediated reductive cyclisation-rearrangement of the nitro-ketone **269** (obtained from the nitro-acid, **268**) gave the dibenzothiazepine derivative **270** in good overall yield; mechanistically it was proposed that a hydroxylamine intermediate leads to the rearrangement after intramolecular nucleophilic addition to the ketone [02JOC8662].

268 269 270

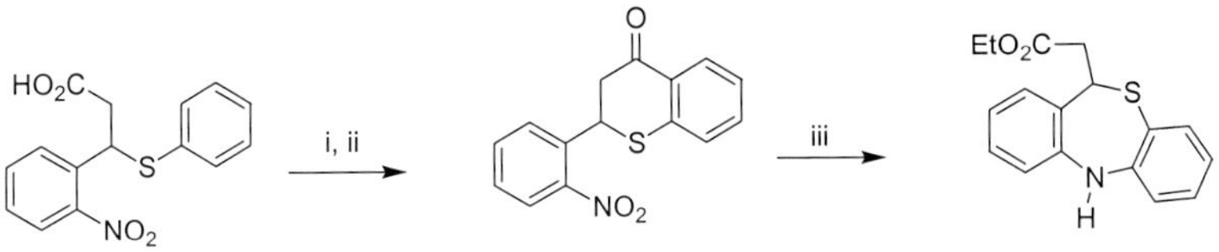

Reagents: (i), $(COCl)_2$, CH_2Cl_2; (ii), $SnCl_4$, CH_2Cl_2 (70%, 2 steps); (iii), 5 equiv of $SnCl_2$, EtOH, reflux (>97%).

Benzotriazole-mediated reactions feature in synthetic routes to the isoindolo- and pyrrolo-fused benzothiazole derivatives **272** (e.g. $R^1 = CH_3$, $R^2 = R^3 = H$) and **273** (e.g. $R^1 = R^3 = H$, $R^2 = Cl$) in good overall yields; the common precursors were the thioamines **271**

[01JOC5590]. Photoinduced decarboxylative cyclisation has been applied to the preparation of the isoindolo-fused thiazepines **275** (m = 3; R = H, CH$_3$) from the salts **274** [01EJO1831].

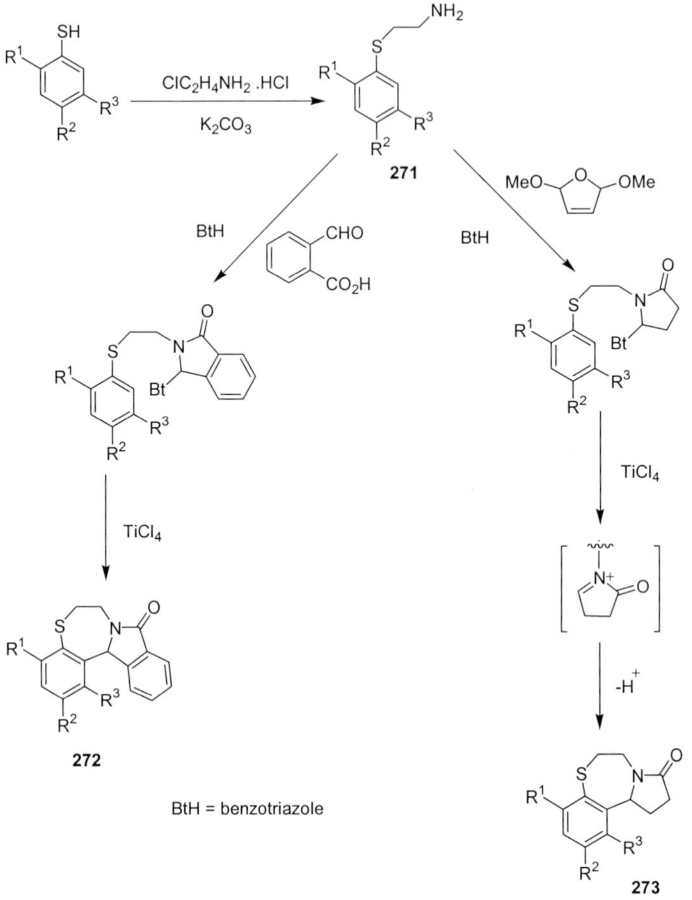

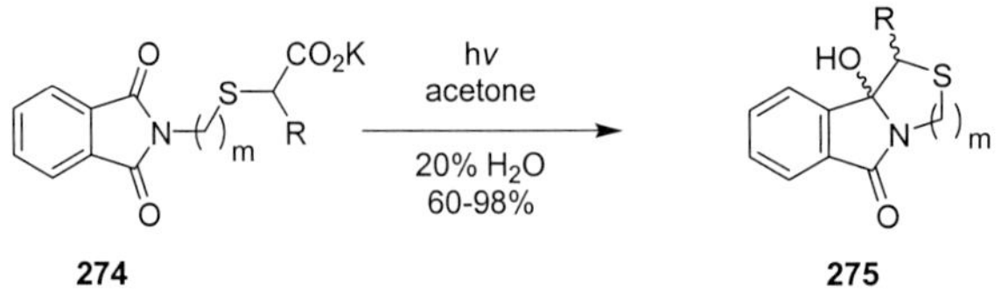

BtH = benzotriazole

Ring-closing metathesis methodology has been used to access 7-membered ring sultones (e.g. **278**, n = 1, m = 1) efficiently from the acyclic diene precursor **277**, which could readily be made in turn fom the appropriate olefinic sulfonyl chloride and alcohol [02SL2019].

276 277 278

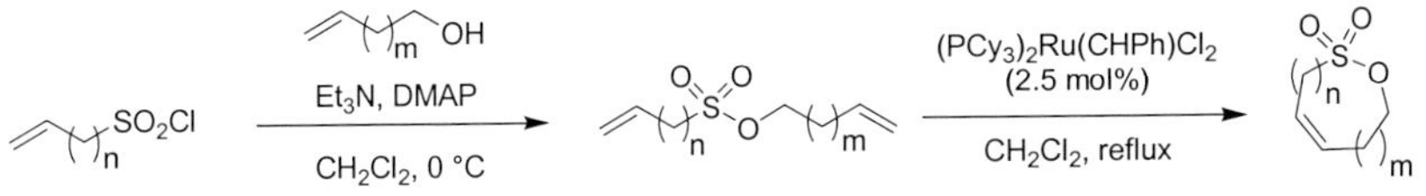

A new approach to dihydro-1,4-oxathiepines **280** (e.g. Ar = Ph, R^1 = Ph, R^2 = Me) has been reported, involving a high yielding base-induced cyclisation of the imides **279** in the key step [01TL4637].

279 280

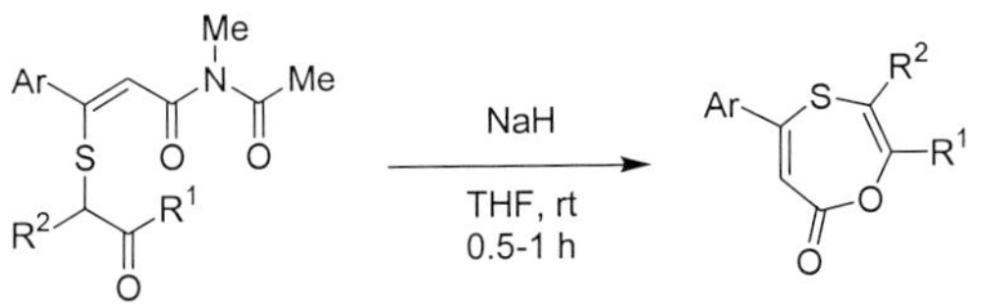

The photoinduced 1,3-proton shift in methyldithiepines (e.g. **281** to **282** and the reverse) has been highlighted as a potential means of modulating hyperpolarizabilities in organic electronic materials [01JOC1894].

281 282

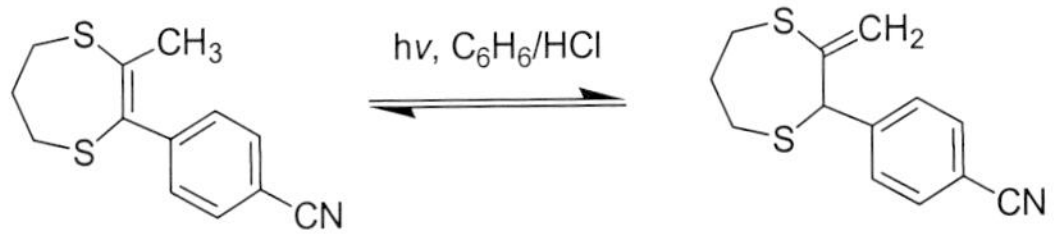

Oxidation of the dihydro-2,4-benzothiepines **283** (R = CH_3, Ph, *t*-Bu) with *m*-chloroperbenzoic acid has been shown to be highly diastereoselective; the resultant *trans*-sulfoxides **284** exist as an equilibrium mixture (observed at –60 °C in $CDCl_3$) of the *chair* and *boat* forms with the substituents equatorially disposed [01JGU960].

283 284

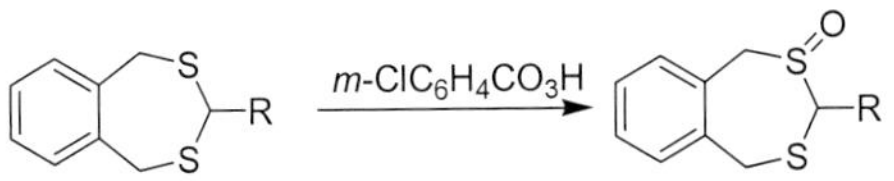

7.4 SEVEN-MEMBERED SYSTEMS CONTAINING THREE OR MORE HETEROATOMS

7.4.1 Systems with N, S and/or O.

In some elegant experimental and interpretative work, Wentrup and co-workers have shown that the triazacycloheptatetraene **287** is a key intermediate in the photolysis (λ >

260nm) in an Ar matrix at cryogenic temperatures of tetrazolopyrazine **285** / 2-azidopyrazine **286**; the final products are the imidazole **288** and the carbodiimide **289** [02JOC8538].

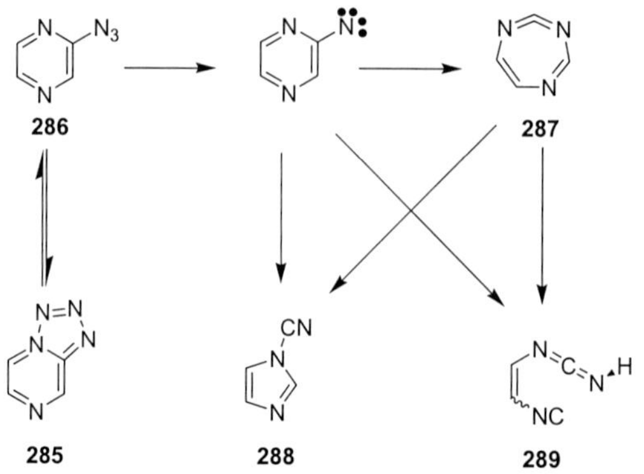

A group of novel triazepine-ruthenium(II) complexes have been described, based on the ligands **290a,b** [01JOM265].

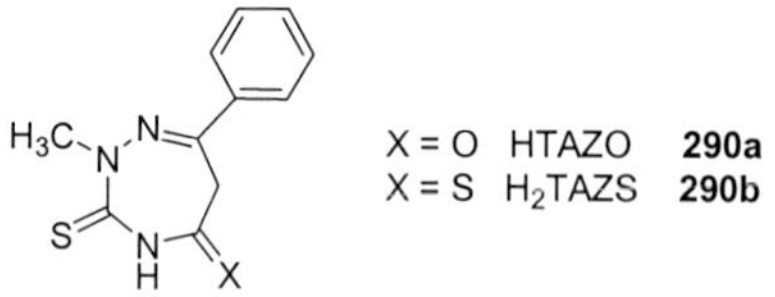

X = O   HTAZO   **290a**
X = S   H$_2$TAZS   **290b**

Parallel solid-phase synthesis has been used to generate a range of 1,7-disubstituted – 1,3,5-triazepane-2,4-diones **291** from resin-bound amino acids; the ring forming step involved reaction of phenyl isocyanato formate with a resin-bound diamine in this efficient sequence [01OL2797].

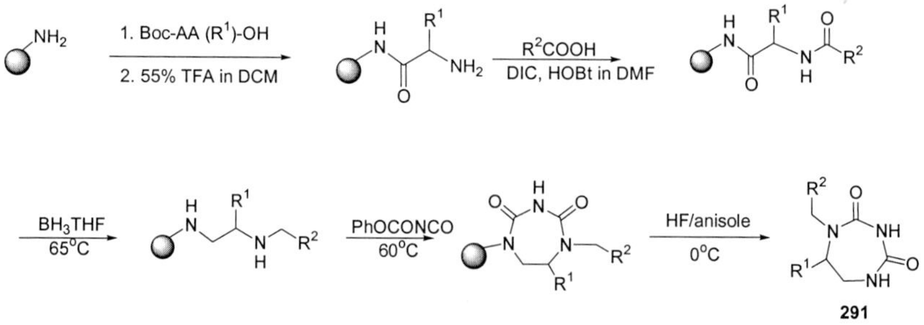

A new efficient synthesis of dihydro-1,3,4-benzotriazepin-5-ones is referred to in a report by Al-Abadelah *et al.* The strategy is based on ring construction starting with the reaction of anthranilic acid and the hydrazonoyl chlorides **292** to give compounds **293**, followed by CDI-mediated cyclisation to **294**. While **294** could also exist as the tautomer **295**, single crystal X-ray crystallographic analysis indicated that tautomer **294** (X = Me) is preferred at least in the solid state [02H2365].

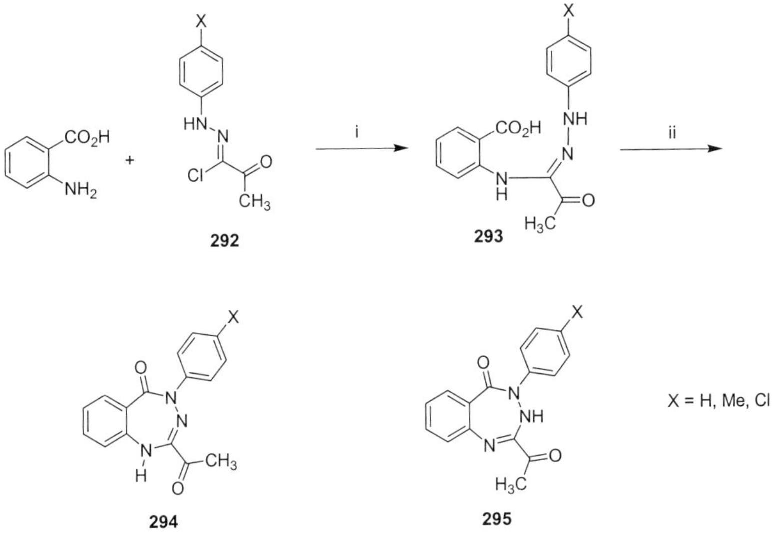

Reagents: (i), aq. MeOH, THF, NEt_3 / 0 °C to 20 °C, 2-3 h; (ii), 1,1'-Carbonyldiimidazole, THF / 20 °C, 1-2 h.

The synthesis and anti-tumour promoting activity of a fused 1,2,4-triazepinone derivative has also been reported by Nagai *et al.* [01JHC1097].

Reaction of perfluoro-5-azanon-4-ene with the amino alcohols **296** in the presence of base affords a convenient route to the oxadiazepine derivatives **297a,b** [01JFC11].

297a X = H, Y = C_2H_5
297b X = Y = CH_3

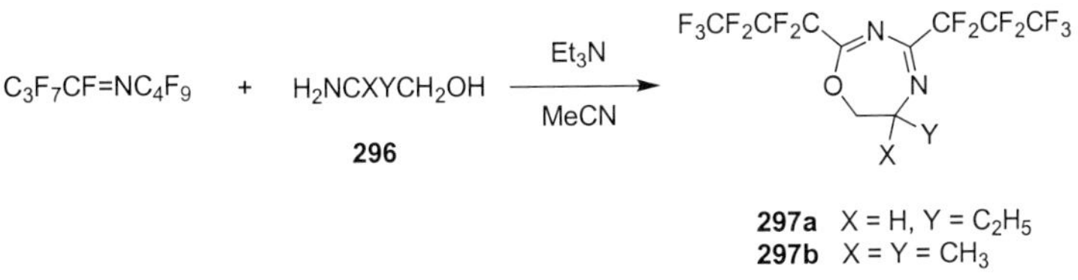

The unstable fused 1,2,6-oxadiazepine intermediate **299** has been reported as resulting from 1,7-electrocyclisation of the non-stabilised azomethine ylide **298** on to the nitro group; subsequent ring contraction afforded the indazole-*N*-oxide **300** [01TL5081].

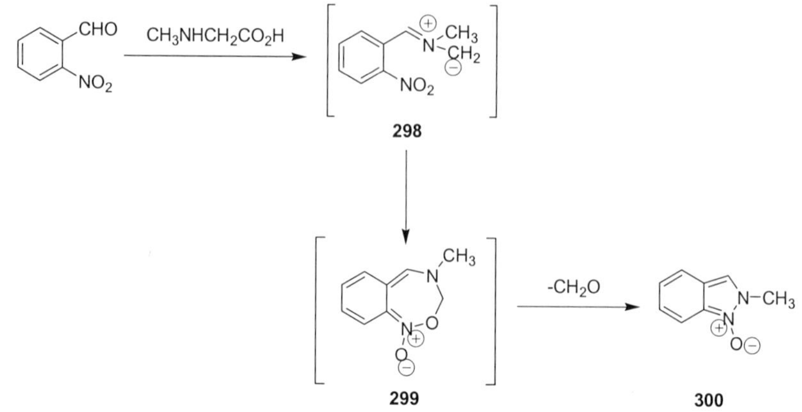

A facile synthesis of the bis-fused oxadiazepines **302** by condensation of the *bis*-aminophenols **301** with ethanedial has been reported by Ochoa and co-workers; yields however were only modest [01T55].

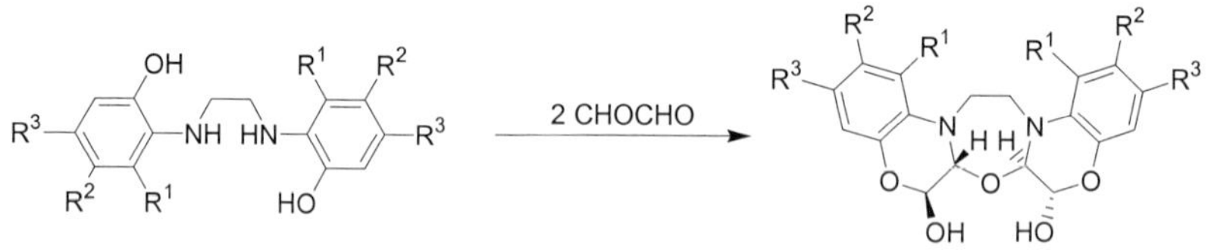

The synthesis of a series of symmetric cyclic sulfamides **303** by a ring construction methodology has been reported; these compounds are of interest as HIV-1 protease inhibitors and formed part of SAR studies in this area [01JMC155]. Single crystal X-ray crystallographic studies have been reported on the substituted 1,4,5-thiadiazepine **304**, together with its corresponding 1,1-dioxide derivative; both compounds possess exact C_2 symmetry and the 7-membered ring conformations (*twist-boat*) are similar in both cases [01AX431].

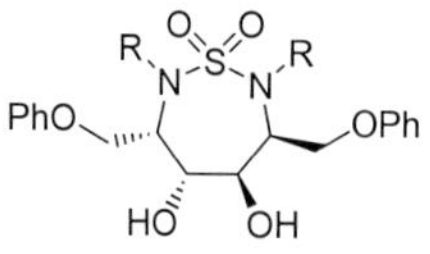

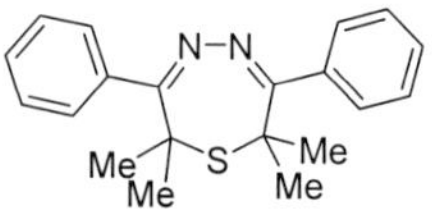

Another example of microwave-assisted condensation involves the formation of the antimicrobial triazolo-1,3,4-thiadiazepines **307** from **305** and **306** in the presence of basic alumina but in the absence of solvent; yields were very good (e.g. **307**, R = 1-tetrazolylmethyl, R' = 4-MeOC$_6$H$_4$, X = H, 90%) [01BMC217].

305 306 307

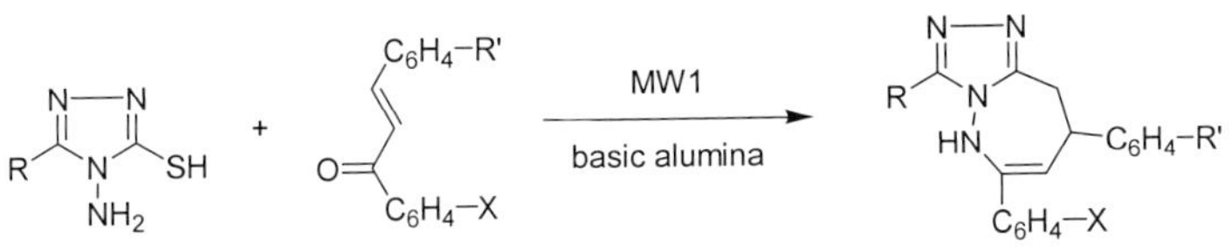

The acid-catalysed reaction of acetonitrile with either the benzo-1,2,3-dioxathepane 2-oxide **308** [01JGU150] or the dihydro-2,3,4-benzodioxasilepine **308a** [01JGU295] affords the benzoxazepine derivative **309**, although in very low yield from the latter; alkaline hydrolysis then yields the amino alcohol **310**.

308 308a 309 310

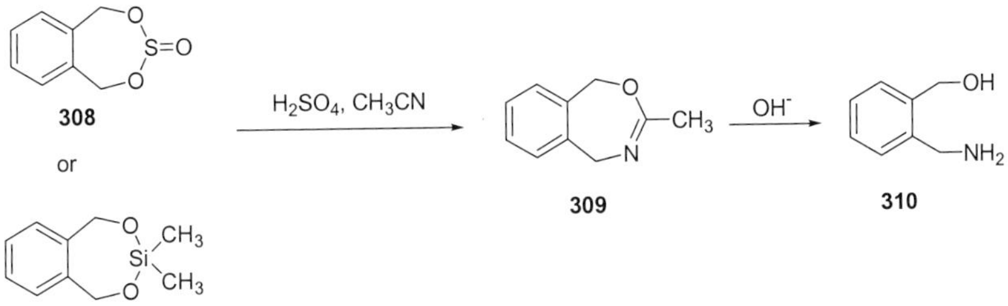

Other examples of 7-membered ring heterocycles with three heteroatoms in the ring include the chiral titanium complex **311** [01SL1889], the novel phosphoramidite ligand **312** [01SL1375], and the cyclipostin natural products from a *Streptomyces* species (e.g. Cyclipostin N, **313**), which is a hormone-sensitive lipase inhibitor [02JAN480].

311 312 313

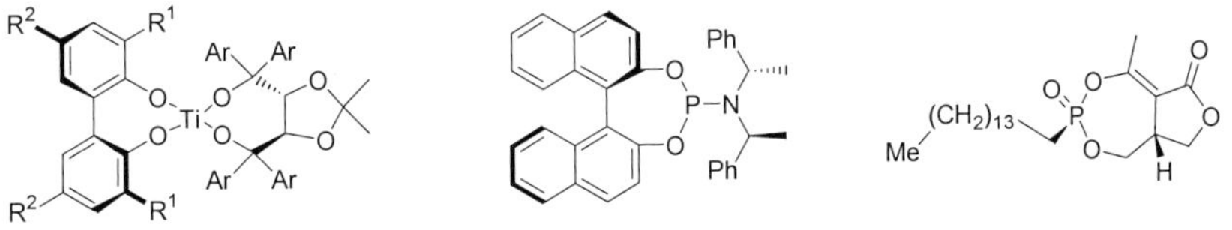

7.4.2 Miscellaneous systems

The novel *bis*-fused 7-membered ring derivatives **316** have been prepared in high yields by reacting, in a 1:2 stoichiometric ratio in toluene or acetonitrile, the salen ligands **314a,b** with the arylboronic acids **315**; reaction in ethanol however gave mainly open bimetallic boronic esters [01IC6405].

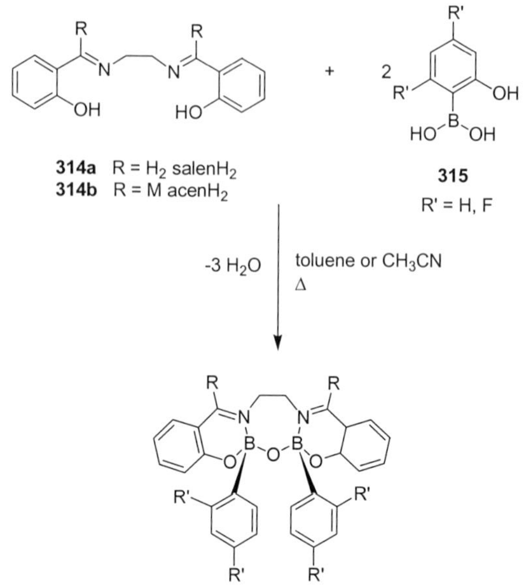

314a R = H₂ salenH₂
314b R = M acenH₂

315 R' = H, F

−3 H₂O | toluene or CH₃CN, Δ

316

Interest continues to be shown in the naturally occurring benzopentathiepine, varacin, in view of its cytotoxicity [01JA10379]. Analogues have also been made, for example, **319**, by reaction of the dithiol, **317**, with S₈, and subsequent removal of the Boc protecting group from **318** [01H145].

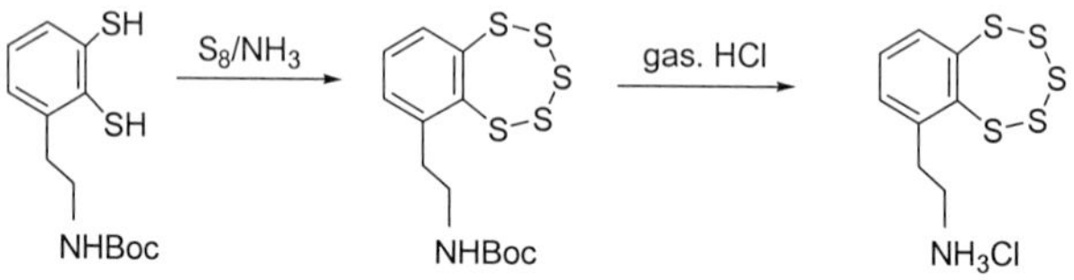

317 → (S₈/NH₃) → **318** → (gas. HCl) → **319**

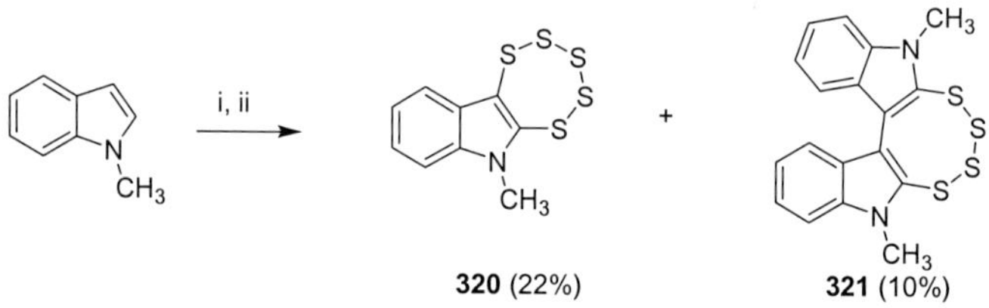

320 (22%) + **321** (10%)

Reagents: (i), n-BuLi, THF, −78 to −10 °C; (ii), S₈ (20 equiv.), −10 °C, 30 min, AcOH.

Rewcastle and co-workers have also reported a concise synthesis of an indole-fused analogue **320**, together with some of the dimeric derivative **321** [01T7185], while Rees *et al.* have described a novel oxime **322** to pentathiepine **324** cascade reaction induced by S_2Cl_2; the 1,2,3-dithiazole **323** is also formed [01CC403].

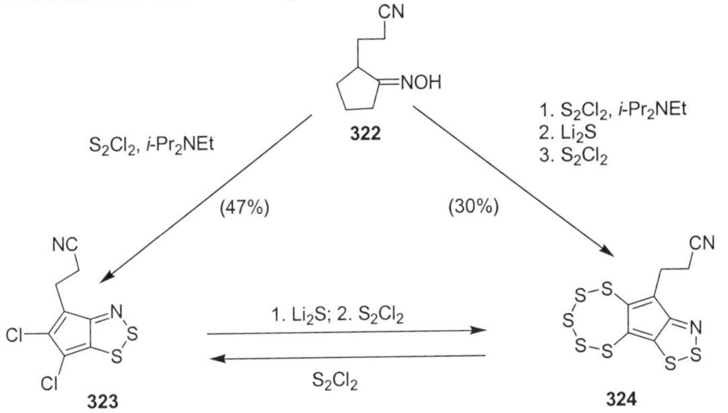

Metal complexes of functionalised sulfur-containing ligands, including the 1,2,3,5,6-pentathiepane **326**, have been described; compound **326** was made by oxidation of **325** with *m*-chloroperbenzoic acid [01ZAAC1518].

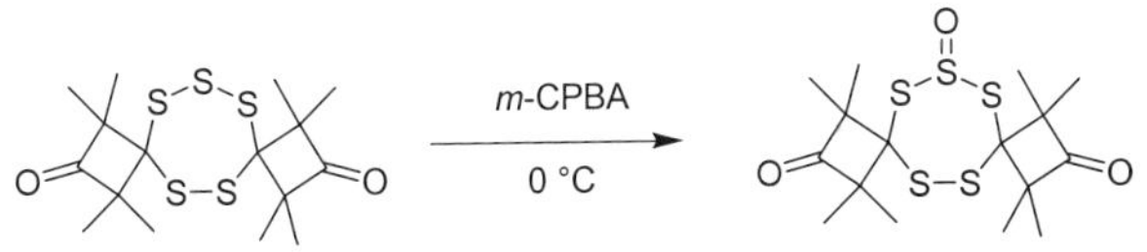

7.5 SEVEN-MEMBERED SYSTEMS OF PHARMACOLOGICAL SIGNIFICANCE

A review on the antidepressant tianeptine, a dibenzo[*c,f*][1,2]thiazepine derivative, has appeared [01CNS231].

The antimalarial, artemisinin, which has an embedded 1,2-dioxapane moiety within its structure, together with a range of related synthetic peroxides, is treated in a detailed review by Jefford [01MI1803]; structure-activity and mode of action considerations are also reviewed.

Amongst other examples of pharmacologically active seven-membered heterocyclic ring compounds the following are of interest: 5-(2-morpholin-4-yl-ethoxy)-benzofuran-2-carboxylic acid (*S*)-3-methyl-1-{3-oxo-1-[2-(3-pyridin-2-yl-phenyl)-ethenoyl]azepan-4-ylcarbanoyl}-butyl-amide (SB331750) which is a potent non-peptide inhibitor of rat cathepsin K (and hence of interest in the control of osteoclast-mediated bone resorption) [02MI746], *N*-substituted azepanes as non-peptide inhibitors of caspases 3 and 7 [01JMC2015] and as estrogens [01JMC1654], a 7*H*-oxepin-2-one (klaivanolide, from the plant *Uvaria klaineana*) as a potent antileishmanial agent [02P885], AS-8112 (**327**), a novel antagonist at dopamine D_2/D_3 and 5-HT_3 receptors [01BJP253], [01EJP361], and [02CPB941], a benzo[*e*][1,4]diazepine derivative (**328**) as a γ-secretase inhibitor [01JBC45394], an imidazo-diazepine nucleosides as AMP deaminase inhibitors [01JMC613], a thiazolo-azepine as a weak modulator of ligand response in $G_α$-protein coupled $α_{2A}$-adrenoceptors [01BP1079] and an oxazolo-azepine (BHT933) in adrenoceptor identification [01LS143], a pyrido[2,3-*b*]azepine derivative as a potent integrin receptor antagonist [02PCT], quaternary ammonium salts with lateral dibenz[*b*,*f*]azepine substituents as allosteric modulators of muscarinic receptors [01AP121], dibenz[*b*,*f*]azepines as new GABA uptake inhibitors [01JMC2152], dialkyldipyridazinodiazepinones as potential HIV-1 reverse transcriptase inhibitors related to nevirapine [01JHC125], pharmacological effects *in vitro* and *in vivo* of dibenzo[*c*,*f*]pyrazino[1,2-*a*]azepine (**329**) (6-methoxymianserin) [02JMC3280], triazolothiadiazepines as potential anti-inflammatory, analgesic and anthelmintic agents [02IJC(B)1712], the reversal of cisplatin resistance by the 1,4-benzothiazepine derivative, JTV-519 (**330**) [01MI597], the pharmacology and associated properties of the antiallergic drug olopatadine hydrochloride, a dibenz[*b*,*e*]oxepine derivative [02MI379], and the anti-angionic effect of ozonides, including the ANO compound (**331**) [02MI220].

7.6 FUTURE DIRECTIONS

While further examples of the powerful ring closing metathesis methodology appeared in 2001/2, the full potential of this ring forming technique is yet to be realised in heterocyclic synthesis. Microwave assisted syntheses of seven-membered heterocyclic systems are making more of an impact, and this is predicted to grow with the availability of custom-designed laboratory microwave irradiation units; cost, however, is still a limitation on widespread adoption.

The use of classical heterocyclic scaffolds (e.g. dibenzazepines) in new medicinal agent design is likely to continue to grow in the context of combinational synthetic methods. The appearance of new spirocyclic seven-membered heterocyclic systems is still not large, yet the scope here continues to be significant particularly with respect to novel molecular entities for specific pharmacological activity. The discovery of a range of new, efficient methodologies for the synthesis of such spiro systems is a challenge for the future.

7.7 REFERENCES

01AG(E)177	H. Frauenrath, D. Brethauer, S. Reim, M. Maurer, G. Raabe, *Angew. Chem. Int. Ed.* **2001**, *40*, 177.
01AP121	R. Li, C. Tränkle, K. Mohr, A. Holzgrabe, *Arch. Pharm. Pharm. Med. Chem.* **2001**, *334*, 121.
01AX431	E. Cuthbertson, C.S. Frampton, D.D. MacNicol, *Acta Crystallogr.* **2001**, *C57*, 431.
01BJP253	T. Yoshikawa, N. Yoshida, M. Oka, *Br. J. Pharmacol.* **2001**, *133*, 253.
01BMC217	M. Kidwai, P. Sapra, P. Misra, R.K. Saxena, M. Singh, *Bioorg. Med. Chem.* **2001**, *9*, 217.
01BMCL1009	S. Pikul, K.M. Dunham, N.G. Almstead, B. De, M.G. Natchus, Y.O. Taiwo, L.E. Williams, B.A. Hynd, L.C. Hsieh, M.J. Janusz, F. Gu, G.E. Mieling, *Bioorg. Med. Chem. Lett.* **2001**, *11*, 1009.
01BP1079	P.J. Pauwels, S. Tardif, F.C. Colpaert, T. Wurch, *Biochem. Pharmacol.* **2001**, *61*, 1079.
01CC403	S. Macho, C.W. Rees, T. Rodríguez, T. Torroba, *Chem. Commun.* **2001**, 403.
01CC2080	B.C.H. May, A.D. Abell, *Chem. Commun.* **2001**, 2080.
01CNS231	A.J. Wagstaff, D. Ormrod, C.M. Spencer, *CNS Drugs.* **2001**, *15*, 231.
01EJO1525	J-M. Coustard, *Eur. J. Org. Chem.* **2001**, 1525.
01EJO1831	A.G. Griesbeck, M. Oelgemöller, J. Lex, A. Haeuseler, M. Schmittel, *Eur. J. Org. Chem.* **2001**, 1831.
01EJO3313	K. Knobloch, M. Keller, W. Eberbach, *Eur. J. Org. Chem.* **2001**, 3313.
01EJP361	T. Yoshikawa, N. Yoshida, M. Oka, *Eur. J. Pharmacol.* **2001**, *431*, 361.
01GC143	S.M. Patel, U. van Chudasama, P.A. Ganeshpure, *Green Chem.* **2001**, *3*, 143.
01H145	R. Sato, T. Ohyama, T. Kawagoe, M. Baba, S. Nakajo, T. Kimura, S. Ogawa, *Heterocycles* **2001**, *55*, 145.
01H567	K. Saito, Y. Emoto, *Heterocycles* **2001**, *54*, 567.
01H855	Y. Sugita, C. Kimura, I. Yokoe, *Heterocycles* **2001**, *55*, 855.
01H967	Y. Horiguchi, T. Saitoh, S. Terakado, K. Honda, T. Kimura, J. Toda, T. Sano, *Heterocycles* **2001**, *54*, 967.
01H1057	A. Kotali, E. Lazaridou, V.P. Papageorgiou, P.A. Harris, *Heterocycles* **2001**, *55*, 1057.
01H1443	B. Kaboudin, K. Navaee, *Heterocycles* **2001**, *55*, 1443.
01H1451	M. Karikomi, K. Ayame, T. Toda, *Heterocycles* **2001**, *55*, 1451.
01H1455	M. Decker, M. Faust, M. Wedig, M. Nieger, U. Holzgrabe, J. Lehmann, *Heterocycles* **2001**, *55*, 1455.
01H1519	J.Y. Lee, N.J. Baek, S.J. Lee, H. Park, Y.S. Lee, *Heterocycles* **2001**, *55*, 1519.
01H1919	M. Kawase, H. Koiwai, T. Tanaka, S. Tani, H. Miyamae, *Heterocycles* **2001**, *55*, 1919.
01HCA1976	J-R. Weng, M-I. Chung, M-H. Yen, C-N. Lin, *Helv. Chim. Acta* **2002**, *84*, 1976.
01IC6405	M. Sánchez, H. Höpfl, M-E. Ochoa, N. Farfán, R. Santillan, R. Sojas, *Inorg. Chem.* **2001**, *40*, 6405.
01JA1017	B.G. Van den Hoven, H. Alper, *J. Am. Chem. Soc.* **2001**, *123*, 1017.
01JA10379	A. Greer, *J. Am. Chem. Soc.* **2001**, *123*, 10379.
01JBC45394	D. Beher, J.D.J. Wrigley, A. Nadin, G. Evin, C.L. Masters, T. Harrison, J.L. Castro, M.S. Shearman, *J. Biol. Chem.* **2001**, *276*, 45394.
01JCO224	F. Micheli, F. Degiorgis, A. Feriani, A. Paio, A. Pozzan, P. Zarantonello, P. Seneci, *J. Comb. Chem.* **2001**, *3*, 224.
01JCS(P1)279	P. Wyatt, S. Warren, M. McPartlin, T. Woodroffe, *J. Chem. Soc. Perkin Trans. 1* **2001**, 279.
01JFC1	J. Kvíčala, P. Hovorková, O. Paleta, *J. Fluorine Chem.* **2001**, *108*, 1.
01JFC11	K-W. Chi, H-A. Kim, G.G. Furin, E.L. Zhuzhgov, N. Protzuk, *J. Fluorine Chem.* **2001**, *110*, 11.
01JGU150	V.V. Kuznetsov, Y.E. Brusilovskii, A.V. Mazepa, S.P. Krasnoshchekaya, *J. Gen. Chem. USSR* **2001**, *71*, 150.
01JGU295	V.V. Kuznetsov, S.A. Bochlor, Y.E. Brusilovskii, A.V. Mazepa, *J. Gen. Chem. USSR* **2001**, *37*, 295.
01JGU960	P.A. Kikilo, B.I. Khairutdinov, R.A. Shaikhutdinov, Y.G. Shtyrlin, V.V. Klochkov, E.N. Klimovitskii, *J. Gen. Chem. USSR* **2001**, *71*, 960.
01JHC89	Y.K. Koh, K-H. Bang, H-S. Kim, *J. Heterocycl. Chem.* **2001**, *38*, 89.
01JHC125	G. Heinisch, B. Matuszczak, E. Spielmann, M. Witvrouw, C. Pannecouque, E. De Clercq, *J. Heterocycl. Chem.* **2001**, *38*, 125.
01JHC383	B.C. Sekhar, D.V. Ramana, S.R. Ramadas, *J. Heterocycl. Chem.* **2001**, *38*, 383.
01JHC561	Q-Y. Xing, H-Z. Wang, X. Zhou, S. Jin, Y-M. Li, A.S.C. Chan, *J. Heterocycl. Chem.* **2001**, *38*, 561.

01JHC579	J.K. Gallos, C.C. Dellios, *J. Heterocycl. Chem.* **2001**, *38*, 579.
01JHC895	M.E. Hedrera, I.A. Perillo, B. Fernández, *J. Heterocycl. Chem.* **2001**, *38*, 895.
01JHC1011	A. Lévai, *J. Heterocycl. Chem.* **2001**, *38*, 1011.
01JHC1031	H-Z. Wang, X. Zhou, J-X. Xu, S. Jin, Y-M. Li, A.S.C. Chan, *J. Heterocycl. Chem.* **2001**, *38*, 1031.
01JHC1097	S-I. Nagai, S. Takemoto, T. Ueda, K. Mizutani, Y. Uozumi, H. Tokuda, *J. Heterocycl. Chem.* **2001**, *38*, 1097.
01JMC155	W. Schaal, A. Karlsson, G. Ahlsén, J. Lindberg, H.O. Andersson,, U.H. Danielson, B. Classon, T. Unge, B. Samuelsson, J. Hultén, A. Hallberg, A. Karlén, *J. Med. Chem.* **2001**, *44*, 155.
01JMC613	S.R. Kasibhatla, B.C. Bookser, W. Xiao, M.D. Erion, *J. Med. Chem.* **2001**, *44*, 613.
01JMC1654	C.P. Miller, M.D. Collini, B.D. Tran H.A. Harris, Y.P. Kharode, J.T. Marzolf, R.A. Moran, R.A. Henderson, R.H.W. Bender, R.J. Unwalla, L.M. Greenberger, J.P. Yardley, M.A. Abou-Gharbia, C.R. Lyttle, B.S. Komm, *J. Med. Chem.* **2001**, *44*, 1654.
01JMC2015	D. Lee, S.A. Long, J.H. Murray, J.L. Adams, M.E. Nuttall, D.P. Nadeau, K. Kikly, J.D. Winkler, C-M. Sung, M.D. Ryan, M.A. Levy, P.M. Keller, W.E. DeWolf Jr., *J. Med. Chem.* **2001**, *44*, 2015.
01JMC2152	K.E. Andersen, J.L. Sørensen, J. Lau, B.F. Lundt, H. Petersen, P.O. Huusfeldt, P.D. Suzdak, M.D.B. Swedberg, *J. Med. Chem.* **2001**, *44*, 2152.
01JNP85	S-B. Chen, G-Y. Gao, H-W. Leung, H-W. Yeung, J-S. Yang, P-G. Xiao, *J. Nat. Prod.* **2001**, *64*, 85.
01JNP125	J-R. Dai, B.K. Carte, P.J. Sidebottom, A.L.S. Yew, S-B. Ng, Y. Huang, M.S. Butler, *J. Nat. Prod.* **2001**, *64*, 125.
01JNP147	C. Ito, M. Itogawa, Y. Mishina, H. Tomiyasu, M. Litaudon, J-P. Cosson, T. Mukainaka, H. Tokuda, H. Nishino, H. Furukawa, *J. Nat. Prod.* **2001**, *64*, 147.
01JNP308	T. Hu, I.W. Burton, A.D. Cembella, J.M. Curtis, M.A. Quilliam, J.A. Walter, J.L.C. Wright, *J. Nat. Prod.* **2001**, *64*, 308.
01JNP415	M. Dreyer, B.W. Nugroho, F.I. Bohnenstengel, R. Ebel, V. Wray, L. Witte, G. Bringmann, J. Mühlbacher, M. Herold, P.D. Hung, L.C. Kiet, P. Proksch, *J. Nat. Prod.* **2001**, *64*, 415.
01JNP421	H.M. Garraffo, P. Jain, T.F. Spande, J.W. Daly, T.H. Jones, L.J. Smith, V.E.Zottig, *J. Nat. Prod.* **2001**, *64*, 421.
01JOC1733	Z. Thale, F.R. Kinder, K.W. Bair, J. Bontempo, A.M. Cruchta, R.W. Versace, P.E. Phillips, M.L. Sanders, S. Wattanasin, P. Crews, *J. Org. Chem.* **2001**, *66*, 1733.
01JOC1894	Y. Wan, A. Kurchan, A. Kutateladze, *J. Org. Chem.* **2001**, *66*, 1894.
01JOC2429	G-J. ten Brink, J-M. Vis, I.W.C.E. Arends, R.A. Sheldon, *J. Org. Chem.* **2001**, *66*, 2429.
01JOC2784	A. Witt, J. Bergman, *J. Org. Chem.* **2001**, *66*, 2784.
01JOC2881	W-P. Hu, J-J. Wang, F-L. Lin, Y-C. Lin, S-R. Lin, M-H. Hsu, *J. Org. Chem.* **2001**, *66*, 2881.
01JOC3564	L.A. Paquette, C.S. Ra, J.D. Schloss, S.M. Leit, J.C. Gallucci, *J. Org. Chem.* **2001**, *66*, 3564.
01JOC5590	A.R. Katritzky, Y-J. Xu, H-Y. He, S. Mehta, *J. Org. Chem.* **2001**, *66*, 5590.
01JOM149	T. Kondo, T. Okada, T. Suzuki, T. Mitsudo, *J. Organomet. Chem.* **2001**, *622*, 149.
01JOM160	I. Ahmad, M.L. Falck-Pedersen, K. Undheim, *J. Organomet. Chem.* **2001**, *625*, 160.
01JOM265	M. Aitali, M.Y.A. Itto, A. Hasnaoui, A. Riahi, A. Karim, J-C. Daram, *J. Organomet. Chem.* **2001**, *619*, 265.
01LS143	E.W. Willems, L.F. Valdivia, E.R-S. Juan, P.R. Saxena, S.M. Villalón, *Life Sci.* **2001**, *69*, 143.
01MC109	N.B. Chernysheva, A.V. Samet, V.N. Marshalkin, V.A. Polukeev, V.V. Semenov, *Mendeleev Commun.* **2001**, 109.
01MI31	M.M. McGee, G. Campiani, A. Ramunno, C. Fattorusso, V. Nacci, M. Lawler, D.C. Williams, D.M. Zisterer, *J. Pharmacol. Exp. Therap.* **2001**, *296*, 31.
01MI183	N. Kim, A.F. El-Kattan, C.S. Asbill, R.J. Kennette, J.W. Sowell Sr., R. Latour, B.B. Michniak, *Journal of Controlled Release*. **2001**, *73*, 183.
01MI363	P. Coelho, L. Blanco, *Main Group Metal Chem.* **2001**, *24*, 363.
01MI597	T. Nakamura, F. Koizumi, N. Kaneko, T. Tamura, F. Chiwaki, Y. Koh, S. Akutagawa, N. Saijo, K. Nishio, *Jpn. J. Cancer Res.* **2001**, *92*, 597.
01MI1803	C.W. Jefford, *Current Medicinal Chemistry* **2001**, *8*, 1803.
01MRC471	E. Gößnitzer, W. Wendelin, *Magn. Reson. Chem.* **2001**, *39*, 471.
01NAT423	A. Corma, L.T. Nemeth, M. Renz, S. Valencia, *Nature* **2001**, *412*, 423.
01NJC102	R.W. Hoffmann, J. Krüger, D. Brückner, *New J. Chem.* **2001**, *25*, 102.
01OL161	S.R. Gilbertson, Z. Fu, *Org. Lett.* **2001**, *3*, 161.
01OL1053	D.J. Wardrop, A. Basak, *Org. Lett.* **2001**, *3*, 1053.

01OL1089	L. Abrous, J. Hynes Jr., S.R. Friedrich, A.B. Smith III., R. Hirschmann, *Org. Lett.* **2001**, *3*, 1089.
01OL2797	Y. Yu, J.M. Ostrech, R.A. Houghten, *Org. Lett.* **2001**, *3*, 2797.
01OL3741	M.P. Doyle, W. Hu, D.J. Timmons, *Org. Lett.* **2001**, *3*, 3741.
01SC3281	M. Longchar, A. Chetia, S. Ahmed, R. C. Boruah, J. S. Sandhu, *Synth. Commun.* **2001**, *31*, 3281.
01SL757	E. Licandro, S. Maiorana, B. Vandoni, D. Perdicchia, P. Paravidino, C. Baldoli, *Synlett* **2001**, 757.
01SL803	M. Catellani, C. Catucci, G. Celentano, R. Ferraccioli, *Synlett* **2001**, 803.
01SL848	L. Chacun-Lefèvre, B. Joseph, J-Y. Mérour, *Synlett* **2001**, 848.
01SL1375	A. Alexakis, S. Rosset, J. Allamand, S. March, F. Guillen, C. Benhaim, *Synlett* **2001**, 1375.
01SL1423	J-B. Clement, J.F. Hayes, H.M. Sheldrake, P.W. Sheldrake, A.S. Wells, *Synlett* **2001**, 1423.
01SL1506	T. Ito, N. Yamazaki, C. Kibayashi, *Synlett* **2001**, 1506.
01SL1784	M. Moreno-Mañas, R. Pleixats, A. Santamaria, *Synlett* **2001**, 1784.
01SL1889	M. Ueki, Y. Matsumoto, J.J. Jodry, K. Mikami, *Synlett* **2001**, 1889.
01SL1953	B. Zaleska, D. Cież, J. Lech, *Synlett* **2001**, 1953.
01T55	M.E. Ochoa, S. Rojas-Lima, H. Höpfl, P. Rodríguez, D. Castillo, N. Farfán, R. Santillan, *Tetrahedron* **2001**, *57*, 55.
01T3645	T. Nishinaga, Y. Izukawa, K. Komatsu, *Tetrahedron* **2001**, *57*, 3645.
01T3951	M.A. Trujillo, J.A. Gómez, J. Campos, A. Espinosa, M.A. Gallo, *Tetrahedron* **2001**, *57*, 3951.
01T7181	P. Kahnberg, O. Sterner, *Tetrahedron* **2001**, *57*, 7181.
01T7185	G.W. Rewcastle, T. Janosik, J. Bergman, *Tetrahedron* **2001**, *57*, 7185.
01TA975	J.E.G. Kemp, R.A. Bass, J. Bordner, P.E. Cross, R.F. Gammon, J.A. Price, *Tetrahedron: Asymmetry*, **2001**, *12*, 975.
01TA2205	A. Costa, C. Nájera, J.M. Sansano, *Tetrahedron: Asymmetry*, **2001**, *12*, 2205.
01TL141	R.A. Aitken, G.M. Buchanan, N. Karodia, T. Massil, R.J. Young, *Tetrahedron Lett.* **2001**, *42*, 141.
01TL573	A. Chihab-Eddine, A. Daïch, A. Jilale, B. Decroix, *Tetrahedron Lett.* **2001**, *42*, 573.
01TL1095	Y. Sugita, C. Kimura, H. Hosoya, S. Yamadoi, I. Yokoe, *Tetrahedron Lett.* **2001**, *42*, 1095.
01TL1543	R. Matsumara, T. Suzuki, H. Hagiwara, T. Hoshi, M. Ando, *Tetrahedron Lett.* **2001**, *42*, 1543.
01TL2305	B. Villaume, C. Gérardin-Charbonnier, S. Thiébaut, C. Selve, *Tetrahedron Lett.* **2001**, *42*, 2305.
01TL2469	M. Yus, F. Foubelo, *Tetrahedron Lett.* **2001**, *42*, 2469.
01TL4637	B.S. Kim, K. Kim, *Tetrahedron Lett.* **2001**, *42*, 4637.
01TL5081	M. Nyerges, I. Fejas, A. Virányi, P.W. Groundwater, L. Toke, *Tetrahedron Lett.* **2001**, *42*, 5081.
01TL5459	E. Yoneda, S-W. Zhang, K. Onitsuka, S. Takahashi, *Tetrahedron Lett.* **2001**, *42*, 5459.
01TL5749	H. Ovaa, G.A. van der Marel, J.H. van Boom, *Tetrahedron Lett.* **2001**, *42*, 5749.
01TL6219	H. Imai, H. Uehara, M. Inoue, H. Oguri, T. Oishi, M. Hirama, *Tetrahedron Lett.* **2001**, *42*, 6219.
01TL7237	B. Danieli, G. Lesma, D. Passarella, A. Sacchetti, A. Silavani, *Tetrahedron Lett.* **2001**, *42*, 7237.
01ZAAC1518	W. Weigand, R. Wünsch, K. Polborn, G. Mloston, *Z. Anorg. Allg. Chem.* **2001**, *627*, 1518.
02CC634	K. Tuhina, D.R. Bhowmik, R.V. Venkateswaran, *Chem. Commun.* **2002**, 634.
02CC1956	A.P. Henderson, E. Mutlu, A. Leclercq, C. Bleasdale, W. Clegg, R.A. Henderson, B.T. Golding, *Chem. Commun.* **2002**, 1956.
02CPB941	Y. Hirokawa, H. Harada, T. Yoshikawa, N. Yoshida, S. Kato, *Chem. Pharm. Bull.* **2002**, *50*, 941.
02EJO3801	G. Túrós, A. Csámpai, T. Lovász, A. Györfi, H. Wamhoff, P. Sohár, *Eur. J. Org. Chem.* **2002**, 3801.
02H11	T. Yamaguchi, S. Ito, N. Mibu, K. Sumoto, *Heterocycles* **2002**, *57*, 11.
02H85	K. Nakashima, M. Imoto, T. Miki, T. Miyake, N. Fujisaki, S. Fukunaga, R. Mizutani, M. Sono, M. Tori, *Heterocycles* **2002**, *56*, 85.
02H113	Y. Sakamoto, M. Koshizuka, H. Koshino, T. Nakata, *Heterocycles* **2002**, *56*, 113.
02H123	H. Yamashita, T. Ohtani, S. Morita, K. Otsubo, K. Kan, J. Matsubara, K. Kitano, Y. Kawano, M. Uchida, F. Tabusa, *Heterocycles* **2002**, *56*, 123.
02H223	K. Satake, Y. Kubota, H. Okamoto, M. Kimura, *Heterocycles* **2002**, *57*, 223.
02H379	F. Po gan, S. Polanc, M. Kocevar, *Heterocycles* **2002**, *56*, 379.
02H425	N. Nishiwaki, T. Ogihara, M. Tamura, N. Asaka, K. Hori, Y. Tohda, M. Ariga, *Heterocycles* **2002**, *56*, 425.

02H449	A. Chihab-Eddine, A. Daich, A. Jilale, B. Decroix, *Heterocycles* **2002**, *58*, 449.
02H635	T. Ohtani, K. Kitano, J. Matsubara, S. Morita, Y. Kawano, M. Komatsu, M. Bando, M. Kido, M. Uchida, F. Tabusa, *Heterocycles* **2002**, *58*, 635.
02H1063	Y. Horiguchi, T. Saitoh, T. Kashiwagi, L. Katura, M. Itagaki, J. Toda, T. Sano, *Heterocycles* **2002**, *57*, 1063.
02H1501	T-I. Ho, W-S. Chen, Y-M. Tsai, J-M Fang, *Heterocycles* **2002**, *57*, 1501.
02H1997	E-C. Wang, M-K. Hsu, Y-L. Lin, K-S. Huang, *Heterocycles* **2002**, *57*, 1997.
02H2365	J.A. Zahra, B.A.A. Thaher, M.M. El-Abadelah, M. Klinga, *Heterocycles* **2002**, *57*, 2365.
02IJC(B)1712	B. Kalluraya, M.A. Rahiman, D. Banji, *Indian J. Chem., Sect. B.* **2002**, *41B*, 1712.
02JA7395	L.A. Friedman, M. Sabat, W.D. Harman, *J. Am. Chem. Soc.* **2002**, *124*, 7395.
02JAN480	L. Vértesy, B. Beck, M. Brönstrup, K. Ehrlich, M. Kurz, G. Müller, D. Schummer, G. Seibert, *J. Antibiot.* **2002**, *55*, 480.
02JCS(P1)2080	H-S. Chong, B. Ganguly, G.A. Broker, R.D. Rogers, M.W. Brechbiel, *J. Chem. Soc. Perkin Trans. 1* **2002**, 2080.
02JCS(P1)2673	K.A. Carey, W. Clegg, M.R.J. Elsegood, B.T. Golding, M.N.S. Hill, H. Maskill, *J. Chem. Soc. Perkin Trans. 1* **2002**, 2673.
02JCS(P2)388	W.L. Karney, C.J. Kastrup, S.P. Oldfield, H.S. Rzepa, *J. Chem. Soc. Perkin Trans. 2.* **2002**, 388.
02JMC3280	H.V. Wikström, M.M. Mensonides-Harsema, T.I.F.H. Cremers, E.K. Moltzen, J. Arnt, *J. Med. Chem.* **2002**, *45*, 3580.
02JOC1199	O. Baudoin, M. Cesario, D. Guénard, F. Guéritte, *J. Org. Chem.* **2002**, *67*, 1199.
02JOC7144	E.G. Occhiato, C. Prandi, A. Ferrali, A. Guarna, A. Deagostino, P. Venturello, *J. Org. Chem.* **2002**, *67*, 7144.
02JOC8538	C. Addicott, M.W. Wong, C. Wentrup, *J. Org. Chem.* **2002**, *67*, 8538.
02JOC8662	D.K. Bates, K. Li, *J. Org. Chem.* **2002**, *67*, 8662.
02MI220	K. Arakawa, Y. Endo, M. Kimura, T. Yoshida, T. Kitaoka, T. Inakazu, Y. Nonami, M. Abe, A. Masuyama, M. Nojima, T. Sasaki, *Int. J. Cancer.* **2002**, *100*, 220.
02MI379	K. Ohmori, K-I. Hayashi, T. Kaise, E. Ohshima, S. Kobayashi, T. Yamazaki, A. Mukouyama, *Jpn. J. Pharmacol.* **2002**, *88*, 379.
02MI746	M.W. Lark, G.B. Stroup, I.E. James, R.A. Dodds, S.M. Hwang, S.M. Blake, B.A. Lechowska, S.J. Hoffman, B.R. Smith, R. Kapadia, X. Liang, K. Erhard, Y. Ru, X. Dong, R.W. Marquis, D. Veber, M. Gowen, *Bone.* **2002**, *30*, 746.
02MI1181	R. Schobert, G.J. Gordon, *Curr. Org. Chem.* **2002**, *6*, 1181.
02OL2109	R.K. Boeckman Jr., T.J. Clarke, B.C. Shook, *Org. Lett.* **2002**, *4*, 2109.
02P885	B. Akendengue, F. Roblot, P.M. Loiseau, C. Bories, E. Ngou-Milama, A. Laurens, R. Hocquemiller, *Phytochemistry.* **2002**, *59*, 885.
02PCT	R.S. Meissner, W. Xu, (Merck & Co., Inc., USA). *PCT Int. Appl.* **2002**, 27pp, CAN 136:309803.
02S2232	D. Nilov, R. Räcker, O. Reiser, *Synthesis.* **2002**, 2232.
02SL403	J. Ishihara, M. Horie, Y. Shimada, S. Tojo, A. Murai, *Synlett* **2002**, 403.
02SL519	P. Raboisson, B. Norberg, J.R. Casimir, J-J. Bourguignon, *Synlett* **2002**, 519.
02SL580	P.C.B. Page, G.A. Rassias, D. Barros, A. Ardakani, D. Bethell, E. Merifield, *Synlett* **2002**, 580.
02SL1350	M. Pauvert, V. Dupont, A. Guingant, *Synlett* **2002**, 1350.
02SL1827	H. Habib-Zahmani, S. Hacini, E. Charonnet, J. Rodriguez, *Synlett* **2002**, 1827.
02SL1860	R. Ferraccioli, D. Carenzi, M. Catellani, *Synlett* **2002**, 1860.
02SL1987	F. Cachoux, M. Ibrahim-Ouali, M. Santelli, *Synlett* **2002**, 1987.
02SL2019	S. Karsch, P. Schwab, P. Metz, *Synlett* **2002**, 2019.
02TL1595	B-C. Chen, S.T. Chao, J.E. Sundeen, J. Tellew, S. Ahmad, *Tetrahedron Lett.* **2002**, *43*, 1595.
02TL7781	S.K. Chattopadhyay, S. Maity, S. Panja, *Tetrahedron Lett.* **2002**, *43*, 7781.

Chapter 8

Eight-Membered and Larger Rings

George R. Newkome
The University of Akron, Akron, Ohio USA
newkome@uakron.edu

8.1 INTRODUCTION

As we enter the twenty-first century, there has been a continuing trend from synthetic studies of classical "crown ethers" towards polyazamacromolecules and the introduction of multiple heteroatoms, including most recently metal atom centers.

Numerous reviews and perspectives have appeared throughout 2002 that are of interest to the macroheterocyclic scientist and those delving into supramolecular chemistry at the molecular level, as well as those studying supermolecules and crystal engineering: classic annulenes in nonclassical applications <02ACR944>; Lewis acid catalysis <02ACR209>; calixphyrins, a hybrid structure between porphyrins and calixpyrroles <01PAC1041>; calixarenes <01MI01>; calixpyrroles <02H169>; host-guest chemistry <02TCC1>; macrocyclic polyethers, as probes <01CCR127>; "confused" porphyrins <02CC1795>; selenoethers and telluroethers <02CCR159>; one-step macrocyclic construction <01JHC1239>; lariat ether receptor systems <02ACR878>; three-dimensional, coordination-driven, self-assembly <02ACR972, 01CM3113>; C,N,O-macromolecules with a Si-surface <02IRPC137>; metal-assembled and metal cluster architectures <02WHX27>; phosphinines and diphosphaferrocenes <02PSSRE1529>; supramolecular chemistry of macrocyclic polyamine <02YZ219>; azacalixarenes <02JIPMC169>; corroles and core-modified corroles <02EJOC1735>; chiral lanthanide complexes <02CR1807>; gadolinium MRI contrast agents <02TCC103, 02TCC1>; complexes of crown ethers <02RJCC153>; synthesis of macrocycles using metal complexes <02YGKK184>; tetrapyrroles <02HCB39, 02HCB13>; extractions with crown ethers <02RJCC697>; molecular encapsulation <02AC1488>; synthesis of combinatorial libraries <02JCC369>; cucurbituril chemistry <02RCR840, 02CSR96>; P macrocycles <02TCC1>; fluorescent sensors and switches <02CSR116>; macrobicyclic and macrotricyclic, heteroclathrochelate complexes <02CCR255>; bridged fullerenes <02CCR141>; metal-based supramolecular systems <02CCR171>; coordination polygons and polyhedra <02CCR91>; self-complementary assembly <02CCR199>; cavitands and related containers <02H2179>; shape-persistent, nano-sized macrocycles <02EJOC3075>; and assembly of catenanes <02SL1743>.

Because of space limitations, only meso- and macrocycles possessing heteroatoms and/or subheterocyclic rings are reviewed; in general, lactones, lactams, and cyclic imides have been excluded. In view of the delayed availability of some articles appearing in previous years, several have been incorporated.

8.2 CARBON–OXYGEN RINGS

Numerous macrocyclic crown ethers possessing diverse attachments, for example: 4'-(ammoniummethylene)benzo-18-crown-6 <02CC3015>, a [Ru(bpy)$_3$]$^{2+}$ moiety attached to a dibenzo-24-crown-8 unit <02JA12786>, benzo-21-crown-7 moieties on a helical 14-residue peptide <02CC1694>, lariat ethers with two pyrenylmethyl sidearms <02OL2641>, two crown ethers on a phenolphthalein core for the visual recognition of nonprotected dipeptides <02OL2313>, anthraquinone-containing cyclic polyethers as hosts for hydronium ion <02JOC3878>, an N-(aminoalkylaryl)urea attached to a benzo-18-crown-6 as a detector of protons <02OL881>, butadienyl dyes of the benzothiazole series with benzocrown ethers <02HCA60>, a functionalized 4-(anthracene-9'-ylmethyl)benzo-15-crown-5 as a molecular photoionic switch <02CC1360>, and [60]fullerene *bis*adducts possessing a dibenzocrown ether moiety <02CEJ5094, 02JA4329> have been reported. A convenient and effective route to benzocrown ether under phase-transfer catalysis conditions has appeared <02S2266>.

A chiral pseudo-24-crown-8 ring was synthesized and shown to possess good chiral recognition of secondary amines, such as N-α-dimethylbenzylamine and propranol <02TL8539>. Chiral *trans-anti-trans*-dicyclohexano-18-crown-6 isomers were prepared *via* a lipase-catalyzed reaction <02TL5805, 02TL5229, 02TL2153>. The first crown ether derivative, Boc-[18-crown-6]-β3-(*L*)-DOPA-OMe has been prepared by *bis-O*-alkylation of the catechol function with

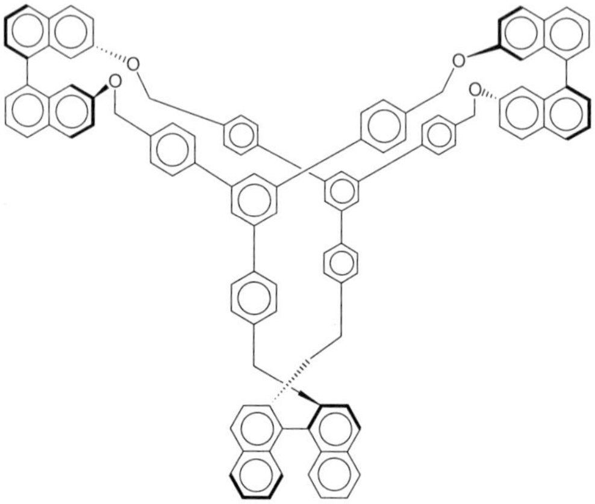

1

cyclization using pentaethyleneglycol ditosylate <02TL8241>. Coupling of an aryl tribromide with binaphthol afforded a novel series of chiral cyclophanes (1) with six coordination sites for complexation <02CEJ5094, 02TL1909>. The 1,4- and 1,6-bridged 20- and 23-membered galactocrown ethers have been prepared by the intramolecular *trans*-glycosidation of an appropriate phenyl 1-thio-D-galactopyranoside <02S1851>.

A chemically addressable, bistable [2]rotaxane that incorporated a dumbbell-shaped component containing both secondary dialkylammonium and 1,2-*bis*(pyridinium)ethane recognition sites for dibenzo-24-crown-8 has been constructed <02JOC9175>; whereas a cyclic dimeric daisy chain compound has been assembled from a difunctional [2]rotaxane in a sequence of noncovalent and covalent synthetic steps <02CC2948>. Two analogous multivalent receptors, each containing either seven dibenzo-24-crown-8 or seven benzometa-phenylene-25-crown-8 moieties appended to a modified β-cyclodextrin core have been generated <02JOC7968>. Post-assembly covalent modification using Wittig chemistry of [2]rotaxane ylides has afforded a [3]catenane and a [3]rotaxane with a precise and synthetically prescribed shortage of dibenzo-24-crown-8 rings <02OL3561>. The novel approach to rotaxanes *via* covalent bond formation has been reported; the rotaxanes were composed of crownophanes having two phenolic hydroxy groups as a molecular rotor and an axle having diamide moieties, prepared in three simple steps <02TL5747>. Protonated 1,2-*bis*(4,4′-bipyridinium)ethane axles and dibenzo-24-crown-8 wheels thread to form [2]pseudorotaxanes, which associate in the solid state to form pseudopolyrotaxanes by *H*-bonding or π-stacking <02CC1282>.

Calixarenes have been a prime core for inclusion in the C,O-macrocyclic family, since they have been monointra- <02JOC7569, 02TL7311, 02OL2129, 02TL3883, 02TL1209, 02JOC684>, *bis*intra- <02JOC6188, 02TL1629, 02TL1225>, and inter-<02JOC7569, 02TL2857, 02JA1341> molecularly bridged with PEG units. Several functional groups were introduced on the upper rim of homooxacalix[3]arene <02JOC8151>. An artificial host consisting of homooxacalix[3]arene and Reichardt dye E_T1 has appeared; the proton-ionizable phenol group acts as a chemical switch to generate a color change with added alkaline metals and various amines <02OL2301>.

Cis- and *trans*-stilbeno*bis*(18-crown-6) were easily obtained by a McMurry reductive coupling of 4′-formylbenzo-18-crown-6 <02JOC521>. The formation of 18-membered macrospirocyclic esters and ethers was readily achieved from tetraene precursors *via* tandem metathesis reactions <02TL7851>. A new intra-annularly linked cyclophane (2) was synthesized in five-steps from *p*-cresol by an intramolecular Eglington coupling of an appropriate precyclophane <02TL7695>. Irradiation of dibenzobarrelene (3) and alkali metal complexes preferentially afforded the dibenzocyclooctatrene in solution but the related dibenzosemibullvalene in the solid state <02OL3247>. A preparative approach to 1,3,5-triaroylbenzene-based functionalized cyclophanes has been reported, based on a regio-selective cross-benzannulation between *bis*(arylethynyl) ketone and enaminone reactants <02JOC4547>. A cyclophane host for anthracene was formed by connecting the oxygen atoms of two α,α′-di(4-hydroxyphenyl)-1,4-diisopropylbenzene with two pentamethylene spacers <02TL5017>. Several examples of nanoscale oxaarenecyclynes have recently appeared <02CEJ5094, 02TL3277>. An interesting incorporation of a hexa-2,5-diyne-1,6-dioxy moiety in the cyclodextrin backbone has been reported <02CEJ5094, 02HCA265>. Ring-closing metathesis procedures have been used on numerous occasions to enable macrocyclization, for example, leading to head-to-head α-cyclodextrins <02TL5533>, as well as the related intramolecular coupling-electrocyclization processes <02CC1842>, and a macrocyclization with $CpCoCl_2$ in which the Co-containing macrocycles were isolated <02JOC6856>.

8.3 CARBON–NITROGEN RINGS

As pointed out by Lash <02JOC4860>, "The porphyrin macrocycle, which serves as one Nature's most versatile ligand systems, attracted considerable attention throughout the 20th century." I t w ould a ppear t hat this interest will continue well into the 21st century, since the trend seems unending. A series of *meso*(8-substituted naphth-1-yl) porphyrins has been designed to create a "tight recognition environment" over the porphyrin plane <02JOC7457>; thus, condensation of *bis*naphthaldehyde with phenyldipyrromethane led to the α,α-5,15 isomer. Generally, one thinks of confused chemists, but in the current context we need to deal with confused porphyrins <02CC1795>. Diels-Alder reaction of Ni(II) *N*-confused tetraarylporphyrins with *o*-benzoquinodimethane afforded the Ni(II) *N*-confused isoquinoporphyrins <02CC1816>. The related Mn(II) adduct has also been prepared <02CC1942>. The *N*-confused porphyrin bearing four crown ethers attached at the *meso*-positions was synthesized and shown to form a *face-to-face* dimer more readily the its porphyrin counterpart <02TL4881>. The condensation of tripyrrinonealdehyde, a red pigment prepared by silver nitrate mediated oxidation of mesobiliverdin XIIIα, with *bis*(2,4-dimethylpyrrole-3-yl)methane yielded (51%) the blue-green tetra-*N*-confused cyclohexa-pyrrole **4** <02JOC4997>. Treatment of 5,10,15,20-tetraphenyl-2-aza-21-carbaporphyrinato-nickel(II) with dihalomethanes in the presence of a proton scavenger afforded a 2,21'-CH$_2$-linked dimer of Ni(II) confused porphyrins in *ca*. 90% yield; the 2,2'-linked isomer was a minor product <02CC92>. The reaction of *N*-confused tetraphenylporphyrin with copper acetate in refluxing toluene and subsequent acid hydrolysis afforded the ring fragmented p roduct, 1 4-benzoyl-5,10-diphenyl-1-tripyrrinone <02OL181>. Syntheses of new oxybenzi-porphyrins, possessing a *meta*-aryl component, obtained by simple [3 + 1]-methodology and a related Pd-complex have appeared <02CC462>. 5,10,15,20-Tetraphenyl-*p*-porphyrin was obtained, albeit in low yield (1%), by from a mixture of pyrrole, the arylaldehyde, and 1,4-di(α-benzyl alcohol)benzene <02JA3838>. The 1,3-diformylindene was condensed with tripyrranes in the presence of TFA followed by DDQ oxidation to give a series of benzo-carbaporphyrins in excellent yields <02JOC4860>; a one-pot Rothemund-type synthesis of *meso*-tetraphenylazuliporphyrin (**5**) has also been developed <02CEJ5397>, also see <02CC1660>. The aza-deficient 5,10,15,20-tetraaryl-21-vacataporphyrin (**6**) has been synthesized by elimination of a tellurium atom from 5,10,15,20-

tetraaryl-21-telluraporphyrin **7** upon treatment with HCl <02CEJ5403>. The synthesis of the heptapyrrolic macrocycle, [30]heptaphyrin(1.1.1.1.1.0.0) has been reported and shown to exhibit a 'figure eight'-like structure in the solid state <02CC328>. The synthesis of nonmetalated triazolephthalo-cyanines has been described for the first time <02JOC1392>. Three types of central free base porphyrins possessing eight pyrazine sites on the periphery have been synthesized and evaluated as light-harvesting assemblies <02JA1182>.

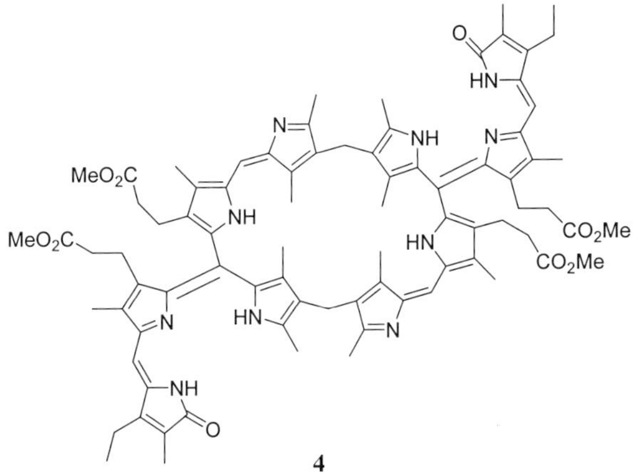

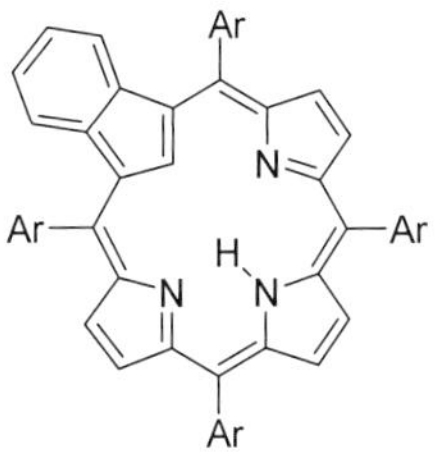

The *meso*-decamethylcalix[5]pyrrole was synthesized from the corresponding furan-based analogue, its binding constant toward chloride ion was found to be lower than the tetrameric counterpart <02OL2695>. Trihalogenated solvents act as effective and selective materials in the template-assisted formation of *meso*-hexaphenylcalix[6]pyrrole <02CC404>. The *meso*-hexamethyl-*meso*-hexaphenylcalix[6]pyrrole assembles into a well-defined dimeric capsules in the crystalline phase; the capsules serve as an efficient host for solvent guests <02CC726>. [1.1](3,3′)-Azobenzenophane, where two azobenzenes are cyclically connected by methylene chains at the meta position, has been reported; crystal structures of all three isomers (*cis, cis-, cis, trans-*, and *trans, trans-*) have been compared <02OL3907>.

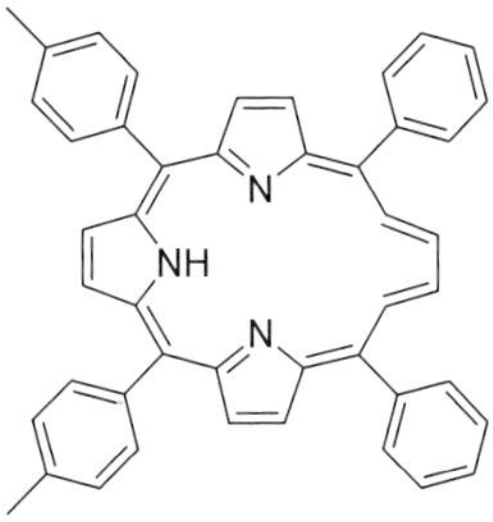

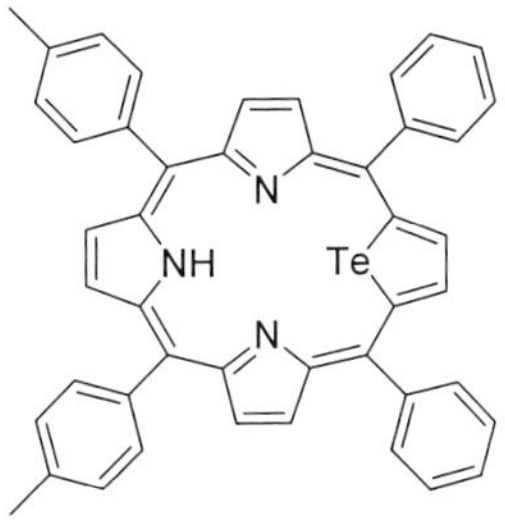

Uracilophane **8** was prepared (29%) by ring-closure of acyclic counterpart, derived from thymine and two 3,6-dimethyluracil components, with paraformaldehyde <02TL9683>. The [3+1] condensation of 2,6-*bis*(imidazolmethyl)pyridine and 2,5-*bis*(trimethylamino-methyl)-pyrrole diiodide afforded (68%) the off-white cyclophane **9** <02TL3423>; the related *bis*pyrrolophane was generated in a similar manner using two 2,5-*bis*methylpyrrole units. A convergent [3 + 1] synthetic strategy was reported for the construction of [1₄]metahetero-phanes composed of both highly π-excessive and π-deficient heterocycles linked in an 1,3-altenating manner <02CEJ474>. Formation of dicationic [14]imidazoliophanes containing two imidazolium rings was shown to be generated by anion-directed through *H*-bonding; the template effect utilized a chloride anion in the ring-closure reaction <02JOC8463>. Thiazole- and oxazole-containing m odified c yclopeptide c ages h ave b een p repared using a trimeric linker to control the assembly and cyclooligomerization of the heterocyclic amino acids <02TL2459>. Macromolecules based on *bis*-pyrazolylpyridine and diethylene-triaminepentaacetic acid have been synthesized and shown to possess luminescence properties upon formation of their lanthanide chelates <02OL213>. [3](3,6)Pyridazino-[3](1,3)indolophane was synthesized from indole by a two-fold sequential hydroboration-/Suzuki-Miyaura cross-coupling procedure; the cyclophane underwent a transannular Diels-Alder reaction to form a pentacyclic precursor to a variety of indole alkaloids <02OL127>.

New twistophanes (**10**), composed of cyclically conjugated dehydrobenzoannulene framework that incorporates the 6,6′-connected-2,2′-bipyridine moieties for metal coordination, have been reported <02CEJ5250>. The synthesis of three new tris-macrocycles containing three [12]aneN₄, [12]aneN₃O, and [14]aneN₄ moieties appended to a trene unit has appeared <02JOC9107>. Chiral calixarenes **11** possessing amino acid residues into macrocyclic rings were prepared by the cyclization of a *bis*(chloromethyl)-phenol-formaldehyde tetramer with an amino acid methyl ester <02JOC7519>. A general, lithium templated/catalyzed *o*-polyazamacrolactamization afforded a route to 12-, 13-, 16-, 17-, and 19-membered N₃₋₆-containing rings <02TL7213>. Receptor **12** was prepared in simple steps in 60% overall yield; the resultant bowl-shaped receptor is well suited to recognize tetrahedral anions <01CC1456>. A new methodology has been reported for the synthesis of β-lactam fused enediynes <02TL4241>. A new and efficient synthesis (three steps, overall 65%) of cyclen utilizing 1,1′-ethylenedi-2-imidazoline with 1,2-

dibromo-ethane has appeared <02JOC4081>. The Pd-catalyzed arylation of cyclam, cyclen, and azacrown ether by an aryl halide was reported <02TL1193>; alternatively, a selective and high yield synthesis of mono N-substituted derivatives of cyclam and cyclen is available <02TL3217>. Treatment of selectively N-tritylated spermidine and spermine derivatives with succinic anhydride, followed by PyBrOP-mediated cyclization and subsequent hydride reduction, gave entry to a series of cyclic polyamines <02TL2593>. Synthesis of 27- and 30-membered rings of the trianglimine and trianglamine type by a [3 + 3] cyclocondensation has been described <02TL3329>. A series of 15-membered triolefinic N-macrocycles containing ferrocenyl groups has been reported <02TL1425>. Efficient syntheses of chiral [26]-N_6, [12]-N_4, [9]-N_3, and [14]-N_4 systems from chiral aziridines as a common building block have appeared <02OL949>. A simple, post-synthetic route to 1,4,7-triazacyclononane-d_{12} has been reported <02TL771>.

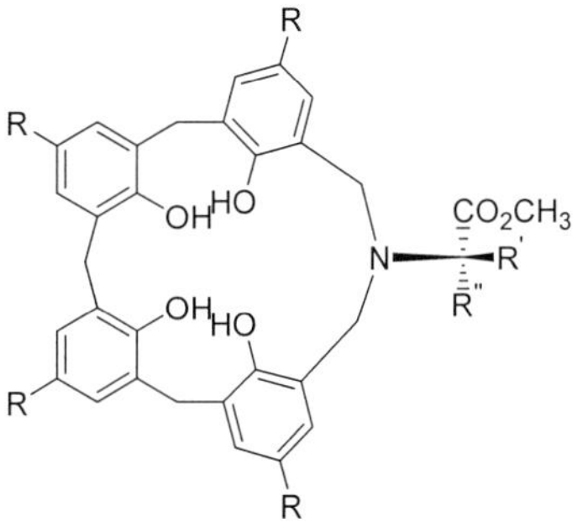

11 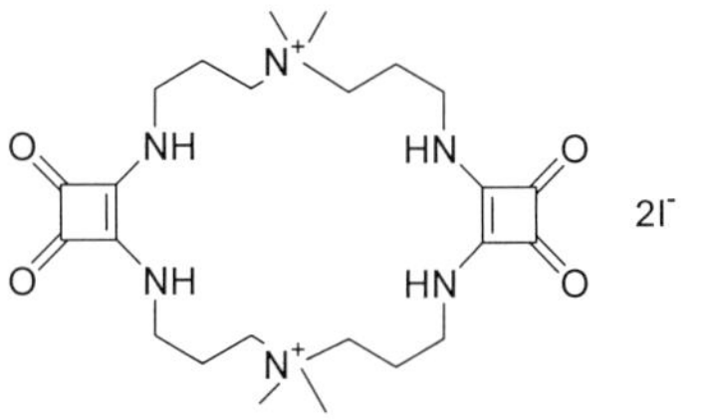 **12**

8.4 CARBON–SULFUR RINGS

The synthesis of a new aromatic cyclic enediyne, 9-thiotribenzo[c,g,j]cycloundeca-3,7,10-triene-1,5-diyne (**13**) using a double Wittig reaction, followed by bromination and di-dehydrobromination, has been reported <02JOC383>. The oxidation of thiols to disulfides with molecular bromine on silica gel solid support afforded high yields of 1,2-dithia-cyclooctane and -decane <02YL6271>. Sulfenylation of 6,6'-dimethoxy-2,2'-dihydroxy-biphenyl with phthalimidesulfenyl chloride gave the 3,3'-N,N'-dithiophthalimide with complete regioselectivity, which when treated with LAH gave the cyclic disulfide **14** in 86% yield <02JOC2019>.

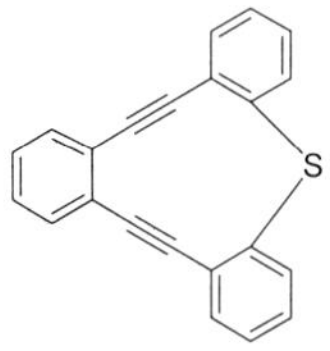

13 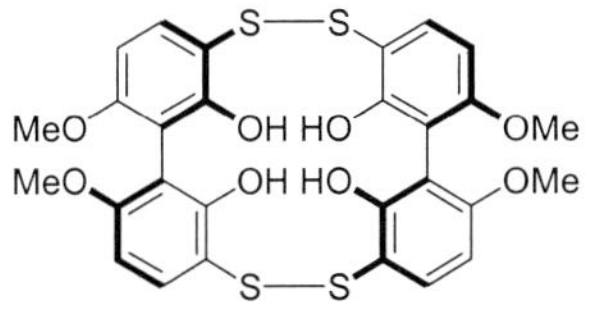 **14** 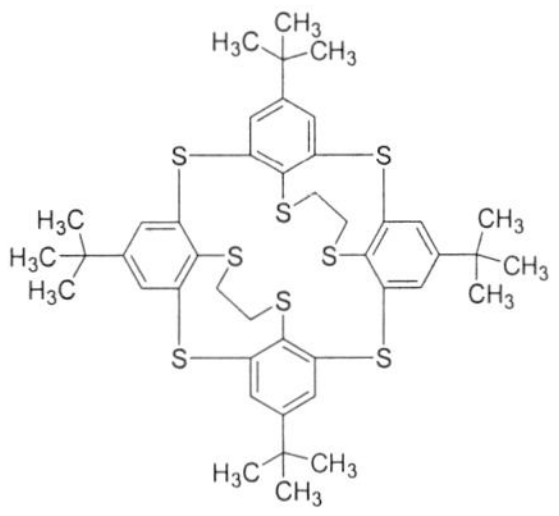 **15**

Treatment of *p-tert*-butylthia- or *p-tert*-butylthiatetramercaptocalix[4]arene with 1,2-dibromoethane in the presence of K_2CO_3 generated two new basket-type materials, e.g. 15 <02TL8975>. Tetra- and hexa-nuclear mercury(II) complexes have been generated from tetrathia- and tetramercaptotetrathia-calix[4]arene, respectively <02CC1042>. *p*-(1-Adamantyl)thiacalix[4]arene was synthesized either by condensation of *p*-(1-adamantyl)-phenol with sulfur in the presence of base or alkylation of *p-H*-thiacalix[4]arene with adamantanol in F_3CCO_2H <02TL5153>. A new macrotricyclic ligand with an N_4S_2 donor set has been synthesized from cyclam; it was shown to encapsulate lithium and transition metal ions <02CC170>.

8.5 CARBON-SILICON RINGS

A novel Si-macrocycle was prepared by the demetalation of the zirconocene-containing dimeric cyclophane intermediate, which was formed by the coupling of dimethyl-*bis*(4-trimethylsilylethynylphenyl)silane with [Zr(Cp)$_2$Cl$_2$] <02CEJ74>. Synthesis, X-ray analysis, and characterization of a dibenzo macrocyclic acetylenic diphenylsilane have been reported <02TL2079>.

8.6 CARBON-SELENIUM RINGS

Cyclic tetraselenadiynes **16** have been synthesized in five-steps from trimethylsilyl-acetylene; the intermolecular Se⋯Se distances were evaluated and interesting columnar structures were noted <02JOC4290>. Related self-organization of cyclic selenaethers into columnar structures has also been reported.<02TL5767, 02OL4503, 02JA10638>; nanotube formation in related S, Se, and Te macrocycles has been shown <02JA10638>.

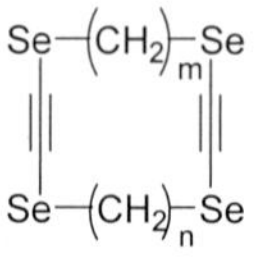

Se—(CH$_2$)$_m$—Se
‖ ‖
Se—(CH$_2$)$_n$—Se
m=n=5

16

8.7 CARBON–NITROGEN-OXYGEN RINGS

Novel 1,3-alternate calix[4]azacrowns having an azo chromophoric pendent moiety, e.g. [4-nitrophenyl(azo)phenyl] or pyrene **17** <02JOC2348> with variable crown ether components have been prepared and confirmed by X-ray analysis <02JOC1372>. The more complicated calix[4]-*bis*-azacrown possessing the iodoaniline chromophore with one methylene spacer attached at both azacrown centers has appeared <02JOC6514>. Numerous aza-crown ethers possessing *N*-substituted groups, such as ethylindole <02CC1806>, ethylbenzene and tyramine <02CC1810>, methoxycoumaryl and 7-acetamidocoumarin <02TL4413> as well as nitroso derivatives <02AL1621>. *N,N'*-Trimethylene-*bis*(benzoaza-15-crown-5) was prepared (21%) from the benzoaza-15-crown-5 with 1,3-propanediol ditosylate <02JOC3533>. The insertion of a 2,6-pyridinedicarbonyl spacer **18** into a permethylated α-cyclodextrin has appeared <02CC1596>.

17   **18**

Monoalkylation of diaza-18-crown-6 with 2-bromoethylbenzene gave *N*-(2-phenylethyl)-4,13-diaza-18-crown-6, which when heated with MeOCH$_2$CH$_2$OMs gave rise to the unsymmetrical *bis*substituted crown <02CC1808>. Eleven monoanthracylmethyl *bis*-azacrown ethers have been synthesized and evaluated as fluorescence sensors for the marine toxin saxitoxin <02JA13448>. Multiply linked diazacrown ethers have been shown to transport Na$^+$ through phospholipid bilayers at a rate of *ca*. 27% that of gramicidin and *N,N'*-*bis*(dodececyl)-4,13-diaza-18-crown-6 fails to transport Na$^+$ at a rate detectable by ^{23}Na NMR; the spacer length was demonstrated to be critical to efficient cation transport <02JA9022>. New metal-free phthalocyanines 9,10-fused symmetrically in the peripheral positions with diazahexaoxamacrobicycles have been prepared by bicyclotetramerization of isoindolinediimine derivatives of the appropriately substituted cryptand <02TL5343>. Ring-closing metathesis of suitable 1,ω-dienes led to efficient atom economic synthetic approaches towards azacrown ether derivatives with eight- to twenty four-membered ring sizes <02TL4207>. A crown ether-containing macrobicycle was used as a wheel component in a templated synthesis of a [2]rotaxane with an acetal-containing axle <02JOC1436>. A *bis*(thiourea) receptor based on a highly flexible dibenzo-diaza-30-crown-10 scaffold represented a K$^+$-organized microenvironment to markedly accelerate the cleavage of the phosphodiester linkage of 2-hydroxypropyl-*p*-nitrophenylphosphate, a RNA model substrate <02TL3455>. Modified tripyrranes incorporating furan (and thiophene) were found to condense with benzene, pyridine, indene dialdehydes to give a series of porphyrin analogues such as oxa-carbaporphyrin <02CC2426>.

The reaction of tripyrranes with pyrrolecarboxaldehyde in the presence of TFA catalyst followed by oxidation with chloranil results in a simultaneous oxidative coupling and condensation to generate a meso-free corrole <02OL4233>. A fluorescence receptor based on triaza-18-crown-6 ether combined with two *N*-guanidinium groups (and one *N*-anthracen-9-ylmethyl moiety) could bind several biologically important amino acids in aqueous methanol

<02TL7243>. [2 + 2]-, [3 + 3], and [4 + 4]-Macrocyclocondensation was demonstrated by the Schiff-base reaction with methylenebis(4,4'-methyl-6,6'-salicylaldehyde) and 1,2-bis(2-aminoethoxy)ethane under high dilution conditions, the imines were readily reduced to the corresponding multiamino, multiphenolic macrocycles <02TL8261>. Ring-closure metathesis of 1,5-bis-o-allyloxyphenylformazans with the suitably located 1,ω-dienes leads to an efficient, highly stereoselective, synthetic route to Z-olefinic 15-membered ring crown-formazans <02TL6971>. An oxa-analogue of 5,10,15-triarylcorrole where two pyrrole rings were replaced by furan moieties has been prepared by condensation of 2,5-bis(arylhydroxymethyl)furan, 2-phenylhydroxymethylfuran, and pyrrole <02JOC5644>.

8.8 CARBON–SULFUR–OXYGEN RINGS

The synthesis of 20,28-dimethyl-24,33-dithia-1,4,7,10,13,16-hexaoxa[16.3.3](1,2,6)-cyclophane <02TL9199> and a related oligomer <02OL3211> was demonstrated by $CsCO_3$-assisted high dilution macrocyclization with the appropriate tetrabromide in the presence of $Na_2S \cdot 9H_2O$. In a related system, inner bridging was accomplished by using interconvertible crown ethers bearing two thiol groups or a disulfide moiety by redox reactions <02TL5719>. A new molecular system, 2,11-dithio[4,4]metametaquinocyclophane, containing a quinone moiety, was created and shown to act as a "molecular flapper" under redox conditions <02OL3971>. Another approach to a molecular switch was demonstrated by the self-assembly of monolayers of tetrathiafulvalene-based redox-switchable ligands attached to a Au(III) surface <02NJC1320>. The first representatives of 1,3-alt-thiacalix[4]arene bis(crown-5 and -6) ethers were prepared by cyclocondensation of thiacalix[4]arene with tetraethylene glycol ditosylate and 1,14-diiodo-3,6,9,12-tetraoxatetradecane <02TL4153>. A series of related $\beta\beta$-1,6-thio-linked cycloglucopyranosides was reported, in which they suggest that the crucial macrocyclization step was achieved in high yields and well-controlled stereochemistry by base-promoted intramolecular S_N2 glycosylation <02OL4503>. A very novel spherical host **19** with D_{2d} symmetry consisted of a tetrathia-[3.3.3.3]paracyclophane and two 18-crown-6 ether units <02OL3911>.

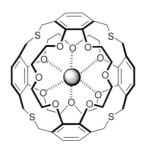

19

8.9 CARBON-PHOSPHORUS-OXYGEN

Condensation of a dichlorophosphite with different *bis*phenols generated macrocyclic phosphates in different sizes dependent on the constraints imposed by the *bis*phenolic moiety <02TL5245>. The macrocyclization of 1,1,7,7-tetrabenzyl<P.O.P-podand-7> with 1,1'-oxy-*bis*[2-bromoethane] afforded the 12-membered 4,4,10,10-tetrabenzyl-12<O.P.O.P-coronand-4>-4,10-diium dibromide in 50 - 60% yields; other related coronands were similarly prepared <02HCA1714>.

8.10 CARBON–NITROGEN–SULFUR RINGS

The syntheses of new aromatic 30-π-heptaphyrins either *via* a [5 + 2] or a [4 + 3] acid-catalyzed condensation and oxidative coupling reactions of easily available, air-stable precursors have been presented <02JOC6309>. Calixpyrrole-like mixed cyclic oligomers containing bithiophanes and pyrroles have been generated by condensation of hybrid oligopyrrolic precursors; whereas, condensation of 1,7-*bis*[(α-hydroxy-α,α-diethyl)-methyl]-bithiophene with oligopyrroles afforded a mixture of large cyclic oligomers <02TL9493>. *Meso*-furyl porphyrins with N_4, N_3S, and N_2S_2 cores were prepared and their spectra compared <02TL9453>. 21,23-Dithiaporphyrins with *bis*-alkoxy substituted on the β-thiophenes were synthesized; X-ray data gives insight to the planarity imposed by the butoxy moieties <02CC2642>.

A convenient one-pot reaction of 2,6-dibromopyridine with NaHS gave rise to thiacalix-[3]-, [4]-, and [6]pyridine; the X-ray structure of the thiacalix[4]pyridine was presented <02CC1686>.

8.11 CARBON-OXYGEN-SELENIUM RING

Two calix[4](diseleno)crown ethers were synthesized by the treatment of the disodium salt of 1,3-propanediselenol with preorganized 1,3-dibromoethoxycalixarenes; the 1,3-Se-bridged calixarene forms an infinite sheet aggregate by means of self-inclusion and intermolecular Se⋯Se interactions in the solid state <02TL131>.

8.12 CARBON-SULFUR-TELLURIUM RING

The preparation of the first examples of mixed S,Te-macrocycles, such as 1,4-dithia-7-telluracyclononane, 14-dithia-8-telluracycloundecane, 1,5-dithia-11-tellurocyclotetradecane, and 1,4,7-trithia-11-telluracyclotetradecane, *via* a "disguised dilution" procedure (not therein defined) was reported <02CC427>.

8.13 CARBON–NITROGEN–SULFUR–OXYGEN RINGS

The *N,N'*-difunctionalization of *bis*(pyrrolo[3,4-*d*]tetrathiafulvalene with ω-polyethylene glycol) derivatives gave rise to *bis*(pyrrolo)tetrathiafulvalene macrocycles <02OL2461>; whereas, the use of the related *bis*-cyanoethylthiolate protected monopyrrolo-TTF monomer gave rise to a tetrathiafulvalene-belt **20** <02OL1327> or the novel tetrathiafulvalene-cage **21** <02OL4189>. Macrocyclic aromatic ether-imide-sulfones were prepared by direct [1 + 1] cyclocondensation of a *bis*-aryl amine with pyromellitic or 1,4,5,8-naphthalenetetra-carboxylic dianhydride in refluxing *N,N*-dimethylacetamide under pseudo-high-dilution conditions; these macrocycles have extremely high (493 and 547 °C, respectively) melting points and show no thermo-oxidative decomposition in air at temperatures up to 400 °C as well as act as supramolecular receptors <02JA13346>. The fairly rigid calix[4]furan was treated with thiazyl chloride to give *bis*-, *tris*- or tetra*kis*-isothiazole macrocycles depending on the reaction conditions <02CC232>.

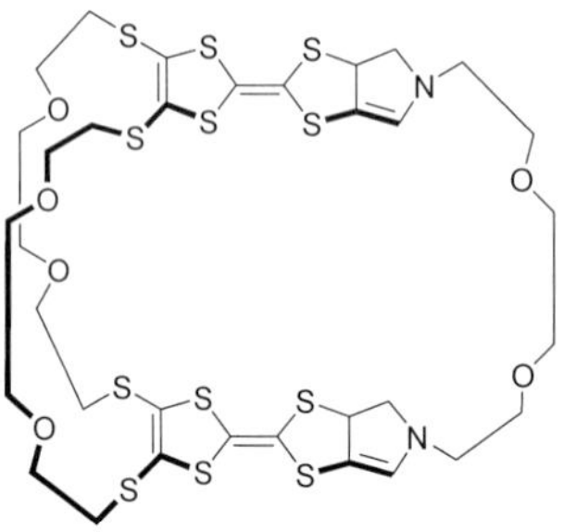

20 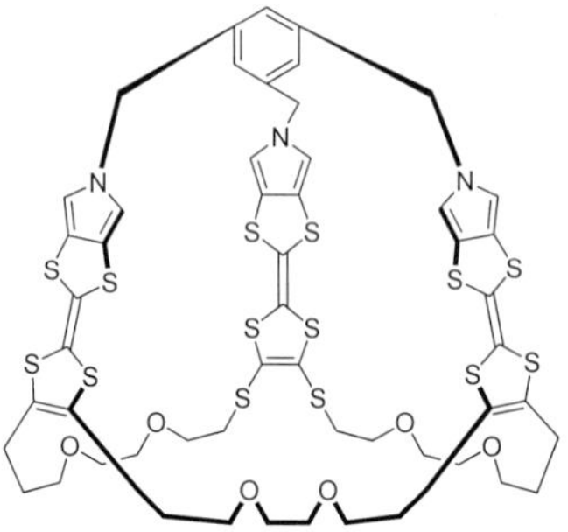   21

8.14 CARBON–METAL RINGS

Treatment of 6,6′-dichloro-2,2′-diethoxy-1,1′-binaphthyl-4,4′-*bis*(acetylene) with one equivalent *cis*-Pt(PEt$_3$)$_2$Cl$_2$ in the presence of CuI in a mixture of diethylamine and solvent at room temperature gave *ca*. 40% yields of a chiral organometallic triangle Pt-complex for asymmetric catalysis <02JA12948>.

8.15 CARBON–NITROGEN–METAL RINGS

Molecular directivity is an important component to the supramolecular construction of two- and three-dimensional nanoscopic metal assemblies. In most cases, such frameworks are macroheterocyclic in nature, and thus are herein considered; also see for a recent review of this topic <02ACR972>. A 4- or 3-pyridylmethyl moiety is attached to the ends of either 1,4-diaminobutane or 1,5-diaminopentane to produce a short "string", which reacts with cucurbituril thus inserting the bead on the chain, then treatment with Pt(en)(NO$_3$)$_2$ (en = ethylenediamine) produced the desired necklace with 3 **22** or 4 beads, respectively <02JA2140>. Putting together the properly directed components in the appropriate ratios gives rise to the three-dimensional construct, for example the molecular cage **23** for a 3:2 ratio of the directed components **24** and **25** <02OL913>. Interestingly, the supramolecular self-assembly was shown to be accelerated under solvent-free conditions affording a higher-order fabrication of two- and three-dimensional topology and even double helicates <02CC1362>. A hollow and roughly spherical cage of ca. 2 nm in diameter was self-assembled from 2,4,6-*tris*(4-pyridyl)-1,3,5-triazine and Pd(diamine)(ONO$_2$)$_2$; this cage enclathrated a variety of neutral organic molecules in the aqueous phase <02JA13576>. Complexation of 1,3,5-*tris*(4-picolyl)benzene (L) with Pd(NO$_3$)$_2$ quantitatively self-assembled to a Pd$_6$L$_8$ molecular sphere <02CC2486>. The rhombic-shaped macrocycle [Cu$_2$(bpa)$_2$(phen)$_2$(H$_2$O)$_2$]·H$_2$O was assembled (48%) by the reaction of phenanthrolines Cu(NO$_3$)$_2$ in water with biphenyl-4,4'-dicarboxylic acid (bpa) and Et$_3$N <02CC1442>.

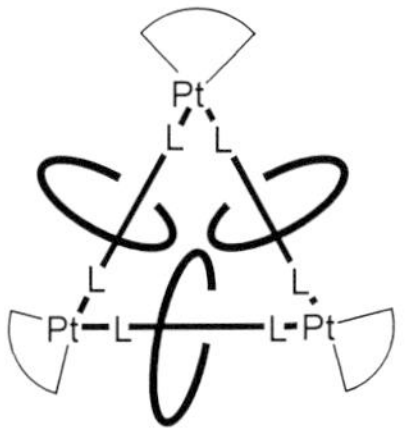

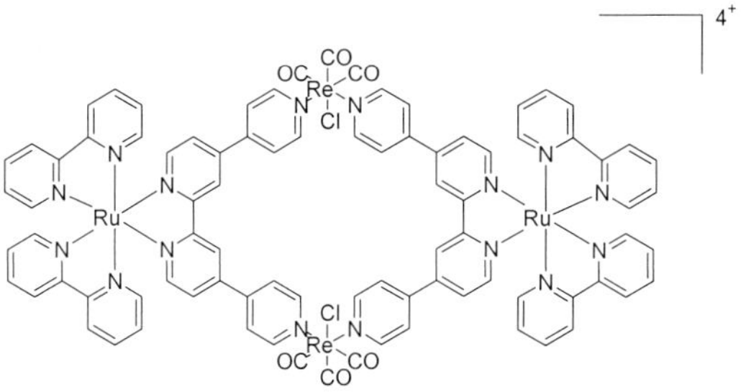

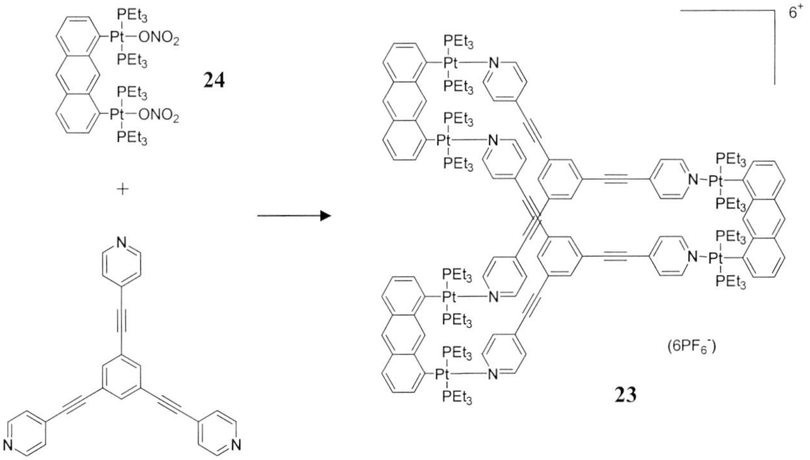

Treatment of [{MCl$_2$(Cp*)}], where M = Ir or Rh, with bidentate ligand, such as 1,4-diisocyano-2,5-dimethylbenzene, 1,4-diisocyano-2,3,5,6-tetramethylbenzene, pyrazine, or 4,4'-dipyridine gave the corresponding dinuclear [{MCl$_2$(Cp*)}$_2$(L)], which were converted into the tetranuclear [MCl$_2$(μ-Cl)$_2$(Cp*)$_2$}$_2$(L)$_2$](OTf)$_4$ upon addition of Ag(OTf) <02CEJ372>. The reaction of [Ru(bpy)$_2$(L)]$^{2+}$, where L = 2,2':4,'':4,4''-quaterpyridinyl, with Pd(en)NO$_3$)$_2$] or [Ru(CO)$_5$Cl] gave the respective heterometallomacrocyclic complex **26** <02CC2540>.

8.16 CARBON-PHOSPHORUS-METAL RINGS

Reaction of equimolar amounts of *bis*(dimethylphosphino)methane (dmpm) with PdMe$_2$ (tmeda) in benzene generated a 91% yield of **27**, as a thermally stable complex that does not eliminate ethane in solution <01JA4081>. New types of topologically chiral [2]catenanes generated from the self-assembly of [{4-BrC$_6$H$_4$CH(4-C$_6$H$_4$OCH$_2$CCAu)$_2$}$_n$] with diphosphine ligands <02JA3959>. The *cis*-configured PtI$_2$ center in **28** rearranged over 4 weeks to the unexpected "ring-fused" system **29** with an extended BINAP-like ligand <02CEJ4622>.

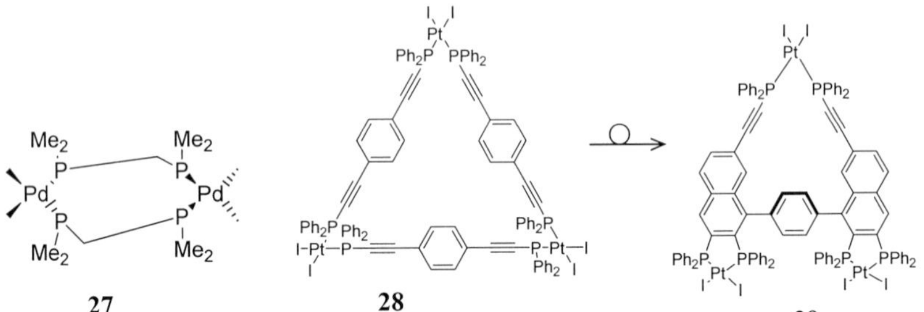

8.17 CARBON-OXYGEN-NITROGEN-METAL RINGS

Three metallosupramolecular ring-within-a-ring structures **30** have been assembled using the HETPHEN concept in the presence of Cu(II) with macrocyclic phenanthrolines exhibiting exotopic and endotopic coordination motifs <02OL2289>. The use of this technique was applied to the construction of metallosupramolecular assemblies of linear and macrocyclic *bis*phenanthrolines in the presence of Cu(II) to provide "nanoboxes" with the internal volumes greater than 5,000 Å^3 <02CC2566>.

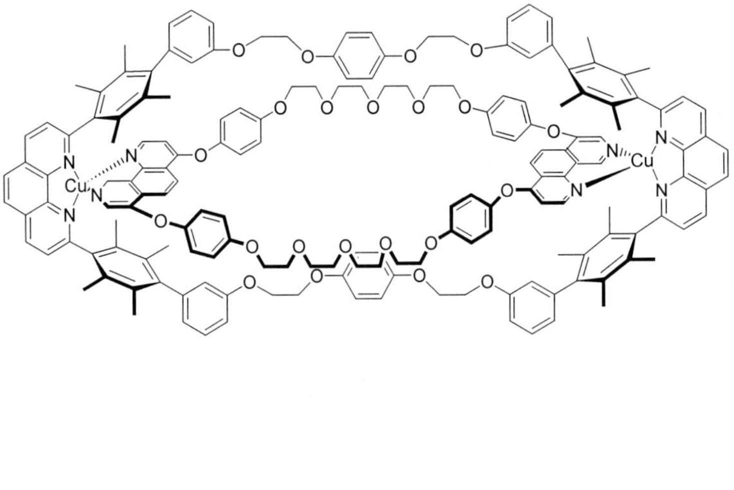

8.18 REFERENCES

01CC1456	R. Prohens, G. Martorell, P. Ballester, A. Costa, *Chem. Commun.* **2001**, 1456.
01CCR127	G.W. Gokel, L.J. Barbour, S.L. De Wall, E.S. Meadows, *Coord. Chem. Rev.* **2001**, *222*, 127.
01CM3113	P.H. Dinolfo, J.T. Hupp, *Chem. Mater.* **2001**, *13*, 3113.
01JHC1239	K.E. Krakowiak, R.M. Izatt, J.S. Bradshaw, *J. Heterocycl. Chem.* **2001**, *38*, 1239.
01PAC1041	J.L. Sessler, R.S. Zimmerman, C. Bucher, V. Kral, B. Andrioletti, *Pure Appl. Chem.* **2001**, *73*, 1041.
01TCC1	*Calixarenes 2001*, (Eds.: Z. Asfari, V. Bohmer, J. Harrowfield, J. Vicens) Kluwer, Dordrecht, The Netherlands **2001**.
02AC1488	F. Hof, S.L. Craig, C. Nuckolls, J. Rebek, *Angew. Chem. Int. Ed.* **2002**, *41*, 1488.
02ACR209	S. Kobayashi, K. Manabe, *Acc. Chem. Res.* **2002**, *35*, 209.
02ACR878	G.W. Gokel, L.J. Barbour, R. Ferdani, J. Hu, *Acc. Chem. Res.* **2002**, *35*, 878.
02ACR944	M.J. Marsella, *Acc. Chem. Res.* **2002**, *35*, 944.
02ACR972	S.R. Seidel, P.J. Stang, *Acc. Chem. Res.* **2002**, *35*, 972.
02CC92	I. Schmidt, P.J. Chmielewski, *Chem. Commun.* **2002**, 92.
02CC170	T.M. Barclay, A. McAuley, S. Subramanian, *Chem. Commun.* **2002**, 170.
02CC232	J. Guillard, O. Meth-Cohn, C.W. Rees, A.J.P. White, D.J. Williams, *Chem. Commun.* **2002**, 232.
02CC328	C. Bucher, D. Seidel, V. Lynch, J.L. Sessler, *Chem. Commun.* **2002**, 328.
02CC404	B. Turner, A. Shterenberg, M. Kapon, K. Suwinska, Y. Eichen, *Chem. Commun.* **2002**, 404.
02CC427	W. Levason, S.D. Orchard, G. Reid, *Chem. Commun.* **2001**, 427.
02CC462	S. Venkatraman, V.G. Anand, S.K. Pushpan, J. Sankar, T.K. Chandrashekar, *Chem. Commun.* **2002**, 462.
02CC726	B. Turner, A. Shterenberg, M. Kapon, M. Botoshansky, K. Suwinska, Y. Eichen, *Chem. Commun.* **2002**, 726.
02CC1042	H. Akdas, E. Graf, M.W. Hosseini, A. De Cian, A. Bilyk, B.W. Skelton, G.A. Koutsantonis, I. Murray, J.M. Harrowfield, A.H. White, *Chem. Commun.* **2002**, 1042.
02CC1282	J. Tiburcio, G.J.E. Davidson, S.J. Loeb, *Chem. Commun.* **2002**, 1282.
02CC1360	S.A. de Silva, B. Amorelli, D.C. Isidor, K.C. Loo, K.E. Crooker, Y.E. Pena, *Chem. Commun.* **2002**, 1360.
02CC1362	A. Orita, L. Jiang, T. Nakano, N. Ma, J. Otera, *Chem. Commun.* **2002**, 1362.
02CC1442	G.-F. Liu, B.-H. Ye, Y.-H. Ling, X.-M. Chen, *Chem. Commun.* **2002**, 1442.
02CC1596	T. Kida, T. Michinobu, W. Zhang, Y. Nakatsuji, I. Ikeda, *Chem. Commun.* **2002**, 1596.
02CC1660	S. Venkatraman, V.G. Anand, V.P. Raja, H. Rath, J. Sankar, T.K. Chandrashekar, W. Teng, K.R. Senge, *Chem. Commun.* **2002**, 1660.
02CC1686	R. Tanaka, T. Yano, T. Nishioka, K. Nakajo, B.K. Breedlove, K. Kimura, I. Kinoshita, K. Isobe, *Chem. Commun.* **2002**, 1686.
02CC1694	Y.R. Vandenburg, D.B. Smith, E. Biron, N. Voyer, *Chem. Commun.* **2002**, 1694.
02CC1795	H. Furuta, H. Maeda, A. Osuka, *Chem. Commun.* **2002**, 1795.
02CC1806	J. Hu, L.J. Barbour, R. Ferdani, G.W. Gokel, *Chem. Commun.* **2002**, 1806.
02CC1808	J. Hu, L.J. Barbour, G.W. Gokel, *Chem. Commun.* **2002**, 1808.
02CC1810	J. Hu, L.J. Barbour, R. Ferdani, G.W. Gokel, *Chem. Commun.* **2002**, 1810.
02CC1816	Z. Xiao, B.O. Patrick, D. Dolphin, *Chem. Commun.* **2002**, 1816.
02CC1842	M.A. Sierra, J.C. del Amo, M.J. Mancheño, M. Gómez-Gallego, M.R. Torres, *Chem. Commun.* **2002**, 1842.
02CC1942	J.D. Harvey, C.J. Ziegler, *Chem. Commun.* **2002**, 1942.
02CC2426	D. Liu, T.D. Lash, *Chem. Commun.* **2002**, 2426.
02CC2486	D.K. Chand, K. Biradha, M. Fujita, S. Sakamoto, K. Yamaguchi, *Chem. Commun.* **2002**, 2486.
02CC2540	P. de Wolf, S.L. Heath, J.A. Thomas, *Chem. Commun.* **2002**, 2540.
02CC2566	M. Schmittel, H. Ammon, V. Kalsani, A. Wiegrafe, C. Michel, *Chem. Commun.* **2002**, 2566.
02CC2642	N. Agarwal, S.P. Mishra, A. Kumar, C.-H. Hung, M. Ravikanth, *Chem. Commun.* **2002**, 2642.
02CC2948	S.-H. Chiu, S.J. Rowan, S.J. Cantrill, J.F. Stoddart, A.J.P. White, D.J. Williams, *Chem. Commun.* **2002**, 2948.
02CC3015	O.P. Kryatova, S.V. Kryatov, R.J. Staples, E.V. Rybak-Akimova, *Chem. Commun.* **2002**, 3015.

02CCR91	G.F. Swiegers, T.J. Malefetse, *Coord. Chem. Rev.* **2002**, *225*, 91.
02CCR141	M.D. Meijer, G.P. M. van Klink, G. van Koten, *Coord. Chem. Rev.* **2002**, *230*, 141.
02CCR159	W. Levason, S.D. Orchard, G. Reid, *Coord. Chem. Rev.* **2002**, *225*, 159.
02CCR171	S.-S. Sun, A.J. Lees, *Coord. Chem. Rev.* **2002**, *230*, 171.
02CCR199	Y. Sunatsuki, Y. Motoda, N. Matsumoto, *Coord. Chem. Rev.* **2002**, *226*, 199.
02CCR255	A. Ingham, M. Rodopoulos, K. Coulter, T. Rodopoulos, S. Subramanian, A. McAuley, *Coord. Chem. Rev.* **2002**, *233-234*, 255.
02CEJ74	L.L. Schafer, J.R. Nitschke, S.S.H. Mao, F.-Q. Liu, G. Harder, M. Haufe, T. D. Tilley, *Chem. Eur. J.* **2002**, *8*, 74.
02CEJ372	Y. Yamamotoa, H. Suzuki, N. Tajima, K. Tatsumi, *Chem. Eur. J.* **2002**, *8*, 372.
02CEJ474	E. Alcalde, N. Mesquida, L. Pérez-García, S. Ramos, M. Alemany, M.L. Rodríguez, *Chem. Eur. J.* **2002**, *8*, 474.
02CEJ4622	T. Baumgartner, K. Huynh, S. Schleidt, A.J. Lough, I. Manners, *Chem. Eur. J.* **2002**, *8*, 4622.
02CEJ5094	A. Godt, S. Duda, Ö. Ünsal, J. Thiel, A. Härter, M. Roos, C. Tschierske, S. Diele, *Chem. Eur. J.* **2002**, *8*, 5094.
02CEJ5250	P.N.W. Baxter, *Chem. Eur. J.* **2002**, *8*, 5250.
02CEJ5397	D.A. Colby, T.D. Lash, *Chem. Eur. J.* **2002**, *8*, 5397.
02CEJ5403	E. Pacholska, L. Latos-Grazynski, Z. Ciunik, *Chem. Eur. J.* **2002**, *8*, 5403.
02CR1807	H.C. Aspinall, *Chem. Rev.* **2002**, *102*, 1807.
02CSR96	K. Kim, *Chem. Soc. Rev.* **2002**, *31*, 96.
02CSR116	K. Rurack, U. Resch-Genger, *Chem. Soc. Rev.* **2002**, *31*, 116.
02EJOC1735	D.T. Gryko, *Eur. J. Org. Chem.* **2002**, 1735.
02EJOC3075	C. Grave, A.D. Schlüter, *Eur. J. Org. Chem.* **2002**, 3075.
02H169	W. Sliwa, *Heterocycles* **2002**, *57*, 169.
02H2179	W. Sliwa, G. Matusiak, M. Deska, *Heterocycles* **2002**, *57*, 2179.
02HCA60	S.P. Gromov, A.I. Vedernikov, E.N. Ushakov, L.G. Kuz'mina, A.V. Feofanov, V.G. Avakyan, A.V. Churakov, Y.S. Alaverdyan, E.V. Malysheva, M.V. Alfimov, J.A.K. Howard, B. Eliasson, U.G. Edlund, *Helv. Chim. Acta* **2002**, *85*, 60.
02HCA265	B. Hoffmann, B. Bernet, A. Vasella, *Helv. Chim. Acta* **2002**, *85*, 265.
02HCA1714	G. Märkl, J. Reisinger, P. Kreitmeier, J. Langer, H. Nöth, *Helv. Chim. Acta* **2002**, *85*, 1714.
02HCB13	K.M. Smith, *Heme, Chlorophyll, and Bilins* **2002**, 13.
02HCB39	J.C. Bommer, P. Hambright, *Heme, Chlorophyll, and Bilins* **2002**, 39.
02JA1182	K. Sugou, K. Sasaki, K. Kitajima, T. Iwaki, Y. Kuroda, *J. Am. Chem. Soc.* **2002**, *124*, 1182.
02JA1341	S.E. Matthews, P. Schmitt, V. Felix, M.G.B. Drew, P.D. Beer, *J. Am. Chem. Soc.* **2002**, *124*, 1341.
02JA2140	K.-M. Park, S.-Y. Kim, J. Heo, D. Whang, S. Sakamoto, K. Yamaguchi, K. Kim, *J. Am. Chem. Soc.* **2002**, *124*, 2140.
02JA3838	M. Stepien, L. Latos-Grazynski, *J. Am. Chem. Soc.* **2002**, *124*, 3838.
02JA3959	C.P. McArdle, S. Van, M.C. Jennings, R.J. Puddephatt, *J. Am. Chem. Soc.* **2002**, *124*, 3959-.
01JA4081	S.M. Reid, J.T. Mague, M.J. Fink, *J. Am. Chem. Soc.* **2001**, *123*, 4081-4082.
02JA4329	Y. Nakamura, A. Asami, T. Ogawa, S. Inokuma, J. Nishimura, *J. Am. Chem. Soc.* **2002**, *124*, 4329.
02JA9022	W.E. Leevy, G.M. Donato, R. Ferdani, W.E. Goldman, P.H. Schlesinger, G.W. Gokel, *J. Am. Chem. Soc.* **2002**, *124*, 9022.
02JA10638	D.B. Werz, R. Gleiter, F. Rominger, *J. Am. Chem. Soc.* **2002**, *124*, 10638.
02JA12786	R. Ballardini, V. Balzani, M. Clemente-León, A. Credi, M.T. Gandolfi, E. Ishow, J. Perkins, J.F. Stoddart, H.-R. Tseng, S. Wenger, *J. Am. Chem. Soc.* **2002**, *124*, 12786.
02JA12948	S.J. Lee, A. Hu, W. Lin, *J. Am. Chem. Soc.* **2002**, *124*, 12948.
02JA13346	H.M. Colquhoun, D.J. Williams, Z. Zhu, *J. Am. Chem. Soc.* **2002**, *124*, 13346.
02JA13448	R.E. Gawley, S. Pinet, C.M. Cardona, P.K. Datta, T. Ren, W.C. Guida, J. Nydick, R.M. Leblanc, *J. Am. Chem. Soc.* **2002**, *124*, 13448.
02JA13576	T. Kusukawa, M. Fujita, *J. Am. Chem. Soc.* **2002**, *124*, 13576.
02JCC369	R.E. Dolle, *J. Combin. Chem.* **2002**, *4*, 369.
02JIPMC169	H. Takemura, *J. Inclusion Phenom. Macrocycl. Chem.* **2002**, *42*, 169.
02JOC383	H. Wandel, O. Wiest, *J. Org. Chem.* **2002**, *67*, 383.
02JOC521	R. Cacciapaglia, S. Di Stefano, L. Mandolini, *J. Org. Chem.* **2002**, *67*, 521.

02JOC684	G. De Salvo, G. Gattuso, A. Notti, M.F. Parisi, S. Pappalardo, *J. Org. Chem.* **2002**, *67*, 684.
02JOC1372	J.S. Kim, O.J. Shon, J.K. Lee, S.H. Lee, J.Y. Kim, K.-M. Park, S.S. Lee, *J. Org. Chem.* **2002**, *67*, 1372.
02JOC1392	M. Nicolau, S. Esperanza, T. Torres, *J. Org. Chem.* **2002**, *67*, 1392.
02JOC1436	J.M. Mahoney, R. Shukla, R.A. Marshall, A.M. Beatty, J. Zajicek, B.D. Smith, *J. Org. Chem.* **2002**, *67*, 1436.
02JOC2019	G. Capozzi, G. Delogu, D. Fabbri, M. Marini, S. Menichetti, C. Nativi, *J. Org. Chem.* **2002**, *67*, 2019.
02JOC2348	J.S. Kim, O.J. Shoen, J.A. Rim, S.K. Kim, J. Yoon, *J. Org. Chem.* **2002**, *67*, 2348.
02JOC3533	M. Nakamura, H. Yokono, K. Tomita, M. Ouchi, M. Miki, R. Dohno, *J. Org. Chem.* **2002**, *67*, 3533.
02JOC3878	B. Kampmann, Y. Lian, K.L. Klinkel, P.A. Vecchi, H.L. Quiring, C.C. Soh, A.G. Sykes, *J. Org. Chem.* **2002**, *67*, 3878.
02JOC4081	P.S. Athey, G.E. Keifer, *J. Org. Chem.* **2002**, *67*, 4081.
02JOC4290	C.B. Werz, R. Gleiter, F. Rominger, *J. Org. Chem.* **2002**, *67*, 4290.
02JOC4547	F.C. Pigge, F. Ghasedi, N.P. Rath, *J. Org. Chem.* **2002**, *67*, 4547.
02JOC4860	T.D. Lash, M.J. Hayes, J.D. Spence, M.A. Muckey, G.M. Ferrence, L.F. Szczepura, *J. Org. Chem.* **2002**, *67*, 4860.
02JOC4997	Y. Zhang, Q. Chen, Z. Wang, C. Yan, G. Li, J.S. Ma, *J. Org. Chem.* **2002**, *67*, 4997.
02JOC5644	M. Pawlicki, L. Latos-Grazynski, L. Szterenberg, *J. Org. Chem.* **2002**, *67*, 5644.
02JOC6188	A. Arduini, E. Brindani, G. Giorgi, A. Pochini, A. Secchi, *J. Org. Chem.* **2002**, *67*, 6188.
02JOC6309	V.G. Anand, S.K. Pushpan, S. Venkatraman, S.J. Narayanan, A. Dey, T.K. Chandrashekar, R. Roy, B.S. Joshi, S. Deepa, G.N. Sastry, *J. Org. Chem.* **2002**, *67*, 6309.
02JOC6514	J.S. Kim, O.J. Shon, S.H. Yang, J.Y. Kim, M.J. Kim, *J. Org. Chem.* **2002**, *67*, 6514.
02JOC6856	G.A. Virtue, N.E. Coyne, D.G. Hamilton, *J. Org. Chem.* **2002**, *67*, 6856.
02JOC7457	A.N. Cammidge, O. Öztürk, *J. Org. Chem.* **2002**, *67*, 7457.
02JOC7519	K. Ito, M. Noike, A. Kida, Y. Ohba, *J. Org. Chem.* **2002**, *67*, 7519.
02JOC7569	A. Notti, S. Occhipinti, S. Pappalardo, M.F. Parisi, I. Pisagatti, A.J.P. White, D.J. Williams, *J. Org. Chem.* **2002**, *67*, 7569.
02JOC7968	D.A. Fulton, S.J. Cantrill, J.F. Stoddart, *J. Org. Chem.* **2002**, *67*, 7968.
02JOC8151	K. Tsubaki, T. Otsubo, T. Morimoto, H. Maruoka, M. Furukawa, Y. Momose, M. Shang, K. Fuji, *J. Org. Chem.* **2002**, *67*, 8151.
02JOC8463	S. Ramos, E. Alcalde, G. Doddi, P. Mencarelli, L. Pérez-García, *J. Org. Chem.* **2002**, *67*, 8463.
02JOC9107	C. Bazzicalupi, A. Bencini, E. Berni, A. Bianchi, S. Ciattini, C. Giorgi, S. Maoggi, P. Paoletti, B. Valtancoli, *J. Org. Chem.* **2002**, *67*, 9107.
02JOC9175	A.M. Elizarov, S.-H. Chiu, J.F. Stoddart, *J. Org. Chem.* **2002**, *67*, 9175.
02NJC1320	G. Trippé, M. Oçafrain, M. Besbes, V. Monroche, J. Lyskawa, F. Le Derf, M. Sallé, J. Becher, B. Colonna, L. Echegoyen, *New J. Chem.* **2002**, 1320.
02OL181	H. Furuta, H. Maeda, A. Osuka, *Org. Lett.* **2002**, *4*, 181.
02OL127	G.J. Bodwell, J. Li, *Org. Lett.* **2002**, *4*, 127.
02OL213	E. Brunet, O. Juanes, R. Sedano, J.-C. Rodríguez-Ubis, *Org. Lett.* **2002**, *4*, 213.
02OL881	M.H. Al-Sayah, N.R. Branda, *Org. Lett.* **2002**, *4*, 881.
02OL913	C.J. Kuehl, T. Yamamoto, S.R. Seidel, P.J. Stang, *Org. Lett.* **2002**, *4*, 913.
02OL949	B.M. Kim, S.M. So, H.J. Choi, *Org. Lett.* **2002**, *4*, 949.
02OL1327	K. Nielsen, J.O. Jeppesen, N. Thorup, J. Becher, *Org. Lett.* **2002**, *4*, 1327.
02OL2129	D. Jokic, Z. Asfari, J. Weiss, *Org. Lett.* **2002**, *4*, 2129.
02OL2289	M. Schmittel, A. Ganz, D. Fenske, *Org. Lett.* **2002**, *4*, 2289.
02OL2301	K. Tsubaki, T. Morimoto, T. Otsubo, K. Fuji, *Org. Lett.* **2002**, *4*, 2301.
02OL2313	K. Tsubaki, T. Kusumoto, N. Hayashi, M. Nuruzzaman, K. Fuji, *Org. Lett.* **2002**, *4*, 2313.
02OL2461	G. Trippé, E. Levillain, F. Le Derf, A. Gorgues, M. Sallé, J.O. Jeppesen, K. Nielsen, J. Becher, *Org. Lett.* **2002**, *4*, 2461.
02OL2641	Y. Nakahara, Y. Matsumi, W. Zhang, T. Kida, Y. Nakatsuji, I. Ikeda, *Org. Lett.* **2002**, *4*, 2641.
02OL2695	G. Cafeo, F.H. Kohnke, M.F. Parisi, R.P. Nascone, G.L. La Torre, D.J. Williams, *Org. Lett.* **2002**, *4*, 2695.

02OL3211	J. Xu, Y.-H. Lai, *Org. Lett.* **2002**, *4*, 3211.
02OL3247	H. Ihnels, M. Schneider, M. Waidelich, *Org. Lett.* **2002**, *4*, 3247.
02OL3561	S.-H. Chiu, A.M. Elizarov, P.T. Glink, J.F. Stoddart, *Org. Lett.* **2002**, *4*, 3561.
02OL3907	Y. Norikane, K. Kitamoto, N. Tamaoki, *Org. Lett.* **2002**, *4*, 3907.
02OL3911	J. Xu, Y.-H. Lai, C. He, *Org. Lett.* **2002**, *4*, 3911.
02OL3971	H.G. Kim, C.-W. Lee, S. Yun, B.H. Hong, Y.-O. Kim, D. Kim, H. Ihm, J.W. Lee, E.C. Lee, P. Tarakeshwar, S.-M. Park, K.S. Kim, *Org. Lett.* **2002**, *4*, 3971.
02OL4189	K.A. Nielsen, J.O. Jeppesen, E. Levillain, N. Thorup, J. Becher, *Org. Lett.* **2002**, *4*, 4189.
02OL4233	J. Sankar, V.G. Anand, S. Venkatraman, H. Rath, T.K. Chandrashekar, *Org. Lett.* **2002**, *4*, 4233.
02OL4503	L. Fan, O. Hindsgaul, *Org. Lett.* **2002**, *4*, 4503.
02PSS1529	P. Rosa, X. Sava, N. Mezailles, M. Melaimi, L. Ricard, F. Mathey, P. Le Floch, *Phosphorus, Sulfur Silicon Relat. Elem.* **2002**, *177*, 1529.
02RCR840	O.A. Geras'ko, D.G. Samsonenko, V.P. Fedin, *Russ. Chem. Rev.* **2002**, *71*, 840.
02RJCC153	V.P. Barannikov, S.S. Guseinov, A.I. V'ugin, *Russ. J. Coord. Chem.* **2002**, *28*, 153.
02RJCC697	V.V. Yakshin, *Russ. J. Coord. Chem.* **2002**, *28*, 697.
02IRPC137	X. Lu, M.C. Lin, *Int. Rev. Phys. Chem.* **2002**, *21*, 137.
02S1851	F. Faltin, V. Fehring, R. Miethchen, *Synthesis* **2002**, 1851.
02S2266	T. Bogaschenko, S. Basok, C. Kulygina, A. Lyapunov, N. Lukyanenko, *Synthesis* **2002**, 2266.
02SL1621	M.A. Zolfigol, A. Bamoniri, *Synlett* **2002**, 1621.
02SL1743	L. Raehm, D.G. Hamilton, J.K.M. Saunders, *Synlett* **2002**, 1743.
02TCC1	S. Penades, *Host-Guest Chemistry. Mimetic Approaches to Study Carbohydrate Recognition*, **2002**, pp. 1-241.
02TCC2	*New Aspects in Phosphorus Chemistry I*, Topics in Current Chemistry ed. (Ed.: J.-P. Majoral) **2002**, pp. 1-244.
02TCC3	H. Gries, *Top. Curr. Chem.* **2002**, *221*, 1-24.
02TCC4	E. Brucher, *Top. Curr. Chem.* **2002**, *221*, 103-122.
02TL131	X. Zeng, X. Han, L. Chen, Q. Li, F. Xu, X. He, Z.-Z. Zhang, *Tetrahedron Lett.* **2002**, *43*, 131.
02TL771	M. Pacchioni, A. Bega, A.C. Fabretti, D. Rovai, A. Cornia, *Tetrahedron Lett.* **2002**, *43*, 771.
02TL1193	I.P. Beletskaya, A.D. Averin, A.G. Bessmertnykh, F. Denat, R. Guilard, *Tetrahedron Lett.* **2002**, *43*, 1193.
02TL1209	G.M.L. Consoli, F. Cunsolo, C. Geraci, E. Gavuzzo, P. Neri, *Tetrahedron Lett.* **2002**, *43*, 1209.
02TL1225	A. Mathieu, Z. Asfari, J. Vicens, *Tetrahedron Lett.* **2002**, *43*, 1225.
02TL1425	A. Llobet, E. Masllorens, M. Moreno-Mañas, A. Pla-Quintana, M. Rodríguez, A. Roglans, *Tetrahedron Lett.* **2002**, *43*, 1425.
02TL1629	V.S. Talanov, G.G. Talanova, M.G. Gorbunova, R.A. Bartsch, *Tetrahedron Lett.* **2002**, *43*, 1629.
02TL1909	P. Rajakumar, M. Srisailas, *Tetrahedron Lett.* **2002**, *43*, 1909.
02TL2079	H. Zhang, K.T. Lam, Y.L. Chen, T. Mo, C.C. Kwok, W.Y. Wong, M.S. Wong, A.W.M. Lee, *Tetrahedron Lett.* **2002**, *43*, 2079.
02TL2153	K. Yamato, R.A. Bartsch, M.L. Dietz, R.D. Rogers, *Tetrahedron Lett.* **2002**, *43*, 2153-2156.
02TL2459	G. Pattenden, T. Thompson, *Tetrahedron Lett.* **2002**, *67*, 2459.
02TL2593	M. Militsopoulou, N. Tsiakopoulos, C. Chochos, G. Magoulas, D. Papaioannou, *Tetrahedron Lett.* **2002**, *43*, 2593.
02TL2857	J. Budka, P. Lhoták, I. Stibor, V. Michlová, J. Sykora, I. Cisarová, *Tetrahedron Lett.* **2002**, *43*, 2857.
02TL3217	C. Li, W.-T. Wong, *Tetrahedron Lett.* **2002**, *43*, 3217.
02TL3277	Y. Yamaguchi, S. Kobayashi, N. Amita, T. Wakamiya, Y. Matsubara, K. Sugimoto, Z. Yoshida, *Tetrahedron Lett.* **2002**, *43*, 3277.
02TL3329	N. Kuhnert, A.M. Lopez-Periago, *Tetrahedron Lett.* **2002**, *43*, 3329.
02TL3423	R.S. Simons, J.C. Garrison, W.G. Kofron, C.A. Tessier, W.J. Youngs, *Tetrahedron Lett.* **2002**, *43*, 3423.
02TL3455	T. Tozawa, S. Tokita, Y. Kubo, *Tetrahedron Lett.* **2002**, *43*, 3455.
02TL3883	Y.H. Kim, N.R. Cha, S.-K. Chang, *Tetrahedron Lett.* **2002**, *43*, 3883.
02TL4153	A. Grün, V. Csokai, G. Parlagh, I. Bitter, *Tetrahedron Lett.* **2002**, *43*, 4153.
02TL4207	Y.A. Ibrahim, H. Behbehani, M.R. Ibrahim, *Tetrahedron Lett.* **2002**, *43*, 4207.

02TL4241	A. Basak, S. Mandal, *Tetrahedron Lett.* **2002**, *43*, 4241.
02TL4413	P. Kele, J. Orbulescu, T.L. Calhoun, R.E. Gawley, R.M. Leblanc, *Tetrahedron Lett.* **2002**, *43*, 4413.
02TL4881	H. Shinmori, H. Furuta, A. Osuka, *Tetrahedron Lett.* **2002**, *43*, 4881.
02TL5017	R.A. Bartsch, P. Kus, N.K. Dalley, X. Kou, *Tetrahedron Lett.* **2002**, *43*, 5017.
02TL5153	E. Shokova, V. Tafeenko, V. Kovalev, *Tetrahedron Lett.* **2002**, *43*, 5153.
02TL5229	K. Yamato, F.A. Fernandez, H.F. Vogel, R.A. Bartsch, M.L. Dietz, *Tetrahedron Lett.* **2002**, *43*, 5229.
02TL5343	A. Bilgin, Y. Gök, *Tetrahedron Lett.* **2002**, *43*, 5343.
02TL5245	I. Bauer, W.D. Habicher, *Tetrahedron Lett.* **2002**, *43*, 5245.
02TL5533	T. Lecourt, J.-M. Mallet, P. Sinaÿ, *Tetrahedron Lett.* **2002**, *43*, 5533.
02TL5719	T. Nabeshima, D. Nishida, *Tetrahedron Lett.* **2002**, *43*, 5719.
02TL5747	K. Hiratani, J. Suga, Y. Nagawa, H. Houjou, H. Tokuhisa, M. Numata, K. Watanabe, *Tetrahedron Lett.* **2002**, *43*, 5747.
02TL5767	D.B. Werz, B.J. Rausch, R. Gleiter, *Tetrahedron Lett.* **2002**, *43*, 5767.
02TL5805	K. Yamato, R.A. Bartsch, G.A. Broker, R.D. Rogers, M.L. Dietz, *Tetrahedron Lett.* **2002**, *43*, 5805.
02TL6271	M.H. Ali, M. McDermott, *Tetrahedron Lett.* **2002**, *43*, 6271.
02TL6971	Y.A. Ibrahim, H. Behbehani, M.R. Ibrahim, N.M. Abrar, *Tetrahedron Lett.* **2002**, *43*, 6971.
02TL7213	K. Drandarov, M. Hesse, *Tetrahedron Lett.* **2002**, *43*, 7213.
02TL7243	S. Sasaki, A. Hashizume, D. Citterio, E. Fujii, K. Suzuki, *Tetrahedron Lett.* **2002**, *43*, 7243.
02TL7311	A. Casnati, C. Massera, N. Pelizzi, I. Stibor, E. Pinkassik, F. Ugozzoli, R. Ungaro, *Tetrahedron Lett.* **2002**, *43*, 7311.
02TL7695	P. Rajakumar, V. Murali, *Tetrahedron Lett.* **2002**, *43*, 7695.
02TL7851	R.A.J. Wybrow, L.A. Johnson, B. Auffray, W.J. Moran, H. Adams, J.P.A. Harrity, *Tetrahedron Lett.* **2002**, *43*, 785.
02TL8241	A. Gaucher, O. Barbeau, W. Hamchaoui, L. Vandromme, K. Wright, M. Wakselman, J.-P. Mazaleyrat, *Tetrahedron Lett.* **2002**, *43*, 8241.
02TL8261	H. Shimakoshi, T. Kai, I. Aritome, Y. Hisaeda, *Tetrahedron Lett.* **2002**, *43*, 8261.
02TL8539	K. Hirose, A. Fujiwara, K. Matsunaga, N. Aoki, Y. Tobe, *Tetrahedron Lett.* **2002**, *43*, 8539.
02TL8975	H. Akdas, L. Bringel, V. Bulach, E. Graf, M.W. Hosseini, A. De Cian, *Tetrahedron Lett.* **2002**, *43*, 8975.
02TL9199	J. Xu, Y.-H. Lai, *Tetrahedron Lett.* **2002**, *43*, 9199.
02TL9453	I. Gupta, M. Ravikanth, *Tetrahedron Lett.* **2002**, *43*, 9453.
02TL9493	E.C. Lee, Y.-K. Park, J.-H. Kim, H. Hwang, Y.-R. Kim, C.-H. Lee, *Tetrahedron Lett.* **2002**, *43*, 9493.
02TL9683	V.E. Semenov, V.D. Akamsin, V.S. Reznik, A.V. Chernova, G.M. Dorozhkina, Y.Y. Efremov, A.A. Nafikova, *Tetrahedron Lett.* **2002**, *43*, 9683.
02WHX27	Y. Gao, M. Du, J. Li, X. Bu, *Wuji Huaxue Xuebao* **2002**, *18*, 27.
02YGKK184	T. Nabeshima, T. Saiki, S. Akine, *Yuki Gosei Kagaku Kyokaishi* **2002**, *60*, 184.
02YZ219	E. Kimura, *Yakugaku Zasshi* **2002**, *122*, 219.

INDEX

Ab initio methods, 308
Aceanthryleno[1,2-*e*][1,2,4]triazines, 353
Aceanthryleno[1,2-*g*]pteridines, 353
2-Acetamido-2-deoxy-β-D-glucopyranosides, 275
Acetogenins, 184
4-Acetoxydihydrobenzo[*b*]furan, 192
2-Acetyl-1,4-naphthoquinone, 172
3-Acetyl-1-phenylsulfonylindole, 354
N-Acetyl-2-azetine, 100
1-Acetyl-2-methoxyazulene, 210
Acetylcholine receptor ligands, 263
Acetylcholinesterase inhibitors, 127
1-Acetylisatin, 109
2-Acetyloxyacrylate, 266
N-Acryloylbenzoxazol-2(3*H*)-ones, 273
N-Acyl-1*H*-benzotriazole-1-carboximidamides, 279
2-Acyl-3-phenyl-*L*-menthopyrazoles, 207
3-Acyl-4-alkylthiazolidine-2-thiones, 243
Acylating agents, 315
5-Acylazirines, 39
1-Acylbenzotriazoles, 219
N-Acyliminium ions, 13, 395
3-Acylisoxazoles, 261
N-Acyloxazolidinones, 278
5-Acyloxypyrazoles, 207
Acylsilanes, 110
5-Acyltetrazoles, 223
N-Acylthiazolidine-4-carboxylic acids, 237, 272
5-Acylzirconocene chloride, 186
Adaline, 25
p-(1-Adamantyl)phenol, 438
p-(1-Adamantyl)thiacalix[4]arene, 438
Adenosine deaminase, 348
Adriatoxinm 361
Aerobic oxidation, 292
Aglaforbesin, 401
Akashins, 67
Aldose reductase inhibitors, 306
α,ω-Alkenediols, 186
2-Alkenylfurans, 182
6-Alkoxy-8,9-dialkylpurines, 348
3-Alkoxyazetidin-2-ones, 104
3-Alkoxycarbonyl-2-methylisoxazolidines, 269
4-Alkoxypyrazoles, 207
6-Alkyl-2*H*-pyran-2-ones, 369
1-Alkyl-4-polyfluoroalkyl-1,2,4-triazolium salts, 220
3-Alkylamino-1,2,4-oxadiazoles, 279
3-Alkylamino-1,2,4-triazoles, 222
5-Alkylidenefuran-2(5*H*)-ones, 369
5-Alkylideneselenazoline-2-ones, 247
2-Alkylidenetetrahydrofurans, 187, 188
2-(Alkylthio)isoflavones, 373
Alkynyliodonium salts, 262

Alkynyltungsten complexes, 187
3,4-Allen-1-ols, 362
Allenes, 109,143
2-Allyl-1,2,3-triazoles, 218
1,5-*bis*-*o*-Allyloxyphenylformazans, 440
Allylsamarium bromide, 187
Allyltributyl stannane, 248
Allyltrimethylsilane lithium, 2780
Aminations, 214
Aminoalkynes, 247
Amino 2,6-dideoxyazasugars, 271
Amino acids, 233, 234
α-Amino camphor sultam, 233
5-Amino-1,2,4-triazoles, 347
Amino-1,4,6,7-tetrahydroimidazo[4,5-*b*]pyridin-5-ones, 215
R-(−)-2-Amino-1-butanol, 243
N-(1-Amino-2,2-dichloroethyl)benzamides, 274
4-Amino-2-aryl-2-oxazolines, 274
2-Amino-4-(3,5,5-trimethyl-2-pyrazolino)-1,3,5-triazines, 342
3-Amino-4-arylfuran-2-carboxylates, 181
5-Amino-4-chloro-6-(alkylamino)pyrimidines, 348
2-Amino-5-(ethylsulphonyl)thiadiazole, 246
2-Amino-5-acylthiazoles, 242
4-Amino-5-substituted-1,2,4-triazole-3-thiones, 348
4-Aminobenzopyran, 366
1,2-*bis*(2-Aminoethoxy)ethane, 440
2-Aminoguanidine, 353
2-Aminoimidazo[1,2-*a*]pyridines, 215, 216
2-Aminoimidazole, 210
2-Aminoimidazolones, 217
5-(Aminomethyl)-2-furancarboxylic acid, 180
2-Amino-*N*,*N*,*N*-trimethyl-9*H*-purin-6-ylammonium chloride, 349
α-Aminonitriles, 223
Aminopalladation, 140
2-Aminopyrido[1,2-*a*][1,3,5]triazin-4-ones, 351
Aminotetrahydrofuran, 183
2-Aminothiazoles, 237, 239
Aminothietanes, 272
Aminothiophenes, 117
6-Aminouracils, 351
Amphidinolide, 24
Anabasine, 297
Ancorinolates, 63
Angiotensin II receptor antagonist, 309
Angustmycin C, 128
4-Anilinofuro[2,3-*b*]quinolines, 178
Anthocyanin, 360
4-(Anthracene-9'-ylmethyl)benzo-15-crown-5, 432
Anthrathiophene, 127
Antiangiogenics, 316

Antibacterials, 345
Anticancer agents, 328
Antiferromagnetic exchange coupling, 309
Antifungal agents, 328
Antiinflammatory, 64
Antimetabolites, 64
Antimicrobial activity, 69
Antioxidants, 63
Antiprotozoal agents, 128
Antituberculosis agents, 328
Antitumor agents, 64, 128, 316
Antiviral activity, 69
Aplysinopsin, 64
Apoptosis, 68
Appel's salt, 235
Aquiledine, 407
Arisugacin A, 361
Artemisinin derivatives, 377, 378
1-Aryl-1,2-diazabuta-1,3-dienes, 209
5-Aryl-2-furaldehydes, 180
1-Aryl-3-methylazuleno[1,2-d]pyrazoles, 210
1-Aryl-4,6-diamino-1,2-dihydro-1,3,5-triazines, 341
1-Aryl-4,6-dinitro-1H-indazoyl-3-methylcarboxylates, 208
3-Aryl-5-alkylisoxazoles, 262
3-Aryl-5-isoxazolecarboxaldehydes, 263
2-(Arylamino)benzimidazoles, 212, 216
3-Arylcarbonyl-benzo[b]furans, 197
Aryldiazoacetates, 194
1,3,5-$tris$-Aryl-hexahydro-1,3,5-triazines, 341
2-Aryliminoimidazolidines, 217
3-Arylmethyl-5-(methylthio)isoxazoles, 263
5-Aryloxy-1,2,3-thiadiazoles, 244
Arylsulfinyl-5-alkoxyfuran-2(5H)-ones, 262
1-Arylsulfonyl-5-hydroxy-1H-1,2,3-triazoles, 218
Arylsulfonylimidazolium salts, 213
5-Aryltetrazoles, 223
3-(Arylthio)isothiazoles, 235
5-Arylthio-1,2,3-thiadiazoles, 244
(+)-Aspicilin, 28
Asymmetric epoxidation, 78, 394
Asymmetric transfer hydrogenation, 96
ATPH, 268
Atropisomerism, 148
Attenol A, 20
Australine, 26
Auxarconjugatins, 60
Axinellamine, 60
1-Aza-2-azoniaallene cation, 206
1-Azabicyclo[1.1.0]butane, 102
Azabicyclo[n.3.1]alkenes, 14
Azacalixarenes, 431
Azacrown ethers, 436
bis-Azacrown ethers, 439
bis(Azaferrocene) ligand, 105
2-Azafluorenones, 343
Azaporphyrin, 125
Azathiabicyclo[2.1.0]pentene, 50
Azepines, 10
Azepino[4,5-b]indoles, 394
Azepino[5,4,3-cd]indoles, 121
Azepinone derivatives, 387
Azetidin-2-ones, 102–108
Azetidin-3-ones, 101
Azetidine 2-phosphonic acids, 101
Azetidine-2,3-diones, 102, 103
Azetidines, 2-acyl, 100
Azetidines, 2-imino, 102
Azetidines, 3,3-dichloro-, 100
Azetidines, bicyclic, 101
Azetidines, hydroxylation, 101
Azetidinium salts,101
Azetine, 49
Azeto[1,2-a]benzimidazoles, 102, 212
β-Azido-α,β-unsaturated ketones, 262
Azidolysis, 93
2-Azidopyrazines, 419
N-Azinylpyridinium-N-aminides, 209
Aziridinium ions, 95, 388, 393
Azirine, thioformyl, 54
2H-Azirines, 108, 270
Azirines, 40
[1.1](3,3')-Azobenzenophane, 435
Azomethine ylide, 108, 151, 152, 157, 242, 291
Azoniabicyclo[2.2.1]heptane, 349
Azulene-substituted thiophenes, 121
Azuleno[1,2-c]thiophenes, 116
Azuleno[1,2-d]pyrazoles, 209
Azuliporphyrins, 129
Bacteriochlorins, 267
Baeyer-Villiger oxidation, 80, 378
Baker-Venkataraman rearrangement, 373
Bartoli reaction, 148
Barton-Zard reaction, 145
Baylis-Hillman reaction, 101, 207, 208, 214, 215, 263, 291, 365, 413
BE-10988, 157
Bengamides, 388
Benz[b]indeno[1,2-d]thiophene, 123
Benz[b]indeno[2,1-e]pyrandiones, 373
Benzazepine, 23
2-Benzazepines, 390
Benzazocine, 23
Benzene-oxides, 397
Benzimidazolo-1,3,5,2-triazaphosphorine 2-oxides, 353
1,2-Benzisothiazolin-3-ones, 231
Benzo[1,2,3,4]tetrazine 1,3-dioxides, 351
Benzo[1,2-b:5,4-b']difuran, 197
Benzo[3,2-c]isoquinoline, 198
Benzo[4,5]furopyridines, 198
Benzo[a] xanthene, 374

Benzo[*b*]azepin-2-one, 392
Benzo[*b*]furan-3-triflates, 177
Benzo[*b*]furancarbaldehyde, 178
Benzo[*b*]furylmethanols, 178
Benzo[*b*]oxepines, 401
Benzo[*b*]tellurophene, 131
Benzo[4,5]thieno[2,3-*c*]pyridines, 119
Benzo[*b*]thieno[2,3-*b*]pyridines, 124
Benzo[*b*]thiophenes, 116, 117, 119, 121, 122
Benzo[*b*]thiopheno[2,3-*f*]indoles, 127
Benzo[*b*]xanthene-6,11,12-triones, 374
Benzo[*c*]xanthene. 374
Benzo[*c*]chromene, 366
Benzo[*c*]furan, 199
Benzo[*c*]selenophene, 130
Benzo[*c*]thiophenes, 116
Benzo[*e*][1,4]diazepin-5-ones, 411
Benzo-1,2,3-dioxathepane 2-oxide, 423
Benzo-18-crown-6, 432
Benzo-21-crown-7, 432
Benzoaza-15-crown-5, 439
Benzocarbaporphyrins, 434
2,4-Benzodiazepin-1-ones, 219
Benzodiazepine receptors, 354
1,5-Benzodiazepine, 408
1,4-Benzodiazepines, 218, 409
2,3,4-Benzodioxasilepine, 423
1,3-Benzodioxoles, 249
Benzodiselenatelluroles, 257
Benzodithiatelluroles, 257
1,3-Benzodithioles, 252,256
Benzometaphenylene-25-crown-8, 433
Benzopentathiepine, 424
2*H*-[1]Benzopyrans, 364, 366
4*H*-[1]Benzopyrans, 364
1*H*-[2]Benzopyrans, 368
o-Benzoquinodimethane, 374
Benzothiaselenatelluroles, 257
1,4-Benzothiazepines, 218
1,5-Benzothiazepines, 408, 416
Benzothieno[3,2-*b*]pyridines, 127
1,3,4-Benzotriazepin-5-ones, 419
1,2,3-Benzotriazinones, 352
5-*N*-(Benzotriazol-1-ylmethyl)amino-3-*tert*-butyl-1-phenylpyrazole, 209
Benzotriazole, photochemistry, 218
Benzotriazole chemistry, 219, 417, 352
bis(Benzotriazolylmethyl)amines, 219
Benzotrithiole 1-oxides, 257
1,4-Benzoxathiins, 244, 380
1,4-Benzoxazepines, 218, 385
1,5-Benzoxazepines, 385
4,1-Benzoxazepines. 385
1,4-Benzoxazinone, 196
Benzoxazol-2(3*H*)-ones, 273
3-Benzoyl-2-methylbenzo[*b*]furan, 178

14-Benzoyl-5,10-diphenyl-1-tripyrrinone, 434
1-Benzoyl-*cis*-1-buten-3-yne, 179
1-Benzyl-5-nitroimidazol, 215, 410
9-Benzyl-6-chloropurine, 349
Benzylated 3,4,5-tribromopyrazole-1-oxides, 209
3-(Benzyloxy)isothiazole, 230
Benzynes, 119, 150
Bicyclotetraamerization, 439
Biginelli reaction, 317, 321
2,2'-Biimidazoles, 210
Bioelectronic sensors, 130
Biphenyl-4,4'-dicarboxylic acid, 443
Bipyrroles, 61, 62
Birch reduction, 178
Bischler-Napieralski cyclization, 215, 294
Bisindoles, 155
Bisoxazolines, 274
Bispropargyloxynorbornene, 363
2',4-Bithiazoles, 236
Bithieno[3,2-*b*][1,4]oxathiine, 130
2,2'-Bithienyl, 120
Bohlmann-Rahtz reactions, 285
(+)-Brefelin A, 266
Brevetoxin, 24
(*E*)-2-Bromo-2-en-4-ynoic acids, 369
2-Bromo-4,4-dimethyloxazoline, 276
5-Bromo-4-bromomethyloxazole, 271
3-Bromo-4*H*-pyran, 361
2-Bromo-5-chlorothiazole, 243
1-(Bromoacetyl)azulenes, 196
6-Bromogranulatimide, 65
3-Bromoindole, 62
4-Bromothiazole, 236
6-Bromotryptophan, 64
Buchwald-Hartwig amination, 288, 306
Butadienyl dyes, 432
2,3-Butanedione, 407
4-*tert*-Butoxy-4*H*-pyran, 361
(*E*)-*tert*-Butoxymethylenepyruvate, 362
1-Butyl-3-methylimidazolium hexafluorophosphorate, 215
N-*tert*-Butylbenzenesulphonamide, 231
3(5)-*tert*-Butylpyrazole, 206
p-*tert*-Butylthiatetramercaptocalix[4]arene, 438
C_{60}, 339
(+)-Calanolide A, 364
Calcitriol, 183
Calcium channel blocker UK-74,756, 404
Calix[4](diseleno)crown ethers, 441
Calix[4]azacrowns, 438
Calix[4]-*bis*-azacrown, 438
Calix[4]furan, 442
Calix[4]naphthalene, 367
Calixarenes, 431
Calixphyrins, 431
Calixpyrrole, 431, 441

Callipeltoside A, 360
(-)-Callystatin A, 361
Camphorselenyl sulphate, 241
Camphorsultam derivative, 234
Candida cylindracea lipase, 176
ε-Caprolactone, 378
Carbazoles, 158
Carbene complexes, 143, 276
Carbene chemistry, 107, 145
Carbenoid centers, 91
Carbocationic cascade rearrangement, 160
Carbolines, 158
Carbonylation reactions, 105, 109
Carbopalladation, 151
3,5-*bis*(5-Carboxy-6-azauracil-1-yl)aniline, 340
1,3,5-*tris*(5-Carboxy-6-azauracil-1-yl)benzene, 340
3-Carboxyphenyltrioxanes, 378
Catalytic kinetic resolution, 86
Catenanes, 431, 433, 444
Cavitands, 431
Cetiedil, 123
Chagas' disease, 324
Chartellamide B, 64
Chartelline A, 64
Chemiluminescence, 263
Chemiluminophors, 312
Chimonanthines, 160
Chiral epoxidation, 77
Chiral Lewis acids, 206
Chiral nitridomanganese complex, 89
Chiral nitrogen transfer, 89
Chiral oxazolidinone ketone catalyst, 78
Chiral porphyrins, 77
Chiral *trans-anti-trans*-dicyclohexano-18-crown-6, 432
Chloptosin, 68
Chlorins, 267
2-Chloro-1,3,5-triazines, 344
2-Chloro-3-nitropyridines, 210
3-Chlorocarbazole, 69
Chlorocarbenes, 275
Chlorohyellazole, 69
3-Chloroindole, 62
4-Chloroindoleacetic acid, 66
Chlorooximes, 261
3-Chloropentafluoropropane-1,2-oxide, 411
6-Chloropurine, 243
Chloroxidation, 313
Chromenes, 17
Chromium complexes, 143
Chromium salen complexes, 75
Chromone-3-carboxaldehyde, 374
Ciguatoxins, 24, 399
Cinoxacin, 309
Circumdatins, 411
Claisen rearrangement, 18

Clathria sp, 350
Cobalt nanoparticles, 191
Coelenterazine, 329
Coerulescine, 152
Coleophomone, 27
Colombian poison frog, 390
Colubricidin A, 60
Combinatorial synthesis, 212, 431
'Confused' tetraarylporphyrins, 431, 434
ß-Conhydrine, 9
Coniceine, 28
Conjugate addition, 156
Conotoxins, 64
Contrast agents, 431
Convolutamydine E, 63
Convolutindole A, 63
Coordination polygons, 431
Copper catalysis, 105, 144, 219, 293, 381, 409
Copper iodide Michael reactions, 189
Copper *N*-arylation, 145
Copper phosphoramidite catalysis, 176
Corroles, 431
Corticotropin-releasing factor, 348
Coumarins, 195
Croalbinecine, 28
Cross-coupling, 154
Cross-metathesis, 213
Crown ethers, 431
Cryptand, 439
Crystal engineering, 431
Cucurbituril, 431. 442
CuI–catalyzed cycloisomerization, 180
Curtius rearrangement, 157
Cuscohygrine, 4
5-Cyano-1,2,4-triazines, 343
3-Cyano-1,5-dimethylpyrazole, 42
2-Cyano-azetidines. 100
1-Cyanoindoles, 151
2-Cyanophenanthroline, 345
2-Cyanopyrroles, 42
Cyanosulfides, 53
2-Cyanothiazolecarbazole, 235
Cyanothiophenes, 42
Cyanuric chloride, 344
Cyclam, 436
Cyclen, 436
Cyclic amino acids, 311
Cyclic peptides, 65
Cyclipostin natural products, 423
5-*endo-trig* Cyclization, 149
[4+2]Cycloaddition reactions, 285
Cycloaddition reactions, 294. 296
Cycloaddition strategies, 87
[4+2]/[3+2] Cycloaddition, 269
[3+3]Cycloaddition, 297
Cycloalkenopyridines, 342

N-(1-Cycloalkenyl)pyrazoles, 207
2H-Cyclobuta[c]chroman-4-ols, 366
Cyclobutanone, 366
Cyclocinamide A, 65
Cyclocondensation, 319
[3 + 3]Cyclocondensation, 437
[1 + 1]Cyclocondensation, 441
Cyclohexadienone, 172
Cyclohexyl isocyanide, 182, 372
Cyclopent[b]indole, 160
Cyclopenta[b]benzofuran, 195
Cyclopenta[c]carbazoles, 160
Cyclopenta[c]thiophenes, 124
Cyclopeptide cages, 436
Cyclopropanation, 143
Cyclopropapyranones, 397
Cyclopropylideneacetic acids, 370
o-Cyclopropylphenols, 177
Cyclothiazomycin, 236
Cyclotriphosphazane, 245, 340
Cyclotriveratrylene, 339
L-Cysteine, 238
Cystothiazole A, 244
Danishefsky's diene, 106, 371
DDQ oxidation, 434
5-Deazapterin, 351
Debromostevensine, 60
meso-Decamethylcalix[5]pyrrole, 173, 435
Decatromicins, 60
Dehydrobenzoannulene, 436
Demetalation, 438
Dendritic oxazoline ligand, 277
Density functional calculation, 206
Deoxy-azasugars, 8
Deoxyepolone B, 366
Deprotonated photocleavage, 52
Depsidones, 412
Dermal penetration enhancers, 387
Desmotropes, pyrazoles, 206
Dess-Martin oxidation, 157, 400
Desulfurization, 126
Desymmetrization process, 2, 26, 297
2-Deuterio-1,5-dimethylimidazole, 47
5-Deuterio-1-methyl-4-phenylimidazole, 38
5-Deuterio-1-methyl-4-phenylpyrazole, 38
5-Deuterio-4-phenylisothiazole, 50
4-Deuterio-5-phenylisothiazole, 55
4-Deuterio-5-phenylthiazole, 50
Deuterium labeling, 45–46
Dewar pyridines, 286
2,2-Di(hydroxymethyl)chroman, 368
3,4-Diacetoxydihydrobenzo[b]furan, 192
1,3-Dialkylimidazolium tetrafluoroborate, 215
p-Dialkynylarenes, 198
1,3-Diamines, 208
3,5-Diamino-1,2,4-triazoles, 221

5,7-Diamino-1,2,4-triazolo[1,5-a][1,3,5]triazines, 347
2,6-Diamino-1,3,5-triazine, 339, 340
3,4-Diaminothiophene, 120
2,3-Diarylbenzo[b]thiophenes, 122
Diaryltetrazoles, 223
1,2-Diaza-1,3-butadienes, 248
Diaza-18-crown-6, 439
1,2-Diaza-2,3-butadienes, 308
2,5-Diazabicyclo[2.1.0]pentene, 40
1,5-Diazabicyclo[2.1.0]pentene, 40, 44, 45
2,5-Diazabicyclopentene, 51
Diazabutadienes, 142
Diazacrown ethers, 439
Diazahexaoxamacrobicycles, 439
Diaza-metallacycles, 307
3,4-Diazanorcaradienes, 347
2,6-Diazatricyclo[4.2.0.0^{2,4}]octan-7-one, 108
1,2-Diazetines, 101
Diazinium ylides, 307
Diazo decomposition, 144
Diazonamides, 193, 272
Diazo-thioketone coupling, 374
Dibenz[b,e]azepin-11-one, 393
Dibenzazepinium salts, 394
Dibenzo[1,4]dithiins, 379
Dibenzo[1,4]oxathiins, 379
Dibenzo[a,c]xanthene, 374
5H-Dibenzo[a,d]cycloheptene, 363
Dibenzo[b,f]-1,4-oxazepinones, 414
Dibenzo[c,e]oxepines, 401
Dibenzo-24-crown-8, 432
Dibenzodiaza-30-crown-10, 439
Dibenzo-dithiepine, 379
Dibenzo-oxathiepine, 379
Dibenzothiazepines, 417
Dibenzothiophenes, 117
1,5-Dibenzoyl-2,4-dialkoxybenzene, 197
Dibenzoylacetylene, 183
3,6-Dibromoindole, 63
2,4-Dibromothiazole, 236
4,5-Dichloro-1,2,3-dithiazolium chloride, 235
6,6'-Dichloro-2,2'-diethoxy-1,1'-binaphthyl-4,4'-bis(acetylene), 442
(Z)-1,4-Dichloro-2-butene, 389
3,3-Dichloro-3H-benz[c][1,2]oxathiazol-1,1-dioxide, 232
2,4-Dichloro-6-alkylamino-1,3,5-triazine, 344
Dichloro-p-toluenesulfonamide, 210
Dicobalt octacarbonyl, 266
Dicyanomethanides, 314
2,6-Dideoxysugars, 17
Dieckmann cyclization, 117
Diels–Alder reactions, 3, 16,103, 106, 108, 120, 124,126, 153, 159, 170-172, 176, 187, 198, 232, 272, 280, 288, 291, 297, 308, 311, 314, 319. 320, 328,

343, 344, 347, 362, 366, 367, 371, 380, 387, 397, 434,
4,5-Diethylidene-oxazolidinone, 278
2,3-Dihydro[1,2,4]triazino[6,5-*f*]quinoline-4-oxides, 353
2,5-Dihydro[*b*]oxepins, 18
3,5-Dihydro-1,2-dioxins, 116
4,7-Dihydro-1,3-dioxepine, 412
Dihydro-1,4-oxathiepines, 418
4,5-Dihydro-1*H*-1,2,3-triazoles, 218
4,7-Dihydro-1*H*-imidazo[4,5-*e*]-1,2,4-triazepin-8-one, 215
3,6-Dihydro-2*H*-pyran-2-one, 371
5,6-Dihydro-2*H*-pyrans, 362
3,4-Dihydro-2*H*-pyrans, 363
2,3-Dihydro-benzo[1,4]diazepines, 411
2,3-Dihydrobenzo[*b*]furans, 193, 194
Dihydrobenzo[*b*]oxepines, 400
Dihydrobenzo[*c*]furans, 199
spiro-2,3-Dihydrobenzo[*d*]isothiazole-1,1-diones, 231
Dihydro-2,4-benzothiepines, 418
4,5-Dihydrobenzo[*e*][1,2,4]triazolo[5,1-*c*][1,4,2]diazaphosphinines, 4,5-Dihydrocorynantheol, 26
Dihydrodibenzo[*b*,*f*]phosphine 5-oxides, 405
Dihydrodibenzothiepines, 404
2,3-Dihydrofurans, 175, 176, 191
2,5-Dihydrofurans, 191
2,3-Dihydroimidazo[2,1-*b*][1,3]oxazoles, 216
6,7-Dihydroimidazo[4,5-*d*][1,3]diazepin-8(3*H*)-one, 215
Dihydroselenophenes, 248
Dihydrotetrathiafulvalenes, 255
Dihydrotetrazine, 346
2,5-Dihydrothiophenes, 246
(*Z*)-1,4-Dihydroxy-2-alkenes,191
2,2-Dihydroxystilbenes, 194
1,14-Diiodo-3,6,9,12-tetraoxatetradecane, 440
o,*o'*-Diiodo-4,5-diarylisoxazoles, 264
Diiron(II) complexes, 315
1,4-Diisocyano-2,3,5,6-tetramethylbenzene, 443
1,4-Diisocyano-2,5-dimethylbenzene, 443
1,4-Dilithio-1,3-dienes, 192
1,3-Dimesityl-4,5-dihydroimidazol-2-ylidene ligand, 214
4,6-Dimethoxy-1,3,5-triazinyl carbonate, 344
6,6'-Dimethoxy-2,2'-dihydroxybiphenyl, 437
α,α-Dimethoxyketones, 181
4,6-Dimethyl-1,3,5-triazines, 344
4,4-Dimethyl-2-(*o*-tolyl)oxazoline, 275
Dimethyl-3-arylsulfonyl-4,5-dihydroimidazolium, 211
1,5-Dimethyl-3-trifluoromethylpyrazole, 43
1,2-Dimethyl-4-trifluoromethylimidazole, 43
3-(*N*,*N*-Dimethylamino)-2-isocyanoacrylate, 211

N,*N'*-Dimethylbarbituric acid, 182
Dimethyl-*bis*(4-trimethylsilylethynylphenyl)silane, 438
Dimethyldioxirane, 109
2,3-Dimethylidenechroman-4-ones. 181
1,2-Dimethylimidazole, 40
4,5-Dimethylisothiazolium salts, 232
bis(Dimethylphosphine)methane, 443
1,4-Dimethylpyrazole, 38
1,3-Dimethylpyrazole, 40
1,5-Dimethylpyrazole, 42
2,3-Dioxabicyclo[3.3.1]nonane, 375
1,5-Dioxaspiro[2.4]heptanes. 184
1,7-Dioxaspiro[5.5]undecanes, 26
1,4-Dioxepin-5-ones, 412
Dioxetanes, 111–112, 263
1,2-Dioxins, 376
1,3-Dioxolan-2-ones, 249,250
1,3-Dioxolan-4-ones, 251
1,3-Dioxolane-2-thiones, 250
1,3-Dioxolanes, 249–252, 2255
1,3-Dioxoles, 249
1,5-Diphenyl-2,4-dithiobiuret, 340
Diphosphaferrocenes, 431
1,3-Dipolar cycloadditions, 118, 145, 152, 157, 159, 206, 208, 217, 221, 222, 237, 238, 242, 261-268, 297, 308, 345
Directed *ortho* metalation, 23, 292
Dirhodium tetrakis[*N*-phthaloyl-(*S*)-*tert*-leucinate], 194
Diselenaditellurafulvalenes, 253
Diselenadithiafulvalenes, 253
1,3-Diselenoles, 254
3,4-Disubstituted pyrazolin-5-ones, 207
trans-5,6-Disubstituted trioxanes, 378
1,3-Ditelluroles, 254
1,3-Di-*tert*-butylimidazol-2-ylidene, 210
1,5-Dithia-11-tellurocyclotetradecane, 441
1,4-Dithia-7-telluracyclononane, 441
14-Dithia-8-telluracycloundecane, 441
Dithiaporphyrins, 129, 441
1,2-Dithiin, 379
2,11-Dithio[4,4]metametaquinocyclophane, 440
1,3-Dithiolane 1,3-dioxides, 252
1,3-Dithiolane 1-oxides, 252
1,3-Dithiolanes, 252,253
1,2-Dithiolanes, 256
1,3-Dithiole-2-thiones, 252,253
1,2-Dithiole-3-thiones, 256
1,3-Dithioles, 252
1,2-Dithioles, 256
[1,2]Dithiolo[3,4-*b*]pyrroles, 147
3,3'-*N*,*N'*-Dithiophthalimide, 437
DNA alkylating agents, 210
DNA cleaving agents, 162, 328
DNA intercalators, 328

DNA, 316, 323
N,N'-bis(Dodececyl)-4,13-diaza-18-crown-6, 439
1,5-Doxaspiro[3,2]hexane, 109
Dragmacidins, 65, 327
Dynamic kinetic resolution, 297
Ebelactones, 110
Echinosulfone A, 67
Echinosulfonic acids, 67
Eglington coupling, 433
Electrochemical oxidation, 320
Electrocyclization, 145, 221, 286, 292
Electroluminescence, 207
Electroluminescent materials, 128
Electron-transfer, 129
Ellipticine, 235
Enaminoisocyanide, 38, 39
Enaminones, 344
Enaminonitriles, 38, 39
Enantioenrichment, 75
Enantioselectivity, 76
Endothelin antagonists, 128
(±)-Epiasarinin, 191
Episulfonium ion, 363
Epothilone, 24, 244
Epoxydibenz[b,f]oxepinol, 400
Epoxyimonobactams, 108
Epoxysilane, 82
5-Ethoxy-5,6-diphenyl-4,5-dihydro-2H-1,2,4-triazine, 344
C-Ethoxycarbonyl N-methyl nitrene, 266
Ethyl 3-methoxy-1-methyl-1H-pyrazol-4-carboxylate, 207
1,1'-Ethylenedi-2-imidazoline, 436
Ethylideneisopropylamine, 341
Eudistomins, 69
Excited states, 40
Factor Xa, 11
Feist–Bénary condensation, 191
Ferrier reaction, 107
Fiscalin B, 327
Fischer carbene complexes, 124, 361, 379, 386
Flash vacuum pyrolysis, 206, 386
Flavanones, 194
Flower pigments, 360
Fluorescence, 315
Fluorescent dyes, 328
Fluorination, 145
3-Fluoro-3-imidazolylpropenoic acids, 213
Fluoroionophores, 128
Fluoropyrazoles, 42
Folic acid, 352
β-Formyl enamides, 413
4'-Formylbenzo-18-crown-6, 433
Fosfomycin, 80
Friedel-Crafts reactions, 120, 156, 294
(−)-Frondosin, 177

Fullerenes, 431
2-Furaldehyde diethyl acetal, 180
Furan, natural products, 167–170
Furan-3,4-dicarboxylic acid, 180
Furano lignans, 186
Furano[2,3-c]pyrans, 179
Furanomycin, 16
Furans, cycloadditions, 170–171
Furazano[3,4-b]-pyrazines, 348
Furo[2,3-b]pyridines, 197
Furo[2,3-c][1]benzopyrylium, 369
Furo[2,3-c]pyridines. 197
Furo[2,3-d]pyrimidinediones, 182
Furo[3,2-b]pyridines, 197
Furo[3,4-b][1]benzopyran-9-ones, 373
Furo-[3,4-b]indole, 154, 181
Furo[3,4-b]pyrrole, 147, 171
Furoanthocyanidin, 196
Furocoumarins, 311
Furoic acids, 173
Furoquinolinone, 195
1-(3-Furyl)-1H-imidazoles, 213
2-Furylmethanol, 109
Galactocrown ethers, 432
Galanthamine, 193
Garner's aldehyde, 7, 106
Gaussian-3 study, 308
Geldanamycin, 11
Gewald reaction, 117
α-Glucosidase inhibitor, 127
Glycosphingolipids, 20
Glycosyl nitrones, 267
Glycosyl oxazolines, 275
N-Glyoxyloylborane-10,2-sultam, 233
Gramicidin, 439
Grandberg indole synthesis, 153
Green chemistry, 212, 307
Grignard reagents, 148, 218, 290, 293, 300
Griseoviridin, 24
Growth hormone secretagogue, 152
Grubbs metathesis catalyst, 1, 176, 380, 386, 388, 398. 400. 404
N-Guanidinium, 440
N-Gulosyl-nitrones, 267
Halichlorine, 9
Halichondrin B. 186
Haliclamine A, 126
Halipeptin C, 244
2-Halo-2H-azirines, 263
5-[2-(Halogenobenzylthio)-ethyl]thio-6-azauracil, 343
4-Haloisoxazoles, 263
α-Haloketones, 237
α-Halomethyl ketones, 242
Hamacanthin, 67

Index

Hantzsch syntheses, 118, 242, 243, 317
Heats of formation, 308
Heck reactions, 14, 107, 123, 145, 157, 158, 175, 193, 198, 214, 293, 366, 368, 374, 412,
Heck–type cobalt–catalysis, 190
Hemetsberger reaction, 153
Heptaphyrin macrocycles, 131
30-π-Heptaphyrins, 441
Heptapyrrolic macrocycle, 435
Herbarumin, 4
N-(Hetero)arylation, 343
Heteroaryl thioamides, 240
Heterocalixarenes, 161, 178
Hetero-clathrochelate, 431
N-Heterocyclic carbenes, 214
Hetero-o-quinodimethanes, 171
HETPHEN concept, 444
Hexa-aryloxy-cyclotriphosphazenes, 345
Hexahydro-1,3,5-triazine, 341
Hexahydro-8H-pyrido[1,2-a]pyrrolo[2,1-c]pyrazines, 219
Hexahydroimidazo[1,5-b]isoquinolines, 219
$meso$-Hexamethyl-$meso$-hexaphenylcalix[6]pyrrole, 435
$meso$-Hexaphenylcalix[6]pyrrole, 435
High dilution macrocyclization, 440
Hinckdentine A, 64
(-)-Histrionicotoxin, 267
anti-HIV agents, 11, 128, 422
D-Homo-10-epi-adrenosterone, 171
Homooxacalix[3]arene, 433
Hormone-sensitive lipase inhibitor, 423
Host-guest chemistry, 431
6-Hydrazino-uracils, 348
Hydrazonoyl chlorides, 340, 419
Hydroamination, 386
Hydrolytic kinetic resolution, 75
2-Hydroxy-1-naphthaldehyde, 365
2-Hydroxy-4,6-dimethoxy-1,3,5-triazine, 344
Hydroxyacrylonitrile, 181
o-Hydroxyaldimines, 175
2-Hydroxybenzaldehydes, 365
2-Hydroxychalcones, 373
7-Hydroxychromenes, 367
Hydroxycoumarins, 372
4-Hydroxycyclopentenones, 173
4-Hydroxydihydrobenzo[b]furan, 192
1-Hydroxyindoles, 150
α-Hydroxylactam, 395
9[(Hydroxymethyl)phenyl]adenines, 348
2-Hydroxymethyl-2-methyl-2H-[1]benzopyrans, 365
N-Hydroxyphthalimide catalysis, 108
2-Hydroxypropyl-p-nitrophenylphosphate, 439
1-Hydroxypyrazoles, 207, 209
Hydroxypyrrolizidines, 8

1,7-bis[α-Hydroxy-α,α-diethyl)-methyl]bithiophene, 441
Hypertricone, 399
Ibotenic acid, 102
Imidazo[1,2-a][1,3]diazepin-3-ones, 410
Imidazo[1,2-a]pyridines, 215
Imidazo[1,2-c]pyrimido[5,4-e]pyrimidinones, 354
Imidazo[1,5-b]isoquinolines, 218
Imidazo[1,5-c]pyrimidinones, 215, 216
Imidazo[1,5-d]thieno[2,3-b][1,4]thiazines, 127
Imidazo[2,1-b][1,3]oxazoles, 215
Imidazo[2,1-c]purinones, 354
Imidazo[4,5-b]pyridinones, 216
Imidazo[4,5-c]pyridinones, 215
2H,4H-Imidazo[4,5-d][1,2,3]triazoles, 219
Imidazo[4,5-e][1,2,4]triazepin-8-one, 353
Imidazo[4,5-h]isoquinolin-9-ones, 215
Imidazo[5,1-f][1,2,4]triazin-4(3H)-ones, 347
Imidazol[1,2-a]pyridines, 215
Imidazole-4,5-dicarboxylic acid, 213
Imidazolidin-2-ones, 211
[1$_4$]Imidazoliophanes, 210
[14]Imidazoliophanes, 436
Imidazolium ionic liquids, 214
2,6-bis(Imidazolmethyl)pyridine, 436
2-Imidazolones, 217
1-(2-Imidazolonyl)-1,2,4-triazoles, 221
Imidazopyridodiazepines, 216
Imidazotetrazines, 349
Iminium ion, 146
Imino-1,2,3-dithiazole, 235
2-Imino-1,3-selenazolidin-4-one, 247
Iminoazirines, 37, 39, 40
Immunosensors, 130
IMPDH inhibitor, 271
Indazole-N-oxide, 420
Indeno[4,5-b]thiophene, 124
Indenopyrrolocarbazoles, 162
Indium-mediated reactions, 234, 290
Indolo[1,2-b]indazoles, 209
Indolo-1,2,4-triazine derivatives, 353
Indole-2,3-quinodimethane, 181
Indole-benzo-fused azepinones, 394
Indole-fused azepinones, 394
Indolizidines, 3, 12, 13, 28, 146
Indolo-2,3-quinodimethanes, 159
Indolocarbazoles, 140, 162
Indolophane, 159
Inosine monophosphate dehydrogenase, 342
Intramolecular amination, 151
Intramolecular copper-catalyzed aziridination, 90
Intramolecular Michael addition, 146
Iodocyclization reaction, 188
α-Iodocyclohexenones, 367
Iodoenolcyclization, 191

Iodoetherification, 265
5-Iodopyran-2-ones, 373
o-Iodoxybenzoic acid, 271
Ionic liquids, 214
Iridinium-catalysis, 104, 210, 291, 293
Isatin imines, 222
Isatins, 153, 157
Isatin-3-thiosemicarbazone, 342
Isatogens, 157
Isatoic anhydride, 409
(+)-Isoaltholactone, 173
Isochromanones, 373
Isochromanquinone, 369
Isochromans, 369
Isocyanates, 145
α-Isocyanoacetamides, 270
Isocyanosulfides, 53
Isoindolo-fused benzothiazoles, 417
Isoindolo-fused thiazepines, 418
Isoindolone, 146
Isolaurallene, 28
Isoquinoline, 13
Isothianaphthene, 119
Isothiazole, 48
Isothiochromans, 374
Isoxazoles, 370
Isoxazolidine-4-carbaldehyde, 268
Isoxazolidinyl-indolizidines, 267
Isoxazolidinyl-quinolines, 267
Isoxazolidinyl-β-lactams, 266
Isoxazolin-5-ones, 266
Isoxazoline N-oxides, 265
Isoxazolinocyclobutenones, 264
Isoxazolopyridazinones, 262
Jacobsen's catalyst, 76
JandaJel resins, 271
Japp-Klingemann reaction, 220
Jaspamides, 65
Juliá-Colonna epoxidation, 80
Katsuki-type salen ligand, 76
Keramamides, 65, 66
2-Ketocarbacepham, 106
Ketorolac, 120
Kinugasa reaction, 105
Kornfeld's ketone, 171
Kottamides, 64
Krapcho dealkoxycarbonylation, 314
Kumata-type cross-coupling, 288
β-Lactams, 3-hydroxy, 103
β-Lactams, biological studies 106
β-Lactams, enantiopure, 106
β-Lactams, polymer-bound, 105
β-Lactones, 108–110
Lariat ethers, 431, 432
Laulimalide, 24, 360

Lawesson's reagent, 116, 241, 307
Leptomycin, 361
S-(+)-Leucinol, 243
Lewis acid catalyst, 83, 86
Light-harvesting assemblies, 435
Lipase-catalysed, 393
Liquid crystal, 128, 313
Lithiation, 153
Lotrafiban, 409
Luche reduction, 173
Luminescence, 341
Lunarine, 193
Macrocyclization, 433
Macrolactonisation, 360
Macrospirocyclic esters, 433
Madelung procedure, 148, 153
Magrastatin, 28
(-)-Malyngolide, 361
Mannich bases, 120, 327
Mannich reactions, 160, 218
α-D-C-Mannosyl-(R)-alanine, 273
Manzamine A, 24
Matemone, 63
Matrix metalloproteinase inhibitors, 406
McMurry reductive coupling, 433
Medermycin, 172
Meldrum's acid, 377
Menthofuran, 181
4-Mercapto-1,3-diols, 380
2-Mercaptoimidazoles, 213
Merrifield's resin, 76
1,3-bis(Mesityl)imidazolinium chloride, 192
Mesobiliverdin XIIIa, 434
Meso-free corrole, 439
Mesoionic heterocycles, 218
[14]Metaheterophanes, 436
Metal-assisted chemistry, 325
Metal-centered catalysts, 77
Metalla-perhydro-1,3,5-triazines, 339
Metallated indoles, 153
Metallic crystals, 130
Metallosupramolecular assemblies, 444
Metallosupramolecular, 444
Metathesis, 214
[3 + 1]-Methodology, 434
N,N'-bis(4-Methoxybenzylidene)ethane-1,2-diamine, 211
1-Methoxycyclohexene, 362
Methyl malonylchloride, 409
3-Methyl-2-furoic acids, 173
1-Methyl-2-thiaadamantane, 374
1-Methyl-4-nitroimidazole, 213
1-Methyl-4-phenylpyrazole, 38
2-Methyl-5-lithiofuran, 180
1-Methyl-5-phenylpyrazole, 43

2-Methyl-5-vinyl-3-furoate, 170
2-Methylbenzo[*b*]furan, 178
Methyldifluorodiethoxyphosphonodithioacetate, 243
α-Methylene tetrazoles, 222
4-Methylene-2,6-disubstituted tetrahydropyrans, 364
3-Methylene-4-vinyltetrahydrofurans, 190
Methylene*bis*(4,4'-methyl-6,6'-salicylaldehyde), 440
exo-Methyleneprolinate, 264
2-Methylenetetrahydrofuran, 175
3-Methylenetetrahydrofuran, 188, 189, 190
Methylfurolabdane, 179
1-Methylimidazoles, 38, 39, 45, 47
2-(*N*-Methylimino)-2*H*-azirine, 37
Methylisothiazoles, 48
Methylisothiazolium ions, 48
O-Methylisourea, 210
Methyllycaconitine, 19
2-Methyloxazolo[5,4-*b*]pyridine, 272
1-Methylpyrazoles, 37, 38, 45, 51
Methylsulfanylimidazo[4,5-*c*]pyrazoles, 209
Methylthiazoles, 48
2-Methylthio-4-nitrothiophene, 262
2,3,4,5-Tetrabromopyrrole, 58
Michael additions, 90, 56, 161, 206, 214, 286, 298, 300
Microcarpalide, 28
Microsclerodermins, 66
Microwave conditions, 158, 207, 210, 212, 214, 217, 285, 288, 292, 293, 317, 318, 324
Mint perfume, 181
Mitosene, 12
Mitozolomide, 349
Mitsunobu reactions, 155, 243, 296, 348, 360, 406
Möbius-like conformations, 402
Mo-catalysed, 370
Molecular directivity, 442
Molecular encapsulation, 431
Molecular flapper, 440
Molecular rotor, 433
Molecular switch, 440
Molecular wires, 128
Monoxime oxidase inhibitor, 69
(−)-Morphine, 193
α-Morpholinoamides, 219
Morulin Pm, 64
Mukaiyama reagent, 106
Mukaiyama's reagent, 216
Mukanadin B, 59
Multi-coordination modes, 345
Munchnones, 237
Muscarinic M_1 receptor antagonist, 407
Mycoepoxydiene, 27
Nakamuric acid, 60
Nanoboxes, 444
Nanoscopic metal assemblies, 442

1,4-Naphtho[1,8a,8-*ef*]diazepin-2(1*H*)-one, 411
Naphtho[2,1-*b*]thiopyran-1'-ylidene-9*H*-thioxanthenes, 374
Naphtho[2,3-*b*]thiophenequinones, 127
Naphthodioxoles, 256
Naphthodiselenoles, 256
Naphthoditelluroles, 256
Naphthodithioles, 256
Naphthooxaselenoles, 256
Naphthooxatelluroles, 256
Naphthooxathioles, 256
Naphthoselenatelluroles, 256
Naphthothiaselenoles, 256
Naphthothiatelluroles, 256
Naproxen, 220
Neber rearrangement, 95
Negishi couplings, 116, 155, 177, 236, 271, 288, 325
Neopyrrolomycin, 59
Ni(II) *N*-confused isoquinoporphyrins, 434
Ni-catalyzed cross-coupling, 307, 370
Nickel–catalyzed [2+2+2] cycloaddition, 198
Nicotinic acetylcholine receptor ligands, 308
Ningalin B, 141
Ninhydrin, 191
Nitrenes, 54, 387
Nitric oxide synthase inhibitors, 121
Nitric oxide synthase, 350
Nitrile oxides, 261, 262, 264
Nitrilimine, 340
Nitrilimines *N*-azinylpyridinium-*N*-aminides, 208
6-Nitro-1*H*-[1,4]-diazepines, 406
Nitroalkenes, 268
3-Nitrochromenes, 365
3-Nitrocoumarins, 367
Nitrogen transfer, 88
β-Nitro-*meso*-tetraphenylporphyrin, 207
Nitrones, 104, 107, 172, 267, 269, 390
N-2-Nitrophenylimidates, 212
6-Nitroquinoline, 353
N-Nitrosobenzotriazole, 218
β-Nitrostyrenes, 365
NMR methods, 310, 324
(+)-Nonactic acid, 184
Norephedrine, 7
Noyori catalyst, 4
Nucleoside, 243, 318
4-Iodoisocoumarins, 372
5-Iodopyran-2-ones, 369
Oligo-*p*-phenylenes, 198
Oligopyrroles, 441
Thiophenes, 116
Oligothiophenes, 121, 122
Olivacine, 311
Oppolzer's sultam, 268
Organocuprate reagents, 300

Osmium complexes, 148
Oxaarenecyclynes, 433
Oxabicyclo[2.2.1]heptadienes, 171
Oxabicyclo[3.2.1]octenes, 176
Oxa-carbaporphyrin, 439
endo-Oxacyclization, 85
3-Oxacycloalkenes, 15
Oxadiazepine derivatives, 420
1,2,6-Oxadiazepine, 420
1,3,4-Oxadiazole, 187, 280
1,2,5-Oxadiazolo[3,4-c]pyridines, 130
Oxa–di–π–methane rearrangement, 174
Oxalopropionate, 110
1-Oxapentadienyl cations, 181
Oxa–Pictet–Spengler reaction, 179
1,2-Oxaselenolanes, 257
Oxaspiro[5.5]undecane, 364
Oxasqualenoids. 184
Oxatelluretanes, 111-112
1,3-Oxathiane, 380
1,3-Oxathiolan-5-ones, 255
1,2-Oxathiolane 2,2-dioxides, 257
1,2-Oxathiolane 2-oxides, 257
1,3-Oxathiolanes, 255,256
1,2-Oxathiolanes, 256
1,2-Oxathiole 2,2-dioxides, 256
1,3-Oxathioles, 255
1,4-Oxazepin-7-ones, 413
1,4-Oxazepines, 413
1,3-Oxazepines, 413
Oxazepino-indolo[2,3-a]quinolizines, 415
[1,2]-Oxazine, 265
3-(Oxazol-5-yl)-indoles, 271
Oxazoline ligands, 275-277
4-Oxazolinyloxazole, 271
Oxazolinyloxirane, 275
Oxazolo-azepinone, 393
Oxazololactams, 277
Oxepine, 3
2-Oxetanones, 108–110
Oxetanones, biological activity, 110
Oxidative couplings, 439, 441
Oxidative cyclization, 150
Oxindoles, 152
Oxo-1,2,3-benzotriazine, 351
N-(2-Oxoalkyl)oxazolinium salts, 210, 275
α-Oxoamides, 103
4-Oxoazetidine-2-carbaldehydes, 104
Oxocanes, 175
Oxone, 78, 80
Oxyallyl cation, 16
Oxybenziporphyrins, 434
(+)-Oxybiotin, 187
1,1'-Oxy-bis[2-bromoethane], 441
Palladacycles, 366

Palladium-catalysed reactions, 6, 122, 123, 124, 129, 143, 151, 152, 158, 207, 209, 212, 214, 217, 222, 223, 243, 268, 293, 295, 307, 314, 318, 327, 392
Palladium(II)-pyrazolato cyclic trimer, 206
(+)-Pamamycin-607, 184, 186
Para[2.2]cyclophane, 130
(+)-(R)-[2.2]Paracyclophane[4,5-d]oxazol-2(3H)-one, 272
Paraherquamide alkaloids, 364
Parallel synthesis, 212
Passerini reactions, 177, 222, 223, 270
Paterno–Büchi reaction, 109
Pauson–Khand reactions, 17, 272, 298
Pd(diamine)(ONO$_2$)$_2$, 442
Pechmann reaction, 371
Pectenotoxins, 361
Penitrems, 66
Pent-4-yn-1-ol, 363
Pentafluorophenyl diphenyl phosphinate, 272
Pentathiaheptaphyrins, 129
Pentathiepane, 425
Pentathiepines, 147, 425
Pentathiepino[2,3-b]thiophene, 121
Pentathiepino[6,7-b]benzo[d]thiophene, 121
Perfluorohexylation, 174
Pericyclic reactions, 124
Permethylated α-cyclodextrin, 439
Pharmacologically active seven-membered heterocycles, 425-426
Phase-transfer catalysis, 432
2-Phenacyl thiazolidine, 238
Phenanthro[9,10-d]isoxazoles, 264
4-Phenyl-1,2,4-triazoline-3,5-dione, 397
3-Phenylazetine, 50
Phenyldipyrromethane, 434
N-(2-Phenylethyl)-4,13-diaza-18-crown-6, 439
Phenylisothiazoles, 49, 55
1-Phenylpyrazole, 46
Phenylsulfanylacrylamides, 392
[1,2]-Phenylsulfonyl migrations, 183
Phenylthiazoles. 49, 50, 55
5-Phenylthiomethyltropone, 366
Phomactin A, 181
Phosphinines, 431
Phosphonoacrolein, 362
Phosphonosugars, 23
Phosphorus 4-membered heterocycles. 112–113
Phosphorylase inhibitors, 318
Photocatalyzed cycloaddition, 284
Photochemical mechanism, 44, 45
Photochemical cleavage, 37–57
Photochemical isomerizations, 37–57
Photochemical mechanism, 40, 42, 49, 51
Photochemical transpositions, 37–57
Photochromic systems, 125

Index 461

Photocleavage, 39
Photocyclization reactions, 291
Photocyclizations, 38, 107, 221
Photocycloaddition reactions, 124, 143, 157, 217, 272, 292
Photoinduced decarboxylations, 418
Photoionic switch, 432
Photoisomerization, 280
Photolithography, 130
Photoluminescence, 128, 207
Photo–oxidation, 144,174
Phototransposition, 39
Phthalans, 373, 379
1,4-Phthalazinedione, 172
Phthalimidesulfenyl chloride, 437
Phthalocyanines, 439
Pibocin, 65
Pinnaic acid, 9
Pinnatoxin A, 361
Pinolidoxin, 28
Piperidines, 6, 7, 8, 13
Plakohypaphorines, 64
Plakortamines, 69
Plakortic acid, 376
Platelet aggregation inhibitor, 316, 409
Platelet-derived growth factor receptor kinase, 154
Poly(fluorinated oxetane)s, 110
Poly[1-(2,4,6-trichlorophenyl)-1H-1,2,4-triazol-5-yl]alkanes, 220
o-Polyazamacrolactamization, 436
Polyazamacromolecules, 431
Polyheterocyclic frameworks, 322
Polyhydroxyindolizidines, 267
Polymer supported reagents, 218, 221
Polymer-bound glycerine, 217
Polymer-supported chalcones, 219
Polystyrene-carbamyl chloride, 217
Polythiophenes, 130
Porphyrins, 431
Potassium channel blocker, 123
Prelaureatin, 28
Prins reactions, 298, 364
N-Propargyl pyrimidines, 262
Propargylamines, 278
3-Propargylmercapto-1,2,4-triazin-5-one, 343
Propargyltrimethylsilane, 104
Propynyltrimethylsilane lithium, 278
Protein phosphatase inhibitor, 126
Protein tyrosine phosphatase inhibitors, 128
Pseudo-24-crown-8, 432
Pseudoceratidine, 60
Pseudopolyrotaxanes, 433
[2]Pseudo-rotaxanes, 433
Psoralen, 195
Pteridines, 351, 352
Pterulone, 401

Pummerer reactions, 160, 218, 390
Pyramidalized nitrogen, 108
2H-Pyran-2-ones. 182
Pyrano azepinediones, 393
Pyrano[2,3-b]chromones, 372
Pyrano[3,2-c]benzopyrans, 365, 372
Pyrano[3,2-c]coumarins, 372
Pyrano[4,3,2-de] [1]benzopyrylium pigments, 369
Pyranoanthocyanidin, 196
Pyranocoumarin, 372
Pyrano-fused oxepanes, 400
Pyranopyrandiones, 370
Pyranopyrans, 362
1-[1-(Pyrazol-1-yl)cycloalkyl]benzotriazoles, 207
Pyrazole thioamides, 210
Pyrazole-related C-nucleosides, 206
Pyrazoles, 206–210
Pyrazolino[60]fullerenes, 206
Pyrazolo[1,5-a]pyridines, 209
Pyrazolo[1,5-a]pyrimidines, 209
Pyrazolo[1,5-d][1,2,4]triazin-7-ones, 347
Pyrazolo[1,5-e][1,3,5]benzoxadiazocines, 209
Pyrazolo[1,5-e][1,3,5]benzoxadiazocines, 209
Pyrazolo[3,4-a]quinoline-1-oxides, 209
1H-Pyrazolo[3,4-b]quinoxalines, 207
Pyrazolo[3,4-b]quinoxalines, 348
Pyrazolo[3,4-b]thieno[2,3-d]pyridines, 117, 127
Pyrazolo[3,4-d]pyrimidines, 348
Pyrazolo[4,3-d]pyrimidines, 348
Pyridazine furocoumarins, 347
Pyridazino[3,4-b]quinoxalin-4-ones, 306
[3](3,6)Pyridazino[3](1,3)indolophane, 436
Pyridazinopyrrolocoumarins, 347
8H-Pyrido[1,2-a]pyrrolo[2,1-c]pyrazines, 218
Pyrido[2,3-b]pyrazines, 351
Pyrido[2,3-b]thieno[3,2-d]pyrimidines, 118, 352
Pyrido[2,3-d][1,2,4]triazolo[4,3-a]pyrimidine-5,6-dione, 354
Pyrido[2,3-d]pyrimidines, 351, 352
Pyrido[4,3-b]carbazole, 311
2-Pyridylpyrroles, 142
Pyrimido[1,2-a][1,3,5]triazines, 350
Pyrimido[1,2-b][1,2,4,5]tetrazines , 351
Pyrimido[3,4-a][1,3,5]-triazines , 350
Pyrimido[4,5-d]pyridazines, 351
Pyrimido[5',4':5,6]-pyrido[2,8-d]pyrimidine, 352
Pyrimido[5,4-d]pyrimidines, 352
bis-Pyrrazolylpyridine, 436
Pyrrolecarboxaldehyde, 439
Pyrrolidines, 4, 6, 7
Pyrrolines, 141
Pyrrolizidine, 28
Pyrrolnitrin, 59
bis(Pyrrolo)tetrathiafulvalene, 441
Pyrrolo[1,2-a]indoles, 159
Pyrrolo[1,2-a]pyrazines, 218

Pyrrolo[1,2-c]thiazoles, 238, 272
Pyrrolo[2,1-a]isoquinolines, 295
Pyrrolo[2,3-b]indoles, 153
Pyrrolo[3,2-c]quinolines, 147
Pyrrolo[3,2-c]quinolones, 291
Pyrrolo[3,4-b]hexahydro-1H-1,5-benzodiazepines, 411
Pyrrolo[3,4-b]indoles, 159
Pyrrolo[3,4-d]pyrimidines, 351
bis(Pyrrolo[3,4-d]tetrathiafulvalene, 441
Pyrrolobenzo[d]thiepines, 404
Pyrazolino[60]fullerenes, 206
Pyrrolo-benzodiazepinone natural product, 411
Pyrrolo-fused benzothiazoles, 417
Pyrrolomycin A, 59
Pyrroloquinazoline alkaloid, 323
Quaterbenzimidazole, 215
2,2':4,":4,4"-Quaterpyridinyl, 443
Quaterthiophenes, 129
1-(2-Quinazolonyl)-1,2,4-triazoles, 221
ortho-Quinodimethane, 125
Quinolizidines, 12
o-Quinone isochromenes, 368
o-Quinone methide, 126, 129, 365, 366
Quinoxalinium dichromate, 328
Radical conditions, 144, 147, 160
Radicicol, 25
Ramberg-Bäcklund, 21
Reductive amination, 298
Reductive deoxygenation, 87
Reichardt dye $E_T 1$, 433
Retro-Mannich reaction, 154
Reverse transcriptase inhibitor, 121, 219
(-)-Rhazinilam, 276
Rhinovirus 3C-protease, 369
Rhizopus arrhizus lipase, 176
Rhodium catalysis, 88, 144, 150, 157, 176, 186, 190, 191, 210, 212, 362
Rhombic-shaped macrocycle, 443
Rhopaladins A-D, 159
Riboflavin synthase, 351
5-exo Ring closure in $S_{RN}1$, 195
Ring expansion, 146
Ring rearrangement metatheses, 3
Ring-closing metathesis, 1, 2, 159, 192, 213, 298, 363, 399, 403, 404, 418, 433, 439
Ring-opening metathesis, 1, 213
Robinson-Gabriel synthesis, 270
(+)-Rogioloxepane A, 398
Rosenmund-von Braun reaction, 214, 215
Roseophilin, 24
[2]Rotaxane, 433, 439
[3]Rotaxane, 433
Rotemund-type synthesis, 434
Ru(II) complexes, 341

Ruthenium carbene complexes, 5
Ruthenium catalysis, 4, 124, 157, 192, 213, 289, 290, 363, 374, 398,
Saccharosamine, 363
Sakurai reaction, 360
Salacinol, 127
Salen complexes, 76
Salicylaldehydes, 365
Salicylihalamide A, 24
Samarium diiodide, 290
Samarium, 145
Saxitoxin, 439
Scandium perchlorate, 155
Scandium(III) triflate, 175
Schrock metathesis catalyst, 1
Securamines, 65
1,2,3-Selenadiazoles, 130, 248
Selenaethers, 438
2-Selenazolin-4-one, 248
Selenazolone, 247
Selenoethers, 431
Selenoureas, 247, 248
Self-assembly, 431, 440, 442
Semi-bullvalenes, 347
Sensors, 431
Serine proteases inhibitors, 121
Sexithiophenes, 129
Sharpless dihydroxylation, 79, 173, 365
Sialic acid, 26
[2,3]Sigmatropic rearrangement, 118
[3,3] Sigmatropic rearrangement, 118
[1,3]-Sigmatropic rearrangement, 143
Silatropylium ion, 405
Silicon 4-membered heterocycles, 112–113
Silthiofam, 117
Silylalkynones, 208
6-Silylpyrazoles, 207
Siphonarin B, 361
Skraup reaction, 292
Slagenin A, 59
Smiles rearrangement, 220
cis-Solamin, 183
Solanapyrones, 361
Solid phase approaches, 122, 128, 145, 151, 153, 215, 216, 238, 288, 297, 317, 326
Sonagashira couplings, 119, 122,158, 177, 197, 214, 368, 372
Sorbitol dehydrogenase inhibitor, 342
Spiro(3H-indole-3,2'-oxetane)s, 109
Spiro-[3H-indole-3,3-[1,2,4]triazolidine]-2-ones, 222
Spiroannulation, 185
Spiroazepinone, 387
Spirocyclic piperidines, 9
Spirocyclisation, 361
Spirodihydrofurans, 191
Spiroisoxazolinoprolinates, 264

Spiroketals, 361
Spirooxindoles, 147
Spiropiperidines, 10
Spiro-β-lactams. 103
Spongistatin, 17
Spongistatins, 361
π-Stacking, 116
(-)-Statine, 277
Staudinger cycloaddition, 104
Staudinger reaction, enantioselective, 105
Staudinger reaction, polymer-supported, 105
Stilbenobis(18-crown-6), 433
Stille reactions, 122, 151, 162, 236, 238, 261, 271, 306, 349, 369,
Streptopyrroles, 60
Strychnine, 311
Styryldiazoacetates, 385
meso-(8-Substituted naphtha-1-yl) porphyrins, 434
2-Substituted-2,5-dimethoxy-2,5-dihydrofurans, 180
Sulfamides, 421
4(5)-Sulfanyl-1H-imidazoles, 211
5-Sulfanylthiazoles, 239
Sulfenylation, 155, 437
Sulfinylaziridine, 92
N-Sulfinylimines, 8
3-Sulfolenes, 126
N-Sulfonyliminium ions, 8, 13
5-Sulfonyltetrazoles, 223
Sulfur ylide, 81
Sultones, 380
Superbase, 287
Supermolecules, 431
Supramolecular chemistry, 431
Supramolecular construction, 442
Suzuki couplings, 122, 130, 177, 214, 215, 271, 288, 306, 307, 312, 318, 325, 350, 361, 371,
Suzuki-Miyaura couplings, 141, 214, 215, 436
Swainsonine, 3
Switches, 431
C_3 Symmetrical ligands, 276
Takai's titanium protocol, 15
Tambjamines, 61
Taurodispacamide A, 59
Te macrocycles, 438
Telluro[3,4-c]thiophene, 131
Telluroethers, 431
Temozolomide, 347, 349
5,10,15,20-Tetraaryl-21-telluraporphyrin, 435
5,10,15,20-Tetraaryl-21-vacataporphyrin, 434
Tetra-azaacenaphthene-3,6-diones, 354
Tetra-azaphenalene 1,7-dione, 354
1,2,4,5-Tetraazaspiro-cycloalkane-3-thiones, 346
Tetrachloropyrimido[4,5-d]pyrimidine, 352
Tetrahydro-1,4-benzodiazepines, 219
Tetrahydro-1,4-benzothiazepines, 219
Tetrahydro-1,4-benzoxazepines, 219

Tetrahydro-1H-imidazole-2,4-diones, 210
Tetrahydroazepines, 10
Tetrahydrobenzo[b]xanthones, 374
(-)-Δ^9-Tetrahydro-cannabivarins, 367
Tetrahydroimidazo[1,5-b]isoquinolin-1(5H)-ones, 219
Tetrahydropyrano[3,2-b]pyran, 362
1,2,3,4-Tetrahydropyrrolo[1,2-a]pyrazines, 219
Tetrahydrotetrazines, 347
Tetramercaptotetrathiacalix[4]arene, 438
Tetra-N-confused cyclohexapyrrole, 434
7,8,15,16-Tetraoxadispiro[5.2.5.2]hexadecane, 378
1,2,4,5-Tetraoxanes, 378
5,10,15,20-Tetraphenyl-2-aza-21-carbaporphyrinatonickel(II), 434
meso-Tetraphenylazuliporphyrin, 434
5,10,15,20-Tetraphenyl-p-porphyrin, 434
Tetrapyrroles, 431
Tetraselenadiynes, 438
Tetraselenafulvalenes, 253
Tetrathia[3.3.3.3]paracyclophane, 440
Tetrathiafulvalenes, 128, 253–256
Tetrathiocin, 128
1,3,6,8-Tetrazatricyclo[4.4.1.1(3.8)]dodecane, 341
1,5-bis(Tetrazol-5-yl)-3-oxapentane, 222
Tetrazoles, 222–223
Tetrazolo[1,5-d][1,4]diazepin-6-one, 411
Tetrazolopyrazines, 419
Tetrazolopyrimidines, 222, 223, 348
TFA catalyst, 439
Δ^9-THC, 367
Thermal cyclization, 149
2-Thia-3,5,6,7,9-pentaazabenz[cd]azulenes , 354
p-H-Thiacalix[4]arene, 438
Thiacalix[4]arene, 440
Thiacalix[4]pyridine, 441
Thiadiazacenaphthylenes, 215
1,4,5-Thiadiazepine, 422
1,3,4-Thiadiazole, 118
1,2,3-Thiadiazole-4-carbohydrazides, 245
1,3,4-Thiadiazolines, 246
1,3,4-Thiadiazolo[2,3-b]quinazoline, 246
1,3,4-Thiadiazolo[2,3-c][1,2,4]triazinones, 347
Thiaheterohelicenes, 129
Thianaphth[4,5-e][1,4]oxazine, 127
Thiaporphyrins, 121, 129
9-Thiatricyclo[4.3.1.1]undecane, 374
2-(Z)-6-(E)-4H-[1,4]-Thiazepin-5-ones, 416
1,2-Thiazepines, 416
Thiazepines, 416
Thiazine, 238
Thiazole, 48
Thiazolines, 240, 241
Thiazolo[2,3-c][1,2,4]triazinones. 343
Thiazolo[3',2':2,3][1,2,4]triazino[5,6-b]indoles, 353

Thiazolo[3,2-*b*]-1,2,4-thiadiazine-1,1-dioxide, 240
Thieno[2,3-*b*][1,5]benzoxazepine, 121, 127
Thieno[2,3-*b*]pyrazines, 127
Thieno[2,3-*c*]pyridines, 117, 127
Thieno[2,3-*d*]-1,3-dithiole-2-thione, 128
Thieno[2,3-*d*]pyrimidines, 118, 127
Thieno[3',2':4,5]thieno[2,3-*c*]quinolines, 124
Thieno[3,2-*b*]thiophenes, 122
Thieno[3,2-*c*]benzazepines, 127
Thieno[3,2-*c*]pyridines, 117, 127
Thieno[3,2-*e*]-1,2,4-thiadiazines, 127
Thieno[3,2-*e*]-1,2-thiazines, 127
Thieno[3,4-*b*]-1,4-oxathiazine, 120
Thieno[3,4-*b*]indolizine, 128
Thieno[3,4-*b*]pyrazine, 120
Thieno[3,4-*c*]thiophenes, 116
Thieno[*b*]azepinediones, 118, 127
Thietanes, 111
Thiirene-1-oxides, 271
5-Thio-7*H*-pyrazolo[5,1-*b*][1,3]thiazine-2,7-diones, 245
Thiocarbonyl ylide, 118
Thiocarbonylation, 278
Z,Z-3,3'-Thiodi(1-*p*-chlorophenyl-3-phenylprop-2-en-1-one), 342
5-Thio-furans, 182
Thioibotenic acid, 230
Thioindoxyl, 119
b,1,6-Thio-linked cycloglucopyranosides, 440
6-Thiomethylidene-1,3,4-oxadiazin-5-ones, 245
Thiopeptide antibiotics, 285
Thiophen-2-one, 120
Thiophene 1,1-oxides, 116
Thiophene-1,1-oxides, 120
Thiophene-1-oxides, 120
Thiopyran-2-thiones, 374
5-Thiopyrazolones, 245
o-Thioquinones, 380
9-Thiotribenzo[*c,g,j*]cycloundeca-3,7,10-triene-1,5-diyne, 437
Thiourazoles, 222
Thiourea, 242
2-Thioxohydantoin ketene dithioacetals, 209
5-Thioxoperhydroimidazo[4,5-*d*]imidazol-2-ones, 215
5-Thioxoperhydroimidazo[4,5-*d*]imidazol-2-ones, 215
Thioxo-pyrido[2,3-*d*]pyrimidinones, 352
Thomitrem A, 66
Three-component coupling, 100, 171, 223
Threonine β-lactone, 110
Thrombin inhibitors, 128
Thymidine dimers, 222
(1*R*,3*S*)-Thysanone, 369
Titanium complex, 295
Titanium induced process, 210, 284

Titanocene reagents, 189
Titanocene(III) chloride, 108
α-Tocopherol, 366
(*S*)-α-Tocotrienol, 368
Topoisomerase II inhibitor, 157
Topsentin, 67
4-Tosylcoumarin, 371
Tosylmethyl isocyanide, 141, 271
Traceless linking, 145
Transition metal catalyzed processes, 150
Transition structures, 78
1,3,5-Tri(4-picolyl)benzene, 442
2,4,6-Tri(4-pyridyl)-1,3,5-triazine, 442
1,2,3-Tri(oxazolinyl)cyclopropane, 275
N-Trialkyl borazines, 345
4,5,6-Triamino-pyrimidine, 353
Triaminotriazine, 344
Trianglamine, 437
5,10,15-Triarylcorrole, 440
5,10,15-Triarylcorrole, oxa-analogue, 180
Triaza-18-crown-6, 440
Triazacycloheptatetraene, 418
1,4,7-Triazacyclonone-d_{12}, 437
1,3,5-Triazepane-2,4-diones, 419
Triazepine-ruthenium(II) complexes, 419
1,2,4-Triazepinones, 420
N-(Triazin-1,3,5-yl)phenylalanine, 342
N-[*as*-Triazin-3-yl]nitrilimines, 221
1,2,4-Triazin-5(4*H*)-ones, 343
1,2,4-Triazin-6-one oxime, 339
1,3,5-Triazine-2,4-diones, 342
1,3,5-Triazine-2,4-dithione, 340
1,2,4-Triazine-4-oxides, 342
1,3,5-Triazine-6-thione, 344
1,3,5-Triazine-based linker, 339
1,2,4-Triazines, 343
[1,3,5]Triazino[1,2-*a*]benzimidazole-2,4-(3*H*,10*H*)-diones, 215
[1,2,4]Triazino[6,5-*f*]quinoline
dihydrodipyridopyrazines , 353
[1,2,4]Triazino[6,5-*f*]quinoline 4-oxides, 343
1,2,4-Triazole-3-thiones, 221
Triazolephthalocyanines, 435
1,2,4-Triazoles, 220–222
Triazolidinediones, 220
1,2,4-[1,2,4]Triazolo[1,5-*a*]azepine, 221
[1,2,3]Triazolo[1,5-*a*][1,4]benzodiazepine-6-(4*H*)-ones, 354
[1,2,3]Triazolo[1,5-*a*]benzotriazines, 219
[1,2,3]Triazolo[1,5-*a*]pyrazines, 219
1,2,3-Triazolo[1,5-*a*]pyrazines, 348
1,2,3-Triazolo[1,5-*a*]pyrazinium-3-olate, 348
1,2,3-Triazolo[1,5-*a*]pyrimidines, 219, 354
[1,2,3]Triazolo[1,5-*a*]quinazoline, 219, 353
1,2,3-Triazolo[1,5-*a*]quinazolines, 219

[1,2,3]Triazolo[1,5-*a*]quinolines, 218, 219, 353
[1,2,3]Triazolo[1,5-*a*]quinoxalines, 219, 221
1,2,4-Triazolo[1,5-*a*]quinoxalines, 222
[1,2,3]Triazolo[1,5-*c*]benzo[1,2,4]triazine, 353
[1,2,4]Triazolo[3,2-*d*][1,5]benzoxazepines, 221
1,2,4-Triazolo[3,4-*b*][1,3,4]thiadiazine, 350
7*H*-1,2,4-Triazolo[3,4-*b*][1,3,4]thiadiazines, 221
1,2,4-Triazolo[3,4-*b*][1,3,4]thiadiazines, 222
1,2,4-Triazolo[3,4-*b*]-1,3-(4*H*)-benzothiazines, 221
1,2,4-Triazolo[4,3-*a*][1,8]naphthyridines, 219, 220
1,2,4-Triazolo[4,3-*a*]5(1*H*)-pyrimidinones, 220
s-Triazolo[4,3-*b*]-*as*-triazin-7(8*H*)-ones, 221, 343
1,2,3-Triazolo[4,5-*d*]pyrimidines, 348, 350
1,2,3-Triazolo[4,5-*e*]1,2,4]triazolo[4,3-*c*]pyrimidine, 350
1,2,4-Triazolo[4'.3':2,3][1,2,4]triazino[5,6-*b*]indoles, 354
1,2,4-Triazolo[5,1-*c*][1,2,4]triazin-4-ones, 349
1,2,4-Triazolo[5,4-*b*][1,3,4]benzotriazepin-6-ones, 221
Triazolo-1,3,4-thiadiazepines, 422
Triazolopyrimidines, 348
Triazolotriazines, 340
1,3,5-Tribromo-2,4,6-tris(bromomethyl)benzene, 368
2,3,5-Tribromobenzofuran, 177
2,3,4-Tribromopyrrole, 58
Tributyl(1-ethoxyvinyl)tin, 238
2,4,6-Trichloro-1,3,5-triazine, 339
1,1,1-Trichloroethyl propargyl ethers, 178
Trichloroisocyanuric acid, 220
α-Triethylsiloxy aldehydes, 185
bis(Triethylsilyl)dienyl ether, 188
Trifloyl oxazoles, 271
3,6-*bis*(Trifluoromethyl)-1,2,4,5-tetrazine, 198
Trifluoromethyl-2-arylbenzimidazoles, 212
2-Trifluoromethyl-4*H*-chromen-4-imines, 373
4-Trifluoromethylfurans, 181
2-Trifluoromethylimidazo[1,2-*a*]pyridines, 215
Trifluoromethyl-pyrazoles, 208
2,5-*bis*(Trimethylaminomethyl)pyrrole diiodide, 436
N,N'-Trimethylene-*bis*(benzoaza-15-crown-5), 439
1,1,1-Trimethylhydrazinium iodide, 213
1,2,5-Trimethylimidazole, 40
1,2,4-Trimethylimidazole, 41
1,3,5-Trimethylpyrazole, 40
1,3-*bis*(Trimethylsiloxy)-1,3-butadienes, 188
2-Trimethylsiloxyfuran, 171
Trimethylsilyl azide. 223
5,6-*bis*(Trimethylsilyl)benzo[*c*]furan, 199
Trimethylsilylacetylene, 438
2,4,5-Triols, 363
3,7,11-Trioxa-1,2,3,4,5,6,7,8,9,10,11,12-dodecahydro-triphenylenes, 368
1,2,4-Trioxanes, 180, 377, 378
1,2,3-Trioxolanes, 257
1,2,4-Trioxolanes, 257

Triphenylphosphine/hexachloroethane, 270
1-(Triphenylphosphoroylideneaminoalkyl)benzo
Triphosgene, 103
Triptycene cyclopentenedione, 265
Tripyrranes, 439
Tripyrrinonealdehyde, 434
Triselenatellurafulvalenes, 253
2,6,9-Trisubstituted purines, 350
1,3,5-Trisubstituted pyrazolines, 207
Trisulcusine, 323
1,4,7-Trithia-11-telluracyclotetradecane, 441
Trithiahexaphyrins, 129
Trithiophenes, 129
Tröger's base and analogues, 120, 209, 321
Tropanes, 9
Tropolone *o*-quinone methide, 366
Trypanocidal phenazine, 324
Tubastrindole A, 68
Tuberostemonine, 25
Tumor necrosis factor-a, 117
Tungsten-catalysed, 363
Tyrian Purple, 67
Ugi four-center three-component reaction, 106
Ugi reactions, 171, 222, 270, 326
Ugibohlin, 60
'Ümpolung' nitration, 307
Uracilophane, 435
Urazoles, 220, 222
Urokinase inhibitors, 128
Cyclophane, 440
Vanadium induced reaction, 145
Vancosamine, 363
Varacin, 424
Vasopressin V₂ receptor agonist, 393
Vicarious nucleophilic substitution reactions, 121
Vilsmeier reagent, 215, 286, 320
Vinoxine, 289
Vinyl epoxides, 7, 84
Vinyl nitrenes, 39, 45
Vinyl phosphonium, 240
Vinylpyrroles, 140
4-Vinyl-2,3-dihydrofuran, 176
3-Vinylbenzo[*b*]furans, 196
3-Vinylfuropyridines, 196
4-Vinyloxetan-2-ones, 371
Wacker oxidation, 188
Wang resin, 214
Wasabi phytoalexin, 161
Watasenia preluciferin (coelenterazine), 329
Weinreb amides, 8, 270
Williamson cycloetherization, 183
Wittig chemistry, 12, 123, 141, 437
aza-Wittig reactions, 124
Woodrosin I, 24
Xanthates, 221
Xanthines, 348, 350

Xantphos, 327, 343
X-ray crystal structures, 161, 234, 308, 309, 316, 324
Yamaguchi cyclisation, 360
Yessotoxin, 361
Yingzhaosu, 376
Ylide, 90

Ynamines, 6
Ynolate anions, 110
Ytterbium triflate reaction, 155
Zoapatanol, 19
[Zr(Cp)$_2$Cl$_2$], 438